Fu
D

# Fundamentals of Digital Electronics

**Robert K. Dueck**
*Seneca College of Applied Arts and Technology*

**West Publishing Company**
Minneapolis/St. Paul • New York
Los Angeles • San Francisco

*Production Services and Text Design:* The Book Company
*Copyedit:* Rebecca Pepper
*Illustration:* Fog Press, Impact Publications, Interactive Composition Corporation
*Composition:* The Clarinda Company
*Index:* Sonsie Conroy

**West's Commitment to the Environment**

In 1906, West Publishing Company began recycling materials left over from the production of books. This began a tradition of efficient and responsible use of resources. Today, up to 95% of our legal books and 70% of our college and school texts are printed on recycled, acid-free stock. West also recycles nearly 22 million pounds of scrap paper annually—the equivalent of 181,717 trees. Since the 1960s, West has devised ways to capture and recycle waste inks, solvents, oils, and vapors created in the printing process. We also recycle plastics of all kinds, wood, glass, corrugated cardboard, and batteries, and have eliminated the use of styrofoam book packaging. We at West are proud of the longevity and the scope of our commitment to our environment.

Printing and Binding by West Publishing Company.

Copyright ©1994   By West Publishing Company
610 Opperman Drive
P.O. Box 64526
St. Paul, MN 55164-0526

**Library of Congress Cataloging-in-Publication Data**

Dueck, Robert K.
    Fundamentals of digital electronics/Robert K. Dueck.
        p.     cm.
    Includes index.
    ISBN 0-314-02424-7 (alk. paper)
    1. Digital electronics.   I. Title.
TK7868.D5D84   1994
621.39'5--dc20                                                        93-33038
                                                                     CIP

*To four great teachers:*

Don Trim and Wally Lewin, who taught me how to think and
Pat Fenske and Stephanie Purvis, who taught me how to write

# • B R I E F    C O N T E N T S •

**Preface**

| | | |
|---|---|---|
| **Chapter 1** | Basic Principles of Digital Systems | 1 |
| **Chapter 2** | Logic Functions and Gates | 19 |
| **Chapter 3** | Logic Gate Networks and Boolean Algebra | 63 |
| **Chapter 4** | Analysis, Design, and Troubleshooting of Logic Gate Networks | 103 |
| **Chapter 5** | Number Systems and Codes | 157 |
| **Chapter 6** | Latches and Flip-Flops | 179 |
| **Chapter 7** | Multivibrators | 239 |
| **Chapter 8** | Introduction to Programmable Logic | 263 |
| **Chapter 9** | Digital Arithmetic and Arithmetic Circuits | 321 |
| **Chapter 10** | MSI Combinational Logic | 373 |
| **Chapter 11** | Introduction to Counters | 455 |
| **Chapter 12** | MSI Counters and Synchronous Counter Design | 503 |
| **Chapter 13** | Shift Registers | 561 |
| **Chapter 14** | Electrical Characteristics of Logic Gates | 601 |
| **Chapter 15** | Logic Gate Circuitry | 637 |
| **Chapter 16** | Interfacing Analog and Digital Circuits | 641 |
| **Chapter 17** | Memory | 735 |
| **Chapter 18** | RAM Devices and Memory Systems | 767 |
| **Appendix A** | Selected Data Sheets: 74F00, 74ALS00 | 809 |
| **Appendix B** | CMOS Handling Precautions for CMOS | 832 |
| **Appendix C** | EPROM Data for a Digital Function Generator | 834 |
| **Appendix D** | Pinouts of Common LSTL and CMOS Logic Gate ICs | 843 |
| | Answers to Problems | 845 |
| | Index | 901 |

# • C O N T E N T S •

**Chapter 1** Basic Principles of Digital
▼ Systems 1

1.1 Digital Versus Analog Electronics 2

1.2 Digital Logic Levels 4

1.3 The Binary Number System 5
Positional Notation 5; Binary Inputs 6

1.4 Digital Waveforms 9
Periodic Waveforms 10; Aperiodic
Waveforms 11; Pulse Waveforms 12

**Chapter 2** Logic Functions and
▼ Gates 19

2.1 Basic Logic Functions 20
NOT, AND, and OR Functions 21; Active
Levels 23

2.2 Derived Logic Functions 25
NAND/NOR Functions 25; Exclusive OR and
Exclusive NOR Functions 27

2.3 Truth Table of a Generalized Gate 29
Generalized Multiple-Input Gates 32

2.4 DeMorgan's Theorems and Gate Equivalence 33
Changing a Gate to Its DeMorgan Equivalent 35

2.5 Enable and Inhibit Properties of Logic Gates 37
AND and OR Gates 38; NAND and NOR
Gates 39; Exclusive OR and Exclusive NOR
Gates 40

2.6 Integrated Circuit Logic Gates 41

2.7 Introduction to Troubleshooting 44
Test Equipment for Troubleshooting Digital
Circuits 47; Possible Faults in Logic Gates 52

**Chapter 3** Logic Gate Networks and
▼ Boolean Algebra 63

3.1 Combinational Logic and Boolean Algebra 64

3.2 Boolean Expressions from Logic Diagrams 65
Order of Precedence 67

3.3 Logic Diagrams from Boolean Expressions 69

3.4 Evaluating Boolean Expressions 73

3.5 Truth Tables From Logic Diagrams 74

3.6 Truth Tables From Boolean Expressions 77

3.7 Boolean Expressions From Truth Tables 78

3.8 Exclusive OR and Exclusive NOR Circuits 84

3.9 Theorems of Boolean Algebra 85
Commutative, Associative, and Distributive
Properties 85; Single Variable Theorems 87;
Multivariable Theorems 91

**Chapter 4** Analysis, Design, and
▼ Troubleshooting of Logic
Gate Networks 103

4.1 Simplifying SOP and POS Expressions 104

4.2 Simplification by the Karnaugh Map Method 109
Two-Variable Map 110; Three- and Four-Variable
Maps 111; Grouping Cells Along Outside
Edges 112; Loading a K-Map From a Truth
Table 113; Multiple Groups 114; Overlapping
Groups 115; Conditions for Maximum
Simplification 116; Using K-Maps for Partially
Simplified Circuits 117; Don't Care States 120;
POS Simplification 123

4.3 Simplification Using DeMorgan Equivalent
Gates 125

4.4   Universal Property of NAND/NOR Gates   128
         All-NAND Forms 128;  All-NOR Forms 129

4.5   Practical Circuit Implementation   135

4.6   Pulsed Operation of Logic Circuits   139

4.7   Designing Logic Circuits from Word
       Problems   141

4.8   Troubleshooting Combinational Logic
       Circuits   143
         General Approach to Troubleshooting 143

## Chapter 5   Number Systems and Codes   157

5.1   Positional Notation   158
         Binary-to-decimal conversion 159

5.2   Binary Numbers   159
         Counting in Binary 159;  Decimal to Binary
         Conversion 160;  Fractional Binary Numbers 163

5.3   Hexadecimal Numbers   165
         Counting in Hexadecimal 165;  Hexadecimal to
         Decimal Conversion 166;  Decimal to Hexadecimal
         Conversion 166;  Conversions Between Hexadecimal
         and Binary 167

5.4   Octal Numbers   168
         Counting in Octal 168;  Conversions Between Octal
         and Binary, Octal and Decimal 169

5.5   BCD Codes   170
         8421 Code 171;  Excess-3 Code 171

5.6   Gray Code   172

5.7   ASCII Code   172

## Chapter 6   Latches and Flip-Flops   179

6.1   Latches   180

6.2   NAND/NOR Latches   183
         R=0, S=0 (no change condition) 184
         S=0, R=1 (reset condition) 185
         S=1, R =0 (set condition) 186
         S=1, R=1 (forbidden condition) 186
         Latch as a switch debouncer 188

6.3   Gated Latches   190
         Gated SR Latch 191;  Transparent Latch (Gated D
         Latch) 192

6.4   Edge Triggered SR and D Flip-Flops   198
         SR Flip-Flops 198;  D Flip-Flops 201;  Applications
         of D Flip-Flops 203

6.5   Edge Triggered JK Flip-Flops   212
         Asynchronous Inputs (Preset and Clear) 215

6.6   Timing Parameters   218

6.7   Troubleshooting Latch Circuits   221

## Chapter 7   Multivibrators   239

7.1   Multivibrators   240
         Bistable Multivibrators 240;  Monostable
         Multivibrators 240;  Astable Multivibrators 242

7.2   IC Monostables   243

7.3   The 555 Timer   247
         Internal Configuration 248

7.4   The 555 Timer as a Monostable
       Multivibrator   250

7.5   The 555 Timer as an Astable Multivibrator   253
         Calculating Timing Component Values 256;
         Minimum Value of $R_a$ 258

## Chapter 8   Introduction to Programmable Logic   263

8.1   Introduction to Programmable Logic   264
         PLD Architectures 266

8.2   PAL Fuse Matrix and Combinational Outputs   270

8.3   PAL Outputs with Programmable Polarity   273

8.4   PAL and GAL Devices with Registered
       Outputs   277
         Generic Array Logic (GAL) 283

8.5   UV-Erasable Programmable Logic Devices
       (EPLD)   291
         85C220/224 292;  85C22V10 292;
         85C060/090 295

8.6   Development Software   295
         Software Vendors 299;  Using PLDasm 301

## Chapter 9   Digital Arithmetic and Arithmetic Circuits   321

9.1   Digital Arithmetic   322
         Unsigned Binary Arithmetic 323

9.2   Representing Signed Binary Numbers   325
         True-Magnitude Form 326;  1's Complement Form
         326;  2's Complement Form 327

9.3   Signed Binary Arithmetic   327
      Signed Addition 328;  Subtraction 328;  Negative
      Sum or Difference 329;  Range of Signed Numbers
      330;  Sign-Bit Overflow 331

9.4   Hexadecimal Arithmetic   334

9.5   Binary Adders and Subtractors   337
      Half and Full Adders 337;  Parallel Binary
      Adder/Subtractor 342;  Overflow Detection 350

9.6   Register Arithmetic Circuits   352

9.7   BCD Adders   357
      Carry Output 358;  Sum Correction 359;
      Multiple-Digit BCD Adders 359

9.8   Programmable Logic Implementation of Arithmetic
      Circuits   361
      PLD Implementation of a Parallel Adder 361;
      Adder/Subtractor with Overflow Indication 364;
      BCD Adder 365

**Chapter 10**   MSI Combinational
▼                Logic   373

10.1  Decoders   374
      Single-Gate Decoders 375;  MSI Decoders 377;
      BCD-to-Seven-Segment Decoders 386

10.2  Encoders   394
      Priority Encoder 394

10.3  Magnitude Comparators   401

10.4  Parity Generators and Checkers   408

10.5  Multiplexers   415
      MSI Multiplexers 417;  Single-Channel Data
      Selection 417;  Boolean Function Generator 418;
      Time-Dependent Multiplexer Applications 422;
      Multi-Channel Data Selection 426

10.6  Demultiplexers   428
      Demultiplexing a TDM Signal 428;  CMOS Analog
      Multiplexer/Demultiplexer 429

10.7  Combinational Logic Applications of PLDs   432

**Chapter 11**   Introduction to
▼                Counters   455

11.1  Digital Counters   456
      Basic Concepts of Digital Counters 458

11.2  Asynchronous Counters   463
      Asynchronous UP Counters and DOWN
      Counters 465;  Truncated Sequence Counters 468

11.3  Decoding a Counter   472
      Propagation Delay, Glitches, and Strobing 473;
      Maximum frequency of an asynchronous
      counter 479

11.4  Synchronous Counters   481
      Full Sequence Counters 482;  Truncated Sequence
      Synchronous Counters 491

**Chapter 12**   MSI Counters and
▼                Synchronous Counter
                 Design   503

12.1  Asynchronous MSI Counters   505

12.2  Synchronous Presettable MSI Counters   509
      Bidirectional Synchronous Counters 509;  Presettable
      MSI Synchronous Counters 513;  Cascading MSI
      Counters 516;  Mod-n Counters 521;  Multi-Chip
      mod-n Counters 523

12.3  IEEE/ANSI Notation   525
      74LS93 and 74LS90 Asynchronous Counters 525;
      74LS161A/163A Synchronous Presettable
      Counters 526;  74LS191 Bidirectional Synchronous
      Presettable Counter 527

12.4  Design of Synchronous Sequential Circuits   528
      Design Procedure 530;  Design of a Synchronous
      mod-12 Counter 530;  Counters with Nonstandard
      Sequences 535;  Use of Control Inputs 541

12.5  Sequential Logic Applications of PLDs   542
      Two 8-Bit Counters (Boolean Equation
      Method) 544;  4-Bit Counter (State Table
      Method) 548

**Chapter 13**   Shift Registers   561
▼

13.1  Basic Shift Register Configurations   562
      Serial Shift Registers 563;  Parallel-Load Shift
      Registers 569;  Applications of Parallel-Load Shift
      Registers 572

13.2  Bidirectional Shift Registers   577
      74LS194A Universal Shift Register 580

13.3  Shift Register Counters   583
      Ring Counters 585;  Johnson Counters 589

13.4  IEEE/ANSI Notation   595
      74LS91 Serial-In-Serial-Out Shift Register 595
      74LS164 Serial-In-Parallel-Out Shift Register;  595
      74LS95B Parallel-Access Shift Register;  596
      74LS194A Universal Shift Register;  596

# Chapter 14 ▼ Electrical Characteristics of Logic Gates   601

14.1  Electrical Characteristics of Logic Gates   602

14.2  Propagation Delay   603
Propagation Delay in Logic Circuits 607

14.3  Fanout   611

14.4  Power Dissipation   615
Power Dissipation in TTL Devices 616;  Power Dissipation in CMOS Devices 620

14.5  Noise Margin   624

14.6  Interfacing TTL and CMOS Gates   627
CMOS Driving LSTTL 628;  LSTTL Driving 74HCT CMOS 629;  LSTTL Driving 74HC or 4000B CMOS 629;  Interfacing Devices with Different Power Supplies 629

# Chapter 15 ▼ Logic Gate Circuitry   637

15.1  Internal Circuitry of TTL gates   638
Bipolar Transistors as Logic Devices 638;  TTL Open Collector Inverter and NAND Gate 641;  Open Collector Applications 647;  Totem Pole Outputs 653;  Tristate Gates 658;  Other Basic TTL Gates 660

15.2  Internal Circuitry of NMOS and CMOS Gates   663
MOSFET Structure 664;  Bias Requirement for MOS Transistors 665;  NMOS Inverter 669;  CMOS Inverter 669;  CMOS NAND/NOR Gates 671;  CMOS AND/OR Gates 674;  CMOS Transmission Gates 675

15.3  TTL and CMOS Variations   667
TTL Logic Families 667;  CMOS Logic Families 680

15.4  Other Logic Families   682
Emitter Coupled Logic 682;  BiCMOS Logic 684

# Chapter 16 ▼ Interfacing Analog and Digital Circuits   691

16.1  Analog and Digital Signals   692
Sampling an analog voltage 693

16.2  Digital-to-Analog Conversion   697
Weighted Resistor D-to-A Converter 699;  R-2R Ladder D-to-A Converter 701;  MC1408 D-to-A Converter 704;  DAC Performance Specifications 714

16.3  Analog-to-Digital Conversion   716
Flash A-to-D Conversion 716;  Successive Approximation A-to-D 718;  Dual Slope A-to-D 722;  Sample and Hold Circuit 727

16.4  Data Acquisition   728

# Chapter 17 ▼ Memory   735

17.1  Basic Memory Concepts   736
Address and Data 736;  RAM and ROM 739;  Memory Capacity 739;  Control Signals 741

17.2  Random Access Read/Write Memory (RAM)   742
Static RAM Cells 743;  Dynamic RAM Cells 749

17.3  Read Only Memory (ROM)   752
Mask-Programmed ROM 752;  Fusible-Link PROM 754;  EPROM 757;  EEPROM 761

17.4  Sequential Memory: FIFO and LIFO   762

# Chapter 18 ▼ RAM Devices and Memory Systems   767

18.1  Static RAM Timing Cycles   768
Timing Parameters 768

18.2  SRAM Read and Write Cycles   771
SRAM Read Cycle 771;  SRAM Write Cycles 775

18.3  Dynamic Ram Timing Cycles   776
Normal Read Cycle 777;  Write Cycles 777;  Read-Write Cycles 778;  Fast Access Cycles 778;  Refresh Cycles 783

18.4  Dynamic RAM Modules   785

18.5  Memory Systems   785
Memory Mapping and Address Decoding 789;  Decoding With MSI Chips 794

**Appendix A**  Selected Data Sheets:   809

**Appendix B**  CMOS Handling Precautions   832

**Appendix C**  Function Generator: EPROM Contents and QuickBASIC Program to Generate Data   834

**Appendix D**  Pinouts of Common LSTTL and CMOS Logic Gate ICs   843

**Answers to Odd-Numbered Problems**

**Index**

## Intended Audience

This book is intended as a textbook for an introductory course in digital electronics in an electronics technician or technologist program. There is also sufficient material for a second course that expands upon the principles of an introductory course. Others who may find this book useful include electronics professionals who want a basic reference text with a practical slant and hobbyists desiring a book for self-study.

No prior knowledge of digital systems is assumed. Prerequisite or corequisite courses in basic DC circuits and introductory college algebra, while not strictly necessary, allow the student to derive maximum benefit from the course of study laid out by this book. A working knowledge of transistors and operational amplifiers is very helpful for understanding the material in Chapters 15 and 16, which would usually be covered in a second course in digital electronics.

## Pedagogical Features

Special features of this book include:

- An outline and list of objectives at the beginning of each chapter.
- Definitions of key terms at the beginning of each new topic section. (First use in context is also indicated in boldface.)
- Early introduction of application examples for each new topic.
- About 300 solved problem examples.
- About 800 end-of-chapter problems.
- Section review problems to assist in retention of recently learned material.
- An introduction to troubleshooting techniques and the concept of troubleshooting as thinking.
- An entire chapter on programmable logic devices (PLDs) that examines PLD hardware *and software* (unique to this text, as far as I know).
- Sections in three chapters that use PLDs in various applications.

## General Outline

The overall plan of the book is as follows:

Introductory concepts (Chapter 1)

Building blocks:

   SSI combinational logic (Chapters 2–4)

   Number systems and codes (Chapter 5)

   SSI sequential logic (Chapters 6–7)

Programmable logic (Chapter 8)

Applications:

   Digital arithmetic (Chapter 9)

   MSI combinational logic (Chapter 10)

   MSI sequential logic (Chapters 11–13)

Electrical characteristics of logic gates (Chapters 14–15)

A/D and D/A conversion (Chapter 16)

Memory (Chapters 17–18)

## About This Book

The problem with most introductory digital textbooks is that they do not tell students why they are studying these curious new gadgets they have never heard of. Unless they are keen enough to have read ahead, they won't have any clue. For instance, in my first or second year of teaching, one of my brighter students asked, after several lectures on flip-flops, "Are these things used much?" (This is the sort of thing that keeps teachers awake at night.)

To address this concern, this book illustrates simple applications of devices almost as soon as they are introduced. Often these examples are somewhat simplistic, in order to speak to the students at their level, but I hope that they will at least tell the students that, indeed, these devices are used for something. Maybe even lots of things.

This book has more word problems than you find in many texts of this sort. The majority of problems that confront the working technician or technologist are really word problems. A customer or employer will not ask a technician to solve a Boolean equation; rather, he or she will ask for a working circuit with a particular verbal specification. I hope that these word problems will help students to extract pertinent information from verbal descriptions and make them realize that Boolean algebra can talk about the world around them in a meaningful way.

I have delayed the discussion of number systems and codes (Chapter 5) until after the material on logic gate networks (Chapters 2 to 4). Until that point, all students really need is a basic knowledge of the binary system (i.e., how to count from 0000 to 1111). The first chapter has a small amount of material on binary numbers to get them up to speed.

I prefer to pique my students' interest by introducing logic gates as soon as possible. The topic of number systems, necessary as it is in later sections, is fairly tedious stuff for the average student. Too much early attention to this topic could lose the students before they knew there was anything better coming. Instructors who disagree with this approach should feel free to introduce the chapter on number systems earlier in the course, as it does not depend on any previous material. If so desired, this could be followed immediately by a unit on digital arithmetic (Sections 9.1 to 9.4).

This book refers extensively to actual devices, both TTL and CMOS (most texts concentrate on only one of these families, usually TTL). There are many references to various data books, mostly Texas Instruments, Motorola, and Intel. The book assumes that students have access only to a good TTL or High-Speed CMOS data book, so that they won't need to have a stack of data books just to read this text. A limited number of data sheets are included in Appendix A.

A feature of this text that sets it apart from similar books is the extent to which it covers programmable logic devices (PLDs). An entire chapter (Chapter 8) is dedicated to the fundamentals of this exciting new technology, including a basic introduction to PLD-shell Plus/PLDasm, Intel's programmable logic design software. Several subsequent chapters (9, 10, and 12) have sections that contain PLD applications, including portions of the PLDasm source files for each one. The complete source files and corresponding JEDEC files are supplied to instructors on a separate diskette. The JEDEC files enable instructors to use the example files even if they have programming software other than Intel's.

One caution: It is beyond the scope of this book to teach a particular software package (PLDasm) in any great detail. The material in this text is supplementary to the manual supplied with PLDshell Plus. As of this writing, this software is available from Intel, free of charge. Call their Literature Hotline (1-800-548-4725).

## How To Use This Book

Very early in the review process of this book, I learned that everybody wants something different from a textbook and that nobody (including me) is likely to use the entire thing. How much or how little of this book you use depends entirely on your own needs. I have tried to make it as modular as possible, so that you can skip sections you don't need.

**Using This Book for Two Courses.** If you have the luxury of two courses in digital logic, the obvious approach is to go straight through the book, from front to back. You may wish to rearrange the material to suit your course. Here are some possibilities:

1. Teach number systems and digital arithmetic early in the course by following Chapter 1 with Chapter 5 and Sections 9.1 to 9.4 in Chapter 9.
2. Follow a complete sequence in combinational logic before studying sequential logic. An example of such a sequence would be Chapters 2 through 4, 9, and 10 (combinational) and Chapters 6, 7, and 11 through 13 (sequential).
3. Delay the study of programmable logic (Chapter 8) until after the chapters on combinational and sequential logic, listed above.
4. Use Chapter 14 (Electrical Characteristics of Logic Gates) any time after Chapter 2 (Logic Functions and Gates). This choice requires that students have a reasonably good background in DC circuit theory, including Ohm's Law and Kirchhoff's Voltage and Current Laws.

**Using This Book for One Course.** A minimal course in digital logic, with some emphasis in troubleshooting, could be derived from this book as follows:

Chapters 1 through 3

Sections 4.1 through 4.5, 4.6, 4.8

Sections 5.1 through 5.3, 5.5, 5.7

Chapter 6

Sections 10.1, 10.2, 10.5, 10.6

Chapter 11

The material in Chapter 5 could come immediately after Chapter 1, if desired. If more time is available, you could add material from Chapters 7, 9, 10, 12, and 13, as required.

I suppose it is theoretically possible for one person to write a book of this scope without any outside help at all, but in my opinion, it wouldn't be much of a book. A large number of people supported me during this project, in greater or lesser degree. I owe each one of them a vote of thanks.

First and foremost, I want to acknowledge the help of my two acquiring editors, Chris Conty and his predecessor, Tom Tucker, without whom this book would never have been started, or finished. Each of these stellar gentlemen acted as editor, project manager, coach, psychologist, and sounding board throughout the "long middle" of this project. Liz Riedel, my developmental editor, did a superb job of lining up reviewers and arranging for ancillaries to this package. In addition to these editorial duties, Liz was always ready to listen when I was convinced that this book would never be finished or that the only person to buy it would be my mother.

My production editor, Mary Verrill, was always on top of the production details and more than ready to bring them to my attention. Mary combines an incredible competence with great reserves of kindness and sympathy. Without her, this book would still be a 6-inch stack of typewritten pages.

Special thanks are due to the production staff at The Book Company: George Calmenson, who managed the editorial side of this project; Wendy Calmenson, who did the text design; and Rebecca Pepper, my copy editor, who crossed out every "which" that I ever wrote.

Thanks to the many reviewers whose comments and helpful suggestions greatly contributed to the quality of the final manuscript. These reviewers include:

- John Blankenship, DeVry Institute of Technology, Decatur, GA
- John Dunbar, DeVry Institute of Technology, Decatur, GA
- James Emerson, Wentworth Institute of Technology, Boston, MA
- Michael R. Gale, Red River Community College, Manitoba
- Frank Grimsley, Jr., San Antonio College, San Antonio, TX
- Ronald Harris, Weber State University, Ogden, UT
- Stephen Harsany, Mt. San Antonio College, Walnut, CA
- David Hata, Portland Community College, Portland, OR
- Jerry Humphrey, Tulsa Junior College, Tulsa, OK
- Bruce Hutchinson, Mohawk College, Ontario

- Edward Kaufenberg, Blackhawk Technical College, WI
- Vladya Kosiba, Seneca College of Applied Arts and Technology, Ontario
- Samuel Kraemer, Oklahoma State University, OK
- Robert Martin, Northern Virginia Community College, Annandale, VA
- Vicki A. Mee, DeVry Institute of Technology, Columbus, OH
- Alan Moltz, Waterbury State Technical College, Waterbury, CT
- Paul Papaioannou, DeVry Institute of Technology, Chicago, IL
- Karl Perusich, Purdue University at South Bend, IN
- Arthur Roitstein, NYC Technical College, New York, NY
- James J. Schreiber, DeVry Institute of Technology, Phoenix, AZ
- George Shaiffer, Pikes Peak Community College, Colorado Springs, CO
- Ernest K. Sharp, Oklahoma State University, Okmulgee, OK
- Roy Siegel, DeVry Institute of Technology, City of Industry, CA
- Diane D. Snyder, Springfield Technical Community College, Springfield, MA
- Wayne Vyrostek, Westark Community College, Fort Smith, AR
- Jean Walls, Mohawk Valley Community College, Utica, NY
- Morrie Walworth, Indiana University at South Bend, IN
- Steven I. Yelton, Cincinnati Technical College, Cincinnati, OH

Special thanks to Ed Kaufenberg and Alan Moltz, the authors of the lab manuals that accompany this text, both for their work on the manuals and for their thorough and helpful reviews. Also to Michael Gale, John Blankenship, Bruce Hutchinson, and Arthur Roitstein, for extra effort in the review process. The criticism and suggestions that they offered could always be counted on to improve the quality of the text.

I gratefully acknowledge the help of my former student Marzena Bachtin, for checking problems in the first third of the manuscript. She made me see many things from a student's point of view that were not immediately obvious. Thanks are due to my colleague Ben Shefler, for his careful checking of the chapter problems.

I was given a lot of moral support throughout the writing of this book by family, friends, and colleagues. In particular, I want to acknowledge the support of my friend and colleague Len Klochek, who was there literally from the beginning of this project and continued to give encouragement throughout. Thanks to my friends at Toronto United Mennonite Church, many of whom have promised to use this book as a doorstop or a coaster, especially to Judi Bergen and Lydia Harder, for reasons best known to themselves. Thanks to my parents, Peter and Esther Dueck, for continued support throughout the years and to my sister, Janice Friesen, for saying, "You've never actually finished anything that took this long, have you?"

*Bob Dueck*
*Toronto, Ontario, Canada*
*June 1993*

# Basic Principles of Digital Systems

## CHAPTER CONTENTS

**1.1** Digital Versus Analog Electronics

**1.2** Digital Logic Levels

**1.3** The Binary Number System

**1.4** Digital Waveforms

## CHAPTER OBJECTIVES

Upon successful completion of this chapter, you will be able to:

- Describe some differences between analog and digital electronics.

- Understand the concept of HIGH and LOW logic levels.

- Explain the basic principles of a positional notation number system.

- Calculate the decimal equivalent of a binary number.

- Translate logic HIGHs and LOWs into binary numbers.

- Generate a sequence of binary numbers.

- Distinguish between the most significant bit and least significant bit of a binary number.

- Describe the difference between periodic, aperiodic, and pulse waveforms.

- Calculate the frequency, period, and duty cycle of a periodic digital waveform.

- Calculate the pulse width, rise time, and fall time of a digital pulse.

## INTRODUCTION

Digital electronics is the branch of electronics based on the switching of voltages called logic levels. Any quantity in the outside world, such as temperature, pressure, or voltage, can be symbolized inside a digital circuit by a number made up of logic levels.

Each logic level (logic HIGH or logic LOW) corresponds to a digit in the binary (base 2) number system. The *binary digits,* or bits, 0 and 1, are sufficient to write any number, given enough places. (For example, 100 in binary is 4 in the decimal system; 1100100 in binary is 100 in decimal.)

Inputs and outputs are not always static. Often they vary with time. Time-varying digital waveforms can have three forms:

1. Periodic waveforms, which repeat a pattern of logic HIGHs and LOWs
2. Aperiodic waveforms, which do not repeat
3. Pulse waveforms, which produce a momentary variation from a constant logic level

## 1.1

# Digital Versus Analog Electronics

**Continuous**  *Smoothly connected. An unbroken series of consecutive values with no instantaneous changes.*

**Discrete**  *Separated into distinct segments or pieces. A series of discontinuous values.*

**Analog**  *A way of representing some physical quantity, such as temperature or velocity, by a proportional continuous voltage or current. An analog voltage or current can have any value within a defined range.*

**Digital**  *A way of representing a physical quantity by a series of binary numbers. A digital representation can have only specific discrete values.*

The study of electronics often is divided into two basic areas: **analog** and **digital** electronics. Analog electronics has a longer history and can be regarded as the "classical" branch of electronics. Digital electronics, although newer, has achieved greater prominence through the advent of the computer age. The modern revolution in microcomputer chips, as part of everything from personal computers to cars and coffee makers, is founded almost entirely on digital electronics.

The main difference between analog and digital electronics can be stated simply. Analog voltages or currents are **continuously** variable between defined values, and digital voltages or currents can vary only by distinct, or **discrete,** steps.

Some keywords highlight the differences between digital and analog electronics:

| Analog | Digital |
|---|---|
| Continuously variable | Discrete steps |
| Amplification | Switching |
| Voltages | Numbers |

An example often used to illustrate the difference between analog and digital devices is the comparison between a light dimmer and a light switch. A light dimmer is an analog device, since it can make the light it controls vary in brightness anywhere within a defined range of values. The light can be fully on, fully off, or at some brightness level in between. A light switch is a digital device, since it can turn the light on or off, but there is no value in between those two states.

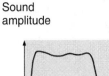

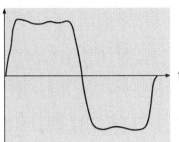

a. **Original audio source**

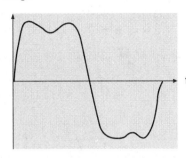

b. **Analog reproduction (shows distortion)**

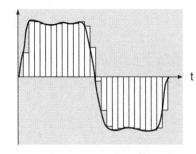

c. **Digital reproduction (simplified)**

**Figure 1.1**

**Digital and Analog Sound Reproduction**

The light switch/light dimmer analogy, although easy to understand, does not show any particular advantage to the digital device. If anything, it makes the digital device seem limited.

One modern application in which a digital device is clearly superior to an analog one is digital audio reproduction. Compact disc players have achieved their high level of popularity because of the accurate and noise-free way in which they reproduce recorded music. This high quality of sound is possible because the music is stored, not as a mechanical copy of the sound vibrations, as in analog records and tapes, but as a series of numbers that represent amplitude steps in the sound waves.

Figure 1.1 shows a sound waveform and its representation in both analog and digital forms.

The analog voltage, shown in Figure 1.1b, is a copy of the original waveform and introduces distortion both in the storage and playback processes. (Think of how a photocopy deteriorates in quality if you make a copy of a copy, then a copy of the new copy, and so on. It doesn't take long before you can't read the fine print.)

A digital audio system doesn't make a copy of the waveform, but rather stores a code (a series of amplitude numbers) that tells the compact disc player how to re-create the original sound every time a disc is played. During the recording process, the sound waveform is "sampled" at precise intervals. The recorder transforms each sample into a digital number corresponding to the amplitude of the sound at that point.

The "samples" (the voltages represented by the vertical bars) of the digitized audio waveform shown in Figure 1.1c are much more widely spaced than they would be in a real digital audio system. They are shown this way to give the general idea of a digitized waveform. In real digital audio systems, each amplitude value can be indicated by a number having as many as 16,000 to 65,000 possible values. Such a large number of possible values means the voltage difference between any two consecutive digital numbers is very small. The numbers can thus correspond extremely closely to the actual amplitude of the sound waveform. If the spacing between the samples is made small enough, the reproduced waveform is almost exactly the same as the original.

---

**Section Review Problem for Section 1.1**

1.1. What is the basic difference between analog and digital audio reproduction?

## 1.2

## Digital Logic Levels

> **Logic level**  *A voltage level that represents a defined digital state in an electronic circuit.*
>
> **Logic HIGH**  *The higher of two voltages in a digital system with two logic levels.*
>
> **Logic LOW**  *The lower of two voltages in a digital system with two logic levels.*
>
> **Positive logic**  *A system in which logic LOW represents binary digit 0 and logic HIGH represents binary digit 1.*
>
> **Negative logic**  *A system in which logic LOW represents binary digit 1 and logic HIGH represents binary digit 0.*

Digitally represented quantities, such as the amplitude of an audio waveform, are usually represented by binary, or base 2, numbers. When we want to describe a digital quantity electronically, we need to have a system that uses voltages or currents to symbolize binary numbers.

The binary number system has only two digits, 0 and 1. Each of these digits can be denoted by a different voltage called a **logic level.** For a system having two logic levels, the lower voltage (usually 0 volts) is called a **logic LOW** and represents the digit 0. The higher voltage (usually 5 V, but in some systems a specific value between 3 V and 18 V) is called a **logic HIGH,** which symbolizes the digit 1. Except for some allowable tolerance, as shown in Figure 1.2, the range of voltages between HIGH and LOW logic levels is undefined.

---

For the voltages in Figure 1.2:

$$+5 \text{ V} = \text{Logic HIGH} = 1$$
$$0 \text{ V} = \text{Logic LOW} = 0$$

---

The system assigning the digit 1 to a logic HIGH and digit 0 to logic LOW is called **positive logic.** Throughout the remainder of this text, logic levels will be referred to as HIGH/LOW or 1/0 interchangeably.

**Figure 1.2**

**Logic Levels Based on +5 V and 0 V**

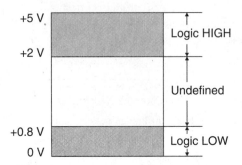

(A complementary system, called **negative logic,** also exists that makes the assignment the other way around, but as it is seldom used in practice, we will not concern ourselves with it except to note that it exists.)

## 1.3

## The Binary Number System

> **Binary number system** *A number system used extensively in digital systems, based on the number 2. It uses two digits, 0 and 1, to write any number.*
>
> **Positional notation** *A system of writing numbers where the value of a digit depends not only on the digit, but also on its placement within a number.*
>
> **Bit** B*inary dig*it. *A 0 or a 1.*

### *Positional Notation*

The **binary number system** is based on the number 2. This means that we can write any number using only two binary digits (or **bits**), 0 and 1. Compare this to the decimal system, which is based on the number 10, where we can write any number with only ten decimal digits, 0 to 9.

The binary and decimal systems are both **positional notation** systems; the value of a digit in either system depends on its placement within a number. In the decimal number 845, the digit 4 really means 40, whereas in the number 9426, the digit 4 really means 400 (845 = 800 + 40 + 5; 9426 = 9000 + 400 + 20 + 6). The value of the digit is determined by *what* the digit is as well as *where* it is.

In the decimal system, a digit in the position immediately to the left of the decimal point is multiplied by 1 ($10^0$). A digit two positions to the left of the decimal point is multiplied by 10 ($10^1$). A digit in the next position left is multiplied by 100 ($10^2$). The positional multipliers, as you move left from the decimal point, are ascending powers of 10.

The same idea applies in the binary system, except that the positional multipliers are powers of 2 ($2^0 = 1$, $2^1 = 2$, $2^2 = 4$, $2^3 = 8$, $2^4 = 16$, $2^5 = 32$, . . .). For example, the binary number 101 has the decimal equivalent:

$$(1 \times 2^2) + (0 \times 2^1) + (1 \times 2^0)$$
$$= (1 \times 4) + (0 \times 2) + (1 \times 1)$$
$$= \quad 4 \quad + \quad 0 \quad + \quad 1$$
$$= \quad 5$$

**EXAMPLE 1.1**     Calculate the decimal equivalents of the binary numbers 1010, 111, and 10010.

**Solutions**
$$1010 = (1 \times 2^3) + (0 \times 2^2) + (1 \times 2^1) + (0 \times 2^0)$$
$$= (1 \times 8) + (0 \times 4) + (1 \times 2) + (0 \times 1)$$
$$= 8 + 2 = 10$$

$$111 = (1 \times 2^2) + (1 \times 2^1) + (1 \times 2^0)$$
$$= (1 \times 4) + (1 \times 2) + (1 \times 1)$$
$$= 4 + 2 + 1 = 7$$
$$10010 = (1 \times 2^4) + (0 \times 2^3) + (0 \times 2^2) + (1 \times 2^1) + (0 \times 2^0)$$
$$= (1 \times 16) + (0 \times 8) + (0 \times 4) + (1 \times 2) + (0 \times 1)$$
$$= 16 + 2 = 18$$

## Binary Inputs

**Most significant bit** *The leftmost bit in a binary number. This bit has the number's largest positional multiplier.*

**Least significant bit** *The rightmost bit of a binary number. This bit has the number's smallest positional multiplier.*

A major class of digital circuits, called combinational logic, operates by accepting logic levels at one or more input terminals and producing a logic level at an output. In the analysis and design of such circuits, it is frequently necessary to find the output logic level of a circuit for all possible combinations of input logic levels.

The digital circuit in the black box in Figure 1.3 has three inputs. Each input can have two possible states, LOW or HIGH, which can be represented by positive logic as 0 or 1. The number of possible input combinations is $2^3 = 8$. (In general, a circuit with $n$ binary inputs has $2^n$ possible input combinations, having decimal equivalents ranging from 0 to $2^n - 1$.) Table 1.1 shows a list of these combinations, both as logic levels and binary numbers, and their decimal equivalents.

A list of output logic levels corresponding to all possible input combinations, applied in ascending binary order, is called a truth table. This is a standard form for showing the function of a digital circuit.

The input bits on each line of Table 1.1 can be read from left to right as a series of 3-bit binary numbers. The numerical values of these eight input combinations range from 0 to 7 (0 to $2^n - 1$) in decimal.

Bit $A$ is called the **most significant bit** (MSB), and bit $C$ is called the **least significant bit** (LSB). As these terms imply, a change in bit $A$ is more significant, since it has the greatest effect on the number of which it is part.

**Figure 1.3**

**3-Input Digital Circuit**

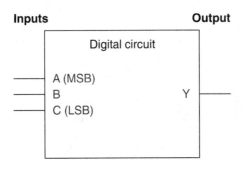

**Table 1.1** Possible Input Combinations for a 3-Input
Digital Circuit

| Logic Level | | | Binary Value | | | Decimal Equivalent |
|---|---|---|---|---|---|---|
| *A* | *B* | *C* | *A* | *B* | *C* | |
| L | L | L | 0 | 0 | 0 | 0 |
| L | L | H | 0 | 0 | 1 | 1 |
| L | H | L | 0 | 1 | 0 | 2 |
| L | H | H | 0 | 1 | 1 | 3 |
| H | L | L | 1 | 0 | 0 | 4 |
| H | L | H | 1 | 0 | 1 | 5 |
| H | H | L | 1 | 1 | 0 | 6 |
| H | H | H | 1 | 1 | 1 | 7 |

**Table 1.2** Effect of Changing the LSB and MSB of a Binary
Number

| | *A* | *B* | *C* | Decimal | |
|---|---|---|---|---|---|
| **Original** | 0 | 1 | 1 | 3 | |
| **Change MSB** | 1 | 1 | 1 | 7 | Difference = 4 |
| **Change LSB** | 0 | 1 | 0 | 2 | Difference = 1 |

Table 1.2 shows the effect of changing each of these bits in a 3-bit binary number and compares the changed number to the original by showing the difference in magnitude. A change in the MSB of any 3-bit number results in a difference of 4. A change in the LSB of any binary number results in a difference of 1. (Try it with a few different numbers.)

▼

**EXAMPLE 1.2**

Figure 1.4 shows a 4-input digital circuit. List all the possible binary input combinations to this circuit and their decimal equivalents. What is the value of the MSB?

**Solution**

Since there are four inputs, there will be $2^4 = 16$ possible input combinations, ranging from 0000 to 1111 (0 to 15 in decimal). Table 1.3 shows the list of all possible input combinations.

The MSB has a value of 8 (decimal).

**Figure 1.4**

**Example 1.2**
**4-Input Digital Circuit**

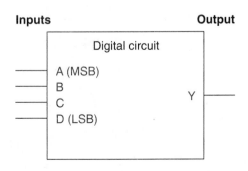

**Table 1.3** Possible Input Combinations for a 4-Input Digital Circuit

| A | B | C | D | Decimal |
|---|---|---|---|---------|
| 0 | 0 | 0 | 0 | 0 |
| 0 | 0 | 0 | 1 | 1 |
| 0 | 0 | 1 | 0 | 2 |
| 0 | 0 | 1 | 1 | 3 |
| 0 | 1 | 0 | 0 | 4 |
| 0 | 1 | 0 | 1 | 5 |
| 0 | 1 | 1 | 0 | 6 |
| 0 | 1 | 1 | 1 | 7 |
| 1 | 0 | 0 | 0 | 8 |
| 1 | 0 | 0 | 1 | 9 |
| 1 | 0 | 1 | 0 | 10 |
| 1 | 0 | 1 | 1 | 11 |
| 1 | 1 | 0 | 0 | 12 |
| 1 | 1 | 0 | 1 | 13 |
| 1 | 1 | 1 | 0 | 14 |
| 1 | 1 | 1 | 1 | 15 |

There are two ways to remember how to construct a binary sequence:

**1.** *Learn to count in binary.* You should know all the binary numbers from 0000 to 1111 and their decimal equivalents (0 to 15). *Make this your first goal in learning the basics of digital systems.*

Each binary number is a unique representation of its decimal equivalent. You can work out the decimal value of a binary number by adding the weighted values of all the bits.

For instance, the binary equivalent of the decimal sequence 0, 1, 2, 3 can be written using two bits: the 1's bit and the 2's bit. The binary count sequence is:

$$00 \ (= 0 + 0)$$
$$01 \ (= 0 + 1)$$
$$10 \ (= 2 + 0)$$
$$11 \ (= 2 + 1)$$

To count beyond this, you need another bit: the 4's bit. The decimal sequence 4, 5, 6, 7 has the binary equivalents:

$$100 \ (= 4 + 0 + 0)$$
$$101 \ (= 4 + 0 + 1)$$
$$110 \ (= 4 + 2 + 0)$$
$$111 \ (= 4 + 2 + 1)$$

The two least significant bits of this sequence are the same as the bits in the 0 to 3 sequence; a repeating pattern has been generated.

The sequence from 8 to 15 requires yet another bit: the 8's bit. The three LSBs of this sequence repeat the 0 to 7 sequence. The binary equivalents of 8 to 15 are:

$$1000 \ (= 8 + 0 + 0 + 0)$$

$$1001 \ (= 8 + 0 + 0 + 1)$$

$$1010 \ (= 8 + 0 + 2 + 0)$$

$$1011 \ (= 8 + 0 + 2 + 1)$$

$$1100 \ (= 8 + 4 + 0 + 0)$$

$$1101 \ (= 8 + 4 + 0 + 1)$$

$$1110 \ (= 8 + 4 + 2 + 0)$$

$$1111 \ (= 8 + 4 + 2 + 1)$$

Practice writing out the binary sequence until it becomes familiar. In the 0 to 15 sequence, it is standard practice to write each number as a 4-bit value, as in Example 1.2, so that all numbers have the same number of bits. Numbers up to 7 have leading zeros to pad them out to 4 bits.

This convention has developed because each bit has a physical location in a digital circuit; we know a particular bit is logic 0 because we can measure 0 V at a particular point in a circuit. A bit with a value of 0 doesn't go away just because there is not a 1 at a more significant location.

While you are still learning to count in binary, you can use a second method.

**2.** *Follow a simple repetitive pattern.* Look at Tables 1.1 and 1.3 again. Notice that the least significant bit follows a pattern. The bits alternate with every line, producing the pattern 0, 1, 0, 1, . . . . The next bit to the left alternates every two lines: 0, 0, 1, 1, 0, 0, 1, 1, . . . . The third bit to the left alternates every four lines: 0, 0, 0, 0, 1, 1, 1, 1, . . . . This pattern can be expanded to cover any number of bits, with the number of lines between alternations doubling with each bit to the left.

---

**Section Review Problems for Section 1.3**

1.2.  How many different binary numbers can be written with 6 bits?

1.3.  How many can be written with 7 bits?

1.4.  Write the sequence of 7-bit numbers from 1010000 to 1010111.

1.5.  Write the decimal equivalents of the numbers written for Problem 1.4.

---

## 1.4

## Digital Waveforms

**Digital waveform**  *A series of logic 1s and 0s plotted as a function of time.*

$\mathsf{T}$he inputs and outputs of digital circuits often are not fixed logic levels but **digital waveforms,** where the input and output logic levels vary with time. There are three possible types of digital waveform. *Periodic* waveforms repeat the same pattern of logic levels over a specified period of time. *Aperiodic* waveforms do not repeat. *Pulse* waveforms follow a HIGH-LOW-HIGH or LOW-HIGH-LOW pattern and may be periodic or aperiodic.

## Periodic Waveforms

**Periodic waveform**  *A time-varying sequence of logic HIGHs and LOWs that repeats over a specified period of time.*

**Period *(T)***  *Time required for a periodic waveform to repeat. Unit: seconds (s).*

**Frequency *(f)***  *Number of times per second that a periodic waveform repeats.* f = 1/T *Unit: Hertz (Hz).*

**Time HIGH *($t_h$)***  *Time during one period that a waveform is in the HIGH state. Unit: seconds (s).*

**Time LOW *($t_l$)***  *Time during one period that a waveform is in the LOW state. Unit: seconds (s).*

**Duty cycle *(DC)***  *Fraction of the total period that a digital waveform is in the HIGH state. DC = $t_h$/T (often expressed as a percentage: %DC = $t_h$/T × 100%).*

**Periodic waveforms** repeat the same pattern of HIGHs and LOWs over a specified period of time. The waveform may or may not be symmetrical, that is, it may or may not be HIGH and LOW for equal amounts of time.

**EXAMPLE 1.3**

Calculate the **time LOW, time HIGH, period, frequency,** and **percent duty cycle** for each of the periodic waveforms in Figure 1.5.

How are the waveforms similar? How do they differ?

**Figure 1.5**

**Example 1.3**
**Periodic Digital Waveforms**

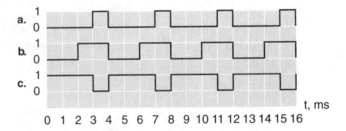

**Solutions**

a. Time LOW: $t_l$ = 3 ms
   Time HIGH: $t_h$ = 1 ms
   Period: $T = t_l + t_h$ = 3 ms + 1 ms = 4 ms
   Frequency: $f = 1/T$ = 1/(4 ms) = 0.25 kHz = 250 Hz

Duty cycle: $\%DC = (t_h/T) \times 100\% = (1 \text{ ms}/ 4 \text{ ms}) \times 100\%$
$$= 25\%$$
(1 ms = 1/1000 second; 1 kHz = 1000 Hz.)

**b.** Time LOW: $t_l = 2$ ms
Time HIGH: $t_h = 2$ ms
Period: $T = t_l + t_h = 2 \text{ ms} + 2 \text{ ms} = 4$ ms
Frequency: $f = 1/T = 1/(4 \text{ ms}) = 0.25 \text{ kHz} = 250$ Hz
Duty cycle: $\%DC = (t_h/T) \times 100\% = (2 \text{ ms}/ 4 \text{ ms}) \times 100\%$
$$= 50\%$$

**c.** Time LOW: $t_l = 1$ ms
Time HIGH: $t_h = 3$ ms
Period: $T = t_l + t_h = 1 \text{ ms} + 3 \text{ ms} = 4$ ms
Frequency: $f = 1/T = 1/(4 \text{ ms}) = 0.25 \text{ kHz} = 250$ Hz
Duty cycle: $\%DC = (t_h/T) \times 100\% = (3 \text{ ms}/ 4 \text{ ms}) \times 100\%$
$$= 75\%$$

The waveforms all have the same period but different duty cycles. A square waveform, shown in Figure 1.5b, has a duty cycle of 50%.

## Aperiodic Waveforms

**Aperiodic waveform**  *A time-varying sequence of logic HIGHs and LOWs that does not repeat.*

An **aperiodic waveform** does not repeat a pattern of 0s and 1s. Thus, the parameters of time HIGH, time LOW, frequency, period, and duty cycle have no meaning for an aperiodic waveform. Most waveforms of this type are one-of-a-kind specimens. (It is also worth noting that most digital waveforms are aperiodic.)

Figure 1.6 shows some examples of aperiodic waveforms.

**Figure 1.6**
**Aperiodic Digital Waveforms**

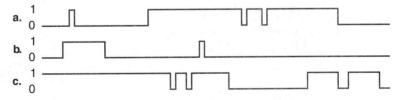

**EXAMPLE 1.4**    A digital circuit generates the following strings of 0s and 1s:

**a.** 001111110110101101000011 0000

**b.** 0011001100110011001100110011

**c.** 00000000111111110000000001111

**d.** 101110111011101110111 0111011

**Figure 1.7**
**Example 1.4**
**Waveforms**

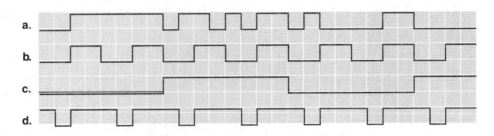

The time between two bits is always the same. Sketch the resulting digital waveforms. Which waveforms are periodic and which are aperiodic?

**Solution** Figure 1.7 shows the waveforms corresponding to the strings of bits above. The waveforms are easier to draw if you break up the bit strings into smaller groups of, say, 4 bits each. For instance:

**a.** 0011 1111 0110 1011 0100 0011 0000

All of the waveforms except Figure 1.7a are periodic.

## Pulse Waveforms

**Pulse** *A momentary variation of voltage from one logic level to the opposite level and back again.*

**Amplitude** *The instantaneous voltage of a waveform. Often used to mean maximum amplitude, or peak voltage, of a pulse.*

**Edge** *The transitional part of a pulse.*

**Rising edge** *The part of a pulse where the logic level is in transition from a LOW to a HIGH.*

**Falling edge** *The part of a pulse where the logic level is in transition from a HIGH to a LOW.*

**Leading edge** *The edge of a pulse that occurs earliest in time.*

**Trailing edge** *The edge of a pulse that occurs latest in time.*

**Pulse width** $(t_w)$ *Elapsed time from the 50% point of the leading edge of a pulse to the 50% point of the trailing edge.*

**Rise time** $(t_r)$ *Elapsed time from the 10% point to the 90% point of the rising edge of a pulse.*

**Fall time** $(t_f)$ *Elapsed time from the 90% point to the 10% point of the falling edge of a pulse.*

Figure 1.8 shows the forms of both an ideal and a nonideal **pulse.** The **rising and falling edges** of an ideal pulse are vertical. That is, the transitions between logic HIGH and

**Figure 1.8**
**Ideal and Nonideal Pulses**

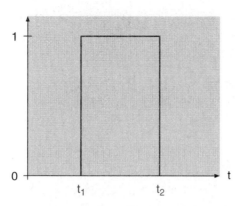 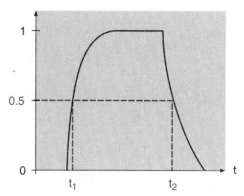

**a. Ideal pulse (instantaneous transitions)**    **b. Nonideal pulse**

**Figure 1.9**
**Pulse Edges**

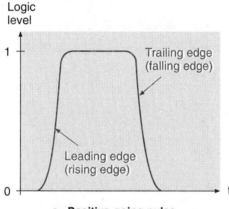 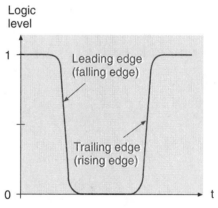

**a. Positive-going pulse**    **b. Negative-going pulse**

LOW levels are instantaneous. There is no such thing as an ideal pulse in a real digital circuit. Circuit capacitance and other factors make the pulse more like the nonideal pulse in Figure 1.8b.

Pulses can be either positive-going or negative-going, as shown in Figure 1.9. In a positive-going pulse, the measured logic level is normally LOW, goes HIGH for the duration of the pulse, and returns to the LOW state. A negative-going pulse acts in the opposite direction.

Nonideal pulses are measured in terms of several timing parameters. Figure 1.10 shows the 10%, 50%, and 90% points on the rising and falling edges of a nonideal pulse. (100% is the maximum **amplitude** of the pulse.)

The 50% points are used to measure **pulse width** because the edges of the pulse are not vertical. Without an agreed reference point, the pulse width is indeterminate. The 10% and 90% points are used as references for the **rise and fall times,** since the edges of a nonideal pulse are nonlinear. Most of the nonlinearity is below the 10% or above the 90% point.

**Figure 1.10**

**Pulse Width, Rise Time, Fall Time**

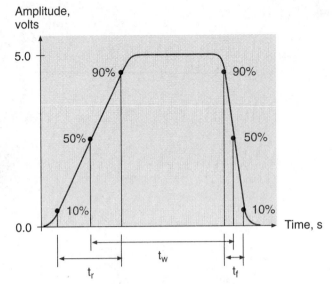

---

**EXAMPLE 1.5**    Calculate the pulse width, rise time, and fall time of the pulse shown in Figure 1.11.

**Figure 1.11**

**Example 1.5 Pulse**

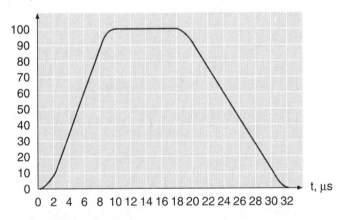

**Solution**    From the graph in Figure 1.11, read the times corresponding to the 10%, 50%, and 90% values of the pulse on both the **leading and trailing edges.**

| *Leading edge:* | 10%: | 2 µs | *Trailing edge:* | 90%: | 20 µs |
|---|---|---|---|---|---|
| | 50%: | 5 µs | | 50%: | 25 µs |
| | 90%: | 8 µs | | 10%: | 30 µs |

*Pulse width:* 50% of leading edge to 50% of trailing edge.

$$t_w = 25 \ \mu s - 5 \ \mu s = 20 \ \mu s$$

*Rise time:* 10% of rising edge to 90% of rising edge.

$$t_r = 8 \ \mu s - 2 \ \mu s = 6 \ \mu s$$

*Fall time:* 90% of falling edge to 10% of falling edge.

$$t_f = 30 \ \mu s - 20 \ \mu s = 10 \ \mu s$$

---

▲

**Section Review Problems for Section 1.4**

A digital circuit produces a waveform that can be described by the following periodic bit pattern: 0011001100110011.

1.6. What is the duty cycle of the waveform?

1.7. Write the bit pattern of a waveform with the same duty cycle and twice the frequency of the original.

1.8. Write the bit pattern of a waveform having the same frequency as the original and a duty cycle of 75%.

## ● GLOSSARY

**Amplitude**  The instantaneous voltage of a waveform. Often used to mean maximum amplitude, or peak voltage, of a pulse.

**Analog**  A way of representing some physical quantity, such as temperature or velocity, by a proportional continuous voltage or current. An analog voltage or current can have any value within a defined range.

**Aperiodic waveform**  A time-varying sequence of logic HIGHs and LOWs that does not repeat.

**Binary number system**  A number system used extensively in digital systems, based on the number 2. It uses two digits to write any number.

**Bit**  *Bi*nary dig*it*. A 0 or a 1.

**Continuous**  Smoothly connected. An unbroken series of consecutive values with no instantaneous changes.

**Digital**  A way of representing a physical quantity by a series of binary numbers. A digital representation can have only specific discrete values.

**Digital waveform**  A series of logic 1s and 0s plotted as a function of time.

**Discrete**  Separated into distinct segments or pieces. A series of discontinuous values.

**Duty cycle (DC)**  Fraction of the total period that a digital waveform is in the HIGH state. $DC = t_h/T$ (often expressed as a percentage: $\%DC = t_h/T \times 100\%$).

**Edge**  The transitional part of a pulse.

**Fall time** $(t_f)$  Elapsed time from the 90% point to the 10% point of the falling edge of a pulse.

**Falling edge**  The part of a pulse where the logic level is in transition from a HIGH to a LOW.

**Frequency** $(f)$  Number of times per second that a periodic waveform repeats. $f = 1/T$ Unit: Hertz (Hz).

**Leading edge**  The edge of a pulse that occurs earliest in time.

**Least significant bit (LSB)**  The rightmost bit of a binary number. This bit has the number's smallest positional multiplier.

**Logic HIGH**  The higher of two voltages in a digital system with two logic levels.

**Logic level**  A voltage level that represents a defined digital state in an electronic circuit.

**Logic LOW**  The lower of two voltages in a digital system with two logic levels.

**Most significant bit (MSB)**  The leftmost bit in a binary number. This bit has the number's largest positional multiplier.

**Negative logic**  A system in which logic LOW represents binary digit 1 and logic HIGH represents binary digit 0.

**Period** $(T)$  Time required for a periodic waveform to repeat. Unit: seconds (s).

**Periodic waveform**  A time-varying sequence of logic HIGHs and LOWs that repeats over a specified period of time.

**Positional notation**  A system of writing numbers in which the value of a digit depends not only on the digit, but also on its placement within a number.

**Positive logic** A system in which logic LOW represents binary digit 0 and logic HIGH represents binary digit 1.

**Pulse** A momentary variation of voltage from one logic level to the opposite level and back again.

**Pulse width** $(t_w)$ Elapsed time from the 50% point of the leading edge of a pulse to the 50% point of the trailing edge.

**Rise time** $(t_r)$ Elapsed time from the 10% point to the 90% point of the rising edge of a pulse.

**Rising edge** The part of a pulse where the logic level is in transition from a LOW to a HIGH.

**Time HIGH** $(t_h)$ Time during one period that a waveform is in the HIGH state. Unit: seconds (s).

**Time LOW** $(t_l)$ Time during one period that a waveform is in the LOW state. Unit: seconds (s).

**Trailing edge** The edge of a pulse that occurs latest in time.

## ● P R O B L E M S

### Section 1.1 Digital Versus Analog Electronics

**1.1** You have two voltmeters, one digital and one analog. The digital meter has a three-digit display. The analog meter displays a voltage by the angular position of needle, much like an automotive speedometer. Briefly state the difference between the meter displays when each meter measures an exact voltage of 15.34 V.

**1.2** What is the difference between the way an analog and a digital voltmeter displays a voltage in terms of the values each can represent?

**1.3** Which of the following quantities is analog in nature and which digital? Explain your answers.

   **a.** Water temperature at the beach

   **b.** Weight of a bucket of sand

   **c.** Grains of sand in a bucket

   **d.** Waves hitting the beach in one hour

   **e.** Height of a wave

   **f.** People in a square mile

### Section 1.3 The Binary Number System

**1.4** Calculate the decimal values of each of the following binary numbers:

   **a.** 100     **f.** 11101

   **b.** 1000   **g.** 111011

   **c.** 11001  **h.** 1011101

   **d.** 110     **i.** 100001

   **e.** 10101  **j.** 10111001

**1.5** Translate each of the following combinations of HIGH (H) and LOW (L) logic levels to binary numbers using positive logic:

   **a.** H H L H

   **b.** L H L H

   **c.** H L H L

   **d.** L L L H

   **e.** H L L L

**1.6** List the sequence of binary numbers from 101 to 1000.

**1.7** List the sequence of binary numbers from 10000 to 11111.

**1.8** What is the decimal value of the most significant bit for the numbers in Problem 1.7?

### Section 1.4 Digital Waveforms

**1.9** Calculate the time LOW, time HIGH, period, frequency, and percent duty cycle for the waveforms shown in Figure 1.12. How are the waveforms similar? How do they differ?

**1.10** Which of the waveforms in Figure 1.13 are periodic and which are aperiodic? Explain your answers.

**1.11** Sketch the pulse waveforms represented by the following strings of 0s and 1s. State which waveforms are periodic and which are aperiodic.

   **a.** 11001111001110110000000110110101

   **b.** 11100011100011100011100011100011

Figure 1.12
**Problem 1.9
Waveforms**

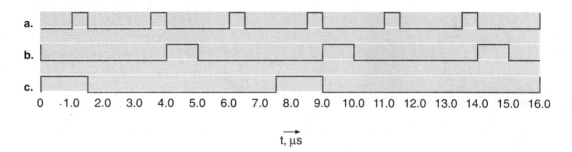

Figure 1.13
**Problem 1.10
Waveforms**

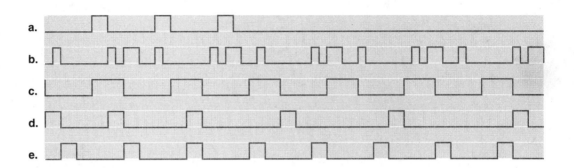

**c.** 1111111100000000111111111111111

**d.** 011001100110011001100110011001100110

**e.** 0111011010011010010110100111011101110

**1.12** Calculate the pulse width, rise time, and fall time of the pulse shown in Figure 1.14.

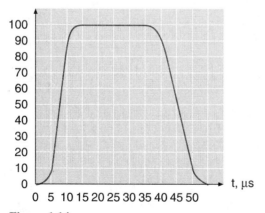

Figure 1.14
**Problem 1.12
Pulse**

**1.13** Repeat Problem 1.12 for the pulse shown in Figure 1.15.

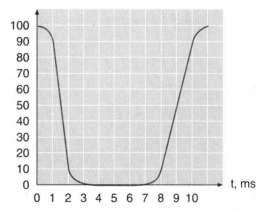

Figure 1.15
**Problem 1.13
Pulse**

## ▼ *Answers to Section Review Questions*

### Section 1.1

**1.1.** An analog audio system makes a direct copy of the recorded sound waves. A digital system stores the sound as a series of binary numbers.

### Section 1.3

**1.2.** 64; **1.3.** 128; **1.4.** 1010000, 1010001, 1010010, 1010011, 1010100, 1010101, 1010110, 1010111; **1.5.** 80, 81, 82, 83, 84, 85, 86, 87.

### Section 1.4

**1.6.** 50%; **1.7.** 0101010101010101; **1.8.** 0111011101110111.

# Logic Functions and Gates

2

● CHAPTER CONTENTS

**2.1** Basic Logic Functions
**2.2** Derived Logic Functions
**2.3** Truth Table of a Generalized Gate
**2.4** DeMorgan's Theorems and Gate Equivalence

**2.5** Enable and Inhibit Properties of Logic Gates
**2.6** Integrated Circuit Logic Gates
**2.7** Introduction to Troubleshooting

● CHAPTER OBJECTIVES

Upon successful completion of this chapter, you will be able to:

- Describe the basic logic functions: AND, OR, and NOT.
- Describe those logic functions derived from the basic ones: NAND, NOR, Exclusive OR, and Exclusive NOR.
- Explain the concept of active levels and identify active LOW and HIGH terminals of logic gates.
- Choose appropriate logic functions to solve simple design problems.
- Draw the truth table of any logic gate.
- Draw any logic gate, given its truth table.

- Draw the DeMorgan equivalent form of any logic gate.
- Determine when a logic gate will pass a digital waveform and when it will block the signal.
- Describe several types of integrated circuit packaging for digital logic gates.
- Describe and predict the effects of basic logic circuit faults.
- Explain the use of basic test equipment, such as a logic probe, logic pulser, and oscilloscope, to detect simple logic circuit faults.

● INTRODUCTION

All digital logic functions can be synthesized by various combinations of the three basic logic functions: AND, OR, and NOT. These so-called Boolean functions are the basis for all further study of combinational logic circuitry. (Combinational logic circuits are digital circuits whose outputs are functions of their inputs, regardless of the order the inputs are applied.) Standard circuits, called logic gates, have been developed for these and for more complex digital logic functions.

Logic gates can be represented in various forms. A standard set of distinctive-shape symbols has evolved as a universally understandable means of representing the various functions in a circuit. A useful pair of mathematical theorems, called DeMorgan's Theorems, enables us to draw these gate symbols in different ways to represent different aspects of the same function. A newer way of representing standard logic gates is outlined in IEEE/ANSI Standard 91-1984, a standard copublished by the Institute of Electrical and Electronic Engineers and the American National Standards Institute. It uses a set of symbols called rectangular-outline symbols.

Logic gates can be used as electronic switches to block or allow passage of digital waveforms. Each logic gate has a different set of properties for enabling (passing) or inhibiting (blocking) digital waveforms.

## 2.1

# Basic Logic Functions

> **Boolean variable**  *A variable having only two possible values, such as HIGH/LOW, 1/0, On/Off, or True/False.*
>
> **Boolean algebra**  *A system of algebra that operates on Boolean variables. The binary (two-state) nature of Boolean algebra makes it useful for analysis, simplification, and design of combinational logic circuits.*
>
> **Boolean expression**  *An algebraic expression made up of Boolean variables and operators, such as AND (•), OR (+), or NOT ( ̄). Also referred to as a* **Boolean function** *or a* **logic function.**
>
> **Logic gate**  *An electronic circuit that performs a Boolean algebraic function.*

At its simplest level, a digital circuit works by accepting logic 1s and 0s at one or more inputs and producing 1s or 0s at one or more outputs. A branch of mathematics known as **Boolean algebra** (named after 19th-century mathematician George Boole) describes the relation between inputs and outputs of a digital circuit. We call these input and output values **Boolean variables** and the functions **Boolean expressions, logic functions,** or **Boolean functions.** The distinguishing characteristic of these functions is that they are made up of variables and constants that can have only two possible values: 0 or 1.

All possible operations in Boolean algebra can be created from three basic logic functions: AND, OR, and NOT.[1] Electronic circuits that perform these logic functions are called **logic gates.** When we are analyzing or designing a digital circuit, we usually don't concern ourselves with the actual circuitry of the logic gates, but treat them as black boxes that perform specified logic functions. We can think of each variable in a logic function as a circuit input and the whole function as a circuit output.

In addition to gates for the three basic functions, there are also gates for compound functions that are derived from the basic ones. NAND gates combine the NOT and AND

---

[1]Words in uppercase letters represent either logic functions (AND, OR, NOT) or logic levels (HIGH, LOW). The same words in lowercase letters represent their conventional nontechnical meanings.

functions in a single circuit. Similarly, NOR gates combine the NOT and OR functions. Gates for more complex functions, such as Exclusive OR and Exclusive NOR, also exist. We will examine all these devices later in the chapter.

## NOT, AND, and OR Functions

> **Truth table**  *A list of all possible input values to a digital circuit, listed in ascending binary order, and the output response for each input combination.*
>
> **Inverter**  *Also called a NOT gate or an inverting buffer. A logic gate that changes its input logic level to the opposite state.*
>
> **Bubble**  *A small circle indicating logical inversion on a circuit symbol.*
>
> **Distinctive-shape symbols**  *Graphic symbols for logic circuits that show the function of each type of gate by a special shape.*
>
> **IEEE/ANSI Standard 91-1984**  *A standard format for drawing logic circuit symbols as rectangles with logic functions shown by a standard notation inside the rectangle for each device.*
>
> **Rectangular-outline symbols**  *Rectangular logic gate symbols that conform to IEEE/ANSI Standard 91-1984.*
>
> **Qualifying symbol**  *A symbol in IEEE/ANSI logic circuit notation, placed in the top center of a rectangular symbol, that shows the function of a logic gate. Some of the qualifying symbols include: 1 = "buffer"; & = "AND"; ≥1 = "OR"; =1 = "Exclusive OR."*
>
> **Buffer**  *An amplifier that acts as a logic circuit. Its output can be inverting or noninverting.*

**Table 2.1** NOT Function Truth Table

| A | Y |
|---|---|
| 0 | 1 |
| 1 | 0 |

### NOT Function

The NOT function, the simplest logic function, has one input and one output. The input can be either HIGH or LOW (1 or 0), and the output is always the opposite logic level. We can show these values in a **truth table,** a list of all possible input values and the output resulting from each one. Table 2.1 shows a truth table for a NOT function.

The NOT function is represented algebraically by the Boolean expression:

$$Y = \overline{A}$$

This is pronounced *"Y equals NOT A"* or *"Y equals A bar."* We can also say *"Y is the complement of A."*

The circuit that produces the NOT function is called the NOT gate or, more usually, the **inverter.** Several possible symbols for the inverter, all performing the same logic function, are shown in Figure 2.1.

The symbols shown in Figure 2.1a are the standard **distinctive-shape symbols** for the inverter. The triangle represents an amplifier circuit, and the **bubble** (the small circle on the input or output) represents inversion. There are two symbols because sometimes it is convenient to show the inversion at the input and sometimes it is convenient to show it at the output.

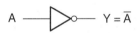

**a. Distinctive-shape**

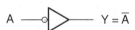

**b. Rectangular-outline**
(IEEE std. 91-1984)

**Figure 2.1**

**Inverter Symbols**

**Table 2.2** 2-input AND
Function Truth Table

| A | B | Y |
|---|---|---|
| 0 | 0 | 0 |
| 0 | 1 | 0 |
| 1 | 0 | 0 |
| 1 | 1 | 1 |

**a. Distinctive-shape**

**b. Rectangular-outline**

**Figure 2.2**

**2-Input AND Gate Symbols**

**Table 2.3** 3-input AND
Function Truth Table

| A | B | C | Y |
|---|---|---|---|
| 0 | 0 | 0 | 0 |
| 0 | 0 | 1 | 0 |
| 0 | 1 | 0 | 0 |
| 0 | 1 | 1 | 0 |
| 1 | 0 | 0 | 0 |
| 1 | 0 | 1 | 0 |
| 1 | 1 | 0 | 0 |
| 1 | 1 | 1 | 1 |

A
B
C — $Y = ABC$

**a. Distinctive-shape**

A
B   &
C — $Y = ABC$

**b. Rectangular-outline**

**Figure 2.3**

**3-Input AND Gate Symbols**

Figure 2.1b shows the **rectangular-outline** inverter symbol specified by **IEEE/ ANSI Standard 91-1984.** This standard is the latest attempt to unify logic circuit symbology. It is most useful for specifying the symbols for the more complex digital devices known as medium- and large-scale integration (MSI and LSI) devices. We will show the basic gates in both distinctive-shape and rectangular-outline symbols, although most examples will use the distinctive-shape symbols.

The "1" in the top center of the IEEE symbol is a **qualifying symbol,** indicating the logic gate function. In this case, it shows that the circuit is a **buffer,** a circuit whose output depends on only one input. The arrows at the input and output of the two IEEE symbols show inversion, like the bubbles in the distinctive-shape symbols.

## AND Function

> **AND gate** *A logic circuit whose output is HIGH when all inputs* (e.g., *A* AND *B* AND *C*) *are HIGH.*
>
> **Logical product** *AND function.*

The AND function combines two or more input variables so that the output is HIGH only if *all* the inputs are HIGH. The truth table for a 2-input AND function is shown in Table 2.2.

Algebraically, this is written:

$$Y = A \cdot B$$

Pronounce this expression *"Y equals A* AND *B."* The AND function is equivalent to multiplication in linear algebra and thus is sometimes called the **logical product.** The dot between variables may or may not be written, so it is equally correct to write $Y = AB$. The logic circuit symbol for an **AND gate** is shown in Figure 2.2 in both distinctive-shape and IEEE/ANSI rectangular-outline form. The qualifying symbol in IEEE/ANSI notation is the ampersand (&).

Table 2.3 shows the truth table for a 3-input AND function. Each of the three inputs can have two different values, which means the inputs can be combined in $2^3 = 8$ different ways. In general, *n* binary (i.e., two-valued) variables can be combined in $2^n$ ways.

Figure 2.3 shows the logic symbols for the device. The output is HIGH only when all inputs are HIGH.

## OR Function

> **OR gate** *A logic circuit whose output is HIGH when at least one input (e.g., A* OR *B* OR *C) is HIGH.*
>
> **Logical sum** *OR function.*

The OR function combines two or more input variables in such a way as to make the output variable HIGH if *at least one* input is HIGH. Table 2.4 gives the truth table for the 2-input OR function.

Table 2.4  2-input OR
Function Truth Table

| A | B | Y |
|---|---|---|
| 0 | 0 | 0 |
| 0 | 1 | 1 |
| 1 | 0 | 1 |
| 1 | 1 | 1 |

**a. Distinctive-shape**

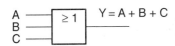

**b. Rectangular-outline**

**Figure 2.4**

**2-Input OR Gate Symbols**

Table 2.5  3-input OR
Function Truth Table

| A | B | C | Y |
|---|---|---|---|
| 0 | 0 | 0 | 0 |
| 0 | 0 | 1 | 1 |
| 0 | 1 | 0 | 1 |
| 0 | 1 | 1 | 1 |
| 1 | 0 | 0 | 1 |
| 1 | 0 | 1 | 1 |
| 1 | 1 | 0 | 1 |
| 1 | 1 | 1 | 1 |

A —⟩ Y = A + B + C
B
C

**a. Distinctive-shape**

A —[ ≥ 1 ]— Y = A + B + C
B
C

**b. Rectangular-outline**

**Figure 2.5**

**3-Input OR Gate Symbols**

The algebraic expression for the OR function is:

$$Y = A + B$$

which is pronounced *"Y equals A OR B."* This is similar to the arithmetic addition function, but it is not the same. The last line of the truth table tells us that $1 + 1 = 1$ (pronounced "1 OR 1 equals 1"), which is not what we would expect in standard arithmetic. The similarity to the addition function leads to the name **logical sum.** (This is different from the "arithmetic sum," where, of course, $1 + 1$ *does not* equal 1.)

Figure 2.4 shows the logic circuit symbols for an **OR gate.** The qualifying symbol for the OR function in IEEE/ANSI notation is "≥1," which tells us that *one or more* inputs must be HIGH to make the output HIGH.

Like AND gates, OR gates can have several inputs, such as the 3-input OR gates shown in Figure 2.5. Table 2.5 shows the truth table for this gate. Again, three inputs can be combined in eight different ways. The output is HIGH when at least one input is HIGH.

## Active Levels

**Active level**  *A logic level defined as the "ON" state for a particular circuit input or output. The active level can be either HIGH or LOW.*

**Active HIGH**  *An active-HIGH terminal is considered "ON" when it is in the logic HIGH state. Indicated by the absence of a bubble at the terminal in distinctive-shape symbols.*

**Active LOW**  *An active-LOW terminal is considered "ON" when it is in the logic LOW state. Indicated by a bubble at the terminal in distinctive-shape symbols.*

An **active level** of a gate input or output is the logic level, either HIGH or LOW, of the terminal when it is performing its designated function. An **active LOW** is shown by a bubble or an arrow symbol on the affected terminal. If there is no bubble or arrow, we assume the terminal is **active HIGH.**

The AND function has active-HIGH inputs and an active-HIGH output. To make the output HIGH, inputs A AND B must *both* be HIGH. The gate performs its designated function only when *all* inputs are HIGH.

The OR gate requires input A OR input B to be HIGH for its output to be HIGH. The HIGH active levels are shown by the absence of bubbles or arrows on the terminals.

**Active Levels of AND and OR Functions**

AND function: *All* inputs must be HIGH to make the output HIGH.
OR function: *At least one* input must be HIGH to make the output HIGH.

Active levels are useful in the analysis of digital circuits. They will tell us what logic level is required to turn a circuit input ON and what response to expect when the output is ON.

**EXAMPLE 2.1**

A bank has set up a new investment program, and it wants to offer this program first to people over 55 years of age who have had accounts with the bank for more than 20 years. Somebody at the bank writes a computer program that looks through the bank's customer file for each of these conditions and arranges to print the customer's name and address if both these conditions are met. The program has three variables assigned as follows:

$A = 1$ if the customer is over 55 years old,

$B = 1$ if the customer has had an account at the bank for more than 20 years, and

$Y = 1$ means "print the customer's name and address."

Use a logic function to accomplish the above task and describe the function in terms of its active levels.

**Solution**

Variable $A$ is HIGH when it performs its designated function: to decide whether a customer is over 55 years old. Thus, $A$ is active HIGH. If the customer is 55 or younger, $A$ is LOW, which is its inactive state. Variables $B$ and $Y$ are also active HIGH since they are HIGH when they are performing their designated functions.

$Y$ is HIGH only if both $A$ and $B$ are HIGH, that is, if both conditions for eligibility (55+ years old, account open for 20+ years) are met.

Table 2.6 shows a truth table that describes this situation. It is exactly the same as the AND function truth table and equation given earlier.

The Boolean expression for this application is $Y = A \cdot B$, the AND logic function.

One way of activating the printer in this example would be to get the computer to set an $A$ output to logic HIGH and a $B$ output to logic HIGH when the appropriate conditions were encountered. These outputs could be inputs to an AND gate whose output could activate the printer. Figure 2.6 shows a simplified block diagram.

**Table 2.6** Truth Table for Example 2.1

| A | B | Y |
|---|---|---|
| 0 | 0 | 0 |
| 0 | 1 | 0 |
| 1 | 0 | 0 |
| 1 | 1 | 1 |

**Figure 2.6**
**AND Gate Application**

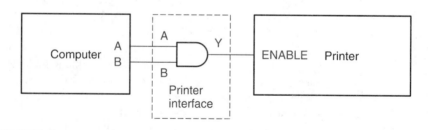

A truth table and Boolean expression are just compact substitutes for words describing some kind of decision process. Instead of saying, "Print a customer's name and address if the customer is over 55 years old AND has had an account for more than 20 years," we just say $Y = A \cdot B$.

---

**Section Review Problem for Section 2.1**

A 4-input gate has input variables $A$, $B$, $C$, and $D$ and output $Y$. Write a descriptive sentence for the active output state(s) if the gate is

2.1. AND;

2.2. OR.

## 2.2

# Derived Logic Functions

**NAND gate**  *A logic circuit whose output is LOW when all inputs are HIGH.*

**NOR gate**  *A logic circuit whose output is LOW when at least one input is HIGH.*

**Exclusive OR gate**  *A 2-input logic circuit whose output is HIGH when one input (but not both) is HIGH.*

**Exclusive NOR gate**  *A 2-input logic circuit whose output is the complement of an Exclusive OR gate.*

**Coincidence gate**  *An Exclusive NOR gate.*

The basic logic functions, AND, OR, and NOT, can be combined to make any other logic function. Special logic gates exist for several of the most common of these derived functions. In fact, for reasons we will discover later, two of these derived-function gates, NAND and NOR, are the most common of all gates, and *each* can be used to create any logic function.

## NAND and NOR Functions

**Table 2.7**  NAND Function Truth Table

| A | B | Y |
|---|---|---|
| 0 | 0 | 1 |
| 0 | 1 | 1 |
| 1 | 0 | 1 |
| 1 | 1 | 0 |

The names NAND and NOR are contractions of NOT AND and NOT OR, respectively. The NAND is generated by inverting the output of an AND function. The symbols for the **NAND gate** and its equivalent circuit are shown in Figure 2.7.

**a. Distinctive-shape**      **b. Rectangular-outline**      **c. Equivalent circuit**

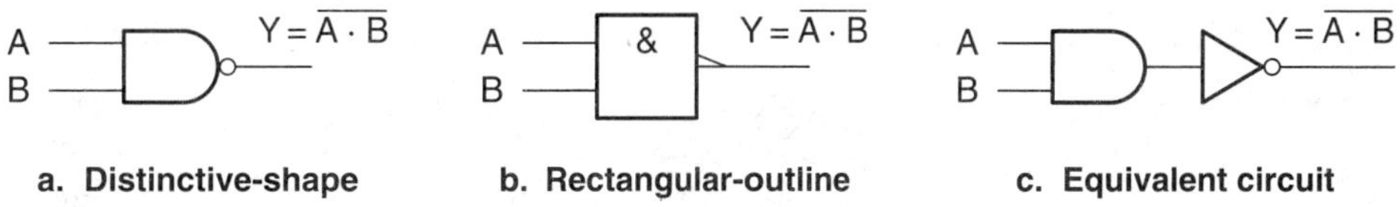

**Figure 2.7**
**NAND Gate Symbols**

The algebraic expression for the NAND function is:

$$Y = \overline{A \cdot B}$$

The entire function is inverted because the bubble is on the NAND gate output.

Table 2.7 shows the NAND gate truth table. The output is LOW when $A$ AND $B$ are HIGH.

We can generate the NOR function by inverting the output of an OR gate. The NOR function truth table is shown in Table 2.8. The truth table tells us that the output is LOW when $A$ OR $B$ is HIGH.

**Table 2.8**  NOR Function Truth Table

| A | B | Y |
|---|---|---|
| 0 | 0 | 1 |
| 0 | 1 | 0 |
| 1 | 0 | 0 |
| 1 | 1 | 0 |

Figure 2.8 shows the logic symbols for the **NOR gate.**

The algebraic expression for the NOR function is:

$$Y = \overline{A + B}$$

**Figure 2.8**
**NOR Gate Symbols**

| a. Distinctive-shape | b. Rectangular-outline | c. Equivalent circuit |

$Y = \overline{A + B}$   $Y = \overline{A + B}$   $Y = \overline{A + B}$

The entire function is inverted because the bubble is on the gate output.

---

**Active Levels of NAND and NOR Functions**

NAND Function: *All* inputs must be HIGH to make the output LOW.
NOR Function: *At least one* input must be HIGH to make the output LOW.

---

We know that the outputs of both gates are active LOW because of the bubbles on the output terminals. The inputs are active HIGH because there are no bubbles on the input terminals.

---

**EXAMPLE 2.2**

Modify the problem of Example 2.1 as follows. When the conditions for printout are satisfied (i.e., the customer file shows the customer is over 55 and has had an account for more than 20 years), the output variable *Y* activates the printer by going LOW.

In other words, we have specified that the printer interface output must now be *active LOW.* Find the truth table for this new function. What logic function does this represent?

**Solution**   As in Example 2.1, both input conditions must be satisfied (both HIGH) to activate the interface output. The output of the interface is LOW only if *A* AND *B* are HIGH. This yields the same truth table as that of Table 2.7, the NAND function table.

---

**EXAMPLE 2.3**

For safety, a punch press requires two-handed operation. Figure 2.9 shows the block diagram of a possible control system for the press. When either one or both of two hand switches is released, the switch opens. A switch-monitoring circuit sees a HIGH input and produces a HIGH output. This HIGH output energizes a relay, which cuts off power to the press.

Decide which logic function is best suited to implement the switch-control logic. Draw the logic symbol for the gate required and write its truth table and Boolean expression.

**Figure 2.9**

**Example 2.3**
**Punch Press Block Diagram**

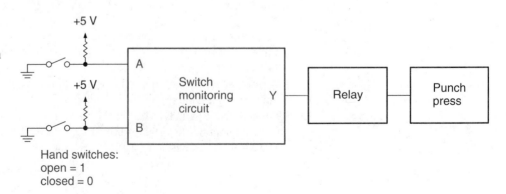

Hand switches:
open = 1
closed = 0

**Solution**

**Figure 2.10**

**Example 2.3
OR Gate**

The switch-monitoring circuit has an active-HIGH output and two active-HIGH inputs. A HIGH state at either input activates the output. This is the same as an OR function. The gate is shown in Figure 2.10. Table 2.9 shows the gate truth table.

The Boolean expression of the switch-monitoring circuit is given by $Y = A + B$.

**Table 2.9** OR Truth Table

| A | B | Y |
|---|---|---|
| 0 | 0 | 0 |
| 0 | 1 | 1 |
| 1 | 0 | 1 |
| 1 | 1 | 1 |

The hand switches in Example 2.3 are configured as simple logic switches. We will see this configuration often in future examples. When the switch is closed, it switches a ground (logic 0) to the input of a logic circuit (A or B in Figure 2.9). The resistor limits current from +5 V to ground, preventing damage to the power supply.

When the switch is open, no current flows from the +5 V supply to ground. Since the input resistance of any logic circuit is very high, very little current flows into the logic circuit input. By Ohm's Law, a small current through a resistor implies that the voltage is about the same at both ends of the resistor. Thus, an open logic switch supplies about +5 V (logic 1) to a circuit input.

> Here is a simple way to remember the logic switch operation: A closed switch connects a logic input to ground, which is logic LOW. An open switch is opposite, and therefore HIGH.

**EXAMPLE 2.4**

Repeat the problem in Example 2.3 for the case where the output of the switch-monitoring circuit produces a logic LOW to cut the power to the punch press.

**Solution**

**Figure 2.11**

**Example 2.4
NOR Gate**

The solution is similar to that of Example 2.3, except this time a HIGH at either input to the monitor circuit makes its output LOW. This is a NOR function (OR with an active-LOW output). Figure 2.11 shows the gate symbol, and the truth table is given in Table 2.10.

The Boolean expression of the monitor circuit is $Y = \overline{A + B}$.

**Table 2.10** NOR Truth Table

| A | B | Y |
|---|---|---|
| 0 | 0 | 1 |
| 0 | 1 | 0 |
| 1 | 0 | 0 |
| 1 | 1 | 0 |

## Multiple-Input NAND and NOR Gates

Table 2.11 shows the truth tables of the 3-input NAND and NOR functions. The logic circuit symbols for these gates are shown in Figure 2.12.

The truth tables of these gates can be generated by understanding the active levels of the gate inputs and outputs. The NAND has a LOW output if *all* inputs are HIGH (the last line in the truth table). The NOR has a LOW output if *at least one* input is HIGH (all but the first line in the truth table).

### *Exclusive OR and Exclusive NOR Functions*

The Exclusive OR function (sometimes abbreviated XOR) is a special case of the OR function. The output of a *2-input* XOR gate is HIGH when *one and only one* of the inputs is HIGH. (Multiple-input XOR circuits do not expand as simply as other functions. As we will see in a later chapter, an XOR output is HIGH when an *odd number* of inputs is HIGH.)

**Table 2.11**  3-input NAND and NOR
Function Truth Tables

| $A$ | $B$ | $C$ | $\overline{A \cdot B \cdot C}$ | $\overline{A + B + C}$ |
|---|---|---|---|---|
| 0 | 0 | 0 | 1 | 1 |
| 0 | 0 | 1 | 1 | 0 |
| 0 | 1 | 0 | 1 | 0 |
| 0 | 1 | 1 | 1 | 0 |
| 1 | 0 | 0 | 1 | 0 |
| 1 | 0 | 1 | 1 | 0 |
| 1 | 1 | 0 | 1 | 0 |
| 1 | 1 | 1 | 0 | 0 |

**Figure 2.12**

**3-input NAND and NOR
Gates**

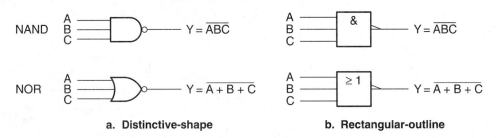

a. **Distinctive-shape**          b. **Rectangular-outline**

a. **Distinctive-shape**

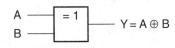

b. **Rectangular-outline**

**Figure 2.13**

**Exclusive OR Gate**

**Table 2.12**  Exclusive OR
Function Truth Table

| $A$ | $B$ | $Y$ |
|---|---|---|
| 0 | 0 | 0 |
| 0 | 1 | 1 |
| 1 | 0 | 1 |
| 1 | 1 | 0 |

Unlike the OR gate, which is sometimes called an Inclusive OR, a HIGH at both inputs makes the output LOW. (We could say that the case in which both inputs are HIGH is excluded.)

The gate symbol for the **Exclusive OR** gate is shown in Figure 2.13.

Table 2.12 shows the truth table for the XOR function.

Another way of looking at the Exclusive OR gate is that its output is HIGH when the inputs are different and LOW when they are the same. This is a useful property in some applications, such as error detection in digital communication systems. (Transmitted data can be compared with received data. If they are the same, no error has been detected.)

The XOR function is expressed algebraically as:

$$Y = A \oplus B$$

The Exclusive NOR function is the inversion of the Exclusive OR function and shares some of the same properties. The symbol, shown in Figure 2.14, is an XOR gate with a bubble on the output, implying that the entire function is inverted. Table 2.13 shows the Exclusive NOR truth table.

The algebraic expression for the Exclusive NOR function is:

$$Y = \overline{A \oplus B}$$

The output of the **Exclusive NOR gate** is HIGH when the inputs are the same and LOW when they are different. For this reason, the XNOR gate is also called a **coincidence gate.** This same/different property is similar to that of the Exclusive OR gate, only opposite in sense. Many of the applications that make use of this property can use either the XOR or the XNOR gate.

**Figure 2.14**
**Exclusive NOR Gate**

A
B
Y = $\overline{A \oplus B}$

**a. Distinctive-shape**

A
B
= 1
Y = $\overline{A \oplus B}$

**b. Rectangular-outline**

**Table 2.13** Exclusive NOR Function Truth Table

| A | B | Y |
|---|---|---|
| 0 | 0 | 1 |
| 0 | 1 | 0 |
| 1 | 0 | 0 |
| 1 | 1 | 1 |

---

**Section Review Problems for Section 2.2**

The output of a logic gate turns on a light when it is HIGH. The gate has two inputs, each of which is connected to a logic switch.

2.3.   What type of gate will turn on the light when the switches are in opposite positions?

2.4.   Which gate will turn off the light only when both switches are HIGH?

2.5.   What type of gate turns on the light only when both switches are LOW?

2.6.   Which gate turns on the light when the switches are in the same position?

---

## 2.3

# Truth Table of a Generalized Gate

If we are going to design or troubleshoot digital circuitry, it is essential that we have the truth table of any logic gate at our fingertips; we can't tell whether a gate is working if we don't know what it is supposed to do.

On the other hand, there is no point in memorizing the truth table for every gate we are likely to encounter. It is more useful to develop a system that will tell us how to get the truth table of any AND- or OR-shaped gate. We can do this by asking three questions about the gate:

1. What is its shape? (AND/OR?)
2. What are the active levels of the gate inputs? (active HIGH or active LOW?)
3. What is the active level of the gate output? (active HIGH or active LOW?)

If the gate is AND-shaped, the function implied is, "*All* inputs must be active to make the output active." If the gate is OR-shaped, the function implied is, "*At least one* input must be active to make the output active." The active levels are shown by the presence (active LOW) or absence (active HIGH) of a bubble.

An example of an AND-shaped gate and of an OR-shaped gate are shown in Figures 2.15 and 2.16. The truth tables for both sets of gates are shown in Tables 2.14 and 2.15. Let us see how the truth table for each gate is developed.

In Figure 2.15, *A* AND *B* must *both* be LOW to make *Y* HIGH. This describes the first line of the truth table and is thus the active state of the gate output. The remaining lines show the gate output in the opposite state (LOW).

**Figure 2.15**
**AND-Shaped Gate**

A —○⌐\
B —○⌐⟩— $\overline{A} \cdot \overline{B}$

**a. AND-shaped gate**

A —▷○— $\overline{A}$\
B —▷○— $\overline{B}$ —⟩— $\overline{A} \cdot \overline{B}$

**b. Logic equivalent**

**Table 2.14** Truth Table for Gate in Figure 2.15

| A | B | Y |
|---|---|---|
| 0 | 0 | 1* |
| 0 | 1 | 0 |
| 1 | 0 | 0 |
| 1 | 1 | 0 |

*Output is in its active state

**Figure 2.16**
**OR-Shaped Gate**

A —⟩○\
B —○⟩— $Y = \overline{\overline{A} + B}$

**a. OR-shaped gate**

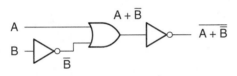

A —⟩— $A + \overline{B}$ —▷○— $\overline{A + \overline{B}}$\
B —▷○— $\overline{B}$

**b. Logic equivalent**

**Table 2.15** Truth Table for Gate in Figure 2.16

| A | B | Y |
|---|---|---|
| 0 | 0 | 0* |
| 0 | 1 | 1 |
| 1 | 0 | 0* |
| 1 | 1 | 0* |

*Output is in its active state

In Figure 2.16, *A* must be HIGH OR *B* must be LOW to make *Y* LOW. *At least one* of these conditions is satisfied on the first, third, and fourth lines of Table 2.15. The remaining state is inactive (HIGH).

The output active states are marked on each truth table with an asterisk. Note that the AND-shaped gate has only one *active* state and the OR-shaped gate has only one *inactive* state. This is generally true for all AND- and OR-shaped gates.

**EXAMPLE 2.5**

Write a descriptive statement of the operation for each of the gates in Figure 2.17. Find the truth table and Boolean expression for *Y* in each case.

**Figure 2.17**
**Example 2.5**
**Logic Gates**

A —⟩○— Y    A —○⟩○— Y    A —○⟩○— Y    A —○⟩○— Y
B —          B —○          B —○         B —

a.          b.          c.          d.

**Solutions**    For each gate, examine the shape, input active levels, and output active levels.

    **a. 1.** Shape: AND (= "all")

      **2.** Inputs: active HIGH

      **3.** Output: active LOW

Descriptive statement: "All[1] inputs must be HIGH[2] to make the output LOW[3]."

**Table 2.16** Truth Table for Gate in Figure 2.17a

| A | B | Y |
|---|---|---|
| 0 | 0 | 1 |
| 0 | 1 | 1 |
| 1 | 0 | 1 |
| 1 | 1 | 0* |

*Output is in its active state

**Table 2.17** Truth Table for Gate in Figure 2.17b

| A | B | Y |
|---|---|---|
| 0 | 0 | 1* |
| 0 | 1 | 0 |
| 1 | 0 | 0 |
| 1 | 1 | 0 |

*Output is in its active state.

The resulting truth table is shown in Table 2.16.

Boolean expression: $Y = \overline{A \cdot B}$

(The function is inverted because the gate output is active LOW.)

**b. 1.** Shape: AND (= "all")
   **2.** Inputs: active LOW
   **3.** Output: active HIGH

Descriptive statement: "All[1] inputs must be LOW[2] to make the output HIGH[3]."

The resulting truth table is shown in Table 2.17.

Boolean expression: $Y = \overline{A} \cdot \overline{B}$

(Variables $A$ and $B$ are inverted because the inputs are active LOW.)

**c. 1.** Shape: OR (= "at least one")
   **2.** Inputs: active LOW
   **3.** Output: active HIGH

Descriptive statement: "At least one[1] input must be LOW[2] to make the output HIGH[3]."

The resulting truth table is shown in Table 2.18.

Boolean expression: $Y = \overline{A} + \overline{B}$

**d. 1.** Shape: AND (= "all" or "both")
   **2.** Inputs: $A$ is active LOW
        $B$ is active HIGH
   **3.** Output: active HIGH

Descriptive statement: "Both[1] $A$ must be LOW AND $B$ must be HIGH[2] to make the output HIGH[3]."

The resulting truth table is shown in Table 2.19.

Boolean expression: $Y = \overline{A} \cdot B$

(Variable $A$ is inverted because its input is active LOW. $B$ is not inverted because it is active HIGH.)

**Table 2.18** Truth Table for Gate in Figure 2.17c

| A | B | Y |
|---|---|---|
| 0 | 0 | 1* |
| 0 | 1 | 1* |
| 1 | 0 | 1* |
| 1 | 1 | 0 |

*Output is in its active state.

**Table 2.19** Truth Table for Gate in Figure 2.17d

| A | B | Y |
|---|---|---|
| 0 | 0 | 0 |
| 0 | 1 | 1* |
| 1 | 0 | 0 |
| 1 | 1 | 0 |

*Output is in its active state.

## Generalized Multiple-Input Gates

Logic gates with more than two inputs can also be analyzed by the system described in the previous section.

**EXAMPLE 2.6**

Determine the truth table and output Boolean expression for each of the 3-input gates in Figure 2.18.

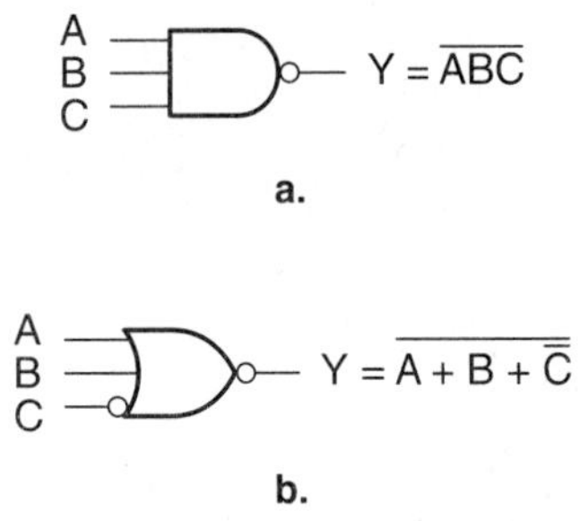

**Table 2.20** Truth Table for Gate in Figure 2.18a

| A | B | C | Y |
|---|---|---|---|
| 0 | 0 | 0 | 1 |
| 0 | 0 | 1 | 1 |
| 0 | 1 | 0 | 1 |
| 0 | 1 | 1 | 1 |
| 1 | 0 | 0 | 1 |
| 1 | 0 | 1 | 1 |
| 1 | 1 | 0 | 1 |
| 1 | 1 | 1 | 0* |

*Output is in its active state.

$Y = \overline{ABC}$

a.

$Y = \overline{A + B + \overline{C}}$

b.

**Figure 2.18**

**Example 2.6 Logic Gates**

**Table 2.21** Truth Table for Gate in Figure 2.18b

| A | B | C | Y |
|---|---|---|---|
| 0 | 0 | 0 | 0* |
| 0 | 0 | 1 | 1 |
| 0 | 1 | 0 | 0* |
| 0 | 1 | 1 | 0* |
| 1 | 0 | 0 | 0* |
| 1 | 0 | 1 | 0* |
| 1 | 1 | 0 | 0* |
| 1 | 1 | 1 | 0* |

*Output is in its active state.

### Solutions

a. 1. Shape: AND (= "all")

 2. Inputs: active HIGH

 3. Output: active LOW

Descriptive statement: "All[1] inputs must be HIGH[2] to make the output LOW[3]."

The resulting truth table is shown in Table 2.20.

Boolean expression: $Y = \overline{A \cdot B \cdot C}$

b. 1. Shape: OR (= "at least one")

 2. Inputs: $A$ is active HIGH

 $B$ is active HIGH

 $C$ is active LOW

 3. Output: active LOW

Descriptive statement: "At least one[1] of the following must be true to make the output LOW[3]: $A$ is HIGH OR $B$ is HIGH OR $C$ is LOW[2]."

The resulting truth table is shown in Table 2.21.

Boolean expression: $Y = \overline{A + B + \overline{C}}$

(Note that we invert variable $C$ (bubble on input $C$) and the entire function (bubble on output).)

---

**Section Review Problem for Section 2.3**

2.7. The gate symbol in Figure 2.18a changes so that input $A$ is active LOW. All other active levels remain unchanged. Write a descriptive statement, the output Boolean expression, and the truth table of the new gate.

## 2.4

# DeMorgan's Theorems and Gate Equivalence

**DeMorgan's Theorems** *Two theorems in Boolean algebra that allow us to transform any gate from an AND-shaped to an OR-shaped gate and vice versa.*

**DeMorgan equivalent forms** *Two gate symbols, one AND-shaped and one OR-shaped, that are equivalent according to DeMorgan's Theorems.*

**Table 2.22** NAND Truth Table

| A | B | Y |
|---|---|---|
| 0 | 0 | 1 |
| 0 | 1 | 1 |
| 1 | 0 | 1 |
| 1 | 1 | 0 |

Recall the truth table (repeated in Table 2.22) and description of a NAND gate.

NAND function: "*All* inputs must be HIGH to make the output LOW."

Another way of saying this is, "*At least one* input must be LOW to make the output HIGH." (Prove this to yourself by looking at the first three lines of Table 2.22.) We can implement this function with an OR gate that has two active-LOW inputs and an active-HIGH output, as shown in Figure 2.19.

The algebraic expression for the OR-shaped gate is:

$$Y = \overline{A} + \overline{B}$$

The input variables, not the function, are complemented, because the inputs to the gate are active LOW and the output is active HIGH. The gates shown in Figure 2.19 are called **DeMorgan equivalent forms.**

The AND-shaped and the OR-shaped gates are the same gate. Both gates have the same truth table, but each symbol represents a different aspect or a different way of looking at the gate. We can extend this observation to state that *any* gate (except XOR/XNOR) has two equivalent forms, one AND, one OR.

**Figure 2.19**

**NAND Gate and DeMorgan Equivalent**

A
B ——[ $\overline{A \cdot B}$ ]  $\Longleftrightarrow$  A
B ——[ $\overline{A} + \overline{B}$ ]

**Table 2.23** Attributes of a NAND Gate and Its DeMorgan Equivalent Form

| Shape | AND | OR |
|---|---|---|
| **Inputs** | Active HIGH | Active LOW |
| **Output** | Active LOW | Active HIGH |

Table 2.23 summarizes the attributes of the gates in Figure 2.19.

**EXAMPLE 2.7**

To open a bank vault door, an electric motor is turned on by a relay, which is activated when a logic LOW is applied to its coil from a controller output. The output of the controller, shown in Figure 2.20, goes LOW when a HIGH is applied to one input from a key-operated switch AND a HIGH is applied to another input from a timer.

**Figure 2.20**
**Example 2.7**
**Bank Vault Controller**

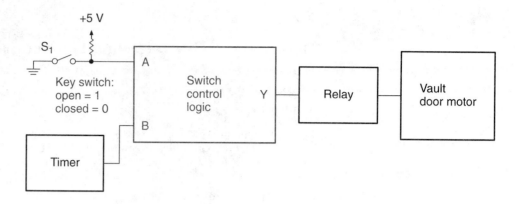

Draw the logic gate that represents the operation of the controller. Write the truth table of the gate.

**Solution** The output is active LOW. *All* inputs must be HIGH to make the output LOW. This is a NAND function. Figure 2.21 shows the NAND gate.

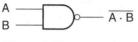

**Figure 2.21**
**Example 2.7**
**NAND Gate**

Table 2.24  Truth Table for Gate in Figure 2.21

| *A* | *B* | *Y* |
|-----|-----|-----|
| 0 | 0 | 1 |
| 0 | 1 | 1 |
| 1 | 0 | 1 |
| 1 | 1 | 0 |

The NAND truth table is shown in Table 2.24.

**EXAMPLE 2.8**

An electric lamp in a storeroom turns on when either one of two doors into the room is open. When a door opens, a switch closes and sends a LOW to the lamp control circuit, shown in Figure 2.22. The control circuit responds by making its output HIGH and turning on the lamp.

**Figure 2.22**
**Example 2.8**
**Lamp Control Circuit**

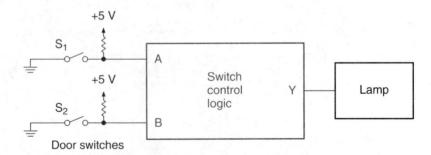

Draw the logic gate that represents the operation of the controller. Write the truth table of the gate. Compare the truth table with the one derived in Example 2.7.

**Solution** The controller output is active HIGH. The output goes HIGH when *at least one* input is LOW. This can be done with an OR-shaped gate having active-LOW inputs and an active-

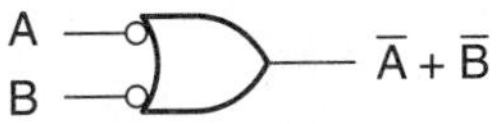

A ─o⊐ ─ $\overline{A} + \overline{B}$
B ─o

**Figure 2.23**

**Example 2.8**
**Logic Gate for Controller**

**Table 2.25** Truth Table for
Gate in Figure 2.23

| A | B | Y |
|---|---|---|
| 0 | 0 | 1 |
| 0 | 1 | 1 |
| 1 | 0 | 1 |
| 1 | 1 | 0 |

**Table 2.26** NOR Gate Truth
Table

| A | B | Y |
|---|---|---|
| 0 | 0 | 1 |
| 0 | 1 | 0 |
| 1 | 0 | 0 |
| 1 | 1 | 0 |

HIGH output, as shown in Figure 2.23. (This is the DeMorgan equivalent of a NAND gate.)

The truth table is the same as for the gate in Example 2.7 (see Table 2.25).

The equivalence of the two gates in the above examples implies that $\overline{A \cdot B} = \overline{A} + \overline{B}$. This algebraic statement is the first of **DeMorgan's Theorems.**

> The OR-shaped gate in Figures 2.19 and 2.23 is *not* the same as a NOR gate. (Compare its truth table to that of the NOR gate to find out why.) This implies that $\overline{A + B} \neq \overline{A} + \overline{B}$.

There is also a DeMorgan equivalent form of the NOR gate. Recall the truth table (Table 2.26) and description of the NOR gate.

NOR function: "*At least one* input must be HIGH to make the output LOW."

Another way of looking at it would be, "*All* inputs must be LOW for the output to be HIGH." This function can be achieved by using an AND-shaped gate ("all inputs") with active-LOW inputs and an active-HIGH output, as shown in Figure 2.24.

The output of the AND-shaped gate is expressed algebraically as:

$$Y = \overline{A} \cdot \overline{B}$$

which implies that $\overline{A + B} = \overline{A} \cdot \overline{B}$, the second of DeMorgan's Theorems. It is important to note that $\overline{A \cdot B} \neq \overline{A} \cdot \overline{B}$. (Prove this to yourself by comparing the NAND truth table to that of the AND-shaped DeMorgan equivalent of the NOR.)

**Figure 2.24**
**DeMorgan Equivalent**
**Forms of NOR Gate**

A ─o⊐ ─ $\overline{A} \cdot \overline{B}$   ⟺   A ─⊐o ─ $\overline{A + B}$
B ─o                         B

> DeMorgan's Theorems:  $\overline{A + B} = \overline{A} \cdot \overline{B}$
> $\overline{A \cdot B} = \overline{A} + \overline{B}$
> Inequalities:  $\overline{A \cdot B} \neq \overline{A} \cdot \overline{B}$
> $\overline{A + B} \neq \overline{A} + \overline{B}$
>    You can remember how to use DeMorgan's Theorems by a simple rhyme: "Break the line and change the sign."

The AND-shaped and OR-shaped versions of a gate have different, mutually exclusive active states on the same truth table. AND-shaped gates have one active output state. OR-shaped gates have only one output state that is not active.

## Changing a Gate to Its DeMorgan Equivalent

We can use our system of gate analysis outlined in Section 2.3 to change any gate to its DeMorgan equivalent. Just remember two words: *Change everything.*

For example, the NOR gate in Figure 2.24 has the following attributes:

Shape: OR

Inputs: active HIGH

Output: active LOW

Descriptive statement: "At least one input must be HIGH to make the output LOW."

When we apply the dictum "change everything," we get the following new attributes:

Shape: AND

Inputs: active LOW

Outputs: active HIGH

Descriptive statement: "All inputs must be LOW to make the output HIGH."

The two descriptive statements are different ways of saying the same thing, a sure indicator that you have the DeMorgan equivalent gate.

---

**EXAMPLE 2.9**

Convert each gate in Figure 2.25 to its DeMorgan equivalent form and write the Boolean expression for the new gate.

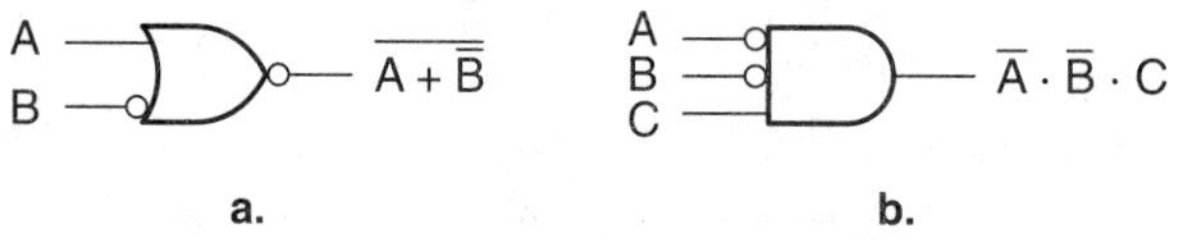

**Figure 2.25**
**Example 2.9**
**Logic Gates**

a.  $\overline{A + \overline{B}}$

b.  $\overline{A} \cdot \overline{B} \cdot C$

**Solution**   Change everything.

| a. | Original form: | DeMorgan equivalent form: |
|---|---|---|
| Shape | OR | AND |
| Inputs | $A$ = active HIGH | $A$ = active LOW |
| | $B$ = active LOW | $B$ = active HIGH |
| Output | Active LOW | Active HIGH |

The DeMorgan equivalent gate is shown in Figure 2.26a. Its Boolean expression is:

$$Y = \overline{A} \cdot B$$

| b. | Original form: | DeMorgan equivalent form: |
|---|---|---|
| Shape | AND | OR |
| Inputs | $A$ = active LOW | $A$ = active HIGH |
| | $B$ = active LOW | $B$ = active HIGH |
| | $C$ = active HIGH | $C$ = active LOW |
| Output | Active HIGH | Active LOW |

**Figure 2.26**

**Example 2.9**
**DeMorgan Equivalent Gates**

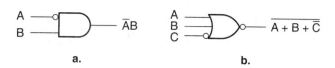

a.    b.

The DeMorgan equivalent gate is shown in Figure 2.26b. Its Boolean expression is:

$$Y = \overline{A + B + \overline{C}}$$

---

**Section Review Problem for Section 2.4**

2.8. The output of a gate is described by the following Boolean expression:

$$Y = \overline{A} + B + C + \overline{D}$$

Write the Boolean expression for the DeMorgan equivalent form of this gate.

---

### 2.5

# Enable and Inhibit Properties of Logic Gates

**Digital signal (or pulse waveform)** *A series of 0s and 1s plotted over time.*

**True form** *Not inverted.*

**Complement form** *Inverted.*

**Enable** *A logic gate is enabled if it allows a digital signal to pass from an input to the output in either true or complement form.*

**Inhibit (or disable)** *A logic gate is inhibited if it prevents a digital signal from passing from an input to the output.*

**In phase** *Two digital waveforms are in phase if they are always at the same logic level at the same time.*

**Out of phase** *Two digital waveforms are out of phase if they are always at opposite logic levels at any given time.*

In Chapter 1 (Example 1.4), we saw that a **digital signal** is just a string of bits (0s and 1s) generated over time. A major task of digital circuitry is the direction and control of such signals. Logic gates can be used to **enable** (pass) or **inhibit** (block) these signals. (The word "gate" gives a clue to this function; the gate can "open" to allow a signal through or "close" to block its passage.)

## AND and OR Gates

The simplest case of the enable and inhibit properties is that of an AND gate used to pass or block a logic signal. Figure 2.27 shows the output of an AND gate under different conditions of input A when a digital signal (an alternating string of 0s and 1s) is applied to input B.

**Figure 2.27**

**Enable/Inhibit Properties of an AND Gate**

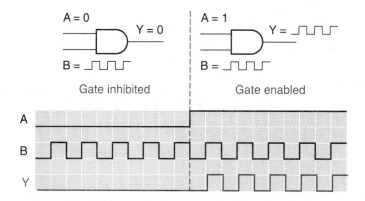

Recall the properties of an AND gate: both inputs must be HIGH to make the output HIGH. Thus, if input *A* is LOW, the output must always be LOW, regardless of the state of input *B*. The digital signal applied to *B* has no effect on the output, and we say that the gate is inhibited or disabled. This is shown in the first half of the timing diagram in Figure 2.27.

If *A* AND *B* are HIGH, the output is HIGH. When *A* is HIGH and *B* is LOW, the output is LOW. Thus, output *Y* is the same as input *B* if input *A* is HIGH; that is, *Y* and *B* are **in phase** with each other. The input waveform is passed to the output in **true form,** and we say the gate is enabled. The last half of the timing diagram in Figure 2.27 shows this waveform.

It is convenient to define terms for the *A* and *B* inputs. Since we apply a digital signal to *B*, we will call it the Signal input. Since input *A* controls whether or not the signal passes to the output, we will call it the Control input. These definitions are illustrated in Figure 2.28.

Each type of logic gate has a particular set of enable/inhibit properties that can be predicted by examining the truth table of the gate. Let us examine the truth table of the AND gate to see how the method works.

**Figure 2.28**

**Control and Signal Inputs of an AND Gate**

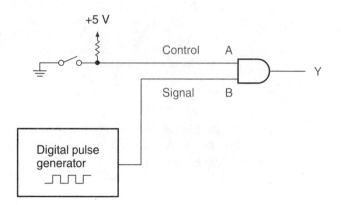

**Table 2.27**  AND Truth Table Showing Enable/Inhibit Properties

| A | B | Y | |
|---|---|---|---|
| 0 | 0 | 0 | (Y = 0) |
| 0 | 1 | 0 | Inhibit |
| 1 | 0 | 0 | (Y = B) |
| 1 | 1 | 1 | Enable |

Divide the truth table in half, as shown in Table 2.27. Since we have designated $A$ as the Control input, the top half of the truth table shows the inhibit function ($A = 0$), and the bottom half shows the enable function ($A = 1$). To determine the gate properties, we compare input $B$ (the Signal input) to the output in each half of the table.

*Inhibit mode:* If $A = 0$ and $B$ is pulsing ($B$ is continuously going back and forth between the first and second lines of the truth table), output $Y$ is always 0. Since the Signal input has no effect on the output, we say that the gate is disabled or inhibited.

*Enable mode:* If $A = 1$ and $B$ is pulsing ($B$ is going continuously between the third and fourth lines of the truth table), the output is the same as the Signal input. Since the Signal input affects the output, we say that the gate is enabled.

**EXAMPLE 2.10**

Use the method just described to draw the output waveform of an OR gate if the input waveforms of $A$ and $B$ are the same as in Figure 2.27. Indicate the enable and inhibit portions of the timing diagram.

**Solution**   Divide the OR gate truth table in half. Designate input $A$ the Control input and input $B$ the Signal input.

As shown in Table 2.28, when $A = 0$ and $B$ is pulsing, the output is the same as $B$ and the gate is enabled. When $A = 1$, the output is always HIGH. (At least one input HIGH makes the output HIGH.) Since $B$ has no effect on the output, the gate is inhibited. This is shown in Figure 2.29 in graphical form.

**Table 2.28**  OR Truth Table Showing Enable/Inhibit Properties

| A | B | Y | |
|---|---|---|---|
| 0 | 0 | 0 | (Y = B) |
| 0 | 1 | 1 | Enable |
| 1 | 0 | 1 | (Y = 1) |
| 1 | 1 | 1 | Inhibit |

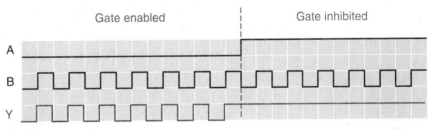

**Figure 2.29**
**Example 2.10**
**OR Gate Enable/Inhibit Waveform**

Example 2.10 shows that a gate can be in the inhibit state even if its output is HIGH. It is natural to think of the HIGH state as "ON," but this is not always the case. Enable or inhibit states are determined by the effect the Signal input has on the gate's output. If an input signal does not affect the gate output, the gate is inhibited. If the Signal input does affect the output, the gate is enabled.

## NAND and NOR Gates

When inverting gates, such as NAND and NOR, are enabled, they will invert an input signal before passing it to the gate output. In other words, they transmit the signal in **complement form.** Figures 2.30 and 2.31 show the output waveforms of a NAND and a NOR gate when a square waveform is applied to input $B$ and input $A$ acts as a Control input.

**Figure 2.30**

**Enable/Inhibit Properties of a NAND Gate**

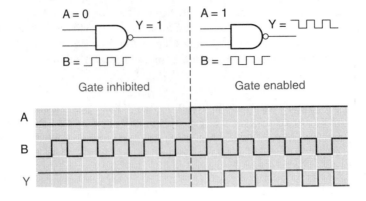

Gate inhibited     Gate enabled

**Figure 2.31**

**Enable/Inhibit Properties of a NOR Gate**

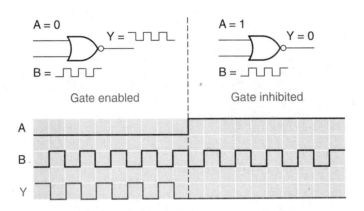

Gate enabled     Gate inhibited

**Table 2.29** NAND Truth Table Showing Enable/Inhibit Properties

| A | B | Y | |
|---|---|---|---|
| 0 | 0 | 1 | $(Y = 1)$ |
| 0 | 1 | 1 | Inhibit |
| 1 | 0 | 1 | $(Y = \overline{B})$ |
| 1 | 1 | 0 | Enable |

**Table 2.30** NOR Truth Table Showing Enable/Inhibit Properties

| A | B | Y | |
|---|---|---|---|
| 0 | 0 | 1 | $(Y = \overline{B})$ |
| 0 | 1 | 0 | Enable |
| 1 | 0 | 0 | $(Y = 0)$ |
| 1 | 1 | 0 | Inhibit |

The truth table for the NAND gate is shown in Table 2.29, divided in half to show the enable and inhibit properties of the gate.

Table 2.30 shows the NOR gate truth table, divided in half to show its enable and inhibit properties.

Figures 2.30 and 2.31 show that when the NAND and NOR gates are enabled, the Signal and output waveforms are opposite to one another; we say that they are **out of phase.**

Compare the enable/inhibit waveforms of the AND, OR, NAND, and NOR gates. Gates of the same shape are enabled by the same Control level. AND and NAND gates are enabled by a HIGH on the Control input and inhibited by a LOW. OR and NOR are the opposite. A HIGH Control input inhibits the OR/NOR; a LOW Control input enables the gate.

## Exclusive OR and Exclusive NOR Gates

*Neither the XOR nor the XNOR gate has an inhibit state.* The Control input on both of these gates acts only to determine whether the output waveform will be in or out of phase with the input signal. Figure 2.32 shows the dynamic properties of an XOR gate.

The truth table for the XOR gate, showing the gate's dynamic properties, is given in Table 2.31.

Notice that when $A = 0$, the output is in phase with $B$ and when $A = 1$, the output is out of phase with $B$. A useful application of this property is to use an XOR gate as a

**Figure 2.32**

**Dynamic Properties of an Exclusive OR Gate**

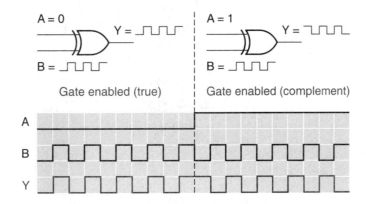

**Table 2.31** XOR Truth Table Showing Dynamic Properties

| A | B | Y | |
|---|---|---|---|
| 0 | 0 | 0 | $(Y = B)$ |
| 0 | 1 | 1 | Enable |
| 1 | 0 | 1 | $(Y = \overline{B})$ |
| 1 | 1 | 0 | Enable |

programmable inverter. When $A = 1$, the gate is an inverter; when $A = 0$, it is a noninverting buffer.

The XNOR gate has properties similar to the XOR gate. That is, an XNOR has no inhibit state, and the Control input switches the output in and out of phase with the Signal waveform, although not the same way as an XOR gate does. You will derive these properties in one of the end-of-chapter problems.

Table 2.32 summarizes the enable/inhibit properties of the six gates examined above.

**Table 2.32** Summary of Enable/Inhibit Properties

| Control | AND | OR | NAND | NOR | XOR | XNOR |
|---|---|---|---|---|---|---|
| $A = 0$ | $Y = 0$ | $Y = B$ | $Y = 1$ | $Y = \overline{B}$ | $Y = B$ | $Y = \overline{B}$ |
| $A = 1$ | $Y = B$ | $Y = 1$ | $Y = \overline{B}$ | $Y = 0$ | $Y = \overline{B}$ | $Y = B$ |

---

**Section Review Problem for Section 2.5**

2.9. Briefly explain why an AND gate is inhibited by a LOW Control input and an OR gate is inhibited by a HIGH Control input.

---

## 2.6

# Integrated Circuit Logic Gates

**Integrated circuit (IC)** *An electronic circuit having many components, such as transistors, diodes, resistors, and capacitors, in a single package.*

**Small-scale integration (SSI)** *An integrated circuit having 12 or fewer gates in one package.*

**Transistor-transistor logic (TTL)** *A family of logic devices used in digital circuits.*

**Complementary metal-oxide semiconductor (CMOS)** *A family of logic devices used in digital circuits.*

**Chip** *An integrated circuit. Specifically, a chip of silicon on which an integrated circuit is constructed.*

**Dual in-line package (DIP)** *A type of IC with two parallel rows of pins for the various circuit inputs and outputs.*

**Breadboard** *A circuit board for wiring temporary circuits, usually used for prototypes or laboratory work.*

**Surface-mount technology (SMT)** *A system of mounting and soldering integrated circuits on the surface of a circuit board.*

**Small outline (SO)** *An IC package similar to a DIP, but smaller, which is designed for automatic placement and soldering on the surface of a circuit board.*

**Ceramic leaded chip carrier (CLCC)** *A square IC package with leads on all four sides designed for surface mounting on a circuit board.*

**Plastic leaded chip carrier (PLCC)** *An IC package like the CLCC, but made of plastic.*

All the logic gates we have looked at so far are available in **integrated circuit** (IC) form. Most of these **small-scale integration** (SSI) functions are available in either **transistor-transistor logic** (TTL) or **complementary metal-oxide semiconductor** (CMOS) technologies. TTL and CMOS devices differ not in their logic functions, but in their construction and electrical characteristics.

TTL and CMOS **chips** are designated by an industry-standard numbering system. One of the most popular TTL families, Low Power Schottky TTL, is designated 74LS*XX*, where *XX* is a different number for each logic function (e.g., 74LS08 designates a package with four 2-input AND gates). A common CMOS family is the 4000B series. Different logic functions in this family are designated by the last three digits (e.g., 4011B designates a package with four 2-input NAND gates).

Newer developments in CMOS manufacturing have produced the 74HC line of High-Speed CMOS, which is intended as a best-of-both-worlds type of technology, combining the relatively higher switching speed of TTL with the low power consumption of CMOS. High-Speed CMOS is designed to be compatible with TTL and has the same numerical designations for many logic functions.

We will look at the electrical properties of these families in more detail in a later chapter, but for now, just recognize that different logic families exist.

The most common way to package logic gates is in a plastic or ceramic **dual in-line package,** or DIP, which has two parallel rows of pins. The standard spacing between pins in one row is 0.1 inch. For packages having fewer than 28 pins, the spacing between rows is 0.3 inch. For larger packages, the rows are spaced by 0.6 inch.

This type of package is designed to be inserted in a printed circuit board in one of two ways: (1) The pins are inserted through holes in the circuit board and soldered in place; or (2) a socket is soldered to the circuit board and the IC is placed in the socket. The latter method is more expensive but makes chip replacement much easier. A socket can occasionally cause its own problems by making a poor connection to the pins of the IC.

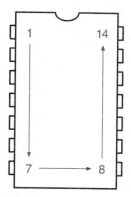

**Figure 2.33**

**14-pin DIP (Top View)**

The DIP is also convenient for laboratory and prototype work, since it can be inserted easily into a **breadboard,** a special type of temporary circuit board with internal connections between holes of a standard spacing.

The outline of a 14-pin DIP is shown in Figure 2.33. There is a notch on one end to show the orientation of the pins. When the IC is oriented as shown and viewed from above, pin 1 is at the top left corner and the pins are numbered counterclockwise from that point.

Besides DIP packages, there are numerous other types of packages for digital ICs, including, among others, SO (**small outline**) and PLCC or CLCC (**plastic** or **ceramic leaded chip carrier**) packages. They are used mostly in applications where circuit board space is at a premium and in manufacturing processes relying on **surface-mount technology** (SMT).

SMT is a sophisticated technology that relies on automatic placement of chips and soldering of pins onto the surface of a circuit board, not through holes in the circuit board. This technique allows a manufacturer to mount components on both sides of a circuit board.

Figure 2.34a shows the DIP, PLCC, and CLCC packages. Figure 2.34b shows the SO packaging.

Logic gates come in packages containing several gates. Common groupings available in DIP packages are six 1-input gates, four 2-input gates, three 3-input gates, or two 4-input gates, although other arrangements are available. The usual way of stating the number of logic gates in a package is to use the numerical prefixes hex (6), quad or quadruple (4), triple (3), or dual (2).

Some common gate packages are listed in Table 2.33.

Information about pin configurations of specific gates is available in any TTL or CMOS data book. If you plan to do any work in digital electronics, it is essential to get a good data book that shows the pinouts of logic devices, as well as electrical characteristics, logic functions, and often some applications. If you're going to get just one book, get a TTL or High-Speed CMOS book by a major manufacturer, such as Texas Instruments, National Semiconductor, or Motorola. If you get a TTL book, make sure that it has as wide a range of logic families as possible. It is especially important to have information about LS series logic. (This may not be true in a few years, due to the rising prominence of more advanced TTL logic families.)

**Figure 2.34**

**(a) Plastic Leaded Chip Carrier (PLCC) and (b) Small Outline Packaging for Integrated Circuits (photo courtesy of Texas Instruments Incorporated)**

**Table 2.33** Some Common Logic Gate Packages

| Gate | Family | Function |
|------|--------|----------|
| 74LS00 | TTL | Quad 2-input NAND gate |
| 74LS02 | TTL | Quad 2-input NOR gate |
| 74LS04 | TTL | Hex inverter |
| 74LS11 | TTL | Triple 3-input AND gate |
| 4011B | CMOS | Quad 2-input NAND gate |
| 4001B | CMOS | Quad 2-input NOR gate |
| 4069UB | CMOS | Hex inverter |
| 4073B | CMOS | Triple 3-input AND gate |
| 74HC00A | High-Speed CMOS | Quad 2-input NAND gate |
| 74HC02A | High-Speed CMOS | Quad 2-input NOR gate |
| 74HC04A | High-Speed CMOS | Hex inverter |
| 74HC11 | High-Speed CMOS | Triple 3-input AND gate |

Figure 2.35 shows the internal diagrams of gates listed in Table 2.32. Notice that the gates can be oriented inside a chip in a number of ways and that gates with the same function are not necessarily oriented in the same way (e.g., 74LS00 and 4011B NANDs; 74LS02 and 4001B NORs; 74LS11 and 4073B ANDs). That's why it is important to confirm pin connections with a data book.

In addition to the gate inputs and outputs, there are two more connections to be made on every chip: the power and ground connections. In TTL, connect $V_{CC}$ to +5 volts and GND to ground. In CMOS, connect $V_{DD}$ to the supply voltage (+3 V to +18 V) and $V_{SS}$ to ground. The gates won't work without these connections.

*Every chip requires power and ground.* This might seem obvious, but it's surprising how often it is forgotten. With beginning digital students, as well as with some who are more experienced, the lack of this knowledge accounts for numerous, maybe even most, laboratory wiring errors. This is probably because most digital circuit diagrams don't show the power connections but assume that you know enough to make them.

A chip gets its required power only through the $V_{CC}$ or $V_{DD}$ pin. Even if the power supply is connected to a logic input as a logic HIGH, you still need to connect it to the power supply pin.

Even more important is a good ground connection. A circuit with no power connection will probably not work at all. A circuit without a ground may appear to work, but it will often produce bizarre errors that are very difficult to detect and repair.

---

**Section Review Problem for Section 2.6**

2.10. How are the pins numbered in a dual in-line package?

---

## 2.7

# Introduction to Troubleshooting

**Troubleshooting** *Detection and correction of wiring errors and component failures in a circuit.*

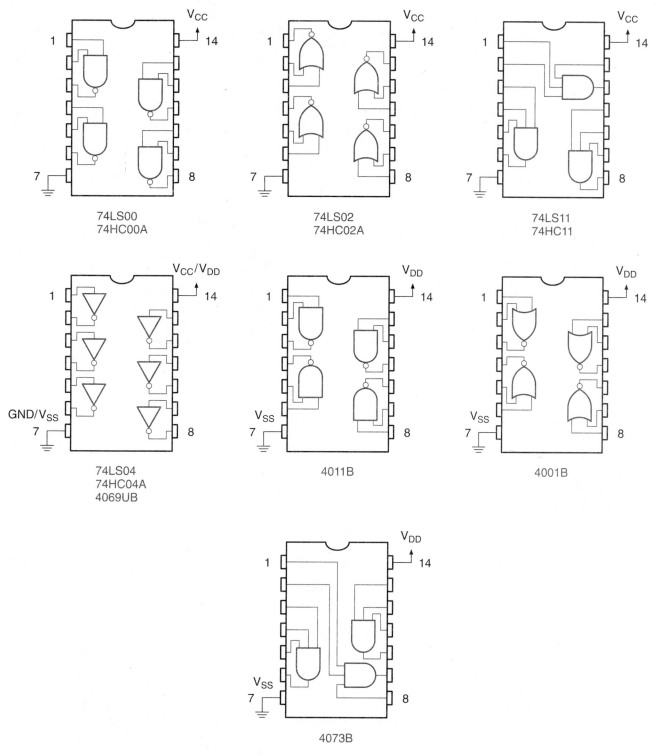

**Figure 2.35**
**Pinouts of Some Common Logic Gate ICs**

As an electronics technician or technologist, chances are very good that much of your working life will consist of **troubleshooting** circuits, that is, finding and repairing wiring errors and component failures. These two types of troubleshooting are fundamentally different.

Wiring errors (or printed circuit board errors) occur mainly in new designs under development, in prototypes, and in student laboratory work. The main difficulty with this type of error is that available schematics do not correspond to the circuit as it has actually been built. The main remedies are visual inspection and measurement of logic levels at various points in the circuit to find where the schematic and circuit do not agree and correction of the fault.

Component failure can occur in any circuit but is the main source of errors in mature designs in industrial settings. Learning to detect these failures in a student laboratory is fairly difficult. Common types of component failure can be simulated, but these simulated faults never seem to be as ingenious as those cooked up by Mother Nature. Experience really is the best teacher.

The first principle of troubleshooting is: *Troubleshooting is thinking!* By far the best troubleshooting tool you carry is the one between your ears—your brain.

Richard Feynman, one of the foremost American physicists of the 20th century, illustrates this point in an anecdote he tells about fixing radios when he was a boy, during the Depression.

> One job was really sensational. I was working at the time for a printer, and a man who knew that printer knew I was trying to get jobs fixing radios, so he sent a fellow around to the print shop to pick me up. The guy is obviously poor—his car is a complete wreck—and we go to his house which is in a cheap part of town. On the way, I say, "What's the trouble with the radio?"
>
> He says, "When I turn it on it makes a noise, and after a while the noise stops and everything's all right, but I don't like the noise at the beginning."
>
> I think to myself: "What the hell! If he hasn't got any money, you'd think he could stand a little noise for a while."
>
> And all the time, on the way to his house, he's saying things like, "Do you know anything about radios? How do you know about radios—you're just a little boy!"
>
> He's putting me down the whole way, and I'm thinking, "So what's the matter with him? So it makes a little noise."
>
> But when we got there I went over to the radio and turned it on. Little noise? *My God!* No wonder the poor guy couldn't stand it. The thing began to roar and wobble—WUH BUH BUH BUH BUH—A *tremendous* amount of noise. Then it quieted down and played correctly. So I started to think: "How can that happen?"
>
> I start walking back and forth, thinking, and I realize that one way it can happen is that the tubes are heating up in the wrong order—that is, the amplifier's all hot, the tubes are ready to go, and there's nothing feeding in, or there's some back circuit feeding in, or something wrong in the beginning part—the RF part—and therefore it's making a lot of noise, picking up something. And when the RF circuit's finally going, and the grid voltages are adjusted, everything's all right.
>
> So the guy says, "What are you doing? You come to fix the radio, but you're only walking back and forth!"
>
> I say, "I'm thinking!" Then I said to myself, "All right, take the tubes out and reverse the order completely in the set." (Many radio sets in those days used the same tubes in different places—212's, I think they were, or 212-A's.) So I changed the tubes around, stepped to the front of the radio, turned the thing on, and it's quiet as a lamb: it waits until it heats up, and then plays perfectly—no noise.
>
> When a person has been negative to you, and then you do something like that, they're usually a hundred per cent the other way, kind of to compensate. He got me

other jobs, and kept telling everybody what a tremendous genius I was, saying, "He fixes radios by *thinking!*" The whole idea of thinking, to fix a radio—a little boy stops and thinks, and figures out how to do it—he never thought that was possible.

*Reprinted from "SURELY YOU'RE JOKING, MR. FEYNMAN!" Adventures of a Curious Character, by Richard P. Feynman, as told to Ralph Leighton, edited by Edward Hutchings, by permission of W. W. Norton & Company, Inc. Copyright (c) 1985 by Richard P. Feynman and Ralph Leighton.

## Test Equipment for Troubleshooting Digital Circuits

**Logic probe**  *A hand-held test instrument that will indicate logic HIGH and LOW levels, open circuits, and single or continuous digital pulses.*

**Floating**  *An undefined logic state, neither HIGH nor LOW.*

**Logic pulser**  *A test instrument used as a source of single or multiple digital pulses.*

**Clock generator**  *A circuit that generates a periodic digital waveform.*

**Oscilloscope**  *A test instrument that displays a plot of input voltage over time on a cathode ray tube (CRT) screen.*

**Trace**  *A line produced by a beam of electrons that periodically sweeps across an oscilloscope screen to display the waveform of the input voltage.*

**Time base**  *A control on an oscilloscope that sets the speed at which a trace sweeps across the display screen.*

**Trigger**  *An oscilloscope control that starts a trace when the input voltage has a particular magnitude and direction.*

A number of test instruments have been developed for troubleshooting digital circuits. Three of the most common are the **logic probe, logic pulser,** and **oscilloscope.** The logic probe and oscilloscope, shown in Figure 2.36 and 2.37, are indicating instruments; they

**Figure 2.36**

**An oscilloscope is used to measure steady-state or periodic voltages by showing the signal as a function of time (photo courtesy of Hewlett-Packard Company)**

**Figure 2.37**

**A logic probe measures logic HIGH and LOW levels and indicates open-circuit conditions. It can also detect a digital pulse (photo courtesy of Tektronix, Inc.)**

**Figure 2.38**

**A logic pulser generates a single pulse or a continuous train of pulses that can be traced through a logic circuit to test its operations (photo courtesy of Tektronix, Inc.)**

indicate steady or time-varying logic states at various points in a circuit. The logic pulser, shown in Figure 2.38, is used to apply a single pulse or a pulse train to a circuit. If a logic pulser is not available, a **clock generator,** either as part of a digital trainer or a stand-alone unit, can be used to produce a train of pulses. These pulses can be traced through a circuit with a logic probe or oscilloscope.

### Logic Probe and Pulser

Figure 2.39 shows how a logic probe indicates various logic states: HIGH, LOW, open circuit (or **floating**), and pulsing. When a logic state is steady, the corresponding light is ON and the other indicator is OFF. There is a separate indicator light for HIGH and LOW states so that we can distinguish between logic LOW and floating, a state that is neither HIGH nor LOW. (If there is only a HIGH indicator, we may incorrectly assume that a terminal is LOW when it is actually floating. It's a bit like the difference between "zero" and "nothing." Zero is a numerical digit that can indicate positive information, such as "zero in the 8's position"; nothing is, well . . ., nothing—a blank space with no meaning.)

The gate output in Figure 2.39c is open circuited because of the open switch at the terminal. Since the terminal cannot be pulled HIGH or LOW by the gate output, we say that it floats at an unknown level. In such a case neither indicator on the logic probe is ON.

One input of the gate in Figure 2.39d is forced HIGH by a resistor to the supply voltage. The other is pulsed by a logic pulser. This instrument can produce a train of pulses, as shown, or a single pulse. The pulser has a trigger switch to control when the pulses are generated. The logic probe at the gate output will see the output pulsing between HIGH and LOW. Therefore, both indicators will flash alternately. Some logic probes have circuitry that slows down a fast pulse train so that it can be observed visually. This is necessary if the pulse train is faster than about 30 Hz, the fastest speed that the human eye can follow.

### Oscilloscope

An oscilloscope is, in effect, a graphical voltmeter; it shows the measured voltage at its inputs as a function of time. It is most useful when its input voltage either is at a constant level or is a periodic function; it is not especially effective for measuring aperiodic waveforms or single pulses.

**Figure 2.39**
**Logic Probe Indications**

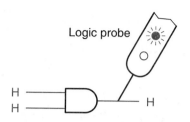

**a. Output constant HIGH (HIGH indicator on)**

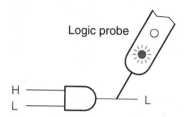

**b. Output constant LOW (LOW indicator on)**

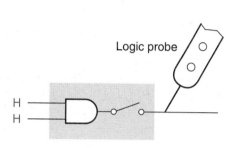

**c. Output open circuit (neither indicator on)**

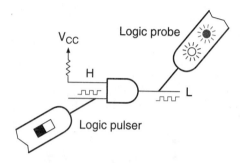

**d. Pulsed input (HIGH and LOW indicators alternately flashing)**

Most oscilloscopes in use today are of a type called "triggered-sweep" oscilloscopes. A beam of electrons generated inside a cathode ray tube (CRT—the part of the scope that looks like a television picture tube) moves across the screen at a regular interval, making a **trace** on the screen.

The beam starts at the left side of the screen, moves to the far right, and then returns to the left to begin the sweep again. The beam sweeps along at a known rate set by a control called the **time base.** The time base sets the number of cycles of the input waveform that can be displayed on the screen and is calibrated in units of time per division (e.g., 2 ms/div, 5 μs/div). One "division" is one square on the oscilloscope screen; most screens are ten divisions wide and eight divisions high, each being further subdivided into five small divisions.

During the sweep, the trace deviates from its baseline position proportionally to the oscilloscope input voltage. The range of this deviation (called the vertical deflection) is controlled by a dial marked off in volts per division (e.g., 1 V/div, 2 mV/div).

We use two controls to set the baseline (0 V reference): (1) a zero-reference (usually marked "0" or "GND"), which shorts the input to ground; and (2) a vertical position adjustment (usually marked "Y position").

The procedure is:

1. Set the input to zero-reference (press "0").
2. Select a horizontal line on the screen for the 0 V line (often, but not always, the center of the screen).
3. Use the "Y position" control to line up the trace with the selected 0 V line.
4. Release the zero-reference input.

**Figure 2.40**

**DC Voltage on an Oscilloscope**

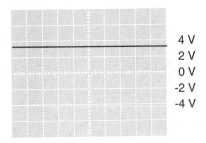

4 V
2 V
0 V
-2 V
-4 V

Vertical: 2 V / div

We can read the amplitude of a waveform directly from the screen; we can calculate the frequency of a periodic waveform from the horizontal dimension (since $f = 1/T$) if we know the time base setting.

The **trigger** is an adjustable circuit that starts the trace when the measured input voltage has a particular magnitude and direction. When the trace reaches the right side of the screen, the electron beam returns to the left side, but it does not start a new sweep right away. It waits for the input voltage to be the same as the trigger voltage (i.e., the correct magnitude and direction) before it begins a new sweep. If the input is periodic, the same line is traced over and over again, thus appearing to stand still.

Figure 2.40 shows a steady DC voltage displayed on an oscilloscope screen. The voltage does not change with time and is thus shown as a straight line. The vertical scale is 2 V/div. If we set the center line as the zero-reference, the voltage is +3.6 V DC. (To make this measurement, the scope *must* be set for DC coupling. This is true for all digital circuit measurements. DC coupling is usually selected by a switch marked "AC/DC" or "AC/GND/DC".)

> Although an oscilloscope can be used to measure steady-state logic levels, it cannot distinguish between logic LOW (0 V) and an open circuit (no input). Thus, you cannot determine the integrity of a ground connection with an oscilloscope. Use a logic probe for this job.

Figure 2.41 shows a clock generator (i.e., a periodic pulse source) connected to one input of the oscilloscope. Figure 2.42 shows the displayed waveform. From the given vertical and horizontal scales, we see that the pulse amplitude is 4 V (assuming a centered zero-reference) and the period is 15 μs. The frequency is $f = 1/T = 1/15$ μs $= 66.7$ kHz.

In practice, the waveform displayed in Figure 2.42 will look like the one in Figure 2.43. The rise and fall times of a digital pulse are too fast compared to the sweep rates of most oscilloscope settings to be properly displayed. The trace appears to jump instantaneously between logic HIGH and LOW levels.

Most oscilloscopes have two or more input channels, each one controlling the vertical deflection of a trace. All channels are controlled by the same time base so that they can be compared in a known timeline. Figure 2.44 shows how the input and output of a NAND gate can be monitored at the same time. The traces are shown in Figure 2.45.

Each trace has its own adjustable zero-reference. In this case, we have offset the traces so that they won't overlap. The input (Channel A) zero-reference is one division

**Figure 2.41**

**Oscilloscope with Periodic Pulse Input**

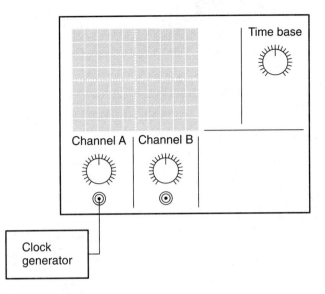

**Figure 2.42**

**Displayed Pulse Waveform**

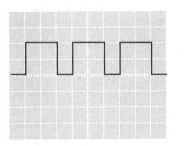

Vertical: 2 V / div
Horizontal: 5 μs / div

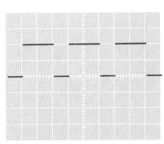

**Figure 2.43**

**Actual Display Pattern**

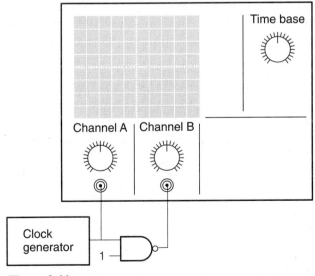

**Figure 2.44**

**Comparing NAND Input and Output Waveforms**

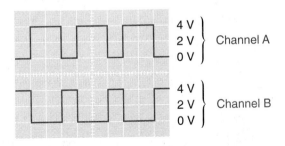

Vertical: 2 V / div (both channels)
Horizontal: 5 μs / div

**Figure 2.45**

**Displayed NAND Waveforms**

above center; the output (Channel B) reference is three divisions below center. The vertical scales on both channels are set to 2 V/div. As we would expect, the two signals are exactly out of phase (same amplitude, opposite duty cycle).

---

**Section Review Problem for Section 2.7a**

2.11. Redraw the displayed oscilloscope waveforms under the following changed conditions: (a) the gate is changed to an OR gate; and (b) the time base is changed to 2 µs/div. (See Figure 2.46 on page 62.)

---

## Possible Faults in Logic Gates

Logic gates are generally tested in a logic circuit, not on their own. It is occasionally necessary to test an individual gate outside of a circuit, so we will examine some of the most likely faults. Gates can be checked by taking truth tables and by examining their enable/inhibit behavior, in other words by noting how they respond to pulsed inputs.

---

There are differences in behavior between TTL and CMOS gates. Because of the way they are constructed, open-circuit TTL inputs act as logic HIGHs; open-circuit CMOS inputs are floating. This causes different behavior under certain fault conditions.

---

We can use our basic troubleshooting tools to test the operation of SSI logic gates. The most common failures are: fault in internal logic; open circuit; terminal stuck HIGH or LOW; internal short circuit (other than HIGH or LOW).

### Fault in Internal Logic Circuit

This type of flaw is relatively uncommon. It can be detected by taking the truth table of the gate under test and comparing it to the desired truth table. If the truth tables are different, the gate should be replaced.

### Open Circuit

An open-circuit output will be neither HIGH nor LOW; it will float regardless of the states of the inputs. Figure 2.39c shows how this can be detected with a logic probe. Connect logic switches to the inputs and go through all input combinations to take the truth table. If the output is open circuited, neither the HIGH nor the LOW indicator will be on for any input combination.

Figure 2.47 shows an open-circuit input under test. A logic pulser applies a pulse to the input. A logic probe monitors the output.

In Figure 2.47a, the good input of a TTL AND gate is held HIGH. The pulse applied to the open input cannot get through the gate. Since an open TTL input is HIGH, the output is at a permanent HIGH state. When the inputs are reversed, the open HIGH enables the gate, allowing the pulse through.

In Figure 2.47b, the CMOS AND gate needs both inputs HIGH to make the output HIGH, so the output will be LOW regardless of which way we apply the inputs.

We could make similar measurements with other logic gates, keeping in mind the idea that an open TTL input is HIGH and an open CMOS input is floating.

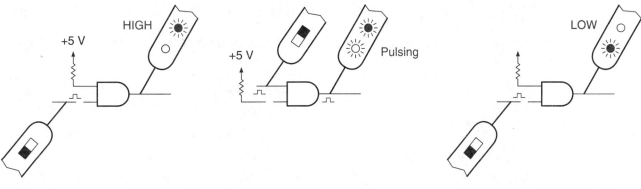

a. Testing an open circuit (TTL  AND)

b. Testing an open circuit (CMOS  AND)

**Figure 2.47**
**Testing an Open Input**

## Input or Output Stuck HIGH or LOW

It is possible for an input or output of a gate to short to ground or the supply voltage, causing the terminal to get stuck in either the HIGH or the LOW state. In such a case, the terminal will ignore any attempts to switch states.

Figure 2.48 shows an AND gate with an input or output stuck HIGH or LOW and the resulting indications on a logic probe. Since an AND gate needs two HIGH inputs to make a HIGH output, the gate in Figure 2.48a will always be LOW. The TTL AND gate with the input stuck HIGH (Figure 2.48b) will have either a HIGH output or a pulsing

**Figure 2.48**

**AND Gate With Input or Output Stack HIGH or LOW**

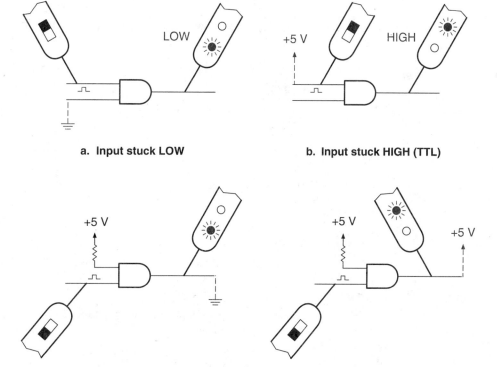

a. Input stuck LOW

b. Input stuck HIGH (TTL)

c. Output stuck LOW

d. Output stuck HIGH

output, depending on whether we apply the logic pulser to the good input or the shorted one.

If the output is stuck LOW or HIGH, as shown in Figures 2.48c and d, the inputs will have no effect on the output.

If a short circuit is external to the gate, it may be repairable, depending on whether an input or output terminal is shorted. A gate will probably work normally if an external input short is removed. A gate with an externally shorted output will probably be permanently damaged.

## Internal Short Circuit (Other Than HIGH or LOW)

Figure 2.49 shows an AND gate with its inputs internally short-circuited. When this happens, the output operates correctly, *provided* A *and* B *are at the same logic level.*

Figure 2.50 shows how the output looks on an oscilloscope, compared to various combinations of input levels. (Three traces are shown for convenience, even though many scopes can only display two.)

The shading highlights the times when the input logic levels are opposite one another. Under these conditions, the output is in an indeterminate state, neither HIGH nor LOW.

**Figure 2.49**

**AND Gate With Shorted Inputs**

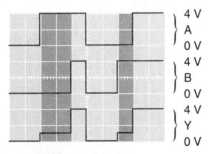

Vertical: 2 V / div

**Figure 2.50**

**Oscilloscope Display: AND With Shorted Inputs**

---

**Section Review Problems for Section 2.7b**

2.12. What would the logic probe in Figure 2.48b indicate if the gate under test was CMOS instead of TTL?

2.13. What would the probe indicate if the pulser was applied to the good input rather than the shorted input? Would it make a difference whether the gate was TTL or CMOS?

# ● GLOSSARY

**Active HIGH**  An active-HIGH terminal is considered "ON" when it is in the logic HIGH state. Indicated by the absence of a bubble at the terminal in distinctive-shape symbols.

**Active level**  A logic level defined as the "ON" state for a particular circuit input or output. The active level can be either HIGH or LOW.

**Active LOW**  An active-LOW terminal is considered "ON" when it is in the logic LOW state. Indicated by a bubble at the terminal in distinctive-shape symbols.

**AND gate**  A logic circuit whose output is HIGH when *all* inputs (e.g., *A* AND *B* AND *C*) are HIGH.

**Boolean algebra**  A system of algebra that operates on Boolean variables. The binary (two-state) nature of Boolean algebra makes it useful for analysis, simplification, and design of combinational logic circuits.

**Boolean expression**  An algebraic expression made up of Boolean variables and operators, such as AND ($\cdot$), OR ($+$), or NOT ($\bar{\phantom{x}}$). Also referred to as a **Boolean function** or a **logic function.**

**Boolean variable**  A variable having only two possible values, such as HIGH/LOW, 1/0, On/Off, or True/False.

**Breadboard**  A circuit board for wiring temporary circuits, usually used for prototypes or laboratory work.

**Bubble**  A small circle indicating logical inversion on a circuit symbol.

**Buffer**  An amplifier that acts as a logic circuit. Its output can be inverting or noninverting.

**Ceramic leaded chip carrier (CLCC)**  A square IC package with leads on all four sides designed for surface mounting on a circuit board.

**Chip**  An integrated circuit. Specifically, a chip of silicon on which an integrated circuit is constructed.

**Clock generator**  A circuit that generates a periodic digital waveform.

**Coincidence gate**  An Exclusive NOR gate.

**Complement form**  Inverted.

**Complementary metal-oxide semiconductor (CMOS)**  A family of logic devices used in digital circuits.

**DeMorgan equivalent forms**  Two gate symbols, one AND-shaped and one OR-shaped, that are equivalent according to DeMorgan's Theorems.

**DeMorgan's Theorems**  Two theorems in Boolean algebra that allow us to transform any gate from an AND-shaped to an OR-shaped gate and vice versa.

**Digital signal (or pulse waveform)**  A series of 0s and 1s plotted over time.

**Distinctive-shape symbols**  Graphic symbols for logic circuits that show the function of each type of gate by a special shape.

**Dual in-line package (DIP)**  A type of IC with two parallel rows of pins for the various circuit inputs and outputs.

**Enable**  A logic gate is enabled if it allows a digital signal to pass from an input to the output in either true or complement form.

**Exclusive NOR gate**  A two-input logic circuit whose output is the complement of an Exclusive OR gate.

**Exclusive OR gate**  A two-input logic circuit whose output is HIGH when one input (but not both) is HIGH.

**Floating**  An undefined logic state, neither HIGH nor LOW.

**IEEE/ANSI Standard 91-1984**  A standard format for drawing logic circuit symbols as rectangles with logic functions shown by a standard notation inside the rectangle for each device.

**In phase**  Two digital waveforms are in phase if they are always at the same logic level at the same time.

**Inhibit (or disable)**  A logic gate is inhibited if it prevents a digital signal from passing from an input to the output.

**Integrated circuit (IC)**  An electronic circuit having many components, such as transistors, diodes, resistors, and capacitors, in a single package.

**Inverter**  Also called a NOT gate or an inverting buffer. A logic gate that changes its input logic level to the opposite state.

**Logic function**  See Boolean expression.

**Logic gate**  An electronic circuit that performs a Boolean algebraic function.

**Logic probe**  A hand-held test instrument that will indicate logic HIGH and LOW levels, open circuits, and single or continuous digital pulses.

**Logic pulser**  A test instrument used as a source of single or multiple digital pulses.

**Logical product** AND function.

**Logical sum** OR function.

**NAND gate** A logic circuit whose output is LOW when *all* inputs are HIGH.

**NOR gate** A logic circuit whose output is LOW when *at least one* input is HIGH.

**OR gate** A logic circuit whose output is HIGH when *at least one* input (e.g., *A* OR *B* OR *C*) is HIGH.

**Oscilloscope** A test instrument that displays a plot of input voltage over time on a cathode ray tube (CRT) screen.

**Out of phase** Two digital waveforms are out of phase if they are always at opposite logic levels at any given time.

**Plastic leaded chip carrier (PLCC)** An IC package like the CLCC, but made of plastic.

**Qualifying symbol** A symbol in IEEE/ANSI logic circuit notation, placed in the top center of a rectangular symbol, that shows the function of a logic gate. Some qualifying symbols include: 1 = "buffer"; & = "AND"; ≥1 = "OR"; =1 = "Exclusive OR."

**Rectangular-outline symbols** Rectangular logic gate symbols that conform to IEEE/ANSI Standard 91-1984.

**Small outline (SO)** An IC package similar to a DIP, but smaller, which is designed for automatic placement and soldering on the surface of a circuit board.

**Small-scale integration (SSI)** An integrated circuit having 12 or fewer gates in one package.

**Surface-mount technology (SMT)** A system of mounting and soldering integrated circuits on the surface of a circuit board.

**Time base** A control on an oscilloscope that sets the speed at which a trace sweeps across the display screen.

**Trace** A line produced by a beam of electrons that periodically sweeps across an oscilloscope screen to display the waveform of the input voltage.

**Transistor-transistor logic (TTL)** A family of logic devices used in digital circuits.

**Trigger** An oscilloscope control that starts a trace when the input voltage has a particular magnitude and direction.

**Troubleshooting** Detection and correction of wiring errors and component failures in a circuit.

**True form** Not inverted.

**Truth table** A list of all possible input values to a digital circuit, listed in ascending binary order, and the output response for each input combination.

## ● P R O B L E M S

*Problems with asterisks are more challenging. Number of asterisks indicates level of difficulty.*

### Section 2.1 Basic Logic Functions

**2.1** Draw the logic gate symbols for a 4-input AND gate in both distinctive-shape and rectangular-outline forms. How many different combinations of input states are possible with this gate?

**2.2** Repeat Problem 2.1 for an 8-input AND gate. How many possible combinations of input states does this gate allow?

**2.3** Write the truth table for a 4-input AND gate.

**2.4** Repeat Problems 2.1, 2.2, and 2.3 for an OR gate.

**2.5** State the active level of each input and output for the gates in Figure 2.51.

### Section 2.2 Derived Logic Functions

**2.6** For a 4-input NAND gate with inputs *A*, *B*, *C*, and *D* and output *Y:*

  **a.** Write the truth table and a descriptive sentence.

  **b.** Write the Boolean expression.

  **c.** Draw the logic circuit symbol in both distinctive-shape and rectangular-outline symbols.

**2.7** Repeat Problem 2.6 for a 4-input NOR gate.

**2.8** State the active levels of the inputs and outputs of a NAND gate and a NOR gate.

**2.9** Write a descriptive sentence of the operation of a 5-input NAND gate with inputs *A, B, C, D,* and *E* and output *Y.* How many lines would the truth table of this gate have?

**2.10** Repeat Problem 2.9 for a 5-input NOR gate.

**2.11** A police radio dispatch system has an automatic signaling scheme to keep track of the location of all its cars. A radio transmitter in each police car sends an identification signal to the system dis-

Figure 2.51

**Problem 2.5
Logic Gates**

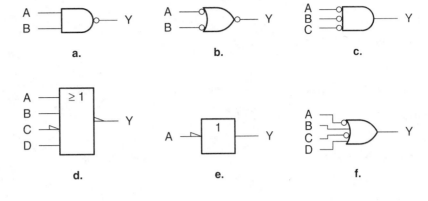

Figure 2.52

**Problem 2.11
Block Diagram of Transmit
Circuitry**

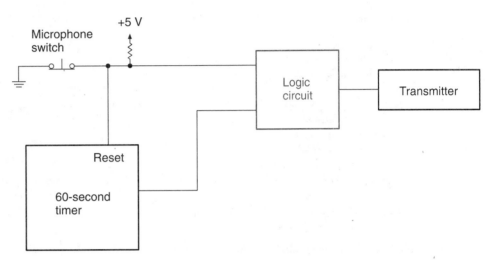

patcher every time a call is made from the car. If a call is not made within 60 seconds of the last call, the ID is transmitted automatically.

Figure 2.52 shows a block diagram of the transmit circuitry. Each radio has a timer whose output goes HIGH every 60 seconds and a microphone switch that produces a HIGH when the operator presses it to transmit. (The microphone switch also resets the timer; the timer output will not go HIGH for the next 60 seconds after the microphone switch is pressed.) A logic circuit output produces a logic LOW to make the radio transmit. Draw the symbol and truth table of the gate that corresponds to the action of the logic circuit.

**2.12** A pump motor in an industrial plant will start only if the temperature and pressure of liquid in a tank exceed a certain level. The temperature sensor and pressure sensor, shown in Figure 2.53, each produce a logic HIGH if the measured quan-

tities exceed this value. The logic circuit interface produces a HIGH output to turn on the motor. Draw the symbol and truth table of the gate that corresponds to the action of the logic circuit.

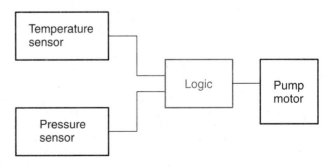

Figure 2.53

**Problem 2.12
Temperature and Pressure Sensors**

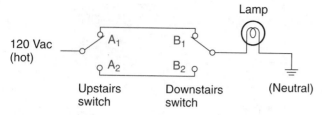

Figure 2.54
**Problem 2.14**
**Circuit for Two-Way Switch**

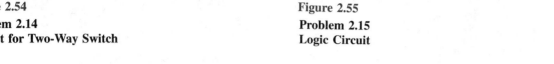

**Figure 2.55**
**Problem 2.15**
**Logic Circuit**

**2.13** Repeat Problem 2.12 for the case in which the motor is activated by a logic LOW.

*2.14 Figure 2.54 shows a circuit for a two-way switch for a stairwell. This is a common circuit that allows you to turn on a light from either the top or the bottom of the stairwell and off at the other end. The circuit also allows anyone coming along after you to do the same thing, no matter which direction they are coming from.

The lamp is ON when the switches are in the same positions and OFF when they are in opposite positions. What logic function does this represent? Draw the truth table of the function and use it to explain your reasoning.

**2.15** Find the truth table for the logic circuit shown in Figure 2.55.

**2.16** Recall the description of a 2-input Exclusive OR gate: "Output is HIGH if one input is HIGH, but not both." This is not the best statement of the operation of a multiple-input XOR gate. Look at the truth table derived in Problem 2.15 and write

a more accurate description of n-input XOR operation.

## Section 2.3 Truth Table of a Generalized Gate
## Section 2.4 DeMorgan's Theorems and Gate Equivalence

**2.17** For each of the gates in Figure 2.56:

a. Write the truth table.

b. Indicate with an * which lines on the truth table show the gate output in its active state.

c. Convert the gate to its DeMorgan equivalent form.

d. Rewrite the truth table and indicate which lines on the truth table show output active states for the DeMorgan equivalent form of the gate.

**2.18** Draw the logic gate symbols for each of the following truth tables and specified input/output active levels.

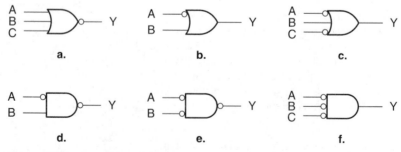

**Figure 2.56**
**Problem 2.17**
**Logic Gates**

| A | B | Y |
|---|---|---|
| 0 | 0 | 0 |
| 0 | 1 | 0 |
| 1 | 0 | 1 |
| 1 | 1 | 0 |

B: active HIGH

A: active LOW

| A | B | C | Y |
|---|---|---|---|
| 0 | 0 | 0 | 1 |
| 0 | 0 | 1 | 1 |
| 0 | 1 | 0 | 1 |
| 0 | 1 | 1 | 0 |
| 1 | 0 | 0 | 1 |
| 1 | 0 | 1 | 1 |
| 1 | 1 | 0 | 1 |
| 1 | 1 | 1 | 1 |

Y: active LOW

| A | B | C | Y |
|---|---|---|---|
| 0 | 0 | 0 | 0 |
| 0 | 0 | 1 | 0 |
| 0 | 1 | 0 | 0 |
| 0 | 1 | 1 | 1 |
| 1 | 0 | 0 | 0 |
| 1 | 0 | 1 | 0 |
| 1 | 1 | 0 | 0 |
| 1 | 1 | 1 | 0 |

Y: active LOW

| A | B | C | Y |
|---|---|---|---|
| 0 | 0 | 0 | 0 |
| 0 | 0 | 1 | 0 |
| 0 | 1 | 0 | 0 |
| 0 | 1 | 1 | 0 |
| 1 | 0 | 0 | 0 |
| 1 | 0 | 1 | 0 |
| 1 | 1 | 0 | 1 |
| 1 | 1 | 1 | 0 |

A: active HIGH

C: active LOW

| A | B | Y |
|---|---|---|
| 0 | 0 | 0 |
| 0 | 1 | 1 |
| 1 | 0 | 1 |
| 1 | 1 | 1 |

Y: active LOW

| A | B | Y |
|---|---|---|
| 0 | 0 | 1 |
| 0 | 1 | 1 |
| 1 | 0 | 1 |
| 1 | 1 | 0 |

Y: active LOW

**2.19** Repeat Problem 2.11 for the case where the timer and microphone switch have logic LOW outputs. Write a sentence describing the action of the new gate.

**2.20** Draw the DeMorgan equivalent of the gate derived in Problem 2.19 and write a sentence describing its action. Which form of the gate makes the system function easiest to understand? Why?

**2.21** An elevator motor will not turn on unless the door is closed and one of six floor destination switches has been pushed. The system is shown in Figure 2.57.

Each destination switch produces a logic LOW when it is pushed. When a switch is pressed, the output of logic circuit 1 produces a HIGH, which is stored in a latch circuit. (A latch

**Figure 2.57**

**Problem 2.21**
**Elevator Circuits**

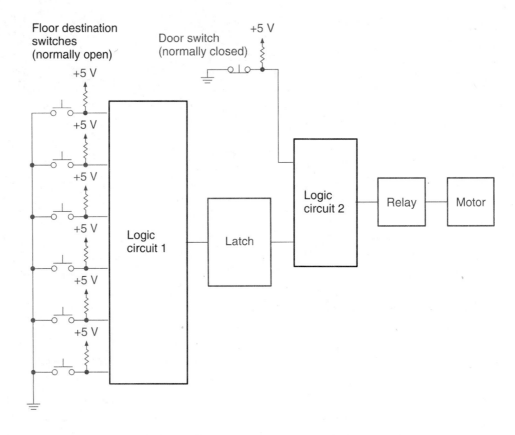

circuit stores a logic 1 or 0 indefinitely. This is necessary since the pushbuttons are momentary action.)

The elevator door switch produces a logic HIGH when the door closes. When the door is closed and a floor destination switch has been pushed, logic circuit 2 produces a LOW output to activate the elevator motor via the relay.

Draw the logic symbols of the gates corresponding to logic circuits 1 and 2. Write a sentence describing the action of each gate. How many lines would the truth table of logic circuit 1 have? Which form of the gate for circuit 1 best shows its function?

**2.22** Write a sentence describing the combined action of logic circuits 1 and 2 in Problem 2.21.

## Section 2.5 Enable and Inhibit Properties of Logic Gates

**2.23** Draw the output waveform of the Exclusive NOR gate when a square waveform is applied to one input and

**a.** The other input is held LOW

**b.** The other input is held HIGH

How does this compare to the waveform that would appear at the output of an Exclusive OR gate under the same conditions?

**2.24** Sketch the input waveforms represented by the following 32-bit sequences (use ¼-inch graph paper, 1 square per bit):

*A:* 00000000000011111111111111100000
*B:* 10100111001010110101001110011011

Assume that these waveforms represent inputs to a logic gate. Sketch the waveform for gate output *Y* if the gate function is:

**a.** AND

**b.** OR

**c.** NAND

**d.** NOR

**e.** XOR

**f.** XNOR

**Figure 2.58**

**Problem 2.25**

**Waveform**

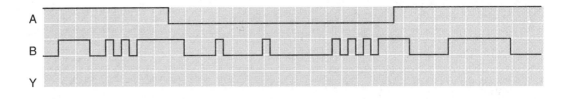

**Figure 2.59**

**Problem 2.26**

**Waveforms**

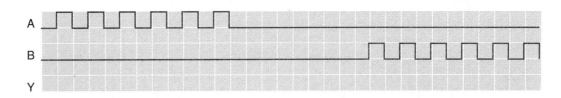

**2.25** Repeat Problem 2.24 for the waveforms shown in Figure 2.58.

**2.26** The *A* and *B* waveforms shown in Figure 2.59 are inputs to an OR gate. Complete the sketch by drawing the waveform for output *Y*.

**2.27** Repeat Problem 2.26 for a NOR gate.

**2.28** Figure 2.60 shows a circuit that will make a lamp flash at 3 Hz when the gasoline level in a car's gas tank drops below a certain point. A float switch in the tank monitors the level of gasoline. What logic level must the float switch produce to make the light flash when the tank is approaching empty? Why?

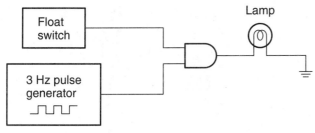

**Figure 2.60**

**Problem 2.28**

**Gasoline Level Circuit**

**2.29** Repeat Problem 2.28 for the case where the AND gate is replaced by an OR gate.

**2.30** Will the circuit in Figure 2.60 work properly if the AND gate is replaced by an Exclusive OR gate? Why or why not?

## Section 2.6 Integrated Circuit Logic Gates

**2.31** Name two logic families used to implement digital logic functions.

**2.32** List the industry-standard numbers for a Quadruple 2-input NAND gate in Low Power Schottky TTL, CMOS, and High-Speed CMOS technologies.

**2.33** Repeat Problem 2.32 for a Quadruple 2-input NOR gate. How does each numbering system differentiate between the NAND and NOR functions?

**2.34** List four types of packaging that a logic gate could come in.

## Section 2.7 Introduction to Troubleshooting

**2.35** Calculate the frequency and duty cycle of the waveform shown on the oscilloscope display of Figure 2.61.

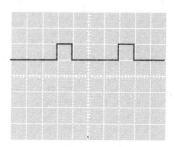

Vertical: 5 V / div
Horizontal: 10 ms / div

**Figure 2.61**

**Problem 2.35**

**Waveform**

**2.36** Redraw the oscilloscope waveform of Figure 2.61 for a new horizontal scale of 20 ms/div.

**2.37** The waveform of Figure 2.61 is applied to one input of a 2-input NOR gate. The other input is held LOW. Draw the waveform of the NOR output if the vertical deflection is the same for both channels.

**2.38** Suppose the LOW input of the NOR gate in Problem 2.37 becomes open circuited. Describe what a logic probe at the gate output would indicate if the gate was TTL.

**2.39** Repeat Problem 2.38 for a CMOS gate.

**2.40** The pulsed input of the NOR gate in Problem 2.37 becomes shorted to ground. Describe what a logic probe will show if it monitors the gate output. Does it matter if the gate is TTL or CMOS? Why or why not?

**2.41** A CMOS NAND gate is tested as shown in Figure 2.62. The gate output is always HIGH regardless of the pulsing at one input. List the possible faults that might cause this behavior.

**Figure 2.62**

**Problem 2.41**
**Testing a CMOS NAND Gate**

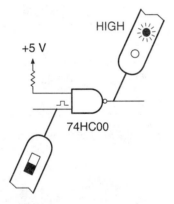

## ▼ Answers to Section Review Questions

### Section 2.1
**2.1** AND: "*A* AND *B* AND *C* AND *D* must be HIGH to make *Y* HIGH."   **2.2.** OR: "*A* OR *B* OR *C* OR *D* must be HIGH to make *Y* HIGH."

### Section 2.2
**2.3.** XOR;   **2.4.** NAND;   **2.5.** NOR;   **2.6.** XNOR.

### Section 2.3
**2.7.** "*A* must be LOW AND *B* AND *C* must be HIGH to make *Y* LOW."

$$Y = \overline{A} \, B \, C$$

| A | B | C | Y |
|---|---|---|---|
| 0 | 0 | 0 | 1 |
| 0 | 0 | 1 | 1 |
| 0 | 1 | 0 | 1 |
| 0 | 1 | 1 | 0 |
| 1 | 0 | 0 | 1 |
| 1 | 0 | 1 | 1 |
| 1 | 1 | 0 | 1 |
| 1 | 1 | 1 | 1 |

### Section 2.4
**2.8.** $Y = \overline{A \, \overline{B} \, \overline{C} \, D}$

### Section 2.5
**2.9.** An AND needs two HIGH inputs to make a HIGH output. If the Control input is LOW, the output can never be HIGH; the output remains LOW. An OR output is HIGH if one input is HIGH. If the Control input is HIGH, the output is always HIGH, regardless of the level at the Signal input. In both cases, the output is "stuck" at one level, signifying that the gate is inhibited.

### Section 2.6
**2.10.** Viewed from above, with the notch in the package away from you, pin 1 is on the left side at the far end. The pins are numbered counterclockwise from that point.

### Section 2.7a
**2.11.** Figure 2.46 shows the modified traces.

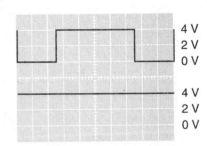

Vertical: 2 V / div (both channels)
Horizontal: 2 μs / div

**Figure 2.46**

**OR Gate Input and Output (One Input HIGH)**

### Section 2.7b
**2.12.** LOW;   **2.13.** Pulsing. Same for TTL and CMOS.

# Logic Gate Networks and Boolean Algebra

## ● CHAPTER CONTENTS

**3.1** Combinational Logic and Boolean Algebra

**3.2** Boolean Expressions From Logic Diagrams

**3.3** Logic Diagrams From Boolean Expressions

**3.4** Evaluating Boolean Expressions

**3.5** Truth Tables From Logic Diagrams

**3.6** Truth Tables From Boolean Expressions

**3.7** Boolean Expressions From Truth Tables

**3.8** Exclusive OR and Exclusive NOR Circuits

**3.9** Theorems of Boolean Algebra

## ● CHAPTER OBJECTIVES

Upon successful completion of this chapter, you will be able to:

- Derive the unsimplified version of a Boolean expression from a logic circuit diagram.

- Explain the relationship between Boolean expression, truth table, and logic circuit and be able to derive any one from either of the other two.

- Describe the equivalent networks for Exclusive OR and Exclusive NOR functions.

- Use the rules of Boolean algebra to simplify Boolean expressions derived from logic circuits and truth tables.

## ● INTRODUCTION

Single logic gates, such as we studied in the previous chapter, are of limited use for the generation of any but the most basic logic functions. A truly useful digital technology requires us to combine many gates to produce the required logic functions required by most applications.

More-complex digital functions are performed using combinational logic. The term "combinational logic" refers to digital circuits whose outputs are derived from combinations of input logic levels independently of the sequence in which they are applied to the circuit. (For example, an AND gate has a HIGH output regardless of whether input $A$ goes HIGH first and then $B$ or input $B$ goes HIGH first and then $A$.)

Boolean algebra is a powerful mathematical tool that can be used for the analysis and design of combinational logic circuits. We can use it to analyze the expected behavior of a circuit and thus determine if it is operating correctly. We can also use it to design a circuit to perform a function specified by a word problem, a truth table, or a Boolean algebraic equation.

## 3.1

# Combinational Logic and Boolean Algebra

> **Logic gate network**  *A digital circuit consisting of two or more logic gates.*
>
> **Combinational logic circuit**  *A digital logic circuit whose output depends only on the combination of input logic states. The output does not depend on the order in which the inputs are applied.*
>
> **Logic diagram**  *A circuit diagram of a digital logic circuit.*

In the examples in Chapter 2, we described various design situations by using basic logic functions and generated each function with a single logic gate. Most logic functions cannot be synthesized with a single gate; they must be produced by a **logic gate network,** a combination of two or more gates. This type of digital circuit is also called a **combinational logic circuit** because its output is a function of the combination of input states. Any combinational logic function can be implemented using the AND, OR, and NOT functions studied in the previous chapter.

(Another type of digital logic, called sequential logic, also exists. The output of a sequential circuit depends on the circuit's past history as well as its present input values. We will study this type of circuit in future chapters.)

Understanding the intended operation of a circuit is essential for the service technician as well as the circuit designer. In plain terms, you can't fix it if you don't know what it's supposed to do.

Combinational logic circuits can be described in three ways:

1. By a **logic diagram,** which is more or less a schematic diagram of the logic circuit. (It is not a true schematic, since it does not show power connections.)
2. By a truth table.
3. By a Boolean expression.

Any of these descriptions can be derived from either of the other two. Figure 3.1 shows this relationship.

Sometimes the Boolean expression of a logic circuit is not in its simplest form. If this is the case, we will often use theorems of Boolean algebra to simplify the expression. This makes it easier to derive a truth table or logic diagram from the expression.

**Figure 3.1**

**Three Descriptions of a Combinational Logic Circuit**

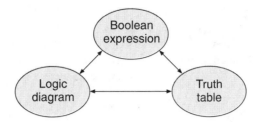

## 3.2

# Boolean Expressions From Logic Diagrams

The basic logic gates and their DeMorgan equivalents are shown in Figure 3.2. These gates form the basis for logic circuit analysis. The $x$ and $y$ input variables can be fixed logic levels, single Boolean variables, or larger terms that are combinations of Boolean variables.

For example, the output of the AND gate shown is given by $x \cdot y$. If $x = 0$ and $y = 1$, the gate output is 0; each input is a fixed logic level and the output is determined by the gate's truth table. If $x = A$ and $y = B$, then the output is $A \cdot B$; each input is a single Boolean variable. Inputs $x$ and $y$ can each be more-complex Boolean expressions. For example, if $x = A \cdot B + C \cdot D$ and $y = E \cdot F$, the output of the AND gate is $(A \cdot B + C \cdot D) \cdot (E \cdot F)$.

**Figure 3.2**

**Basic Logic Gates and Functions**

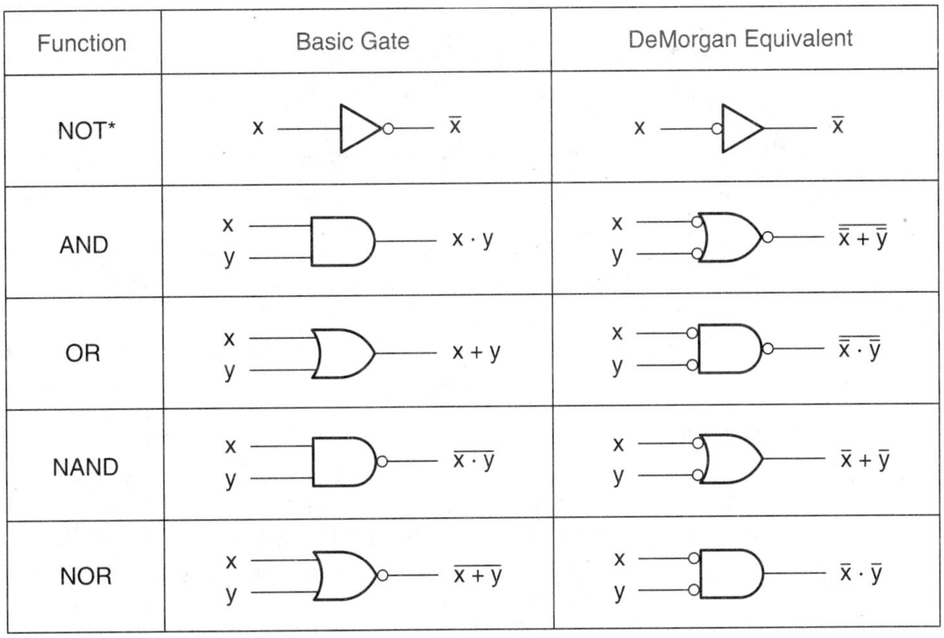

| Function | Basic Gate | DeMorgan Equivalent |
|---|---|---|
| NOT* | $x$ —▷∘— $\bar{x}$ | $x$ —∘▷— $\bar{x}$ |
| AND | $x$, $y$ —D— $x \cdot y$ | $x$, $y$ —D∘— $\overline{\bar{x} + \bar{y}}$ |
| OR | $x$, $y$ —D— $x + y$ | $x$, $y$ —D∘— $\overline{\bar{x} \cdot \bar{y}}$ |
| NAND | $x$, $y$ —D∘— $\overline{x \cdot y}$ | $x$, $y$ —D— $\bar{x} + \bar{y}$ |
| NOR | $x$, $y$ —D∘— $\overline{x + y}$ | $x$, $y$ —D— $\bar{x} \cdot \bar{y}$ |

* NOT, being a single-variable function, is not subject to DeMorgan's Theorems. It is included in this figure to show two senses of the function.

**EXAMPLE 3.1**

Figure 3.3 shows a combinational logic circuit that might be used as part of a cruise control circuit (i.e., automatic speed control) in a car with a manual transmission. The circuit is designed to release the cruise control function and return control to the accelerator pedal. If *RELEASE* = 1, the circuit sends a signal to disengage the cruise control.

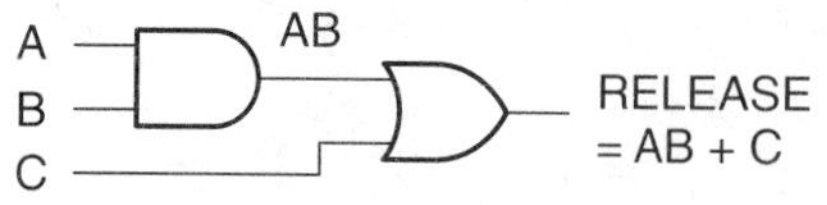

**Figure 3.3**

**Example 3.1**

**Cruise Control RELEASE Circuit**

The input variables are assigned as follows:

$A = 1$ if the clutch pedal is depressed.

$B = 1$ if the shift lever is shifted out of its present position.

$C = 1$ if the brake pedal is depressed.

Find the Boolean expression of the output and briefly describe the operation of the circuit.

**Solution**    The Boolean expression of the AND gate output is $A \cdot B$, as we would expect. The OR gate function, according to Figure 3.2, is $x + y$. In this case,

$$x = A \cdot B \text{ and}$$

$$y = C$$

The function for the circuit is the same as the OR gate output:

$$RELEASE = A \cdot B + C$$

All variables are active HIGH. *RELEASE* is HIGH when $A$ AND $B$ are HIGH OR when $C$ is HIGH. The definitions of the input and output variables thus imply that the cruise control circuit will disengage if (the clutch pedal is depressed AND the shift lever is moved) OR (if the brake pedal is depressed).

The circuit operation in Example 3.1 is defined equally well by the Boolean expression or by the verbal description. Our preference is to use the Boolean expression, since it is more compact and we can manipulate it with Boolean algebra.

**EXAMPLE 3.2**

Figure 3.4 shows a control circuit for a large radio transmitter's cooling fan. The transmitter is used only for short periods of time, thus requiring only occasional cooling. There is a temperature sensor monitoring the transmitter's power amplifier, a timer that measures the time since the fan switched off, and a "Push to Talk" line that is active when the transmitter is on. The control circuit turns on the fan with a LOW at its output. The other variables are defined as follows:

$A = 0$ when temperature > 75°C.

$B = 0$ when timer > 3 minutes.

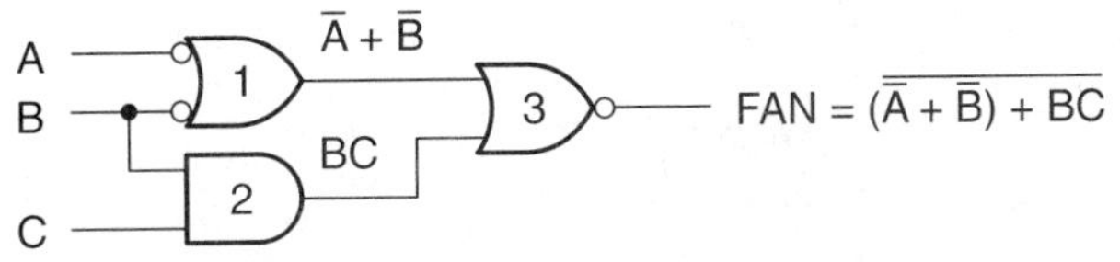

**Figure 3.4**

**Example 3.2**

**Fan Control Circuit**

$B = 1$ when timer $\leq 3$ minutes.

$C = 1$ when the transmitter is on.

Write the Boolean expression for the fan control circuit specified above and a short verbal description of its operation.

**Solution** Look at the output of each gate separately. Gate 1 is a NAND gate in its DeMorgan equivalent form, and gate 2 is an AND gate.

$$\text{Gate 1:} \quad \overline{A} + \overline{B}$$

$$\text{Gate 2:} \quad B \cdot C$$

The output of gate 3 is given by the NOR function, with Boolean expressions for inputs, rather than single variables.

$$\text{Gate 3:} \quad \text{FAN} = \overline{x + y} \text{ where } x = \overline{A} + \overline{B}, \text{ and}$$

$$y = B \cdot C$$

$$\text{FAN} = \overline{(\overline{A} + \overline{B}) + B \cdot C}$$

For the defined variables and the active levels shown in the logic diagram:

1. The output of gate 1 is HIGH if temperature > 75°C OR the fan has been off for 3 minutes.

2. The output of gate 2 is HIGH if the fan has been off for less than 3 minutes AND the transmitter is ON.

3. The output of gate 3 goes LOW and turns on the fan if either of the previous conditions are satisfied.

Therefore, the fan will operate if the temperature of the transmitter power amplifier exceeds 75°C OR if the fan has been off for 3 minutes OR if (the fan has been off for less than 3 minutes AND the transmitter is ON).

(There is a redundancy in the definition of this problem. Can you think of a simpler circuit that will do the same job?)

## Order of Precedence

> **Order of precedence** *The standard order in which Boolean functions (AND, OR, NOT) are performed when evaluating a Boolean expression.*

When we have a Boolean expression made up of several terms, we must have a way of knowing which operations are performed first. In other words, we must establish an **order of precedence** for Boolean functions. For instance, if we have the expression $Y =$

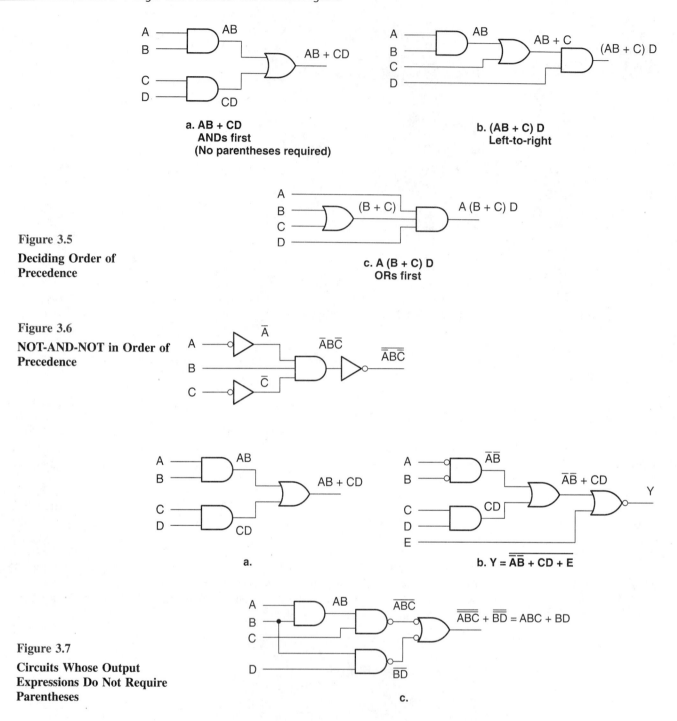

**Figure 3.5**

**Deciding Order of Precedence**

a. AB + CD
ANDs first
(No parentheses required)

b. (AB + C) D
Left-to-right

c. A (B + C) D
ORs first

**Figure 3.6**

**NOT-AND-NOT in Order of Precedence**

a.

b. Y = $\overline{\overline{AB} + CD + E}$

**Figure 3.7**

**Circuits Whose Output Expressions Do Not Require Parentheses**

c.

*AB + CD,* do we perform the AND operations first: *(AB) + (CD);* the operations from left to right: *((AB) + C)D;* or the ORs first: *A(B + C)D?* This is an important question, because each grouping gives a very different circuit, as shown in Figure 3.5.

The usual convention is to choose the order of evaluation shown in Figure 3.5a: ANDs first unless otherwise indicated by parentheses. Table 3.1 shows the priority of the basic logic functions in Boolean algebra.

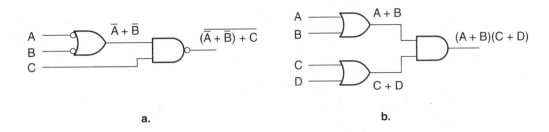

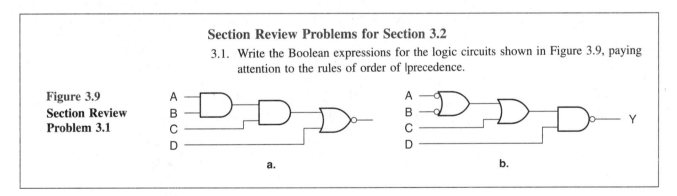

**Figure 3.8**
**Circuits Whose Output Expressions Require Parentheses**

**Table 3.1** Order of Precedence for Boolean Functions

| Priority | Function |
|----------|----------|
| 1 | NOT (variable) |
| 2 | AND |
| 3 | OR |
| 4 | NOT (function) |

The NOT function has a different priority, depending on whether it operates on a function or a variable. For example, in the function $Y = \overline{\overline{A}\ B\ \overline{C}}$, the order of precedence is NOT (variables $A$ and $C$), AND, then NOT (complete function), which gives us the logic diagram shown in Figure 3.6. The idea behind this priority is that individual inputs are inverted before combined outputs.

Figure 3.7 shows several circuits whose output Boolean expressions require no parentheses. In each circuit, the order of gates is the same as the standard order of precedence. The main clue is that all ANDs are closer to the circuit input than any of the ORs.

Figure 3.8 shows several circuits whose output expressions *do* require parentheses. In each of these circuits, there is at least one place where the standard order of precedence (AND before OR) is reversed and so must be indicated by parentheses.

---

**Section Review Problems for Section 3.2**

3.1. Write the Boolean expressions for the logic circuits shown in Figure 3.9, paying attention to the rules of order of |precedence.

**Figure 3.9**
**Section Review Problem 3.1**

---

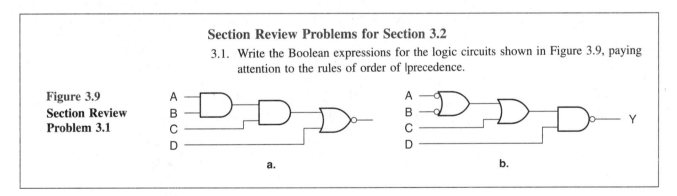

## 3.3

# Logic Diagrams From Boolean Expressions

**Level of gating**  *All gates in a logic circuit required to supply the inputs of the next gate or set of gates. Level 1 is the circuit output.*

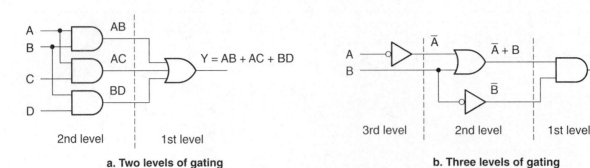

a. Two levels of gating

b. Three levels of gating

**Figure 3.10**

**Levels of Gating**

**Table 3.2** Order of Precedence for Figure 3.10a

| Level | Gate | Output |
|-------|------|--------|
| 1 | OR | $AB + AC + BD$ |
| 2 | AND | $AB$ |
|  | AND | $AC$ |
|  | AND | $BD$ |

**Table 3.3** Order of Precedence for Figure 3.10b

| Level | Gate | Output |
|-------|------|--------|
| 1 | AND | $(\overline{A} + B) \cdot \overline{B}$ |
| 2 | OR | $\overline{A} + B$ |
|  | NOT | $\overline{B}$ |
| 3 | NOT | $\overline{A}$ |

The concept of **levels of gating** can help us draw logic diagrams from Boolean expressions. This is a method of drawing a logic diagram starting from the output and working back to the inputs. The circuit output gate is at level 1. All gates needed to supply the level 1 inputs are at level 2. The gates required to supply all level 2 inputs are at level 3, and so on. As before, each input can be a logic level, a variable, or a Boolean expression.

The Boolean expression

$$Y = AB + AC + BD$$

has two levels of gating, as shown in Figure 3.10a. The order of precedence rules, listed in Table 3.2, tell us to perform AND functions before OR functions. Thus, the level 1 function (the last performed) will be an OR. The AND gates are all at level 2 because all their outputs are needed to supply the inputs for level 1, the OR gate.

The expression

$$Y = (\overline{A} + B) \cdot \overline{B}$$

requires three levels of gating, as shown in Figure 3.10b. Order of precedence rules, shown in Table 3.3, tell us to invert $A$ and $B$, perform the parenthesized OR, and then perform the AND.

Both inverters in Figure 3.10b seem to have the same function, so why is inverter $A$ a third-level gate, while inverter $B$ is at the second level? Inverter $B$ supplies the output OR, making it a level 2 gate. Inverter $A$, on the other hand, supplies a level 2 OR, thus placing it in level 3.

**EXAMPLE 3.3**

Draw the logic circuit for the following Boolean expression:

$$Y = (A + \overline{B})(\overline{A} + C)$$

**Solution**    Order of precedence:    1. Invert $A$ and $B$
                                        2. Parenthesized OR
                                        3. AND

This order of precedence is matched with gating levels in Table 3.4. Figure 3.11 shows how we build the circuit a step at a time.

**Table 3.4**  Order of Precedence for Figure 3.11

| Level | Gate | Output | Figure |
|-------|------|--------|--------|
| 1 | AND | $(A + \overline{B})(\overline{A} + C)$ | 3.11a |
| 2 | OR | $A + \overline{B}$ | 3.11b |
|   | OR | $\overline{A} + C$ | |
| 3 | NOT | $\overline{A}$ | 3.11c |
|   | NOT | $\overline{B}$ | |

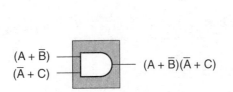

a. 1st level

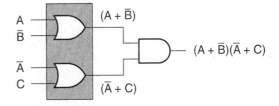

b. 2nd level

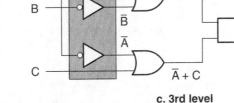

c. 3rd level

**Figure 3.11**

**Example 3.3**
**Building the Logic Circuit**
**for Y = (A + B̄) (Ā + C)**

**EXAMPLE 3.4**

Repeat the problem of Example 3.3 for the expression:

$$Y = A + \overline{B}\,\overline{A} + C$$

**Solution**    Order of precedence:    1. Invert $A$ and $B$
                                        2. AND function
                                        3. OR function

Table 3.5 shows the corresponding gating levels. Figure 3.12 shows the step-by-step construction of the network.

**Table 3.5** Order of Precedence for Figure 3.12

| Level | Gate | Output | Figure |
|-------|------|--------|--------|
| 1 | OR | $A + \overline{B}\,\overline{A} + C$ | 3.12a |
| 2 | AND | $\overline{B}\,\overline{A}$ | 3.12b |
| 3 | NOT | $\overline{A}$ | 3.12c |
| 3 | NOT | $\overline{B}$ | 3.12c |

**Figure 3.12**

**Example 3.4**
**Building the Logic Circuit**
**for Y = A + $\overline{B}\,\overline{A}$ + C**

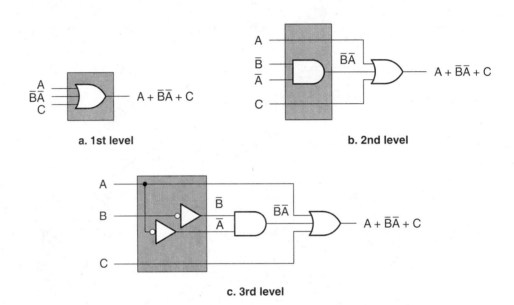

a. 1st level

b. 2nd level

c. 3rd level

We can invert a logic function with an inverting gate, such as a NAND or a NOR, or an inverter. The inverter, if used, adds another level of gating, whereas the NAND and NOR do not. A NAND function has the same priority as an AND function, and a NOR has the same priority as an OR.

**EXAMPLE 3.5**

Draw the logic diagram corresponding to the following Boolean expression:

$$Y = \overline{AB + \overline{(C + D)}} \cdot \overline{ACD}$$

**Table 3.6** Order of Precedence for Figure 3.13

| Level | Gate | Output | Figure |
|-------|------|--------|--------|
| 1 | AND | $\overline{AB + \overline{(C + D)}} \cdot \overline{ACD}$ | 3.13a |
| 2 | NOR | $\overline{AB + (C + D)}$ | 3.13b |
|  | NAND | $\overline{ACD}$ |  |
| 3 | AND | $AB$ | 3.13c |
|  | NOR | $\overline{C + D}$ |  |

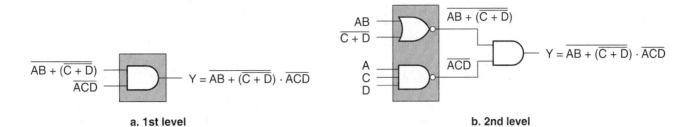

a. 1st level          b. 2nd level

c. 3rd level

**Figure 3.13**
**Example 3.5**
**Building the Logic Circuit for Y = $\overline{AB + \overline{(C + D)}} \cdot \overline{ACD}$**

**Solution**    Table 3.6 shows the order of precedence and gating levels for the required Boolean expression. Figure 3.13 shows the step-by-step construction of the logic diagram.

In Example 3.5, the AND is lower in the order of precedence than either of the OR functions. This is because both ORs are under bars (i.e., NOR functions) and are thus treated as though they were in parentheses.

---

**Section Review Problems for Section 3.3**

3.2.  Draw the logic diagrams corresponding to the following Boolean expressions.

a.  $Y = (AB + C)BC$

b.  $Y = \overline{\overline{AB} + \overline{(B + C)}}$

---

## 3.4

# Evaluating Boolean Expressions

A Boolean expression is a shorthand way of listing the responses of a digital system to all possible combinations of inputs. A truth table of a circuit is a list of these responses. To evaluate an expression for a particular set of inputs, we substitute the numerical values into the expression and apply the rules for the NOT, AND, and OR functions, shown in Table 3.7.

**Table 3.7** Numerical Values of NOT, AND, and OR Functions

| NOT | AND | OR |
|---|---|---|
| $\overline{0} = 1$ | $0 \cdot 0 = 0$ | $0 + 0 = 0$ |
| | $0 \cdot 1 = 0$ | $0 + 1 = 1$ |
| $\overline{1} = 0$ | $1 \cdot 0 = 0$ | $1 + 0 = 1$ |
| | $1 \cdot 1 = 1$ | $1 + 1 = 1$ |
| (Output opposite of input) | (Output HIGH if both inputs HIGH) | (Output HIGH if either input HIGH) |

**EXAMPLE 3.6**

Evaluate the expression:

$$Y = \overline{\overline{A} \cdot B} \cdot \overline{(A \cdot C + \overline{B} \cdot C)}$$

for $A = 1$, $B = 0$, $C = 1$.

**Solution**

$$Y = \overline{\overline{1} \cdot 0} \cdot \overline{(1 \cdot 1 + \overline{0} \cdot 1)}$$

$$= \overline{\overline{1} \cdot 0} \cdot \overline{(1 \cdot 1 + 1 \cdot 1)}$$

$$= \overline{0} \cdot \overline{(1 + 1)}$$

$$= \overline{0} \cdot \overline{1}$$

$$= 1 \cdot 0$$

$$= 0$$

---

**Section Review Problems for Section 3.4**

3.3.  Evaluate the same expression in Example 3.6 if:

a.  $A = B = C = 0$

b.  $A = B = 1$, $C = 0$

---

## 3.5

## Truth Tables From Logic Diagrams

Figure 3.15 shows a simple logic circuit. To get a truth table for this circuit, we need to find out how the different combinations of input values affect not only the output, but various intermediate points within the circuit.

Since this circuit is so simple, there is only one such intermediate point: the output of the AND gate, which is designated as (1) in Table 3.8 and Figure 3.15. Each entry under column (1) is the output of the AND gate for the corresponding values of the $A$ and $B$ inputs. We can find the output $Y$ by ORing the values in column (1) with $C$. Thus, $Y = 1$ if $(A \cdot B = 1)$ OR $(C = 1)$.

For more complex circuits, we would need a column for every gate output in the circuit.

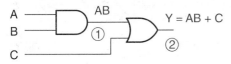

**Figure 3.15**
**Logic Circuit**

**Table 3.8** Truth Table for Figure 3.15

| A | B | C | (1) A · B | Y = (1) + C | |
|---|---|---|---|---|---|
| 0 | 0 | 0 | 0 | 0 | |
| 0 | 0 | 1 | 0 | 1 | (C = 1) |
| 0 | 1 | 0 | 0 | 0 | |
| 0 | 1 | 1 | 0 | 1 | (C = 1) |
| 1 | 0 | 0 | 0 | 0 | |
| 1 | 0 | 1 | 0 | 1 | (C = 1) |
| 1 | 1 | 0 | 1 | 1 | (A · B = 1) |
| 1 | 1 | 1 | 1 | 1 | (A · B = 1, C = 1) |

▼

**EXAMPLE 3.7**    Find the truth table for the circuit shown in Figure 3.16.

**Solution**    The circuit shown has the following intermediate points marked:

$(1) = \overline{A}$

$(2) = \overline{C}$

$(3) = \overline{A}B$

$(4) = B\overline{C}$

Construct the truth table by finding the logic level at each designated point for each possible input combination. Each logic level is determined by an AND, an OR, a NOT, a NAND, or a NOR function operating on the logic levels of one or two previous columns. Table 3.9 shows the truth table for the circuit.

**Figure 3.16**

**Example 3.7**
**Logic Circuit**

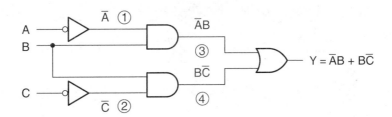

**Table 3.9** Truth Table for Figure 3.16

| A | B | C | $\overline{A}$ (1) | $\overline{C}$ (2) | (3) = $\overline{A}B$ (1) · B | (4) = $B\overline{C}$ (2) · B | Y = (3) + (4) |
|---|---|---|---|---|---|---|---|
| 0 | 0 | 0 | 1 | 1 | 0 | 0 | 0 |
| 0 | 0 | 1 | 1 | 0 | 0 | 0 | 0 |
| 0 | 1 | 0 | 1 | 1 | 1 | 1 | 1 |
| 0 | 1 | 1 | 1 | 0 | 1 | 0 | 1 |
| 1 | 0 | 0 | 0 | 1 | 0 | 0 | 0 |
| 1 | 0 | 1 | 0 | 0 | 0 | 0 | 0 |
| 1 | 1 | 0 | 0 | 1 | 0 | 1 | 1 |
| 1 | 1 | 1 | 0 | 0 | 0 | 0 | 0 |

**EXAMPLE 3.8**     Find the truth table for the logic diagram in Figure 3.17.

**Figure 3.17**

**Example 3.8**
**Logic Circuit**

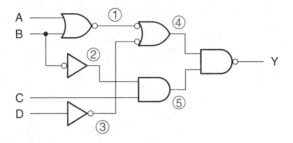

**Solution**     Designate the following intermediate points in the circuit:

$(1) = \overline{A + B}$                 (LOW if $A$ OR $B$ HIGH)

$(2) = \overline{B}$

$(3) = \overline{D}$

$(4) = \overline{(1)} + \overline{(3)} = (A + B) + D$     (HIGH if (1) OR (3) LOW) (Alternate interpretation:
                                                         HIGH if $A$ OR $B$ OR $D$ HIGH)

$(5) = (2) \cdot C = \overline{B} \cdot C$

Use the logic levels and functions at each intermediate point to construct the truth table. The complete truth table, including intermediate points, is shown in Table 3.10.

**Table 3.10** Truth Table for Figure 3.17

| A | B | C | D | $\overline{A + B}$ (1) | $\overline{B}$ (2) | $\overline{D}$ (3) | $\overline{(1)} + \overline{(3)}$ (4) | $(2) \cdot C$ (5) | $Y = \overline{(4) \cdot (5)}$ |
|---|---|---|---|---|---|---|---|---|---|
| 0 | 0 | 0 | 0 | 1 | 1 | 1 | 0 | 0 | 1 |
| 0 | 0 | 0 | 1 | 1 | 1 | 0 | 1 | 0 | 1 |
| 0 | 0 | 1 | 0 | 1 | 1 | 1 | 0 | 1 | 1 |
| 0 | 0 | 1 | 1 | 1 | 1 | 0 | 1 | 1 | 0 |
| 0 | 1 | 0 | 0 | 0 | 0 | 1 | 1 | 0 | 1 |
| 0 | 1 | 0 | 1 | 0 | 0 | 0 | 1 | 0 | 1 |
| 0 | 1 | 1 | 0 | 0 | 0 | 1 | 1 | 0 | 1 |
| 0 | 1 | 1 | 1 | 0 | 0 | 0 | 1 | 0 | 1 |
| 1 | 0 | 0 | 0 | 0 | 1 | 1 | 1 | 0 | 1 |
| 1 | 0 | 0 | 1 | 0 | 1 | 0 | 1 | 0 | 1 |
| 1 | 0 | 1 | 0 | 0 | 1 | 1 | 1 | 1 | 0 |
| 1 | 0 | 1 | 1 | 0 | 1 | 0 | 1 | 1 | 0 |
| 1 | 1 | 0 | 0 | 0 | 0 | 1 | 1 | 0 | 1 |
| 1 | 1 | 0 | 1 | 0 | 0 | 0 | 1 | 0 | 1 |
| 1 | 1 | 1 | 0 | 0 | 0 | 1 | 1 | 0 | 1 |
| 1 | 1 | 1 | 1 | 0 | 0 | 0 | 1 | 0 | 1 |

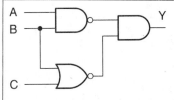

**Section Review Problem for Section 3.5**

3.4. Find the truth table of the logic diagram shown in Figure 3.18.

◀ **Figure 3.18**
**Section Review Problem 3.4**

### 3.6

## Truth Tables From Boolean Expressions

We can find a truth table directly from the Boolean expression of a circuit output. The technique is similar to that for finding the truth table from a logic circuit diagram.

Let us find the truth table of the expression $Y = B(A + C)$. Designate one column for the term $(A + C)$. According to order of precedence rules, we must evaluate this term first. The output column is filled by evaluating the AND function $(1) \cdot B = (A + C) \cdot B$ for each line. Table 3.11 shows the resulting truth table.

**Table 3.11** Truth Table for $Y = B(A + C)$

| A | B | C | $(A + C)$ (1) | $Y = (1) \cdot B$ |
|---|---|---|---|---|
| 0 | 0 | 0 | 0 | 0 |
| 0 | 0 | 1 | 1 | 0 |
| 0 | 1 | 0 | 0 | 0 |
| 0 | 1 | 1 | 1 | 1 |
| 1 | 0 | 0 | 1 | 0 |
| 1 | 0 | 1 | 1 | 0 |
| 1 | 1 | 0 | 1 | 1 |
| 1 | 1 | 1 | 1 | 1 |

**EXAMPLE 3.9**

Find the truth table for the expression

$$Y = (B \cdot C + \overline{B \cdot C} \cdot \overline{D}) \cdot (A + C \cdot D)$$

**Solution**    Designate a column in the truth table for each of the following intermediate steps in the equation.

$$(1) = B \cdot C$$

$$(2) = \overline{B \cdot C}$$

$$(3) = \overline{D}$$

$$(4) = (2) \cdot (3) = \overline{B \cdot C} \cdot \overline{D}$$

$$(5) = (1) + (4) = B \cdot C + (\overline{B \cdot C} \cdot \overline{D})$$

$$(6) = CD$$

$$(7) = A + (6)$$

$$Y = (5) \cdot (7)$$

**Table 3.12**  Truth Table for Example 3.9

| A | B | C | D | $BC$ (1) | $\overline{BC}$ (2) | $\overline{D}$ (3) | (2) · (3) (4) | (1) + (4) (5) | $CD$ (6) | $A$ + (6) (7) | (5) · (7) $Y$ |
|---|---|---|---|---|---|---|---|---|---|---|---|
| 0 | 0 | 0 | 0 | 0 | 1 | 1 | 1 | 1 | 0 | 0 | 0 |
| 0 | 0 | 0 | 1 | 0 | 1 | 0 | 0 | 0 | 0 | 0 | 0 |
| 0 | 0 | 1 | 0 | 0 | 1 | 1 | 1 | 1 | 0 | 0 | 0 |
| 0 | 0 | 1 | 1 | 0 | 1 | 0 | 0 | 0 | 1 | 1 | 0 |
| 0 | 1 | 0 | 0 | 0 | 1 | 1 | 1 | 1 | 0 | 0 | 0 |
| 0 | 1 | 0 | 1 | 0 | 1 | 0 | 0 | 0 | 0 | 0 | 0 |
| 0 | 1 | 1 | 0 | 1 | 0 | 1 | 0 | 1 | 0 | 0 | 0 |
| 0 | 1 | 1 | 1 | 1 | 0 | 0 | 0 | 1 | 1 | 1 | 1 |
| 1 | 0 | 0 | 0 | 0 | 1 | 1 | 1 | 1 | 0 | 1 | 1 |
| 1 | 0 | 0 | 1 | 0 | 1 | 0 | 0 | 0 | 0 | 1 | 0 |
| 1 | 0 | 1 | 0 | 0 | 1 | 1 | 1 | 1 | 0 | 1 | 1 |
| 1 | 0 | 1 | 1 | 0 | 1 | 0 | 0 | 0 | 1 | 1 | 0 |
| 1 | 1 | 0 | 0 | 0 | 1 | 1 | 1 | 1 | 0 | 1 | 1 |
| 1 | 1 | 0 | 1 | 0 | 1 | 0 | 0 | 0 | 0 | 1 | 0 |
| 1 | 1 | 1 | 0 | 1 | 0 | 1 | 0 | 1 | 0 | 1 | 1 |
| 1 | 1 | 1 | 1 | 1 | 0 | 0 | 0 | 1 | 1 | 1 | 1 |

Table 3.12 shows the resulting truth table.

---

**Section Review Problem for Section 3.6**

3.5.  Find the truth table for the following Boolean expression:

$$Y = \overline{BC} \cdot (\overline{\overline{A} + D}) \cdot AC$$

## 3.7

# Boolean Expressions From Truth Tables

**Product term**  *A term in a Boolean expression where one or more true or complement variables are ANDed (e.g., $\overline{A}\,\overline{C}$).*

**Minterm**  *A product term in a Boolean expression where all possible variables appear once in true or complement form (e.g., $\overline{A}\,\overline{B}\,C$; $A\,\overline{B}\,\overline{C}$).*

**Sum term**  *A term in a Boolean expression where one or more true or complement variables are ORed (e.g., $\overline{A} + B + \overline{D}$).*

**Maxterm**  *A sum term in a Boolean expression where all possible variables appear once, in true or complement form (e.g., $(\overline{A} + \overline{B} + C)$; $(A + \overline{B} + C)$).*

**Sum-of-products (SOP)**  *A type of Boolean expression where several product terms are summed (ORed) together (e.g., $\overline{A}\,B\,\overline{C} + \overline{A}\,\overline{B}\,C + A\,B\,C$).*

> **Product-of-sums (POS)** *A type of Boolean expression where several sum terms are multiplied (ANDed) together (e.g., $(\overline{A} + \overline{B} + C)(A + \overline{B} + \overline{C})(\overline{A} + \overline{B} + \overline{C})$).*
>
> **Bus form** *A way of drawing a logic diagram so that each true and complement input variable is available along a continuous conductor called a bus.*

$\mathsf{S}$uppose we have an unknown digital circuit, represented by the block in Figure 3.19. All we know is which terminals are inputs, which are outputs, and how to connect the power. Given only that information, we can find the Boolean expression of the output.

The first thing to do is find the circuit truth table by applying all possible input combinations in binary order and reading the output for each one. Suppose the unknown circuit in Figure 3.19 yields the truth table shown in Table 3.13.

The truth table output is HIGH for three conditions:

1. When *A* AND *B* AND *C* are all LOW, OR
2. When *A* is LOW AND *B* AND *C* are HIGH, OR
3. When *A* is HIGH AND *B* AND *C* are LOW.

**Figure 3.19**

**Digital Circuit With Unknown Function**

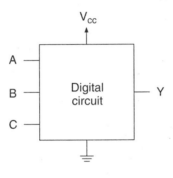

**Table 3.13**  Truth Table for Figure 3.19

| *A* | *B* | *C* | *Y* |
|-----|-----|-----|-----|
| 0 | 0 | 0 | 1 |
| 0 | 0 | 1 | 0 |
| 0 | 1 | 0 | 0 |
| 0 | 1 | 1 | 1 |
| 1 | 0 | 0 | 1 |
| 1 | 0 | 1 | 0 |
| 1 | 1 | 0 | 0 |
| 1 | 1 | 1 | 0 |

Each of those conditions represents a **minterm** in the output Boolean expression. (A minterm is a **product term** (AND term) that includes all variables *(A, B, C)* in true or complement form.) The minterms are:

1. $\overline{A}\,\overline{B}\,\overline{C}$
2. $\overline{A}\,B\,C$
3. $A\,\overline{B}\,\overline{C}$

Since condition 1 OR condition 2 OR condition 3 produces a HIGH output from the circuit, the Boolean function *Y* consists of all three minterms summed (ORed) together, as follows:

$$Y = \overline{A}\,\overline{B}\,\overline{C} + \overline{A}\,B\,C + A\,\overline{B}\,\overline{C}$$

This expression is in a standard form called **sum-of-products** (SOP) form. Figure 3.20 shows the equivalent logic circuit.

**Figure 3.20**

**Logic Circuit for Y = $\overline{A}\,\overline{B}\,\overline{C}$ + $\overline{A}BC$ + $A\overline{B}\,\overline{C}$**

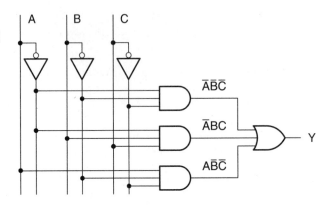

The inputs *A, B,* and *C* and their complements are shown in **bus form.** Each variable is available, in true or complement form, at any point along a conductor. This is a useful, uncluttered notation for circuits that require all of the input variables several times.

> We can derive an SOP expression from a truth table as follows:
>
> 1. Every line on the truth table that has a HIGH output corresponds to a minterm in the truth table's Boolean expression.
>
> 2. Write all truth table variables for every minterm in true or complement form. If a variable is 0, write it in complement form (with a bar over it); if it is 1, write it in true form (no bar).
>
> 3. Combine all minterms in an OR function.

We can also find the Boolean function of a truth table in **product-of-sums** (POS) form. The product-of-sums form of a Boolean expression consists of a number of **maxterms** (i.e., **sum terms** (OR terms) containing all variables in true or complement form) that are ANDed together. To find the POS form of *Y,* we will find the SOP expression for $\overline{Y}$ and apply DeMorgan's Theorems.

Recall DeMorgan's Theorems:

$$\overline{x + y + z} = \overline{x}\,\overline{y}\,\overline{z}$$

$$\overline{x\,y\,z} = \overline{x} + \overline{y} + \overline{z}$$

When the theorems were introduced, they were presented as two-variable theorems, but in fact they are valid for any number of variables.

Let's reexamine Table 3.13. To find the sum-of-products expression for *Y,* we wrote a minterm for each line where $Y = 1$. To find the SOP expression for $\overline{Y}$, we must write a minterm for each line where $Y = 0$. Variables *A, B,* and *C* must appear in each minterm, in true or complement form. A variable is in complement form (with a bar over the top) if its value is 0 in that minterm, and it is in true form (no bar) if its value is 1 in that minterm.

We get the following minterms for $\overline{Y}$:

$$\overline{A}\,\overline{B}\,C$$
$$\overline{A}\,B\,\overline{C}$$

$$A\,\overline{B}\,C$$
$$A\,B\,\overline{C}$$
$$A\,B\,C$$

Thus, the SOP form of $\overline{Y}$ is:

$$\overline{Y} = \overline{A}\,\overline{B}\,C + \overline{A}\,B\,\overline{C} + A\,\overline{B}\,C + A\,B\,\overline{C} + A\,B\,C$$

To get $Y$ in POS form, we must invert both sides of the above expression and apply DeMorgan's Theorems to the righthand side.

$$Y = \overline{\overline{Y}} = \overline{\overline{A}\,\overline{B}\,C + \overline{A}\,B\,\overline{C} + A\,\overline{B}\,C + A\,B\,\overline{C} + A\,B\,C}$$

$$= (\overline{\overline{A}\,\overline{B}\,C})(\overline{\overline{A}\,B\,\overline{C}})(\overline{A\,\overline{B}\,C})(\overline{A\,B\,\overline{C}})(\overline{A\,B\,C})$$

$$= (A + B + \overline{C})(A + \overline{B} + C)(\overline{A} + B + \overline{C})(\overline{A} + \overline{B} + C)(\overline{A} + \overline{B} + \overline{C})$$

This Boolean expression can be implemented by the logic circuit in Figure 3.21.

**Figure 3.21**

**Logic Circuit for**
**Y = (A + B + C̄)**
**(A + B̄ + C) (Ā + B + C̄)**
**(Ā + B̄ + C) (Ā + B̄+ C̄)**

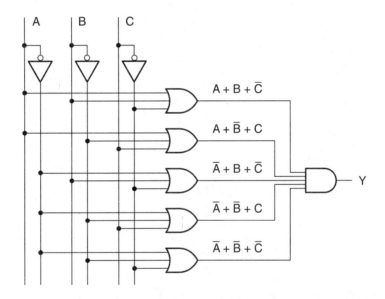

We don't have to go through the whole process outlined above every time we want to find the POS form of a function. We can find it directly from the truth table, following the procedure summarized below. Use this procedure to find the POS form of the expression given by Table 3.13. The terms in this expression are the same as those derived by DeMorgan's Theorem.

Deriving a POS expression from a truth table:

1. Every line on the truth table that has a LOW output corresponds to a maxterm in the truth table's Boolean expression.

2. Write all truth table variables for every maxterm in true or complement form. If a variable is 1, write it in complement form (with a bar over it); if it is 0, write it in true form (no bar).

3. Combine all maxterms in an AND function.

Note that these steps are all opposite to those used to find the SOP form of the Boolean expression.

---

**EXAMPLE 3.10**

Find the Boolean expression, in both SOP and POS forms, for the logic function represented by Table 3.14. Draw the logic circuit for each form.

**Table 3.14** Truth Table for Example 3.10 (with minterms and maxterms)

| A | B | C | D | Y | Minterms | Maxterms |
|---|---|---|---|---|---|---|
| 0 | 0 | 0 | 0 | 1 | $\overline{A}\,\overline{B}\,\overline{C}\,\overline{D}$ | |
| 0 | 0 | 0 | 1 | 1 | $\overline{A}\,\overline{B}\,\overline{C}\,D$ | |
| 0 | 0 | 1 | 0 | 0 | | $A + B + \overline{C} + D$ |
| 0 | 0 | 1 | 1 | 1 | $\overline{A}\,\overline{B}\,C\,D$ | |
| 0 | 1 | 0 | 0 | 0 | | $A + \overline{B} + C + D$ |
| 0 | 1 | 0 | 1 | 0 | | $A + \overline{B} + C + \overline{D}$ |
| 0 | 1 | 1 | 0 | 0 | | $A + \overline{B} + \overline{C} + D$ |
| 0 | 1 | 1 | 1 | 0 | | $A + \overline{B} + \overline{C} + \overline{D}$ |
| 1 | 0 | 0 | 0 | 1 | $A\,\overline{B}\,\overline{C}\,\overline{D}$ | |
| 1 | 0 | 0 | 1 | 0 | | $\overline{A} + B + C + \overline{D}$ |
| 1 | 0 | 1 | 0 | 1 | $A\,\overline{B}\,C\,\overline{D}$ | |
| 1 | 0 | 1 | 1 | 0 | | $\overline{A} + B + \overline{C} + \overline{D}$ |
| 1 | 1 | 0 | 0 | 1 | $A\,B\,\overline{C}\,\overline{D}$ | |
| 1 | 1 | 0 | 1 | 1 | $A\,B\,\overline{C}\,D$ | |
| 1 | 1 | 1 | 0 | 1 | $A\,B\,C\,\overline{D}$ | |
| 1 | 1 | 1 | 1 | 0 | | $\overline{A} + \overline{B} + \overline{C} + \overline{D}$ |

**Solution**  All minterms (for SOP form) and maxterms (for POS form) are shown in the last two columns of Table 3.14.

**Boolean Expressions:**

*SOP form:*

$$Y = \overline{A}\,\overline{B}\,\overline{C}\,\overline{D} + \overline{A}\,\overline{B}\,\overline{C}\,D + \overline{A}\,\overline{B}\,C\,D + A\,\overline{B}\,\overline{C}\,\overline{D} + A\,\overline{B}\,C\,\overline{D} + A\,B\,\overline{C}\,\overline{D}$$

$$+ A\,B\,\overline{C}\,D + A\,B\,C\,\overline{D}$$

*POS form:*

$$Y = (A + B + \overline{C} + D)\,(A + \overline{B} + C + D)\,(A + \overline{B} + C + \overline{D})\,(A + \overline{B} + \overline{C} + D)$$

$$(A + \overline{B} + \overline{C} + \overline{D})\,(\overline{A} + B + C + \overline{D})\,(\overline{A} + B + \overline{C} + \overline{D})\,(\overline{A} + \overline{B} + \overline{C} + \overline{D})$$

The logic circuits are shown in Figures 3.22 and 3.23.

**Figure 3.22**
**Example 3.10**
**SOP Form**

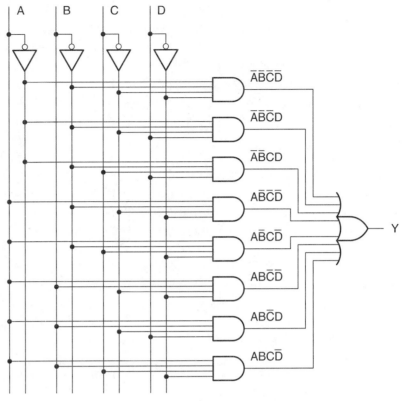

**Figure 3.23**
**Example 3.10**
**POS Form**

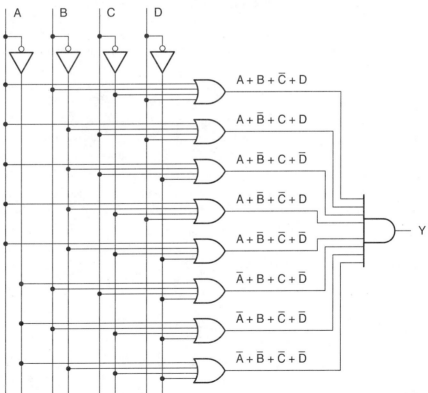

**Section Review Problems for Section 3.7**

3.6. Find the SOP and POS forms of the Boolean functions represented by the following truth tables.

| a. | A | B | C | Y |
|---|---|---|---|---|
| | 0 | 0 | 0 | 0 |
| | 0 | 0 | 1 | 0 |
| | 0 | 1 | 0 | 0 |
| | 0 | 1 | 1 | 0 |
| | 1 | 0 | 0 | 1 |
| | 1 | 0 | 1 | 1 |
| | 1 | 1 | 0 | 0 |
| | 1 | 1 | 1 | 0 |

| b. | A | B | C | Y |
|---|---|---|---|---|
| | 0 | 0 | 0 | 1 |
| | 0 | 0 | 1 | 0 |
| | 0 | 1 | 0 | 0 |
| | 0 | 1 | 1 | 0 |
| | 1 | 0 | 0 | 1 |
| | 1 | 0 | 1 | 1 |
| | 1 | 1 | 0 | 1 |
| | 1 | 1 | 1 | 0 |

## 3.8

# Exclusive OR and Exclusive NOR Circuits

The Exclusive OR and Exclusive NOR functions studied in Chapter 2 can be implemented easily in either SOP or POS form. The truth tables for these functions are shown in Tables 3.15 and 3.16.

The SOP form of the Exclusive OR function is:

$$\text{SOP: } Y = \overline{A}B + A\overline{B}$$

The SOP form of the Exclusive NOR function is:

$$\text{SOP: } Y = \overline{A}\,\overline{B} + A\,B$$

Figure 3.24 shows the circuits for these expressions. You will be asked to find the POS forms of these expressions in the problems at the end of the chapter.

**Table 3.15** Truth Table for Exclusive OR

| A | B | Y |
|---|---|---|
| 0 | 0 | 0 |
| 0 | 1 | 1 |
| 1 | 0 | 1 |
| 1 | 1 | 0 |

$$Y = A \oplus B$$

**Table 3.16** Truth Table for Exclusive NOR

| A | B | Y |
|---|---|---|
| 0 | 0 | 1 |
| 0 | 1 | 0 |
| 1 | 0 | 0 |
| 1 | 1 | 1 |

$$Y = \overline{A \oplus B}$$

**Figure 3.24**

**SOP Forms of XOR and XNOR Circuits**

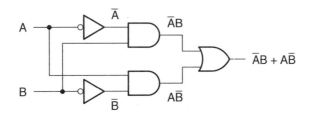

a. SOP form of XOR  function

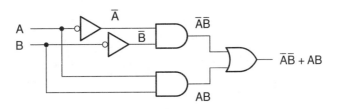

b. SOP form of XNOR  function

## 3.9

# Theorems of Boolean Algebra

At this point you might well ask, "Is there a simpler way to implement a logic expression that doesn't use all the gates required in an SOP or POS circuit?" The circuit derived from a truth table in SOP or POS form is seldom the most efficient way to implement a logic function; it can often be done with at least a few less gates. However, the price we pay for this simplification is the necessity of learning a few rules of Boolean algebra.

Boolean algebra was originally a philosophical tool. It is named for George Boole, a 19th-century English mathematician, who developed a system of algebra to determine the truth or falsehood of logical propositions. Since Boole was looking for a way to solve questions with yes/no answers, his algebra assumes that any variable can have only two possible values.

In technical applications, we use Boolean algebra in much the same way as Boole did to solve true/false philosophical questions. However, the questions we want to answer are of a different sort. Instead of asking, "Is it true what Socrates says about truth and beauty?" we ask, "Should this machine turn on?"

Variables used in stating the rules of Boolean algebra will be lowercase letters, such as *w, x, y,* and *z.* By using lowercase letters, we imply that any valid Boolean constant (1 or 0); any single Boolean variable *(A, B, C),* which would normally be capitalized; or any group of variables *(AC; AB + BD; (C + D)(A + C))* can be substituted into a general rule.

## *Commutative, Associative, and Distributive Properties*

Some of the basic theorems of Boolean algebra are so obvious that one feels they hardly need to be stated. We have, in fact, assumed some of these properties in previous sections of this chapter. However, when learning a new system of mathematics, it is best to formalize our assumptions.

The Commutative, Associative, and Distributive Properties are the same in Boolean algebra as they are in linear algebra.

---

**Commutative Property of Addition** *Two or more Boolean variables can be summed (ORed) in any order.*

1. $x + y = y + x$
   $x + y + z = z + y + x = y + x + z$, etc.

**Commutative Property of Multiplication** *Two or more Boolean variables can be multiplied (ANDed) in any order.*

2. $x \cdot y = y \cdot x$
   $x \cdot y \cdot z = z \cdot y \cdot x = x \cdot z \cdot y$, etc.

---

Figure 3.25 shows the **Commutative Properties** represented as logic gates.

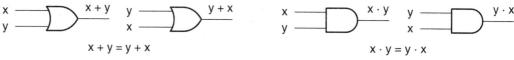

**Figure 3.25**
**Commutative Properties**

$x + y = y + x$

**a. Addition**

$x \cdot y = y \cdot x$

**b. Multiplication**

---

**Associative Property of Addition** *Two or more OR operations can be performed in any order.*

3. $x + (y + z) = (x + y) + z$

**Associative Property of Multiplication** *Two or more AND operations can be performed in any order.*

4. $x(yz) = (xy)z$

---

The rules of order of precedence say that Boolean operations in parentheses are performed first. The circuits in Figure 3.26 reflect this aspect of the **Associative Properties.**

---

**Distributive Properties** *AND operations can be "multiplied through," or distributed over several OR operations.*

5. $x(y + z) = xy + xz$
6. $(x + y)(w + z) = xw + xz + yw + yz$

---

The **Distributive Properties** change a product-of-sums circuit to a sum-of-products circuit. Figure 3.27 shows logic circuits for both sides of the preceding equations.

Figure 3.26
**Associative Properties**

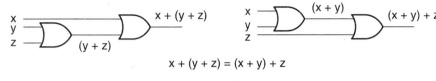

$$x + (y + z) = (x + y) + z$$

**a. Addition**

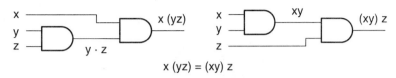

$$x (yz) = (xy) z$$

**b. Multiplication**

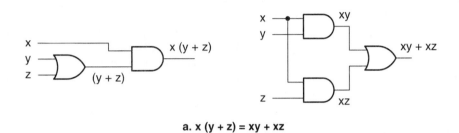

**a. x (y + z) = xy + xz**

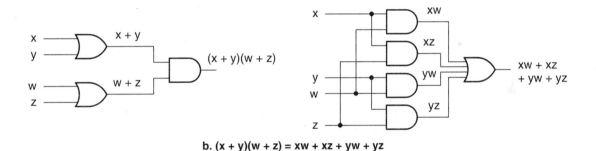

**b. (x + y)(w + z) = xw + xz + yw + yz**

Figure 3.27
**Distributive Properties**

## *Single-Variable Theorems*

There are nine theorems that can be used to manipulate a single variable in a Boolean expression. An easy way to remember these theorems is to divide them into three groups:

1. Four theorems: $x$ AND/OR 0/1
2. Four theorems: $x$ AND/OR $x/\bar{x}$
3. One theorem: Double Inversion

**Figure 3.28**
**X AND/OR 0/1**

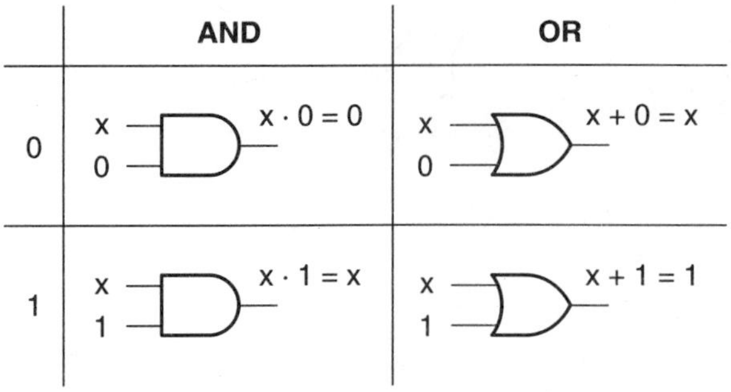

|   | AND | OR |
|---|-----|-----|
| 0 | $x \cdot 0 = 0$ | $x + 0 = x$ |
| 1 | $x \cdot 1 = x$ | $x + 1 = 1$ |

## $x$ AND/OR 0/1

The theorems in the first group can be generated by asking what happens when $x$, a Boolean variable or expression, is at one input of an AND or an OR gate and a 0 or a 1 is at the other.

Examine the truth table of the gate in question. Hold one input of the gate constant and find the effect of the other on the output. This is the same procedure we used in Chapter 2 to examine the enable/inhibit properties of logic gates.

Each of these four theorems can be represented by a logic gate, as shown in Figure 3.28.

**$x \cdot 0$:**

| A | x | Y |
|---|---|---|
| 0 | 0 | 0 |
| 0 | 1 | 0 |
| ~~1~~ | ~~0~~ | ~~0~~ |
| ~~1~~ | ~~1~~ | ~~1~~ |

If $x = 0$, $Y = 0$

If $x = 1$, $Y = 0$

> 7. $x \cdot 0 = 0$

(Can never have both inputs HIGH, therefore output is always LOW.)

**$x + 0$:**

| A | x | Y |
|---|---|---|
| 0 | 0 | 0 |
| 0 | 1 | 1 |
| ~~1~~ | ~~0~~ | ~~1~~ |
| ~~1~~ | ~~1~~ | ~~1~~ |

If $x = 0$, $Y = 0$

If $x = 1$, $Y = 1$

> 8. $x + 0 = x$

(LOW input enables OR gate.)

$x \cdot 1$:

| A | x | Y |
|---|---|---|
| ~~0~~ | ~~0~~ | ~~0~~ |
| ~~0~~ | ~~1~~ | ~~0~~ |
| 1 | 0 | 0 |
| 1 | 1 | 1 |

If $x = 0$, $Y = 0$
If $x = 1$, $Y = 1$
(HIGH input enables AND gate.)

9. $x \cdot 1 = x$

$x + 1$:

| A | x | Y |
|---|---|---|
| ~~0~~ | ~~0~~ | ~~0~~ |
| ~~0~~ | ~~1~~ | ~~1~~ |
| 1 | 0 | 1 |
| 1 | 1 | 1 |

If $x = 0$, $Y = 1$
If $x = 1$, $Y = 1$

10. $x + 1 = 1$

(One input always HIGH, therefore output is always HIGH.)

## $x$ AND/OR $x/\overline{x}$

Four theorems are generated by combining a Boolean variable or expression, $x$, with itself or its complement in an AND or in an OR function.

Again, we can use the AND and OR truth tables. For the first two theorems, we look only at the lines where both inputs are the same. For the other two, we use the lines where the inputs are different.

Figure 3.29 shows the logic gates that represent these theorems.

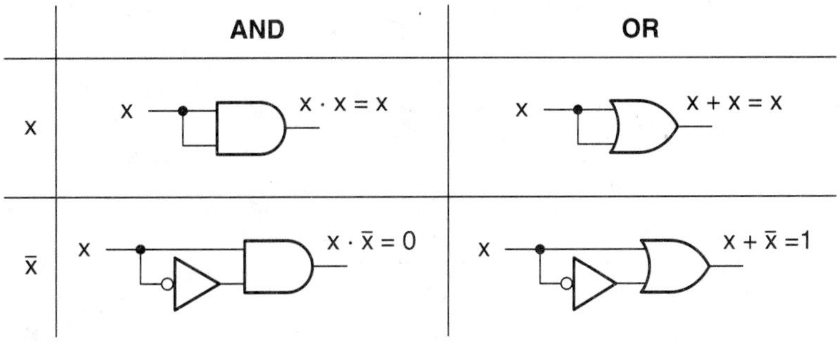

**Figure 3.29**
**X AND/OR X/$\overline{\text{X}}$**

*x · x:*

| A | x | Y |
|---|---|---|
| 0 | 0 | 0 |
| ~~0~~ | ~~1~~ | ~~0~~ |
| ~~1~~ | ~~0~~ | ~~0~~ |
| 1 | 1 | 1 |

If $x = 0$, $Y = 0$

If $x = 1$, $Y = 1$

$$\boxed{11. \ x \cdot x = x}$$

*x + x:*

| A | x | Y |
|---|---|---|
| 0 | 0 | 0 |
| ~~0~~ | ~~1~~ | ~~1~~ |
| ~~1~~ | ~~0~~ | ~~1~~ |
| 1 | 1 | 1 |

If $x = 0$, $Y = 0$

If $x = 1$, $Y = 1$

$$\boxed{12. \ x + x = x}$$

*x · x̄:*

| A | x | Y |
|---|---|---|
| ~~0~~ | ~~0~~ | ~~0~~ |
| 0 | 1 | 0 |
| 1 | 0 | 0 |
| ~~1~~ | ~~1~~ | ~~1~~ |

If $x = 0$, $Y = 0$

If $x = 1$, $Y = 0$

(Since inputs are opposite, can never have both HIGH. Output always LOW.)

$$\boxed{13. \ x \cdot \bar{x} = 0}$$

*x + x̄:*

| A | x | Y |
|---|---|---|
| ~~0~~ | ~~0~~ | ~~0~~ |
| 0 | 1 | 1 |
| 1 | 0 | 1 |
| ~~1~~ | ~~1~~ | ~~1~~ |

If $x = 0$, $Y = 1$

If $x = 1$, $Y = 1$

(Since inputs are opposite, one input always HIGH. Therefore, output is always HIGH.)

$$\boxed{14. \ x + \bar{x} = 1}$$

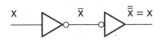

**Figure 3.30**
**Double inversion**

## Double Inversion

The ninth single-variable theorem is just common sense. It states that a variable or expression inverted twice is the same as the original variable or expression. It is given by:

$$15.\ \bar{\bar{x}} = x$$

This theorem is illustrated by the two inverters in Figure 3.30.

## *Multivariable Theorems*

There are numerous multivariable theorems we could learn, but we will look only at three of the most useful.

$$16.\ x + xy = x$$

**Proof:**

$$x + xy = x\,(1 + y) \quad \text{(Distributive Property)}$$
$$= x \cdot 1 \quad\quad (1 + y = 1)$$
$$= x$$

Figure 3.31 illustrates the circuit in this theorem. Note that the equivalent is not a circuit at all, but a single, unmodified variable. Thus, the circuit shown need never be built.

**Figure 3.31**
**Theorem 16**

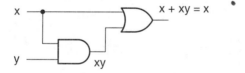

---

**EXAMPLE 3.11**

Simplify the following Boolean expressions, using Theorem 16 and other rules of Boolean algebra. Draw the logic circuits of the unsimplified and simplified expressions.

**a.** $H = K\bar{L} + K$
**b.** $Y = (A + B)CD + (\overline{A + B})$
**c.** $W = (PQR + \bar{P}\,\bar{Q})(S + T) + (\bar{P} + \bar{Q})(S + T) + (S + T)$

**Solution**    Figure 3.32 shows the logic circuits for the unsimplified and simplified versions of the above expressions.

**a.** Let $x = K$, let $y = \bar{L}$:

$$H = x + xy = K + K\bar{L}$$

**Unsimplified** $\Longleftrightarrow$ **Simplified**

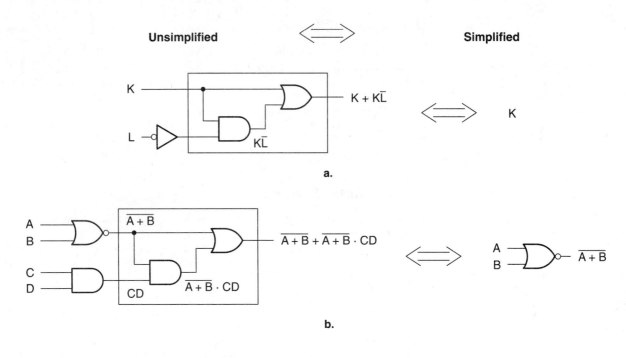

a.

b.

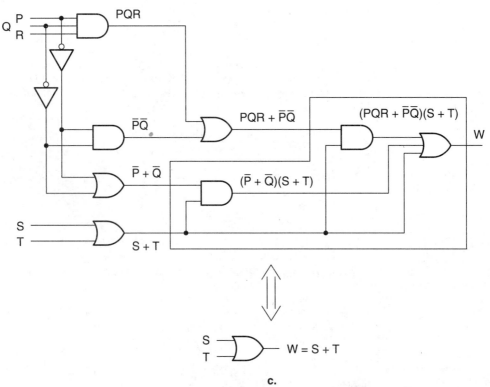

c.

**Figure 3.32**

**Example 3.11**
**Logic Circuits for Unsimplified and Simplified Expressions**

Theorem 16 states $x + xy = x$. Therefore $K + K\overline{L} = K$.

**b.** Let $x = (\overline{A + B})$, let $y = CD$:

$$Y = x + xy = x = \overline{A + B}$$

**c.** Let $x = S + T$, let $y = (\overline{P} + \overline{Q})$:

Since $x + xy = x$, $(\overline{P} + \overline{Q})(S + T) + (S + T) = (S + T)$.

$$W = (PQR + \overline{P}\,\overline{Q})(S + T) + (S + T)$$

Let $x = S + T$, let $y = (PQR + \overline{P}\,\overline{Q})$:

$$W = x + xy = x = S + T$$

**Alternate method:**

$$W = (PQR + \overline{P}\,\overline{Q})(S + T) + (\overline{P} + \overline{Q})(S + T) + (S + T)$$

By the Distributive Property:

$$W = ((PQR + \overline{P}\,\overline{Q}) + (\overline{P} + \overline{Q}))(S + T) + (S + T)$$

Let $x = S + T$, let $y = ((PQR + \overline{P}\,\overline{Q}) + (\overline{P} + \overline{Q}))$:

$$W = x + xy = x = S + T$$

---

17. $(x + y)(x + z) = x + yz$

**Proof:**

$$
\begin{aligned}
(x + y)(x + z) &= xx + xz + xy + yz && \text{(Distributive Property)} \\
&= (x + xy) + xz + yz && (xx = x;\ \text{Associative Property}) \\
&= x + xz + yz && (x + xy = x\ \text{(Theorem 15)}) \\
&= (x + xz) + yz && \text{(Associative Property)} \\
&= x + yz && \text{(Theorem 15)}
\end{aligned}
$$

Figure 3.33 shows the logic circuits for the left and right sides of the equation for Theorem 17. This theorem is a special case of one of the Distributive Properties, Theorem 6, where $w = x$.

**Figure 3.33**

**Theorem 17**

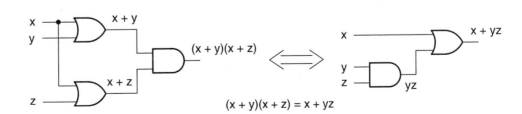

$(x + y)(x + z) = x + yz$

**EXAMPLE 3.12**

Simplify the following Boolean expressions, using Theorem 17 and other rules of Boolean algebra. Draw the logic circuits of the unsimplified and simplified expressions.

  **a.** $L = (M + \overline{N})(M + \overline{P})$
  **b.** $Y = (\overline{A + B} + AB)(\overline{A + B} + C)$

**Solution**

Figure 3.34 shows the logic circuits for the unsimplified and simplified versions of the above expressions.

$$\text{Theorem 17: } (x + y)(x + z) = x + yz$$

  **a.** Let $x = M$, let $y = \overline{N}$, let $z = \overline{P}$:

$$L = (x + y)(x + z) = x + yz = M + \overline{N}\,\overline{P}$$

  **b.** Let $x = \overline{A + B}$, let $y = AB$, let $z = C$:

$$Y = (x + y)(x + z) = x + yz = \overline{A + B} + ABC = \overline{A}\,\overline{B} + ABC$$

**Figure 3.34**

**Example 3.12
Logic Circuits for
Unsimplified and
Simplified Expressions**

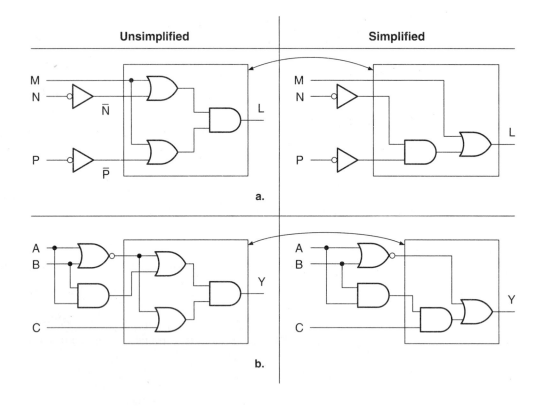

18. $x + \overline{x}y = x + y$

**Proof:**    $x + \bar{x}y = (x + \bar{x})(x + y)$    $(x + y)(x + z) = x + yz$
(Theorem 17)

$$= 1 \cdot (x + y) \qquad (x + \bar{x} = 1)$$

$$= x + y$$

Figure 3.35 illustrates Theorem 18 with a logic circuit.

**Figure 3.35**

**Theorem 18**

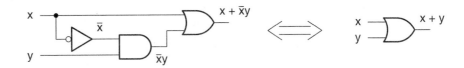

Here is another way to remember Theorem 18:

If a variable $(x)$ is ORed with a term consisting of a different variable (y), which is ANDed with the first variable's complement $(\bar{x})$, the complement disappears.

$$x + \bar{x}y = x + y$$

**EXAMPLE 3.13**

Simplify the following Boolean expressions, using Theorem 18 and other rules of Boolean algebra. Draw the logic circuits of the unsimplified and simplified forms of the expressions.

    **a.** $W = \bar{U} + U\bar{V}$

    **b.** $P = Q\bar{R}S + (\bar{Q} + R + \bar{S})\,T$

    **c.** $J = \overline{KM}\,(\bar{K} + L + M) + KM$

**Solution**    Figure 3.36 shows the circuits for the unsimplified and simplified expressions.

Theorem 18: $x + \bar{x}y = x + y$

**a.** Let $x = \bar{U}$, let $y = \bar{V}$:

$$W = x + \bar{x}y = x + y = \bar{U} + \bar{V}$$

**b.**    $P = Q\bar{R}S + (\bar{Q} + R + \bar{S})\,T$

$$= Q\bar{R}S + \overline{Q\bar{R}S}\,T \qquad \text{(DeMorgan's Theorem)}$$

Let $x = Q\bar{R}S$, let $y = T$:

$$P = x + \bar{x}y = x + y = Q\bar{R}S + T$$

**c.** Let $x = KM$, let $y = (\bar{K} + L + M)$:

$$J = x + \bar{x}y = x + y = KM + \bar{K} + L + M$$

$$= \bar{K} + L + (M + KM) \qquad \text{(Associative Property)}$$

$$= \bar{K} + L + M \qquad \text{(Theorem 16)}$$

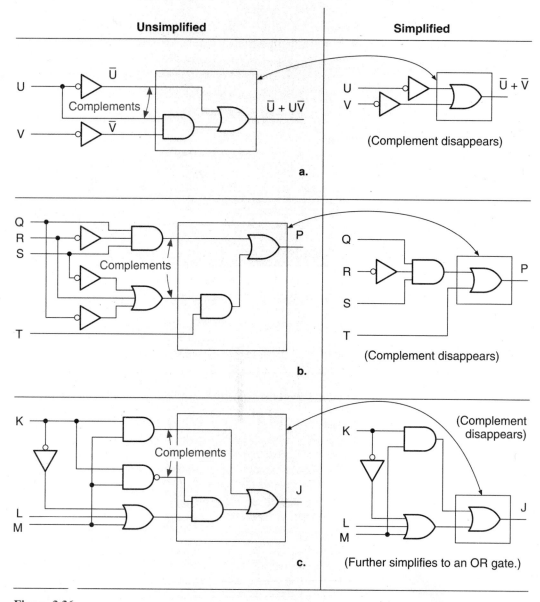

**Figure 3.36**
**Example 3.13**
**Logic Circuits for Unsimplified and Simplified Expressions**

The rules of Boolean algebra are summarized in Table 3.17. Don't try to memorize all these rules. The Commutative, Associative, and Distributive Properties are the same as their counterparts in ordinary algebra. The single-variable theorems can be reasoned out by your knowledge of logic gate operation. That leaves only three multivariable theorems, which you can memorize, if you insist. It is preferable to understand where they come from.

**Table 3.17** Rules of Boolean Algebra

**Commutative Properties**

1. $x + y = y + x$
2. $x \cdot y = y \cdot x$

**Associative Properties**

3. $x + (y + z) = (x + y) + z$
4. $x(yz) = (xy)z$

**Distributive Properties**

5. $x(y + z) = xy + xz$
6. $(x + y)(w + z) = xw + xz + yw + yz$

**x AND/OR 0/1**

7. $x \cdot 0 = 0$
8. $x + 0 = x$
9. $x \cdot 1 = x$
10. $x + 1 = 1$

**x AND/OR x/$\bar{x}$**

11. $x \cdot x = x$
12. $x + x = x$
13. $x \cdot \bar{x} = 0$
14. $x + \bar{x} = 1$

**Double Inversion**

15. $\bar{\bar{x}} = x$

**Multivariable Theorems**

16. $x + xy = x$
17. $(x + y)(x + z) = x + yz$ (special case of Theorem 6 where $w = x$)
18. $x + \bar{x}y = x + y$

---

### Section Review Problems for Section 3.9

3.7. Use theorems of Boolean algebra to simplify the following Boolean expressions.

a. $Y = \overline{AC} + (\overline{A} + \overline{C})D$
b. $Y = \overline{A} + \overline{C} + ACD$
c. $Y = (A\overline{B} + \overline{B}C)(A\overline{B} + \overline{C})$

## ● GLOSSARY

**Associative Property of Addition** Two or more OR operations can be performed in any order.

**Associative Property of Multiplication** Two or more AND operations can be performed in any order.

**Bus form** A way of drawing a logic diagram so that each true and complement input variable is available along a continuous conductor called a bus.

**Combinational logic circuit**  A digital logic circuit whose output depends only on the combination of input logic states. The output does not depend on the order in which the inputs are applied.

**Commutative Property of Addition**  Two or more Boolean variables can be summed (ORed) in any order (i.e., $x + y = y + x$; $x + y + z = z + y + x$; etc.).

**Commutative Property of Multiplication**  Two or more Boolean variables can be multiplied (ANDed) in any order (i.e., $x \cdot y = y \cdot x$; $x \cdot y \cdot z = z \cdot y \cdot x$; etc.).

**Distributive Properties**  AND operations can be "multiplied through," or distributed over several OR operations.

**Level of gating**  All gates in a logic circuit required to supply the inputs of the next gate or set of gates. Level 1 is the circuit output.

**Logic diagram**  A circuit diagram of a digital logic circuit.

**Logic gate network**  A digital circuit consisting of two or more logic gates.

**Maxterm**  A sum term in a Boolean expression where all possible variables appear once, in true or complement form (e.g., $(\overline{A} + \overline{B} + C)$; $(A + \overline{B} + C)$).

**Minterm**  A product term in a Boolean expression where all possible variables appear once in true or complement form (e.g., $\overline{A}\,\overline{B}\,C$; $A\overline{BC}$).

**Order of precedence**  The standard order in which Boolean functions (AND, OR, NOT) are performed when evaluating a Boolean expression.

**Product-of-sums (POS)**  A type of Boolean expression where several sum terms are multiplied (ANDed) together (e.g., $(\overline{A} + \overline{B} + C)(A + \overline{B} + \overline{C})(\overline{A} + \overline{B} + \overline{C})$).

**Product term**  A term in a Boolean expression where one or more true or complement variables are ANDed (e.g., $\overline{A}\,\overline{C}$).

**Sum-of-products (SOP)**  A type of Boolean expression where several product terms are summed (ORed) together (e.g., $\overline{ABC} + \overline{A}\overline{B}C + ABC$).

**Sum term**  A term in a Boolean expression where one or more true or complement variables are ORed (e.g., $\overline{A} + B + \overline{D}$).

---

## ● PROBLEMS

### Section 3.1 Combinational Logic and Boolean Algebra

### Section 3.2 Boolean Expressions From Logic Diagrams

**3.1**  Write the Boolean unsimplified expression for each logic diagram shown in Figure 3.37.

**3.2**  Write the Boolean unsimplified expression for each logic diagram shown in Figure 3.38.

**3.3**  Write the Boolean unsimplified expression for the circuit shown in Figure 3.39.

**3.4**  Write the Boolean unsimplified expressions for the circuits shown in Figure 3.40. Which expression requires parentheses?  Why?

**3.5**  The circuit shown in Figure 3.41 is called a majority vote circuit. It will turn on an indicator lamp (active HIGH) only if a majority of inputs are HIGH. Write the Boolean expression for the circuit.

### Section 3.3 Logic Diagrams From Boolean Expressions

**3.6**  Draw a majority vote circuit (see Problem 3.5) for four input sensors and write the Boolean expression of its output. (Assume that three out of four constitutes a majority.)

**3.7**  Draw the logic circuit for the following Boolean expressions:

**a.** $Y = A B + B C$

**b.** $Y = A C D + B C D$

**c.** $Y = (B + D)(C + A)$

**d.** $Y = B + D C + A$

**e.** $Y = \overline{A C} + \overline{B + C}$

**f.** $Y = \overline{A C} + B + C$

**g.** $Y = \overline{\overline{\overline{A B D}} + \overline{B} C} + \overline{A + C}$

**h.** $Y = \overline{\overline{\overline{A} B} + \overline{A} C + B C}$

**i.** $Y = \overline{\overline{A} B} + \overline{\overline{A} C} + B C$

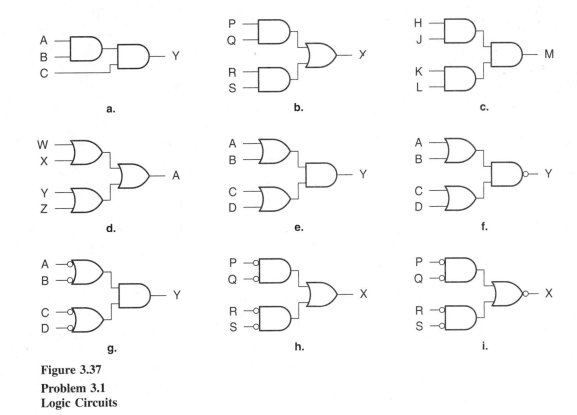

**Figure 3.37**
**Problem 3.1**
**Logic Circuits**

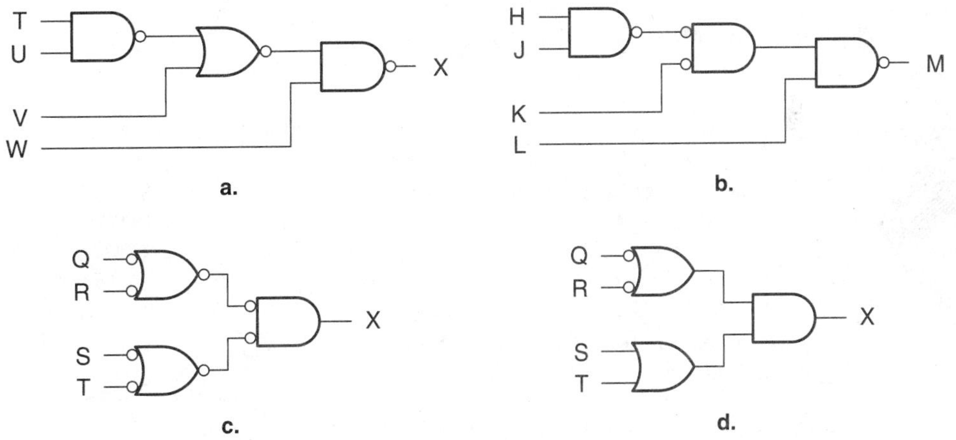

**Figure 3.38**
**Problem 3.2**
**Logic Circuits**

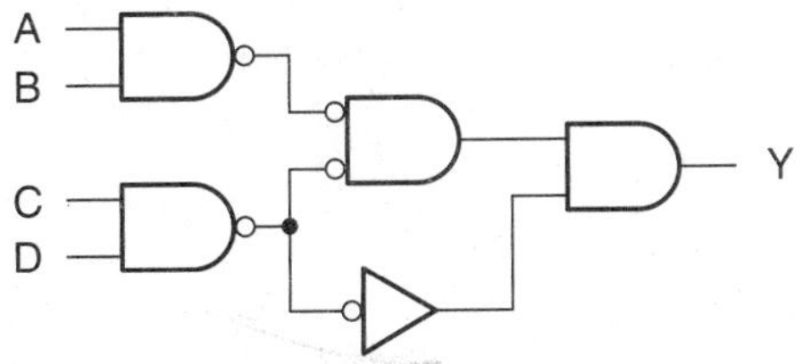

**Figure 3.39**
**Problem 3.3**
**Logic Circuit**

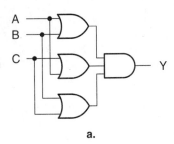

a.

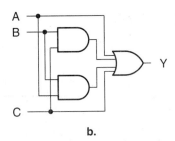

b.

**Figure 3.40**

**Problem 3.4**
**Logic Circuits**

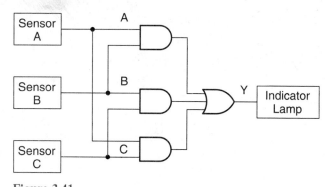

**Figure 3.41**

**Problem 3.5**
**Majority Vote Circuit**

## Section 3.4 Evaluating Boolean Expressions

**3.8** Evaluate the Boolean expression in Problem 3.7, part a, for the following cases:

a. $A = 0, B = 1, C = 1$

b. $A = 1, B = 0, C = 1$

**3.9** Evaluate the Boolean expressions in Problem 3.7, parts c and d, for $A = 0, B = 0, C = 1, D = 1$.

**3.10** Evaluate the Boolean expressions in Problem 3.7, parts e and f, for $A = 1, B = 0, C = 0$.

**3.11** Evaluate the Boolean expression in Problem 3.7, part g, for $A = 0, B = 1, C = 0, D = 0$.

**3.12** Evaluate the Boolean expressions in Problem 3.7, parts h and i, for $A = 1, B = 1, C = 0$.

## Section 3.5 Truth Tables From Logic Diagrams

## Section 3.6 Truth Tables From Boolean Expressions

**3.13** Find the truth tables for the circuits in Problems 3.2 to 3.4.

**3.14** Find the truth tables for the Boolean expressions in Problem 3.7, parts a, g, h, and i.

## Section 3.7 Boolean Expressions From Truth Tables

**3.15** Find the Boolean expression, in both sum-of-products and product-of-sums forms, for the logic function represented by the following truth table. Draw the logic diagram for each form.

| A | B | C | Y |
|---|---|---|---|
| 0 | 0 | 0 | 1 |
| 0 | 0 | 1 | 1 |
| 0 | 1 | 0 | 1 |
| 0 | 1 | 1 | 1 |
| 1 | 0 | 0 | 0 |
| 1 | 0 | 1 | 0 |
| 1 | 1 | 0 | 0 |
| 1 | 1 | 1 | 0 |

**3.16** Repeat Problem 3.15 for the following truth table.

| A | B | C | Y |
|---|---|---|---|
| 0 | 0 | 0 | 0 |
| 0 | 0 | 1 | 1 |
| 0 | 1 | 0 | 0 |
| 0 | 1 | 1 | 0 |
| 1 | 0 | 0 | 1 |
| 1 | 0 | 1 | 0 |
| 1 | 1 | 0 | 1 |
| 1 | 1 | 1 | 0 |

**3.17** Repeat Problem 3.15 for the following truth table.

| A | B | C | Y |
|---|---|---|---|
| 0 | 0 | 0 | 0 |
| 0 | 0 | 1 | 1 |
| 0 | 1 | 0 | 1 |
| 0 | 1 | 1 | 0 |
| 1 | 0 | 0 | 0 |
| 1 | 0 | 1 | 1 |
| 1 | 1 | 0 | 1 |
| 1 | 1 | 1 | 1 |

**3.18** Repeat Problem 3.15 for the following truth table.

| A | B | C | D | Y |
|---|---|---|---|---|
| 0 | 0 | 0 | 0 | 1 |
| 0 | 0 | 0 | 1 | 1 |
| 0 | 0 | 1 | 0 | 1 |
| 0 | 0 | 1 | 1 | 0 |
| 0 | 1 | 0 | 0 | 0 |
| 0 | 1 | 0 | 1 | 0 |
| 0 | 1 | 1 | 0 | 1 |
| 0 | 1 | 1 | 1 | 0 |
| 1 | 0 | 0 | 0 | 1 |
| 1 | 0 | 0 | 1 | 1 |
| 1 | 0 | 1 | 0 | 1 |
| 1 | 0 | 1 | 1 | 0 |
| 1 | 1 | 0 | 0 | 1 |
| 1 | 1 | 0 | 1 | 0 |
| 1 | 1 | 1 | 0 | 1 |
| 1 | 1 | 1 | 1 | 0 |

## Section 3.8 Exclusive OR and Exclusive NOR Circuits

**3.19** Figure 3.24 shows the sum-of-products circuits for the 2-input XOR and XNOR functions. Find the product-of-sums forms of the Boolean ex-

pressions for these functions ($A \oplus B$ and $\overline{A \oplus B}$). Draw the logic circuit of each expression and compare them to the circuits for the SOP forms.

## Section 3.9 Theorems of Boolean Algebra

**3.20** Use the rules of Boolean algebra to simplify the following expressions as much as possible.

**a.** $Y = A A B + C$

**b.** $Y = A \overline{A} B + C$

**c.** $J = K + L \overline{L}$

**d.** $S = (T + U) V \overline{V}$

**e.** $S = T + V \overline{V}$

**f.** $Y = (A \overline{B} + \overline{C})(B \overline{D} + F)$

**\*3.21** Use the rules of Boolean algebra to simplify the following expressions as much as possible.

**a.** $M = P Q + \overline{P Q} R$

**b.** $M = P Q + P Q \overline{R}$

**c.** $S = (\overline{T + U}) V + (T + U)$

**d.** $Y = (\overline{A} + B + \overline{D}) A C + A \overline{B} D$

**e.** $Y = (\overline{A} + B + \overline{D}) A C + \overline{A \overline{B} D}$

**f.** $P = (\overline{Q R} + S T)(\overline{Q R} + Q)$

**g.** $U = (X + \overline{Y} + W Z)(\overline{W} Y + \overline{Y} + W Z)$

**\*\*3.22** Use the rules of Boolean algebra to simplify the following expressions as much as possible.

**a.** $Y = \overline{A B} C D + (\overline{A} + \overline{B}) \overline{C + D} + \overline{A} + \overline{B}$

**b.** $Y = \overline{A B} C D + (\overline{A} + \overline{B}) \overline{C + D} + \overline{\overline{A} + \overline{B}}$

**c.** $K = (L \overline{M} + L \overline{M})(M \overline{N} + L M N) + M(\overline{N} + L)$

**d.** $W = ((\overline{X + \overline{Z}}) + (X + \overline{Z})(X\overline{Y} + \overline{Z}))$
$\phantom{W=} (\overline{(\overline{X\overline{Y} + Z})} \, (\overline{X} + X\overline{Y}) + (\overline{X} + X\overline{Y}))$

**e.** $W = \overline{\overline{((X + \overline{Z})} + \overline{(X + \overline{Z})(X\overline{Y} + \overline{Z})})}$
$\phantom{W=} \overline{\overline{((\overline{X\overline{Y} + Z})(\overline{X} + X\overline{Y})} + \overline{(\overline{X} + X\overline{Y})})}$

## ▼ Answers to Section Review Questions

### Section 3.2
**3.1a.** $Y = \overline{ABC + D}$  **3.1b.** $Y = \overline{(\overline{A} + \overline{B} + C) \cdot D}$

### Section 3.3
**3.2.** Figure 3.14 shows the final logic diagrams corresponding to the Boolean expressions.

### Section 3.4
**3.3a.** $Y = 1$  **3.3b.** $Y = 0$

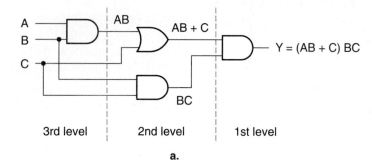

**3rd level**    **2nd level**    **1st level**

**a.**

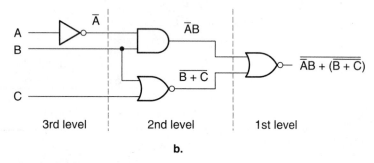

**3rd level**    **2nd level**    **1st level**

**b.**

**Figure 3.14**
**Section Review Problem 3.2**

**Section 3.6**
**3.5.**

| c | A | B | C | D | Y |
|---|---|---|---|---|---|
| | 0 | 0 | 0 | 0 | 0 |
| | 0 | 0 | 0 | 1 | 0 |
| | 0 | 0 | 1 | 0 | 0 |
| | 0 | 0 | 1 | 1 | 0 |
| | 0 | 1 | 0 | 0 | 0 |
| | 0 | 1 | 0 | 1 | 0 |
| | 0 | 1 | 1 | 0 | 0 |
| | 0 | 1 | 1 | 1 | 0 |
| | 1 | 0 | 0 | 0 | 0 |
| | 1 | 0 | 0 | 1 | 0 |
| | 1 | 0 | 1 | 0 | 1 |
| | 1 | 0 | 1 | 1 | 0 |
| | 1 | 1 | 0 | 0 | 0 |
| | 1 | 1 | 0 | 1 | 0 |
| | 1 | 1 | 1 | 0 | 0 |
| | 1 | 1 | 1 | 1 | 0 |

**Section 3.5**
**3.4.**

| A | B | C | Y |
|---|---|---|---|
| 0 | 0 | 0 | 1 |
| 0 | 0 | 1 | 0 |
| 0 | 1 | 0 | 0 |
| 0 | 1 | 1 | 0 |
| 1 | 0 | 0 | 1 |
| 1 | 0 | 1 | 0 |
| 1 | 1 | 0 | 0 |
| 1 | 1 | 1 | 0 |

**Section 3.7**

**3.6a.** SOP: $Y = A \overline{B} \overline{C} + A \overline{B} C$
POS: $Y = (A + B + C)(A + B + \overline{C})$
$(A + \overline{B} + C)(A + \overline{B} + \overline{C})(\overline{A} + \overline{B} + C)(\overline{A} + \overline{B} + \overline{C})$
**3.6b.** SOP: $Y = \overline{A} \overline{B} \overline{C} + A \overline{B} \overline{C} + A \overline{B} C + A B \overline{C}$
POS: $Y = (A + B + \overline{C})(A + \overline{B} + C)$
$(A + \overline{B} + \overline{C})(\overline{A} + \overline{B} + \overline{C})$

**Section 3.9**

**3.7a.** $Y = \overline{AC}$  **3.7b.** $Y = \overline{AC} + D$ or $Y = \overline{A} + \overline{C} + D$  **3.7c.**
$Y = A\overline{B}$

# Analysis, Design, and Troubleshooting of Logic Gate Networks

● CHAPTER CONTENTS

**4.1** Simplifying SOP and POS Expressions

**4.2** Simplification by the Karnaugh Map Method

**4.3** Simplification Using DeMorgan Equivalent Gates

**4.4** Universal Property of NAND/NOR Gates

**4.5** Practical Circuit Implementation

**4.6** Pulsed Operation of Logic Circuits

**4.7** Designing Logic Circuits from Word Problems

**4.8** Troubleshooting Combinational Logic Circuits

● CHAPTER OBJECTIVES

Upon successful completion of this chapter, you will be able to:

- Apply the rules of Boolean algebra to simplify Boolean expressions derived from logic circuits and truth tables.

- Apply the Karnaugh map method to reduce Boolean expressions and logic circuits to their simplest forms.

- Simplify the Boolean expression of a circuit by converting some of its logic gates to their DeMorgan equivalent forms.

- Draw the circuit for any logic function using only NAND or only NOR gates.

- Design a combinational logic circuit, using commercially available gates, and account for any unused gates in the IC packages.

- Describe the pulsed operation of simple logic circuits by drawing appropriate timing diagrams.

- Derive simple logic circuits, their truth tables, and Boolean expressions from word problems.

- Troubleshoot combinational logic circuits.

As we saw in Chapter 3, a Boolean expression for a combinational logic circuit can easily be derived from the circuit truth table. It is seldom desirable, from the point of view of either analysis or design, to leave the expression in unsimplified form. We can use several techniques for simplifying a Boolean expression and therefore its corresponding circuit. Three possible techniques are Boolean algebra, Karnaugh mapping (a graphical tool for Boolean simplification), and redrawing a circuit using DeMorgan equivalent gates.

Any combinational logic circuit can be built using only NAND or only NOR gates. We will examine several techniques for building such circuits.

Logic circuits must be built from devices that are actually available on the market. This may seem obvious, but in practice it means that we are limited to certain forms of logic. Fortunately, the range of available devices is sufficiently broad that this does not present a real problem. We will learn how to construct circuits with commercially available gates.

As we can with single gates, we can use logic circuits to enable and inhibit pulsed waveforms. We will examine several such applications.

It is important for us to learn how to translate word problems into digital circuits to solve the problems. In industry, all problems are ultimately word problems. We will look at simple design techniques, both using fixed logic (gates) and, in a later chapter, programmable logic devices (PLDs).

Learning to recognize basic circuit faults is essential to the electronics student. Although troubleshooting, the detection and repair of circuit faults, is best learned by experience, there are some standard procedures that are helpful in the troubleshooting process. We will learn about some common faults in logic gate networks and techniques we can use to find them.

### 4.1

# Simplifying SOP and POS Expressions

**Maximum SOP simplification** *The form of an SOP Boolean expression that cannot be further simplified by canceling variables in the product terms. It may be possible to get a POS form of the expression with fewer terms or variables.*

**Maximum POS simplification** *The form of a POS Boolean expression that cannot be further simplified by canceling variables in the sum terms. It may be possible to get an SOP form of the expression with fewer terms or variables.*

In Chapter 3, we discovered that we can generate a Boolean equation from a truth table and express it in sum-of-products (SOP) or product-of-sums (POS) form. From this equation, we can develop a logic circuit diagram. The next step in the design or analysis of a circuit is to simplify its Boolean expression as much as possible, with the ultimate aim of producing a circuit that has fewer physical components than the unsimplified circuit.

**Table 4.1** Truth Table for the SOP and POS Networks in Figure 4.1

| A | B | C | Y |
|---|---|---|---|
| 0 | 0 | 0 | 1 |
| 0 | 0 | 1 | 0 |
| 0 | 1 | 0 | 0 |
| 0 | 1 | 1 | 1 |
| 1 | 0 | 0 | 1 |
| 1 | 0 | 1 | 0 |
| 1 | 1 | 0 | 0 |
| 1 | 1 | 1 | 0 |

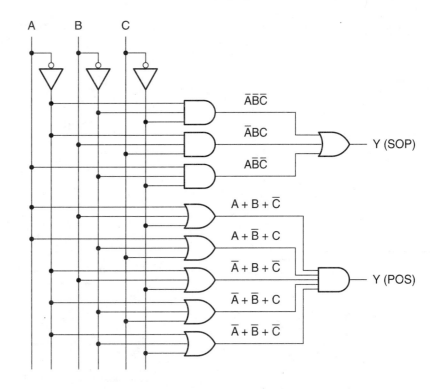

**Figure 4.1**

**Unsimplified SOP and POS Networks**

For example, in Chapter 3 (Section 3.7), we found the SOP and POS forms of the Boolean expression represented by Table 4.1. These forms yield the logic diagrams shown in Figures 3.20 and 3.21. For convenience, the circuits are illustrated again in Figure 4.1. The corresponding algebraic expressions can be simplified by the rules of Boolean algebra to give us a simpler circuit in each case.

The sum-of-products and product-of-sums expressions represented by Table 4.1 are:

$$Y = \overline{A}\,\overline{B}\,\overline{C} + \overline{A}\,B\,C + A\,\overline{B}\,\overline{C} \quad \text{(SOP)}$$

and

$$Y = (A + B + \overline{C})(A + \overline{B} + C)(\overline{A} + B + \overline{C})(\overline{A} + \overline{B} + C)(\overline{A} + \overline{B} + \overline{C}) \quad \text{(POS)}$$

The SOP form is fairly easy to simplify:

$$Y = \overline{A}\,\overline{B}\,\overline{C} + \overline{A}\,B\,C + A\,\overline{B}\,\overline{C}$$
$$= (\overline{A} + A)\,\overline{B}\,\overline{C} + \overline{A}\,B\,C \quad \text{(Distributive Property)}$$
$$= 1 \cdot \overline{B}\,\overline{C} + \overline{A}\,B\,C \quad (x + \overline{x} = 1)$$
$$= \overline{B}\,\overline{C} + \overline{A}\,B\,C \quad (x \cdot 1 = x)$$

Since we cannot cancel any more SOP terms, we can call this final form the **maximum SOP simplification.** The logic diagram for the simplified expression is shown in Figure 4.2.

**Figure 4.2**

**Simplified SOP Circuit**

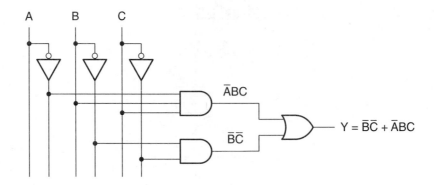

$Y = \overline{B}\overline{C} + \overline{A}BC$

---

Two terms in an SOP expression can be reduced to one if they are identical except for one variable that is in true form in one term and complement form in the other. Such a grouping of a variable and its complement always cancels.

$$x y \overline{z} + x y z = x y (\overline{z} + z) = x y$$

---

There is a similar procedure for the POS form. Examine the following expression:

$$Y = (A + B + \overline{C})(A + B + C)$$

Recall Theorem 17: $(x + y)(x + z) = x + yz$.
Let $x = A + B$, let $y = \overline{C}$, let $z = C$.

$$Y = (A + B) + \overline{C}C \quad \text{(Theorem 17)}$$
$$= (A + B) + 0 \quad (x\overline{x} = 0)$$
$$= (A + B) \quad (x + 0 = x)$$

---

A POS expression can be simplified by grouping two terms that are identical except for one variable that is in true form in one term and complement form in the other.

$$(x + y + \overline{z})(x + y + z) = (x + y) + \overline{z}z = x + y$$

---

Let us use this procedure to simplify the POS form of the previous Boolean expression, shown again below with the terms numbered for our reference.

$$
\begin{array}{ccccc}
(1) & (2) & (3) & (4) & (5) \\
\end{array}
$$
$$Y = (A + B + \overline{C})\,(A + \overline{B} + C)\,(\overline{A} + B + \overline{C})\,(\overline{A} + \overline{B} + C)\,(\overline{A} + \overline{B} + \overline{C})$$

There is usually more than one way to simplify an expression. The following grouping of the numbered POS terms is one possibility.

**Figure 4.3**
**Simplified POS Circuit**

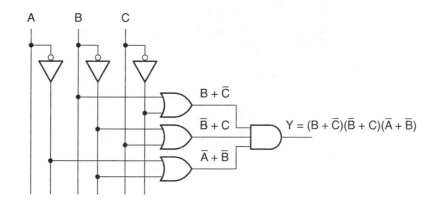

$$(1)(3): (A + B + \overline{C})\,(\overline{A} + B + \overline{C}) = B + \overline{C}$$

$$(2)(4): (A + \overline{B} + C)\,(\overline{A} + \overline{B} + C) = \overline{B} + C$$

$$(4)(5): (\overline{A} + \overline{B} + C)\,(\overline{A} + \overline{B} + \overline{C}) = \overline{A} + \overline{B}$$

Combining the above terms, we get the expression:

$$Y = (B + \overline{C})(\overline{B} + C)(\overline{A} + \overline{B})$$

Figure 4.3 shows the logic diagram for this expression. Compare this logic diagram and that of Figure 4.2 with the unsimplified circuits of Figure 4.1. Since there are no more cancellations of POS terms possible, we can call this the **maximum POS simplification.** We can, however, apply other rules of Boolean algebra and simplify further.

$$
\begin{aligned}
Y &= (B + \overline{C})(\overline{B} + C)(\overline{A} + \overline{B}) \\
&= (\overline{B} + \overline{A}C)(B + \overline{C}) && \text{(Theorem 17)} \\
&= \overline{B}\,B + \overline{B}\,\overline{C} + \overline{A}\,B\,C + \overline{A}\,C\,\overline{C} && \text{(Distributive Property)} \\
&= \overline{B}\,\overline{C} + \overline{A}\,B\,C && (x \cdot \overline{x} = 0)
\end{aligned}
$$

This is the same result we got when we simplified the SOP form of the expression.

To be sure you are getting the maximum SOP or POS simplification, you should be aware of the following guidelines.

1. Each term must be grouped with another, if possible.

2. When attempting to group all terms, it is permissible to group a term more than once, such as term (4) above. The theorems $x \cdot x = x$ (POS forms) and $x + x = x$ (SOP forms) imply that using a term more than once does not change the Boolean expression.

3. Each pair of terms should have at least one term that appears only in that pair. Otherwise, you will have redundant terms that will need to be canceled later. For example, another possible group in the POS simplification above is terms (3) and (5). But since both these terms are in other groups, this pair is unnecessary and would yield a term you would have to cancel.

▼

**EXAMPLE 4.1**

Find the maximum SOP simplification for the Boolean function represented by Table 4.2. Draw the logic diagram for the simplified expression.

**Solution**  SOP form:

$$Y = \overset{(1)}{\overline{A}\,\overline{B}\,\overline{C}\,\overline{D}} + \overset{(2)}{\overline{A}\,\overline{B}\,\overline{C}\,D} + \overset{(3)}{\overline{A}\,B\,\overline{C}\,\overline{D}} + \overset{(4)}{\overline{A}\,B\,\overline{C}\,D} + \overset{(5)}{A\,\overline{B}\,\overline{C}\,\overline{D}} + \overset{(6)}{A\,B\,\overline{C}\,\overline{D}}$$

**Table 4.2** Truth Table for Example 4.1

| A | B | C | D | Y |
|---|---|---|---|---|
| 0 | 0 | 0 | 0 | 1 |
| 0 | 0 | 0 | 1 | 1 |
| 0 | 0 | 1 | 0 | 0 |
| 0 | 0 | 1 | 1 | 0 |
| 0 | 1 | 0 | 0 | 1 |
| 0 | 1 | 0 | 1 | 1 |
| 0 | 1 | 1 | 0 | 0 |
| 0 | 1 | 1 | 1 | 0 |
| 1 | 0 | 0 | 0 | 1 |
| 1 | 0 | 0 | 1 | 0 |
| 1 | 0 | 1 | 0 | 0 |
| 1 | 0 | 1 | 1 | 0 |
| 1 | 1 | 0 | 0 | 1 |
| 1 | 1 | 0 | 1 | 0 |
| 1 | 1 | 1 | 0 | 0 |
| 1 | 1 | 1 | 1 | 0 |

Group the terms as follows:

$$(1) + (5): \overline{A}\,\overline{B}\,\overline{C}\,\overline{D} + A\,\overline{B}\,\overline{C}\,\overline{D} = \overline{B}\,\overline{C}\,\overline{D}$$

$$(2) + (4): \overline{A}\,\overline{B}\,\overline{C}\,D + \overline{A}\,B\,\overline{C}\,D = \overline{A}\,\overline{C}\,D$$

$$(3) + (6): \overline{A}\,B\,\overline{C}\,\overline{D} + A\,B\,\overline{C}\,\overline{D} = B\,\overline{C}\,\overline{D}$$

$$Y = \overline{A}\,\overline{C}\,D + \overline{B}\,\overline{C}\,\overline{D} + B\,\overline{C}\,\overline{D}$$

$$= \overline{A}\,\overline{C}\,D + (\overline{B} + B)\overline{C}\,\overline{D}$$

$$= \overline{A}\,\overline{C}\,D + \overline{C}\,\overline{D}$$

$$= (\overline{A}D + \overline{D})\overline{C}$$

$$= (\overline{A} + \overline{D})\overline{C}$$

$$= \overline{A}\,\overline{C} + \overline{D}\,\overline{C}$$

The logic diagram for the above expression is shown in Figure 4.4.

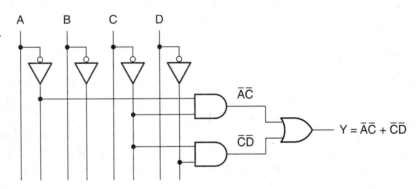

**Figure 4.4**

**Example 4.1 Simplified SOP Circuit**

**Table 4.3** Truth Table for Section Review Problems

| A | B | C | Y |
|---|---|---|---|
| 0 | 0 | 0 | 0 |
| 0 | 0 | 1 | 1 |
| 0 | 1 | 0 | 1 |
| 0 | 1 | 1 | 1 |
| 1 | 0 | 0 | 0 |
| 1 | 0 | 1 | 0 |
| 1 | 1 | 0 | 1 |
| 1 | 1 | 1 | 0 |

**Section Review Problems for Section 4.1**

4.1. Find the maximum SOP and POS simplifications for the function represented by Table 4.3.

## 4.2

# Simplification by the Karnaugh Map Method

**Karnaugh map** *A graphical tool for finding the maximum SOP or POS simplification of a Boolean expression. A Karnaugh map works by arranging the terms of an expression in such a way that variables can be canceled by grouping minterms or maxterms.*

**Cell** *The smallest unit of a Karnaugh map, corresponding to one line of a truth table. The input variables are the cell's coordinates, and the output variable is the cell's contents.*

**Adjacent cell** *Two cells are adjacent if there is only one variable that is different between the coordinates of the two cells. For example, the cells for minterms ABC and $\overline{A}BC$ are adjacent.*

**Pair** *A group of two adjacent cells in a Karnaugh map. A pair cancels one variable in a K-map simplification.*

**Quad** *A group of four adjacent cells in a Karnaugh map. A quad cancels two variables in a K-map simplification.*

**Octet** *A group of eight adjacent cells in a Karnaugh map. An octet cancels three variables in a K-map simplification.*

In Example 4.1, we derived a sum-of-products Boolean expression from a truth table and simplified the expression by grouping minterms that differed by one variable. We made this task easier by breaking up the truth table into groups of four lines. (It is difficult for the eye to grasp an overall pattern in a group of 16 lines.) We chose groups of four because variables *A* and *B* are the same in any one group and variables *C* and *D* repeat the same binary sequence in each group. This allows us to see more easily when we have terms differing by only one variable.

The **Karnaugh map,** or K-map, is a graphical tool for simplifying Boolean expressions that uses a similar idea. A K-map is a square or rectangle divided into smaller squares called **cells,** each of which represents a line in the truth table of the Boolean expression to be mapped. Thus, the number of cells in a K-map is always a power of 2, usually 4, 8, or 16. The coordinates of each cell are the input variables of the truth table. The cell content is the value of the output variable on that line of the truth table. Figure 4.5 shows the formats of Karnaugh maps for Boolean expressions having two, three, and four variables, respectively.

There are two equivalent ways of labeling the cell coordinates: numerically or by true and complement variables. We will use the numerical labeling since it is always the same, regardless of the chosen variables.

The cells in the Karnaugh maps are set up so that the coordinates of any two **adjacent cells** differ by only one variable. By grouping adjacent cells according to specified rules, we can simplify a Boolean expression by canceling variables in their true and complement forms, much as we did algebraically in the previous section.

**Figure 4.5**
**Karnaugh Map Formats**

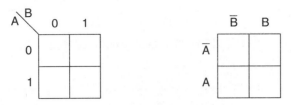

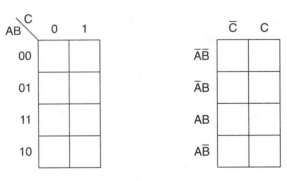

a. Two-variable forms

b. Three-variable forms

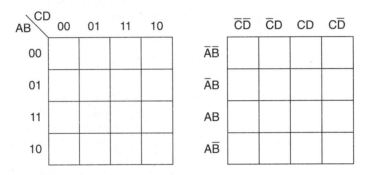

c. Four-variable forms

## Two-Variable Map

**Table 4.4** Truth Table for a Two-Variable Boolean Expression

| A | B | Y |
|---|---|---|
| 0 | 0 | 1 |
| 0 | 1 | 1 |
| 1 | 0 | 0 |
| 1 | 1 | 0 |

Table 4.4 shows the truth table of a two-variable Boolean expression.

The Karnaugh map shown in Figure 4.6 is another way of showing the same information as the truth table. Every line in the truth table corresponds to a cell, or square, in the Karnaugh map.

The coordinates of each cell correspond to a unique combination of input variables (A, B). The content of the cell is the output value for that input combination. If the truth table output is 1 for a particular line, the content of the corresponding cell is also 1. If the output is 0, the cell content is 0.

The SOP expression of the truth table above is

$$Y = \overline{A}\,\overline{B} + \overline{A}\,B$$

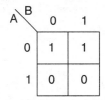

**Figure 4.6**
**Karnaugh Map for Table 4.4**

which can be simplified as follows:

$$Y = \overline{A}\,(\overline{B} + B)$$
$$= \overline{A}$$

We can perform the same simplification by grouping the adjacent **pair** of 1s in the Karnaugh map, as shown in Figure 4.7.

When we circle a pair of 1s in a K-map, we are grouping the common variable in two minterms, then factoring out and canceling the complements.

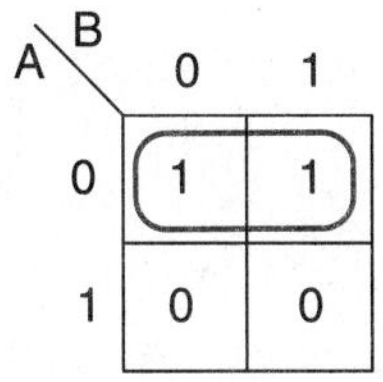

**Figure 4.7**
**Grouping a Pair of Adjacent Cells**

To find the simplified form of the Boolean expression represented in the K-map, we examine the coordinates of all the cells in the circled group. We retain coordinate variables that are the same in all cells and eliminate coordinate variables that are different in different cells.

In this case:

$\overline{A}$ is a coordinate of both cells of the circled pair. (Keep $\overline{A}$.)

$\overline{B}$ is a coordinate of one cell of the circled pair, and $B$ is a coordinate of the other. (Discard $B/\overline{B}$.)

$$Y = \overline{A}$$

## Three- and Four-Variable Maps

Refer to the forms of three- and four-variable Karnaugh maps shown in Figure 4.5. Each cell is specified by a unique combination of binary variables. This implies that the three-variable map has 8 cells (since $2^3 = 8$) and the four-variable map has 16 cells (since $2^4 = 16$).

The variables specifying the row (both maps) or the column (the four-variable map) do not progress in binary order; they advance such that there is only *one change of variable per row or column*. For example, the numbering of the rows is 00, 01, 11, 10, rather than the binary order 00, 01, 10, 11. If we were to use binary order, adjacent cells in rows 2 and 3 or 3 and 4 would differ by two variables, meaning we could not factor out and cancel a pair of complements by grouping these cells. For instance, we cannot cancel complementary variables from the pair $\overline{A}\,B\,C + A\,\overline{B}\,C$, which differs by two variables.

The number of cells in a group must be a power of 2, such as 1, 2, 4, 8, or 16.

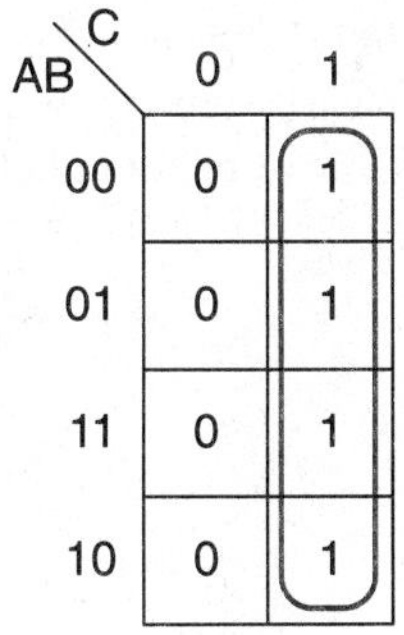

**Figure 4.8**
**Quad**

A group of four adjacent cells is called a **quad.** Figure 4.8 shows a Karnaugh map for a Boolean function whose terms can be grouped in a quad. The Boolean expression displayed in the K-map is:

$$Y = \overline{A}\,\overline{B}\,C + \overline{A}\,B\,C + A\,B\,C + A\,\overline{B}\,C$$

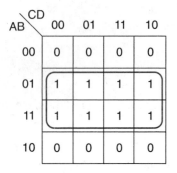

**Figure 4.9**
**Octet**

*A* and *B* are part of the quad coordinates in both true and complement form. (Discard *A* and *B*.)

*C* is a coordinate of *each cell* in the quad. (Keep *C*.)

$$Y = C$$

Grouping cells in a quad is equivalent to factoring two complementary pairs of variables and canceling them.

$$Y = (A + \overline{A})(B + \overline{B})C = C$$

You can verify that this is the same as the original expression by multiplying out the terms.

An **octet** is a group of eight adjacent cells. Figure 4.9 shows the Karnaugh map for the following Boolean expression:

$$Y = \overline{A}\,B\,\overline{C}\,\overline{D} + \overline{A}\,B\,\overline{C}\,D + \overline{A}\,B\,C\,\overline{D} + \overline{A}\,B\,C\,D$$

$$+ A\,B\,\overline{C}\,\overline{D} + A\,B\,\overline{C}\,D + A\,B\,C\,\overline{D} + A\,B\,C\,D$$

Variables *A*, *C*, and *D* are coordinates of the octet cells in both true and complement form. (Discard *A*, *C*, and *D*.)

*B* is a coordinate of *each* cell. (Keep *B*.)

$$Y = B$$

The algebraic equivalent of this octet is an expression where three complementary variables are factored out and canceled.

$$Y = (A + \overline{A})B(C + \overline{C})(D + \overline{D}) = B$$

A Karnaugh map completely filled with 1s corresponds to a Boolean constant. For a Boolean expression *Y*, *Y* = 1.

## Grouping Cells Along Outside Edges

The cells along an outside edge of a three- or four-variable map are adjacent to cells along the opposite edge (only one change of variable). Thus we can group cells "around the outside" of the map to cancel variables. In the case of the four-variable map, we can also group the four corner cells as a quad, since they are all adjacent to one another.

**EXAMPLE 4.2**

Use Karnaugh maps to simplify the following Boolean expressions:

**a.** $Y = \overline{A}\,\overline{B}\,\overline{C} + \overline{A}\,\overline{B}\,C + A\,\overline{B}\,\overline{C} + A\,\overline{B}\,C$

**b.** $Y = \overline{A}\,B\,\overline{C}\,\overline{D} + \overline{A}\,B\,C\,\overline{D} + A\,\overline{B}\,\overline{C}\,\overline{D} + A\,\overline{B}\,C\,\overline{D}$

**Solutions**  Figure 4.10 shows the Karnaugh maps for the Boolean expressions labeled **a** and **b**. Cells in each map are grouped in a quad.

**Figure 4.10**

**Example 4.2**

**K-Maps**

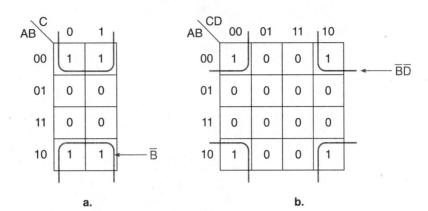

a.

b.

**a.** *A* and *C* are both coordinates of two cells in true form and two cells in complement form. (Discard *A* and *C*.)

$\overline{B}$ is a coordinate of each cell. (Keep $\overline{B}$.)

$$Y = \overline{B}$$

**b.** *A* and *C* are both coordinates of two cells in true form and two cells in complement form. (Discard *A* and *C*.)

$\overline{B}$ and $\overline{D}$ are coordinates of each cell. (Keep $\overline{B}$ and $\overline{D}$.)

$$Y = \overline{B}\,\overline{D}$$

## Loading a K-Map From a Truth Table

> We don't need a Boolean expression to fill a Karnaugh map if we have the function's truth table.

Figures 4.11 and 4.12 show truth table and Karnaugh map forms for three- and four-variable Boolean expressions. The numbers in parentheses show the order of terms in binary sequence for both forms.

The Karnaugh map is not laid out in the same order as the truth table. That is, it is not laid out in a binary sequence. This is due to the criterion for cell adjacency: no more than one variable change between rows or columns.

**Figure 4.11**

**Order of Terms (Three-Variable Function)**

| A | B | C | Y |
|---|---|---|-----|
| 0 | 0 | 0 | (0) |
| 0 | 0 | 1 | (1) |
| 0 | 1 | 0 | (2) |
| 0 | 1 | 1 | (3) |
| 1 | 0 | 0 | (4) |
| 1 | 0 | 1 | (5) |
| 1 | 1 | 0 | (6) |
| 1 | 1 | 1 | (7) |

**a. Truth table**

C
AB      0      1
00    (0)    (1)
01    (2)    (3)
11    (6)    (7)
10    (4)    (5)

**b. K-map**

**Figure 4.12**

**Order of Terms
(Four-Variable Function)**

| A | B | C | D | Y |
|---|---|---|---|---|
| 0 | 0 | 0 | 0 | (0) |
| 0 | 0 | 0 | 1 | (1) |
| 0 | 0 | 1 | 0 | (2) |
| 0 | 0 | 1 | 1 | (3) |
| 0 | 1 | 0 | 0 | (4) |
| 0 | 1 | 0 | 1 | (5) |
| 0 | 1 | 1 | 0 | (6) |
| 0 | 1 | 1 | 1 | (7) |
| 1 | 0 | 0 | 0 | (8) |
| 1 | 0 | 0 | 1 | (9) |
| 1 | 0 | 1 | 0 | (10) |
| 1 | 0 | 1 | 1 | (11) |
| 1 | 1 | 0 | 0 | (12) |
| 1 | 1 | 0 | 1 | (13) |
| 1 | 1 | 1 | 0 | (14) |
| 1 | 1 | 1 | 1 | (15) |

| AB\CD | 00 | 01 | 11 | 10 |
|-------|-----|-----|-----|-----|
| 00 | (0) | (1) | (3) | (2) |
| 01 | (4) | (5) | (7) | (6) |
| 11 | (12) | (13) | (15) | (14) |
| 10 | (8) | (9) | (11) | (10) |

a. Truth table                          b. K-map

Filling in a Karnaugh map from a truth table is easy when you understand a system for doing it quickly. For the three-variable map, fill row 1, then row 2, skip to row 4, then go back to row 3. By doing this, you trace through the cells in binary order. Use the mnemonic phrase "1, 2, skip, back" to help you remember this.

The system for the four-variable map is similar but must account for the columns as well. The rows get filled in the same order as the three-variable map, but within each row, fill column 1, then column 2, skip to column 4, then go back to column 3. Again, "1, 2, skip, back."

The four-variable map is easier to fill from the truth table if we break up the truth table into groups of four lines, as we have done in Figure 4.12. Each group is one row in the Karnaugh map. Following this system will quickly fill the cells in binary order.

Go back and follow the order of terms on the four-variable map in Figure 4.12, using this system. (Remember, for both rows and columns, "1, 2, skip, back.")

## Multiple Groups

If there is more than one group of 1s in a K-map simplification, each group is a term in the maximum SOP simplification of the mapped Boolean expression. The resulting terms are ORed together.

▼

**EXAMPLE 4.3**    Use the Karnaugh map method to simplify the Boolean function represented by Table 4.5.

**Table 4.5** Truth Table for Example 4.3

| A | B | C | D | Y |
|---|---|---|---|---|
| 0 | 0 | 0 | 0 | 1 |
| 0 | 0 | 0 | 1 | 0 |
| 0 | 0 | 1 | 0 | 1 |
| 0 | 0 | 1 | 1 | 0 |
| 0 | 1 | 0 | 0 | 0 |
| 0 | 1 | 0 | 1 | 1 |
| 0 | 1 | 1 | 0 | 0 |
| 0 | 1 | 1 | 1 | 1 |
| 1 | 0 | 0 | 0 | 0 |
| 1 | 0 | 0 | 1 | 0 |
| 1 | 0 | 1 | 0 | 0 |
| 1 | 0 | 1 | 1 | 0 |
| 1 | 1 | 0 | 0 | 0 |
| 1 | 1 | 0 | 1 | 1 |
| 1 | 1 | 1 | 0 | 0 |
| 1 | 1 | 1 | 1 | 1 |

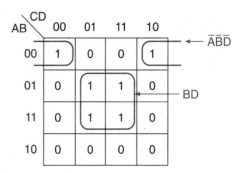

**Figure 4.13**

**Example 4.3 K-Map**

**Solution**    Figure 4.13 shows the Karnaugh map for the truth table in Table 4.5. There are two groups of 1s—a pair and a quad.

**Pair:**

Variables $\overline{A}$, $\overline{B}$, and $\overline{D}$ are coordinates of both cells. (Keep $\overline{A}\,\overline{B}\,\overline{D}$.) $C$ is a coordinate of one cell and $\overline{C}$ is a coordinate of the other. (Discard $C$.)

Term: $\overline{A}\,\overline{B}\,\overline{D}$

**Quad:**

Both $A$ and $C$ are coordinates of two cells in true form and two cells in complement form. (Discard $A$ and $C$.)

$B$ and $D$ are coordinates of all four cells. (Keep $B\,D$.)

Term: $B\,D$

Combine the terms in an OR function:

$$Y = \overline{A}\,\overline{B}\,\overline{D} + B\,D$$

## Overlapping Groups

A cell may be grouped more than once. The only condition is that every group must have at least one cell that does not belong to any other group. Otherwise, redundant terms will result.

**EXAMPLE 4.4**          Simplify the function represented by Table 4.6.

**Solution**    The Karnaugh map for the function in Table 4.6 is shown in Figure 4.14, with two different groupings of terms.

**Figure 4.14**

**Example 4.4**

**K-Maps**

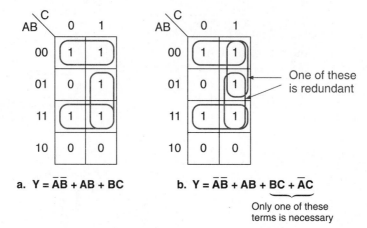

a.  $Y = \overline{A}\,\overline{B} + AB + BC$    b.  $Y = \overline{A}\,\overline{B} + AB + BC + \overline{A}C$

One of these is redundant

Only one of these terms is necessary

**Table 4.6** Truth Table for Example 4.4

| A | B | C | Y |
|---|---|---|---|
| 0 | 0 | 0 | 1 |
| 0 | 0 | 1 | 1 |
| 0 | 1 | 0 | 0 |
| 0 | 1 | 1 | 1 |
| 1 | 0 | 0 | 0 |
| 1 | 0 | 1 | 0 |
| 1 | 1 | 0 | 1 |
| 1 | 1 | 1 | 1 |

**a.** The simplified Boolean expression drawn from the first map has three terms.

$$Y = \overline{A}\,\overline{B} + A\,B + B\,C$$

**b.** The second map yields an expression with four terms.

$$Y = \overline{A}\,\overline{B} + A\,B + B\,C + \overline{A}\,C$$

One of the last two terms is redundant, since neither of the pairs corresponding to these terms has a cell belonging only to that pair. We could retain either pair of cells and its corresponding term, but not both.

We can show algebraically that the last term is redundant and thus make the expression the same as that in part a.

$$Y = \overline{A}\,\overline{B} + A\,B + B\,C + \overline{A}\,C$$
$$= \overline{A}\,\overline{B} + A\,B + B\,C + \overline{A}\,(B + \overline{B})\,C$$
$$= \overline{A}\,\overline{B} + A\,B + B\,C + \overline{A}\,B\,C + \overline{A}\,\overline{B}\,C$$
$$= \overline{A}\,\overline{B}\,(1 + C) + A\,B + B\,C\,(1 + \overline{A})$$
$$= \overline{A}\,\overline{B} + A\,B + B\,C$$

## Conditions for Maximum Simplification

The maximum simplification of a Boolean expression is achieved only if the circled groups of cells in its K-map are as large as possible and there are as few groups as possible.

**EXAMPLE 4.5**     Find the maximum SOP simplification of the Boolean function represented by Table 4.7.

**Solution**     The values of Table 4.7 are loaded into the three K-maps shown in Figure 4.15. Three different ways of grouping adjacent cells are shown. One results in maximum simplification; the other two do not.

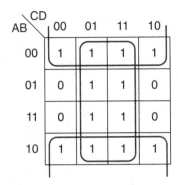

a. **Maximum simplification**
   $Y = \overline{B} + D$

b. **Less than maximum simplification**
   $Y = A\overline{B} + \overline{A}\,\overline{B} + D$

c. **Less than maximum simplification**
   $Y = \overline{B} + BD$

**Figure 4.15**
**Example 4.5**
**K-Maps**

We get the maximum SOP simplification by grouping the two octets shown in Figure 4.15a. The resulting expression is

$$a.\ Y = \overline{B} + D$$

Figures 4.15b and c show two simplifications that are less than the maximum because the chosen cell groups are smaller than they could be. The resulting expressions are:

$$b.\ Y = \overline{A}\,\overline{B} + A\,\overline{B} + D$$

$$c.\ Y = \overline{B} + B\,D$$

Neither of these expressions is the simplest possible, since both can be reduced by Boolean algebra to the form in Figure 4.15a.

**Table 4.7** Truth Table for Example 4.5

| A | B | C | D | Y |
|---|---|---|---|---|
| 0 | 0 | 0 | 0 | 1 |
| 0 | 0 | 0 | 1 | 1 |
| 0 | 0 | 1 | 0 | 1 |
| 0 | 0 | 1 | 1 | 1 |
| 0 | 1 | 0 | 0 | 0 |
| 0 | 1 | 0 | 1 | 1 |
| 0 | 1 | 1 | 0 | 0 |
| 0 | 1 | 1 | 1 | 1 |
| 1 | 0 | 0 | 0 | 1 |
| 1 | 0 | 0 | 1 | 1 |
| 1 | 0 | 1 | 0 | 1 |
| 1 | 0 | 1 | 1 | 1 |
| 1 | 1 | 0 | 0 | 0 |
| 1 | 1 | 0 | 1 | 1 |
| 1 | 1 | 1 | 0 | 0 |
| 1 | 1 | 1 | 1 | 1 |

## Using K-Maps for Partially Simplified Circuits

Figure 4.16 shows a logic diagram that can be further simplified. If we want to use a Karnaugh map for this process, we must do one of two things.

1. Fill in the K-map from the existing product terms. Each product term that is not a minterm will represent more than one cell in the Karnaugh map. When the map is filled, regroup the cells for maximum simplification.

2. Expand the sum-of-products expression of the circuit to get a sum-of-minterms form. Each minterm represents one cell in the K-map. Group the cells for maximum simplification.

**Figure 4.16**

**Logic Diagram That Can Be Further Simplified**

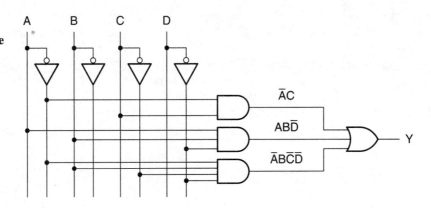

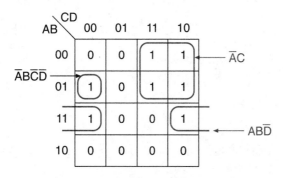

**a. K-map from logic diagram (Figure 4.16)**

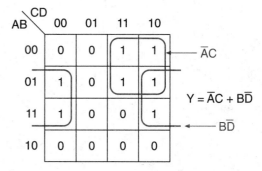

**b. Maximum simplification**

**Figure 4.17**

**Further Simplification of Logic Diagram (Figure 4.16)**

Figure 4.17 shows the K-map derived from the existing circuit and the regrouped cells that yield the maximum simplification.

The algebraic method requires us to expand the existing Boolean expression to get a sum of minterms. The original expression is:

$$Y = \overline{A}\,B\,\overline{C}\,\overline{D} + A\,B\,\overline{D} + \overline{A}\,C$$

The theorem $(x + \overline{x}) = 1$ implies that we can add a variable to a term in true and complement form without changing the term. The expanded expression is:

$$Y = \overline{A}\,B\,\overline{C}\,\overline{D} + A\,B\,(C + \overline{C})\,\overline{D} + \overline{A}\,(B + \overline{B})\,C\,(D + \overline{D})$$

$$= \overline{A}\,B\,\overline{C}\,\overline{D} + A\,B\,C\,\overline{D} + A\,B\,\overline{C}\,\overline{D}$$

$$+ \overline{A}\,B\,C\,D + \overline{A}\,B\,C\,\overline{D} + \overline{A}\,\overline{B}\,C\,D + \overline{A}\,\overline{B}\,C\,\overline{D}$$

The terms of this expression can be loaded into a K-map and simplified, as shown in Figure 4.17b. Figure 4.18 shows the logic diagram for the simplified expression.

**Figure 4.18**

**Simplified Circuit**

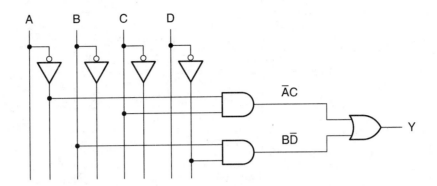

---

**EXAMPLE 4.6**

Use a Karnaugh map to find the maximum SOP simplification of the circuit shown in Figure 4.19.

**Figure 4.19**

**Example 4.6 Circuit to Be Simplified**

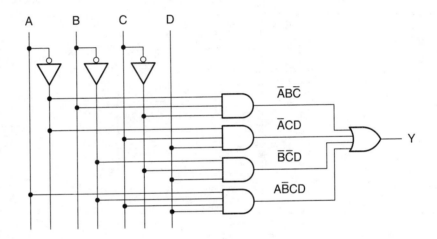

**Solution**  Figure 4.20a shows the Karnaugh map of Figure 4.19 with terms grouped as shown in the original circuit. Figure 4.20b shows the terms regrouped for the maximum simplification, which is given by:

$$Y = \overline{A}\, D + \overline{B}\, D + \overline{A}\, B\, \overline{C}$$

*Alternate method.* The Boolean expression for the circuit in Figure 4.19 is:

$$Y = \overline{A}\, B\, \overline{C} + \overline{A}\, C\, D + \overline{B}\, \overline{C}\, D + A\, \overline{B}\, C\, D$$

This expands to the following expression:

$$Y = \overline{A}\, B\, \overline{C}\,(D + \overline{D}) + \overline{A}\,(B + \overline{B})\, C\, D + (A + \overline{A})\, \overline{B}\, \overline{C}\, D + A\, \overline{B}\, C\, D$$

$$= \overline{A}\, B\, \overline{C}\, D + \overline{A}\, B\, \overline{C}\, \overline{D} + \overline{A}\, B\, C\, D + \overline{A}\, \overline{B}\, C\, D + A\, \overline{B}\, \overline{C}\, D$$

$$+ \overline{A}\, \overline{B}\, \overline{C}\, D + A\, \overline{B}\, C\, D$$

This expression can be loaded directly into the K-map and simplified, as shown in Figure 4.20b. The logic diagram for the simplified expression is shown in Figure 4.21.

**Figure 4.20**

**Example 4.6**

**Maximum Simplification of Figure 4.19**

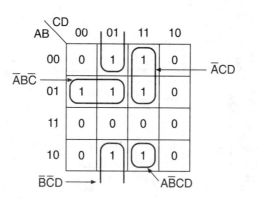

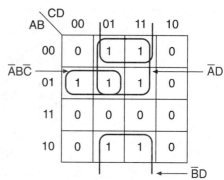

a. **K-map from Figure 4.19**

b. **Maximum simplification**

**Figure 4.21**

**Example 4.6**

**Simplified Circuit**

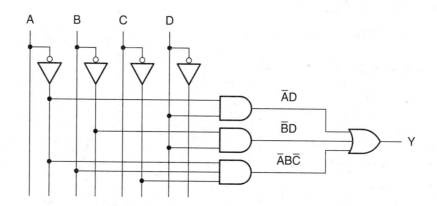

## Don't Care States

**Don't care state** *An output state that can be regarded as either HIGH or LOW, as is most convenient. A don't care state is the output state of a circuit for a combination of inputs that will never occur.*

Sometimes a digital circuit will be intended to work only for certain combinations of inputs; any other inputs will never be applied to the circuit.

In such a case, it may be to our advantage to use so-called **don't care states** to simplify the circuit. A don't care state is shown in a K-map cell as an "X" and can be either a 0 or a 1, depending on which case will yield the maximum simplification.

A common application of the don't care state is a digital circuit designed for binary-coded decimal (BCD) inputs. In BCD, a decimal digit (0–9) is encoded as a 4-bit binary number (0000–1001). This leaves six binary states that are never used (1010, 1011, 1100, 1101, 1110, 1111). In any circuit designed for BCD inputs, these states are don't care states.

All cells containing 1s must be grouped if we are looking for a maximum SOP simplification. (If necessary, a group can contain one cell.) The don't care states can be used to maximize the size of these groups. We need not group all don't care states, only those that actually contribute to a maximum simplification.

**EXAMPLE 4.7**

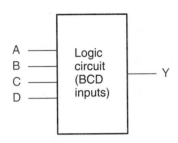

**Figure 4.22**

**Example 4.7**
**Circuit to Be Simplified**

The circuit in Figure 4.22 is designed to accept binary-coded decimal inputs. The output is HIGH when the input is the BCD equivalent of 5, 7, or 9. If the BCD equivalent of the input is not 5, 7 or 9, the output is LOW. The output is not defined for input values greater than 9.

Find the maximum SOP simplification of the circuit.

**Solution**

The Karnaugh map for the circuit is shown in Figure 4.23a.

We can designate three of the don't care cells as 1s—those corresponding to input states 1011, 1101, and 1111. This allows us to group the 1s into two overlapping quads, which yield the following simplification.

$$Y = A D + B D$$

The ungrouped don't care states are treated as 0s. The corresponding circuit is shown in Figure 4.23b.

**Figure 4.23**

**Example 4.7**
**Karnaugh Map and Logic Diagram**

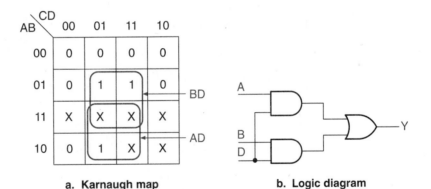

a.  **Karnaugh map**        b.  **Logic diagram**

**EXAMPLE 4.8**

One type of decimal code is called 2421 code, so called because of the positional weights of its bits. (For example, 1011 in 2421 code is equivalent to $2 + 2 + 1 = 5$ in decimal. 1100 is equivalent to decimal $2 + 4 = 6$.) Table 4.8 shows how this code compares to its equivalent decimal digits and to the BCD code used in Example 4.7.

2421 code is sometimes used because it is "self-complementing," a property that BCD code does not have, but that is useful in digital decimal arithmetic circuits. Briefly, the idea of self-complementing code is that when all bits of an encoded number are inverted, the "9's complement" of that number is automatically generated. (9's complement is a way of representing a negative number in a decimal code.) We will examine the self-complementing property in detail in a later chapter on number systems and codes.

The bits of the BCD code are designated $D_4 D_3 D_2 D_1$. The bits of the 2421 code are designated $Y_4 Y_3 Y_2 Y_1$.

Table 4.8  BCD and 2421 Code

| Decimal Equivalent | BCD Code | | | | 2421 Code | | | |
|---|---|---|---|---|---|---|---|---|
| | $D_4$ | $D_3$ | $D_2$ | $D_1$ | $Y_4$ | $Y_3$ | $Y_2$ | $Y_1$ |
| 0 | 0 | 0 | 0 | 0 | 0 | 0 | 0 | 0 |
| 1 | 0 | 0 | 0 | 1 | 0 | 0 | 0 | 1 |
| 2 | 0 | 0 | 1 | 0 | 0 | 0 | 1 | 0 |
| 3 | 0 | 0 | 1 | 1 | 0 | 0 | 1 | 1 |
| 4 | 0 | 1 | 0 | 0 | 0 | 1 | 0 | 0 |
| 5 | 0 | 1 | 0 | 1 | 1 | 0 | 1 | 1 |
| 6 | 0 | 1 | 1 | 0 | 1 | 1 | 0 | 0 |
| 7 | 0 | 1 | 1 | 1 | 1 | 1 | 0 | 1 |
| 8 | 1 | 0 | 0 | 0 | 1 | 1 | 1 | 0 |
| 9 | 1 | 0 | 0 | 1 | 1 | 1 | 1 | 1 |

**Figure 4.24**

**Example 4.8**
**K-Maps: BCD to 2421**

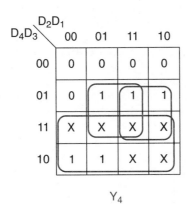

$$Y_4 = D_4 + D_3D_2 + D_3D_1$$
$$Y_3 = D_4 + D_3D_2 + D_3\overline{D}_1$$
$$Y_2 = D_4 + \overline{D}_3D_2 + D_3\overline{D}_2D_1$$
$$Y_1 = D_1$$

Sketch the logic diagram of a circuit with BCD inputs and 2421 outputs, using the Karnaugh map method to obtain the maximum SOP simplification.

**Solution**    The required circuit is called a code converter. Each 4-bit BCD input corresponds to a 4-bit 2421 output. Thus, we must find four Boolean expressions, one for each bit of the 2421 code. We can derive each Boolean expression from a truth table represented by the corresponding output column in Table 4.8.

We can load the 2421 values into four different Karnaugh maps, as shown in Figure 4.24. The cells corresponding to the unused input codes 1010, 1011, 1100, 1101, 1110, and 1111 are don't care states in each map.

The K-maps yield the following simplifications:

$$Y_4 = D_4 + D_3\,D_2 + D_3\,D_1$$

$$Y_3 = D_4 + D_3\,D_2 + D_3\,\overline{D}_1$$

$$Y_2 = D_4 + \overline{D}_3\,D_2 + D_3\,\overline{D}_2\,D_1$$

$$Y_1 = D_1$$

Figure 4.25 shows the logic diagram for these equations.

**Figure 4.25**

**Example 4.9**
**BCD-to-2421 Code Converter**

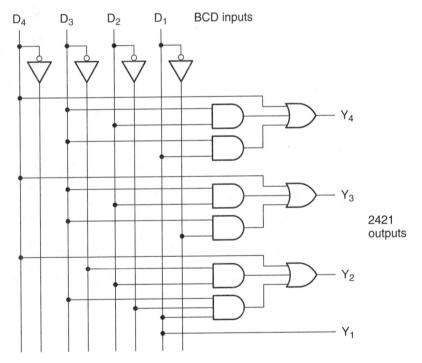

## POS Simplification

(This section can be omitted without loss of continuity.)
Until now, we have looked only at obtaining the maximum SOP simplification from a Karnaugh map. It is also possible to find the maximum POS simplification from the same map.

**Figure 4.26**

**SOP and POS Forms on a K-Map**

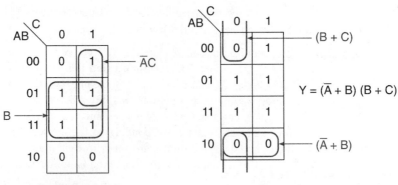

**a. SOP simplification**
$Y = \overline{A}C + B$

**b. POS simplification**
$Y = (\overline{A} + B)\ (B+C)$

Figure 4.26 shows a Karnaugh map with the cells grouped for an SOP simplification and a POS simplification. The SOP simplification is shown in Figure 4.26a and the POS simplification in Figure 4.26b.

When we derive the POS form of an expression from a truth table, we use the lines where the output is 0 and we use the complements of the input variables on these lines as the elements of the selected maxterms. The same principle applies here.

The maxterms are:

$$(A + B + C)\quad \text{Top left cell}$$

$$(\overline{A} + B + C)\quad \text{Bottom left cell}$$

$$(\overline{A} + B + \overline{C})\quad \text{Bottom right cell}$$

The variables are canceled in much the same way as in the SOP form. Remember, however, that the POS variables are the complements of the variables written beside the Karnaugh map.

If there is more than one simplified term, the terms are ANDed together, as in a full POS form.

Cancellations:

*Outside pair:*    $A$ is present in both true and complement form in the pair. (Discard $A$.)

$B$ and $C$ are present in both cells of the pair. (Keep $B$ and $C$.)

Term:   $B + C$

*Bottom pair:*    $\overline{A}$ and $B$ are present in both cells of the pair. (Keep $\overline{A}$ and $B$.)

$C$ is present in both true and complement form in the pair. (Discard $C$.)

Term:   $\overline{A} + B$

Maximum POS simplification:

$$Y = (\overline{A} + B)(B + C)$$

Compare this with the maximum SOP simplification:

$$Y = \overline{A}\,C + B$$

By the Boolean theorem $(x + y)(x + z) = x + yz$, we see that the SOP and POS forms are equivalent.

---

**EXAMPLE 4.9**

Find the maximum POS simplification of the logic function represented by Table 4.9.

**Solution**    Figure 4.27 shows the Karnaugh map from the truth table in Table 4.9. The cells containing 0s are grouped in two quads and there is a single 0 cell left over.

**Table 4.9** Truth Table for Example 4.9

| A | B | C | D | Y |
|---|---|---|---|---|
| 0 | 0 | 0 | 0 | 0 |
| 0 | 0 | 0 | 1 | 0 |
| 0 | 0 | 1 | 0 | 0 |
| 0 | 0 | 1 | 1 | 0 |
| 0 | 1 | 0 | 0 | 1 |
| 0 | 1 | 0 | 1 | 1 |
| 0 | 1 | 1 | 0 | 1 |
| 0 | 1 | 1 | 1 | 1 |
| 1 | 0 | 0 | 0 | 0 |
| 1 | 0 | 0 | 1 | 1 |
| 1 | 0 | 1 | 0 | 0 |
| 1 | 0 | 1 | 1 | 1 |
| 1 | 1 | 0 | 0 | 1 |
| 1 | 1 | 0 | 1 | 0 |
| 1 | 1 | 1 | 0 | 1 |
| 1 | 1 | 1 | 1 | 1 |

*Simplification:*

| | |
|---|---|
| *Corner quad:* | $(B + D)$ |
| *Horizontal quad:* | $(A + B)$ |
| *Single cell:* | $(\overline{A} + \overline{B} + C + \overline{D})$ |

$$Y = (A + B)(B + D)(\overline{A} + \overline{B} + C + \overline{D})$$

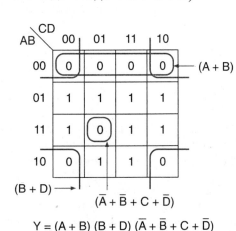

**Figure 4.27**

**Example 4.9**

**POS Simplification of Table 4.9**

$$Y = (A + B)\,(B + D)\,(\overline{A} + \overline{B} + C + \overline{D})$$

---

## 4.3

# Simplification Using DeMorgan Equivalent Gates

If a digital circuit has a lot of inverting gates, such as NANDs, NORs, or inverters, the Boolean expression can get rather cluttered up with bars and become difficult to understand. For instance, look at the circuit in Figure 4.28a. The Boolean expression of the output is given by:

$$Y = \overline{(\overline{A\,B\,\overline{C}})(\overline{\overline{A}\,\overline{B}\,C})}$$

**Figure 4.28**

**Simplification Using DeMorgan Equivalent**

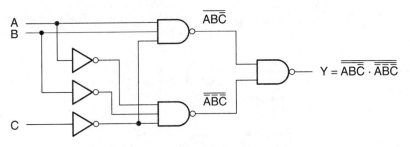

a. Original circuit

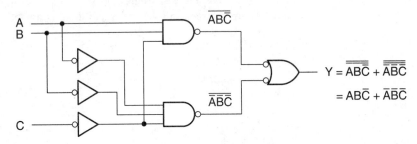

b. DeMorgan equivalent

We can simplify the expression by applying DeMorgan's Theorems:

$$Y = \overline{(\overline{A\,B\,\overline{C}})} + \overline{(\overline{\overline{A}\,\overline{B}\,C})}$$

$$= A\,B\,\overline{C} + \overline{A}\,\overline{B}\,\overline{C}$$

The same simplification can be accomplished by redrawing the circuit as in Figure 4.28b. The output of each gate is at the same active level as the input of the following gate. The only way to do this without changing the logic function of the circuit is to change some of the gates to their DeMorgan equivalent forms.

The output NAND gate changes to its DeMorgan equivalent. The sense of the inverters also changes so that active-HIGH outputs (no bubble) drive active-HIGH inputs. The negations between first- and second-level NANDs cancel, giving us the same result as above without the algebra.

1. When redrawing a circuit with its DeMorgan equivalents, start at the output and work toward the input.

2. Make the circuit output active HIGH so that the Boolean expression doesn't have a bar. Convert the output (first-level) gate to its DeMorgan equivalent if it originally had a bubble on the output.

3. Make the active levels of the second-level gate outputs match the first-level inputs, converting gates to DeMorgan equivalents, if necessary.

4. Repeat the procedure until you reach the circuit inputs. (Generally, this means that you will convert gates on odd-numbered levels (1, 3, 5, . . .) to their DeMorgan equivalents.)

**EXAMPLE 4.10**

a. Write the Boolean expression for the circuit in Figure 4.29.

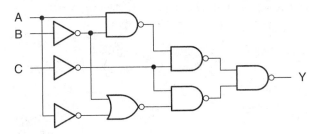

Figure 4.29
Example 4.10
Circuit

b. Redraw the circuit to make the input and output active levels match at each gating level, and write the new Boolean expression.

c. Find the maximum SOP simplification for the new expression, and draw the simplified circuit.

**Solution**    a.

$$Y = \overline{((\overline{A \cdot \overline{B})\overline{C}}) \cdot ((\overline{\overline{A} + \overline{B}}) \cdot \overline{C})}$$

b. The redrawn circuit is shown in Figure 4.30a. The Boolean expression for the output of the new circuit is:

$$Y = (\overline{A} + B)\,\overline{C} + A\,B\,\overline{C}$$
$$= \overline{A}\,\overline{C} + B\,\overline{C} + A\,B\,\overline{C}$$

c. The Boolean expression above can be simplified as follows:

$$Y = \overline{A}\,\overline{C} + B\,\overline{C}\,(1 + A)$$
$$= \overline{A}\,\overline{C} + B\,\overline{C}$$

Figure 4.30b shows the circuit for the simplified expression.

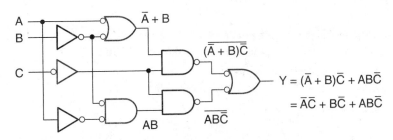

a. DeMorgan equivalent of Figure 4.29

b. Maximum SOP simplification

Figure 4.30
Example 4.10
DeMorgan Equivalent and Maximum SOP Simplification

**Section Review Problem for Section 4.3**

4.2. Redraw the circuit in Figure 4.31 so that active levels are the same on connected inputs and outputs. Write the resulting Boolean expression.

**Figure 4.31**

**Section Review Problem Circuit**

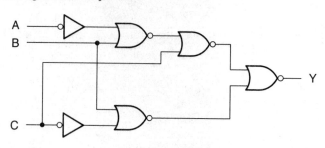

---

## 4.4

# Universal Property of NAND/NOR Gates

At the beginning of Chapter 2 (Logic Functions and Gates), we stated that any Boolean function can be realized by a combination of AND, OR, or NOT functions. We can go one step further and state that any function can be implemented using only NAND functions (NOT and AND) or only NOR functions (NOT and OR).

We can create the basic logic functions from NAND or NOR gates by applying DeMorgan's and other Boolean theorems.

$$\overline{x\,y} = \overline{x} + \overline{y}$$

$$\overline{x + y} = \overline{x}\,\overline{y}$$

### All-NAND Forms

NOT:   An inverter can be constructed from a single NAND gate by connecting both inputs together, as shown in Figure 4.32.

Algebraic expression: $\overline{x \cdot x} = \overline{x}$

**Figure 4.32**

**NOT From NAND**

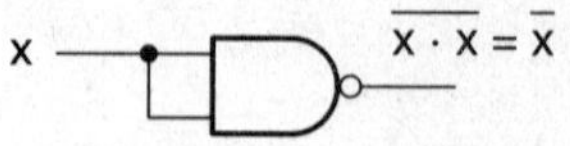

$$x \cdot x = \overline{x}$$

AND:   The AND function is created by inverting a NAND function, as in Figure 4.33.

Algebraic expression: $\overline{\overline{x \cdot y}} = x \cdot y$

**Figure 4.33**

**AND From NANDs**

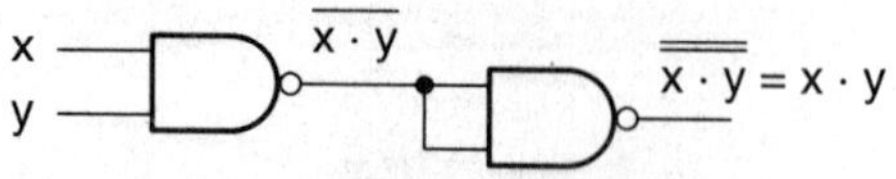

$$\overline{x \cdot y} \qquad \overline{\overline{x \cdot y}} = x \cdot y$$

OR:    The OR function requires three NANDs, as shown in Figure 4.34a. The circuit can be redrawn to get the same algebraic result a little more directly. This equivalent circuit is shown in Figure 4.34b.

Algebraic expressions: **a.** $\overline{\overline{x} \cdot \overline{y}} = \overline{\overline{x + y}} = x + y$ or
**b.** $\overline{\overline{x}} + \overline{\overline{y}} = x + y$

**Figure 4.34**

**OR From NANDs**

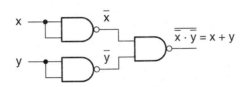

**a. Standard form**

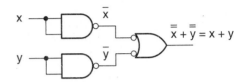

**b. DeMorgan equivalent form**

NOR:    Figure 4.35a and b shows two forms of the NOR circuit, each using four NAND gates. The NOR is implemented by inverting the output of the OR circuit with a NAND inverter. Note that the OR and NOR circuits in Figures 4.34b and 4.35b are drawn with matching input and output active levels in all gates.

Algebraic expressions: **a.** $\overline{\overline{\overline{x} \cdot \overline{y}}} = \overline{\overline{\overline{x + y}}} = \overline{x + y}$ or
**b.** $\overline{\overline{\overline{x}} + \overline{\overline{y}}} = \overline{x + y}$

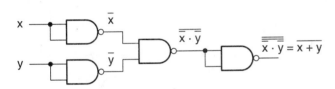

**a. Standard form**

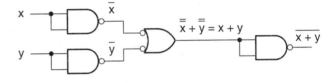

**b. DeMorgan equivalent form**

**Figure 4.35**

**NOR From NANDs**

## All-NOR Forms

NOT:    The NOT function is formed by connecting the inputs of a NOR gate together, as shown in Figure 4.36.

Algebraic expression: $\overline{x + x} = \overline{x}$

**Figure 4.36**

**NOT From NOR**

$$x \multimap\!\!\!\supset\!\!\circ \quad \overline{x + x} = \overline{x}$$

OR:    The OR function is created by inverting the NOR function with a NOR gate inverter. This is shown in Figure 4.37.

Algebraic expression: $\overline{\overline{x + y}} = x + y$

**Figure 4.37**

**OR From NORs**

AND:      The AND function is synthesized from three NOR gates. Two forms of the circuit are shown in Figure 4.38.

Algebraic expressions: **a.** $\overline{\overline{x}+\overline{y}} = \overline{\overline{x}\cdot\overline{y}} = x\cdot y$ or

                  **b.** $\overline{\overline{x}}\cdot\overline{\overline{y}} = x\cdot y$

**Figure 4.38**

**AND From NORs**

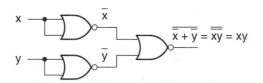
             **a. Standard form**
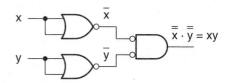
             **b. DeMorgan equivalent form**

NAND:      The NAND function is created by inverting the previously derived AND function. This requires four NOR gates, as shown in Figure 4.39a and b. Note that the AND and NAND circuits in Figures 4.38b and 4.39b have matching input and output active levels in all gates.

Algebraic expressions: **a.** $\overline{\overline{\overline{x}+\overline{y}}} = \overline{\overline{\overline{x}\cdot\overline{y}}} = \overline{x\cdot y}$ or

                  **b.** $\overline{\overline{\overline{x}}\cdot\overline{\overline{y}}} = \overline{x\cdot y}$

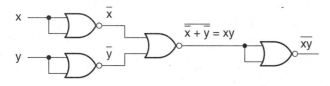
             **a. Standard form**
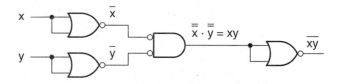
             **b. DeMorgan equivalent form**

**Figure 4.39**

**NAND From NORs**

There are several ways we can make a circuit with only NANDs or only NORs.

**1.** If the circuit is in SOP form, we can change the ANDs and ORs to NANDs directly by drawing a pair of bubbles between each AND output and each OR input. The bubbles will cancel, giving us the original Boolean expression.

If the SOP circuit must be built with only NORs, we convert the AND-OR (SOP) gates to NOR-NOR gates in standard or DeMorgan equivalent form and insert NOR-NOR inverters to adjust the input and output logic levels.

**2.** If the circuit is in POS form, we can replace the ORs and ANDs directly with NOR gates in standard or DeMorgan equivalent form by inserting a pair of bubbles between each OR output and each AND input. We can also replace the OR-AND gates (i.e., the POS function) with NAND gates and use NAND inverters to correct the logic levels of the inputs and output.

**3.** If the circuit is not in SOP or POS form, we can use a "block" approach and replace all the gates with the equivalent all-NAND or all-NOR forms.

**4.** We can apply Boolean algebra to the expression we want to implement to get an equivalent form, which we then use to construct a circuit.

**EXAMPLE 4.11**

Convert the logic diagram of Figure 4.40:

**a.** To an all-NAND circuit

**b.** To an all-NOR circuit

**Figure 4.40**

**Example 4.11**
**SOP Circuit**

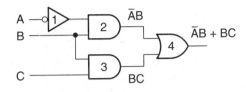

**Solution**

**a.** Figure 4.41a shows the all-NAND equivalent of the SOP circuit of Figure 4.40. All gates are NAND gates in either standard or DeMorgan equivalent form. The inverter is in the same place as in the original circuit.

**b.** Figure 4.41b shows the same circuit in all-NOR form. The circuit is constructed around two levels of AND-shaped and OR-shaped NORs, which preserves the SOP function. We must add three inverters to the circuit to get the Boolean expression back to its original form. The inverters are placed on inputs *B* and *C* and the circuit output. There is no inverter on *A*. Notice that this is the exact opposite of where we placed inverters for the all-NAND form. (Recall that in DeMorgan gate conversions we applied the rule "Change everything." This is similar, although we retain the AND and OR shapes for the SOP function.)

In both circuits the inverters are drawn so that active levels of gate inputs and outputs are matched.

**Figure 4.41**

**Example 4.11**
**All-NAND and All-NOR**
**Forms of an SOP Circuit**

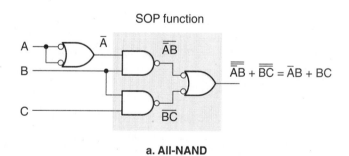

**a. All-NAND**

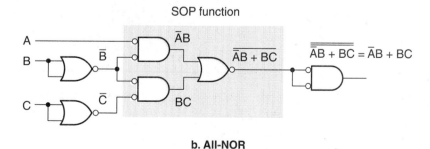

**b. All-NOR**

**EXAMPLE 4.12**    Convert the circuit of Figure 4.42:

    **a.** To an all-NOR circuit

    **b.** To an all-NAND circuit

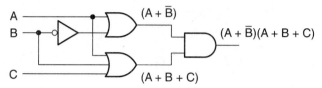

**Figure 4.42**

**Example 4.12**
**POS Circuit**

**Solution**

    **a.** Figure 4.43a shows the all-NOR implementation of the POS circuit in Figure 4.42. We have converted the AND and OR gates to NORs by adding bubbles to the AND inputs and the OR outputs. In the configuration shown, the bubbles cancel, leaving us with the original Boolean expression. The inverter stays at input *B*.

    **b.** Figure 4.43b shows the equivalent all-NAND circuit. We retain the POS function by drawing the NANDs in their OR shape, followed by an AND-shaped gate. The active levels of the circuit must be adjusted by adding NAND inverters as shown. The inverters are at locations complementary to those of the all-NOR circuit.

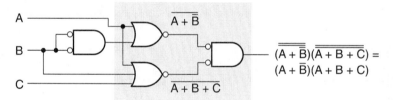

**a. All-NOR circuit**

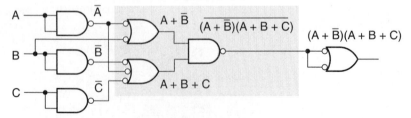

**b. All-NAND circuit**

**Figure 4.43**

**Example 4.12**
**All-NAND and All-NOR Forms of a POS Circuit**

**EXAMPLE 4.13**     Redraw the logic diagram of Figure 4.44 as:

**a.** An all-NAND network

**b.** An all-NOR network

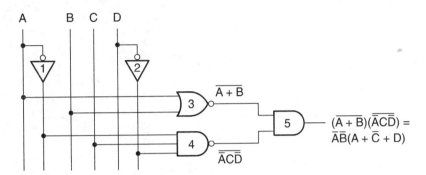

**Figure 4.44**
**Example 4.13**
**Logic Diagram**

**Solution**   The circuit of Figure 4.44 is in neither SOP nor POS form. We can convert this circuit to an all-NAND form by directly substituting the all-NAND equivalent of each gate, as shown in Figure 4.45. The blocks shown are the gate equivalent circuits. We could slightly simplify this circuit by eliminating gate 1 and connecting the topmost input of gate 4 to the output of gate 3a. (Both gates 1 and 3a produce the output $\overline{A}$.)

Figure 4.46 shows the all-NOR form of the circuit in Figure 4.44. Figure 4.46a shows the gates of the original circuit replaced by equivalent NOR gates. We can further simplify the circuit by canceling two pairs of inverters—gates 1–4a and 2–4c. These pairs perform the double inversion function, which cancels out.

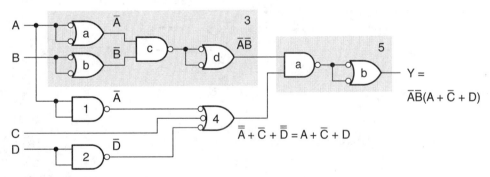

**Figure 4.45**
**Example 4.13**
**All-NAND Form of Circuit in Figure 4.44**

Figure 4.46

Example 4.13
All-NOR Form

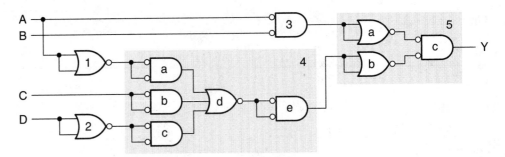

**a. Block circuit**

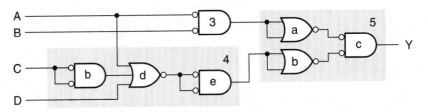

**a. After cancellation of inverter pairs**

**EXAMPLE 4.14**    Implement the circuit in Figure 4.47 in both all-NAND and all-NOR forms. Use Boolean algebra to find the logic diagram of each form. Show the Boolean expression and circuit in each case.

Figure 4.47

Example 4.14
Circuit

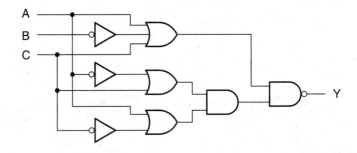

Solutions

All NAND:  $Y = \overline{(A + \overline{B} + C)(\overline{A} + C)(A + \overline{C}))}$

$= \overline{\overline{(A + \overline{B} + C)}\;\overline{(\overline{A} + C)}\;\overline{(A + \overline{C})}}$

$= \overline{(\overline{A\,B\,\overline{C}})(\overline{A\,\overline{C}})(\overline{\overline{A}\,C})}$

Figure 4.48 shows the all-NAND circuit for this expression.

All NOR:  $Y = \overline{(A + \overline{B} + C)(\overline{A} + C)(A + \overline{C})}$

$= \overline{\overline{(A + \overline{B} + C)} + \overline{(\overline{A} + C)} + \overline{(A + \overline{C})}}$

$= \overline{\overline{\overline{(A + \overline{B} + C)} + \overline{(\overline{A} + C)} + \overline{(A + \overline{C})}}}$

The all-NOR circuit is shown in Figure 4.49.

**Figure 4.48**

**Example 4.14**
**All-NAND Implementation of Figure 4.47**

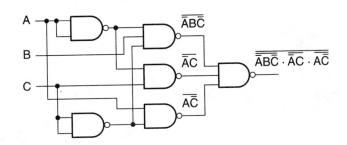

**Figure 4.49**

**Example 4.14**
**All-NOR Implementation of Figure 4.47**

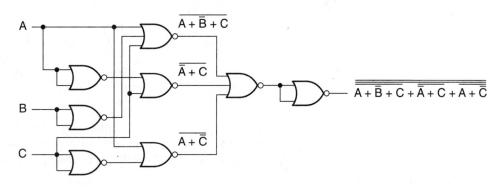

# 4.5

## Practical Circuit Implementation

Until now, we have been drawing logic circuits without regard to whether we can actually build them or not. Even a fairly elementary circuit may be impossible to construct without some modification because the required parts are not available.

For instance, examine the circuit in Figure 4.50. You can't build this circuit from TTL components, because there are no 3-input OR gates available in TTL. The simplest solution is to replace the 3-input OR with two gates of a 74LS32 Quad 2-input OR, as shown in Figure 4.51. The other devices in the circuit are a 74LS11 Triple 3-input AND gate and a 74LS04 Hex Inverter. (Alternatively, we could use 74LS10 Triple 3-input

**Figure 4.50**
**Circuit That Can't Be Built**

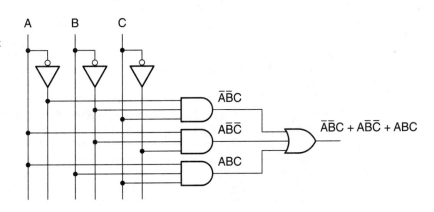

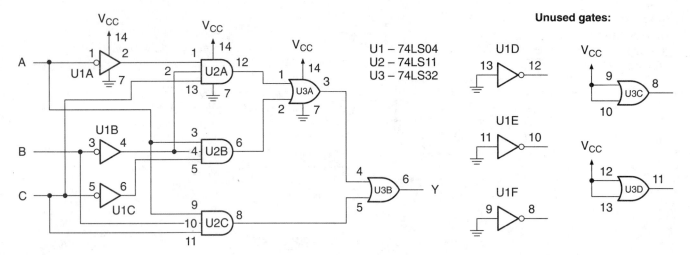

**Figure 4.51**

**Practical Implementation of Figure 4.50**

NAND gates. Refer to Figure 4.41a for the general idea.)

Figure 4.51 shows the pin numbers for each gate, as well as the gate logic functions. The IC packages are designated U1, U2, and U3, with the individual gates shown as U1A, U1B, and so on.

In a practical design, we have to account for the gates left over in each package, since all gates in a package have an effect on various electrical parameters, such as power consumption, whether they are used in a circuit or not. A commercial design would connect the inputs of the unused gates to either a logic HIGH or a logic LOW to protect the other gates in the package from excessive noise. This is particularly important for CMOS components, where unconnected inputs can cause damage by static and overheating.

**Table 4.10**  Available SSI Gates

| Logic Function | Configuration | | Logic Family | | |
|---|---|---|---|---|---|
| | | | LSTTL | CMOS | High-Speed CMOS |
| NAND | Quadruple | 2-input | 74LS00 | 4011B | 74HC00 |
| | Triple | 3-input | 74LS10 | 4023B | 74HC10 |
| | Dual | 4-input | 74LS20 | 4012B | 74HC20 |
| | Single | 8-input | 74LS30 | 4068B | 74HC30 |
| | Single | 13-input | 74LS133 | | 74HC133 |
| NOR | Quadruple | 2-input | 74LS02 | 4001B | 74HC02 |
| | Triple | 3-input | 74LS27 | 4025B | 74HC27 |
| | Dual | 4-input | | 4002B | 74HC4002 |
| | Dual (with strobe) | 4-input | 74LS25 | | |
| | Single | 8-input | | 4078B | |
| AND | Quadruple | 2-input | 74LS08 | 4081B | 74HC08 |
| | Triple | 3-input | 74LS11 | 4073B | 74HC11 |
| | Dual | 4-input | 74LS21 | 4082B | |
| OR | Quadruple | 2-input | 74LS32 | 4071B | 74HC32 |
| | Triple | 3-input | | 4075B | 74HC4075 |
| | Dual | 4-input | | 4072B | |

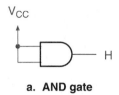

**a. AND gate**

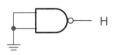

**b. NAND gate**

**Figure 4.52**
**TTL NAND and AND Gates Connected for Minimum Power Consumption**

It doesn't matter at which logic level the input is held for CMOS gates, but generally a TTL gate consumes less power if the output is in the HIGH state. For example, we can connect the inputs of AND and NAND gates as shown in Figure 4.52. The AND inputs are tied HIGH and the NAND inputs are tied LOW. Both connections generate logic HIGH outputs, resulting in lower power consumption than LOW outputs.

Table 4.10 shows the availability of different gates for the basic logic functions in three popular technologies: Low Power Schottky TTL, B-series (metal-gate) CMOS, and High-Speed (poly-gate) CMOS.

The choice of which technology to use will probably depend on factors other than gate availability, since there are a number of ways to produce any logic function with the gates you do have. In fact, we know we can implement any Boolean function with only NAND or NOR gates. Given that, it is interesting to note that the largest selections of gate configurations are for the NAND and NOR functions.

Pin assignments for the 4000B series CMOS gates are shown in Figure 4.53 (see page 138). $V_{DD}$ (power) is connected to pin 14 and $V_{SS}$ (ground) is connected to pin 7 for each of these gates. Pin connections for TTL chips can be found in any good TTL data book. (This text assumes throughout that you have access to a TTL and/or High-Speed CMOS data book.)

Connections for most High-Speed CMOS chips are the same as those for TTL, as this family was designed to supply a CMOS alternative to TTL and thus be completely pin-compatible. Exceptions to this general case are those High-Speed CMOS chips whose functions are not available in TTL, such as 74HC4002 and 74HC4075. These chips have the same pin assignments as their B-series CMOS equivalents.

---

**EXAMPLE 4.15**

The circuit in Figure 4.54 represents the maximum SOP simplification of a Boolean function. Redraw the circuit, including pin numbers and unused gates, for an actual implementation in High-Speed CMOS.

**Solution**

There are no 4-input AND or OR gates available in High-Speed CMOS. One simple solution is to use an all- (well, almost all-) NAND implementation, as shown in Figure 4.55. This requires only three IC packages—a 74HC04 inverter, a 74HC10 3-input NAND, and a 74HC20 4-input NAND. One of the 4-input NANDs (U3B) is drawn in DeMorgan equivalent form.

**Figure 4.54**

**Example 4.15**
**Circuit to Be Redrawn**

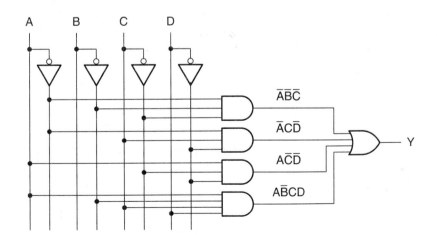

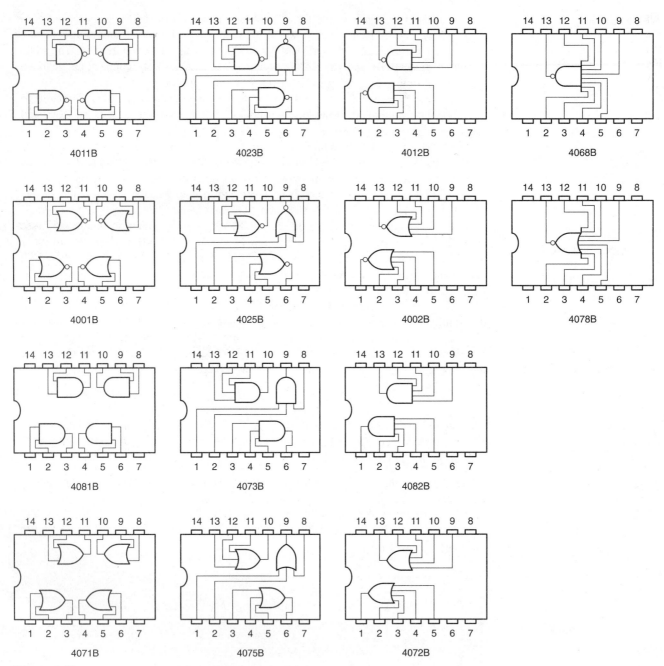

**Figure 4.53**
**Pinouts of CMOS Devices Listed in Table 4.10**

**Figure 4.55**

**Example 4.15**
**Practical Implementation of**
**Figure 4.54 in High-Speed**
**CMOS**

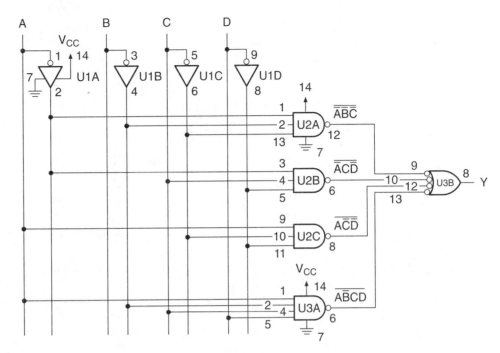

Unused:

U1E                    U1F

# 4.6

# Pulsed Operation of Logic Circuits

In Chapter 2, we looked at the enable and inhibit properties of various individual logic gates. We used these gates to pass or block pulsed digital signals by applying the signal to one input and controlling the logic levels of the remaining inputs.

We can devise more-complex schemes for controlling pulsed signals, using multigate logic circuits. Figure 4.56 shows one such example. The circuit in Figure 4.56 will tell us the period of time, in milliseconds, that one or both of the switches on $A$ and $B$ are open (i.e., in the HIGH state). The counter is a digital circuit whose internal binary value advances by 1 with every cycle of waveform $Y$ (i.e., it counts the number of cycles at output $Y$).

The output $Y$ will be pulsing at 1 kHz, provided the AND gate is in the enable condition, that is, if its Control input is HIGH. This occurs if either $A$ or $B$ is HIGH.

Figure 4.56b shows one possible timing diagram of the circuit. Assume that the counter is reset to zero before either $A$ or $B$ goes HIGH. The third line in the timing diagram, the one showing $(A + B)$, is the enable/inhibit Control input of the AND gate.

**Figure 4.56**
**Multigate Logic Circuit**

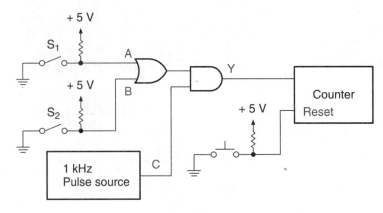

a. Enable circuit

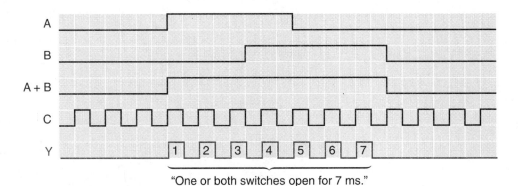

"One or both switches open for 7 ms."

b. Timing diagram

It goes HIGH as a result of either switch *A* or switch *B* opening and enables the pulse waveform for seven cycles, or 7 ms.

---

**EXAMPLE 4.16**

A lamp mounted in a car's instrument panel warns the driver if the engine temperature is too high, the oil pressure is too low, or both. The engine temperature and oil pressure are each monitored by a corresponding sensor, as shown in Figure 4.57.

When the output NAND gate is enabled, it will drive the warning lamp at 3 Hz, alerting the driver to a failure condition. The circuit works only if the ignition switch is closed (logic LOW).

Draw the waveforms, showing a pulsing waveform at *D* and all possible conditions of inputs *A*, *B*, and *C*. Show the resulting waveform of *Y*.

**Solution**

The circuit output is enabled when the Control input of the NAND gate is HIGH. The conditions required are either low oil pressure ($A = 1$) OR high engine temperature ($B = 1$), AND ignition ON ($C = 0$).

The timing diagram in Figure 4.58 shows all possible combinations of *A*, *B*, and *C* and shows the output for all combinations. When *Y* is enabled, it is pulsing out of phase with *D*.

**Figure 4.57**
**Example 4.16**
**Monitoring Sensors**

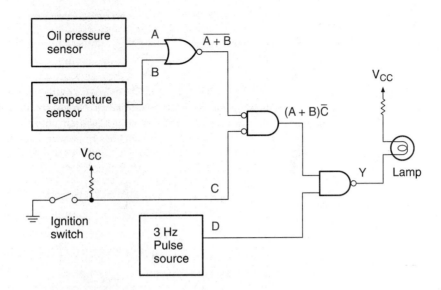

**Figure 4.58**
**Example 4.16**
**Timing Diagram**

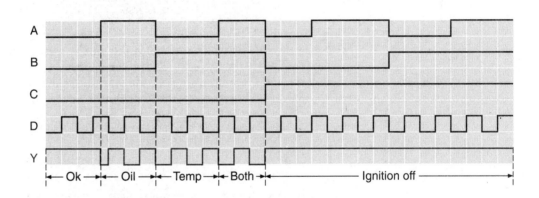

# 4.7

# Designing Logic Circuits From Word Problems

There are several steps involved in the design of digital logic circuits.

1. Define the problem. Make sure you know what it is you are trying to do. Write the problem in words. Have you covered all possibilities?

2. Write a descriptive sentence for each output function, saying what response is required for all possible input conditions. Look for keywords such as "and" and "or," indications of negation, and specified logic levels.

3. Use the description to write a Boolean expression for each circuit output.

4. Simplify the Boolean expression or expressions, if necessary.

5. Draw the circuit for each Boolean expression.

**EXAMPLE 4.17**

An office telephone system is designed to prevent users from making unauthorized long-distance calls. A dialing lockout circuit, shown in block diagram form in Figure 4.59, is installed in each telephone. Once a caller has accessed an outside line and dials either a 0 or a 1 (the long-distance dialing codes), the circuit expects to see a four-digit authorization code. If the access code is then keyed in, the dialing circuit unlocks and the long-distance number can be dialed normally. If, rather than a 0 or 1, a digit from 2 to 9 (i.e., a local call digit) is dialed first, the dialing circuit unlocks and the number can be dialed normally.

The Digit Decoder shown in Figure 4.59 has three outputs, which act as follows:

| Output | State | Comment |
|--------|-------|---------|
| *A* | HIGH | Digit 0 received from telephone keypad. |
| *B* | HIGH | Digit 1 received from telephone keypad. |
| *C* | HIGH | Digit 2 to 9 received from telephone keypad. |

The Access Decoder output acts as follows:

| Output | State | Comment |
|--------|-------|---------|
| *D* | HIGH | Correct access code received from telephone keypad. |

(Assume that the outputs of both decoders are "latching," that is, once made HIGH, an output stays HIGH until the circuit is reset by hanging up the phone.) The *UNLOCK* output is active HIGH.

Write a Boolean expression for the dialing lockout circuit and draw the logic circuit.

**Solution**   Follow the steps for logic circuit design outlined at the beginning of this section.

1. *Define the problem.* This part of the design process has been done for you in the description above.

2. *Write a descriptive sentence for each output.* There is only one output, labeled *UNLOCK*. The dialing lockout circuit will unlock if the first number dialed is a 0 *or* a 1 *and* the correct access code is dialed *or* if the first digit dialed is from 2 to 9. The logic levels corresponding to these conditions are specified in the problem statement.

**Figure 4.59**

**Example 4.17**
**Block Diagram of Dialing Lockout Circuit**

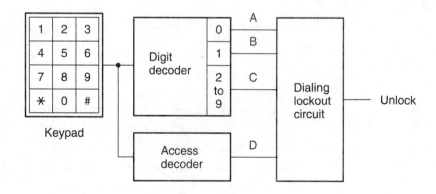

**Figure 4.60**

**Example 4.17**
**Logic Circuit**

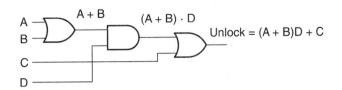

3. *Translate the description into Boolean algebra. UNLOCK* will go HIGH if:

   **a.** The first digit is 0 ($A = 1$) OR 1 ($B = 1$) AND the correct access code is dialed ($D = 1$). Resulting term: $(A + B) \cdot D$
   OR

   **b.** The first digit is from 2 to 9. ($C = 1$)
   This translates into the following Boolean expression:

$$UNLOCK = (A + B) \cdot D + C$$

4. *Simplify the expression.* The expression is already in its simplest form. Alternatively, the expression could be written

$$UNLOCK = A\,D + B\,D + C$$

There is no advantage to one form over the other, as both expressions require the same number of logic gate packages to build.

5. *Draw the corresponding circuit.* The first expression can be implemented with the logic circuit in Figure 4.60. This is the circuit found inside the block labeled "Dialing Lockout Circuit" in Figure 4.59.

# 4.8

# Troubleshooting Combinational Logic Circuits

Combinational logic circuits are subject to the same faults as individual gates: internal logic faults, open circuits, and short circuits to ground, power, and other points in the circuit. These faults may be internal or external to a gate.

## General Approach to Troubleshooting

> When you are troubleshooting a circuit, it is understood that you will be referring to the most up-to-date schematic of the circuit, which *must* include the pin numbers of all gates in the circuit. Without such a diagram, you are working blind.

1. When troubleshooting any circuit, the first thing you should do is make sure that the circuit is getting power. Is the power supply operating properly? (Check with a voltmeter or oscilloscope.) Use a logic probe to check the power and ground pins of the various chips. (An oscilloscope cannot distinguish between a logic LOW (ground) and an open circuit.)

> It is particularly important to check the integrity of the circuit ground connection. Open grounds often produce bizarre faults that cannot be predicted very easily in a pencil-and-paper type of problem.

**2.** Make a quick visual inspection of the circuit, looking for solder bridges, misplaced or loose wires, and obviously damaged components or connectors. Are all input and output cables properly connected?

**3.** Think about possible causes of faulty circuit operation. Ask yourself about proper circuit operation. What is it supposed to do? How does this differ from its observed behavior? Make some notes, listing possible faults. As you check the circuit, refer to your notes, eliminating various possibilities and maybe adding others. Feel free to write notes on your circuit diagrams, if doing so helps you think. After you have located and repaired the circuit fault(s), rewrite your notes neatly so that you or someone else can refer to them in the future.

**4.** Digital circuits can be tested by truth table or pulsed inputs. Here are some possible approaches:

**a.** Take a truth table of the circuit and compare it to the design truth table. Ask yourself what could be responsible for any differences. Test this out by setting the inputs to a combination that causes an error and follow the logic levels through the circuit with a logic probe until you find the source of the error.

**b.** Apply an input pulse to the circuit and follow it through with a logic probe or oscilloscope until you locate a fault.

You can begin the tracing at the input or the output of the circuit, depending on your preference. It often makes sense to start at the output and work back. An output is usually the result of several logic branches, each of which can be tested individually. If one branch is working at the point where it feeds the output, it need not be tested any further, thus eliminating a lot of checking.

---

**EXAMPLE 4.18**

The output of the circuit in Figure 4.61 is always HIGH, regardless of the combination of input logic levels. Power and ground are at proper levels for both chips. Further measurements with a logic probe yield the logic levels shown in Table 4.11.

What are the most likely faults in the circuit? Explain how you would check for these faults.

**Solution**

The circuit output ($U2$–8) is always HIGH because $U2$–10 is always LOW. $U2A$ is working correctly and need not be checked any further. (We know this because $U2$–9 gives the correct logic levels for the term ($\overline{\overline{A}\ B}$.) $U2$–10 could be stuck LOW for a number of reasons, which can be checked in sequence:

**1.** Monitor $U2$–4 and $U2$–5 with a logic probe. If both these pins are stuck HIGH, $U2$–6 will be stuck LOW.

**2.** If $U2$–4 and $U2$–5 are stuck HIGH, disconnect one of these pins from the circuit to see if you can force $U2$–6 HIGH with a pulser on the open input. If so, the inverters ($U1A$ and $U1B$) may not be working properly. Check the inverter output levels and replace, if necessary. Reconnect the gates after testing and/or repairing.

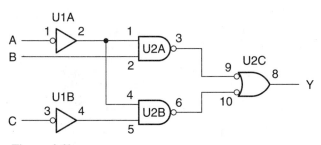

**Figure 4.61**
**Example 4.18**
**Logic Circuit**

**Table 4.11**  Measured Logic Levels

| A | B | C | U2–9 | U2–10 | U2–8 |
|---|---|---|------|-------|------|
| 0 | 0 | 0 | 1 | 0 | 1 |
| 0 | 0 | 1 | 1 | 0 | 1 |
| 0 | 1 | 0 | 0 | 0 | 1 |
| 0 | 1 | 1 | 0 | 0 | 1 |
| 1 | 0 | 0 | 1 | 0 | 1 |
| 1 | 0 | 1 | 1 | 0 | 1 |
| 1 | 1 | 0 | 1 | 0 | 1 |
| 1 | 1 | 1 | 1 | 0 | 1 |

**3.** If the inputs of *U2B* are working properly, the output (*U2–6*) or *U2–10* might be stuck LOW. The question is whether the short is internal or external. Do we have to repair a connection or replace a gate?

   To test this, disconnect *U2–6* and *U2–10*. Monitor *U2–6* with a logic probe to find out if it works properly. Use a pulser on *U2–10* and monitor *U2–8* to check the operation of the output gate. If a gate fails one of these tests, replace it. (Since both gates are part of the same chip, the actual repair—replacing a gate—is the same for both faults.)

**4.** If both *U2B* and *U2C* work properly, the wire or circuit board trace connecting the two pins is shorted to ground. Locate the fault by visual inspection and repair it.

**EXAMPLE 4.19**    The circuit in Figure 4.61 gives the truth table in Table 4.12. What are the possible faults in the circuit? How would you test for them?

**Table 4.12**  Measured Logic Levels

| A | B | C | U2–8 |
|---|---|---|------|
| 0 | 0 | 0 | 0 |
| 0 | 0 | 1 | 0 |
| 0 | 1 | 0 | 1 |
| 0 | 1 | 1 | 1 |
| 1 | 0 | 0 | 0 |
| 1 | 0 | 1 | 0 |
| 1 | 1 | 0 | 0 |
| 1 | 1 | 1 | 0 |

**Table 4.13**  Design Truth Table

| A | B | C | U2–8 |
|---|---|---|------|
| 0 | 0 | 0 | 1 |
| 0 | 0 | 1 | 0 |
| 0 | 1 | 0 | 1 |
| 0 | 1 | 1 | 1 |
| 1 | 0 | 0 | 0 |
| 1 | 0 | 1 | 0 |
| 1 | 1 | 0 | 0 |
| 1 | 1 | 1 | 0 |

**Solution**    Table 4.13 shows the truth table of the correctly operating circuit.

   Figure 4.62 shows the Karnaugh maps for the measured and the design truth tables. By comparing the two K-maps, we see that the circuit is not responding to the input term $\overline{A}\,\overline{C}$. The term $\overline{A}\,\overline{C}$ corresponds to the logic path including *U1A, U1B, U2B,* and *U2C.* This path normally makes the circuit output HIGH by making *U2–10* LOW.

   Possible faults are:

**1.** *U2–10* or *U2–6* could be stuck HIGH. Check this with a logic probe. If this is the case, disconnect one of the pins and determine whether the short is at *U2–6, U2–10,* or in the connection between them.

**Figure 4.62**

**Example 4.19 K-Maps From Tables 4.12 and 4.13**

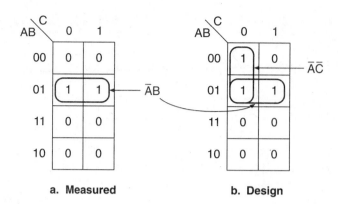

a. Measured

b. Design

2. If *U2–6* is floating, *U2B* probably has an internal open circuit at its output.

3. *U2–10* could have an internal open circuit. Apply a pulse to *U2–10* to see if *U2C* operates correctly. If it does, pulse pin *U2–6* to test the connection between *U2–10* and *U2–6*.

4. If the connection is OK, try to change the state of *U2–6* by pulsing *A* or *C*. (Note: Since $\overline{\overline{A}\,B}$ is working, it is more likely that *C* is faulty or that the connection *U1–2* to *U2–4* is open.) If one or both connections from *U1* to *U2B* are open, *U2–6* will never go LOW. (An open CMOS input is at a floating logic level; *U2–4* and *U2–5* must both be HIGH to make *U2–6* LOW.) You will have to monitor the inverter outputs, the NAND inputs, and the NAND output to determine exactly where such an open circuit might be.

# ● GLOSSARY

**Adjacent cell**   Two cells are adjacent if there is only one variable that is different between the coordinates of the two cells. For example, the cells for minterms $ABC$ and $\overline{A}BC$ are adjacent.

**Cell**   The smallest unit of a Karnaugh map, corresponding to one line of a truth table. The input variables are the cell's coordinates, and the output variable is the cell's contents.

**Don't care state**   An output state that can be regarded either as HIGH or LOW, as is most convenient. A don't care state is the output state of a circuit for a combination of inputs that will never occur.

**Karnaugh map**   A graphical tool for finding the maximum SOP or POS simplification of a Boolean expression. A Karnaugh map works by arranging the terms of an expression in such a way that variables can be canceled by grouping minterms or maxterms.

**Maximum POS simplification**   The form of a POS Boolean expression that cannot be further simplified by canceling variables in the sum terms. It may be possible to get an SOP form of the expression with fewer terms or variables.

**Maximum SOP simplification**   The form of an SOP Boolean expression that cannot be further simplified by canceling variables in the product terms. It may be possible to get a POS form of the expression with fewer terms or variables.

**Octet**   A group of eight adjacent cells in a Karnaugh map. An octet cancels three variables in a K-map simplification.

**Pair**   A group of two adjacent cells in a Karnaugh map. A pair cancels one variable in a K-map simplification.

**Quad**   A group of four adjacent cells in a Karnaugh map. A quad cancels two variables in a K-map simplification.

## ● PROBLEMS

### Section 4.1 Simplifying SOP and POS Expressions

**4.1** Find the maximum SOP and POS simplifications for the function represented by the truth table given in Table 4.14, using the rules of Boolean algebra.

**Table 4.14** Truth Table for Problem 4.1

| A | B | C | Y |
|---|---|---|---|
| 0 | 0 | 0 | 0 |
| 0 | 0 | 1 | 1 |
| 0 | 1 | 0 | 0 |
| 0 | 1 | 1 | 1 |
| 1 | 0 | 0 | 0 |
| 1 | 0 | 1 | 1 |
| 1 | 1 | 0 | 0 |
| 1 | 1 | 1 | 1 |

**4.2** Repeat Problem 4.1 for the truth table given in Table 4.15.

**Table 4.15** Truth Table for Problem 4.2

| A | B | C | Y |
|---|---|---|---|
| 0 | 0 | 0 | 1 |
| 0 | 0 | 1 | 0 |
| 0 | 1 | 0 | 1 |
| 0 | 1 | 1 | 0 |
| 1 | 0 | 0 | 0 |
| 1 | 0 | 1 | 0 |
| 1 | 1 | 0 | 1 |
| 1 | 1 | 1 | 0 |

**4.3** Repeat Problem 4.1 for the truth table given in Table 4.16.

**Table 4.16** Truth Table for Problem 4.3

| A | B | C | D | Y |
|---|---|---|---|---|
| 0 | 0 | 0 | 0 | 0 |
| 0 | 0 | 0 | 1 | 0 |
| 0 | 0 | 1 | 0 | 0 |
| 0 | 0 | 1 | 1 | 0 |
| 0 | 1 | 0 | 0 | 1 |
| 0 | 1 | 0 | 1 | 1 |
| 0 | 1 | 1 | 0 | 0 |
| 0 | 1 | 1 | 1 | 0 |
| 1 | 0 | 0 | 0 | 0 |
| 1 | 0 | 0 | 1 | 1 |
| 1 | 0 | 1 | 0 | 0 |
| 1 | 0 | 1 | 1 | 1 |
| 1 | 1 | 0 | 0 | 1 |
| 1 | 1 | 0 | 1 | 1 |
| 1 | 1 | 1 | 0 | 0 |
| 1 | 1 | 1 | 1 | 1 |

**4.4** Repeat Problem 4.1 for the truth table given in Table 4.17.

**Table 4.17** Truth Table for Problem 4.4

| A | B | C | D | Y |
|---|---|---|---|---|
| 0 | 0 | 0 | 0 | 1 |
| 0 | 0 | 0 | 1 | 0 |
| 0 | 0 | 1 | 0 | 1 |
| 0 | 0 | 1 | 1 | 0 |
| 0 | 1 | 0 | 0 | 1 |
| 0 | 1 | 0 | 1 | 0 |
| 0 | 1 | 1 | 0 | 0 |
| 0 | 1 | 1 | 1 | 0 |
| 1 | 0 | 0 | 0 | 0 |
| 1 | 0 | 0 | 1 | 0 |
| 1 | 0 | 1 | 0 | 0 |
| 1 | 0 | 1 | 1 | 0 |
| 1 | 1 | 0 | 0 | 1 |
| 1 | 1 | 0 | 1 | 1 |
| 1 | 1 | 1 | 0 | 0 |
| 1 | 1 | 1 | 1 | 0 |

**4.5**    Repeat Problem 4.1 for the truth table given in Table 4.18.

**Table 4.18**  Truth Table for Problem 4.5

| A | B | C | D | Y |
|---|---|---|---|---|
| 0 | 0 | 0 | 0 | 0 |
| 0 | 0 | 0 | 1 | 1 |
| 0 | 0 | 1 | 0 | 0 |
| 0 | 0 | 1 | 1 | 1 |
| 0 | 1 | 0 | 0 | 0 |
| 0 | 1 | 0 | 1 | 1 |
| 0 | 1 | 1 | 0 | 1 |
| 0 | 1 | 1 | 1 | 1 |
| 1 | 0 | 0 | 0 | 0 |
| 1 | 0 | 0 | 1 | 1 |
| 1 | 0 | 1 | 0 | 0 |
| 1 | 0 | 1 | 1 | 0 |
| 1 | 1 | 0 | 0 | 0 |
| 1 | 1 | 0 | 1 | 1 |
| 1 | 1 | 1 | 0 | 0 |
| 1 | 1 | 1 | 1 | 1 |

**4.6**    Repeat Problem 4.1 for the truth table given in Table 4.19.

**Table 4.19**  Truth Table for Problem 4.6

| A | B | C | D | Y |
|---|---|---|---|---|
| 0 | 0 | 0 | 0 | 1 |
| 0 | 0 | 0 | 1 | 0 |
| 0 | 0 | 1 | 0 | 1 |
| 0 | 0 | 1 | 1 | 0 |
| 0 | 1 | 0 | 0 | 1 |
| 0 | 1 | 0 | 1 | 0 |
| 0 | 1 | 1 | 0 | 0 |
| 0 | 1 | 1 | 1 | 0 |
| 1 | 0 | 0 | 0 | 1 |
| 1 | 0 | 0 | 1 | 0 |
| 1 | 0 | 1 | 0 | 1 |
| 1 | 0 | 1 | 1 | 1 |
| 1 | 1 | 0 | 0 | 1 |
| 1 | 1 | 0 | 1 | 0 |
| 1 | 1 | 1 | 0 | 1 |
| 1 | 1 | 1 | 1 | 1 |

## Section 4.2 Simplification by the Karnaugh Map Method

**4.7**    Use the Karnaugh map method to reduce the following Boolean expressions to their maximum SOP and POS simplifications (POS optional):

**a.**  $Y = \overline{A}\,\overline{B}\,C + \overline{A}\,B\,C + A\,B\,C$

**b.**  $Y = \overline{A}\,\overline{B}\,C + \overline{A}\,B\,C + A\,B\,\overline{C} + A\,B\,C + A\,\overline{B}\,C$

**c.**  $Y = \overline{A}\,\overline{B}\,\overline{C} + \overline{A}\,B\,C + A\,B\,C + A\,\overline{B}\,C$

**4.8**    Use the Karnaugh map method to reduce the following Boolean expression to its maximum SOP simplification:

$$Y = \overline{A}\,\overline{B}\,\overline{C}\,\overline{D} + \overline{A}\,\overline{B}\,\overline{C}\,D + \overline{A}\,\overline{B}\,C\,D + \overline{A}\,\overline{B}\,C\,\overline{D} + \overline{A}\,B\,C\,\overline{D} + A\,B\,\overline{C}\,D + A\,B\,C\,\overline{D} + A\,\overline{B}\,\overline{C}\,\overline{D} + A\,\overline{B}\,C\,\overline{D}$$

**4.9**    Use a Karnaugh map to reduce the Boolean expression represented by the truth table given in Table 4.20 to its simplest SOP and POS forms (POS optional).

**Table 4.20**  Truth Table for Problem 4.9

| A | B | C | D | Y |
|---|---|---|---|---|
| 0 | 0 | 0 | 0 | 0 |
| 0 | 0 | 0 | 1 | 0 |
| 0 | 0 | 1 | 0 | 0 |
| 0 | 0 | 1 | 1 | 1 |
| 0 | 1 | 0 | 0 | 1 |
| 0 | 1 | 0 | 1 | 1 |
| 0 | 1 | 1 | 0 | 1 |
| 0 | 1 | 1 | 1 | 1 |
| 1 | 0 | 0 | 0 | 0 |
| 1 | 0 | 0 | 1 | 0 |
| 1 | 0 | 1 | 0 | 0 |
| 1 | 0 | 1 | 1 | 1 |
| 1 | 1 | 0 | 0 | 0 |
| 1 | 1 | 0 | 1 | 0 |
| 1 | 1 | 1 | 0 | 0 |
| 1 | 1 | 1 | 1 | 1 |

**4.10** Repeat Problem 4.9 for the truth table given in Table 4.21.

**Table 4.21** Truth Table for Problem 4.10

| A | B | C | D | Y |
|---|---|---|---|---|
| 0 | 0 | 0 | 0 | 0 |
| 0 | 0 | 0 | 1 | 1 |
| 0 | 0 | 1 | 0 | 1 |
| 0 | 0 | 1 | 1 | 0 |
| 0 | 1 | 0 | 0 | 0 |
| 0 | 1 | 0 | 1 | 1 |
| 0 | 1 | 1 | 0 | 1 |
| 0 | 1 | 1 | 1 | 0 |
| 1 | 0 | 0 | 0 | 1 |
| 1 | 0 | 0 | 1 | 0 |
| 1 | 0 | 1 | 0 | 0 |
| 1 | 0 | 1 | 1 | 0 |
| 1 | 1 | 0 | 0 | 0 |
| 1 | 1 | 0 | 1 | 1 |
| 1 | 1 | 1 | 0 | 0 |
| 1 | 1 | 1 | 1 | 1 |

**4.11** Repeat Problem 4.9 for the truth table given in Table 4.22.

**Table 4.22** Truth Table for Problem 4.11

| A | B | C | D | Y |
|---|---|---|---|---|
| 0 | 0 | 0 | 0 | 0 |
| 0 | 0 | 0 | 1 | 0 |
| 0 | 0 | 1 | 0 | 1 |
| 0 | 0 | 1 | 1 | 1 |
| 0 | 1 | 0 | 0 | 0 |
| 0 | 1 | 0 | 1 | 0 |
| 0 | 1 | 1 | 0 | 0 |
| 0 | 1 | 1 | 1 | 0 |
| 1 | 0 | 0 | 0 | 0 |
| 1 | 0 | 0 | 1 | 1 |
| 1 | 0 | 1 | 0 | X |
| 1 | 0 | 1 | 1 | X |
| 1 | 1 | 0 | 0 | X |
| 1 | 1 | 0 | 1 | X |
| 1 | 1 | 1 | 0 | X |
| 1 | 1 | 1 | 1 | X |

$X$ = don't care state

**\*4.12** The circuit in Figure 4.63 represents the maximum SOP simplification of a Boolean function.

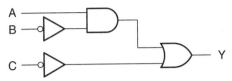

**Figure 4.63**
**Problem 4.12**
**Logic Circuit**

Use a Karnaugh map to derive the circuit for the maximum POS simplification.

**\*4.13** Repeat Problem 4.12 for the circuit in Figure 4.64.

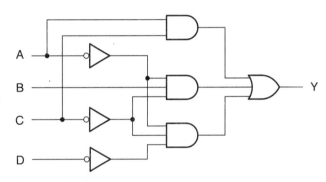

**Figure 4.64**
**Problem 4.13**
**Logic Circuit**

**\*4.14** Refer to the BCD-to-2421 code converter developed in Example 4.8. Use a similar design procedure to develop the circuit of a 2421-to-BCD code converter.

**\*4.15** Excess-3 code is a decimal code that is generated by adding 0011 ($= 3_{10}$) to a BCD code. Table 4.23 shows the relationship between a decimal digital code, natural BCD code, and Excess-3 code. Draw the circuit of a BCD-to-Excess-3 code converter, using the Karnaugh map method to simplify all Boolean expressions.

**Table 4.23** BCD and Excess-3 Code

| Decimal Equivalent | BCD Code | | | | Excess-3 | | | |
|---|---|---|---|---|---|---|---|---|
| | $D_4$ | $D_3$ | $D_2$ | $D_1$ | $E_4$ | $E_3$ | $E_2$ | $E_1$ |
| 0 | 0 | 0 | 0 | 0 | 0 | 0 | 1 | 1 |
| 1 | 0 | 0 | 0 | 1 | 0 | 1 | 0 | 0 |
| 2 | 0 | 0 | 1 | 0 | 0 | 1 | 0 | 1 |
| 3 | 0 | 0 | 1 | 1 | 0 | 1 | 1 | 0 |
| 4 | 0 | 1 | 0 | 0 | 0 | 1 | 1 | 1 |
| 5 | 0 | 1 | 0 | 1 | 1 | 0 | 0 | 0 |
| 6 | 0 | 1 | 1 | 0 | 1 | 0 | 0 | 1 |
| 7 | 0 | 1 | 1 | 1 | 1 | 0 | 1 | 0 |
| 8 | 1 | 0 | 0 | 0 | 1 | 0 | 1 | 1 |
| 9 | 1 | 0 | 0 | 1 | 1 | 1 | 0 | 0 |

**\*4.16** Repeat Problem 4.15 for an Excess-3-to-BCD code converter.

## Section 4.3 Simplification Using DeMorgan Equivalent Gates

**4.17** Redraw the circuit in Figure 4.65 to make the input and output active levels match at each gating level, and write the Boolean expression of the output of the new circuit.

**4.18** **a.** Write the Boolean expression for the circuit shown in Figure 4.66.

**Figure 4.65**

**Problem 4.17 Logic Circuit**

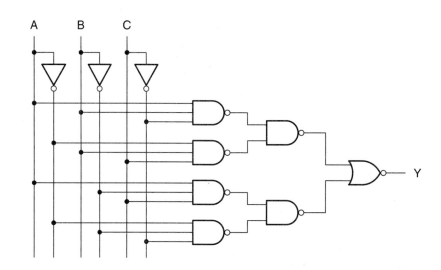

**Figure 4.66**

**Problem 4.18 Logic Circuit**

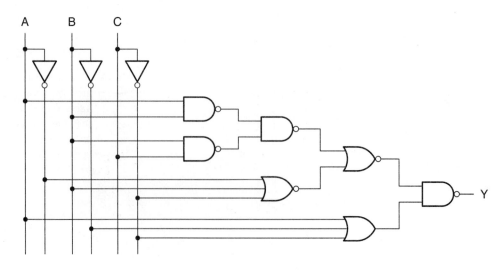

**b.** Redraw the circuit in Figure 4.66 to make the input and output active levels match at each gating level, and write the Boolean expression of the output of the new circuit.

**c.** Simplify the circuit as much as possible.

**4.19** **a.** Draw the circuit represented by the following Boolean expression:

$$Y = (\overline{(\overline{\overline{A} + \overline{B} + C + \overline{A}\,\overline{B}}) + \overline{A} + B + C})\,(A\,B\,C)$$

**b.** Redraw the circuit derived in part a to make the input and output active levels match at each gating level, and write the Boolean expression of the output of the new circuit.

## Section 4.4 Universal Property of NAND/NOR Gates

**4.20** Redraw the circuit in Figure 4.67 as an all-NAND circuit and as an all-NOR circuit. In each case, simplify the circuit as much as possible.

**4.21** Repeat Problem 4.20 for the circuit in Figure 4.68.

**4.22** Repeat Problem 4.20 for the circuit in Figure 4.69.

**Figure 4.67**

**Problem 4.20**
**Logic Circuit**

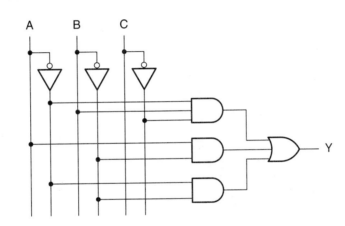

**Figure 4.68**

**Problem 4.21**
**Logic Circuit**

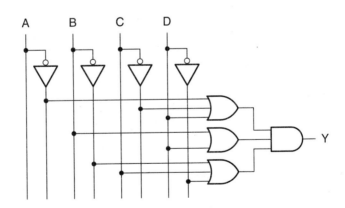

**Figure 4.69**

**Problem 4.22**
**Logic Circuit**

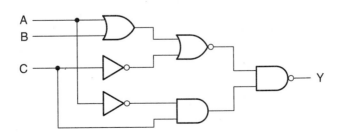

**4.23** Repeat Problem 4.20 for the circuit in Figure 4.70.

## Section 4.5 Practical Circuit Implementation

**4.24** Redraw the circuit shown in Figure 4.71, including pin numbers and unused gates, for an implementation in B-series CMOS. You may use any available CMOS gates. (Refer to Figure 4.53 for pinouts.)

**4.25** Repeat Problem 4.24 for an implementation in Low Power Schottky (LS) TTL.

**4.26** Redraw the circuit in Figure 4.69, including pin numbers and unused gates, for an implementation in High-Speed CMOS.

**4.27** Redraw the circuit in Figure 4.70, including pin numbers and unused gates, for an implementation in Metal-Gate (4000B) CMOS.

**4.28** Repeat Problem 4.27 for an implementation in Low Power Schottky (LS) TTL.

## Section 4.6 Pulsed Operation of Logic Circuits

**4.29** Given inputs $A$, $B$, and $C$ to the logic circuit shown in Figure 4.72 (see page 153), draw the waveform for the circuit output, $Y$.

**4.30** Repeat Problem 4.29 for the circuit in Figure 4.73 (see page 15).

**Figure 4.70**

**Problem 4.23**
**Logic Circuit**

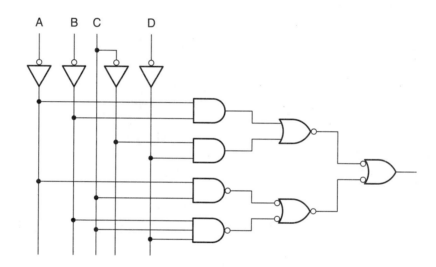

**Figure 4.71**

**Problem 4.24**
**Logic Circuit**

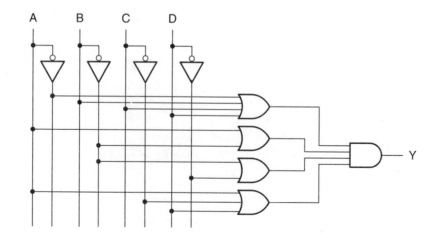

**Figure 4.72**

**Problem 4.29**
**Logic Circuit**

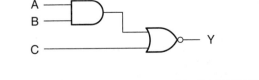

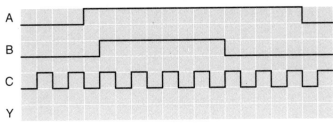

**Figure 4.73**

**Problem 4.30**
**Logic Circuit**

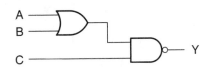

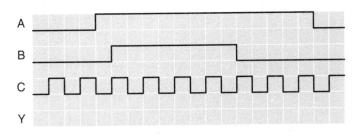

**4.31** Repeat Problem 4.29 for the circuit in Figure 4.74.

**Figure 4.74**

**Problem 4.31**
**Logic Circuit**

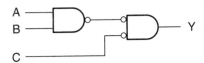

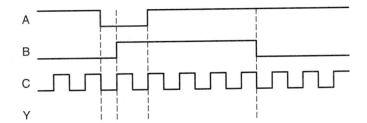

**4.32** Repeat Problem 4.29 for the circuit in Figure 4.75.

**\*4.33** Part of a security system is shown in Figure 4.76. The system sounds a 200-Hz buzzer if certain security conditions are breached. The system is controlled by a conductive window strip, a door switch, and two related timers.

    **a.** When the door opens, the door switch opens, producing a logic HIGH.

    **b.** There are two timers, one for 30 seconds, one for 60 seconds. Each timer starts as soon as

its START input goes HIGH (i.e., when the door opens).

    **c.** The output, A, of the 60-second timer is normally LOW. When this timer starts, A goes HIGH for 60 seconds.

    **d.** The 30-second timer is slightly different. Its output, B, is also normally LOW, but will go HIGH *after* the 30 seconds has elapsed. After this time, it will stay HIGH until the system is reset by pressing the RESET switch.

    **e.** The 30-second timer can be reset any time after the door opens. If it is not reset in the first

**Figure 4.75**

**Problem 4.32
Logic Circuit**

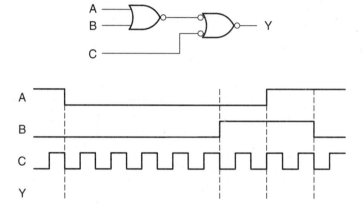

**Figure 4.76**

**Problem 4.33
Security System**

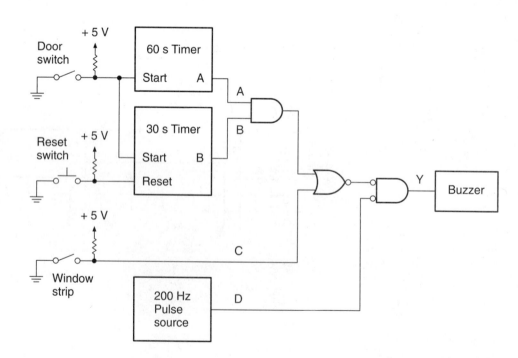

30 seconds, $B$ goes HIGH and a 200-Hz buzzer will sound. (We can prevent the buzzer from sounding by pressing the RESET switch within 30 seconds of opening the door.)

**f.** The buzzer sounds immediately if a conductive strip around a window (represented by a switch on input $C$) opens at any time, making input $C$ HIGH.

Draw the timing diagram for all possible input conditions, and indicate the action taking place for each condition.

### Section 4.7 Designing Logic Circuits From Word Problems

**\*\*4.34** An industrial process involves the mixing of chemicals in a vat with critical restrictions on volume (level), temperature, and pressure of the vat contents. A digitally controlled safety system, shown in Figure 4.77, monitors each of these parameters.

There are three sensors monitoring each parameter, labeled $L_1$, $L_2$, and $L_3$ for level; $T_1$, $T_2$, and $T_3$ for temperature; and $P_1$, $P_2$, and $P_3$ for pressure. Each sensor produces a logic HIGH when its measured parameter is out of the acceptable range.

The output, $Y$, of the safety system goes LOW and switches on an alarm circuit if a majority of parameters (level, temperature, and pressure) give an out-of range indication (logic HIGH). Each parameter will give an out-of-range indication if at least one sensor for that parameter produces a HIGH output.

Write the Boolean expression for the circuit and draw the logic circuit. (Refer to Problem 3.5 (Chapter 3) for the form of a majority vote circuit.)

**\*\*4.35** Redraw the logic circuit derived in Problem 4.34 if a majority of sensors for a parameter must give an out-of-range indication to make the monitored parameter indicate an out-of-range condition.

Write the Boolean expression for the redrawn circuit.

### Section 4.8 Troubleshooting Combinational Logic Circuits

**\*4.36** The circuit in Figure 4.78 yields the truth table shown in Table 4.24. Assuming the power and ground connections are OK, list the possible faults that could cause the response in Table 4.24, and describe procedures you could use to check these possibilities.

**4.37** The circuit in Figure 4.78 yields the truth table shown in Table 4.25 (see page 156). Table 4.25 also shows measurements taken at two other points in the circuit. Based on these readings, describe the possible circuit faults and what you could do to check these possibilities.

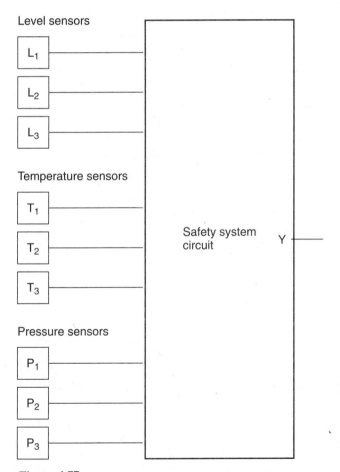

**Figure 4.77**
**Problem 4.34**
**Monitoring System**

**Figure 4.78**

**Problem 4.36**
**Logic Circuit**

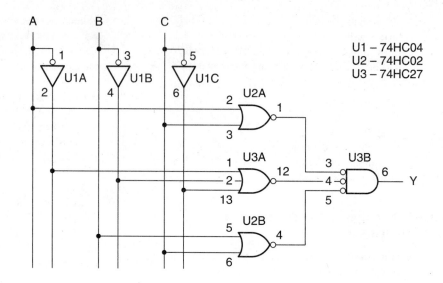

U1 – 74HC04
U2 – 74HC02
U3 – 74HC27

**Table 4.24** Truth Table for Problem 4.36

| A | B | C | Y |
|---|---|---|---|
| 0 | 0 | 0 | 0 |
| 0 | 0 | 1 | 1 |
| 0 | 1 | 0 | 1 |
| 0 | 1 | 1 | 1 |
| 1 | 0 | 0 | 0 |
| 1 | 0 | 1 | 1 |
| 1 | 1 | 0 | 1 |
| 1 | 1 | 1 | 0 |

**Table 4.25** Truth Table for Problem 4.37

| A | B | C | U3–3 | U3–4 | U3–6 |
|---|---|---|------|------|------|
| 0 | 0 | 0 | 1 | 0 | 0 |
| 0 | 0 | 1 | 0 | 0 | 0 |
| 0 | 1 | 0 | 1 | 0 | 0 |
| 0 | 1 | 1 | 0 | 0 | 0 |
| 1 | 0 | 0 | 0 | 0 | 0 |
| 1 | 0 | 1 | 0 | 0 | 0 |
| 1 | 1 | 0 | 0 | 0 | 0 |
| 1 | 1 | 1 | 0 | 1 | 0 |

## ▼ Answers to Section Review Question

### Section 4.1

**4.1.** SOP form: $\overline{A}C + B\overline{C}$
POS form: $(\overline{A} + \overline{C})(B + C)$

### Section 4.3

**4.2.** $Y = (A\overline{B} + C)(B + \overline{C})$

# Number Systems and Codes

## ● CHAPTER CONTENTS

**5.1** Positional Notation
**5.2** Binary Numbers
**5.3** Hexadecimal Numbers
**5.4** Octal Numbers

**5.5** BCD Codes
**5.6** Gray Code
**5.7** ASCII Code

## ● CHAPTER OBJECTIVES

Upon successful completion of this chapter, you will be able to:

- Explain positional notation and write the positional multipliers for any number base.
- Count in binary, octal, decimal, or hexadecimal.
- Convert a number in binary, octal, decimal, or hexadecimal to any of the other number bases.
- Calculate the fractional binary equivalent of any decimal number.

- Write a decimal number in 8421 or Excess-3 codes.
- Generate an $n$-bit Gray code sequence.
- Encode alphanumeric strings into ASCII code and vice versa.

## ● INTRODUCTION

Numbers in digital systems are usually based on a positional notation system, where digits have different values depending on their placement within a number. The most common positional notation systems in digital systems are those based on 2, 8, 10, and 16 (binary, octal, decimal, and hexadecimal, respectively). It is important to understand the different systems of writing numbers for digital systems and how we can convert numbers from one form to another.

Binary numbers are used in nonpositional codes to represent numbers, letters, and various control commands. Binary-coded decimal represents individual decimal numbers as binary numbers. ASCII code represents alphanumeric and control characters in a 7-bit binary format.

## 5.1

# Positional Notation

> **Base** *The number whose powers represent positional multipliers in a number system. Also the number of digits that are used in that system (e.g., ten digits, 0 to 9, are used in the decimal system).*

Positional notation and the decimal number system are so familiar to us that most of the time we don't even realize we are using them. By analyzing these familiar ways of writing numbers, we can get at the principles behind everyday numerical operations and apply them to other number systems that are useful to us in digital systems. Because of the binary, or two-state, nature of digital devices, the most useful numbers in digital systems are those whose **base** is a power of 2, particularly 2, 16, and, less frequently, 8.

A number base is indicated by a subscript, for example, $466_{10}$ (base 10) or $341_8$ (base 8).

Almost all number systems commonly used today are positional. (The best known nonpositional number system is the Roman system (I, II, III, IV, V, etc.). Its nonpositional nature makes it very awkward for any sort of technical calculation.)

In a positional notation system, the value of a digit changes when it appears in a different position within a number. For instance, the digit 8 in the decimal number 2894 really means 800. In the decimal number 1082, the digit 8 means 80.

This is because in the decimal (base 10) system, we must multiply the digit in each position by a power-of-10 multiplier. The value of the positional multiplier is 10 times that of the multiplier for the next position to the right. The multipliers for the first five digit positions are:

$$10^4 \qquad 10^3 \qquad 10^2 \qquad 10^1 \qquad 10^0$$
$$=10000 \quad =1000 \quad =100 \quad =10 \quad =1$$

The number 30428 can be broken down as:

$$
\begin{aligned}
3 \times 10^4 &= 3 \times 10000 = 30000 \\
0 \times 10^3 &= 0 \times 1000 = 0 \\
4 \times 10^2 &= 4 \times 100 = 400 \\
2 \times 10^1 &= 2 \times 10 = 20 \\
8 \times 10^0 &= 8 \times 1 = \underline{\phantom{00}8} \\
&\qquad\qquad\qquad\quad 30428
\end{aligned}
$$

> The values of the positional multipliers of a number system are powers of the system's number base. In the decimal system, the base is 10, so multipliers are powers of 10. In the binary system, the base is 2, so each positional multiplier is a power of 2.

The first five positional multipliers in the binary system are:

$$2^4 \qquad 2^3 \qquad 2^2 \qquad 2^1 \qquad 2^0$$
$$= 16 \qquad = 8 \qquad = 4 \qquad = 2 \qquad = 1$$

The multiplier value for each position is twice that of the multiplier for the next position to the right.

## Binary-to-Decimal Conversion

To convert a binary number to decimal, multiply each positional binary multiplier by the corresponding bit (0 or 1) and add the resultant products.

**EXAMPLE 5.1**

Convert $1011101_2$ to decimal.

**Solution**    The multipliers and the products of their corresponding bits are:

$$1 \times 2^6 = 64$$
$$0 \times 2^5 = 0$$
$$1 \times 2^4 = 16$$
$$1 \times 2^3 = 8$$
$$1 \times 2^2 = 4$$
$$0 \times 2^1 = 0$$
$$1 \times 2^0 = \underline{1}$$
$$93_{10}$$

---

**Section Review Problem for Section 5.1**

5.1. What are the first four positional multipliers for a base 16 number system?

---

## 5.2

# Binary Numbers

Since the binary number system has a base of 2, there are only two digits for writing binary numbers—0 and 1. The value of each bit is determined by a power-of-2 multiplier corresponding to the bit's position within a number.

## Counting in Binary

In Chapter 1, we looked at two ways to generate a binary sequence. We can count in binary by using the weighted values of bits to progress through a count sequence:

$$000 = (0 + 0 + 0) = 0$$
$$001 = (0 + 0 + 1) = 1$$
$$010 = (0 + 2 + 0) = 2$$
$$011 = (0 + 2 + 1) = 3$$
$$100 = (4 + 0 + 0) = 4$$
$$101 = (4 + 0 + 1) = 5$$
$$110 = (4 + 2 + 0) = 6$$
$$111 = (4 + 2 + 1) = 7$$

Alternatively, we can duplicate a pattern in the adjacent numbers.

1. The least significant bit (rightmost bit) alternates 0, 1, 0, 1,. . . with every line.
2. The bit second from the right alternates every two lines: 0, 0, 1, 1, 0, 0, 1, 1,. . ..
3. The third bit from the right alternates every four lines: 0, 0, 0, 0, 1, 1, 1, 1,. . ..
4. The fourth bit from the right alternates every eight lines: 0, 0, 0, 0, 0, 0, 0, 0, 1, 1, 1, 1, 1, 1, 1,. . ..

This pattern can be extended for any number of bits. Each bit alternates every $2^{n-1}$ lines, for the $n$th bit from the right.

The numbers in a binary sequence are often written with a fixed length. The complete 5-bit binary sequence is:

| (0) | 00000 | (8) | 01000 | (16) | 10000 | (24) | 11000 |
|-----|-------|------|-------|------|-------|------|-------|
|     | 00001 |      | 01001 |      | 10001 |      | 11001 |
|     | 00010 |      | 01010 |      | 10010 |      | 11010 |
|     | 00011 |      | 01011 |      | 10011 |      | 11011 |
|     | 00100 |      | 01100 |      | 10100 |      | 11100 |
|     | 00101 |      | 01101 |      | 10101 |      | 11101 |
|     | 00110 |      | 01110 |      | 10110 |      | 11110 |
| (7) | 00111 | (15) | 01111 | (23) | 10111 | (31) | 11111 |

## Decimal-to-Binary Conversion

There are two methods commonly used to convert decimal numbers to binary: sum of powers of 2 and repeated division by 2.

### Sum of Powers of 2

You can convert a decimal number to binary by adding up powers of 2 by inspection, adding bits as you need them to fill up the total value of the number. For example, convert $57_{10}$ to binary.

$$64_{10} > 57_{10} > 32_{10}$$

Thus, 32 ($= 2^5$) is the largest power of 2 that is smaller than 57. The MSB is a 1 in the 32's bit position. The other bits are added as needed to complete the number.

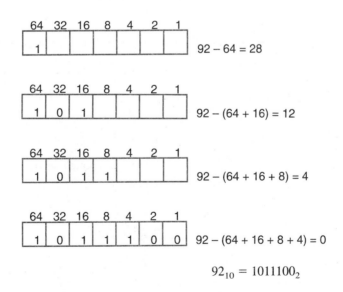

| 32 | 16 | 8 | 4 | 2 | 1 | |
|----|----|----|----|----|----|----|
| 1 |  |  |  |  |  | $57 - 32 = 25$ |

| 32 | 16 | 8 | 4 | 2 | 1 | |
|----|----|----|----|----|----|----|
| 1 | 1 |  |  |  |  | $57 - (32 + 16) = 9$ |

| 32 | 16 | 8 | 4 | 2 | 1 | |
|----|----|----|----|----|----|----|
| 1 | 1 | 1 |  |  |  | $57 - (32 + 16 + 8) = 1$ |

| 32 | 16 | 8 | 4 | 2 | 1 | |
|----|----|----|----|----|----|----|
| 1 | 1 | 1 | 0 | 0 | 1 | $57 - (32 + 16 + 8 + 1) = 0$ |

$$57_{10} = 111001_2$$

**EXAMPLE 5.2**  Convert $92_{10}$ to binary using the sum-of-powers-of-2 method.

**Solution**  $$128 > 92 > 64$$

The 64's ($2^6$) bit is the MSB.

| 64 | 32 | 16 | 8 | 4 | 2 | 1 | |
|----|----|----|----|----|----|----|----|
| 1 |  |  |  |  |  |  | $92 - 64 = 28$ |

| 64 | 32 | 16 | 8 | 4 | 2 | 1 | |
|----|----|----|----|----|----|----|----|
| 1 | 0 | 1 |  |  |  |  | $92 - (64 + 16) = 12$ |

| 64 | 32 | 16 | 8 | 4 | 2 | 1 | |
|----|----|----|----|----|----|----|----|
| 1 | 0 | 1 | 1 |  |  |  | $92 - (64 + 16 + 8) = 4$ |

| 64 | 32 | 16 | 8 | 4 | 2 | 1 | |
|----|----|----|----|----|----|----|----|
| 1 | 0 | 1 | 1 | 1 | 0 | 0 | $92 - (64 + 16 + 8 + 4) = 0$ |

$$92_{10} = 1011100_2$$

## Repeated Division by 2

Any decimal number divided by 2 will leave a remainder of 0 or 1. Repeated division by 2 will leave a string of 0s and 1s that become the binary equivalent of the decimal number. Let us use this method to convert $46_{10}$ to binary.

1. Divide the decimal number by 2 and note the remainder.

$$46/2 = 23 + \text{remainder } 0 \text{ (LSB)}$$

The remainder is the least significant bit of the binary equivalent of 46.

2. Divide the quotient from the previous division and note the remainder. The remainder is the second LSB.

$$23/2 = 11 + \text{remainder } 1$$

3. Continue this process until the *quotient* is 0. The last remainder is the most significant bit of the binary number.

$$11/2 = 5 + \text{remainder } 1$$
$$5/2 = 2 + \text{remainder } 1$$
$$2/2 = 1 + \text{remainder } 0$$
$$1/2 = 0 + \text{remainder } 1 \quad \text{(MSB)}$$

To write the binary equivalent of the decimal number, read the remainders from the bottom up.

$$46_{10} = 101110_2$$

---

**EXAMPLE 5.3**

Use repeated division by 2 to convert $115_{10}$ to a binary number.

**Solution**

$$115/2 = 57 + \text{remainder } 1 \text{ (LSB)}$$
$$57/2 = 28 + \text{remainder } 1$$
$$28/2 = 14 + \text{remainder } 0$$
$$14/2 = 7 + \text{remainder } 0$$
$$7/2 = 3 + \text{remainder } 1$$
$$3/2 = 1 + \text{remainder } 1$$
$$1/2 = 0 + \text{remainder } 1 \text{ (MSB)}$$

Read the remainders from bottom to top: 1110011.

$$115_{10} = 1110011_2$$

---

In any decimal-to-binary conversion, the number of bits in the binary number is the exponent of the smallest power of 2 that is larger than the decimal number.

For example, for the numbers $92_{10}$ and $46_{10}$,

$$2^7 = 128 > 92 \qquad 7 \text{ bits: } 1011100$$
$$2^6 = 64 > 46 \qquad 6 \text{ bits: } 101110$$

## Fractional Binary Numbers

**Radix point**  *The generalized form of a decimal point. In any positional number system, the radix point marks the dividing line between positional multipliers that are positive and negative powers of the system's number base.*

**Binary point**  *A period (".") that marks the dividing line between positional multipliers that are positive and negative powers of 2 (e.g., first multiplier right of binary point = $2^{-1}$; first multiplier left of binary point = $2^0$).*

In the decimal system, fractional numbers use the same digits as whole numbers, but the digits are written to the right of the decimal point. The multipliers for these digits are negative powers of 10—$10^{-1}$ (1/10), $10^{-2}$ (1/100), $10^{-3}$ (1/1000), and so on.

So it is in the binary system. Digits 0 and 1 are used to write fractional binary numbers, but the digits are to the right of the **binary point**—the binary equivalent of the decimal point. (The decimal point and binary point are special cases of the **radix point,** the general name for any such point in any number system.)

Each digit is multiplied by a positional factor that is a negative power of 2. The first four multipliers on either side of the binary point are:

| $2^3$ | $2^2$ | $2^1$ | $2^0$ | binary point . | $2^{-1}$ | $2^{-2}$ | $2^{-3}$ | $2^{-4}$ |
|---|---|---|---|---|---|---|---|---|
| =8 | =4 | =2 | =1 | | =1/2 | =1/4 | =1/8 | =1/16 |

**EXAMPLE 5.4**

Write the binary fraction 0.101101 as a decimal fraction.

**Solution**

$1 \times 1/2 = 1/2$

$0 \times 1/4 = 0$

$1 \times 1/8 = 1/8$

$1 \times 1/16 = 1/16$

$0 \times 1/32 = 0$

$1 \times 1/64 = 1/64$

$1/2 + 1/8 + 1/16 + 1/64 = 32/64 + 8/64 + 4/64 + 1/64$

$= 45/64$

$= 0.703125_{10}$

## Fractional-Decimal-to-Fractional-Binary Conversion

Simple decimal fractions such as 0.5, 0.25, and 0.375 can be converted to binary fractions by a sum-of-powers method. The above decimal numbers can also be written

$0.5 = 1/2$, $0.25 = 1/4$, and $0.375 = 3/8 = 1/4 + 1/8$. These numbers can all be represented by negative powers of 2. Thus, in binary,

$$0.5_{10} \quad = 0.1_2$$

$$0.25_{10} \quad = 0.01_2$$

$$0.375_{10} = 0.011_2$$

The conversion process becomes more complicated if we try to convert decimal fractions that cannot be broken into powers of 2. For example, the number $1/5 = 0.2_{10}$ cannot be exactly represented by a sum of negative powers of 2. (Try it.) For this type of number, we must use the method of repeated multiplication by 2.

**Method:**

1. Multiply the decimal fraction by 2 and note the integer part. The integer part is either 0 or 1 for any number between 0 and 0.999 . . . . The integer part of the product is the first digit to the left of the binary point.

$$0.2 \times 2 = 0.4 \quad \text{Integer part: } 0$$

2. Discard the integer part of the previous product and repeat step 1 until the fraction repeats or terminates.

$$0.4 \times 2 = 0.8 \quad \text{Integer part: } 0$$

$$0.8 \times 2 = 1.6 \quad \text{Integer part: } 1$$

$$0.6 \times 2 = 1.2 \quad \text{Integer part: } 1$$

$$0.2 \times 2 = 0.4 \quad \text{Integer part: } 0$$

(Fraction repeats; product is same as in step 1)

Read the above integer parts from top to bottom to obtain the fractional binary number. Thus, $0.2_{10} = 0.00110011 \ldots_2 = 0.\overline{0011}_2$. The bar shows the portion of the digits that repeats.

---

**EXAMPLE 5.5**

Convert $0.95_{10}$ to its binary equivalent.

**Solution**

$$0.95 \times 2 = 1.90 \quad \text{Integer part: } 1$$

$$0.90 \times 2 = 1.80 \quad \text{Integer part: } 1$$

$$0.80 \times 2 = 1.60 \quad \text{Integer part: } 1$$

$$0.60 \times 2 = 1.20 \quad \text{Integer part: } 1$$

$$0.20 \times 2 = 0.40 \quad \text{Integer part: } 0$$

$$0.40 \times 2 = 0.80 \quad \text{Integer part: } 0$$

$$0.80 \times 2 = 1.60 \quad \text{Fraction repeats last four digits}$$

$$0.95_{10} = 0.11\overline{1100}_2$$

# 5.3

# Hexadecimal Numbers

After binary numbers, hexadecimal (base 16) numbers are the most important numbers in digital applications. Hexadecimal, or hex, numbers are primarily used as a shorthand form of binary notation. Since 16 is a power of 2 ($2^4 = 16$), each hexadecimal digit can be converted directly to four binary digits. Hex numbers can pack more digital information into fewer digits.

Hex numbers have become particularly popular with the advent of small computers, which use binary data having 8, 16, or 32 bits. Such data can be represented by 2, 4, or 8 hexadecimal digits, respectively.

## Counting in Hexadecimal

The positional multipliers in the hex system are powers of sixteen: $16^0 = 1$, $16^1 = 16$, $16^2 = 256$, $16^3 = 4096$, and so on.

We need 16 digits to write hex numbers; the decimal digits 0 through 9 are not sufficient. The usual convention is to use the capital letters A through F, each letter representing a number from $10_{10}$ through $15_{10}$. Table 5.1 shows how hexadecimal digits relate to their decimal and binary equivalents.

**Table 5.1** Hex Digits and Their Binary and Decimal Equivalents

| Hex | Decimal | Binary |
|-----|---------|--------|
| 0 | 0 | 0000 |
| 1 | 1 | 0001 |
| 2 | 2 | 0010 |
| 3 | 3 | 0011 |
| 4 | 4 | 0100 |
| 5 | 5 | 0101 |
| 6 | 6 | 0110 |
| 7 | 7 | 0111 |
| 8 | 8 | 1000 |
| 9 | 9 | 1001 |
| A | 10 | 1010 |
| B | 11 | 1011 |
| C | 12 | 1100 |
| D | 13 | 1101 |
| E | 14 | 1110 |
| F | 15 | 1111 |

**Counting Rules for Hexadecimal Numbers:**

1. Count in sequence from 0 to F in the least significant digit.
2. Add 1 to the next digit to the left and start over.
3. Repeat in all other columns.

For instance, the hex numbers between 19 and 22 are 19, 1A, 1B, 1C, 1D, 1E, 1F, 20, 21, 22. (The decimal equivalents of these numbers are $25_{10}$ through $34_{10}$.)

**EXAMPLE 5.6**

List the hexadecimal digits from $190_{16}$ to $200_{16}$, inclusive.

**Solution**    The numbers follow the counting rules: Use all the digits in one position, add 1 to the digit one position left, and start over.

For brevity, we will list only a few of the numbers in the sequence:

190, 191, 192, . . . , 199, 19A, 19B, 19C, 19D, 19E, 19F,

1A0, 1A1, 1A2, . . . , 1A9, 1AA, 1AB, 1AC, 1AD, 1AE, 1AF,

1B0, 1B1, 1B2, . . . , 1B9, 1BA, 1BB, 1BC, 1BD, 1BE, 1BF,

1CO, . . . , 1CF, 1DO, . . . , 1DF, 1E0, . . . , 1EF, 1F0, . . . , 1FF, 200

**Section Review Problems for Section 5.3a**

5.2. List the hexadecimal numbers from FA9 to FB0, inclusive.

5.3. List the hexadecimal numbers from 1F9 to 200, inclusive.

## Hexadecimal-to-Decimal Conversion

To convert a number from hex to decimal, multiply each digit by its power-of-16 positional multiplier and add the products.

**EXAMPLE 5.7**

Convert $7C6_{16}$ to decimal.

**Solution**

$$7 \times 16^2 = 7_{10} \times 256_{10} = 1792_{10}$$
$$C \times 16^1 = 12_{10} \times 16_{10} = 192_{10}$$
$$6 \times 16^0 = 6_{10} \times 1_{10} = \underline{6_{10}}$$
$$1990_{10}$$

**EXAMPLE 5.8**

Convert $1FD5_{16}$ to decimal.

**Solution**

$$1 \times 16^3 = 1_{10} \times 4096_{10} = 4096_{10}$$
$$F \times 16^2 = 15_{10} \times 256_{10} = 3840_{10}$$
$$D \times 16^1 = 13_{10} \times 16_{10} = 208_{10}$$
$$5 \times 16^0 = 5_{10} \times 1_{10} = \underline{5_{10}}$$
$$8149_{10}$$

---

**Section Review Problem for Section 5.3b**

5.4. Convert the hexadecimal number A30F to its decimal equivalent.

---

## Decimal-to-Hexadecimal Conversion

Decimal numbers can be converted to hex by the sum-of-weighted-hex-digits method or by repeated division by 16. The main difficulty we encounter in either method is remembering to convert decimal numbers 10 through 15 into the equivalent hex digits, A through F.

### Sum of Weighted Hexadecimal Digits

This method is useful for simple conversions (about three digits). For example, the decimal number 35 is easily converted to the hex value 23.

$$35_{10} = 32_{10} + 3_{10} = (2 \times 16) + (3 \times 1) = 23_{16}$$

**EXAMPLE 5.9**

Convert $175_{10}$ to hexadecimal.

**Solution**

$$256_{10} > 175_{10}$$

Since $256 = 16^2$, the hexadecimal number will have two digits.

$$(11 \times 16) > 175 > (10 \times 16)$$

16  1

| A | |

$$175 - (A \times 16) = 175 - 160 = 15$$

16  1

| A | F |

$$175 - ((A \times 16) + (F \times 1))$$
$$= 175 - (160 + 15) = 0$$

### Repeated Division by 16

Repeated division by 16 is a systematic decimal-to-hexadecimal conversion method that is not limited by the size of the number to be converted.

It is similar to the repeated-division-by-2 method used to convert decimal numbers to binary. Divide the decimal number by 16 and note the remainder, making sure to express it as a hex digit. Repeat the process until the quotient is zero. The last remainder is the most significant digit of the hex number.

**EXAMPLE 5.10**

Convert $31581_{10}$ to hexadecimal.

**Solution**

$$31581/16 = 1973 + \text{remainder } 13 \text{ (D) (LSD)}$$
$$1973/16 \ = \ 123 + \text{remainder} \qquad 5$$
$$123/16 \ = \quad 7 + \text{remainder } 11 \text{ (B)}$$
$$7/16 \quad = \quad 0 + \text{remainder} \qquad 7 \text{ (MSD)}$$

$$31581_{10} = 7B5D_{16}$$

---

**Section Review Problem for Section 5.3c**

5.4. Convert the decimal number 8137 to its hexadecimal equivalent.

---

## Conversions Between Hexadecimal and Binary

Table 5.1 shows all 16 hexadecimal digits and their decimal and binary equivalents. Note that for every possible 4-bit binary number, there is a hexadecimal equivalent.

Binary-to-hex and hex-to-binary conversions simply consist of making a conversion between each hex digit and its binary equivalent.

**EXAMPLE 5.11**

Convert $7EF8_{16}$ to its binary equivalent.

**Solution**    Convert each digit individually to its equivalent value:

$$7_{16} = 0111_2$$

$$E_{16} = 1110_2$$

$$F_{16} = 1111_2$$

$$8_{16} = 1000_2$$

The binary number is all the above binary numbers in sequence:

$$7EF8_{16} = 111111011111000_2$$

The leading zero (the MSB of 0111) has been left out.

---

**Section Review Problems for Section 5.3d**

5.5.  Convert the hexadecimal number 934B to binary.

5.6.  Convert the binary number 11001000001101001001 to hexadecimal.

## 5.4

## Octal Numbers

**Table 5.2** Octal Numbers and Their Binary Equivalents

| Octal | Binary |
|-------|--------|
| 0 | 000 |
| 1 | 001 |
| 2 | 010 |
| 3 | 011 |
| 4 | 100 |
| 5 | 101 |
| 6 | 110 |
| 7 | 111 |

Octal numbers are based on powers of 8 and are written using eight digits, from 0 to 7. Like hexadecimal numbers, octal numbers can be converted directly into binary, since 8 is a power of 2. Each octal digit has a 3-bit binary equivalent, as shown in Table 5.2.

Octal numbers are not as widely used in digital systems as hexadecimal numbers. Since the common microprocessor bus sizes of 8, 16, or 32 bits are not evenly divisible by 3, it is not as easy to represent the data in these devices by octal numbers.

### Counting in Octal

Counting in octal is a bit jarring at first; the numbers always seem to be cutting short. We start at 0 and count to 7. Add 1 to the next digit left and start over:

$$10, 11, 12, \ldots, 17$$

Add 1 to the most significant digit and start over in the least significant digit:

$$20, 21, 22, \ldots, 27$$

Eventually, the most significant digit will be 7 (i.e., the highest octal digit):

$$70, 71, 72, \ldots, 77.$$

Add 1 to the next digit and start over:

$$100, 101, 102, \ldots$$

**EXAMPLE 5.12**

a. List the octal numbers from 75 to 112, inclusive.

b. What is the next octal number after 117? After 177? After 777?

**Solution**

a. 75, 76, 77, 100, 101, 102, 103, 104, 105, 106, 107, 110, 111, 112. (Counting rule: Use all possible digits, then add 1 to the next position left and start over.)

b. The next octal number after 117 is 120. (All digits in the least significant digit position are used up.)

   The next octal number after 177 is 200. (All digits are used up in the two least significant digit positions.)

   The next octal number after 777 is 1000.

---

> **Section Review Problem for Section 5.4a**
>
> 5.7.  List the octal numbers in sequence from 567 to 601, inclusive.

## Conversions Between Octal and Binary, Octal and Decimal

Octal can be treated like hexadecimal in conversions to and from binary and decimal. Binary and octal numbers are converted back and forth by direct conversion on a digit-by-digit basis—three binary digits to one octal digit. Octal and decimal numbers are converted back and forth by positionally weighted digits (octal to decimal) or repeated division by 8 (decimal to octal).

---

**EXAMPLE 5.13**

Convert $32175_8$ to its binary equivalent.

**Solution**

Convert each digit individually, then write the resultant bits in sequence. Leading zeros may be omitted in the most significant digit.

$$3_8 = 011$$

$$2_8 = 010$$

$$1_8 = 001$$

$$7_8 = 111$$

$$5_8 = 101$$

$$32175_8 = 11010001111101_2$$

---

**EXAMPLE 5.14**

Convert $10111010110100_2$ to its octal equivalent.

**Solution**

Break the binary number into groups of 3 bits, beginning at the right, and convert each group to an octal digit.

$$10 = 2$$
$$111 = 7$$
$$010 = 2$$
$$110 = 6$$
$$100 = 4$$
$$10111010110100_2 = 27264_8$$

**EXAMPLE 5.15**   Convert $3720_8$ to decimal.

**Solution**

$$3 \times 8^3 = 3 \times 512 = 1536_{10}$$
$$7 \times 8^2 = 7 \times \phantom{0}64 = \phantom{0}448_{10}$$
$$2 \times 8^1 = 2 \times \phantom{00}8 = \phantom{00}16_{10}$$
$$0 \times 8^0 = 0 \times \phantom{00}1 = \underline{\phantom{00000}0}$$
$$\phantom{0 \times 8^0 = 0 \times 1 = }2000_{10}$$

**EXAMPLE 5.16**   Convert $4073_{10}$ to its octal equivalent.

**Solution**

$$4073/8 = 509 + \text{remainder } 1 \quad \text{(LSD)}$$
$$509/8 = \phantom{0}63 + \text{remainder } 5$$
$$63/8 = \phantom{00}7 + \text{remainder } 7$$
$$7/8 = \phantom{00}0 + \text{remainder } 7 \quad \text{(MSD)}$$
$$4073_{10} = 7751_8$$

---

**Section Review Problems for Section 5.4b**

5.8.  Convert $2753_8$ to binary.

5.9.  Convert $10001101011100_2$ to octal.

5.10.  Convert $7541_8$ to decimal.

5.11.  Convert $10764_{10}$ to octal.

**5.5**

# BCD Codes

**Binary-coded decimal (BCD).**   *A code that represents each digit of a decimal number by a binary value.*

**Table 5.3** Decimal Digits and Their 8421 BCD Equivalents

| Decimal Digit | BCD (8421) |
|:---:|:---:|
| 0 | 0000 |
| 1 | 0001 |
| 2 | 0010 |
| 3 | 0011 |
| 4 | 0100 |
| 5 | 0101 |
| 6 | 0110 |
| 7 | 0111 |
| 8 | 1000 |
| 9 | 1001 |

BCD stands for **binary-coded decimal.** As the name implies, BCD is a system of writing decimal numbers with binary digits. There is more than one way to do this, as BCD is a *code,* not a positional number system. That is, the various positions of the bits do not necessarily represent increasing powers of a specified number base.

Two commonly used BCD codes are 8421 code, where the bits for *each decimal digit* are weighted, and Excess-3 code, where each decimal digit is represented by a binary number that is 3 larger than the true binary value of the digit.

## 8421 Code

> **8421 code** *A BCD code that represents each digit of a decimal number by its 4-bit true binary value.*

The most straightforward BCD code is the **8421 code,** also called Natural BCD. Each decimal digit is represented by its 4-bit true binary value. When we talk about BCD code, this is usually what we mean.

This code is called 8421 because these are the positional weights of each digit. Table 5.3 shows the decimal digits and their BCD equivalents.

8421 BCD is not a positional number system, because each decimal digit is encoded separately as a 4-bit number.

---

**EXAMPLE 5.17**

Write $4987_{10}$ in both binary and 8421 BCD.

**Solution**

The binary value of $4987_{10}$ can be calculated by repeated division by 2:

$$4987_{10} = 1\ 0011\ 0111\ 1011_2$$

The BCD digits are the binary values of each decimal digit, encoded separately. We can break bits into groups of 4 for easier reading. Note that the first and last BCD digits each have a leading zero to make them 4 bits long.

$$4987_{10} = 0100\ 1001\ 1000\ 0111_{BCD}$$

---

**Table 5.4** Decimal Digits and Their 8421 and Excess-3 Equivalents

| Decimal Digit | 8421 | Excess-3 |
|:---:|:---:|:---:|
| 0 | 0000 | 0011 |
| 1 | 0001 | 0100 |
| 2 | 0010 | 0101 |
| 3 | 0011 | 0110 |
| 4 | 0100 | 0111 |
| 5 | 0101 | 1000 |
| 6 | 0110 | 1001 |
| 7 | 0111 | 1010 |
| 8 | 1000 | 1011 |
| 9 | 1001 | 1100 |

## Excess-3 Code

> **Excess-3 Code** *A BCD code that represents each digit of a decimal number by a binary number derived by adding 3 to its 4-bit true binary value.*
>
> **9's complement** *A way of writing decimal numbers where a number is made negative by subtracting each of its digits from 9 (e.g., $-726 = 999 - 726 = 273$ in 9's complement).*
>
> **Self-complementing** *A code that automatically generates a negative equivalent (e.g., 9's complement for a decimal code) when all its bits are inverted.*

**Excess-3 code** is a type of BCD code that is generated by adding $11_2$ ($3_{10}$) to the 8421 BCD codes. Table 5.4 shows the Excess-3 codes and their 8421 and decimal equivalents.

The advantage of this code is that it is **self-complementing.** If the bits of the Excess-3 digit are inverted, they yield the **9's complement** of the decimal equivalent.

We can generate the 9's complement of an $n$-digit number by subtracting it from a number made up of $n$ 9s. Thus, the 9's complement of 632 is $999 - 632 = 367$.

The Excess-3 equivalent of 632 is 1001 0110 0101. If we invert all the bits, we get 0110 1001 1010. The decimal equivalent of this Excess-3 number is 367, the 9's complement of 632.

This property is useful for performing decimal arithmetic digitally.

## 5.6

## Gray Code

**Table 5.5** 4-Bit Gray Code

| Decimal | True Binary | Gray Code |
|---------|-------------|-----------|
| 0 | 0000 | 0000 |
| 1 | 0001 | 0001 |
| 2 | 0010 | 0011 |
| 3 | 0011 | 0010 |
| 4 | 0100 | 0110 |
| 5 | 0101 | 0111 |
| 6 | 0110 | 0101 |
| 7 | 0111 | 0100 |
| 8 | 1000 | 1100 |
| 9 | 1001 | 1101 |
| 10 | 1010 | 1111 |
| 11 | 1011 | 1110 |
| 12 | 1100 | 1010 |
| 13 | 1101 | 1011 |
| 14 | 1110 | 1001 |
| 15 | 1111 | 1000 |

> **Gray code** *A binary code that progresses such that only one bit changes between two successive codes.*

Table 5.5 shows a 4-bit **Gray code** compared to decimal and binary values. Any two adjacent Gray codes differ by exactly one bit.

Gray code can be extended indefinitely if you understand the relationship between the binary and Gray digits. Let us name the binary digits $b_3b_2b_1b_0$, with $b_3$ as the most significant bit, and the Gray code digits $g_3g_2g_1g_0$ for a 4-bit code. For a 4-bit code:

$$g_3 = b_3$$

$$g_2 = b_3 \oplus b_2$$

$$g_1 = b_2 \oplus b_1$$

$$g_0 = b_1 \oplus b_0$$

For an $n$-bit code, the MSBs are the same in Gray and binary ($g_n = b_n$). The other Gray digits are generated by the Exclusive OR function of the binary digits in the same position and the next most significant position.

## 5.7

## ASCII Code

> **Alphanumeric code** *A code used to represent letters of the alphabet and numerical characters.*
>
> **ASCII** *American Standard Code for Information Interchange. A 7-bit code for representing alphanumeric and control characters.*
>
> **Case shift** *Changing letters from capitals (uppercase) to small letters (lowercase) or vice versa.*

Digital systems and computers could operate perfectly well using only binary numbers. However, if there is any need for a human operator to understand the input and output data of a digital system, it is necessary to have a system of communication that is understandable to both a human operator and the digital circuit.

A code that represents letters (alphabetic characters) and numbers (numeric characters) as binary numbers is called an **alphanumeric code.** The most commonly used alphanumeric code is **ASCII** ("askey"), which stands for American Standard Code for Information Interchange. ASCII code represents letters, numbers, and other "typewriter characters" in 7 bits. In addition, ASCII has a repertoire of "control characters," codes that are used to send control instructions to and from devices such as video display terminals, printers, and modems.

Table 5.6 shows the ASCII code in both binary and hexadecimal forms. The code for any character consists of the bits in the column heading, then those in the row heading. For example, the ASCII code for "A" is $1000001_2$ or $41_{16}$. The code for "a" is

**Table 5.6** ASCII Code

| LSBs | 000 (0) | 001 (1) | 010 (2) | MSBs 011 (3) | 100 (4) | 101 (5) | 110 (6) | 111 (7) |
|---|---|---|---|---|---|---|---|---|
| **0000 (0)** | NUL | DLE | SP | 0 | @ | P | ' | p |
| **0001 (1)** | SOH | DC1 | ! | 1 | A | Q | a | q |
| **0010 (2)** | STX | DC2 | " | 2 | B | R | b | r |
| **0011 (3)** | ETX | DC3 | # | 3 | C | S | c | s |
| **0100 (4)** | EOT | DC4 | $ | 4 | D | T | d | t |
| **0101 (5)** | ENQ | NAK | % | 5 | E | U | e | u |
| **0110 (6)** | ACK | SYN | & | 6 | F | V | f | v |
| **0111 (7)** | BEL | ETB | ' | 7 | G | W | g | w |
| **1000 (8)** | BS | CAN | ( | 8 | H | X | h | x |
| **1001 (9)** | HT | EM | ) | 9 | I | Y | i | y |
| **1010 (A)** | LF | SUB | * | : | J | Z | j | z |
| **1011 (B)** | VT | ESC | + | ; | K | [ | k | { |
| **1100 (C)** | FF | FS | , | < | L | \ | l | ¦ |
| **1101 (D)** | CR | GS | − | = | M | ] | m | } |
| **1110 (E)** | SO | RS | . | > | N | ^ | n | ~ |
| **1111 (F)** | SI | US | / | ? | O | — | o | DEL |

**Control Characters:**

| | |
|---|---|
| NUL–Null | DLE–Data Link Escape |
| SOH–Start of Header | DC1–Device Control 1 |
| STX–Start Text | DC2–Device Control 2 |
| ETX–End Text | DC3–Device Control 3 |
| EOT–End of Transmission | DC4–Device Control 4 |
| ENQ–Enquiry | NAK–No Acknowledgment |
| ACK–Acknowledge | SYN–Synchronous Idle |
| BEL–Bell | ETB–End of Transmission Block |
| BS–Backspace | CAN–Cancel |
| HT–Horizontal Tabulation | EM–End of Medium |
| LF–Line Feed | SUB–Substitute |
| VT–Vertical Tabulation | ESC–Escape |
| FF–Form Feed | FS–Form Separator |
| CR–Carriage Return | GS–Group Separator |
| SO–Shift Out | RS–Record Separator |
| SI–Shift In | US–Unit Separator |
| SP-Space | DEL–Delete |

$1100001_2$ or $61_{16}$. The codes for capital (uppercase) and lowercase letters differ only by the second most significant bit, for all letters. Thus, we can make an alphabetic **case shift,** like using the Shift key on a typewriter keyboard, by switching just one bit.

The codes in columns 0 and 1 are control characters. They cannot be displayed on any kind of output device, such as a printer or video monitor, although they may be used to control the device. For instance, if the codes $0A_{16}$ (Line Feed) and $0D_{16}$ (Carriage Return) are sent to a printer, the paper will advance by one line and the print head will return to the beginning of the line.

The displayable characters begin at $20_{16}$ ("space") and continue to $7E_{16}$ ( "tilde"). Spaces are considered ASCII characters.

---

**EXAMPLE 5.18**

Encode the following string of characters into ASCII (hexadecimal form). Do not include quotation marks.

"Total system cost: $4,000,000. @ 10%"

**Solution**    Each character, including spaces, is represented by two hex digits as follows:

54 6F 74 61 6C 20 73 79 73 74 65 6D 20 63 6F 73 74 3A 20

T  o  t  a  l  SP s  y  s  t  e  m  SP c  o  s  t  :  SP

24 34 2C 30 30 30 2C 30 30 30 2E 20 40 20 31 30 25

$  4  ,  0  0  0  ,  0  0  0  .  SP @  SP 1  0  %

---

**Section Review Problem for Section 5.7**

5.12.  Decode the following sequence of hexadecimal ASCII codes.

54 72 75 65 20 6F 72 20 46 61 6C 73 65 3A 20 31 2F 34 20 3C 20 31 2F 32

---

## ● GLOSSARY

**Alphanumeric code**    A code used to represent letters of the alphabet and numerical characters.

**ASCII**    American Standard Code for Information Interchange. A 7-bit code for representing alphanumeric and control characters.

**Base**    The number whose powers represent positional multipliers in a number system. Also the number of digits that are used in that system (e.g., 10 digits, 0 to 9, are used in the decimal system).

**Binary-coded decimal (BCD)**    A code that represents each digit of a decimal number by a binary value.

**Binary point**    A period (".") that marks the dividing line between positional multipliers that are positive and negative powers of 2 (e.g., first multiplier right of binary point = $2^{-1}$; first multiplier left of binary point = $2^0$).

**Case shift**    Changing letters from capitals (uppercase) to small letters (lowercase) or vice versa.

**8421 code**    A BCD code that represents each digit of a decimal number by its 4-bit true binary value.

**Excess-3 code**    A BCD code that represents each digit of a decimal number by a binary number derived by adding 3 to its 4-bit true binary value.

**Gray code**    A binary code that progresses such that only one bit changes between two successive codes.

**9's complement**    A way of writing decimal numbers where a number is made negative by subtracting each of its digits from 9 (e.g., $726 = 999 - 726 = 273$ in 9's complement).

**Radix point**  The generalized form of a decimal point. In any positional number system, the radix point marks the dividing line between positional multipliers that are positive and negative powers of the system's number base.

**Self-complementing**  A code that automatically generates a negative-equivalent (e.g., 9's complement for a decimal code) when all its bits are inverted.

## ● P R O B L E M S

### Section 5.1 Positional Notation
### Section 5.2 Binary Numbers

**5.1**  List the first five positional multipliers for a base 8 number system.

**5.2**  Convert the following binary, octal, or hexadecimal numbers to decimal by breaking them into their digits and positional multipliers. For example,

$$174_8 = 1 \times 8^2 = 64_{10}$$
$$7 \times 8^1 = 56_{10}$$
$$4 \times 8^0 = \underline{\quad 4_{10}}$$
$$124_{10}$$

   **a.** $10110_2$       **b.** $11101_2$

   **c.** $110011_2$     **d.** $542_8$

   **e.** $7746_8$       **f.** $3251_8$

   **g.** $914_{16}$       **h.** $8403_{16}$

**5.3**  Complete the binary sequence between 101101 and 110101.

**5.4**  Complete the binary sequence between 1110100 and 1111101.

**5.5**  Convert the following binary numbers to decimal.

   **a.** $01101_2$

   **b.** $110101_2$

   **c.** $110001011011_2$

   **d.** $101101101010_2$

   **e.** $1011010_2$

   **f.** $1111111111_2$

   **g.** $10000000000_2$

**5.6**  Convert the following decimal numbers to binary. Use the sum-of-powers-of-2 method for parts a, c, e, and g. Use the repeated-division-by-2 method for parts b, d, f, and h.

   **a.** $75_{10}$

   **b.** $83_{10}$

   **c.** $237_{10}$

   **d.** $198_{10}$

   **e.** $63_{10}$

   **f.** $64_{10}$

   **g.** $4087_{10}$

   **h.** $8193_{10}$

**5.7**  Convert the following fractional binary numbers to their decimal equivalents.

   **a.** 0.101

   **b.** 0.011

   **c.** 0.1101

**5.8**  Convert the following fractional binary numbers to their decimal equivalents.

   **a.** 0.01

   **b.** 0.0101

   **c.** 0.010101

   **d.** 0.01010101

**\*\*5.9**  The numbers in Problem 5.8 are converging to a closer and closer binary approximation of a simple fraction that can be expressed by decimal integers $a/b$. What is the fraction?

**\*\*5.10**  What is the simple decimal fraction $(a/b)$ represented by the repeating binary number 0.101010 . . .?

**5.11** Convert the following decimal numbers to their binary equivalents. If a number has an integer part larger than 0, calculate the integer and fractional parts separately.

**a.** $0.75_{10}$

**b.** $0.625_{10}$

**c.** $0.1875_{10}$

**d.** $0.65_{10}$

**e.** $1.75_{10}$

**f.** $3.95_{10}$

**g.** $67.84_{10}$

## Section 5.3 Hexadecimal Numbers

**5.12** Write all the hexadecimal numbers in sequence from $308_{16}$ to $321_{16}$, inclusive.

**5.13** Write all the hexadecimal numbers in sequence from $9F7_{16}$ to $A03_{16}$, inclusive.

**5.14** Convert the following hexadecimal numbers to their decimal equivalents.

**a.** $1A0_{16}$

**b.** $10A_{16}$

**c.** $FFF_{16}$

**d.** $1000_{16}$

**e.** $F3C8_{16}$

**f.** $D3B4_{16}$

**g.** $C000_{16}$

**h.** $308AF_{16}$

**5.15** Convert the following decimal numbers to their hexadecimal equivalents.

**a.** $709_{10}$

**b.** $1889_{10}$

**c.** $4095_{10}$

**d.** $4096_{10}$

**e.** $10128_{10}$

**f.** $32000_{10}$

**g.** $32768_{10}$

**5.16** Convert the following hexadecimal numbers to their binary equivalents.

**a.** $F3C8_{16}$

**b.** $D3B4_{16}$

**c.** $8037_{16}$

**d.** $FABD_{16}$

**e.** $30AC_{16}$

**f.** $3E7B6_{16}$

**g.** $743DCF_{16}$

**5.17** Convert the following binary numbers to their hexadecimal equivalents.

**a.** $101111010000110_2$

**b.** $101101101010_2$

**c.** $110001011011_2$

**d.** $110101111000100_2$

**e.** $10101011110000101_2$

**f.** $11001100010110111_2$

**g.** $101000000000000000_2$

## Section 5.4 Octal Numbers

**5.18** What is the next octal number after 5007? After 5077? After 5777? After 7777?

**5.19** List the octal numbers from 367 to 400, inclusive.

**5.20** Convert the following octal numbers to their decimal equivalents.

**a.** $624_8$

**b.** $52_8$

**c.** $177_8$

**d.** $200_8$

**e.** $1750_8$

**f.** $1777_8$

**g.** $2000_8$

**5.21** Convert the following decimal numbers to their octal equivalents.

**a.** $64_{10}$

**b.** $25_{10}$

**c.** $125_{10}$

**d.** $177_{10}$

**e.** $200_{10}$

**f.** $4096_{10}$

**5.22** Convert the binary numbers in Problem 5.17 to their octal equivalents.

**5.23** Convert the octal numbers in Problem 5.20 to their binary equivalents. (Compare the numbers in parts c and d. Also compare the numbers in parts f and g.)

## Section 5.5 BCD Codes

**5.24** Convert the following decimal numbers to true binary, 8421 BCD code, and Excess-3 code.

**a.** $709_{10}$

**b.** $1889_{10}$

**c.** $2395_{10}$

**d.** $1259_{10}$

**e.** $3972_{10}$

**f.** $7730_{10}$

## Section 5.6 Gray Code

**5.25** Make a table showing the equivalent Gray codes corresponding to the range from $0_{10}$ to $31_{10}$.

## Section 5.7 ASCII Code

**5.26** Write your name in ASCII code.

**5.27** Encode the following text into ASCII code:

"10% off purchases over $50. (Monday only)"

**5.28** Decode the following string of ASCII code.

57 41 52 4E 49 4E 47 21 20 54 68 69 73 20 63 6F 6D
6D 61 6E 64 20 65 72 61 73 65 73 20 36 34 30 4B 20
6F 66 20 6D 65 6D 6F 72 79 2E

## ▼ Answers to Section Review Questions

### Section 5.1
**5.1.** $16^0 = 1$, $16^1 = 16$, $16^2 = 256$, $16^3 = 4096$.

### Section 5.3a
**5.2.** FA9, FAA, FAB, FAC, FAD, FAE, FAF, FBO.   **5.3.** 1F9, 1FA, 1FB, 1FC, 1FD, 1FE, 1FF, 200.

### Section 5.3b
**5.4.** $41743_{10}$.

### Section 5.3c
**5.4.** 1FC9.

### Section 5.3d
**5.5.** 1001001101001011.   **5.6.** C8349.

### Section 5.4a
**5.7.** 567, 570, 571, 572, 573, 574, 575, 576, 577, 600, 601.

### Section 5.4b
**5.8.** $10111101011_2$.   **5.9.** $21534_8$.   **5.10.** $3937_{10}$.   **5.11.** $25014_8$.

### Section 5.7
**5.12.** "True or False: 1/4 < 1/2".

# Latches and Flip-Flops

## ● CHAPTER OUTLINE

**6.1** Latches

**6.2** NAND/NOR Latches

**6.3** Gated Latches

**6.4** Edge-Triggered SR and D Flip-Flops

**6.5** Edge-Triggered JK Flip-Flops

**6.6** Timing Parameters

**6.7** Troubleshooting Latch Circuits

## ● CHAPTER OBJECTIVES

Upon successful completion of this chapter, you will be able to:

- Explain the difference between combinational and sequential circuits.

- Define the set and reset functions of an SR latch.

- Draw circuits, function tables, and timing diagrams of NAND and NOR latches.

- Explain the effect of each possible input combination to a NAND and a NOR latch, including set, reset, and no change functions, as well as the ambiguous or forbidden input condition.

- Design circuit applications that employ NAND and NOR latches.

- Describe the use of the *ENABLE* input of a gated SR latch as an enable/inhibit function and as a synchronizing function.

- Outline the problems involved with using a level-sensitive *ENABLE* input on a gated SR latch.

- Explain the concept of edge-triggering and why it is an improvement over level-sensitive enabling.

- Draw circuits, function tables, and timing diagrams of edge-triggered SR, D, and JK flip-flops.

- Explain why D and JK flip-flops are improvements over SR flip-flops.

- Describe the toggle function of a JK flip-flop.

- Describe the operation of the asynchronous preset and clear functions of D and JK flip-flops and be able to draw timing diagrams showing their functions.

- Draw D and JK flip-flops in circuit applications, such as serial and parallel data transfer.

- Explain the operation of pulse-triggered and data-lockout flip-flops.

- Troubleshoot latch and flip-flop circuits and devices.

179

## ● INTRODUCTION

The digital circuits studied to this point have all been combinational circuits, that is, circuits whose outputs are functions only of their present inputs. A particular set of input states will always produce the same output state in a combinational circuit.

This chapter will introduce a new category of digital circuitry: the sequential circuit. The output of a sequential circuit is a function both of the present input conditions and the previous conditions of the inputs and/or outputs. The output depends on the sequence in which the inputs are applied.

We will begin our study of sequential circuits by examining the two most basic sequential circuit elements: the latch and the flip-flop, both of which are part of the general class of circuits called bistable multivibrators. These are similar devices, each being used to store a single bit of information indefinitely. The difference between a latch and a flip-flop is the condition under which the stored bit is allowed to change.

Latches and flip-flops are also used as integral parts of more-complex devices, such as programmable logic devices (PLDs) and medium-scale integration (MSI) devices, usually when an input or output state must be stored.

### 6.1

## Latches

**Sequential circuit** *A digital circuit whose output depends not only on the present combination of inputs, but also on the history of the circuit.*

**Latch** *A sequential circuit with two inputs called SET and RESET, which make the latch store a logic 0 (reset) or 1 (set) until actively changed.*

**SET** *1. The stored HIGH state of a latch circuit.*
     *2. A latch input that makes the latch store a logic 1.*

**RESET** *1. The stored LOW state of a latch circuit.*
        *2. A latch input that makes the latch store a logic 0.*

All the circuits we have seen up to this point have been combinational circuits. That is, their present outputs depend only on their present inputs. The output state of a combinational circuit results from a combination of input logic states.

The other major class of digital circuits is the **sequential circuit.** The present outputs of a sequential circuit depend not only on its present inputs, but also on its past input states.

The simplest sequential circuit is the SR **latch,** whose logic symbol is shown in Figure 6.1a. The latch has two inputs, *SET* and *RESET,* and two complementary outputs, $Q$ and $\overline{Q}$. If the latch is operating normally, the outputs are always in opposite logic states.

The latch operates like a momentary-contact pushbutton with START and STOP functions, shown in Figure 6.2. A momentary-contact switch operates only when it is held down. When released, a spring returns the switch to its rest position.

**Figure 6.1**
**SR Latch (Active HIGH Inputs)**

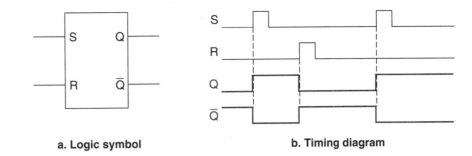

a. Logic symbol          b. Timing diagram

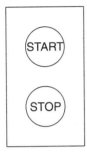

**Figure 6.2**
**Industrial Pushbutton**

Suppose the switch in Figure 6.2 is used to control a motor starter. When you push the START button, the motor begins to run. Releasing the START switch does not turn the motor off; that can be done only by pressing the STOP button. If the motor is running, pressing the START button again has no effect, except continuing to let the motor run. If the motor is not running, pressing the STOP switch has no effect, since the motor is already stopped.

There is a conflict if we press both switches simultaneously. In such a case we are trying to start and stop the motor at the same time. We will come back to this point later.

The latch *SET* input is like the START button in Figure 6.2. The *RESET* input is like the STOP button.

By definition:

A latch is set when $Q = 1$ and $\overline{Q} = 0$.
A latch is reset when $Q = 0$ and $\overline{Q} = 1$.

The latch in Figure 6.1 has active-HIGH *SET* and *RESET* inputs. To set the latch, make $R = 0$ and make $S = 1$. This makes $Q = 1$ until the latch is actively reset, as shown in the timing diagram in Figure 6.1b. To activate the reset function, make $S = 0$ and make $R = 1$. The latch is now reset ($Q = 0$) until the set function is next activated.

Combinational circuits produce an output by combining inputs. In sequential circuits, it is more accurate to think in terms of activating functions. In the latch described, $S$ and $R$ are not *combined* by a Boolean function to produce a particular result at the output. Rather, the set function is *activated* by making $S = 1$, and the reset function is *activated* by making $R = 1$, much as we would activate the START or STOP function of a motor starter by pressing the appropriate pushbutton.

The timing diagram in Figure 6.1b shows that the inputs need not remain active after the set or reset functions have been selected. In fact, the $S$ or $R$ input *must* be inactive before the opposite function can be applied, in order to avoid conflict between the two functions.

**EXAMPLE 6.1**

Latches can have active-HIGH or active-LOW inputs, but in each case $Q = 1$ after the set function is applied and $Q = 0$ after reset. For each latch shown in Figure 6.3, complete the timing diagram shown. $Q$ is initially LOW in both cases.

**Figure 6.3**
**SR Latch**

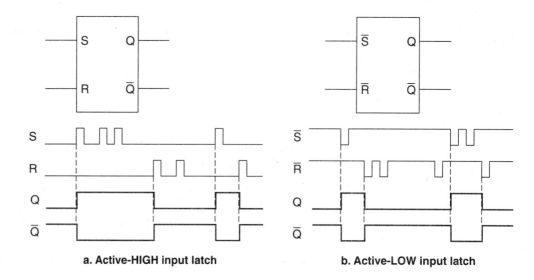

**a. Active-HIGH input latch**          **b. Active-LOW input latch**

Solution    The $Q$ and $\overline{Q}$ waveforms are shown in Figure 6.3. Note that the outputs respond only to the first set or reset command in a sequence of several pulses.

**EXAMPLE 6.2**    Figure 6.4 shows a latching HOLD circuit for an electronic telephone. When HIGH, the *HOLD* output allows you to replace the handset without disconnecting a call in progress.

   The two-position switch is the telephone's hook switch (the switch the handset pushes down when you hang up), shown in the off-hook (in-use) position. The Normally Closed pushbutton is a momentary-contact switch used as a *HOLD* button. The circuit is such that the *HOLD* button does not need to be held down to keep the *HOLD* active. The latch "remembers" that the switch was pressed, until told to "forget" by the reset function.

   Describe the sequence of events that will place a caller on hold and return the call from hold. Also draw timing diagrams showing the waveforms at the *HOLD* input, hook switch inputs, $S$ input, and *HOLD* output for one hold-and-return sequence. (*HOLD* output = 1 means the call is on hold.)

Solution    To place a call on hold, we must set the latch. We can do so if we press and hold the *HOLD* switch, then the hook switch. This combines two HIGHs—one from the *HOLD* switch and one from the on-hook position of the hook switch—into the AND gate, making $S = 1$ and $R = 0$. The handset can be kept on-hook and the *HOLD* button released. The latch stays set, as $S = R = 0$ as long as the handset is on-hook.

**Figure 6.4**
**Example 6.2**
**Latching HOLD Button**

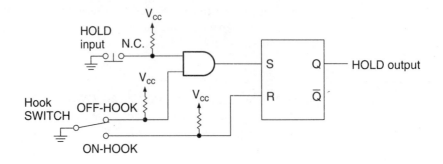

**Figure 6.5**

**Example 6.2**
**HOLD Timing Diagram**

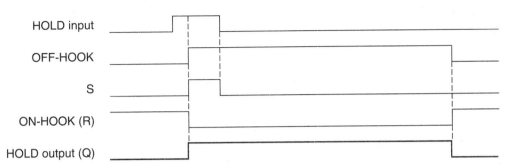

To restore a call, lift the handset. This places the hook switch into the off-hook position and now $S = 0$ and $R = 1$, which resets the latch and turns off the *HOLD* condition.

Figure 6.5 shows the timing diagram for the sequence described.

## 6.2

# NAND/NOR Latches

An SR latch is easy to build with logic gates. Figure 6.6 shows two such circuits, one made from NOR gates and one from NANDs. The NAND gates in the second circuit are drawn in DeMorgan equivalent form.

The two circuits both have the following three features:

1. OR-shaped gates
2. Logic level inversion between the gate input and output
3. Feedback from the output of one gate to an input of the opposite gate

During our examination of the NAND and NOR latches, we will discover why these features are important.

A significant difference between the NAND and NOR latches is the placement of *SET* and *RESET* inputs with respect to the $Q$ and $\overline{Q}$ outputs. Once we define which output is $Q$ and which is $\overline{Q}$, the locations of the *SET* and *RESET* inputs are automatically defined.

In a NOR latch, the gates have active-HIGH inputs and active-LOW outputs. When the input to the $Q$ gate is HIGH, $Q = 0$, since either input HIGH makes the output LOW.

**Figure 6.6**

**SR Latch Circuits**

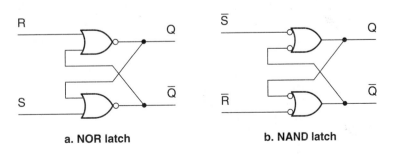

a. NOR latch                    b. NAND latch

**Table 6.1** NOR and NAND Latch Functions

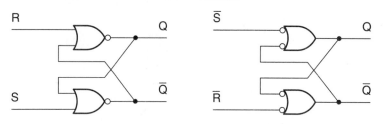

| S | R | Action (NOR Latch) | $\overline{S}$ | $\overline{R}$ | Action (NAND Latch) |
|---|---|---|---|---|---|
| 0 | 0 | Neither *SET* nor *RESET* active; output does not change from previous state | 0 | 0 | Both *SET* and *RESET* active; forbidden condition |
| 0 | 1 | *RESET* input active | 0 | 1 | *SET* input active |
| 1 | 0 | *SET* input active | 1 | 0 | *RESET* input active |
| 1 | 1 | Both *SET* and *RESET* active; forbidden condition | 1 | 1 | Neither *SET* nor *RESET* active; output does not change from previous state |

Therefore, this input must be the *RESET* input. By default, the other is the *SET* input.

In a NAND latch, the gate inputs are active LOW (in DeMorgan equivalent form) and the outputs are active HIGH. A LOW input on the $Q$ gate makes $Q = 1$. This, therefore, is the *SET* input, and the other gate input is *RESET.*

Since the NAND and NOR latch circuits have two binary inputs, there are four possible input states. Table 6.1 summarizes the action of each latch for each input combination. The functions are the same for each circuit, but they are activated by opposite logic levels.

We will examine the NOR latch circuit for each of the input conditions in Table 6.1. The analysis of a NAND latch is similar and will be left as an exercise.

## R = 0, S = 0; No Change Condition

If a NOR latch is set, it will stay set if $S = R = 0$. If the latch is reset, it will stay reset if $S = R = 0$. The inputs are the same in either case, but the outputs depend on the history of the latch (i.e., whether the latch was previously set or reset). Figure 6.7 shows the two possible stable conditions.

The cross-feedback on the gates keeps the circuit in a stable condition. When the latch is set, $Q = 1$, by definition. This 1 is fed back to the gate on the opposite side of the circuit, where it makes $\overline{Q} = 0$. (NOR function: "either input HIGH makes output LOW.") The 0 at $\overline{Q}$ feeds back to the first gate, where, along with the 0 at $R$, it makes $Q = 1$. The outputs and inputs form a closed loop that has no tendency to change.

A similar set of conditions keeps the latch stable in the reset state. If $Q = 0$, the latch is reset. The 0 at $Q$ feeds back to an input on the opposite gate, where, along with the 0 at $S$, it makes $\overline{Q} = 1$. The $\overline{Q}$ output goes to the opposite gate, where it makes $Q = 0$, thus closing the loop.

**Figure 6.7**
**NOR Latch Stable States**

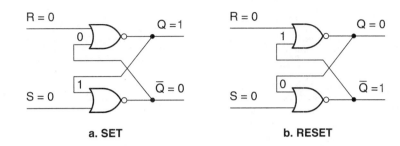

a. SET                    b. RESET

## S = 0, R = 1; Reset Condition

The *SET* and *RESET* inputs can be thought of as "external inputs" to the latch. The two gate inputs that are supplied by feedback connections are connected only to logic gates within the circuit and can thus be referred to as "internal inputs."

Figure 6.8 shows a NOR latch in the transition from the set to the reset state. Figure 6.8a shows the latch in the set condition. To reset the latch, we make $R = 1$ (Figure 6.8b). This makes the $Q$ output go LOW ("at least one input HIGH makes output LOW"), which feeds back to the opposite gate (Figure 6.8c). There, combined with the LOW $S$ input, $\overline{Q} = 1$. The loop closes when the HIGH at $\overline{Q}$ goes back to the "internal" input of the $R$ gate (Figure 6.8d). Since the top NOR gate needs only one HIGH input to make its output LOW, we can make $R = 0$ and still maintain the reset state (Figure 6.8e).

> The OR shape of the latch gates allows the required HIGH input level to be applied externally to the *RESET* input *or* internally from $\overline{Q}$.
>
> The external HIGH *initiates* the change of state; the internal HIGH *maintains* the new state and allows $R$ to become LOW again.

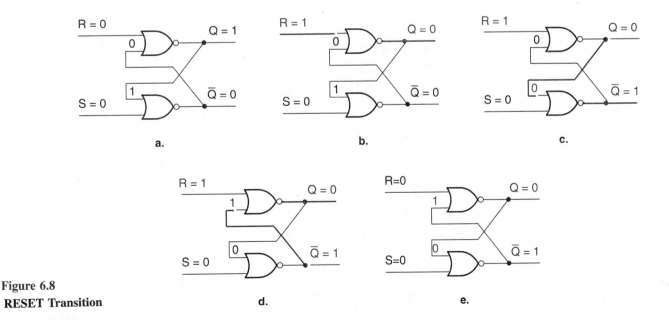

a.                    b.                    c.

d.                    e.

**Figure 6.8**
**RESET Transition**

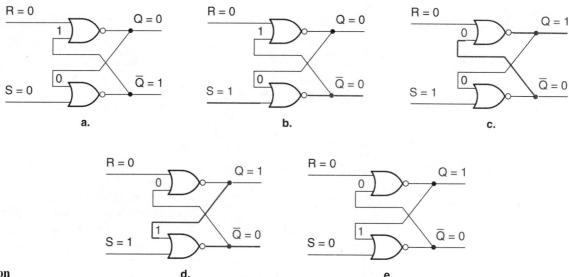

**Figure 6.9**
**SET Transition**

## S = 1, R = 0; Set Condition

Figure 6.9 shows a NOR latch in transition from set to reset. Figure 6.9a shows the latch in the reset state. We initiate the set function when we make $S = 1$ (Figure 6.9b). Since a NOR requires at least one input HIGH to make its output LOW, $\overline{Q} = 0$. Feedback takes this LOW to the opposite side of the latch where, combined with $R = 0$, the $Q$ output goes HIGH (Figure 6.9c). This HIGH is coupled back to the $S$ gate (Figure 6.9d), keeping $\overline{Q} = 0$ via the "internal" input of the $S$ gate. We can remove the HIGH at $S$ and the latch remains stable in the set condition.

---

> The HIGH input level required to keep the latch set can be applied externally to the *SET* input *or* internally from $Q$. The external HIGH *initiates* the change of state; the internal HIGH *maintains* the new state and allows $S$ to become LOW again.

---

## S = 1, R = 1; Forbidden Condition

$S = R = 1$ implies that the latch is being simultaneously set and reset. Since the latch cannot respond to these contradictory instructions, it becomes unstable. Figure 6.10 illustrates why this is so.

When $R = 1$, $Q = 0$. (The NOR function requires one HIGH input to make a LOW output.) For the same reason, when $S = 1$, $\overline{Q} = 0$. This condition (both outputs LOW) is not stable, since the latch won't maintain it if $S$ or $R$ goes back to 0.

Three things can happen, as illustrated by the timing diagrams in Figure 6.10b, c, and d.

1. $R$ goes LOW before $S$, which results in the latch becoming set, as the *SET* input is the last one active.

2. $S$ goes LOW before $R$, causing the latch to reset, $R$ being the last input active.

**Figure 6.10**
**NOR Latch Forbidden State**

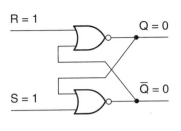

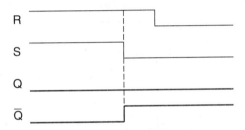

a. Forbidden state (unstable)

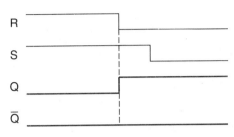

b. Transition from forbidden
(unstable) to SET (stable)

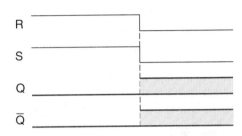

c. Transition from forbidden
(unstable) to RESET (stable)

d. Transition from forbidden to
no change (unknown result)

3. $S$ and $R$ go LOW simultaneously, causing an unknown result. The state illustrated in Figure 6.10a cannot be maintained if both $S$ and $R$ return to 0 at the same time. Both outputs would try to make a LOW-to-HIGH transition, such that $Q = \overline{Q} = 1$, which is impossible because of the feedback connection; a 1 on any internal input would make $Q$ or $\overline{Q} = 0$. The only stable configuration is one where there is only one HIGH internal input in the latch circuit.

In practice, the LOW-to-HIGH output transition always happens slightly faster in one gate than the other, causing the latch to drop into the set or reset condition. The problem is that there is no way of knowing which condition will prevail for a particular pair of gates. Consequently, the $S = R = 1$ state is forbidden for a NOR latch. (We can occasionally use this state to advantage in a design if we know that $S$ and $R$ will never return to LOW simultaneously. This practice is not very common and should generally be avoided unless there is a compelling reason to use it.)

The NOR latch operation can be summarized in a function table, shown in Table 6.2. The designation $Q_0$ indicates the previous value of $Q$ (i.e., $Q$ has not changed).

Table 6.3 shows the function table of a NAND latch.

**Table 6.2** NOR Latch Function Table

| $S$ | $R$ | $Q$ | $\overline{Q}$ | Comment |
|---|---|---|---|---|
| 0 | 0 | $Q_0$ | $\overline{Q}_0$ | No change |
| 0 | 1 | 0 | 1 | Reset |
| 1 | 0 | 1 | 0 | Set |
| 1 | 1 | 0 | 0 | Forbidden |

**Table 6.3** NAND Latch Function Table

| $\overline{S}$ | $\overline{R}$ | $Q$ | $\overline{Q}$ | Comment |
|---|---|---|---|---|
| 0 | 0 | 1 | 1 | Forbidden |
| 0 | 1 | 1 | 0 | Set |
| 1 | 0 | 0 | 1 | Reset |
| 1 | 1 | $Q_0$ | $\overline{Q}_0$ | No change |

**Figure 6.11**
**Switches as Pulse
Generators**

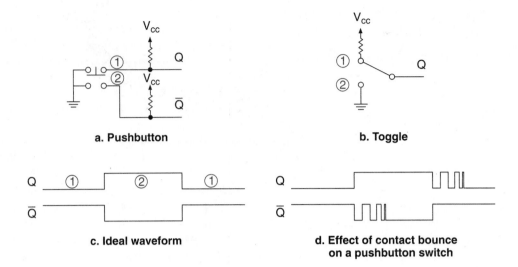

a. Pushbutton

b. Toggle

c. Ideal waveform

d. Effect of contact bounce
on a pushbutton switch

**Figure 6.12**

**NAND Latch as a Switch
Debouncer**

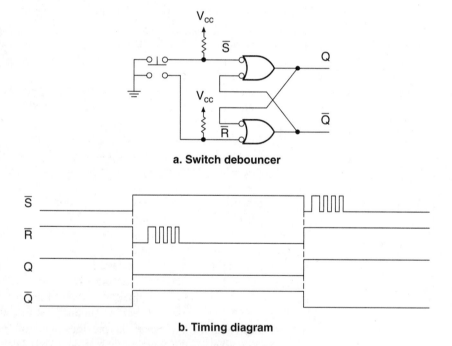

a. Switch debouncer

b. Timing diagram

## Latch as a Switch Debouncer

Pushbutton or toggle switches are sometimes used to generate pulses for digital circuit inputs, as illustrated in Figure 6.11. However, when a switch is operated and contact is made on a new terminal, the contact, being mechanical, will bounce a few times before settling into the new position. Figure 6.11d shows the effect of contact bounce on the waveform for a pushbutton switch. The contact bounce is shown only on the terminal where contact is being made, not broken.

Contact bounce can be a serious problem, particularly when a switch is used as an input to a digital circuit that responds to individual pulses. If the circuit expects to receive one pulse, but gets several from a bouncy switch, it will behave unpredictably.

A latch can be used as a switch debouncer, as shown in Figure 6.12a. When the pushbutton is in the position shown, the latch is set, since $\overline{S} = 0$ and $\overline{R} = 1$. (Recall that the NAND latch inputs are active LOW.) When the pushbutton is pressed, the $\overline{R}$ contact bounces a few times, as shown in Figure 6.12b. However, on the first bounce, the latch is reset. Any further bounces are ignored, since the resulting input state is either $\overline{S} = \overline{R} = 1$ (no change) or $\overline{S} = 1$, $\overline{R} = 0$ (reset).

Similarly, when the pushbutton is released, the $\overline{S}$ input bounces a few times, setting the latch on the first bounce. The latch ignores any further bounces, since they either do not change the latch output ($\overline{S} = \overline{R} = 1$) or set it again ($\overline{S} = 0$, $\overline{R} = 1$). The resulting waveforms at $Q$ and $\overline{Q}$ are free of contact bounce and can be used reliably as inputs to digital sequential circuits.

---

**EXAMPLE 6.3**

A NOR latch can be used as a switch debouncer, but not in the same way as a NAND latch. Figure 6.13 shows two NOR latch circuits, only one of which works as a switch debouncer. Draw a timing diagram for each circuit, showing $R$, $S$, $Q$, and $\overline{Q}$, to prove that the circuit in Figure 6.13b eliminates switch contact bounce but the circuit in Figure 6.13a does not.

**Figure 6.13**

**Example 6.3**
**NOR Latch Circuits**

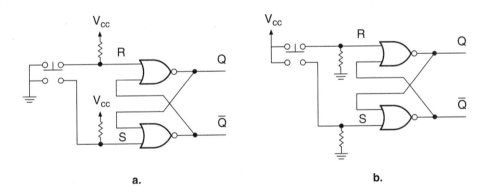

a.

b.

**Solution**    Figure 6.14 shows the timing diagrams of the two NOR latch circuits. In the circuit in Figure 6.13a, contact bounce causes the latch to oscillate in and out of the forbidden state of the latch ($S = R = 1$). This causes one of the two outputs to bounce for each contact closure. (Use the function table of the NOR latch to examine each part of the timing diagram to see that this is so.)

By making the resistors pull down rather than pull up, as in Figure 6.13b, the latch oscillates in and out of the no change state ($S = R = 0$) as a result of contact bounce. The first bounce on the *SET* terminal sets the latch, and other oscillations are disregarded. The first bounce on the *RESET* input resets the latch, and further pulses on this input are ignored.

The principle illustrated here is that a closed switch must present the active input level to the latch, since switch bounce is only a problem on contact closure. Thus, a closed switch must make the input of a NOR latch HIGH or the input of a NAND latch LOW to debounce the switch waveform.

**Figure 6.14**

**Example 6.3**
**NOR Latch Circuits**

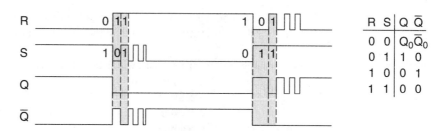

a. Timing diagram for circuit of Figure 6.13a

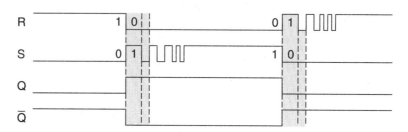

b. Timing diagram for circuit of Figure 6.13b

Note: The NOR latch is seldom used in practice as a switch debouncer. The pull-down resistors need to be about 500 Ω or less to guarantee a logic LOW at the input of a TTL NOR gate. In such a case, a constant current of about 10 mA flows through the resistor connected to the Normally Closed portion of the switch. This value is unacceptably high in most circuits, as it draws too much idle current from the power supply. For this reason, the NAND latch, which uses higher-value pull-up resistors (about 1 kΩ or larger) and therefore draws less idle current, is preferred for a switch debouncer.

---

**Section Review Problem for Section 6.2**

6.1. Why is the input state $S = R = 1$ considered forbidden in the NOR latch? Why is the same state in the NAND latch the no change condition?

## 6.3

# Gated Latches

**Gated SR latch** *An SR latch whose ability to change states is controlled by an extra input called the ENABLE input.*

**Steering gates** *Logic gates, controlled by the ENABLE input of a gated latch, that steer a SET or RESET pulse to the correct input of an SR latch circuit.*

**Transparent latch (gated D latch)** *A latch whose output follows its data input when its ENABLE input is active.*

## Gated SR Latch

It is not always desirable to allow a latch to change states at random times. The circuit shown in Figure 6.15, called a **gated SR latch,** regulates the times when a latch is allowed to change state.

The gated SR latch has two distinct subcircuits. One pair of gates is connected as an SR latch. A second pair, called the **steering gates,** can be enabled or inhibited by a control signal, called *ENABLE*, allowing one or the other of these gates to pass a *SET* or *RESET* signal to the latch gates.

The *ENABLE* input can be used in two principal ways: (1) as an ON/OFF signal, and (2) as a synchronizing signal.

Figure 6.15b shows the *ENABLE* input functioning as an ON/OFF signal. When *ENABLE* = 1, the circuit acts as an active-HIGH latch. The upper gate converts a HIGH

**Figure 6.15**
**Gated SR Latch**

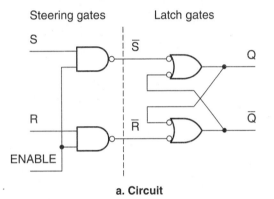

**a. Circuit**

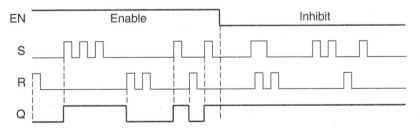

**b. ENABLE used as an ON/OFF signal**

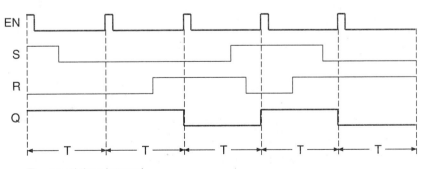

T = equal time interval

**c. ENABLE used as a synchronizing signal**

**Table 6.4** Gated SR Latch Function Table

| EN | S | R | Q | $\overline{Q}$ | Comment |
|----|---|---|---|----|---------|
| 1 | 0 | 0 | $Q_0$ | $\overline{Q}_0$ | No change |
| 1 | 0 | 1 | 0 | 1 | Reset |
| 1 | 1 | 0 | 1 | 0 | Set |
| 1 | 1 | 1 | 0 | 0 | Forbidden |
| 0 | X | X | $Q_0$ | $\overline{Q}_0$ | Inhibited |

at $S$ to a LOW at $\overline{S}$, setting the latch. The lower gate converts a HIGH at $R$ to a LOW at $\overline{R}$, thus resetting the latch.

When $ENABLE = 0$, the steering gates are inhibited and do not allow $SET$ or $RESET$ signals to reach the latch gate inputs. In this condition, the latch outputs cannot change.

Figure 6.15c shows the $ENABLE$ input as a synchronizing signal. A periodic pulse waveform is present on the $ENABLE$ line. The $S$ and $R$ inputs are free to change at random, but the latch outputs will change only when the $ENABLE$ input is active. Since the $ENABLE$ pulses are equally spaced in time, changes to the latch output can occur only at fixed intervals. The outputs can change out of synchronization if $S$ or $R$ change when $ENABLE$ is HIGH. We can minimize this possibility by making the $ENABLE$ pulses as short as possible.

Table 6.4 represents the function table for a gated SR latch.

**EXAMPLE 6.4**

Figure 6.16 shows two gated latches with the same $S$ and $R$ input waveforms but different $ENABLE$ waveforms. $EN_1$ has a 50% duty cycle. $EN_2$ has a duty cycle of 16.67%.

Draw the output waveforms, $Q_1$ and $Q_2$. Describe how the length of the $ENABLE$ pulse affects the output of each latch, assuming that the intent of each circuit is to synchronize the output changes to the beginning of the $ENABLE$ pulse.

**Solution**

Figure 6.16b shows the completed timing diagram. The longer $ENABLE$ pulse at latch 1 allows the output to switch too soon during pulses 1 and 4. ("Too soon" means before the beginning of the next $ENABLE$ pulse.) In each of these cases, the $S$ and $R$ inputs change while the $ENABLE$ input is HIGH. This premature switching is eliminated in latch 2 because the $S$ and $R$ inputs change after the shorter $ENABLE$ pulse is finished. A shorter pulse gives less chance for synchronization error, since the time for possible output changes is minimized.

## Transparent Latch (Gated D Latch)

Two circuits for the gated D ("Data") latch, or **transparent latch,** are shown in Figures 6.17 and 6.18.

Each of these circuits is called a transparent latch because, when the circuit is enabled, the $Q$ output is the same as the $D$ input, thus acting "transparently." When the $ENABLE$ input is not active, the circuit stores the value of $D$ present when the latch was last enabled.

**Figure 6.16**

**Example 6.4
Effect of ENABLE Pulse
Width**

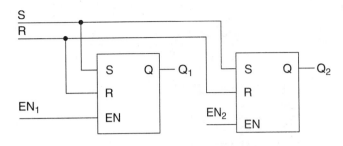

**a. Latches**

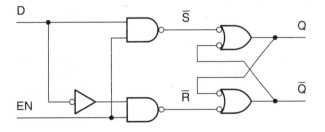

**b. Timing diagram**

**Figure 6.17**

**Transparent Latch (1)**

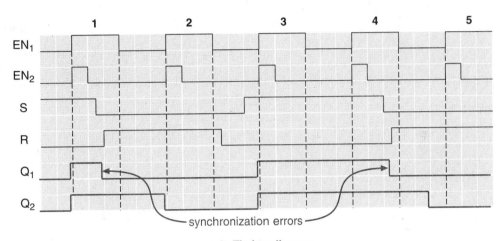

**Figure 6.18**

**Transparent Latch (2)**

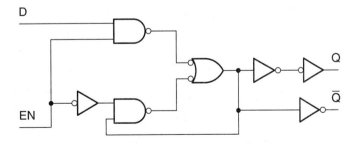

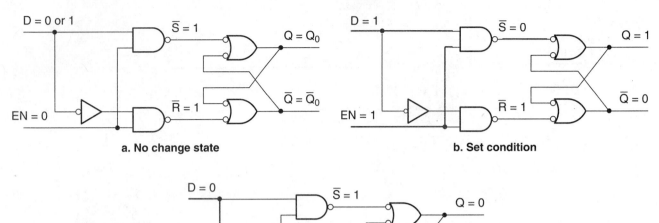

a. No change state

b. Set condition

c. Reset condition

**Figure 6.19**
**Operation of Transparent Latch (1)**

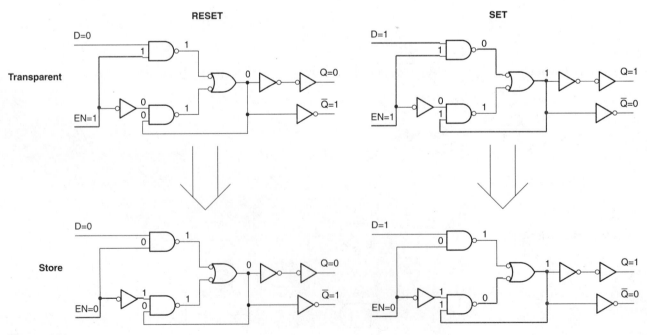

**Figure 6.20**
**Operation of Transparent Latch (2)**

**Table 6.5** Function Table of a Transparent Latch

| EN | D | Q | $\overline{Q}$ | Comment |
|----|---|---|---|---------|
| 0 | X | $Q_0$ | $\overline{Q}_0$ | Store (no change) |
| 1 | 0 | 0 | 1 | Reset |
| 1 | 1 | 1 | 0 | Set |

The latch in Figure 6.17 is a modification of the gated SR latch, configured so that the $S$ and $R$ inputs are always opposite. Under these conditions, the states $S = R = 0$ (no change) and $S = R = 1$ (forbidden) can never occur. However, the equivalent of the no change state happens when the $ENABLE$ input is LOW, when the latch steering gates are inhibited.

Figure 6.19 shows the operation of the first transparent latch (Figure 6.17) in the inhibit (no change), set, and reset states. When the latch is inhibited, the steering gates block any LOW pulses to the latch gates; the latch does not change states, regardless of the logic level at $D$.

If $EN = 1$, $Q$ follows $D$. When $D = 1$, the upper steering gate transmits a LOW to the $SET$ input of the latch and $Q = 1$. When $D = 0$, the lower steering gate transmits a LOW to the $RESET$ input of the output latch and $Q = 0$.

The operation of the second latch circuit (Figure 6.18) is illustrated in Figure 6.20. The upper two diagrams show the latch in its transparent mode, and the lower two show the latch states as it stores a 0 *(RESET)* or 1 *(SET)*.

In the transparent mode, the $EN$ input enables the upper NAND gate path. The output simply follows the $D$ input. In "latch" or "store" mode, $EN = 0$ inhibits the upper NAND gate and the $D$ input has no effect on the output.

If the latch output was 1 (latch set) when $EN$ was last HIGH, the feedback path to the lower NAND gate is active. $EN = 0$ keeps this path active via the inverter, maintaining the set state of the latch.

If $Q$ was 0 when $EN$ was HIGH, the feedback path was never active. Since the upper NAND gate is inhibited, a HIGH at $D$ cannot make it active. The latch remains reset.

The operation of the latch in Figure 6.17 is easier to understand than that in Figure 6.18. However, the latter represents the circuit that is actually used in integrated circuit D latches.

Table 6.5 shows the function table for a transparent latch. This table applies to both latch circuits, since they are functionally identical.

### 74HC75 4-Bit Transparent Latch

The 74HC75 Dual 2-Bit Transparent Latch is a common High-Speed CMOS device containing four transparent latches in two groups of two. (The TTL equivalent, 74LS75, has the same pin configuration.) Each pair of latches has a common $ENABLE$ input. Figures 6.21 and 6.22 show the logic symbol and equivalent logic circuit for this device.

We can use this device to store two 2-bit numbers or a single 4-bit number. In the latter case, the $ENABLE$ inputs are tied together. The circuit can also be used as four semi-independent latches if the common $ENABLE$ is not too important.

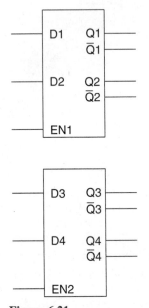

**Figure 6.21**

**74HC75 Latch Logic Symbol**

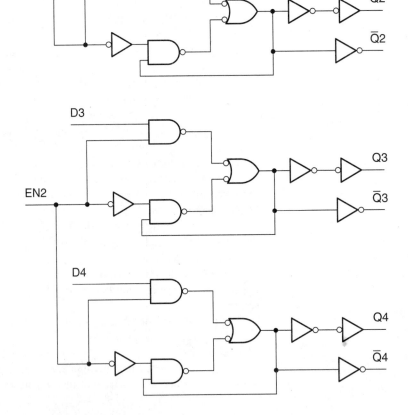

**Figure 6.22**

**74HC75 Latch Logic Diagram**

**EXAMPLE 6.5**

A municipality wants to install a series of traffic lights along a major road. As part of the planning process, the municipal traffic department decides to monitor the traffic flow with a computer and a number of above-ground traffic sensors in the major intersections along the road. A typical intersection, shown in Figure 6.23, has four sensors installed, one for each traffic direction. Each sensor has a circuit interface that produces a logic HIGH when a car passes over it.

The traffic computer cannot monitor all intersections simultaneously, so a 74HC75 transparent latch, as shown in Figure 6.24a, is used to store the status of each intersection at intervals of 5 seconds.

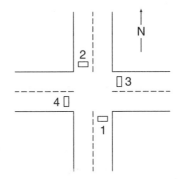

**Figure 6.23**

**Example 6.5**
**Sensor Placement in a**
**Traffic Intersection**

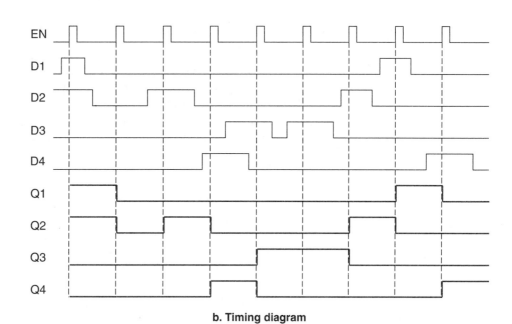

**Figure 6.24**

**Example 6.5**
**Latch Configuration and**
**Timing Diagram**

**b. Timing diagram**

The 5-second interval is set by a systemwide *ENABLE* signal. Some time after each *ENABLE* pulse, the computer reads the status of all intersections, one after the other, and thus sees a "snapshot" of the traffic pattern at the beginning of each 5-second interval.

Figure 6.24b shows the timing diagram of a typical traffic pattern at the intersection. The *D* inputs show the cars passing through the intersection in the various lanes. Complete this timing diagram by drawing the *Q* outputs of the 74HC75 latch.

How should we interpret the *Q* output waveforms?

**Solution**  Figure 6.24b shows the completed timing diagram. The *ENABLE* input synchronizes the random sensor pattern to a 5-second standard interval. The computer can read this pattern at any time within the standard interval. A HIGH on any *Q* output indicates a car over a sensor at the beginning of the interval. For example, at the beginning of the first interval, there is a car in the northbound lane (*Q1*) and one in the southbound lane (*Q2*). Similar interpretations can be made for each interval.

## 6.4

# Edge-Triggered SR and D Flip-Flops

> **Edge** *The HIGH-to-LOW (negative edge) or LOW-to-HIGH (positive edge) transition of a pulse waveform.*
>
> **CLOCK** *An enabling input to a sequential circuit that is sensitive to the positive- or negative-going edge of a waveform.*
>
> **Edge-triggered** *Enabled by the positive or negative edge of a digital waveform.*
>
> **Edge-sensitive** *Edge-triggered.*
>
> **Level-sensitive** *Enabled by a logic HIGH or LOW level.*

In Example 6.4, we saw how a shorter pulse width at the *ENABLE* input of a gated latch increased the chance of the output being synchronized to the *ENABLE* pulse waveform. This is because a shorter *ENABLE* pulse gives less chance for the *SET* and *RESET* inputs to change during the time the latch is enabled.

A logical extension of this idea is to enable the latch for such a small time that the width of the *ENABLE* pulse is almost zero. The best approximation we can make to this is to allow changes to the circuit output only when an enabling, or **CLOCK,** input receives the **edge** of an input waveform. An edge is the part of a waveform that is in transition from LOW to HIGH (positive edge) or HIGH to LOW (negative edge), as shown in Figure 6.25. We can say that a device enabled by an edge is **edge-triggered** or **edge-sensitive.**

Since the *CLOCK* input enables a circuit only while in transition, we can refer to it as a "dynamic" input. This is in contrast to the *ENABLE* input of a gated latch, which is **level-sensitive** or "static," and will enable a circuit for the entire time it is at its active level.

## Edge-Triggered SR Flip-Flops

> **Flip-flop** *A sequential circuit based on a latch whose output changes when its CLOCK input receives either an edge or a pulse, depending on the device.*

The gated SR latch with a *CLOCK* input is called an SR **flip-flop.** The difference between a latch and a flip-flop is not always very clear. We will differentiate between the two devices by their enabling inputs. A flip-flop has an edge-triggered or pulse-triggered *CLOCK* input. (A pulse-triggered input requires a complete pulse cycle, such

CLK \_\_\_\_\_⌐‾‾‾‾        CLK ‾‾‾‾⌐\_\_\_\_

**Figure 6.25**

**Edges of a CLOCK Waveform**

**a. LOW-to-HIGH transition (positive edge)**        **b. HIGH-to-LOW transition (negative edge)**

**Figure 6.26**
**Edge-Triggered SR**
**Flip-Flops**

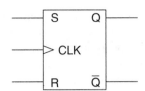

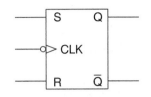

**a. Positive edge-triggered**        **b. Negative edge-triggered**

as LOW-HIGH-LOW, to make an output change.) A latch has either no enabling input or a level-sensitive *ENABLE* input.

Figure 6.26 shows logic symbols for positive and negative edge-triggered SR flip-flops. The triangle at the *CLK* input indicates a dynamic, or edge-triggered, input. The bubble on the *CLK* input of the second device, combined with the triangle, indicates negative edge triggering.

The SR flip-flop acts the same as a gated SR latch, except that output changes occur only on the active *CLOCK* edges. Table 6.6 gives the function tables for the positive and negative edge-triggered flip-flops.

**Table 6.6** Function Tables for Edge-Triggered SR Flip-Flops

| CLK | S | R | Q | $\overline{Q}$ | Comment | CLK | S | R | Q | $\overline{Q}$ | Comment |
|-----|---|---|---|---|---------|-----|---|---|---|---|---------|
| ↑ | 0 | 0 | $Q_0$ | $\overline{Q}_0$ | No change | ↓ | 0 | 0 | $Q_0$ | $\overline{Q}_0$ | No change |
| ↑ | 0 | 1 | 0 | 1 | Reset | ↓ | 0 | 1 | 0 | 1 | Reset |
| ↑ | 1 | 0 | 1 | 0 | Set | ↓ | 1 | 0 | 1 | 0 | Set |
| ↑ | 1 | 1 | 1 | 1 | Forbidden | ↓ | 1 | 1 | 1 | 1 | Forbidden |
| 0 | X | X | $Q_0$ | $\overline{Q}_0$ | Inhibited | 0 | X | X | $Q_0$ | $\overline{Q}_0$ | Inhibited |
| 1 | X | X | $Q_0$ | $\overline{Q}_0$ | Inhibited | 1 | X | X | $Q_0$ | $\overline{Q}_0$ | Inhibited |
| ↓ | X | X | $Q_0$ | $\overline{Q}_0$ | Inhibited | ↑ | X | X | $Q_0$ | $\overline{Q}_0$ | Inhibited |
| | | **Positive Edge-Triggered** | | | | | | **Negative Edge-Triggered** | | | |

**EXAMPLE 6.6**

A positive edge-triggered SR flip-flop has inputs *S*, *R*, and *CLK* as shown in the timing diagram in Figure 6.27. Complete the timing diagram by drawing the lines for *Q* and *Q*. Assume that the flip-flop is initially set.

**Solution**    The output waveforms are shown in Figure 6.27. The *Q* and $\overline{Q}$ waveforms change only when the clock is in transition from LOW to HIGH, regardless of when *S* and *R* change.

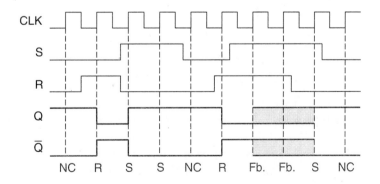

**Figure 6.27**

**Example 6.6**
**Timing Diagram**

The outputs are as specified in Table 6.6. The transition for each clock pulse is shown as set *(S),* reset *(R),* no change *(NC),* or forbidden *(Fb).* Note that when both *S* and *R* are HIGH, the output enters an unknown condition. The outputs will both spike HIGH on each clock pulse, then revert to either a set or reset state, depending on the individual gate speeds. We cannot be certain of the output until the next time a set or reset state is clocked into the flip-flop. (Compare this to the forbidden-state transition shown in Figure 6.10d.)

**EXAMPLE 6.7**

Figure 6.28 shows a positive edge-triggered SR flip-flop and a gated SR latch with the same *S, R,* and *CLK* inputs. A timing diagram shows the input waveforms. Complete the timing diagram by drawing the *Q* outputs for each device. Account for the differences in the outputs.

**Solution** Figure 6.28 shows the completed timing diagram. $Q_1$ changes according to the states of *S* and *R,* but only on the positive edges of the *CLK* waveform. $Q_2$ will change according to *S* and *R* any time that *CLK* is in the HIGH state. Thus, the flip-flop is edge-sensitive and the latch is level-sensitive. This difference results in different transitions on pulses 4/5 and 6/7.

**Figure 6.28**

**Example 6.7**
**Edge- Versus Level-Sensitive**
**Devices**

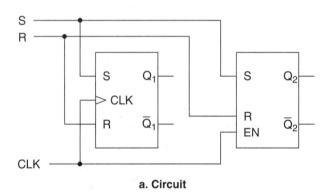

a. Circuit

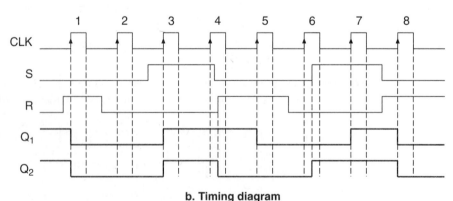

b. Timing diagram

The SR flip-flop is a nice bridge between latch and flip-flop circuits. However, due to improvements in available digital devices, the SR flip-flop as a small-scale integration (SSI) device is pretty much obsolete. (Its forbidden state makes it less than ideal for some applications.) SR flip-flops are sometimes incorporated into medium-scale integration (MSI) logic and programmable logic devices (PLDs), where the *S* and *R* conditions are arranged internally to avoid the use of the forbidden state.

Two SSI devices that can replace the SR flip-flop but do not have the forbidden state are the D flip-flop, which has the same states as a transparent latch, and the JK flip-flop, which replaces the forbidden state with a toggle condition. (Toggle means that the output changes to the opposite logic state with every *CLOCK* pulse.)

The majority of flip-flops in use today are either D or JK flip-flops.

## Edge-Triggered D Flip-Flops

**Edge detector** *A circuit in an edge-triggered flip-flop that converts the active edge of a CLOCK input to an active-level pulse at the internal latch's SET and RESET inputs.*

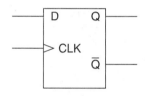

**a. Positive edge-triggered**

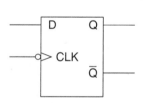

**b. Negative edge-triggered**

**Figure 6.29**

**D Flip-Flop Logic Symbols**

The D, or data, flip-flop has two inputs—*D* and *CLK*—and a pair of complementary outputs—$Q$ and $\overline{Q}$. Figure 6.29a and b shows the logic symbols for the positive and negative edge-triggered versions of the D flip-flop.

As is the case with the transparent D latch, $Q$ follows $D$ when the *CLOCK* is pulsed. Table 6.7 gives the function tables of the D flip-flops.

**Table 6.7** Function Tables for Edge-Triggered D Flip-Flops

| *CLK* | *D* | *Q* | $\overline{Q}$ | Comment | *CLK* | *D* | *Q* | $\overline{Q}$ | Comment |
|---|---|---|---|---|---|---|---|---|---|
| ↑ | 1 | 1 | 0 | Set | ↓ | 1 | 1 | 0 | Set |
| ↑ | 0 | 0 | 1 | Reset | ↓ | 0 | 0 | 1 | Reset |
| 0 | X | $Q_0$ | $\overline{Q}_0$ | Inhibited | 0 | X | $Q_0$ | $\overline{Q}_0$ | Inhibited |
| 1 | X | $Q_0$ | $\overline{Q}_0$ | Inhibited | 1 | X | $Q_0$ | $\overline{Q}_0$ | Inhibited |
| ↓ | X | $Q_0$ | $\overline{Q}_0$ | Inhibited | ↑ | X | $Q_0$ | $\overline{Q}_0$ | Inhibited |
| **Positive Edge-Triggered** | | | | | **Negative Edge-Triggered** | | | | |

Figure 6.30 shows the equivalent circuit of a positive edge-triggered D flip-flop. The circuit is the same as the transparent latch of Figure 6.17, except that the enable input (called *CLK* in the flip-flop) passes through an **edge detector,** a circuit that converts a positive edge to a brief positive-going pulse. (A negative edge detector converts a negative edge to a positive-going pulse.)

Figure 6.31 shows a circuit that acts as a simplified positive edge detector. Edge detection depends on the fact that a gate output does not switch immediately when its

**Figure 6.30**

**D Flip-Flop Equivalent Circuit**

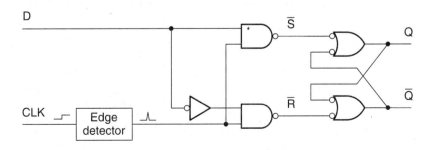

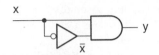

**a. Simplified circuit**

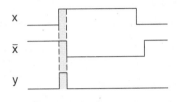

**b. Waveforms**

**Figure 6.31**

**Positive Edge Detector**

input switches. There is a delay of about 3 to 10 ns from input change to output change, called propagation delay.

When input $x$, shown in the timing diagram of Figure 6.31, goes from LOW to HIGH, the inverter output, $\bar{x}$, goes from HIGH to LOW after a short delay. This delay causes both $x$ and $\bar{x}$ to be HIGH for a short time, producing a high-going pulse at the circuit output immediately following the positive edge at $x$.

When $x$ returns to LOW, $\bar{x}$ goes HIGH after a delay. However, there is no time in this sequence when both AND inputs are HIGH. Therefore, the circuit output stays LOW after the negative edge of the input waveform.

Figure 6.32 shows how the D flip-flop circuit operates. When $D = 0$ and the edge detector senses a positive edge at the *CLK* input, the output of the lower NAND gate steers a low-going pulse to the *RESET* input of the latch, thus storing a 0 at $Q$. When $D = 1$, the upper NAND gate is enabled. The edge detector sends a high-going pulse to the upper steering gate, which transmits a low-going *SET* pulse to the output latch. This action stores a 1 at $Q$.

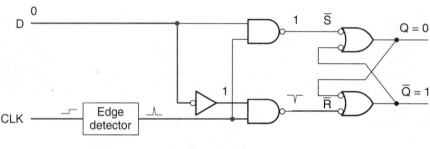

**a. Reset action**

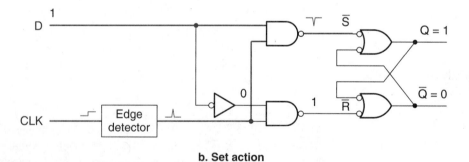

**Figure 6.32**

**Operation of a D Flip-Flop**

**b. Set action**

---

**EXAMPLE 6.8**

Two positive edge-triggered D flip-flops are connected as shown in Figure 6.33a. Inputs $D_A$ and *CLK* are as shown in the timing diagram. Complete the timing diagram by drawing the waveforms for $Q_A$ and $Q_B$, assuming that both flip-flops are initially reset.

**Solution**

Figure 6.33 shows the output waveforms. $Q_A = D_A$ at each point where the *CLK* input has a positive edge. One result of this is that the HIGH pulse on $D_A$ between *CLK* pulses 5 and 6 is ignored, since $D_A = 0$ at positive edges 5 and 6.

Since $D_B = Q_A$ and $Q_B$ follows $D_B$, the waveform at $Q_B$ is the same as at $Q_A$, but delayed by one clock cycle. If $Q_A$ changes, we assume that the value of $D_B$ is the same as $Q_A$ just *before* the *CLK* pulse. This is because delays within the circuitry of the flip-

**Figure 6.33**
**Example 6.8**
**Circuit and Timing Diagram**

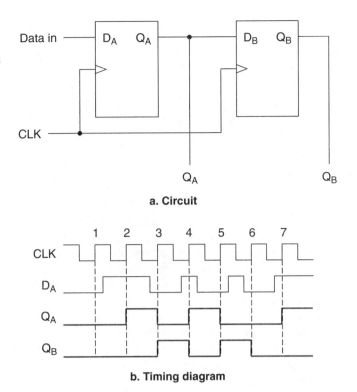

a. Circuit

b. Timing diagram

flops ensure that their outputs will not change for several nanoseconds after an applied clock pulse. Therefore, the level at $D_B$ remains constant long enough for it to be clocked into the second flip-flop.

The data entering the circuit at $D_A$ are moved, or shifted, from one flip-flop to the next. This type of data movement, called "serial shifting," is frequently used in data communication and in computer arithmetic circuits.

## Applications of D Flip-Flops

**Serial transmission**  *A method of sending multibit data between two devices where the bits are sent one after the other along a single line.*

**Parallel transmission**  *A technique of sending multibit data from one device to another where all bits are sent simultaneously over individual parallel lines.*

D flip-flops are often used to store data temporarily and to move them in an orderly fashion from one destination to another. Movement of data one bit after another along a single line, as in Example 6.8, is called **serial transmission.** When data are moved simultaneously along multiple lines, we call this **parallel transmission.**

### Serial-to-Parallel Conversion

Digital data, particularly in communications and computer applications, are frequently sent in multibit packages of fixed size. These packages can represent information such

**Figure 6.34**

**Asynchronous Transmission of an ASCII Character**

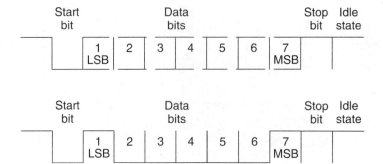

**Figure 6.35**

**Waveform of Serially Transmitted ASCII "A"**

as BCD or hexadecimal digits or ASCII characters. (See Chapter 5, Number Systems and Codes, for information about ASCII code.)

Multibit data within a device, such as a personal computer, often run via parallel transmission, since the device operates faster if all bits are available at the same time. Data are often sent between two devices, such as a computer and a video display terminal (VDT) by serial transmission, that is, along one line with the bits sent one at a time. This saves connector hardware, including cable, which can be expensive, particularly if the connection is fairly long. Serial transmission can also use telephone lines for long-distance communication, whereas parallel transmission cannot.

Communication between a computer and a VDT is achieved by serial transmission of ASCII characters. Figure 6.34 shows a standard serial transmission format for transmitting a single ASCII character (i.e., one letter, number, "typewriter character," or control character).

The transmission line is in a HIGH state when no character is being transmitted. This is called the idle state. The first bit, called the start bit, is always LOW to indicate the beginning of a transmitted character. The data bits follow, LSB to MSB. One or more stop bits, which are always HIGH, signify the end of the character.

For example, the ASCII code for "A" is $41_{16}$ or $100\ 0001_2$. Figure 6.35 shows the waveform for a serially transmitted "A."

---

**EXAMPLE 6.9**

A serially transmitted ASCII character can be converted to parallel form by shifting it into eight D flip-flops, shown in Figure 6.36, and reading the $Q$ outputs of each flip-flop after the entire character has been shifted in.

Figure 6.37 shows a a clock waveform and the bits of ASCII "m," including a start bit and 7 data bits, appearing in sequence at input $D_A$. (ASCII "m" = $6D_{16}$ =

**Figure 6.36**

**Example 6.9**
**D Flip-Flops as a Serial-to-Parallel Converter**

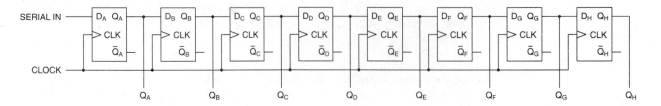

110 1101$_2$). Eight clock pulses are shown. Complete the waveforms for $Q_A$ through $Q_B$ and describe the final resulting output combination. Assume that all flip-flops are initially set.

**Solution**  Figure 6.37 shows the completed timing diagram. The contents of each flip-flop are shifted to the next with each *CLK* pulse.

After eight pulses, the bit that was at $D_A$ on pulse 1 ends up at $Q_H$. The bit at $D_A$ on pulse 2 is at $Q_G$ on pulse 8, and so on. The end result is that the entire ASCII character and the start bit are stored in the flip-flops after eight *CLK* pulses. You can read ASCII "m" in flip-flops *A* through *G*. Outputs $Q_A$ through $Q_G$ are 110 1101 = 6D$_{16}$ = "m." The start bit is in $Q_H$.

**Figure 6.37**

**Example 6.9
ASCII "m" Converted to
Parallel Form**

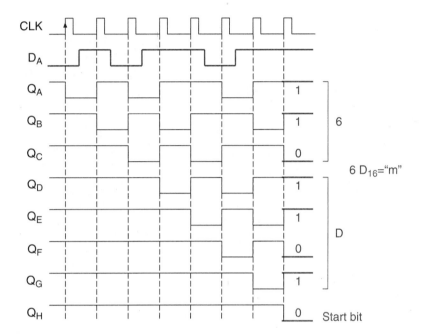

## Parallel Data Transfer

Example 6.9 shows how serially transmitted data can be converted to parallel form. Once in parallel form these data can be moved from one device to another along several parallel lines. Two devices used in this type of transfer are the 74HC373 Octal (8-bit) Transparent Latch and the 74HC374 Octal D Flip-Flop, shown in Figure 6.38. The equivalent LSTTL devices are 74LS373 and 74LS374.

The 74HC373 works the same as a single transparent latch, except that when the device is enabled (*EN* = 1), all eight *Q* outputs follow all eight *D* inputs; *1Q* follows *1D, 2Q* follows *2D,* and so on. The terminal marked $\overline{OC}$ (Output Control) enables the latch outputs when held LOW.

The 74HC374 contains eight individual D flip-flops with a common clock. When the *CLK* input receives a positive edge, all *Q*s follow their respective *D* inputs. As is the case with the 74HC373, the outputs are enabled when $\overline{OC}$ is held LOW.

**Figure 6.38**
**Parallel Devices**

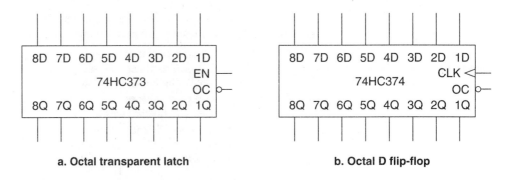

a. Octal transparent latch                    b. Octal D flip-flop

**EXAMPLE 6.10**

Figure 6.39 shows how we can use a pair of '373s or '374s to store an 8-bit number in one of two selected locations. Briefly explain the operation of each circuit. Explain why the transparent latch circuit needs a pair of 2-input AND gates, while the flip-flop circuit requires two 3-input ANDs.

**Solution**    **74HC373 Octal Latches:**

The circuit in Figure 6.39a selects one of two octal latches to store parallel data. All 8 bits are applied to the inputs of both latches simultaneously. The selected latch is transparent when $EN = 1$. It stores the applied parallel data when $EN$ returns to 0.

A latch is selected by two inputs:

1. *LOAD DATA:* This input enables both AND gates, allowing either latch *A* or latch *B* to be selected. When this input is HIGH, one of the two latches is always selected.

2. *WORD SELECT:* This determines which of the two octal latches will receive the parallel data. Since *WORD SELECT* is connected to the inverter input, it places the two *EN* inputs in opposite states: when $EN = 0$ for latch *A*, $EN = 1$ for latch *B*, and vice versa.

Table 6.8 summarizes the operation of the circuit in Figure 6.39a.

**Table 6.8** Truth Table for Latch or Flip-Flop Selection

| *LOAD DATA* | *WORD SELECT* | Selected Latch/Flip-Flop |
|---|---|---|
| 0 | 0 | Neither |
| 0 | 1 | Neither |
| 1 | 0 | Latch/flip-flop *A* |
| 1 | 1 | Latch/flip-flop *B* |

**74HC374 Octal D Flip-Flops:**

The circuit in Figure 6-39b stores 8-bit data in one of two octal D flip-flops. A positive-going pulse must be applied to one of the 74HC374 devices to store the 8-bit parallel data. The *WORD SELECT* and *LOAD DATA* lines are used to enable one of the two AND gates, thus steering a *CLOCK* pulse to either flip-flop *A* or flip-flop *B*. Table 6.8 shows which flip-flop is enabled for each combination of *LOAD DATA* and *WORD SELECT*.

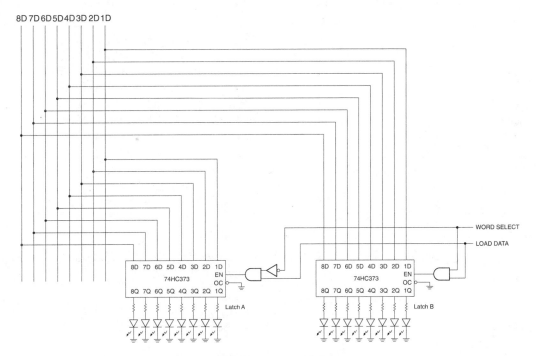

a. Octal transparent latches

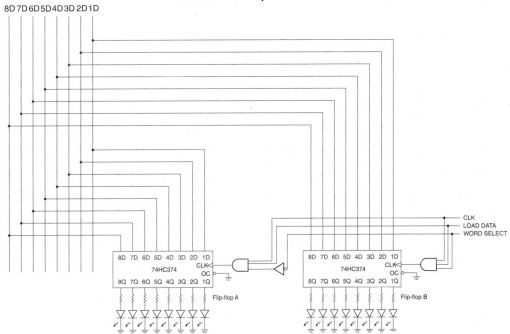

b. Octal D Flip-flops

**Figure 6.39**

**Example 6.10**
**Parallel Data Transfer to One of Two Locations**

**Control Gating:**

The two circuits require different control logic because one is level-sensitive and the other is edge-sensitive. When *LOAD DATA* and *WORD SELECT* enable one of the AND gates in Figure 6.39a, *EN* = 1 for the selected latch. Since the *EN* input of the '373 is level-sensitive, this is sufficient to make the latch accept the parallel data.

The '374 requires a positive edge at its *CLK* input to load the parallel data. Thus, we need an input to enable the load function *(LOAD DATA)*, an input to select which latch to enable *(WORD SELECT)*, and an input to supply the clock pulse.

## Tristate Buffering in Parallel Data Transfer

> **Tristate buffer**  *A gate having three possible output states: logic HIGH, logic LOW, and a high-impedance state.*
>
> **High-impedance state**  *The output state of a tristate buffer that is neither logic HIGH nor logic LOW, but resembles an open circuit.*

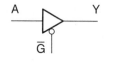

**a. Noninverting tristate buffer**

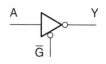

**b. Inverting tristate buffer**

**Figure 6.40**

**Tristate Buffers (Active-LOW Enable)**

Figure 6.40 shows the logic symbols of two **tristate buffers.** These gates are commonly used in parallel data transfer when several device outputs are connected in parallel.

If two standard gate outputs are connected in parallel, they can be damaged if they are at opposite logic levels. A logic LOW is approximately at ground potential. A logic HIGH is about 3.5 V in a TTL device. If two connected outputs are at opposite levels, current from the HIGH output will flow toward the LOW through a relatively low impedance. This current is at a level above the rated value of either gate output and will damage them both over time (a few minutes).

Tristate buffering allows two or more outputs to be connected to a single load without damage, provided only one of the outputs is enabled at a time. The disabled outputs are electrically isolated from the circuit.

The operation of the gates in Figure 6.40 is simple: when $\overline{G} = 0$, the gates act as noninverting or inverting buffers. Logic levels are the usual values of HIGH or LOW, as determined by the gate inputs. $Y = A$ for the gate in Figure 6.40a; $Y = \overline{A}$ for the gate in Figure 6.40b. When $\overline{G} = 1$, the outputs are in the **high-impedance state.** In this condition, the output is neither HIGH nor LOW. It acts as an open circuit, electrically isolating the output from any load it is connected to.

Figure 6.41 shows two devices we can use for tristate buffering of parallel data. The 74HC240 contains eight tristate inverting buffers, arranged in groups of four. Each group has a common *ENABLE*, allowing 4-bit data to be blocked or passed in complement form. The 74HC244 device has the same arrangement, except the buffers are noninverting. The two *ENABLE* $(\overline{G})$ inputs on each chip can be tied together to make an 8-bit inverting or noninverting buffer.

Figure 6.42 illustrates the logic symbol of a 74HC173 4-bit D flip-flop. This device is similar to the 74HC374 flip-flop, except that it has *ENABLE* controls for both the input and output data. Four-bit data are loaded when $\overline{G1} = \overline{G2} = 0$ and a positive clock edge is applied. Data are available at the $Q$ outputs when $\overline{M} = \overline{N} = 0$.

**Figure 6.41**
**Octal Tristate Buffers**

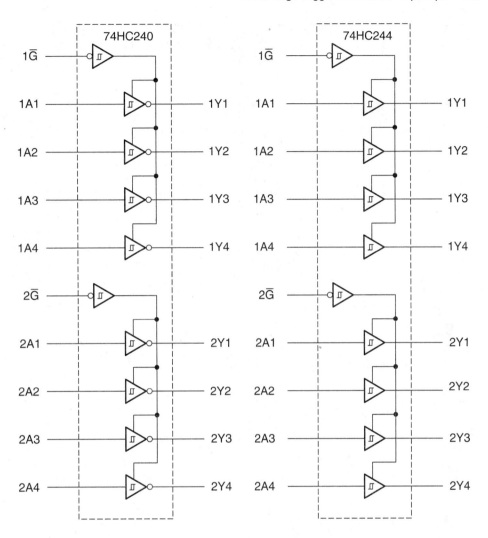

**Figure 6.42**
**74HC173 4-Bit D Flip-Flop**

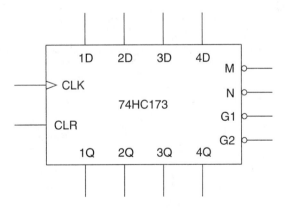

**EXAMPLE 6.11**   Figure 6.43 shows a circuit that can transfer any one of four 4-bit words to either of two 4-bit locations.

**Figure 6.43**

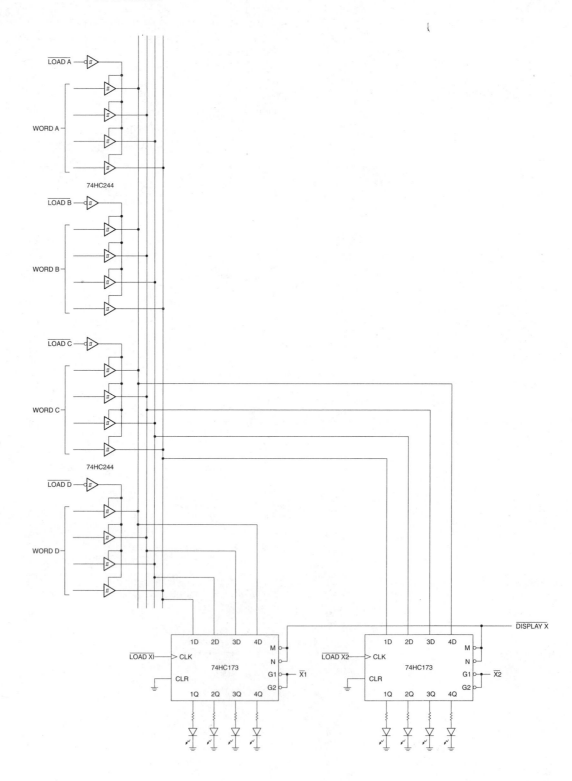

**Figure 6.44**

**Example 6.11**
**Timing Diagram for Parallel Transfer**

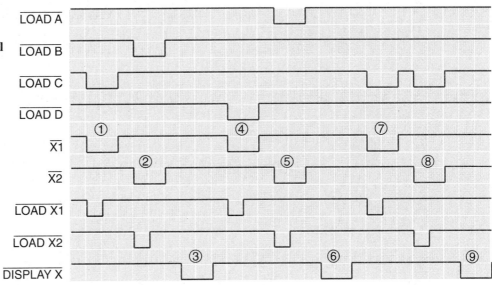

The 4-bit words are:

Word A = 0011

Word B = 1010

Word C = 1110

Word D = 1111

Figure 6.44 shows a timing diagram of the $\overline{LOAD}$, $\overline{X}$, and $\overline{DISPLAY}$ inputs. Write the sequence of bits appearing at the latch outputs at points 3, 6, and 9 in the timing diagram.

**Solution**   Parallel data are transferred from a data input to a latch when the appropriate $\overline{LOAD}$ and $\overline{X}$ lines are LOW (points 1, 2, 4, 5, 7, and 8 in the timing diagram). $\overline{LOAD}$ selects which 4-bit input data appear on the parallel bus. $\overline{X}$ selects which '173 flip-flop will receive the data. The actual transfer occurs on the positive edge of $\overline{LOAD\ X1}$ or $\overline{LOAD\ X2}$. These waveforms are arranged so that each positive edge is on the center of the data-enabling interval. (This practice minimizes the effect of distortions during the rise and fall times of the enabling waveforms.)

The latch contents are displayed when the $\overline{DISPLAY\ X}$ line is LOW (points 3, 6, and 9).

The circuit operation can be summarized as follows:

| Step | Action | Source | Destination | Output |
|------|--------|--------|-------------|--------|
| 1 | Transfer | C | X1 | ——— |
| 2 | Transfer | B | X2 | ——— |
| 3 | Display | — | — | 1110 1010 |
| 4 | Transfer | D | X1 | ——— |
| 5 | Transfer | A | X2 | ——— |
| 6 | Display | — | — | 1111 0011 |

| Step | Action | Source | Destination | Output |
|------|--------|--------|-------------|--------|
| 7 | Transfer | C | X1 | ———— |
| 8 | Transfer | C | X2 | ———— |
| 9 | Display | – | — | 1110 1110 |

## 6.5

# Edge-Triggered JK Flip-Flops

**Toggle** *Change to the opposite binary state with each applied clock pulse.*

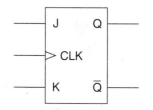

**a. Positive edge-triggered**

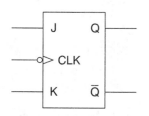

**b. Negative edge-triggered**

**Figure 6.45**

**Edge-Triggered JK Flip-Flops**

The edge-triggered JK flip-flop is one of the most versatile and widely used sequential circuits available. Figure 6.45 shows the logic symbols of the positive and negative edge-triggered versions of this flip-flop. The function tables of these devices are shown in Table 6.9.

$J$ acts as the flip-flop's *SET* input, and $K$ acts as the *RESET* input, both active HIGH. The operation of the JK flip-flop is almost the same as an SR flip-flop, with one important difference: *there is no forbidden state.* Instead, when $J = K = 1$, the output will **toggle** (go to the opposite state) when a *CLK* pulse is applied.

Figure 6.46 shows the simplified circuit of a negative edge-triggered JK flip-flop. The circuit is like that of the SR flip-flop, except that there are two extra feedback lines from the latch outputs to the steering gate inputs. This feedback is responsible for the toggling action.

Figure 6.47 illustrates how the additional two lines cause the flip-flop to toggle. The cross-feedback from $Q$ to $K$ and from $\overline{Q}$ to $J$ enables one, but not both, of the steering gates. The edge detector just after the *CLK* input produces a short positive-going pulse upon detecting a negative edge on the *CLK* waveform. The enabled steering gate complements and transmits this pulse to the latch, activating either the set or reset function. This in turn changes the latch state and enables the opposite steering gate.

Since all inputs of the steering gates must be HIGH to enable one of the latch functions, $J$ and $K$ must both be HIGH to sustain a repeated toggling action. Under these conditions, $\overline{Q}$ and $Q$ alternately enable one of the steering gates.

**Table 6.9**  Function Tables for Edge-Triggered JK Flip-Flops

| CLK | J | K | Q | $\overline{Q}$ | Comment | CLK | J | K | Q | $\overline{Q}$ | Comment |
|-----|---|---|---|---|---------|-----|---|---|---|---|---------|
| ↑ | 0 | 0 | $Q_0$ | $\overline{Q}_0$ | No change | ↓ | 0 | 0 | $Q_0$ | $\overline{Q}_0$ | No change |
| ↑ | 0 | 1 | 0 | 1 | Reset | ↓ | 0 | 1 | 0 | 1 | Reset |
| ↑ | 1 | 0 | 1 | 0 | Set | ↓ | 1 | 0 | 1 | 0 | Set |
| ↑ | 1 | 1 | $\overline{Q}_0$ | $Q_0$ | Toggle | ↓ | 1 | 1 | $\overline{Q}_0$ | $Q_0$ | Toggle |
| 0 | X | X | $Q_0$ | $\overline{Q}_0$ | Inhibited | 0 | X | X | $Q_0$ | $\overline{Q}_0$ | Inhibited |
| 1 | X | X | $Q_0$ | $\overline{Q}_0$ | Inhibited | 1 | X | X | $Q_0$ | $\overline{Q}_0$ | Inhibited |
| ↓ | X | X | $Q_0$ | $\overline{Q}_0$ | Inhibited | ↑ | X | X | $Q_0$ | $\overline{Q}_0$ | Inhibited |
| **Positive Edge-Triggered** | | | | | | **Negative Edge-Triggered** | | | | | |

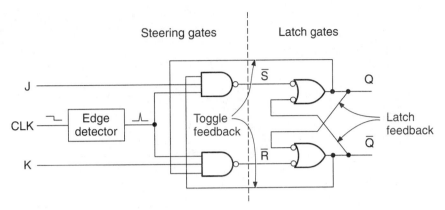

**Figure 6.46**

**JK Flip-Flop Circuit (Simplified)**

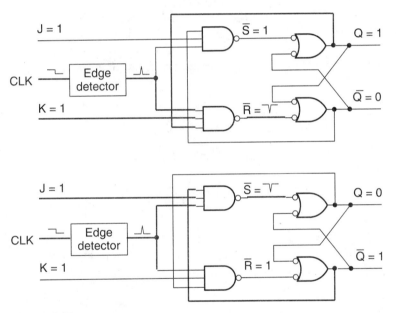

**Figure 6.47**

**Toggle Action of a JK Flip-Flop**

---

**EXAMPLE 6.12**    The *J, K,* and *CLK* inputs of a negative edge-triggered JK flip-flop are as shown in the timing diagram in Figure 6.48. Complete the timing diagram by drawing the waveforms for $Q$ and $\overline{Q}$. Indicate which function (no change, set, reset, or toggle) is performed at each clock pulse. The flip-flop is initially reset.

**Solution**    The completed timing diagram is shown in Figure 6.48. The outputs change only on the negative edges of the *CLK* waveform. Note that the same output sometimes results from different inputs. For example, the function at clock pulse 4 is reset and the function at pulses 5 and 6 is no change, but the $Q$ waveform is LOW in each case.

**Figure 6.48**

**Example 6.12**
**Timing Diagram**
**(Negative-Edge-Triggered**
**JK Flip-Flop)**

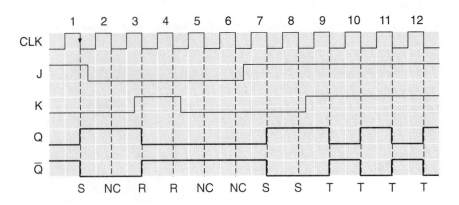

EXAMPLE 6.13

The toggle function of a JK flip-flop is often used to generate a desired output sequence from a series of flip-flops. The circuit shown in Figure 6.49 is configured so that all flip-flops are permanently in toggle mode.

Assume that all flip-flops are initially reset. Draw a timing diagram showing the $CLK$, $Q_A$, $Q_B$, and $Q_C$ waveforms when eight clock pulses are applied. Make a table showing each combination of $Q_C$, $Q_B$, and $Q_A$. What pattern do the outputs form over the period shown on the timing diagram?

**Figure 6.49**

**Example 6.13**
**Circuit**

(circuit diagram with three JK flip-flops: $J_A$ $Q_A$, CLK, $K_A$; $J_B$ $Q_B$, CLK, $K_B$; $J_C$ $Q_C$, CLK, $K_C$; all J and K inputs tied to 1)

**Solution**    The circuit timing diagram is shown in Figure 6.50. All flip-flops are in toggle mode. Each time a negative clock edge is applied to the flip-flop $CLK$ input, the $Q$ output will change to the opposite state.

For flip-flop $A$, this happens with every clock pulse, since it is clocked directly by the CLK waveform. Each of the other flip-flops is clocked by the $Q$ output waveform of the previous stage. Flip-flop $B$ is clocked by the negative edge of the $Q_A$ waveform. Flip-flop $C$ toggles when $Q_B$ goes from HIGH to LOW.

**Table 6.10** Sequence of Outputs for Circuit in Figure 6.49

| Clock Pulse | $Q_C$ | $Q_B$ | $Q_A$ |
|---|---|---|---|
| 0 | 0 | 0 | 0 |
| 1 | 0 | 0 | 1 |
| 2 | 0 | 1 | 0 |
| 3 | 0 | 1 | 1 |
| 4 | 1 | 0 | 0 |
| 5 | 1 | 0 | 1 |
| 6 | 1 | 1 | 0 |
| 7 | 1 | 1 | 1 |
| 8 | 0 | 0 | 0 |

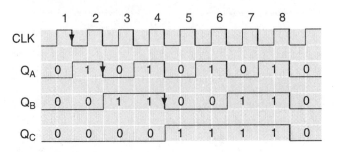

**Figure 6.50**

**Example 6.13**
**Timing Diagram**

Table 6.10 shows the flip-flop outputs after each clock pulse. The outputs form a 3-bit number that counts from 000 to 111 in binary sequence, then returns to 000 and repeats.

This flip-flop circuit is called a 3-bit asynchronous counter. We will study counter circuits in detail in Chapter 11.

▲

## Asynchronous Inputs (Preset and Clear)

**Synchronous inputs** *The inputs of a flip-flop that do not affect the flip-flop's Q outputs unless a clock pulse is applied. Examples include D, J, and K inputs.*

**Asynchronous inputs** *The inputs of a flip-flop that change the flip-flop's Q outputs immediately, without waiting for a pulse at the CLK input. Examples include preset and clear inputs.*

**Preset** *An asynchronous set function.*

**Clear** *An asynchronous reset function.*

The *S, R, D, J,* and *K* inputs of the various flip-flops examined so far are called **synchronous inputs.** This is because any effect they have on the flip-flop outputs is synchronized to the *CLK* input.

Another class of input is also provided on many flip-flops. These inputs, called **asynchronous inputs,** do not need to wait for a clock pulse to make a change at the output. The two functions usually provided are **preset,** an asynchronous set function, and **clear,** an asynchronous reset function. In TTL flip-flops, these functions are generally active LOW, and are abbreviated $\overline{PRE}$ and $\overline{CLR}$.

Figure 6.51 shows a modification to the JK flip-flop of Figure 6.46. The $\overline{PRE}$ and $\overline{CLR}$ inputs have direct access to the latch gates of the flip-flop and thus are not affected by the *CLK* input. They act exactly the same as the *SET* and *RESET* inputs of an SR latch and will override any synchronous input functions currently active.

**Figure 6.51**
**$\overline{PRE}$ and $\overline{CLR}$ Inputs**

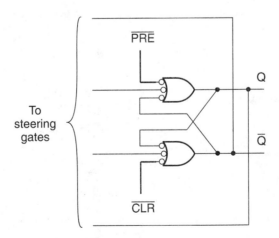

**EXAMPLE 6.14**   The waveforms for the *CLK, J, K, $\overline{PRE}$*, and $\overline{CLR}$ inputs of a 74LS76A negative edge-triggered JK flip-flop are shown in the timing diagram of Figure 6.52. Complete the diagram by drawing the waveform for output *Q*.

**Figure 6.52**

**Example 6.14**
**Waveforms**

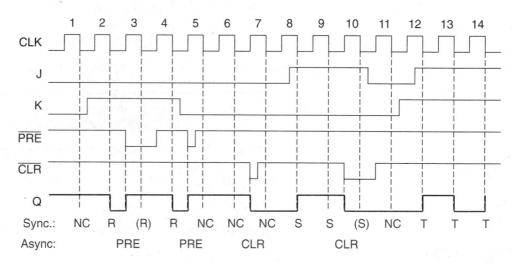

**Solution**   The *Q* waveform is shown in Figure 6.52. The asynchronous inputs cause an immediate change in *Q*, whereas the synchronous inputs must wait for the next negative clock edge. If asynchronous and synchronous inputs are simultaneously active, the asynchronous inputs have priority. This occurs in two places: pulse 3 *(K, $\overline{PRE}$)* and pulse 10 *(J, $\overline{CLR}$)*.

The diagram shows the synchronous functions (no change, reset, set, and toggle) at each clock pulse and the asynchronous functions (preset and clear) at the corresponding transition points.

The function table of a negative edge-triggered JK flip-flop with preset and clear functions is shown in Table 6.11.

**Table 6.11** Function Table of a Negative Edge-Triggered JK Flip-Flop With Preset and Clear Functions

| | $\overline{PRE}$ | $\overline{CLR}$ | *CLK* | *J* | *K* | *Q* | $\overline{Q}$ | Comment |
|---|---|---|---|---|---|---|---|---|
| **Synchronous Functions** | 1 | 1 | ↓ | 0 | 0 | $Q_0$ | $\overline{Q}_0$ | No change |
| | 1 | 1 | ↓ | 0 | 1 | 0 | 1 | Reset |
| | 1 | 1 | ↓ | 1 | 0 | 1 | 0 | Set |
| | 1 | 1 | ↓ | 1 | 1 | $\overline{Q}_0$ | $Q_0$ | Toggle |
| **Asynchronous Functions** | 0 | 1 | X | X | X | 1 | 0 | Preset |
| | 1 | 0 | X | X | X | 0 | 1 | Clear |
| | 0 | 0 | X | X | X | 1 | 1 | Forbidden |
| | 1 | 1 | 0 | X | X | $Q_0$ | $\overline{Q}_0$ | Inhibited |
| | 1 | 1 | 1 | X | X | $Q_0$ | $\overline{Q}_0$ | Inhibited |
| | 1 | 1 | ↑ | X | X | $Q_0$ | $\overline{Q}_0$ | Inhibited |

X = Don't care           ↓ = HIGH-to-LOW transition (negative edge)
$Q_0$ = Previous state of *Q*     ↑ = LOW-to-HIGH transition (positive edge)

If preset and clear functions are not used, they should be disabled by connecting them to logic HIGH (for active-LOW inputs). This prevents them from being activated inadvertently by circuit noise. The synchronous functions of some flip-flops, such as the 74LS76A, will not operate properly unless $\overline{PRE}$ and $\overline{CLR}$ are HIGH.

▼

**EXAMPLE 6.15**    One application of the $\overline{CLR}$ input of a JK flip-flop is to use it to reset a flip-flop circuit upon detection of a particular logic state.

Figure 6.53 shows a modification to the flip-flop circuit of Figure 6.49, using 74HC107 Dual JK Flip-Flops. (This flip-flop has a $\overline{CLR}$ input, but no $\overline{PRE}$ input.) Analyze the operation of the circuit and draw a timing diagram showing the waveforms for $CLK$, $Q_A$, $Q_B$, $Q_C$, and $\overline{CLR}$.

**Figure 6.53**

**Example 6.15**
**Use of $\overline{CLR}$ Inputs**

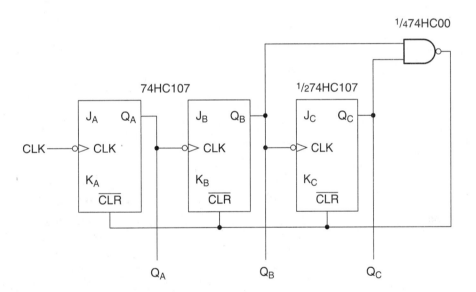

**Solution**    The NAND gate in Figure 6.53 will clear all flip-flops when its output is LOW. This occurs only when $Q_B = Q_C = 1$. As in Example 6.13, the circuit will count in binary sequence, beginning at $Q_C Q_B Q_A = 000$. When the output reaches the state $Q_C Q_B Q_A = 110$, the NAND inputs both go HIGH. This pulls all $\overline{CLR}$ inputs LOW, making $Q_C Q_B Q_A = 000$ after a short delay through the NAND gate and the flip-flops' $\overline{CLR}$ circuitry. (For the sake of simplicity, the delay is not shown in Figure 6.54.)

Since the outputs are now in the state $Q_C Q_B Q_A = 000$, the conditions required to make the $\overline{CLR}$ inputs LOW have disappeared and the outputs will progress to the next binary state, $Q_C Q_B Q_A = 001$, at the next $CLK$ pulse.

Without the NAND gate, the sequence of output states is from 000 to 111. The sequence of the circuit outputs has been truncated; the circuit now counts from 000 to 101. We don't count the state 110, since it is present just long enough to make the flip-flops clear—about 40 ns. Its only effect is to make a short spike on the $Q_B$ output—much shorter than the pulse shown in Figure 6.54.

**Figure 6.54**

**Example 6.15**
**Timing Diagram (Not to Scale)**

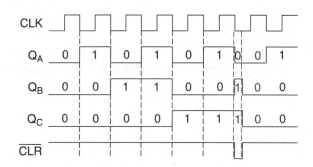

---

**Section Review Problem for Section 6.5**

6.2. Can the JK flip-flops in the circuit of Figure 6.49 be directly replaced by SR flip-flops? Why or why not?

# 6.6

# Timing Parameters

**Setup time** $(t_{su})$   *The time required for the synchronous inputs of a flip-flop to be stable before a CLK pulse is applied.*

**Hold time** $(t_h)$   *The time that the synchronous inputs of a flip-flop must remain stable after the active CLK transition is finished.*

**Pulse width** $(t_W)$   *Minimum time required for an active-level pulse applied to a CLK, $\overline{CLR}$, or $\overline{PRE}$ input, as measured from the midpoint of the leading edge of the pulse to the midpoint of the trailing edge.*

**Recovery time** $(t_{rec})$   *Minimum time from the midpoint of the trailing edge of a $\overline{CLR}$ or $\overline{PRE}$ pulse to the midpoint of an active CLK edge.*

**Propagation delay**   *The time required for the output of a digital circuit to change states after a change at one or more of its inputs.*

Flip-flops are electrical devices with inherent internal switching delays. As such, they have specific requirements for the timing of the input and output waveforms in order for them to operate reliably. Figure 6.55 shows some of the basic timing requirements of a JK flip-flop.

Figure 6.55a illustrates the definitions of **setup time** $(t_{su})$, **hold time** $(t_h)$, and **pulse width** $(t_W)$. The notation used for the "*J or K*" waveform indicates that the *J* or *K* input could be at either logic level and makes a transition to the opposite level at some point. The setup time is measured from the midpoint of the *J* or *K* transition to the midpoint of the active *CLK* edge. The logic level on the *J* or *K* input must be steady for at least this

**Figure 6.55**

**Timing Parameters of a JK Flip-Flop**

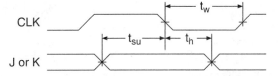
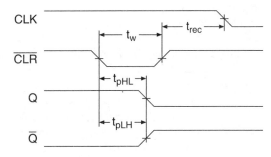

a. Setup, hold, and CLK pulse width

b. $\overline{CLR}$ pulse width, propagation delay, and recovery time

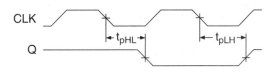

c. Propagation delay from CLK

time for the flip-flop to operate correctly. Setup time for both LSTTL and High-Speed CMOS flip-flops is about 20 ns.

Similarly, the hold time is measured from the midpoint of the *CLK* transition to the midpoint of the next *J* or *K* transition. The *J* or *K* level must be held steady for at least this time to ensure dependable operation. Hold time is 0 for LSTTL and 3 ns for a High-Speed CMOS flip-flop.

The pulse width, $t_w$, shows how long the *CLK* needs to be held LOW after an active *CLK* edge. Although the LOW level does not itself latch data into the flip-flop, internal logic levels must reach a steady state before the device can accept a new clock pulse. This minimum pulse width allows the necessary time for these internal transitions. The data sheet for a 74HC107 flip-flop (High-Speed CMOS) gives the clock pulse width as 16 ns; a data sheet for a 74LS107A device gives the value as 20 ns.

Figure 6.55b shows the pulse width required at the $\overline{CLR}$ input, the **propagation delay** from $\overline{CLR}$ to *Q* and $\overline{Q}$, and the **recovery time** that must be allowed from the end of a $\overline{CLR}$ pulse to the beginning of a *CLK* pulse. These times also apply to a pulse on the $\overline{PRE}$ input of a flip-flop.

Propagation delay is the result of internal electrical delays, primarily the charging and discharging of internal capacitances of the gate transistor junctions. The practical result of this is that a pulse at the $\overline{CLR}$ input makes *Q* go LOW, but not immediately; there is a delay of several nanoseconds between input pulse and output response.

Propagation delay is defined by the direction of the *output* transition. The delay at *Q*, which goes from HIGH to LOW, is called $t_{pHL}$. The delay at $\overline{Q}$, which goes from

LOW to HIGH when cleared, is called $t_{pLH}$. Values for propagation delay from $\overline{CLR}$ to $Q$ or $\overline{Q}$ are about 20 ns for LSTTL and 31 ns for High-Speed CMOS.

The recovery time, $t_{rec}$, allows the internal logic levels of the flip-flop to reach a steady state after a $\overline{CLR}$ pulse. When the internal levels are stable, the device is ready to accept an active $CLK$ edge. The recovery time for High-Speed CMOS is 20 ns and 25 ns for an LSTTL device. (The LSTTL data sheet treats this parameter as a species of setup time; it is shown as setup time after the $\overline{CLR}$ is inactive. Same thing.)

Finally, Figure 6.55c shows the propagation delay from $CLK$ to $Q$. This is the time from the midpoint of an active $CLK$ edge to the midpoint of a transition at $Q$ caused by that $CLK$ edge. The parameters are defined, as before, by the direction of the output transition. Propagation delays $t_{pLH}$ and $t_{pHL}$ are 20 ns, maximum, for a 74LS107A device and 25 ns for a 74HC107 flip-flop.

The timing restrictions of a flip-flop imply that there is a maximum $CLK$ frequency beyond which the device will not operate reliably. Data sheets give these values as about 30 MHz for both LSTTL and High-Speed CMOS devices.

Table 6.12 summarizes the timing parameters of a 74LS107A flip-flop and a 74HC107 device. The values for the latter device are for $V_{cc} = 4.5$ V and a temperature range of $-55°C$ to $25°C$; they increase with a higher temperature range or a lower supply voltage.

**Table 6.12** Timing Parameters of an LSTTL and a High-Speed CMOS Flip-Flop

| Symbol | Parameter | 74LS107A | 74HC107 |
|--------|-----------|----------|---------|
| $t_{su}$ | Setup time | 20 ns | 20 ns |
| $t_h$ | Hold time | 0 ns | 3 ns |
| $t_w$ | $\overline{CLR}$ pulse width | 25 ns | 16 ns |
| | $CLK$ pulse width | 20 ns | 16 ns |
| $t_{rec}$ | Recovery time | 25 ns | 20 ns |
| $t_{pHL,}$ | Propagation delay | | |
| $t_{pLH}$ | (from $\overline{CLR}$) | 20 ns | 31 ns |
| | (from $CLK$) | 20 ns | 25 ns |
| $f_{max}$ | Maximum frequency | 30 MHz | 30 MHz |

**EXAMPLE 6.16**

The timing diagrams in Figure 6.56 represent some of the timing parameters of a JK flip-flop. From these diagrams, determine the setup and hold times and the propagation delays from $CLK$ and $\overline{CLR}$ to $Q$ and $\overline{Q}$.

**Solution** The values are as follows:

Setup time = 15 ns

Hold time = 5 ns

Propagation delays (from $CLK$): 25 ns ($t_{pLH}$ and $t_{pHL}$)
(from $\overline{CLR}$): 20 ns ($t_{pLH}$ and $t_{pHL}$)

**Figure 6.56**

**Example 6.16**
**Timing Parameters**

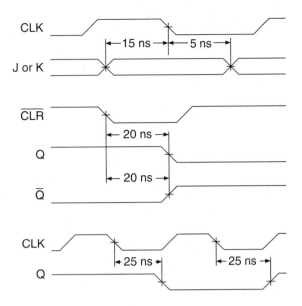

---

## 6.7

## Troubleshooting Latch Circuits

Latches and flip-flops can fail at the component level or the circuit level. Finding faults in either situation demands a good understanding of the normal functions of the circuit or device, applying tests to check the expected functions, and then analyzing the test results to find any reasons for failures.

**EXAMPLE 6.17**

The following behavior is observed at the terminals of a NAND latch, shown in Figure 6.57:

1. The latch will not remain in the set or reset state.

2. When $\overline{S} = 1$ and a pulse waveform is on the $\overline{R}$ input, $\overline{Q}$ pulses out of phase with $\overline{R}$, and $Q$ pulses in phase with $\overline{R}$.

3. When $\overline{R} = 1$, a pulse waveform on the $\overline{S}$ input makes $Q$ pulse out of phase with the waveform. $\overline{Q}$ is always HIGH.

What is the most likely fault in the latch circuit?

**Figure 6.57**

**Example 6.17**
**NAND Latch**

**Solution**

When troubleshooting, it always helps to draw a picture. Partial schematics of the latch, shown in Figure 6.58, are used to help us visualize the troubleshooting test results.

Recall the operation of the NAND latch: The outputs are changed by action at the $\overline{S}$ or $\overline{R}$ input. The latch maintains its new state by feedback from the output of one gate

**Figure 6.58**

**Example 6.17**

**Partial Schematics**

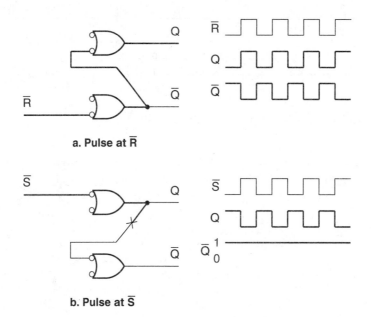

a. Pulse at $\overline{R}$

b. Pulse at $\overline{S}$

to the input of the opposite one. Each of the three tests described above gives information about the latch operation.

*Test 1:* The external inputs, $\overline{S}$ and $\overline{R}$, can make the outputs switch, but they don't stay switched. This means that the problem is in the feedback connections.

*Test 2:* The connection from $\overline{Q}$ to the opposite gate is intact, since the pulse waveform at $\overline{R}$ affects both outputs, as shown in Figure 6.58a.

*Test 3:* This test should produce out-of-phase waveforms at both outputs, but it does not. The likely fault is a broken connection between $Q$ and the opposite gate, as shown in Figure 6.58b, or an open input on that gate.

**EXAMPLE 6.18**

The following behavior is observed at the terminals of the NAND latch in Figure 6.59:

1. The latch reset function works normally.

2. The latch will not set when $\overline{S} = 0$.

3. The latch will set when a negative-going pulse is applied to terminal $a$ of the top gate.

What are the two most likely faults in the circuit?

**Figure 6.59**

**Example 6.18**

**NAND Latch**

**Solution**

Tests 1 and 3 indicate that the latch feedback is working correctly, since the latch can be made to stay in the set or reset condition.

Test 2 indicates that the *SET* pulse is not getting to the input of its corresponding gate. This implies that either the connection is open-circuited or the $\overline{S}$ input of the latch is stuck in the logic HIGH state. Either fault could be inside the chip or in the circuit connection.

**EXAMPLE 6.19**

Figure 6.60 shows a 74LS76A JK flip-flop connected so that it is permanently in toggle mode. When a series of *CLK* pulses is applied, the *Q* output remains in the logic HIGH state. List possible causes of this failure and briefly explain why they result in the permanent HIGH state. How could the fault be further narrowed down?

**Figure 6.60**

**Example 6.19**

**JK Flip-Flop in Toggle Mode**

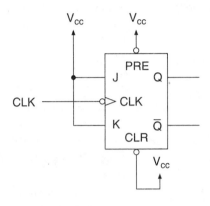

**Solution**    Possible causes:

1.  $\overline{PRE}$ is stuck LOW. This would make the set function of the output latch always active. We could test this assumption by making $\overline{CLR}$ LOW. If $\overline{PRE}$ is stuck LOW, both $Q$ and $\overline{Q}$ will go HIGH. $\overline{Q}$ will return to LOW when $\overline{CLR}$ returns to HIGH.

2.  $K$ is stuck LOW. This results in the input condition $J = 1$, $K = 0$, the synchronous set condition. If this is the fault, you should be able to make $Q = 0$ by pulling $\overline{CLR}$ LOW. $Q$ will go HIGH again on the next *CLK* pulse.

---

## ●GLOSSARY

**Asynchronous inputs**  The inputs of a flip-flop that change the flip-flop's $Q$ outputs immediately, without waiting for a pulse at the *CLK* input. Examples include preset and clear inputs.

**Clear**   An asynchronous reset function.

*CLOCK*   An enabling input to a sequential circuit that is sensitive to the positive- or negative-going edge of a waveform.

**Edge**   The HIGH-to-LOW (negative edge) or LOW-to-HIGH (positive edge) transition of a pulse waveform.

**Edge detector**  A circuit in an edge-triggered flip-flop that converts the active edge of a *CLOCK* input to an active-level pulse at the internal latch's *SET* and *RESET* inputs.

**Edge-sensitive**  Edge-triggered.

**Edge-triggered**   Enabled by the positive or negative edge of a digital waveform.

**Flip-flop**  A sequential circuit based on a latch whose output changes when its *CLOCK* input receives either an edge or a pulse, depending on the device.

**Gated SR latch**  An SR latch whose ability to change states is controlled by an extra input called the *ENABLE* input.

**High-impedance state**  The output state of a tristate buffer that is neither logic HIGH nor logic LOW, but resembles an open circuit.

**Hold time**  $(t_h)$  The time that the synchronous inputs of a flip-flop must remain stable after the active *CLK* transition is finished.

**Latch** A sequential circuit with two inputs called *SET* and *RESET,* which make the latch store a logic 0 (reset) or 1 (set) until actively changed.

**Level-sensitive** Enabled by a logic HIGH or LOW level.

**Parallel transmission** A technique of sending multibit data from one device to another where all bits are sent simultaneously over individual parallel lines.

**Preset** An asynchronous set function.

**Propagation delay** The time required for the output of a digital circuit to change states after a change at one or more of its inputs.

**Pulse width** $(t_w)$ Minimum time required for an active-level pulse applied to a *CLK,* $\overline{CLR}$, or $\overline{PRE}$ input, as measured from the midpoint of the leading edge of the pulse to the midpoint of the trailing edge.

**Recovery time** $(t_{rec})$ Minimum time from the midpoint of the trailing edge of a $\overline{CLR}$ or $\overline{PRE}$ pulse to the midpoint of an active *CLK* edge.

**Reset** 1. The stored LOW state of a latch circuit. 2. A latch input that makes the latch store a logic 0.

**Sequential circuit** A digital circuit whose output depends not only on the present combination of inputs, but also on the history of the circuit.

**Serial transmission** A method of sending multibit data between two devices where the bits are sent one after the other along a single line.

**Set** 1. The stored HIGH state of a latch circuit. 2. A latch input that makes the latch store a logic 1.

**Setup time** $(t_{su})$ The time required for the synchronous inputs of a flip-flop to be stable before a *CLK* pulse is applied.

**Steering gates** Logic gates, controlled by the *ENABLE* input of a gated latch, that steer a *SET* or *RESET* pulse to the correct input of an SR latch circuit.

**Synchronous inputs** The inputs of a flip-flop that do not affect the flip-flop's *Q* outputs unless a clock pulse is applied. Examples include *D, J,* and *K* inputs.

**Toggle** Change to the opposite binary state with each applied clock pulse.

**Transparent latch (gated D latch)** A latch whose output follows its data input when its *ENABLE* input is active.

**Tristate buffer** A gate having three possible output states: logic HIGH, logic LOW, and a high-impedance state.

# ● PROBLEMS

### Section 6.1 Latches

**6.1** Complete the timing diagram in Figure 6.61 for the active-HIGH latch shown. The latch is initially set.

**6.2** Repeat Problem 6.1 for the timing diagram shown in Figure 6.62.

**6.3** Complete the timing diagram in Figure 6.63 (see page 225) for the active-LOW latch shown.

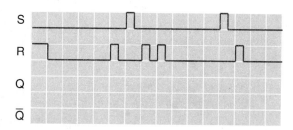

**Figure 6.62**

**Problem 6.2**
**Timing Diagram**

**Figure 6.61**

**Problem 6.1**
**Timing Diagram**

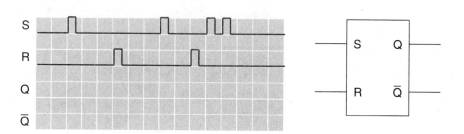

**Figure 6.63**

**Problem 6.3**

**Timing Diagram**

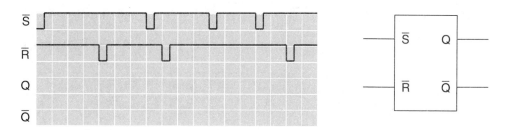

**Figure 6.64**

**Problem 6.4**

**Latch for Motor Starter**

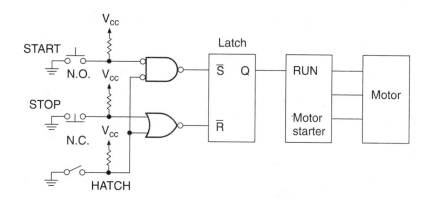

**\*6.4**  Figure 6.64 shows an active-LOW latch used to control a motor starter. The motor runs when $Q = 1$ and stops when $Q = 0$.

   The motor is housed in a safety enclosure that has an access hatch for service. A safety interlock prevents the motor from running when the hatch is open. The *HATCH* switch opens when the hatch opens, supplying a logic HIGH to the circuit. The *START* switch is a Normally Open momentary-contact pushbutton (LOW when pressed). The *STOP* switch is a Normally Closed momentary-contact pushbutton (HIGH when pressed).

   Draw the timing diagram of the circuit, showing *START, STOP, HATCH,* $\overline{S}$, $\overline{R}$, and $Q$ for the following sequence of events:

**a.** *START* is pressed and released.

**b.** The hatch cover is opened.

**c.** *START* is pressed and released.

**d.** The hatch cover is closed.

**e.** *START* is pressed and released.

**f.** *STOP* is pressed and released.

Briefly describe the functions of the three switches and how they affect the motor operation.

## Section 6.2 NAND/NOR Latches

**6.5**  Draw a NAND latch, correctly labeling the inputs and outputs. Describe the operation of a NAND latch for all four possible combinations of $\overline{S}$ and $\overline{R}$.

**6.6**  The timing diagram in Figure 6.65 (see page 226) shows the input waveforms of a NAND latch. Complete the diagram by showing the output waveforms.

**6.7**  Figure 6.66 (see page 226) shows the input waveforms to a NOR latch. Draw the corresponding output waveforms.

**\*6.8**  Figure 6.67 (see page 226) represents two input waveforms to a latch circuit.

**a.** Draw the outputs $Q$ and $\overline{Q}$ if the latch is a NAND latch.

**b.** Draw the output waveforms if the latch is a NOR latch.

(Note that in each case, the waveforms will produce the forbidden state at some point. Even under this condition, it is still possible to produce unambiguous output waveforms. Refer to Figure 6.10 for guidance.)

Figure 6.65
Problem 6.6
Timing Diagram

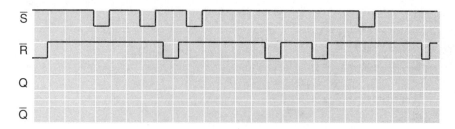

Figure 6.66
Problem 6.7
Input Waveforms to a NOR
Latch

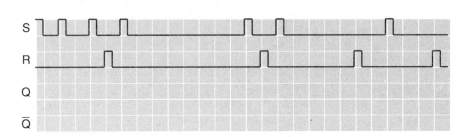

Figure 6.67
Problem 6.8
Input Waveforms to a Latch

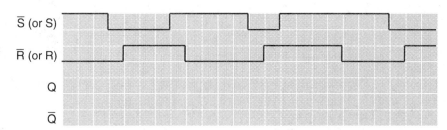

*6.9  a. Draw a timing diagram for a NAND latch showing each of the following sequences of events:
      i. $\overline{S}$ and $\overline{R}$ are both LOW; $\overline{S}$ goes HIGH before $\overline{R}$.
      ii. $\overline{S}$ and $\overline{R}$ are both LOW; $\overline{R}$ goes HIGH before $\overline{S}$.
      iii. $\overline{S}$ and $\overline{R}$ are both LOW; $\overline{S}$ and $\overline{R}$ go HIGH simultaneously.
   b. State why $\overline{S} = \overline{R} = 0$ is a forbidden state for the NAND latch.
   c. Briefly explain what the final result is for each of the above transitions.

6.10  Figure 6.68 shows the effect of mechanical bounce on the switching waveforms of a single-pole double throw (SPDT) switch.
   a. Briefly explain how this effect arises.
   b. Draw a NAND latch circuit that can be used to eliminate this mechanical bounce, and briefly explain how it does so.

## Section 6.3 Gated Latches

6.11  Complete the timing diagram for the gated latch shown in Figure 6.69.

6.12  Complete the timing diagram for the gated latch shown in Figure 6.70.

*6.13  A pump motor can be started at two different locations with momentary-contact pushbuttons $S_1$ and $S_2$. It can be stopped by momentary-contact pushbuttons $ST_1$ and $ST_2$. As in Problem 6.4, a RUN input on the motor controller must be kept HIGH to keep the motor running. After the motor is stopped, a timer prevents the motor from starting for 5 minutes.
       Draw a circuit block diagram showing how an SR latch and some additional gating logic can be used in such an application. The timer can be shown as a block activated by the STOP function. Assume that the timer output goes HIGH for 5 minutes when activated.

**Figure 6.68**
**Problem 6.10**
**Effect of Mechanical Bounce on a SPDT Switch**

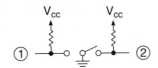

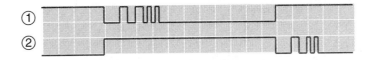

**Figure 6.69**
**Problem 6.11**
**Gated Latch**

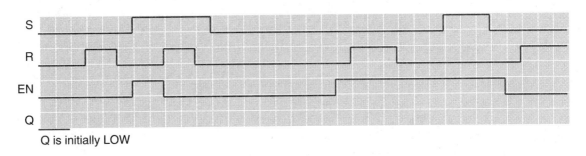

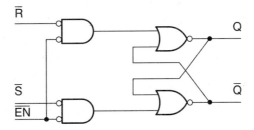

Q is initially LOW

**Figure 6.70**
**Problem 6.12**
**Gated Latch**

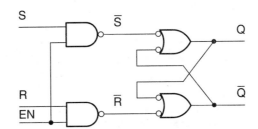

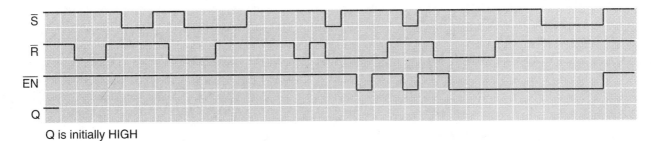

Q is initially HIGH

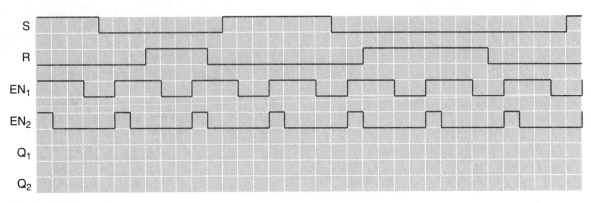

**Figure 6.71**

**Problem 6.14**
**Waveforms**

**6.14** The $S$ and $R$ waveforms in Figure 6.71 are applied to two different gated latches. The *ENABLE* waveforms for the latches are shown as $EN_1$ and $EN_2$. Draw the output waveforms $Q_1$ and $Q_2$, assuming that $S$, $R$, and $EN$ are all active HIGH. Which output is least prone to synchronization errors? Why?

**6.15** Figure 6.72 represents the waveforms of the $EN$ and $D$ inputs of a 74HC75 4-bit transparent latch. Complete the timing diagram by drawing the waveforms for $Q_1$ to $Q_4$.

**\*6.16** An electronic direction finder aboard an aircraft uses a 4-bit number to distinguish 16 different compass points as follows:

| Direction | Degrees | Binary |
|-----------|---------|--------|
| N | 0/360 | 0000 |
| NNE | 22.5 | 0001 |
| NE | 45 | 0010 |
| ENE | 67.5 | 0011 |
| E | 90 | 0100 |
| ESE | 112.5 | 0101 |
| SE | 135 | 0110 |
| SSE | 157.5 | 0111 |
| S | 180 | 1000 |
| SSW | 202.5 | 1001 |
| SW | 225 | 1010 |
| WSW | 247.5 | 1011 |
| W | 270 | 1100 |
| WNW | 295.5 | 1101 |
| NW | 315 | 1110 |
| NNW | 337.5 | 1111 |

The output of the direction finder is stored in a 4-bit latch so that the aircraft flight path can be logged by a computer. The latch is periodically updated by a continuous pulse on the latch enable line.

Figure 6.73 (see page 230) shows a sample reading of the direction finder's output as presented to the latch.

**a.** Complete the timing diagram by filling in the data for the $Q$ outputs.

**b.** Based on the completed timing diagram of Figure 6.73, make a rough sketch of the aircraft's flight path for the monitored time.

## Section 6.4 Edge-Triggered SR and D Flip-Flops

**6.17** The waveforms in Figure 6.74 (see page 230) are applied to the inputs of a positive edge-triggered SR flip-flop. Complete the timing diagram by drawing the waveform for the $Q$ output. $Q$ is initially LOW.

**6.18** Repeat Problem 6.17 for the waveforms in Figure 6.75 (see page 231).

**6.19** The waveforms in Figure 6.76 (see page 231) are applied to the inputs of a positive edge-triggered SR flip-flop and a gated SR latch. Complete the timing diagram where $Q_1$ is the output of the flip-flop and $Q_2$ is the output of the gated latch. Account for any differences between the $Q_1$ and $Q_2$ waveforms.

**Figure 6.72**
**Problem 6.15**
**Waveforms**

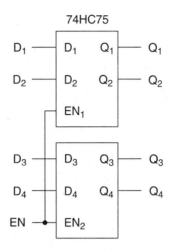

74HC75

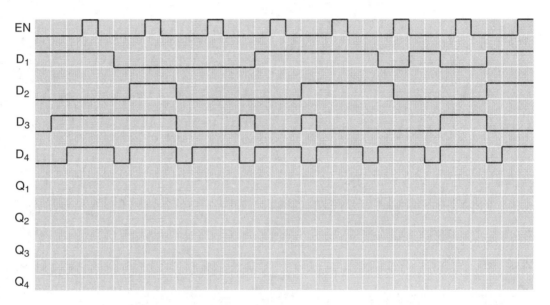

**6.20** Complete the timing diagram for a positive edge-triggered D flip-flop if the waveforms shown in Figure 6.77 (see page 231) are applied to the flip-flop inputs.

**6.21** Repeat Problem 6.20 for the waveforms shown in Figure 6.78 (see page 231).

**6.22** Repeat Problem 6.20 for the waveforms shown in Figure 6.79 (see page 232).

**\*6.23** Draw a logic diagram of a D flip-flop configured for toggle mode. (Hint: The $D$ input must always be the opposite of the $Q$ output.)

**6.24** Briefly describe the difference between serial and parallel transmission.

**\*6.25** The ASCII character "&" is transmitted into an 8-bit serial shift register, shown in Figure 6.36. Draw a timing diagram like the one developed in Example 6.9 showing how the ASCII character is clocked into the circuit in eight pulses. Assume that all flip-flops are initially set. (Refer to the ASCII table in Table 5.6.)

**\*6.26** Repeat Problem 6.25 for ASCII "r".

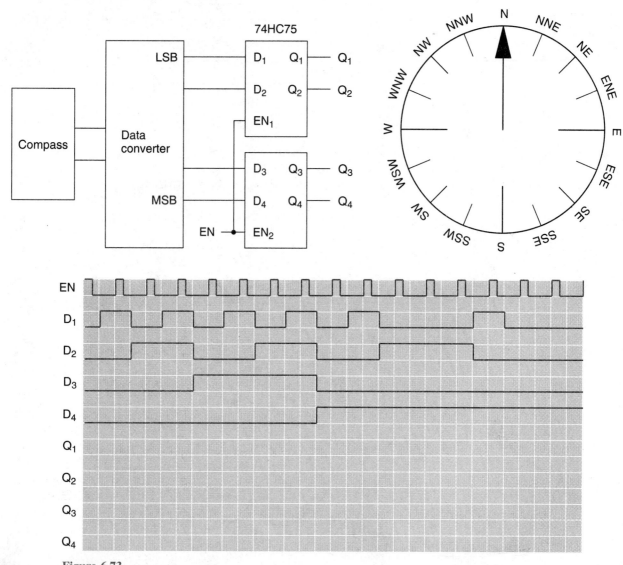

**Figure 6.73**

**Problem 6.16**
**Sample Reading of Direction Finder's Output**

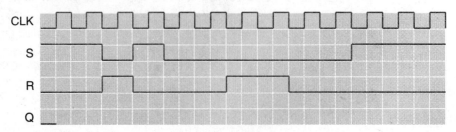

**Figure 6.74**

**Problem 6.17**
**Waveforms**

Figure 6.75
**Problem 6.18**
**Waveforms**

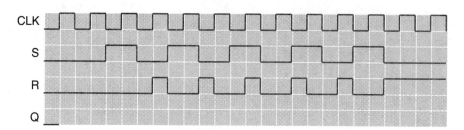

Figure 6.76
**Problem 6.19**
**Waveforms**

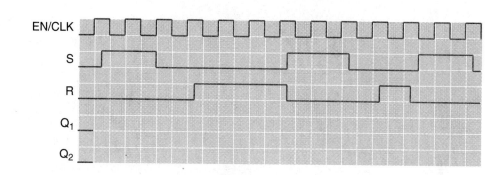

Figure 6.77
**Problem 6.20**
**Waveforms**

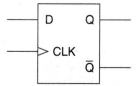

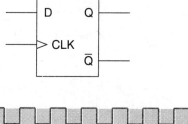

Figure 6.78
**Problem 6.21**
**Waveforms**

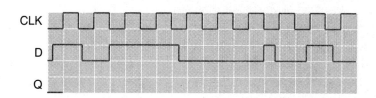

**Figure 6.79**

**Problem 6.22**

**Waveforms**

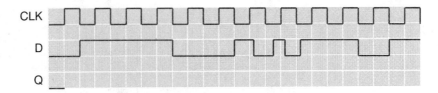

**Figure 6.80**

**Problem 6.27**

**Waveforms**

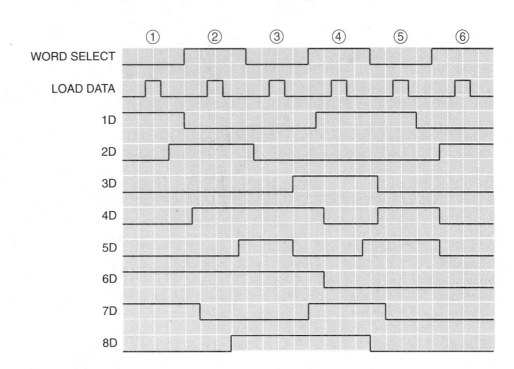

**6.27** The waveforms shown in Figure 6.80 are applied to the circuit in Figure 6.39a. Write a table showing the contents of latch *A* and latch *B* at each numbered point on Figure 6.80 (i.e., the center of each *LOAD DATA* pulse). What is the exact point on the timing diagram where new data are latched into latch *A* or *B?*

**6.28** The waveforms illustrated in Figure 6.81 are applied to the circuit in Figure 6.39b. Write a table showing the contents of flip-flop *A* and flip-flop *B* at each numbered point in Figure 6.81 (i.e., the center of each *LOAD DATA* pulse). What is the exact point on the timing diagram where new data are clocked into flip-flop *A* or *B?*

**6.29** The waveforms illustrated in Figure 6.82 are applied to the 4-bit parallel transfer circuit of Figure 6.43. The four input words are:

Word A = 0011
Word B = 1010
Word C = 1110
Word D = 1111

**a.** Complete the following table:

| Step | Action | Source | Destination | Output |
|------|--------|--------|-------------|--------|
| 1 | Transfer | *D* | *X1* | ——— |
| 2 | Transfer | *B* | *X2* | ——— |
| 3 | Display | – | — | 1111 1010 |
| 4 | | | | |
| 5 | | | | |
| 6 | | | | |
| 7 | | | | |
| 8 | | | | |
| 9 | | | | |

**b.** Use a pen, pencil, or marker on a copy of Figure 6.43 to highlight the path used to transfer the parallel data in step 5 in the timing diagram in Figure 6.82.

**Figure 6.81**
**Problem 6.28**
**Waveforms**

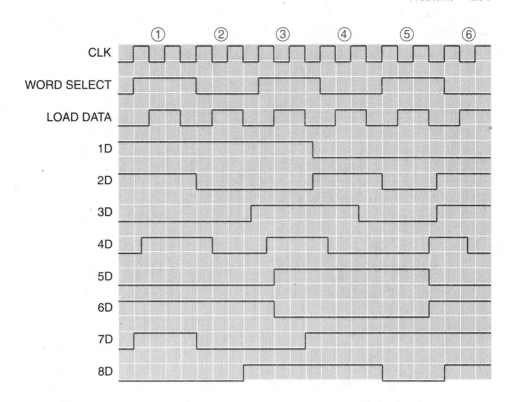

**Figure 6.82**
**Problem 6.29**
**Waveforms**

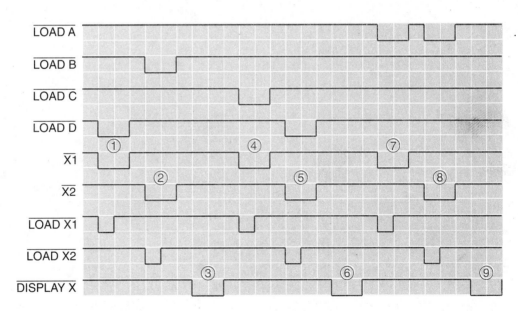

Figure 6.83

Problem 6.30
Waveforms

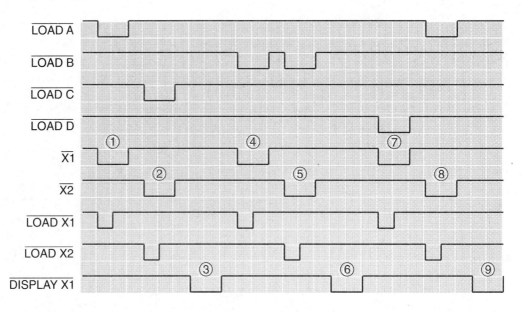

Figure 6.84

Problem 6.32
Waveforms

Figure 6.85

Problem 6.33
Waveforms

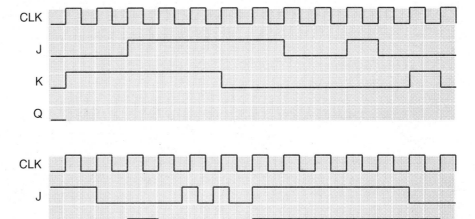

**6.30** Repeat Problem 6.29 for the following input data and the waveforms shown in Figure 6.83.

Word A = 0001
Word B = 0110
Word C = 1011
Word D = 1100

## Section 6.5 Edge-Triggered JK Flip-Flops

**6.31** State the major functional difference between the SR flip-flop and the JK flip-flop.

**6.32** The waveforms in Figure 6.84 are applied to a negative edge-triggered JK flip-flop. Complete the timing diagram by drawing the $Q$ waveform.

**6.33** Repeat Problem 6.32 for the waveforms in Figure 6.85.

**6.34** Given the inputs $x$, $y$, and $z$ to the circuit in Figure 6.86, draw the waveform for output $Q$.

**6.35** Assume that all flip-flops in Figure 6.87 are initially set. Draw a timing diagram showing the $CLK$, $Q_A$, $Q_B$, and $Q_C$ waveforms when eight clock pulses are applied. Make a table showing

**Figure 6.86**
**Problem 6.34**
**Inputs to Circuit**

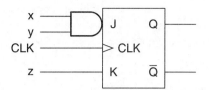

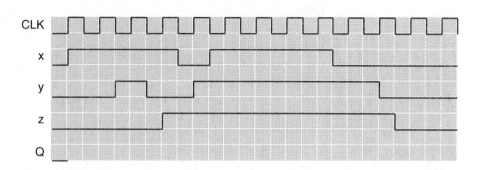

**Figure 6.87**
**Problem 6.35**
**Flip-Flops**

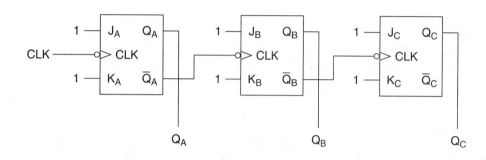

**Figure 6.88**
**Problem 6.36**
**Waveforms**

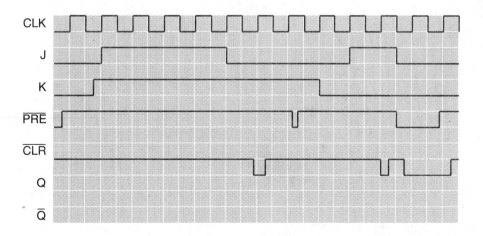

each combination of $Q_C$, $Q_B$, and $Q_A$. What pattern do the outputs form over the period shown on the timing diagram?

**6.36** The waveforms shown in Figure 6.88 are applied to a 74LS76A negative edge-triggered JK flip-flop. The flip-flop's Preset and Clear inputs are active LOW. Complete the timing diagram by drawing the output waveforms.

**6.37** Repeat Problem 6.36 for the waveforms in Figure 6.89.

**Figure 6.89**
**Problem 6.37**
**Waveforms**

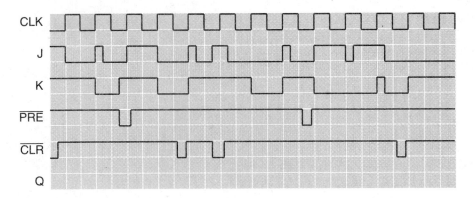

**Figure 6.90**

**Problem 6.38**

**Circuit**

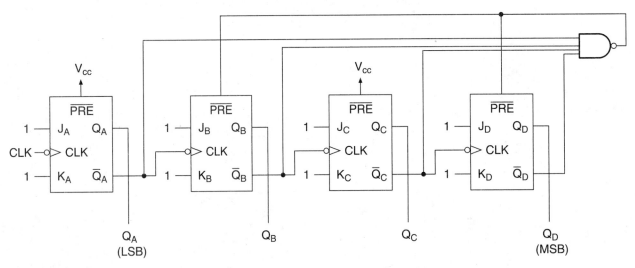

**\*\*6.38** Analyze the operation of the circuit in Figure 6.90 by drawing a timing diagram showing the waveforms for *CLK*, $Q_A$, $Q_B$, $Q_C$, $Q_D$, and $\overline{PRE}$ for a period of 11 clock cycles. Assume that the initial state of the circuit is $Q_D Q_C Q_B Q_A = 1010$. Make a table showing the state $Q_D Q_C Q_B Q_A$ after each clock pulse. What pattern emerges from the data?

## Section 6.6 Timing Parameters

**6.39** Use a TTL or High-Speed CMOS data book, as appropriate, to look up the setup and hold times of the following devices:

**a.** 74LS74A

**b.** 74HC76

**c.** 74LS76A

**d.** 74LS107A

**e.** 74ALS112A

**f.** 74HC112

**6.40** Draw a timing diagram showing the setup and hold times for a 74LS76A flip-flop.

**6.41** Draw timing diagrams (to scale) showing setup and hold times, minimum *CLK* and $\overline{CLR}$ pulse widths, recovery time, and propagation delay

**Figure 6.91**

**Problem 6.42**
**Timing Parameters**

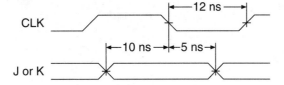

**Figure 6.92**

**Problem 6.43**
**Timing Diagram**

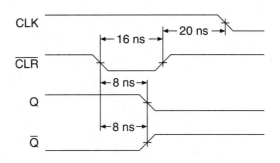

times from *CLK* and $\overline{CLR}$ for both 74LS107A and 74HC107 flip-flops.

**6.42** Write names and values of the JK flip-flop timing parameters illustrated in Figure 6.91.

**6.43** Repeat Problem 6.42 for the timing diagram in Figure 6.92.

### Section 6.7 Troubleshooting Latch Circuits

**6.44** The output of a NOR latch will set properly, but it will not respond to a pulse at its *RESET* input. What are the most likely faults with the circuit? How can you test for these faults?

**6.45** The outputs of a 74HC374 Octal D Flip-Flop are all cleared. The *D* inputs are all made HIGH, and

a *CLK* pulse is applied. None of the outputs change. What are the most likely faults? Explain how you could test for these faults.

**6.46** The same conditions are applied to a 74HC374 Flip-Flop as in Problem 6.45. In this case, all outputs except *5Q* go HIGH. What are the most likely faults? How would you test your reasoning?

**6.47** The outputs of a 74LS76A JK Flip-Flop will set and reset correctly when changed synchronously. However, neither preset nor clear inputs will make the output change. What is the most likely fault?

## ▼ Answers to Section Review Questions

### Section 6.2

**6.1.** The NOR latch has active HIGH inputs. If both inputs are HIGH, you are attempting to activate set and reset functions at the same time, which is contradictory.

The NAND latch has active-LOW inputs. When both inputs are HIGH, you are not activating either set or reset, so the output will not change.

### Section 6.5

**6.2.** No direct replacement is possible, since the circuit uses the toggle state of the JK flip-flops, which is not available in the SR flip-flop.

# Multivibrators

## CHAPTER OUTLINE

**7.1** Multivibrators

**7.2** IC Monostables

**7.3** The 555 Timer

**7.4** The 555 Timer as a Monostable Multivibrator

**7.5** The 555 Timer as an Astable Multivibrator

## CHAPTER OBJECTIVES

Upon successful completion of this chapter, you will be able to:

- Describe the operation of astable, monostable, and bistable multivibrators.

- Draw timing diagrams for retriggerable and nonretriggerable monostable multivibrators.

- Calculate values of external components for circuits using monostable ICs, such as the 4538B Dual Monostable Multivibrator.

- Describe the operation of the 555 timer IC and how it can be connected as an astable or a monostable multivibrator.

- Calculate values of external components for 555 timer circuits.

## INTRODUCTION

The latches and flip-flops studied in Chapter 6 are examples of a larger class of sequential circuits called multivibrators. The output states of multivibrators are functions of their previous history and can be changed either by external control signals or internal design, depending on the specific type of circuit.

The bistable multivibrator (i.e., a latch or flip-flop) derives its name from the fact that its output has two possible states, both of which are stable. There also exist two other types of multivibrator circuits: the monostable multivibrator, whose output has one stable state, and the astable multivibrator, whose output has no stable state. These circuits are used as generators of single pulses (monostable) or as digital oscillators (astable).

## 7.1

# Multivibrators

> **Multivibrator**   *A sequential digital circuit whose output oscillates between logic HIGH and LOW states, either automatically due to the circuit design or under the control of a special input (e.g., SET, RESET, TRIGGER).*
>
> **Stable state**   *An output state of a multivibrator that will not change unless actively changed by an external signal.*
>
> **Unstable state**   *An output state of a multivibrator that will change to the opposite state without requiring a control signal to make it change.*
>
> **Quasistable state**   *A temporarily stable state of a multivibrator output.*

A **multivibrator** is a digital circuit whose output depends on the previous history of the circuit. Depending on the specific type of circuit, a multivibrator can store a specific logic level, produce a single pulse, or generate a continuous series of pulses.

Since they are binary digital circuits, multivibrators have two output states, HIGH and LOW, each of which can be stable, unstable, or quasistable.

A **stable state** has no tendency to change on its own. It requires a specific control signal such as *SET* or *RESET* to make it change.

An **unstable state** will change automatically to the opposite state without a control signal.

A **quasistable state** is a temporarily stable state that will become unstable and revert to the opposite state after a specified time.

Multivibrator circuits are classified by the number of stable states they have (2, 1, or 0), leading to the designations bistable (2), monostable (1), and astable (0).

## Bistable Multivibrators

> **Bistable multivibrator**   *A multivibrator with two stable output states (e.g., a latch or flip-flop).*

Both the HIGH and LOW states of a **bistable multivibrator** are stable. The output changes state only when actively switched by an input such as *SET* or *RESET.* This class of circuits includes all the latches and flip-flops we studied in Chapter 6.

## Monostable Multivibrators

> **Monostable multivibrator (one-shot multivibrator)**   *A multivibrator with one stable state and one quasistable state whose output produces a pulse of fixed length.*
>
> **Trigger**   *The control input of a monostable multivibrator that initiates a timed pulse at the circuit's output.*

> **Nonretriggerable** *A monostable multivibrator in which new trigger pulses are ignored if the output is producing a pulse.*
>
> **Retriggerable** *A monostable multivibrator in which a new trigger pulse restarts the timing of an output pulse, thus increasing the output pulse width.*
>
> **Time out** *Finish a timing cycle.*

A **monostable multivibrator** has one stable state. It can be pushed into a quasistable state by a pulse on a control input called the **trigger.** After a fixed time, the output reverts to the stable state, thus producing a single pulse of specified width.

Monostables can be further classified as **retriggerable** and **nonretriggerable,** depending on how the device responds to new trigger pulses when the output is pulsing. A nonretriggerable monostable simply ignores the new trigger pulses; the output pulse width is always the same. A retriggerable device stretches the output pulse by restarting its timing from when the last trigger pulse was applied.

Figure 7.1 shows the IEEE symbols and functions of a retriggerable and a nonretriggerable monostable multivibrator. The normal rest state (stable state) of each monostable in Figure 7.1 is LOW. The trigger inputs are also active LOW. When a LOW pulse is applied to the trigger input of either circuit, its output goes HIGH (quasistable state) for a predetermined period, after which it returns to the LOW state.

**Figure 7.1**
**Monostable Multivibrator**

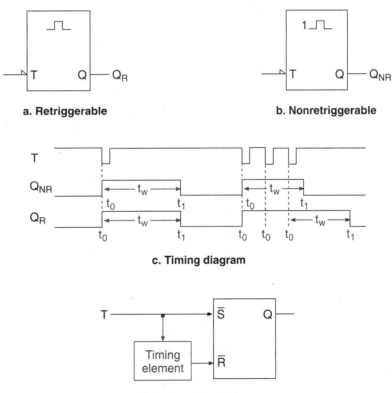

a. Retriggerable

b. Nonretriggerable

c. Timing diagram

d. Functional block diagram

Figure 7.1c demonstrates the difference between the outputs of a retriggerable and a nonretriggerable circuit for two cases:

1. Time between trigger pulses is greater than output pulse width, $t_w$
2. Time between trigger pulses is less than output pulse width, $t_w$

For case 1, the waveforms for nonretriggerable $(Q_{NR})$ and retriggerable $(Q_R)$ monostables are identical. Immediately after the trigger input is pulsed, the output goes HIGH and stays HIGH until time $t_w$ has elapsed. At this point the monostable **times out** and the output reverts to its stable state.

The waveforms for case 2 are different for the two types of devices. The nonretriggerable device *ignores* any further pulses on its trigger if the output is generating a pulse. The output must return to its stable state before the device accepts any new trigger pulses.

The retriggerable device *accepts* new trigger pulses; if the trigger receives a pulse during an output pulse, the timing is restarted from that point.

$Q_{NR}$ times out one pulse width after the *first* trigger pulse. $Q_R$ times out one pulse width after the *last* trigger pulse.

Figure 7.1d shows the functional block diagram of a monostable multivibrator. The trigger acts as a *SET* or *RESET* input to a latch (*SET* in this case) and also activates a timing element such as a resistor/capacitor *(RC)* circuit. After a specified time delay, usually the time it takes a capacitor to charge to a specified voltage, the timing element resets the latch.

## Astable Multivibrators

> **Astable multivibrator**   *A multivibrator with no stable states that acts as a digital oscillator.*

Since the output of an **astable multivibrator** has no stable states, it continually oscillates between HIGH and LOW. Figure 7.2a and b shows the IEEE symbol and the out-

**Figure 7.2**
**Astable Multivibrator**

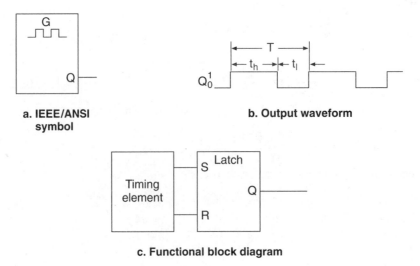

a. IEEE/ANSI
   symbol

b. Output waveform

c. Functional block diagram

put waveform of the astable multivibrator. The astable has no control inputs, only a $Q$ output, which oscillates with a fixed period and duty cycle.

Figure 7.2c shows the function of a common type of astable in which the output of a latch is alternately set and reset. The length of time the output is HIGH or LOW is determined by a timing element, such as a charging or discharging capacitor. We will examine such a circuit in some detail in a later section.

## 7.2

## IC Monostables

A wide variety of integrated circuits have been designed specifically as monostable multivibrators. Some available devices are the TTL 74121/221, 74122, and 74130 devices and the CMOS 4538B/74HC4538 devices. Of these, all but the '121/221 devices are retriggerable.

Retriggerable monostables tend to be more popular than nonretriggerable ones for several reasons: (1) a retriggerable monostable can be configured as a nonretriggerable device by using appropriate feedback; (2) output pulse width, $t_w$, can be extended by pulsing a trigger input before it has timed out; (3) many retriggerable devices are equipped with an external *RESET* input that can be used to shorten the normal pulse width. These features give the designer complete control of output timing.

Figure 7.3 shows the logic circuit and the IEEE/ANSI standard symbol for the High-Speed CMOS 74HC4538 Dual Monostable Multivibrator.

The 74HC4538 monostable is edge-triggered and can be activated by a positive edge at input $A$ or by a negative edge at input $B$. The pulse can be extended by further trigger pulses at $A$ or $B$ or shortened by a LOW at the $\overline{RESET}$ input.

If you wish to use only one of the trigger inputs, you must disable the other one by tying it to its inactive logic level. Since $A$ and $B$ have opposite logic levels, this gives the

**Figure 7.3**

**74HC4538 Dual Monostable Multivibrator**

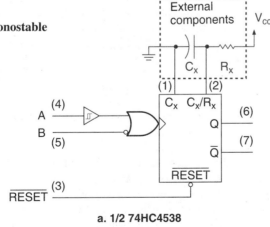

a. 1/2 74HC4538

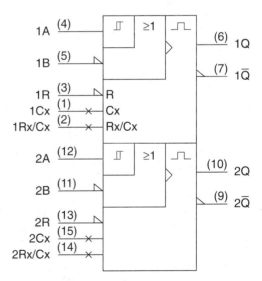

b. IEEE/ANSI symbol for 74HC4538

**Figure 7.4**

**74HC4538 as a Retriggerable Monostable Multivibrator**

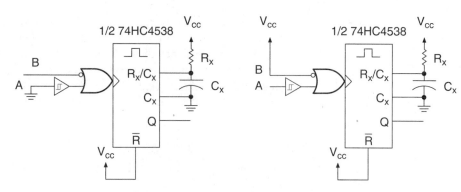

a. Negative edge-triggered          b. Positive edge-triggered

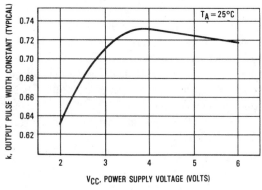

**Figure 7.5**

**Graphical Calculation of Pulse Width for a 74HC4538 Monostable Multivibrator**
(Reprinted with permission of Motorola)

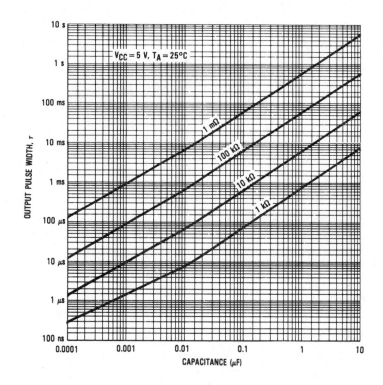

option of making the circuit positive or negative edge-triggered, as illustrated in Figure 7.4.

The IEEE/ANSI symbol shows an "X" on the $R_x/C_x$ input and the $C_x$ input, indicating that these are nonlogic connections. The timing components are connected to these inputs as shown in Figure 7.3a. The output pulse width is a linear function of the $RC$ time constant of the external timing components. That is, $t_w = kR_xC_x$, where constant $k$ depends on the device supply voltage, as represented by the graph in Figure 7.5a.

The monostable monitors the capacitor voltage to set the pulse width, which is the time required to charge from an internally set lower reference voltage to an upper reference voltage.

**EXAMPLE 7.1**

A 74HC4538 monostable multivibrator is connected as a negative edge-triggered device, as in Figure 7.4a. $C_x = 4.7 \ \mu F$, $V_{CC} = +5$ V.

    **a.** Calculate the value of $R$ required to make the circuit output generate a 25-ms pulse.

    **b.** Choose the standard resistor value closest to that calculated in part a.

        The first two significant figures of a standard resistor value can be any of the following:

| 10 | 18 | 33 | 56 | 91 |
|----|----|----|----|----|
| 12 | 22 | 39 | 68 |    |
| 15 | 27 | 47 | 82 |    |

    **c.** Recalculate the *actual* pulse width, $t_w$, based on the standard value. Calculate the percent error of the actual pulse width with respect to the design value.

**Solution**

    **a.** From the graph in Figure 7.5a, for $V_{CC} = +5$ V, $k \approx 0.72$:

$$R_x = t_w/kC_x = 25 \text{ ms}/(0.72)(4.7 \ \mu F) = 7.39 \text{ k}\Omega$$

    **b.** The closest standard resistor value is 6.8 k$\Omega$.

    **c.** Based on the value chosen above, the actual pulse width is given by:

$$t_w = kR_xC_x = (0.72)(6.8 \text{ k}\Omega)(4.7 \ \mu F)$$
$$= 23 \text{ ms}$$

This represents a percent error of:

$$\% \text{ error} = \frac{\text{actual} - \text{design}}{\text{design}} \times 100\%$$

$$= \frac{23 \text{ ms} - 25 \text{ ms}}{25 \text{ ms}} \times 100\% = -8\%$$

(Since standard resistors are generally accurate to within $\pm 5\%$, the error due to component selection is within the same order of magnitude as the error due to component tolerance.)

The graphs in Figure 7.5a and b are taken from a manufacturer's data sheet. The logarithmic graph in Figure 7.5b allows us to select a timing capacitor for a 74HC4538 monostable, given a specified resistor value and pulse width. Alternatively, we can use it to calculate output pulse width from known timing components. (The graph is logarithmic because ratios of 10:1 are shown as equal distances on each axis. The divisions between 1 ms and 10 ms are 1, 1.5, 2, 3, 4, 5, 6, 7, 8, 9, 10. The same proportions apply throughout the graph.)

When we select either a capacitor or a pulse width, we assume that the resistor is one of the values indicated on the heavy lines and select a combination of capacitance and pulse width that intersects the resistor value. (The line indicating "1 m$\Omega$" should read "1 M$\Omega$".)

**EXAMPLE 7.2**

Use the graph in Figure 7.5b to calculate the output pulse width of a 74HC4538 monostable if the timing resistor is 1 kΩ and the timing capacitor is 2 μF. What is the pulse width for $R = 10$ kΩ? For $R = 100$ kΩ?

**Solution**

The line for 1 kΩ intersects the 2-μF line (two divisions to the right of 1 μF) at 1.5 ms (one division above the 1-ms line). $t_w = 1.5$ ms.

For $R = 10$ kΩ, the graph yields a pulse width of about 12 ms. For $R = 100$ kΩ, $t_w \approx 100$ ms. (This implies that the pulse width function is not quite linear, since a linear function would yield pulse widths of 1.5 ms, 15 ms, and 150 ms.)

**EXAMPLE 7.3**

Draw a circuit showing how the 74HC4538 monostable can be configured as a positive edge-triggered nonretriggerable device. Briefly explain how the circuit inhibits trigger pulses when the output is active.

**Solution**

Figure 7.6 illustrates how to connect the 74HC4538 device as a positive edge-triggered nonretriggerable circuit. (We can connect either $\overline{Q}$ to B (for positive edge-triggering) or Q to A (for negative edge-triggering).) When the output, Q, generates a HIGH pulse, $\overline{Q}$ generates a LOW pulse of equal length. When applied to input B, this LOW pulse inhibits the internal OR gate and prevents further trigger pulses from activating a new output pulse before the monostable times out.

**Figure 7.6**

**74HC4538 as a Nonretriggerable Monostable Multivibrator (Positive Edge-Triggered)**

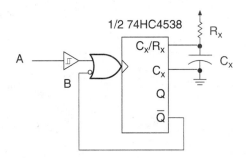

**EXAMPLE 7.4**

Draw a circuit showing how to make an astable multivibrator from two monostable multivibrators. Use a 74HC4538 Dual Monostable. Draw a timing diagram and briefly explain how the circuit works.

**Solution**

Figure 7.7 shows the astable circuit and timing diagram. Each monostable triggers the other one on the trailing edge of its output pulse. Since each monostable generates a HIGH pulse at Q, it must trigger the other one on the negative edge of the output waveform. (We could also choose to connect $\overline{Q}$ of one monostable to A of the other for a positive edge-triggered circuit.)

We must choose one of the monostable outputs as the circuit output. If we choose $Q_2$, then monostable 1 sets time LOW $(t_l)$ and monostable 2 sets the circuit's time HIGH $(t_h)$. These values are given by:

$$t_l = kR_1C_1$$

$$t_h = kR_2C_2$$

The frequency and duty cycle of this circuit are given by:

$$f = \frac{1}{T} = \frac{1}{t_l + t_h} = \frac{1}{k(R_1C_1 + R_2C_2)}$$

$$DC = \frac{t_h}{t_l + t_h} = \frac{kR_2C_2}{k(R_1C_1 + R_2C_2)} = \frac{R_2C_2}{R_1C_1 + R_2C_2}$$

**Figure 7.7**

**Astable Multivibrator From Two Monostables**

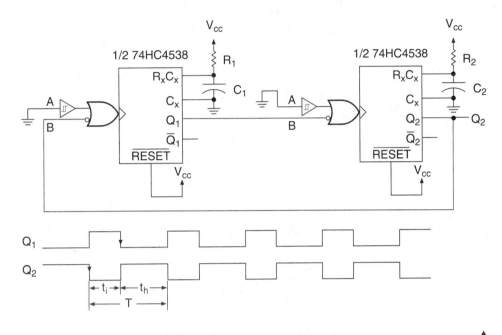

---

**Section Review Problem for Section 7.2**

7.1. Calculate the width of the pulse generated by the circuit in Figure 7.6 if $R_x = 47$ kΩ, $C_x = 0.022$ μF, and $V_{CC} = +5$ V. How can this pulse be lengthened?

## 7.3

# The 555 Timer

**Comparator** *An analog device with two input terminals labeled (+) and (−) that compares the input voltages and produces a logic HIGH output when $v(+) > v(−)$ and a LOW output when $v(+) < v(−)$.*

The 555 timer is a linear integrated circuit that can be used to make an inexpensive astable or monostable multivibrator. It is not a digital device, but it is often used for generating pulses in digital circuits, especially in cases where cost is more important than precision.

The 555 timer can be used with supply voltages ranging from +4.5 V to +16 V and requires few external components. The waveform output times are independent of the supply voltage and depend only on the values of the external components. Two common manufacturers' designations are NE555 and MC1455. The 555 is generally packaged as an 8-pin DIP. Although there are numerous other uses for the 555 timer, we will look only at the multivibrator applications.

## Internal Configuration

Figure 7.8 shows a functional schematic of the 555 timer. The pin assignments are as follows:

1. Ground
2. Trigger
3. Output
4. Reset

8. $V_{CC}$
7. Discharge
6. Threshold
5. Control voltage

The circuit consists of three main sections: (1) a voltage divider and two **comparators** for input voltage sensing, (2) an SR latch, and (3) an output section, consisting of an inverting output buffer and a discharge transistor. Let's look at each of these sections individually.

### Voltage Divider and Comparators

Figure 7.9 shows the input-voltage sensing circuit, consisting of a voltage divider of three 5-kΩ resistors and two comparators. The resistors divide the supply voltage into equal parts, giving values of 1/3 $V_{CC}$ and 2/3 $V_{CC}$ at their junctions. The comparators use these voltages as reference values.

A comparator is an operational amplifier-driven open loop (i.e., with no feedback) so as to give a very high gain. When the (+) input is at a higher voltage than the (−) input, the output goes into positive saturation, or logic HIGH. When the (−) terminal is at a higher voltage than the (+) terminal, the output goes into negative saturation, or logic LOW. (Since there is a single supply voltage, rather than $\pm V_{CC}$, negative satura-

**Figure 7.8**

**Functional Diagram of the 555 Timer**

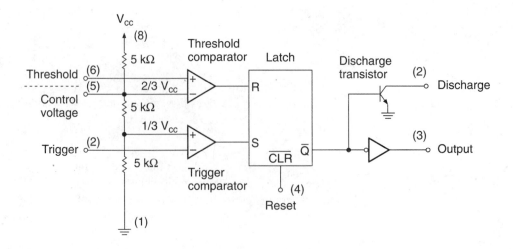

**Figure 7.9**
**Voltage Divider and Comparators**

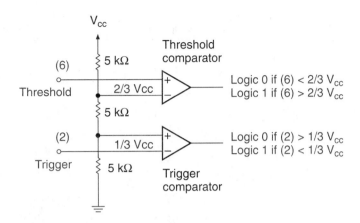

tion is 0 V and positive saturation is $V_{CC}$.) If these op amp terms are unfamiliar, details of comparator operation can be found in any good book on electronic devices. However, it is only important to remember that the output is HIGH if $v(+) > v(-)$ and LOW if $v(+) < v(-)$.

The two comparators are called the Threshold and Trigger comparators. The Threshold comparator has a reference voltage of 2/3 $V_{CC}$ at its $(-)$ input. Its $(+)$ input goes to pin 6. Let us designate the voltage at pin 6 as $V_6$.

---

If $V_6 < 2/3\ V_{CC}$, the Threshold comparator output is LOW.
If $V_6 > 2/3\ V_{CC}$, the Threshold comparator output is HIGH.

---

The Trigger comparator has a reference voltage of 1/3 $V_{CC}$ at its $(+)$ input, and its $(-)$ input connects to pin 2. Let's call the Trigger input voltage $V_2$.

---

If $V_2 > 1/3\ V_{CC}$, the Trigger comparator output is LOW.
If $V_2 < 1/3\ V_{CC}$, the Trigger comparator output is HIGH.

---

## SR Latch

The SR latch is set by the Trigger comparator and reset by the Threshold comparator. Both inputs are active HIGH. Thus:

---

If $V_2 < 1/3\ V_{CC}$, the latch sets.
If $V_6 > 2/3\ V_{CC}$, the latch resets.
If $V_2 > 1/3\ V_{CC}$ and $V_6 < 2/3\ V_{CC}$, the latch neither sets nor resets.

---

The latch can also be cleared directly, by making the *RESET* input, pin 4, LOW.

## Output Section

The output of the 555 timer is the same as the latch's $Q$ output. Thus, the Trigger input makes the output HIGH, and the Threshold input makes the output LOW.

**Figure 7.10**

**Discharge Transistor as a Switch**

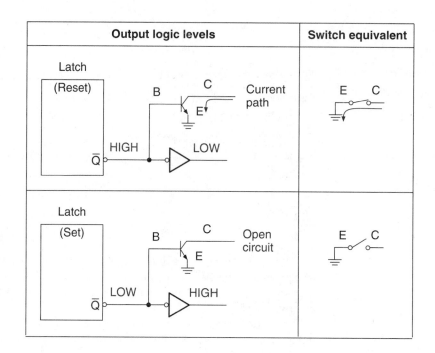

| Output logic levels | Switch equivalent |
|---|---|
| Latch (Reset) ... B C Current path ... E ... $\overline{Q}$ HIGH LOW | E C |
| Latch (Set) ... B C Open circuit ... E ... $\overline{Q}$ LOW HIGH | E C |

The collector of an NPN transistor, called the Discharge transistor, connects to pin 7. The base of the transistor connects to the latch's $\overline{Q}$ output.

The transistor acts as a switch, as shown in Figure 7.10, creating a low-impedance path between collector and emitter when the base is at the logic HIGH state. This happens when the latch is reset. When the latch is set, the transistor base is LOW, making the path from collector to emitter high impedance, thus "opening" the switch.

The transistor is used as a path to discharge a timing capacitor when the 555 timer is used as an astable or monostable multivibrator.

## 7.4

# The 555 Timer as a Monostable Multivibrator

Figure 7.11 shows the pin connections and equivalent circuit of a 555 timer connected as a monostable multivibrator.

The monostable is triggered when pin 2 is pulled to less than 1/3 $V_{CC}$. The output produces a pulse whose width, $t_w$, is determined by the charging $RC$ time constant of the external resistor and capacitor connected to pins 6 and 7.

Pin 5 can be used to set the Threshold and Trigger voltages of the 555 timer by applying a voltage directly to the Threshold comparator reference input. (Threshold voltage is equal to the voltage applied to pin 5. Trigger voltage is half this value.) If we choose not to use this function, the 0.01-$\mu$F capacitor at pin 5 acts as an AC short circuit and a DC open circuit. This shorts to ground unwanted AC signals (which might affect the Threshold and Trigger levels) without changing the DC reference values for the two comparators.

**Figure 7.11**

**555 Timer as a Monostable Multivibrator**

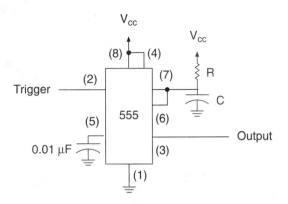

**a. Pin connections**

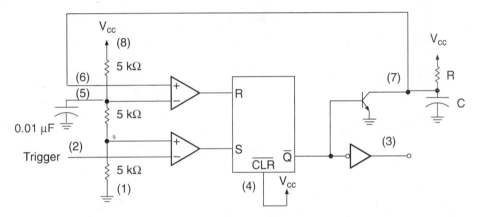

**b. Equivalent circuit**

The operation of the circuit can be broken into two phases: the capacitor charging and discharging cycles. The operation of the charging cycle is shown in Figure 7.12a. The discharging cycle is shown in Figure 7.12b.

A LOW pulse on the Trigger input (pin 2) sets the SR latch and thus makes the output (pin 3) HIGH. At the same time the Discharge transistor turns OFF, since a LOW is applied to its base. Prior to this, the transistor had shorted the capacitor to ground, keeping it completely discharged.

With the base LOW, the collector and emitter of the transistor offer a high-impedance path to ground, effectively removing the transistor from the circuit. This allows the capacitor to begin charging from 0 V, increasing at a rate determined by the *RC* charging equation:

$$v_c(t) = V_{CC}\,(1 - e^{-t/RC})$$

The Threshold comparator monitors the capacitor voltage, $v_c(t)$. When $v_c(t) = 2/3$ $V_{cc}$, the Threshold comparator resets the SR latch, as shown in Figure 7.12b, and the output returns to LOW. The Discharge transistor now turns ON, since its base is HIGH, and provides a low-impedance path from the capacitor to ground. Since there is relatively little resistance in this path, the capacitor discharges almost instantaneously.

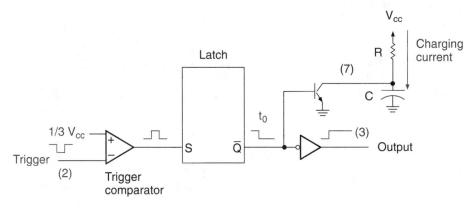

**a. Charging cycle (output HIGH)**

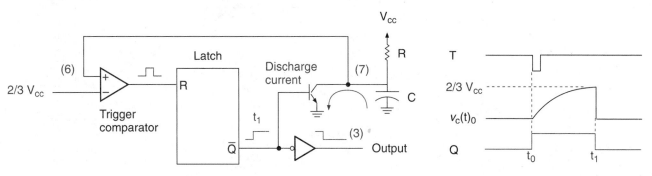

**b. Discharging cycle (output LOW)**

**c. Trigger, capacitor, and output waveforms**

**Figure 7.12**

**Operation of 555 Timer as a Monostable Multivibrator**

Figure 7.12c shows the trigger, capacitor voltage, and output waveforms in relation to one another. The output, $Q$, is HIGH as long as the capacitor is charging. As soon as it discharges, the output goes LOW.

We can solve the charging equation for the time required for the capacitor to charge from 0 V to 2/3 $V_{CC}$. This time is $t_w$, the output pulse width. Thus:

$$2/3 \ V_{CC} = V_{CC} \ (1 - e^{-t/RC})$$

$$2/3 = 1 - e^{-t/RC}$$

$$1 - 2/3 = e^{-t/RC}$$

$$1/3 = e^{-t/RC}$$

$$ln \ (1/3) = -t/RC$$

$$t = -RC \ ln \ (1/3)$$

$$= 1.1 \ RC$$

$$t_w = t = 1.1 \ RC$$

**EXAMPLE 7.5**

A 555 timer IC is connected as a monostable multivibrator, as shown in Figure 7.11. The timing capacitor has a value of 100 pF.

a. Calculate the value of $R$ required to make the circuit output generate a 5-$\mu$s pulse.

b. Choose the standard resistor value closest to that calculated in part a. Recalculate the *actual* pulse width, $t_w$, based on the standard value.

The first two significant figures of a standard resistor value can be any of the following:

| | | | | |
|----|----|----|----|----|
| 10 | 18 | 33 | 56 | 91 |
| 12 | 22 | 39 | 68 | |
| 15 | 27 | 47 | 82 | |

**Solution**

a. The pulse width of the monostable output is given by:

$$t_w = 1.1\ RC$$

The charging resistance is thus given by:

$$R = t_w/(1.1\ C)$$
$$= (5 \times 10^{-6}\ \text{s})/(1.1)(100 \times 10^{-12}\ \text{F})$$
$$= 45.4\ \text{k}\Omega$$

b. The closest standard resistor value is 47 k$\Omega$. This results in a pulse width of:

$$t_w = (1.1)(47\ \text{k}\Omega)(100\ \text{pF})$$
$$= 5.17\ \mu\text{s}$$

This is 3.4% longer than the intended pulse time, but since resistors are generally accurate to within $\pm$5%, it's within an acceptable range.

## 7.5

# The 555 Timer as an Astable Multivibrator

Figure 7.13 shows the 555 timer connected as an astable multivibrator. The circuit generates a periodic digital waveform with the frequency and duty cycle determined by the charging and discharging time constants of $R_a$, $R_b$, and C. As was the case with the monostable circuit, the 0.01-$\mu$F capacitor at pin 5 shorts AC signals to ground without altering the DC values of the Threshold and Trigger voltages. This capacitor has nothing to do with timing.

In the astable circuit, the Trigger and Threshold inputs monitor the time-varying capacitor voltage, $v_c(t)$. These two inputs, working together, keep the capacitor voltage within the range $1/3\ V_{CC} \leq v_c(t) \leq 2/3\ V_{CC}$. The capacitor charging and discharging cycles are shown in Figure 7.14.

When the capacitor discharges to $1/3\ V_{CC}$, the Trigger comparator sets the SR latch, driving the output HIGH and turning the Discharge transistor OFF. The capacitor charges

**Figure 7.13**

**555 Timer as an Astable Multivibrator**

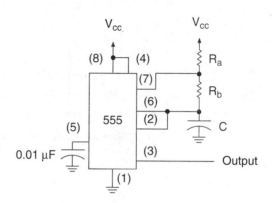

**a. Pin connections**

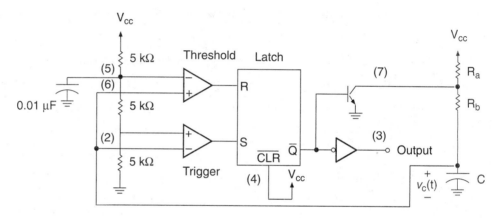

**b. Equivalent circuit**

through $R_a$ and $R_b$, as shown in Figure 7.14a. The capacitor voltage is given by:

$$v_c(t) = V_{0c} + (V_{CC} - V_{0c})(1 - e^{-t/(R_a + R_b)C})$$

where $V_{0c}$ is the capacitor voltage at the beginning of the charging phase ($V_{0c} = 1/3\ V_{CC}$).

When the capacitor voltage reaches $2/3\ V_{CC}$, the Threshold comparator resets the SR latch, making the output LOW and turning ON the Discharge transistor. The capacitor discharges through $R_b$, as shown in Figure 7.14b. During the discharging phase, the capacitor voltage is given by:

$$v_c(t) = V_{0d}e^{-t/(R_bc)}$$

where $V_{0d}$ is the capacitor voltage at the beginning of the discharging phase ($V_{0d} = 2/3\ V_{CC}$).

Figure 7.14c shows the relationship between the output, $Q$, and the capacitor voltage, $v_c(t)$. The output time HIGH, $t_h$, corresponds to the time required for the capacitor to charge from $1/3\ V_{CC}$ to $2/3\ V_{CC}$. Time LOW, $t_l$, is the time required for the capacitor

**Figure 7.14**
**Charging and Discharging**
**Cycles**

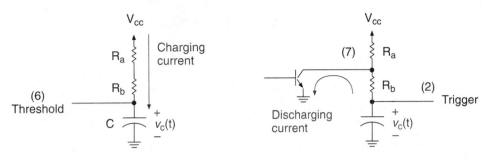

a. Charging cycle (output HIGH)    b. Discharging cycle (output LOW)

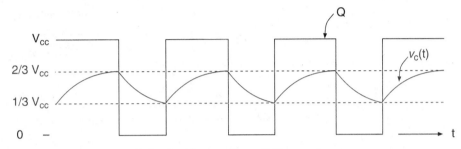

c. Output and capacitor voltage waveforms

to discharge from $2/3\ V_{CC}$ to $1/3\ V_{CC}$. We can calculate these times in terms of the external component values by solving for $t$ in the charging and discharging equations.

**Time HIGH (charging phase):**

$$v_c(t) = V_{0c} + (V_{cc} - V_{0c})(1 - e^{-t/(R_a + R_b)C})$$

Solve for elapsed time, $t$, when $v_c(t) = 2/3\ V_{cc}$.

$$2/3\ V_{CC} = 1/3\ V_{CC} + (V_{CC} - 1/3\ V_{CC})(1 - e^{-t/(R_a + R_b)C})$$

$$2/3\ V_{CC} - 1/3\ V_{CC} = 2/3\ V_{CC}(1 - e^{-t/(R_a + R_b)C})$$

$$1/3\ V_{CC} = 2/3\ V_{CC}(1 - e^{-t/(R_a + R_b)C})$$

$$1/2 = (1 - e^{-t/(R_a + R_b)C})$$

$$e^{-t/(R_a + R_b)C} = (1 - 1/2)$$

$$-t/(R_a + R_b)C = ln\ (1/2)$$

$$t = -(ln\ (1/2))(R_a + R_b)C$$

$$= 0.693(R_a + R_b)C$$

$$t_h = t = 0.693(R_a + R_b)C$$

**Time LOW (discharging phase):**

$$v_c(t) = V_{0d}e^{-t/(R_bC)}$$

Solve for elapsed time, $t$, when $v_c(t) = 1/3\ V_{CC}$.

$$1/3\ V_{CC} = 2/3\ V_{CC}e^{-t/(R_bC)}$$

$$1/2 = e^{-t/(R_bC)}$$

$$ln\ (1/2) = -t/(R_bC)$$

$$t = -R_bC\ ln\ (1/2)$$

$$= 0.693\ R_bC$$

---

$$t_l = t = 0.693\ R_bC$$

---

The values of $t_h$ and $t_l$ can be used to find the frequency and duty cycle of the output waveform, which are generally more useful parameters.

**Frequency:**

Frequency is derived from period, $T$, which is the sum of time LOW and time HIGH.

$$T = t_h + t_l$$

$$= 0.693\ (R_a + R_b)C + 0.693\ R_bC$$

$$= 0.693\ (R_a + 2R_b)C$$

$$f = 1/T$$

---

$$f = \frac{1}{0.693\ (R_a + 2R_b)C} \approx \frac{1.44}{(R_a + 2R_b)C}$$

---

**Duty cycle:**

$$DC = t_h/T$$

$$= \frac{0.693\ (R_a + R_b)C}{0.693\ (R_a + 2R_b)C}$$

---

$$DC = \frac{R_a + R_b}{R_a + 2R_b}$$

---

## Calculating Timing Component Values

The values of two of the three components, $R_a$, $R_b$, and $C$, can be calculated using the above equations, provided the desired frequency and duty cycle of the output waveform are known. The third component must be specified, since only two variables can be determined uniquely from two linear equations.

A common practice is to specify the capacitor value and calculate the resistor values. Any systematic method for solving two linear equations in two unknowns can be used to find the values of the two resistors. One easy method is to use the duty cycle equation to calculate a ratio of the resistors and plug this into the frequency equation to solve for one resistor value. The other resistor is determined from the previously calculated ratio.

---

**EXAMPLE 7.6**

A 555 timer IC is connected as an astable multivibrator, as shown in Figure 7.13. The timing capacitor has a value of 100 pF.

   **a.** Calculate the values of $R_a$ and $R_b$ required to make the timer output oscillate at 72 kHz, with a duty cycle of 75%.

   **b.** Choose the standard resistor values closest to those calculated in part a. Recalculate the *actual* frequency and duty cycle that result from the standard resistors.

**Solution**   **a.** Use the duty cycle equation to calculate a resistor ratio.

$$DC = \frac{R_a + R_b}{R_a + 2R_b} = 0.75$$

$$R_a + R_b = 0.75\,(R_a + 2R_b)$$

$$= 0.75\,R_a + 1.5\,R_b$$

$$R_a - 0.75\,R_a = 1.5\,R_b - R_b$$

$$0.25\,R_a = 0.5\,R_b$$

$$R_a = 2\,R_b$$

Plug this ratio into the frequency equation.

$$f \approx \frac{1.44}{(R_a + 2R_b)C} = \frac{1.44}{(2R_b + 2R_b)C} = \frac{1.44}{4R_bC}$$

Solve for $R_b$:

$$72 \text{ kHz} = 1.44/(4R_bC)$$

$$R_b = 1.44/(4 \times 72 \times 10^3 \text{ Hz} \times 100 \times 10^{-12} \text{ F})$$

$$= 50 \text{ k}\Omega$$

Solve for $R_a$:

$$R_a = 2R_b = 100 \text{ k}\Omega$$

   **b.** Standard values for the resistors calculated in part a are:

$$R_a = 100 \text{ k}\Omega$$

$$R_b = 47 \text{ k}\Omega$$

Actual frequency and duty cycle are:

$$f = 1.44/(100 \text{ k}\Omega + 2\times47 \text{ k}\Omega) \times 100 \text{ pF}$$

$$= 74.2 \text{ kHz} \quad (+3.1\% \text{ from expected})$$

$$DC = (100 \text{ k}\Omega + 47 \text{ k}\Omega)/(100 \text{ k}\Omega + 2\times47 \text{ k}\Omega)$$

$$= 0.757 \quad (+1\% \text{ from expected})$$

## Minimum Value of $R_a$

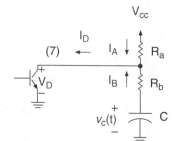

**Figure 7.15**

**Calculation of Minimum $R_a$**

When the 555 timer is in the discharge cycle, current is drawn from the power supply through $R_a$ and the Discharge transistor, as well as from the capacitor through $R_b$. Figure 7.15 shows the discharge circuit and the various currents flowing into pin 7. The manufacturer's specifications indicate that the total current into pin 7 must not exceed 200 mA.

Total current, $I_D$, flowing through the Discharge transistor is given by:

$$I_D = I_A + I_B$$

If we assume that $I_B << I_A$, then $I_D \approx I_A$. (This assumption is made by the manufacturer's data sheet.) Thus, we must choose $R_a$ such that $I_A \leq 200$ mA.

From Figure 7.15, we see that:

$$R_a = \frac{V_{CC} - V_D}{I_A}$$

Assuming that $V_D \approx 0$ and $I_A \leq 200$ mA, then:

$$R_a \geq V_{CC}/200 \text{ mA}$$

and, for $V_{CC} = 5$ V:

$$R_a \geq 5 \text{ V}/200 \text{ mA} = 25 \text{ }\Omega$$

If $R_a$ is at least 25 $\Omega$, it will protect the Discharge transistor from excessive current.

Although $R_a$ cannot be zero, as we have just seen, it can approach zero, relative to $R_b$, and not affect the duty cycle of the output waveform.

When $R_a << R_b$,

$$DC = \frac{R_a + R_b}{R_a + 2R_b} \approx \frac{R_b}{2R_b} = 0.5$$

The charging or discharging time constant of a capacitor is directly proportional to its capacitance and to the resistance through which it charges or discharges. Since the capacitor charges through a resistance of $(R_a + R_b)$ and discharges through a resistance of $R_b$, $t_h$ will always be longer than $t_l$. The two values approach each other as $R_a$ becomes smaller, but they will never be equal, since $R_a$ has a minimum value. Thus, a duty cycle of 50% on the output waveform is not attainable by selection of resistors alone, although it can be quite close.

**EXAMPLE 7.7**

Calculate the frequency and percent duty cycle of a 555 timer astable multivibrator having the following component values:

$$R_a = 27\ \Omega,\ R_b = 5.6\ \text{k}\Omega,\ C = 0.1\ \mu\text{F}$$

(Note that $R_a$ is close to its minimum value of 25 $\Omega$ % $V_{CC} = 5$ V.)

**Solution**

$$f = 1/(0.693(27\Omega + 2\times5.6\times10^3\Omega)0.1\times10^{-6}\text{F})$$

$$= 1.28\ \text{KHz}$$

$$\%DC = ((27\Omega + 5.6\times10^3\Omega)/(27\Omega + 2\times5.6\times10^3\Omega)) \times 100\%$$

$$= 50.12\%$$

If the ratio $R_b:R_a$ is 10:1 or greater, the duty cycle approaches 50%. (For a 10:1 ratio, $\%DC = (11/21) \times 100\% = 52.4\%$.)

A simple way to make the duty cycle of a 555 astable output exactly 50% is to use the output to clock a JK flip-flop in toggle mode and take the desired waveform from the flip-flop output. This works because the negative edges of the 555 output waveform are equally spaced, and each negative edge generates one half-cycle of the flip-flop output waveform. Since the flip-flop divides the astable output frequency by 2, you must compensate by calculating an astable frequency of twice the desired value.

**EXAMPLE 7.8**

Figure 7.16 shows a 555 timer astable multivibrator whose output is connected to a negative edge-triggered JK flip-flop configured for toggle mode.

**Figure 7.16**

**Example 7.8**
**555 Timer Astable**
**Multivibrator**

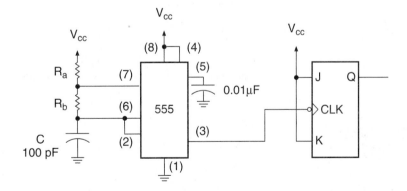

a. Calculate the values of $R_a$ and $R_b$ required to make the flip-flop output oscillate at 100 kHz. Assume that $C = 100$ pF and $R_a = R_b$.

b. Calculate the duty cycle of the astable output.

c. Draw a timing diagram showing the relation between the astable and flip-flop outputs.

**Solution**    a. Solve the frequency equation for $R_a$, assuming $f = 200$ kHz and $R_b = R_a$.

$$f \approx \frac{1.44}{(R_a + 2R_b)C} = \frac{1.44}{(R_a + 2R_a)C} = \frac{1.44}{3R_a C}$$

$$R_a = 1.44/3(200 \times 10^3 \text{Hz})(100 \times 10^{-12}\text{F})$$

$$= 24 \text{ k}\Omega$$

This is not a standard resistor value. The closest standard value is 22 k$\Omega$, which results in an astable output frequency of 218 kHz and a flip-flop output frequency of 109 kHz. (Prove this to yourself by doing the calculation.)

If you require greater accuracy, either $R_a$ or $R_b$ can be replaced by a series combination of a fixed and a variable resistor. (The fixed resistor ensures that the value of $R_b$ never drops below a selected minimum value. For example, replace $R_b$ with an 18-k$\Omega$ resistor in series with a 10-k$\Omega$ potentiometer so that the desired resistance is in the middle range of the pot.) Set the frequency by monitoring the flip-flop output on an oscilloscope.

**b.** $DC = (24\text{k}\Omega + 24\text{k}\Omega)/(24\text{k}\Omega + 2 \times 24\text{k}\Omega) = 0.667$

**c.** Figure 7.17 shows the astable and flip-flop waveforms. Although the astable output does not have a 50% duty cycle, the flip-flop does.

**Figure 7.17**

**Example 7.8**
**Astable and Flip-Flop**
**Waveforms**

Astable output*

Flip-flop output**

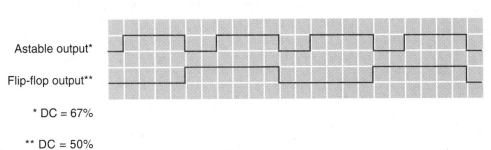

\* DC = 67%

\*\* DC = 50%

---

**Section Review Problem for Section 7.5**

7.2. A 555 timer astable circuit has resistor values $R_a = 1$ k$\Omega$ and $R_b = 2$ k$\Omega$. A similar circuit has resistors $R_a = 2$ k$\Omega$ and $R_b = 1$ k$\Omega$. Will the frequency and duty cycle be the same for each circuit? Explain briefly.

## ● GLOSSARY

**Astable multivibrator** A multivibrator with no stable states that acts as a digital oscillator.

**Bistable multivibrator** A multivibrator with two stable output states. (e.g., a latch or flip-flop).

**Comparator** An analog device with two input terminals labeled (+) and (−) that compares the input voltages and produces a logic HIGH output when $v(+) > v(-)$ and a LOW output when $v(+) < v(-)$.

**Monostable multivibrator (one-shot multivibrator)** A multivibrator with one stable state and one quasistable state whose output produces a pulse of fixed length.

**Multivibrator** A sequential digital circuit whose output oscillates between logic HIGH and LOW states, either automatically, due to the circuit design, or under the control of a special input (e.g., *SET, RESET, TRIGGER*).

**Nonretriggerable** A monostable multivibrator in which new trigger pulses are ignored if the output is producing a pulse.

**Quasistable state** A temporarily stable state of a multivibrator output.

**Retriggerable** A monostable multivibrator in which a new trigger pulse restarts the timing of an output pulse, thus increasing the output pulse width.

**Stable state** An output state of a multivibrator that will not change unless actively changed by an external signal.

**Time out** Finish a timing cycle.

**Trigger** The control input of a monostable multivibrator that initiates a timed pulse at the circuit's output.

**Unstable state** An output state of a multivibrator that will change to the opposite state without requiring a control signal to make it change.

## ● PROBLEMS

### Section 7.1 Multivibrators

### Section 7.2 IC Monostables

**7.1** Briefly explain the difference between a retriggerable and a nonretriggerable monostable multivibrator.

**7.2** Draw one half of a 74HC4538 monostable multivibrator connected as a nonretriggerable circuit with a pulse width of 100 ms ($C = 2.2$ μF, $V_{CC} = +5$ V). Use the $Q$ output to inhibit retriggering. Briefly explain why this connection inhibits retriggering. Is this circuit positive or negative edge-triggered?

**7.3** Use the graph in Figure 7.5b to select a timing capacitor for a 74HC4538 monostable multivibrator with a pulse width of 150 μs. $R = 100$ kΩ.

**7.4** Repeat Problem 7.3 for a pulse width of 1 ms. $R = 100$ kΩ.

**7.5** Draw a circuit showing how to make an astable multivibrator out of two monostable multivibrators. (Use a 74HC4538 Dual Monostable Multivibrator.) Select resistor values such that $f = 100$ kHz and $DC = 40\%$. Assume that each monostable has a timing capacitor of 220 pF.

### Section 7.4 The 555 Timer as a Monostable Multivibrator

**7.6** A 555 timer is connected as a monostable multivibrator, as shown in Figure 7.11. If the timing capacitor, $C$, has a value of 25 μF, calculate the value of the timing resistor, $R$. Assume $t_w = 1$s.
Is the circuit retriggerable or nonretriggerable? Explain your answer.

**7.7** Choose the standard resistor value closest to that calculated in Problem 7.6. Recalculate the *actual* pulse width, $t_w$, based on the standard value.
The first two significant figures of a standard resistor value can be any of the following:

| | | | | |
|----|----|----|----|----|
| 10 | 18 | 33 | 56 | 91 |
| 12 | 22 | 39 | 68 | |
| 15 | 27 | 47 | 82 | |

### Section 7.5 The 555 Timer as an Astable Multivibrator

**7.8** Choose values of $R_a$ and $R_b$ that will make the 555 timer circuit in Figure 7.13 oscillate at 120 kHz with a duty cycle of 60%. The timing capacitor is 0.001 μF.

**7.9** Choose the standard resistor values closest to those calculated in Problem 7.8. Recalculate the *actual* frequency and duty cycle based on the standard values.

**7.10** Calculate the frequency and duty cycle of an astable multivibrator constructed from a 555 timer and the following sets of components:

**a.** $R_a = 1$ kΩ
$R_b = 6.8$ kΩ
$C = 0.01$ μF

**b.** $R_a = 6.8$ kΩ
$R_b = 1$ kΩ
$C = 0.01$ μF

**c.** $R_a = 1$ kΩ
$R_b = 3.3$ kΩ
$C = 0.01$ μF

**d.** $R_a = 1$ kΩ
$R_b = 3.3$ kΩ
$C = 0.1$ μF

Refer to Figure 7.13 when answering Problems 7.11 through 7.13.

**7.11** Comment on the difference in frequency and duty cycle when resistors $R_a$ and $R_b$ are switched around. Why does this make a difference? Use the values from Problem 7.10, parts a and b, as examples.

**7.12** Comment on the effect on frequency and duty cycle of different values of external resistance. Use the values from Problem 7.10, parts a and c, as examples.

**7.13** Comment on the effect on frequency and duty cycle of different values of external capacitance. Use the values from Problem 7.10, parts c and d, as examples.

**7.14** Calculate the frequency and duty cycle of an astable multivibrator constructed from a 555 timer and the following sets of components:

    **a.** $R_a = 10$ kΩ
        $R_b = 2.7$ kΩ
        $C = 0.15$ μF

    **b.** $R_a = 2.7$ kΩ
        $R_b = 10$ kΩ
        $C = 0.15$ μF

    **c.** Explain any differences in duty cycle and frequency in the above.

**\*\*7.15** Recalculate the frequency and duty cycle of the circuit in Problem 7.14b if $R_b$ is bypassed by a diode. (Cathode connects to junction of $R_a$ and $R_b$, anode connects to junction of $R_b$ and $C$.) Redraw the timing components for this case. (Hint: The effect of the diode is to remove $R_b$ from the capacitor charging path, but not the discharging path.)

**\*\*7.16** The bypass diode added in Problem 7.15 removes the restriction of duty cycle being greater than 50% from a 555 timer astable circuit. Why? Write the modified frequency and duty cycle equations for this circuit.

## ▼ Answers to Section Review Questions

### Section 7.2

**7.1.** $t_w = kR_xC_x = 744$ μs. With the circuit as shown, the pulse can be lengthened only by increasing the value of $R_x$ or $C_x$.

### Section 7.5

**7.2.** Different frequency and duty cycle. Both $f$ and $DC$ are inversely proportional to the term $(R_a + 2R_b)$. In the first case, $(R_a + 2R_b) = 5$ kΩ. In the second case, $(R_a + 2R_b) = 4$ kΩ.

Another way to look at it is that the charging time is the same in both circuits since $R_{a1} + R_{b1} = R_{a2} + R_{b2}$, but the discharge time is different, since $R_{b1} \neq R_{b2}$. This changes both frequency and duty cycle.

# Introduction to Programmable Logic

## ● CHAPTER CONTENTS

**8.1** Introduction to Programmable Logic

**8.2** PAL Fuse Matrix and Combinational Outputs

**8.3** PAL Outputs With Programmable Polarity

**8.4** PAL and GAL Devices With Registered Outputs

**8.5** UV-Erasable Programmable Logic Devices (EPLD)

**8.6** Development Software

## ● CHAPTER OBJECTIVES

Upon successful completion of this chapter, you will be able to:

- Draw a diagram showing the basic hardware conventions for a sum-of-products-type programmable logic device.

- Describe the differences in architecture between PROM, FPLA, and PAL.

- Describe the structure of a programmable array logic (PAL) AND matrix.

- Draw fuses on the logic diagram of a PAL to implement simple logic functions.

- Describe the structures of the various types of PAL outputs: combinational, programmable polarity, and registered.

- Explain the structure of an output logic macrocell (OLMC).

- Summarize the features of a UV-erasable PLD, including those features that are supersets of PAL and GAL devices.

- Define software terms, such as source file, JEDEC file, test vector, and simulation, used in the programming of PLDs.

- Explain how a JEDEC file relates to the pattern of intact and blown fuses in a PLD AND matrix.

- Describe how test vectors are used to verify the function of a programmed PLD.

- Write simple source files and simulate PLD designs using Intel's PLDasm Programmable Logic Design Software.

## ● INTRODUCTION

In previous chapters, we have studied logic gates and flip-flops, the basic building blocks of digital logic circuits. Many integrated circuits are available that combine these devices on a single chip to perform some relatively complex logic function.

The problem with these devices is that their functions remain fixed. A particular part might have functions a designer will never need or lack functions required for a specific application.

Programmable logic is a solution to this problem. A programmable logic device (PLD) is an integrated circuit whose function can be programmed by the user, usually in the form of a sum-of-products (SOP) Boolean equation.

The general form of a PLD has a number of input variables (in true and complement form) and AND gate inputs that can be interconnected by a matrix of fuses. Fuses corresponding to desired connections are left intact and the remaining fuses are blown. These connections form product terms that are combined in OR gates to produce SOP functions. These OR gates drive outputs that can be combinational (output buffers) or registered (flip-flops). The more recent types of PLDs have outputs whose configurations are also fuse programmable. Many such devices are also erasable, either electrically or by exposure to ultraviolet radiation.

Historically, fuses were drawn on the logic diagram of a PLD device and blown one at a time. This is never done now. In modern PLD applications, computer software generates the required fuse maps and controls the hardware used to program the fuse matrices of the required PLDs.

### 8.1

# Introduction to Programmable Logic

**Programmable logic device (PLD)**  *A logic device whose function can be programmed by the user, usually in sum-of-products form.*

**Product line**  *A single line on a logic diagram used to represent all inputs to an AND gate (i.e., one product term) in a PLD sum-of-products array.*

**Input line**  *A line that applies the true or complement form of an input variable to the AND matrix of a PLD.*

**PROM**  *Programmable read-only memory. Programmable logic with a fixed AND matrix and a programmable OR matrix.*

**FPLA**  *Field programmable logic array. Programmable logic whose AND and OR matrices are both programmable.*

**PAL**  *Programmable array logic. Programmable logic with a fixed OR matrix and a programmable AND matrix.*

Traditionally, if a manufacturer of electronic equipment wanted to incorporate a nonstandard digital logic function (i.e., one not commercially available in a single device) into a circuit, he would choose one of the following options:

*SSI/MSI logic.* The function would be implemented with standard small-scale integration (SSI) logic and medium-scale integration (MSI) chips. (An MSI chip is a device that performs a fixed logic function that would require from 12 to 100 equivalent gates.) This is a practical solution only if the function to be implemented is fairly simple. Otherwise, circuit boards become too large and complicated. Also, SSI/MSI devices are inflexible; you are stuck with the logic function as defined by the IC manufacturer.

*Custom chips.* A special chip would be designed and manufactured. This is usually a solution to the packaging problem, but is an expensive method unless production volume is high and the design is well established.

A middle way between these methods is programmable logic. The original **programmable logic devices (PLDs)** consisted of a number of AND and OR gates organized in sum-of-products (SOP) arrays in which connections were made or broken by a matrix of fuse links. An intact fuse allowed a connection to be made; a blown fuse would break a connection. Since any combinational logic function can be written in SOP form, any Boolean function could be programmed into these PLDs by blowing selected fuses. The programming was done by special equipment and its associated software. The hardware and software would select each fuse individually and apply a momentary high-current pulse if the fuse was to be blown.

The main problem with fuse-programmable PLDs is that they can be programmed one time only; if there is a mistake in the design and/or programming or if the design is updated, we must program a new PLD. Fuse-programmable PLDs are still in use, but more recent technology has produced several types of erasable PLDs, based not on fuses but on floating-gate metal-oxide-semiconductor transistors. These transistors also form the basis of memory technologies such as UV-erasable programmable read-only memory (UV-EPROM) and electrically erasable programmable read-only memory (EEPROM or $E^2$PROM).

Even though fuse technology is decreasing in importance in PLDs, we will use the fuse metaphor when examining PLD structure, for two reasons. First, modern erasable PLD technology is based on similar structural ideas; the fuse metaphor works well enough to describe the new structures. Second, modern PLDs can be better understood if we understand the models from which they are derived.

A symbol convention, slightly different from the standard logic circuit notation, has been developed for programmable logic. Figure 8.1 shows an example.

The circuit shown in Figure 8.1 is a sum-of-products network whose Boolean expression is given by:

$$F = \overline{A}\ \overline{B}\ C + A\ \overline{B}\ \overline{C}$$

The product terms are accumulated by the AND gates, which are shown as having only one input each, called the **product line.** This shorthand notation greatly reduces the number of AND inputs that must be shown. Each of the ANDs in Figure 8.1 has six inputs when all fuses are intact. This number is usually much greater (typically 32 or more) in actual PLDs.

A buffer having true and complement outputs applies each input variable to the AND matrix, thus producing two **input lines.** Each product line can be joined to any input line by leaving the corresponding fuse intact at the junction between the input and product lines. An intact fuse is shown as an "X" at the appropriate junction. A blown fuse is shown as an unconnected crossover.

If a product line, such as for the third AND gate, has all its fuses intact, we do not show the fuses on that product line. Instead, this condition is indicated by an "X" through

**Figure 8.1**

**PLD Symbology**

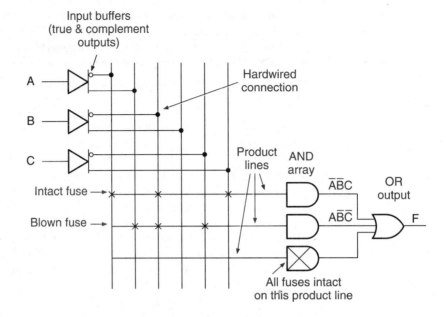

the gate. The output of the third AND gate is a logic 0, since $(\overline{A}\,A\,\overline{B}\,B\,\overline{C}\,C) = 0$. This is necessary to enable the OR gate output:

$$\overline{A}\,\overline{B}\,C + A\,\overline{B}\,\overline{C} + 0 = \overline{A}\,\overline{B}\,C + A\,\overline{B}\,\overline{C}$$

If the unused AND output was HIGH, the function $F$ would be:

$$\overline{A}\,\overline{B}\,C + A\,\overline{B}\,\overline{C} + 1 = 1$$

## PLD Architectures

PLDs are built from arrays of AND and OR gates that can be hardwired (fixed) or fused (programmable). The PLD configuration of Figure 8.1 is the most common of several architectures. The main variations are:

**PROM** (programmable read-only memory):   Fixed AND, programmable OR

**FPLA** (field programmable logic array):   Programmable AND and OR

**PAL** (programmable array logic):   Programmable AND, fixed OR

These differences don't seem very great at first glance, but they do have implications for the flexibility and efficiency of the logic circuits to be implemented. To compare these different architectures properly, let us see how each one can implement the same series of Boolean equations.

$$F1 = \overline{A}$$

$$F2 = \overline{A}\,B + A\,\overline{C}$$

$$F3 = \overline{A}\,B + A\,\overline{B}$$

$$F4 = A\,\overline{C} + \overline{A}\,B\,C$$

## PROM

PROM architecture has a fixed AND matrix and a fuse-programmable OR array. Figure 8.2 shows a PROM programmed with our sample functions.

We don't know what product terms we might need for a general Boolean expression. Since the AND array of the PROM is hardwired, we must make available all possible product terms. The most efficient way to do this is to make a matrix of all possible minterms. The minterms are ORed as needed to make the desired output expressions.

For example, note that the expression $\overline{A}$ cannot be programmed directly; it must be constructed from four minterms:

$$\overline{A} = \overline{A}\,\overline{B}\,\overline{C} + \overline{A}\,\overline{B}\,C + \overline{A}\,B\,\overline{C} + \overline{A}\,B\,C$$

The minterm $A\,B\,C$ is not used in any of the four expressions. In most PLD applications, not all minterms are used. Thus, the AND matrix in a PROM is usually bigger than it needs to be for any specific application. If another variable is needed, the size of the AND matrix must double, even if the variable is used only once. This tends to limit the number of input variables that can be used in a PROM application.

**Figure 8.2**
**PROM Architecture**

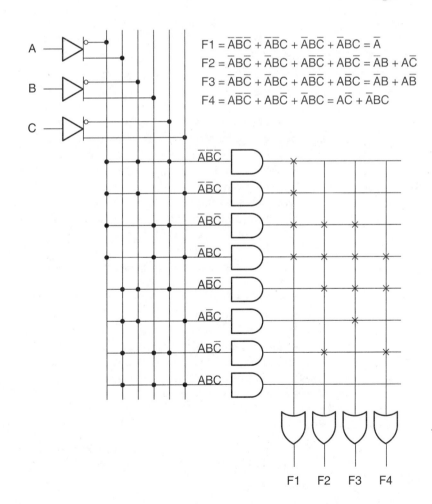

$F1 = \overline{A}\overline{B}\overline{C} + \overline{A}\overline{B}C + \overline{A}B\overline{C} + \overline{A}BC = \overline{A}$

$F2 = \overline{A}\overline{B}C + \overline{A}BC + A\overline{B}\overline{C} + AB\overline{C} = \overline{A}B + A\overline{C}$

$F3 = \overline{A}B\overline{C} + \overline{A}BC + A\overline{B}\overline{C} + A\overline{B}C = \overline{A}B + A\overline{B}$

$F4 = A\overline{B}\overline{C} + AB\overline{C} + \overline{A}BC = A\overline{C} + \overline{A}BC$

**Figure 8.3**
**FPLA Architecture**

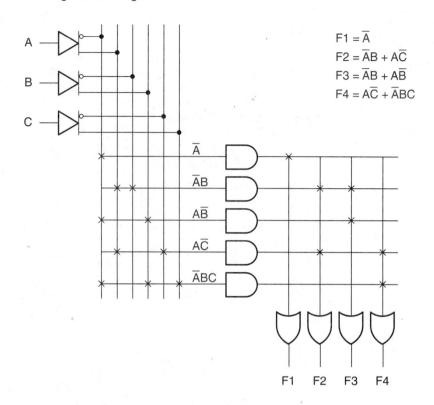

$F1 = \overline{A}$

$F2 = \overline{A}B + A\overline{C}$

$F3 = \overline{A}B + A\overline{B}$

$F4 = A\overline{C} + \overline{A}BC$

## FPLA

Field programmable logic arrays are a great deal more flexible than PROMs because of their programmable AND matrices. This feature implies that any product term can be programmed, not just minterms, and only the product terms actually required by a specific Boolean function need be selected. Figure 8.3 shows an FPLA programmed with the example equations.

The main advantage of FPLA is the reduced size of the AND matrix. The number of AND gates need be no greater than the combined number of product terms in the output expressions, regardless of the number of variables used. In many cases, it can be less, as when two or more outputs use the same product term. In our example, there are two such terms: both *F2* and *F3* contain the term $\overline{A}\,B$, and both *F2* and *F4* use the term $A\,\overline{C}$.

There is an electrical disadvantage to FPLA. All digital circuits experience a certain amount of propagation delay. That is, a circuit's output does not change instantaneously when its input changes; there is a delay of several nanoseconds. This delay limits the speed at which the circuit output can reliably switch from one state to another.

In PLD devices, the propagation delay of a hardwired matrix is less than that of a fused matrix. In FPLA devices, the extra level of fused connections makes the device slower than a comparable PROM.

## PAL

PAL is the most common configuration of programmable logic.[1] It represents a compromise between PROM and FPLA architectures.

---

[1]PAL is a registered trademark of Advanced Micro Devices.

**Figure 8.4**
**PAL Architecture**

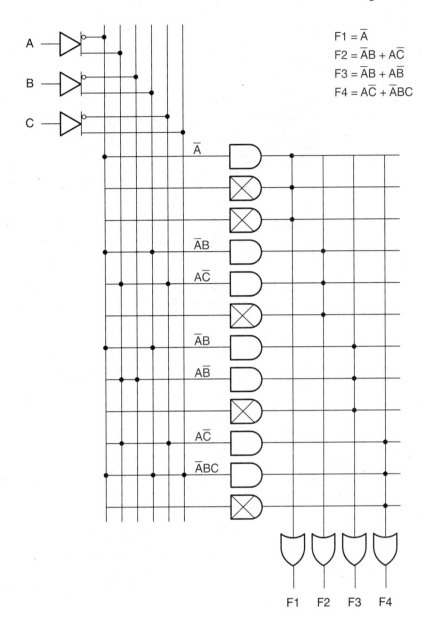

$$F1 = \overline{A}$$
$$F2 = \overline{A}B + A\overline{C}$$
$$F3 = \overline{A}B + A\overline{B}$$
$$F4 = A\overline{C} + \overline{A}BC$$

Since the OR matrix is not programmable, the output ORs usually have the same number of product lines, even though they are probably not all used. The hardwired OR matrix does improve the propagation delay, as compared to FPLA.

Figure 8.4 shows a PAL device programmed with our sample expressions. This PAL has four outputs, each having three product terms. Note that only 7 of the 12 available product lines are used. Also, the fixed OR matrix results in some redundancy; the product terms that are common to more than one function must be programmed separately for each function.

Unlike that of the PROM, the AND matrix of the PAL can easily be expanded to include more variables without doubling the size of the fuse matrix for each new variable.

### 8.2

# PAL Fuse Matrix and Combinational Outputs

> **Cell** *A fuse location in a programmable logic device, specified by the intersection of an input line and a product line.*
>
> **Product line first cell number** *The lowest cell number on a particular product line in a PAL AND matrix where all cells are consecutively numbered.*
>
> **Input line number** *A number assigned to a true or complement input line in a PAL AND matrix.*
>
> **Multiplexer** *A circuit that selects one of several signals to be directed to a single output.*

Figure 8.5 shows the logic diagram of a PAL16L8 PAL circuit. This device can produce up to eight different sum-of-products expressions, one for each group of AND and OR gates. The device has active-LOW tristate outputs, as indicated by the "L" in the part number. Each is controlled by a product line from the related AND matrix.

The pins that can be used only as inputs or outputs are marked "I" or "O," respectively. Six of the pins can be used as inputs or outputs and are marked "I/O." The I/O pins can also feed back a derived Boolean expression into the matrix, where it can be employed as part of another function.

The part number of a PAL device gives the designer information about the number of inputs and outputs and their configurations, as follows:

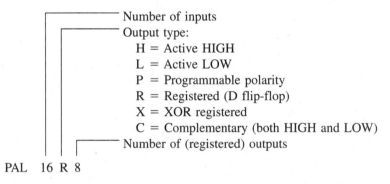

```
              ┌─────────────── Number of inputs
              │  ┌──────────── Output type:
              │  │                H = Active HIGH
              │  │                L = Active LOW
              │  │                P = Programmable polarity
              │  │                R = Registered (D flip-flop)
              │  │                X = XOR registered
              │  │                C = Complementary (both HIGH and LOW)
              │  │  ┌─────────── Number of (registered) outputs
              │  │  │
    PAL   16  R  8
```

The numbering system has some potential ambiguities. For example, it is not possible to use 16 inputs and 8 outputs in a PAL16L8 device at the same time; 6 of the inputs are actually input/output pins. Some possible configurations are as follows:

16 inputs (10 dedicated + 6 I/O) and 2 dedicated outputs

10 inputs and 8 outputs (2 dedicated + 6 I/O)

12 inputs (10 dedicated + 2 I/O) and 6 outputs (2 dedicated + 4 I/O)

Each of the outputs of the PAL16L8 is buffered by a tristate inverter, whose *ENABLE* input is controlled by its own product line. When the *ENABLE* line of the tristate inverter is HIGH, the inverter output is the same as it would normally be—a logic HIGH or LOW, determined by the state of the corresponding OR gate output.

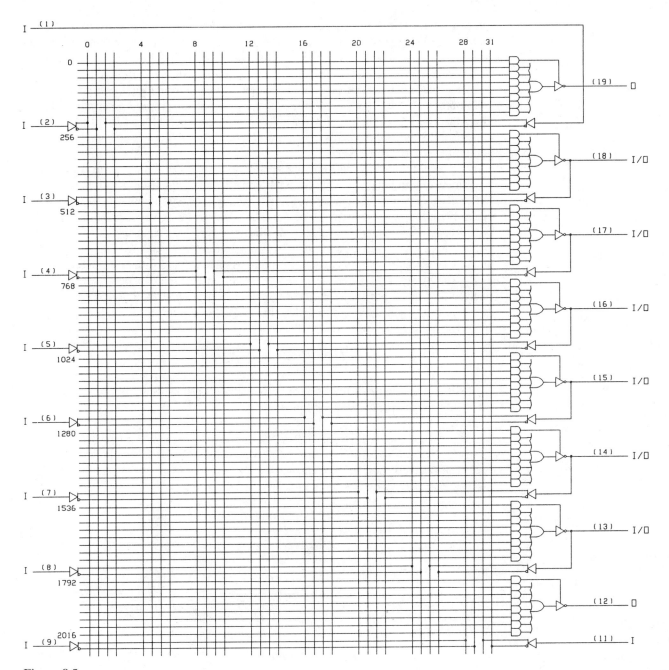

**Figure 8.5**
**Unprogrammed PAL16L8**

When the *ENABLE* line is LOW, the inverter output is in the high-impedance state. The output acts as an open circuit, neither HIGH nor LOW; it is as though the output was completely disconnected from the circuit. The inverter is permanently enabled if all fuses on the *ENABLE* product line are blown, and permanently disabled if these fuses are all intact.

Published logic diagrams of PAL devices generally do not have fuses drawn on them. This allows us to draw fuses for any application. In practice, PLDs have become too complex to manually draw fuse maps for most applications. Fuse assignment is generally done using special software. Historically, PLD programming would begin with fuses drawn on such a diagram, and each fuse would be selected and blown individually by someone operating a hardware device constructed for such a purpose.

Fuse locations, called **cells,** are specified by two numbers: the **product line first cell number,** shown along the left side of the diagram, and the **input line number,** shown along the top. The address of any particular fuse is the sum of its product line first cell number and its input line number. The fuses on the PAL16L8 device are numbered from 0000 to 2047 ($= 2016 + 31$).

In order to examine the general principle of fuse programming, let us develop the programmed logic diagram for a common combinational circuit: a 4-to-1 **multiplexer.** (After developing the fuse maps for several examples, we will not refer to this technique again. From that point on, we will be more concerned with development software.)

This circuit, shown in Figure 8.6a, directs one of four input logic signals, $D_0$ to $D_3$, to output $Y$, depending on the state of two select inputs $S_0$ and $S_1$. The circuit works on the enable/inhibit principle; each AND gate is enabled by a different combination of $S_1 S_0$. The binary state of the select inputs is the same as the decimal subscript of the selected data input. For instance, $S_1 S_0 = 10$ selects data input $D_2$; the AND gate corresponding to $D_2$ is enabled and the other three ANDs are inhibited.

The logic equation for output $Y$ is given by:

$$Y = D_0 \, \overline{S}_1 \, \overline{S}_0 + D_1 \, \overline{S}_1 \, S_0 + D_2 \, S_1 \, \overline{S}_0 + D_3 \, S_1 \, S_0$$

Since the outputs of the PAL16L8 are active LOW, as illustrated in Figure 8.6b, we should rewrite the equation as follows:

$$\overline{Y} = \overline{D}_0 \, \overline{S}_1 \, \overline{S}_0 + \overline{D}_1 \, \overline{S}_1 \, S_0 + \overline{D}_2 \, S_1 \, \overline{S}_0 + \overline{D}_3 \, S_1 \, S_0$$

The $D$ inputs must be complemented to reverse the effect of the active-LOW output. The output is enabled when the $EN$ input is HIGH. Figure 8.7 shows the PAL16L8A logic diagram with fuses for the multiplexer application.

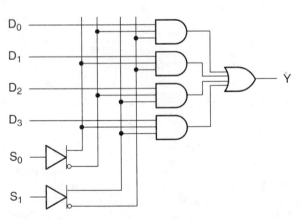

a. Active-HIGH output

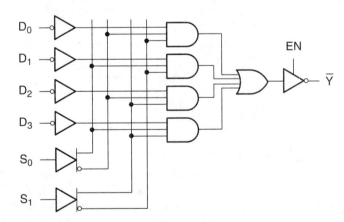

b. Active-LOW output

**Figure 8.6**

**4-to-1 Multiplexer Circuits**

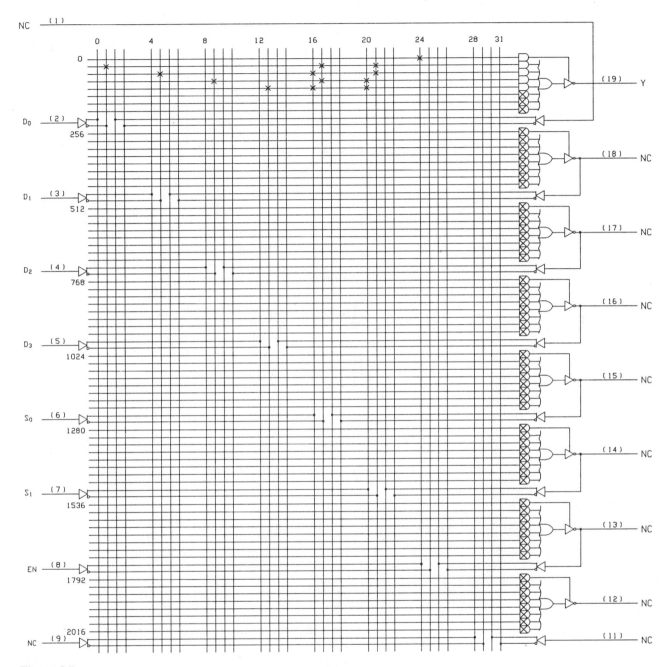

**Figure 8.7**
**Programmed Logic Diagram for a 4-to-1 Multiplexer**

# 8.3
# PAL Outputs With Programmable Polarity

The multiplexer application developed above uses a PAL device whose output is always fixed at the active-LOW polarity. This fixed polarity is suitable for most applications,

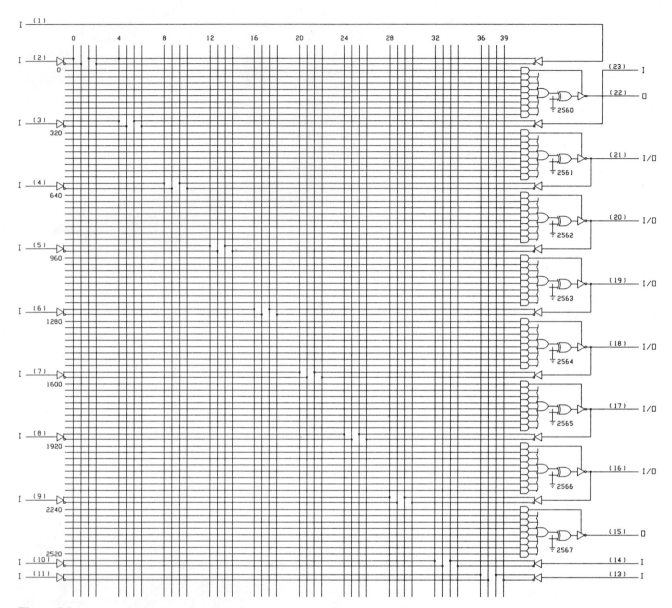

**Figure 8.8**

**PAL20P8 Logic Diagram**

but Boolean functions that would normally have active-HIGH outputs must be implemented in DeMorgan equivalent form, which is not always very straightforward.

Some applications require both active-HIGH and active-LOW outputs. In such cases, it is useful to have a device whose output polarity is fuse programmable.

Figure 8.8 shows the logic diagram of a PAL20P8 PAL device. This device is the same as a PAL16L8, except that there are four more dedicated inputs, and the polarity of each output is programmable. The Exclusive OR gate on each output is programmed to act as either an inverter or a buffer. When its associated fuse is intact, the XOR input is grounded and passes the output of its related SOP network in true form. When combined with the output inverter, this produces an active-LOW output. When the polarity fuse is blown, the fused XOR input floats to the HIGH state, inverting the SOP output; the output pin becomes active HIGH.

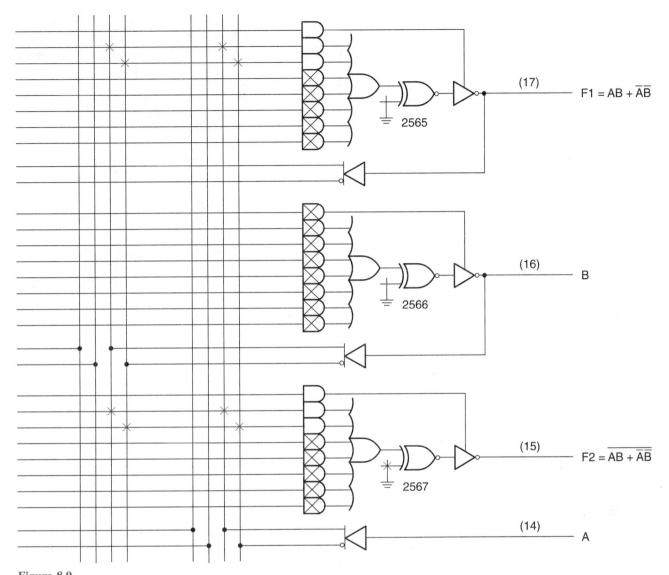

**Figure 8.9**

**PAL Outputs With Programmable Polarity**

The polarity fuses are given numbers higher than those of the main fuse array. In this case, the product line fuses are numbered 0000 to 2559 and the output polarity fuses are numbered 2560 to 2567.

Figure 8.9 illustrates the selection of output polarity. Two Boolean functions, *F1* and *F2*, are programmed into the fuse array, with outputs at pins (17) and (15), respectively. The equations are:

$$F1 = A\,B + \overline{A}\,\overline{B}$$

$$F2 = \overline{A\,B + \overline{A}\,\overline{B}}$$

We could, if we chose, rewrite *F2* to show the output as active LOW:

$$\overline{F2} = A\,B + \overline{A}\,\overline{B}$$

The portion of the PAL20P8 logic diagram shown in Figure 8.9 represents the fuses required to program *F1* and *F2*. Pins (14) and (16) supply inputs *A* and *B* to the matrix. The *ENABLE* lines of the tristate output buffers float HIGH, since all fuses are blown on the corresponding product lines, thus permanently enabling the output buffers.

The fuses numbered 2565 and 2567 select the polarity at pins (15) and (17). Fuse 2565 is blown. The fused input to the corresponding XOR gate floats HIGH, thus making the gate into an inverter. Combined with the tristate buffer, this makes pin (17) active HIGH.

Fuse 2567 is intact. This grounds the input to the corresponding XOR gate, making the gate into a noninverting buffer. Combined with the tristate output buffer, this makes pin (15) active LOW.

---

**EXAMPLE 8.1**

Show how a PAL20P8 device can be used to implement the following logic functions by drawing fuses on the device's logic diagram.

$$\text{NOT:} \quad F_1 = \overline{A}$$
$$\text{AND:} \quad F_2 = BC$$
$$\text{OR:} \quad F_3 = D + E$$
$$\text{NAND:} \quad F_4 = \overline{FG}$$
$$\text{NOR:} \quad F_5 = \overline{H + J}$$
$$\text{XOR:} \quad F_6 = K \oplus L = \overline{K} L + K \overline{L}$$
$$\text{XNOR:} \quad F_7 = \overline{M \oplus N} = \overline{M}\,\overline{N} + M N$$

How would the implementation of these logic functions differ if only active-LOW outputs were available, as in a PAL16L8?

**Solution**   The PAL20P8 has 14 dedicated inputs, 2 dedicated outputs, and 6 lines that can be used as inputs or outputs. Our functions need 13 input variables and 7 output variables. We will use six I/O pins (pins (16) through (21)) and one dedicated output (pin (15)) for the output variables.

All functions must be in SOP form. Outputs for NOT, AND, OR, Exclusive OR, and Exclusive NOR are active HIGH. Therefore, polarity fuses on the outputs for *F1, F2, F3, F6,* and *F7* are blown. NAND and NOR outputs are active LOW; the polarity fuses for *F4* and *F5* remain intact.

Figure 8.10 shows the logic diagram of the programmed PAL. If only active-LOW outputs were available, we would need to rewrite some of the equations to make the outputs correspond to their DeMorgan equivalent forms, as follows:

$$\text{AND:} \quad F_2 = \overline{\overline{B} + \overline{C}}$$
$$\text{OR:} \quad F_3 = \overline{\overline{D}\,\overline{E}}$$
$$\text{XOR:} \quad F_6 = K \oplus L = \overline{\overline{K}\,\overline{L} + K L}$$
$$\text{XNOR:} \quad F_7 = \overline{M \oplus N} = \overline{\overline{M} N + M \overline{N}}$$

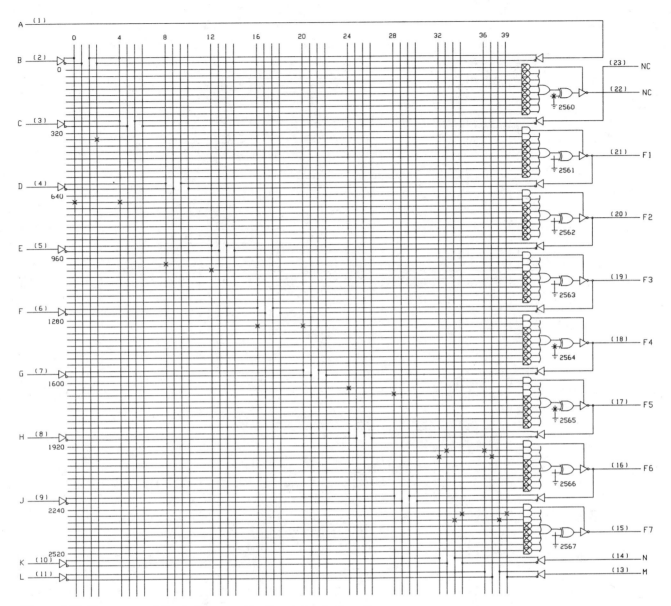

**Figure 8.10 Programmed Logic Diagram for Seven Logic Functions**

## 8.4

# PAL and GAL Devices With Registered Outputs

**Register** *A digital circuit such as a flip-flop or array of flip-flops that stores one or more bits of digital information.*

**Registered output** *An output of a programmable array logic (PAL) device having a flip-flop (usually D-type) that stores the output state.*

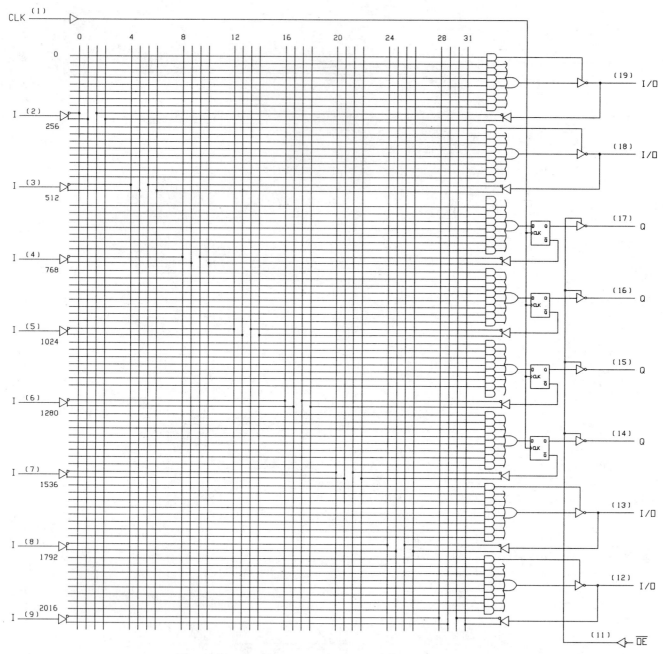

**Figure 8.11**
**PAL16R4 Logic Diagram**

Flip-flops are generally found in programmable logic devices as **registered outputs.** A **register** is one or more flip-flops used to store data. Registered outputs in programmable array logic (PAL) devices can be used for the same functions as individual flip-flops.

Figure 8.11 shows the logic diagram of a PAL device with four registered outputs: a PAL16R4. The fuse matrix is identical to that of a PAL16L8 device; the differences between the two devices are the registered outputs, a dedicated clock input (pin 1), and a pin for enabling all registered outputs (pin 11).

A PAL device with eight registered outputs, a PAL16R8, is shown in Figure 8.12. The fuse matrix is also the same as for a PAL16L8. The *CLOCK* and registered *OUTPUT ENABLE* pins are the same as those of the PAL16R4.

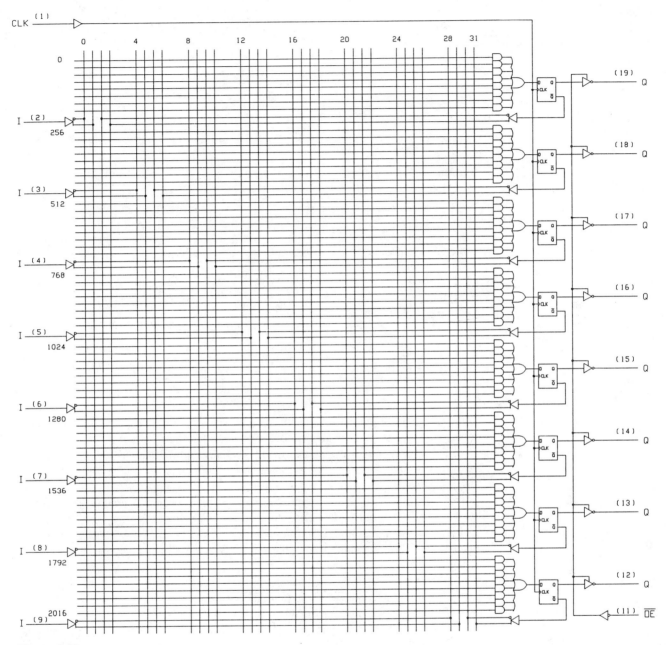

**Figure 8.12**

**PAL16R8 Logic Diagram**

With Registered PAL, the number of outputs shown in the part number indicates the number of registered outputs. For example, a PAL16R4 device has four registered outputs and four combinational I/O pins, a PAL16R6 device (not shown) has six registered outputs and two combinational I/O pins, and a PAL16R8 has eight registered outputs.

---

**EXAMPLE 8.2**

A common data operation is that of "rotation." Figure 8.13 illustrates how a 4-bit number can be rotated to the right by 0, 1, 2, or 3 places. To rotate the data, move all bits the

**Figure 8.13**
**Example 8.2**
**Rotation to the Right of 4-Bit Data**

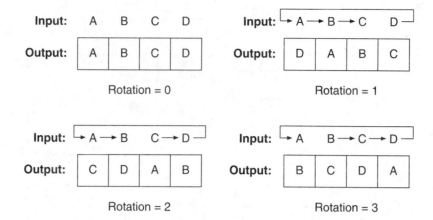

**Table 8.1** Rotation to the Right by a Selectable Number of Bits

| $S_1$ | $S_0$ | $Q_A$ | $Q_B$ | $Q_C$ | $Q_D$ | Rotation |
|-------|-------|-------|-------|-------|-------|----------|
| 0 | 0 | A | B | C | D | 0 |
| 0 | 1 | D | A | B | C | 1 |
| 1 | 0 | C | D | A | B | 2 |
| 1 | 1 | B | C | D | A | 3 |

required number of places to the right. As data reach the rightmost position, move them to the beginning so that they are transferred in a closed loop.

This operation is usually performed by serially shifting the data the required number of places and feeding back the last output to the first input of a serial shift register. (See Examples 6.8 and 6.9 for a description of serial shifting.)

Rotation can also be accomplished by a parallel transfer operation. We can load the bits of the input into four D flip-flops in the order determined by two select inputs, $S_1$ and $S_0$. Assume that the binary number $S_1 S_0$ is the same as the Rotation number in Figure 8.13. Table 8.1 summarizes the contents of the circuit after one clock pulse is applied.

Sketch a circuit, using gates and flip-flops, that can accomplish this rotation as a parallel transfer function. Briefly explain its operation.

Write the Boolean expression(s) for the circuit.

Show how the circuit can be implemented by a PAL16R4 device by drawing fuses on its logic diagram.

**Solution**

Figure 8.14 shows a parallel transfer circuit that will perform the specified rotation. The circuit works by enabling one AND gate in each group of four for each combination of $S_1$ and $S_0$. For example, when $S_1 S_0 = 00$, the rotation is 0 and the leftmost AND gate of each group is enabled, transferring the parallel data into the flip-flops so that $D_A = A$, $D_B = B$, $D_C = C$, and $D_D = D$. After one clock pulse, $Q_A Q_B Q_C Q_D = ABCD$.

Similarly, if $S_1 S_0 = 10$, we select a rotation of 2. The third AND gate from the left is selected in each group of four. This makes the data $D_A = C$, $D_B = D$, $D_C = A$, and $D_D = B$ appear at the flip-flop inputs. After one clock pulse, $Q_A Q_B Q_C Q_D = CDAB$.

The same principle governs the circuit operation for the other two select codes.

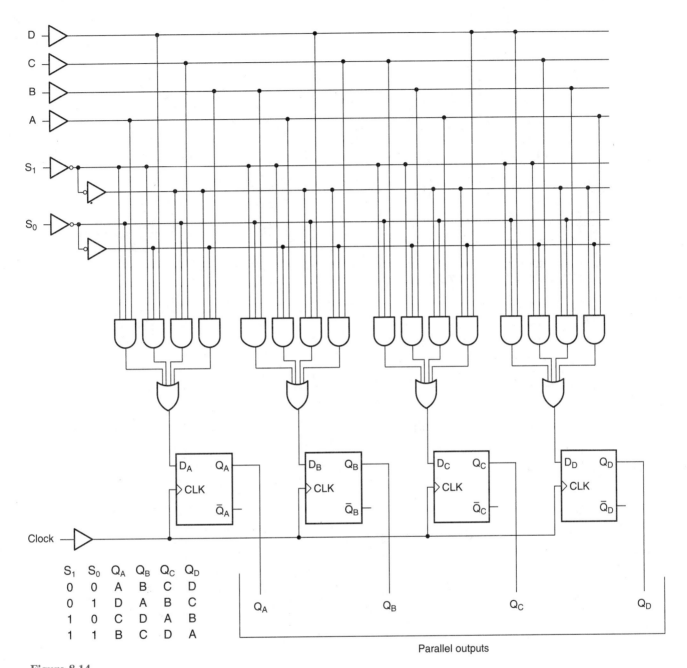

**Figure 8.14**

**Example 8.2**
**Rotation by Parallel Transfer**

When writing Boolean equations for registered PLD outputs, we use the symbol := rather than = as an assignment operator. (In other words, the := symbol assigns the Boolean expression to the variable on the lefthand side of the equation.) The Boolean equations for the circuit are:

$$Q_A := \overline{S}_1\,\overline{S}_0\,A + \overline{S}_1\,S_0\,D + S_1\,\overline{S}_0\,C + S_1\,S_0\,B$$

$$Q_B := \overline{S}_1\,\overline{S}_0\,B + \overline{S}_1\,S_0\,A + S_1\,\overline{S}_0\,D + S_1\,S_0\,C$$

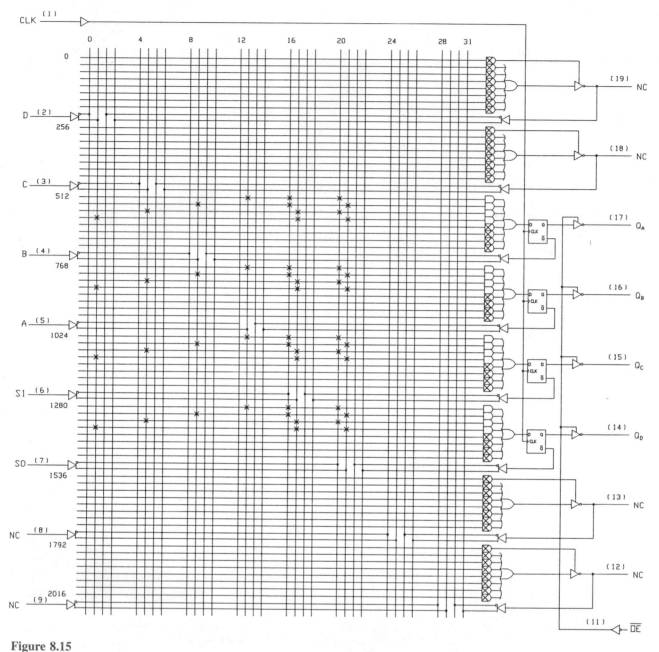

**Figure 8.15**

**Example 8.2**
**Programmed PLD for Selectable Bit Rotation**

$$Q_C := \overline{S}_1 \, \overline{S}_0 \, C + \overline{S}_1 \, S_0 \, B + S_1 \, \overline{S}_0 \, A + S_1 \, S_0 \, D$$

$$Q_D := \overline{S}_1 \, \overline{S}_0 \, D + \overline{S}_1 \, S_0 \, C + S_1 \overline{S}_0 \, B + S_1 \, S_0 \, A$$

These equations imply that each registered output requires us to use four product lines, one for each product term. The programmed logic diagram is shown in Figure 8.15.

## Generic Array Logic (GAL)

**Generic array logic (GAL)**  *A type of programmable logic whose outputs can be configured in a variety of different ways (e.g., registered, combinational, or tristate), depending on the design requirements.*

**Output logic macrocell (OLMC)**  *A programmable output cell used as the output of a generic array logic device that can be configured as a registered, combinational, or tristate output of either polarity, with or without feedback, or as an input.*

**Architecture cell**  *A programmable cell within a generic array logic device that, along with several other such cells, sets the configuration of an output logic macrocell.*

**Global architecture cell**  *An architecture cell that affects all OLMCs within a GAL device.*

**Local architecture cell**  *An architecture cell that affects only one OLMC within a GAL device.*

A refinement of the idea of programmable array logic is **generic array logic (GAL)**[2]. A GAL is a programmable device configured by a sum-of-products fuse matrix, much like PAL. There are two main differences between PAL and GAL. First, the output configurations of a GAL device are programmable, so that they can be combinational or registered, according to design needs. Second, GALs are electrically erasable; if a mistake is found in the fuse matrix or in the fundamental design of a device or if the design version is updated, all cells in the same device can be "unprogrammed" and reprogrammed as required.

There are two basic GAL designs. One type simply emulates a variety of different standard PAL devices. The device to be emulated is selected by the GAL development software. A second type can be configured in any way that the design requires, without reference to any standard PAL device. Thus, the latter type introduces one further degree of flexibility into design.

### GAL16V8

Figure 8.16 shows a GAL of the first type—the GAL16V8. The "V" in the GAL part number indicates a variable-output architecture, meaning it can be programmed so that the output is combinational or registered. The GAL16V8 emulates a fairly large variety of standard PAL devices in three distinct categories: Small PAL, Medium PAL, and Registered PAL. Small and Medium PAL are devices with combinational outputs only. The designations Small and Medium refer to the complexity of their fuse matrices.

Table 8.2 shows the PAL devices in each category that are replaceable by a GAL16V8 device. Each of the emulated PAL devices has a characteristic set of inputs and outputs. The output types are selected by programming an **output logic macrocell (OLMC)** for each output pin.

---

[2]GAL is a registered trademark of Lattice Semiconductor Inc.

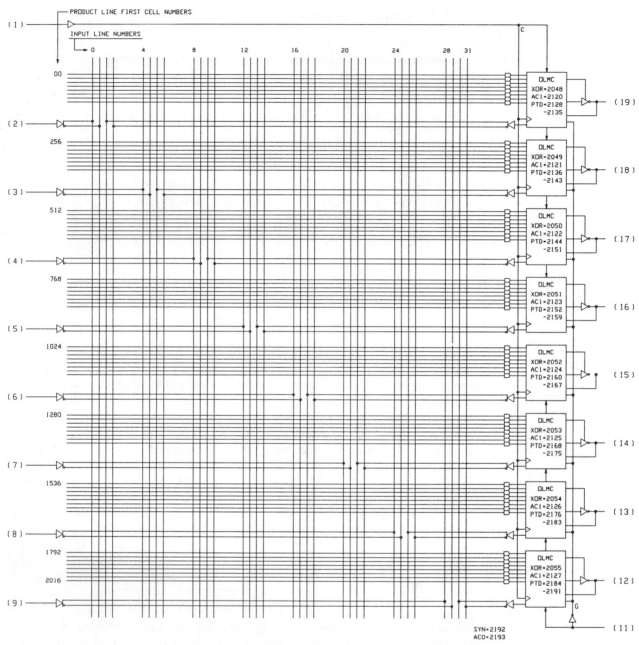

**Figure 8.16**

**Logic Diagram of a GAL16V8**

Figure 8.17 shows the types of outputs that each OLMC can emulate. Figure 8.17a, b, and c are combinational outputs with or without feedback and/or tristate output control. (The functions of the output configurations in Figure 8.17a and b can be performed by that of Figure 8.17c. Three configurations are available for historical reasons; Small PAL devices have the first two types, and Medium PAL the third type of output.)

Figure 8.17d is a registered output with tristate control. All configurations have programmable polarity controlled by the fused XOR gate. Which configuration is assigned to which output depends primarily on the type of PAL that the GAL emulates.

**Table 8.2**  PAL Devices Replaceable by GAL16V8

| | **Small PAL** | | |
|---|---|---|---|
| PAL10L8 | PAL12L6 | PAL14L4 | PAL16L2 |
| PAL10H8 | PAL12H6 | PAL14H4 | PAL16H2 |
| PAL10P8 | PAL12P6 | PAL14P4 | PAL16P2 |
| | **Medium PAL** | | |
| PAL16L8 | | | |
| PAL16H8 | | | |
| PAL16P8 | | | |
| | **Registered PAL** | | |
| PAL16R8 | PAL16R6 | PAL16R4 | |
| PAL16RP8 | PAL16RP6 | PAL16RP4 | |

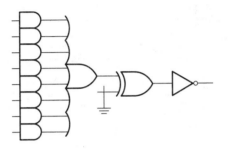

**a. Active combinatorial output**

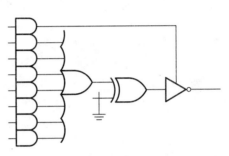

**b. Tristate combinatorial output**

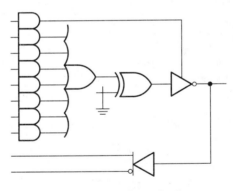

**c. Combinatorial input/output**

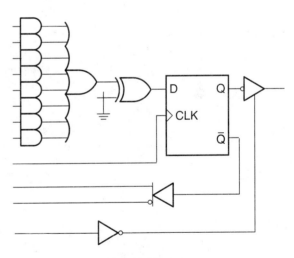

**d. Registered output**

**Figure 8.17**

**OLMC Configurations for a GAL16V8**

Table 8.3 lists the configuration of each pin of a GAL16V8 according to the type of device it emulates. Pins 2 to 9 are not shown on the table; these pins are dedicated inputs. Also not shown are pin 10 (ground) and pin 20 $(V_{CC})$.

Table 8.3 shows that for a PAL12L6, for example, pins 1, 11, 12, and 19 are configured as dedicated inputs; pins 13 through 18 are outputs of the type shown in Figure 8.17a.

**Table 8.3** GAL16V8 Output Configurations

| Pin | Small PAL | | | | Medium PAL | Registered PAL | | |
|---|---|---|---|---|---|---|---|---|
| 1 | IN | IN | IN | IN | IN | CLK | CLK | CLK |
| 11 | IN | IN | IN | IN | IN | OE | OE | OE |
| 12 | OUT | IN | IN | IN | TRI | REG | I/O | I/O |
| 13 | OUT | OUT | IN | IN | I/O | REG | REG | I/O |
| 14 | OUT | OUT | OUT | IN | I/O | REG | REG | REG |
| 15 | OUT | OUT | OUT | OUT | I/O | REG | REG | REG |
| 16 | OUT | OUT | OUT | OUT | I/O | REG | REG | REG |
| 17 | OUT | OUT | OUT | IN | I/O | REG | REG | REG |
| 18 | OUT | OUT | IN | IN | I/O | REG | REG | I/O |
| 19 | OUT | IN | IN | IN | TRI | REG | I/O | I/O |
| | 10L8 | 12L6 | 14L4 | 16L2 | 16L8 | 16R8 | 16R6 | 16R4 |
| | 10H8 | 12H6 | 14H4 | 16H2 | 16H8 | 16RP8 | 16RP6 | 16RP4 |
| | 10P8 | 12P6 | 14P4 | 16P2 | 16P8 | | | |

IN: Dedicated input
OE: Output Enable pin
CLK: Clock input
OUT: Combinational output (Figure 8.17a)
TRI: Tristate combinational output (Figure 8.17b)
I/O: Input/output (Figure 8.17c)
REG: Registered output (Figure 8.17d)

**Table 8.4** GAL16V8 Architecture Cells

**Global Architecture Cells:**

$ACO = 0$ for Small PAL mode
$ACO = 1$ for Medium/Registered PAL mode
$SYN = 0$ for registered output (p1 = $CLK$, p11 = $OE$)
$SYN = 1$ for combinational only (p1, p11 dedicated inputs)

**Local Architecture Cells:**

$PTD$ selects an active product line
Small PAL
    $AC1 = 0$ selects pin as output
    $AC1 = 1$ selects pin as input
Medium/Registered PAL
    $AC1 = 0$ selects pin as registered
    $AC1 = 1$ selects pin as combinational

The table also shows that for a PAL16R6, pin 1 is the clock input, pin 11 is *OUTPUT ENABLE,* pins 12 and 19 are tristate input/output pins of the type shown in Figure 8.17c, and pins 13 through 18 are registered outputs, as shown in Figure 8.17d.

Similar interpretations can be made for the other devices in the table.

The output types are selected internally by the status of several **architecture cells,** designated *SYN, ACO, AC1, PTD$_n$,* and *XOR*. These are cells with addresses as indicated on the GAL logic diagram and automatically programmed by GAL development software. *ACO* and *SYN* are called **global architecture cells** since they affect the configuration of the entire device. *AC1, PTD,* and *XOR* are **local architecture cells** that affect only the configuration of the OLMC to which they belong.

Table 8.4 shows the states of these architecture cells for the various output programming options.

EXAMPLE 8.3    Figure 8.18 (see page 288) shows a 6-bit data transfer circuit that combines the features of a parallel transfer circuit and a serial shift register. Briefly explain the circuit operation and write the Boolean equations required so that a GAL16V8 device can be programmed to implement this function.

Briefly explain why a GAL16V8 cannot be programmed to make an 8-bit serial/parallel shift register.

Solution    When the *SHIFT/LOAD* input is HIGH, the upper AND gate of each pair is enabled, connecting data from the *SERIAL IN* terminal to $D_A$ and also connecting the $Q$ output of each flip-flop to the $D$ input of the next. Thus, data shift serially from one flip-flop to the next and move over one place with each applied clock pulse.

When *SHIFT/LOAD* is LOW, the lower AND gate of each pair is enabled and data at *A, B, C, D, E,* and *F* are loaded in parallel into the flip-flops when a clock pulse is applied.

The Boolean equations required to program this shift register are:

$$Q_A := SL \cdot SI + A \cdot \overline{SL}$$

$$Q_B := SL \cdot Q_A + B \cdot \overline{SL}$$

$$Q_C := SL \cdot Q_B + C \cdot \overline{SL}$$

$$Q_D := SL \cdot Q_C + D \cdot \overline{SL}$$

$$Q_E := SL \cdot Q_D + E \cdot \overline{SL}$$

$$Q_F := SL \cdot Q_E + F \cdot \overline{SL} \quad \text{where } SI = \text{Serial Input and } SL = SHIFT/\overline{LOAD}$$

The GAL16V8 cannot be used to make an 8-bit serial/parallel shift register, because the device does not have enough inputs. We need 8 parallel inputs, *CLK,* and a *SHIFT/LOAD* input, a total of 10 inputs. There are only 9 available.

## GAL22V10

Figure 8.19 shows the logic diagram of a GAL22V10 generic array logic device. This industry-standard device is possibly the most widely used PLD in the world today. Unlike the GAL16V8, the 22V10 can be programmed any way your design requires, without reference to any standard PAL configuration. The GAL22V10 has a number of other features that make it superior to the GAL16V8.

1. There are more outputs (10 as opposed to 8 for the 16V8).

2. There are more inputs (11 dedicated inputs, plus any I/O lines used as inputs).

3. The output logic macrocells are of different sizes, allowing expressions with larger numbers of product terms in some OLMCs than others. There are two OLMCs with each of the following numbers of product lines: 8, 10, 12, 14, and 16. This allows more flexibility in design, while minimizing the number of product lines.

4. OLMC configuration is much simpler than that of a GAL16V8. Two architecture cells per macrocell, $S_0$ and $S_1$, select the output type, as shown in Figure 8.20.

**Figure 8.18**
**Example 8.3**
**Parallel/Serial Shift Register**

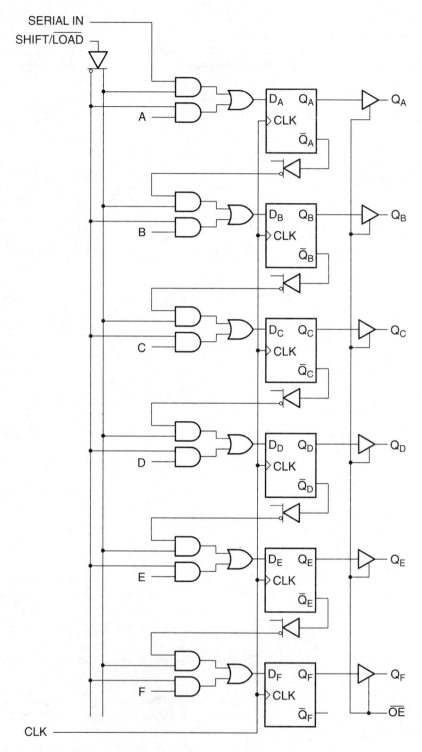

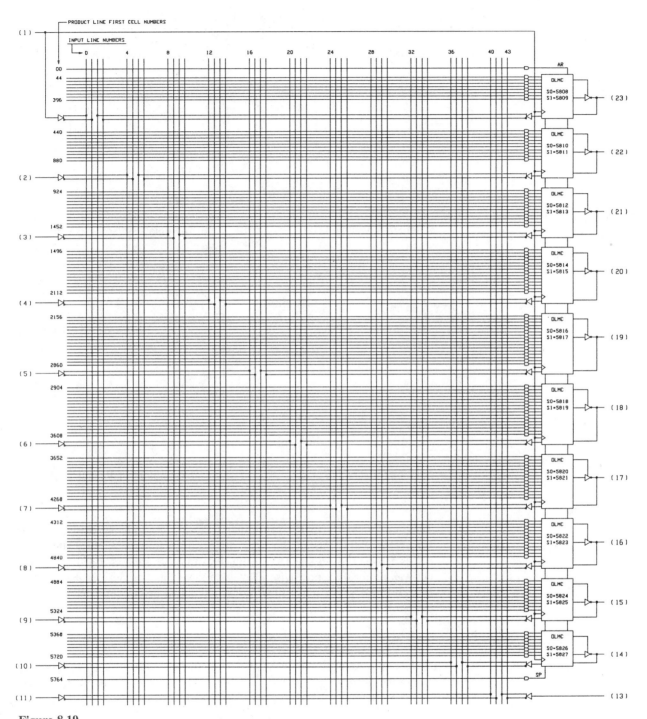

**Figure 8.19**

**GAL22V10 Logic Diagram**

**Figure 8.20**

**GAL22V10 OLMC Configurations**

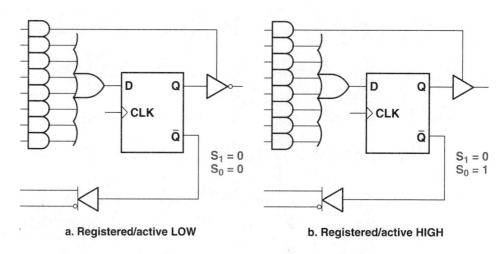

a. Registered/active LOW     b. Registered/active HIGH

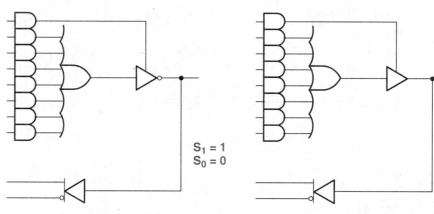

c. Combinatorial/active LOW     d. Combinatorial/active HIGH

5. There are product lines for Synchronous Preset *(SP)* and Asynchronous Reset *(AR)*. The *SP* line sets all flip-flops HIGH on the first clock pulse after it becomes active. The *AR* line sets all flip-flops LOW as soon as it activates, without waiting for the clock pulse. (Note that these lines set or reset the *Q* output of each flip-flop. An active-LOW registered output inverts this state at the output pin.)

**EXAMPLE 8.4**

In Example 8.3, we were unable to make an 8-bit parallel/serial shift register because a GAL16V8 does not have enough inputs. Write the Boolean equations required to make an 8-bit parallel/serial shift register from a GAL22V10.

**Solution**

$$Q_A := SL \cdot SI + A \cdot \overline{SL}$$

$$Q_B := SL \cdot Q_A + B \cdot \overline{SL}$$

$$Q_C := SL \cdot Q_B + C \cdot \overline{SL}$$

$$Q_D := SL \cdot Q_C + D \cdot \overline{SL}$$

$$Q_E := SL \cdot Q_D + E \cdot \overline{SL}$$

$$Q_F := SL \cdot Q_E + F \cdot \overline{SL}$$

$$Q_G := SL \cdot Q_F + G \cdot \overline{SL}$$

$$Q_H := SL \cdot Q_G + H \cdot \overline{SL}$$

## 8.5

# UV-Erasable Programmable Logic Devices (EPLD)

**Superset** *Containing all features of another device plus some additional features not contained in the original device.*

**Erasable programmable logic device (EPLD)** *A PLD with programmable input/output macrocells whose AND matrix can be erased by exposure to ultraviolet light of a specified wavelength.*

**Windowed CERDIP** *A ceramic dual in-line package having a quartz window that passes ultraviolet light.*

**OTP** *One-time programmable. Refers to a programmable device, such as a PLD, that cannot be erased after being programmed.*

The latest generation of PLDs offered by some manufacturers, such as Intel, incorporates several upgrade, or **superset,** features over earlier types of PLDs.

One superset feature is erasability. These **erasable PLDs (EPLDs)** can be erased by exposure to ultraviolet (UV) light. (GALs are also erasable, but by electrical means.) EPLDs packaged in ceramic DIP packages with a quartz window allow the UV light to fall on the die (the silicon chip). High-intensity UV radiation applied at a distance of about 2.5 cm (1 in.) and having a wavelength of 2537 angstroms will erase an EPLD in about 2 hours (as opposed to about 20 minutes for other types of UV-eraseable devices, such as EPROMs). Sunlight and fluorescent lighting will also erase these devices over a period of weeks or years, depending on the source, so the quartz window of a programmed device should be covered with an opaque label.

Other superset features depend on the particular device.

Note that not all EPLDs are packaged in the **windowed CERDIP.** System speeds of modern microprocessor and high-performance digital systems are such that DIP packages are not suitable for many designs. This is because the $V_{CC}$ and ground pins in a DIP are at the corners of the package and thus have the longest bonding wires to the silicon die. This is enough of a difference to degrade the performance of the chip at high clock speeds due to increased propagation delay and inductance.

Some EPLDs are now available only in package types other than DIPs, such as PLCC (plastic leaded chip carrier). These types of packaging do not allow device erasure by UV light, because they cannot be manufactured with a quartz window. These are known as **OTP,** or one-time programmable, devices. The next generation of PLCC devices will likely be electrically erasable, as GALs are.

## 85C220/224

Figure 8.21 shows the architecture of an Intel 85C220 EPLD. This device, which has ten dedicated inputs and eight I/O macrocells, is designed as a superset direct replacement for a number of standard 20-pin PAL and GAL devices. The 85C220 will replace any standard PAL or GAL having one of the suffixes listed in Table 8.5.

The 85C220 has the following superset features over the PAL and GAL devices listed in Table 8.5.

1. *Each macrocell* can be configured as an active-HIGH or active-LOW combinational or registered output. Feedback on each macrocell can be combinational, registered, or directly from the pin (i.e., an input).

2. The SOP network of each macrocell has *eight product terms,* rather than seven.

3. A ninth product line is supplied for an *individual Output Enable at each macrocell.* Registered PALs have only a global Output Enable; *OE* is not controllable for individual registered outputs.

4. *Feedback is available on all eight macrocells,* rather than six in a device such as the 16L8.

The global clock goes to all macrocells to provide synchronous clocking.

Figure 8.22 shows the architecture of the 85C224 EPLD. This device has all the superset features of the 85C220 and directly replaces those 24-pin PAL and GAL devices listed in Table 8.6.

## 85C22V10

The 85C22V10 EPLD has the same logic diagram as other 22V10 devices, but contains two superset features.

1. A *clock invert* feature allows the OLMC register to be clocked on a rising or falling edge.

2. Each macrocell has an additional architecture cell that allows *four more OLMC configurations* than does the standard 22V10. In other 22V10s, the feedback configuration is always the same as the output; registered output means registered feedback, and combinational output means feedback from an I/O pin. The extra architecture cell allows other combinations, such as combinational output with registered feedback. Table 8.7 summarizes these combinations.

**Table 8.5** 20-Pin PAL/GAL Devices Replaceable by 85C220

| 10H8 | 12H6 | 14H4 | 16H2 | 16R8 | 16V8 |
|------|------|------|------|------|------|
| 10L8 | 12L6 | 14L4 | 16L2 | 16RP8 | 18P8 |
|      |      |      | 16L8 | 16R4 | 18V8 |
|      |      |      | 16P8 | 16RP4 |      |
|      |      |      |      | 16RP6 |      |

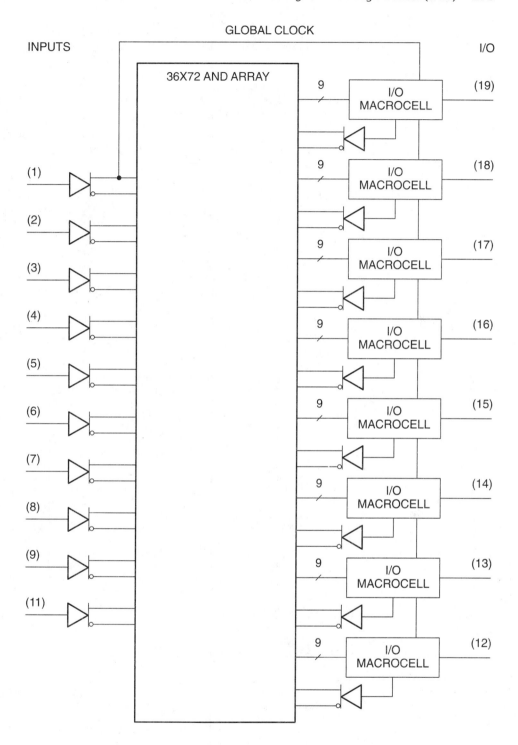

**Figure 8.21**
**85C220 Logic Diagram**

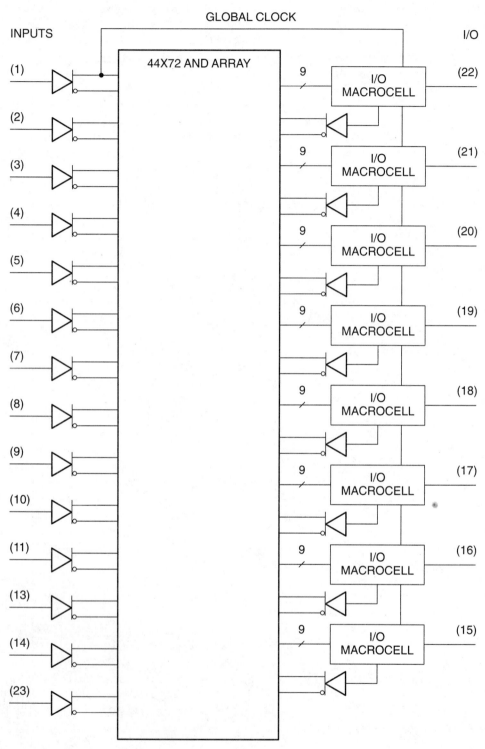

Figure 8.22
85C224 Logic Diagram

Table 8.6  24-Pin PAL/GAL Devices Replaceable by 85C224

| | |
|---|---|
| 14L8 | 20R4 |
| 16L6 | 20R6 |
| 18L4 | 20R8 |
| 20L2 | 20V8 |
| 20L8 | |

Table 8.7  OLMC Configurations for an 85C22V10 EPLD

| S2 | S1 | S0 | Output Type | Feedback |
|----|----|----|-------------|----------|
| 0 | 0 | 0 | Registered/active LOW | Registered |
| 0 | 0 | 1 | Registered/active HIGH | Registered |
| 0 | 1 | 0 | Combinational/active LOW | Pin |
| 0 | 1 | 1 | Combinational/active HIGH | Pin |
| 1 | 0 | 0 | Registered/active LOW | Pin |
| 1 | 0 | 1 | Registered/active HIGH | Pin |
| 1 | 1 | 0 | Combinational/active LOW | Registered |
| 1 | 1 | 1 | Combinational/active HIGH | Registered |

## 85C060/090

Figure 8.23 shows the architecture of an 85C060 EPLD. This is a very flexible device, having 16 I/O macrocells and 4 dedicated inputs.

Each macrocell can be configured as a combinational or registered output of either polarity. If registered, the output can be configured as a D, T, SR, or JK flip-flop. Each macrocell is fed by ten lines from the AND matrix: eight product terms, an Asynchronous Reset *(AR)*, and a selectable Output Enable/Asynchronous Clock *(OE/CLK)*.

There are two synchronous clock inputs, *CLK1* and *CLK2,* each of which clocks eight macrocells. Each macrocell can be configured so that it is clocked synchronously or asynchronously. If synchronous, it is clocked by either *CLK1* or *CLK2,* simultaneous to all other synchronously clocked macrocells in the same group. In this mode, the *OE* product term enables and disables the macrocell tristate output.

If asynchronous, the macrocell output is permanently enabled and the *OE/CLK* line provides independent clocking for that macrocell.

Figure 8.24 shows the architecture of an 85C090 EPLD. This device has the same features as the 85C060, but has more inputs and outputs. There are 24 I/O macrocells and 12 dedicated inputs. Each macrocell can be configured in the same way as those in the 85C060.

**8.6**

# Development Software

**JEDEC**  *Joint Electron Device Engineering Council.*

**JEDEC file**  *An industry-standard form of text file indicating which fuses are blown and which are intact in a programmable logic device.*

**Text file**  *An ASCII-coded document stored on a magnetic disk.*

**Checksum**  *An error-checking code derived from the accumulated sum of the data being checked.*

**Test vector**  *A list of inputs and expected outputs used to test the function of a programmed PLD.*

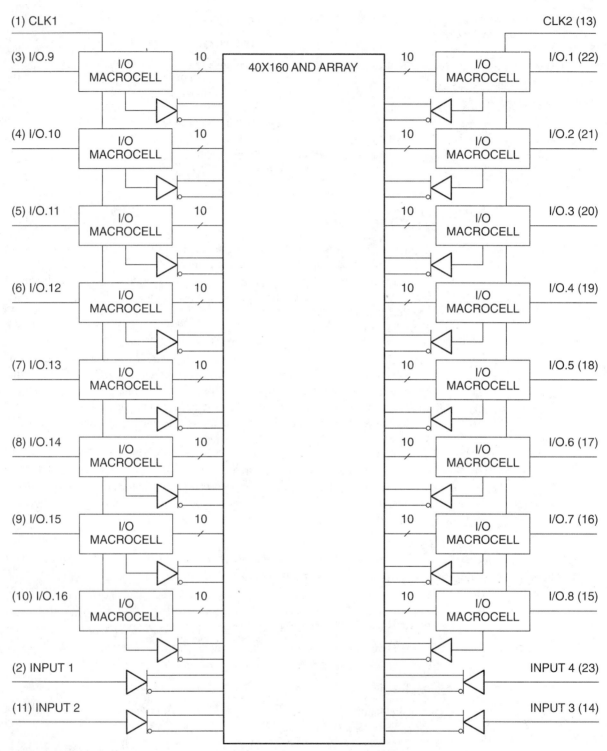

**Figure 8.23**
**85CO60 Logic Diagram**

**Figure 8.24**
**85CO90 Logic Diagram**

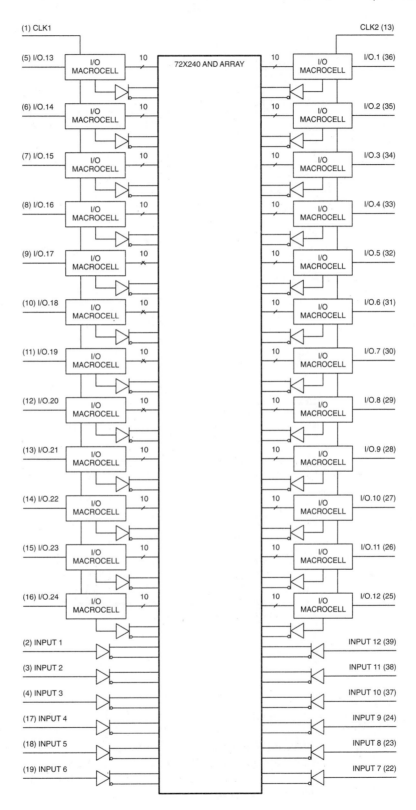

**Figure 8.25**

**Sample JEDEC File (1)**

```
^B
PAL16L8
*
QF2048*QP20*F0*
L0000
1111 1111 1111 1111 1111 1111 1111 1111
1111 1111 0111 1111 1111 1111 0111 1111
1111 1111 0111 1111 1111 1111 1101 1111
1111 1111 1111 1111 1111 1111 0101 1111*
L0256
1111 1111 1111 1111 1111 1111 1111 1111
1110 1111 1111 1011 1111 1111 1111 0111
1110 1111 1111 0111 1111 1111 1111 1011
1101 1111 1111 1011 1111 1111 1111 1011
1101 1111 1111 0111 1111 1111 1111 0111*
L0512
1111 1111 1111 1111 1111 1111 1111 1111
1101 1111 1111 0111 1111 1111 1111 1111
1101 1111 1111 1111 1111 1111 1110 0111
1111 1111 1111 0111 1111 1111 1111 0111*
L0768
1111 1111 1111 1111 1111 1111 1111 1111
1011 1111 1101 1111 1011 1111 1111 1111
1011 1111 1110 1111 0111 1111 1111 1111
0111 1111 1110 1111 1011 1111 1111 1111
0111 1111 1101 1111 0111 1111 1111 1111*
L1024
1111 1111 1111 1111 1111 1111 1111 1111
0111 1111 1111 1111 0111 1111 1111 1111
0111 1111 1101 1111 1111 1111 1111 1111
1111 1111 1101 1111 0111 1111 1111 1111*
L1280
1111 1111 1111 1111 1111 1111 1111 1111
1111 1011 1111 1111 1101 1011 1111 1111
1111 1011 1111 1111 1110 0111 1111 1111
1111 0111 1111 1111 1110 1011 1111 1111
1111 0111 1111 1111 1101 0111 1111 1111*
L1536
1111 1111 1111 1111 1111 1111 1111 1111
1111 0111 1111 1111 1111 0111 1111 1111
1111 0111 1111 1111 1101 1111 1111 1111
1111 1111 1111 1111 1101 0111 1111 1111*
L1792
1111 1111 1111 1111 1111 1111 1111 1111
1111 1111 1011 1111 1111 1111 1001 1111
1111 1111 1011 1111 1111 1111 0110 1111
1111 1111 0111 1111 1111 1111 1010 1111
1111 1111 0111 1111 1111 1111 0101 1111*
L2048
1111 1111*
C8DCF*
^C0000
```

Programming individual fuses in a PAL device is very tedious and error prone for all but the very simplest applications. This task is nearly always done with the help of a computer. A number of computer programs are available to help the digital designer produce the required fuse pattern for any PLD application. Many of these will take inputs such as Boolean equations, truth tables, or other forms and produce the simplest SOP solution to the particular problem.

The end result of such software is a **JEDEC file,** an industry-standard way of listing which fuses in the PLD should remain intact and which should be blown. The JEDEC file is stored as an ASCII **text file** on a magnetic disk. Most PLD programmers will accept the JEDEC file and use it as a template for blowing fuses in the target device.

Figure 8.25 shows an example of a JEDEC file for a PAL16L8 application. (Don't worry about the actual function performed by this PAL; the JEDEC file is shown simply as an example of a standard file format.)

The file starts with an ASCII "Start Text" character (^B). Next is some information required by the PAL programmer about the type of device (PAL16L8), number of fuses (2048), and so forth. The fuse information starts with the line L0000, which is the first product line. The 1s and 0s that follow show the programmed state of each cell in each product line; a 1 is a blown fuse and a 0 is an intact fuse. In other words, each 0 in the JEDEC file represents an X in the same position on the PAL logic diagram.

The product terms for first sum-of-products output are set by the states of fuses 0000 to 0255 (eight product lines). In the file shown, all fuses are blown in the first product line, the second product line shows two intact fuses, and so forth. Since all fuses are intact in the last three of these eight lines, they need not be shown in the JEDEC file.

Whenever some unprogrammed product lines are omitted from the fuse map, the last fuse line shown ends with an asterisk (*). The next line with programmed fuses is indicated by a new fuse number. For example, the second group of fuses (0256 to 0511) in Figure 8.25 begins after the line marked L0256 in the JEDEC file. The remaining fuse lines are similarly indicated.

The JEDEC file in Figure 8.25 ends with a hexadecimal **checksum** (C8DCF), an error-checking code derived from the programming data, and an ASCII "End Text" code (^C).

Some software packages create a JEDEC file that lists all product lines in the PLD matrix, even if they are not all programmed. Figure 8.26 shows such an example. This indicates that JEDEC files can vary somewhat within a standard format, depending on the software used to create it.

Figures 8.25 and 8.26 represent different PLD applications and therefore will have different fuse maps. In Figure 8.26, the fuse map is shown beginning at line 0 and proceeding in sequence without a break. The end of the fuses is indicated by a final asterisk.

At the end of the JEDEC file in Figure 8.26 are a number of **test vectors.** These vectors, shown separately in Figure 8.27, provide a way to test the function of a PLD after it has been programmed, but before it is installed in a circuit board. The PLD programming hardware uses the vectors as a guide to apply specified input levels to the PLD and measure corresponding output levels.

Each vector lists the logic level of each pin in the PLD. The first character in each vector corresponds to pin 1, the second character to pin 2, and so on to pin 20. Thus, Figure 8.27 represents the test vectors for a 20-pin EPLD. (The application is the 4-bit rotation circuit of Example 8.2, fitted for an 85C220 EPLD.)

A 0 or 1 in a test vector represents an input logic level. An H or L specifies an expected HIGH or LOW output level. Pins marked X indicate "don't care" conditions. Pins designated N ($V_{CC}$ and ground) are not tested. We will examine how PLD development software can generate these test vectors in a later part of this chapter.

## Software Vendors

PLD programming software is available from a number of suppliers. One of the first PLD programming languages was PALASM, developed by Advanced Micro Devices, the company holding the trademark for PAL. The early versions of PALASM allowed entry of Boolean equations in sum-of-products forms only. This is not very convenient for even moderately complex designs.

**Figure 8.26**
**Sample JEDEC**
**File (2)**

```
^BTITLE    4-BIT DATA ROTATION
REVISION   1.0
AUTHOR     R.DUECK
COMPANY    SENECA COLLEGE

; 4-BIT PARALLEL LOAD CIRCUIT FOR ROTATE RIGHT FUNCTION
;
;   THIS CIRCUIT ROTATES INCOMING DATA TO THE RIGHT BY THE
; NUMBER OF BITS SPECIFIED BY THE BINARY VALUE OF S1 S0.
; FOR INPUT ABCD, THE ROTATION IS AS FOLLOWS.
;
;           S1 S0   QA QB QC QD  ROTATION
;
;           0 0     A B C D      0
;           0 1     D A B C      1
;           1 0     C D A B      2
;           1 1     B C D A      3
;

OPTIONS: TURBO = ON
SECURITY = OFF
LOC Release [ 2.1 ] SID [ 2.84 ]
*
N PART: 85c220 *
N@ Vcc @20 *
N@ GND @10 *
N@ S1 @2 *
N@ S0 @3 *
N@ A @4 *
N@ B @5 *
N@ C @6 *
N@ D @7 *
N@ CLOCK @1 *
N@ QA @19 *
N@ QB @18 *
N@ QC @17 *
N@ QD @16 *
QF2916*
QP20*
QV14*
X0*
F0*
G0*
L0
11010110111111111111111111111111111
11100111111011111111111111111111111
00000000000000000000000000000000000
00000000000000000000000000000000000
11011011111110111111111111111111111
11101011101111111111111111111111111
00000000000000000000000000000000000
00000000000000000000000000000000000
11111111111111111111111111111111111
11010111101111111111111111111111111
11100111111011111111111111111111111
00000000000000000000000000000000000
00000000000000000000000000000000000
11011010111111111111111111111111111
11101011111011111111111111111111111
00000000000000000000000000000000000
00000000000000000000000000000000000
11111111111111111111111111111111111
11010111110111111111111111111111111
11100110111111111111111111111111111
00000000000000000000000000000000000
00000000000000000000000000000000000
11011011101111111111111111111111111
11101011111110111111111111111111111
00000000000000000000000000000000000
00000000000000000000000000000000000
11111111111111111111111111111111111
11010111111110111111111111111111111
11100111101111111111111111111111111
```

```
00000000000000000000000000000000000
00000000000000000000000000000000000
11011011111101111111111111111111111
11101010111111111111111111111111111
00000000000000000000000000000000000
00000000000000000000000000000000000
11111111111111111111111111111111111
00000000000000000000000000000000000
00000000000000000000000000000000000
00000000000000000000000000000000000
00000000000000000000000000000000000
00000000000000000000000000000000000
00000000000000000000000000000000000
00000000000000000000000000000000000
00000000000000000000000000000000000
00000000000000000000000000000000000
00000000000000000000000000000000000
00000000000000000000000000000000000
00000000000000000000000000000000000
00000000000000000000000000000000000
00000000000000000000000000000000000
00000000000000000000000000000000000
00000000000000000000000000000000000
00000000000000000000000000000000000
00000000000000000000000000000000000
00000000000000000000000000000000000
00000000000000000000000000000000000
00000000000000000000000000000000000
00000000000000000000000000000000000
00000000000000000000000000000000000
00000000000000000000000000000000000
00000000000000000000000000000000000
00000000000000000000000000000000000
00000000000000000000000000000000000
00000000000000000000000000000000000
00000000000000000000000000000000000
11111111111111111111111111111111111
11111111111111111111111111111111111
11111111111111111111111111111111111
11111111111111111111111111111111111
11111111111111111111111111111111111
11111111111111111111111111111111111
11111111111111111111111111111111111
11111111111111111111111111111111111
11111111111111100000000000000000011
*
V0001 XXXXXXXXXNXXXXXXXXXN*
V0002 0001000XXNXXXXXXXXXN*
V0003 0001000XXNXXXXXXXXXN*
V0004 1001000XXNXXXXXLLLHN*
V0005 0001000XXNXXXXXLLLHN*
V0006 0011000XXNXXXXXLLLHN*
V0007 1011000XXNXXXXXLLHLN*
V0008 0011000XXNXXXXXLLHLN*
V0009 0101000XXNXXXXXLLHLN*
V0010 1101000XXNXXXXXLHLLN*
V0011 0101000XXNXXXXXLHLLN*
V0012 0111000XXNXXXXXLHLLN*
V0013 1111000XXNXXXXXHLLLN*
V0014 0111000XXNXXXXXHLLLN*
C78E8*
^C4753
```

**Figure 8.27**

**PLD Test Vectors**

```
V0001 XXXXXXXXXXNXXXXXXXXXN*
V0002 0001000XXNXXXXXXXXXN*
V0003 0001000XXNXXXXXXXXXN*
V0004 1001000XXNXXXXXLLLHN*
V0005 0001000XXNXXXXXLLLHN*
V0006 0011000XXNXXXXXLLLHN*
V0007 1011000XXNXXXXXLLHLN*
V0008 0011000XXNXXXXXLLHLN*
V0009 0101000XXNXXXXXLLHLN*
V0010 1101000XXNXXXXXLHLLN*
V0011 0101000XXNXXXXXLHLLN*
V0012 0111000XXNXXXXXLHLLN*
V0013 1111000XXNXXXXXHLLLN*
V0014 0111000XXNXXXXXHLLLN*
```

PALASM itself has improved and is still the basis for other PLD programming languages. But other programming software now allows programming in *any* Boolean format (not just SOP), in truth table form, or in state table form (for sequential circuits).

These software packages will expand a Boolean equation to its full SOP form and then simplify it, using a logic minimizer program, such as ESPRESSO, a public domain package developed by the University of California at Berkeley. PLD software can often simulate device operation and display the simulated waveforms. Finally, many packages can generate test vectors so that the device can be tested immediately after programming.

Some popular PLD assemblers are ABEL (Data I/O), CUPL (Logical Devices), OPAL (National Semiconductor), and PLDshell Plus/PLDasm (Intel). We will use PLDshell Plus/PLDasm for the PLD examples in this text, since it is easy to use and incorporates all the features listed above. As of this writing, PLDshell is available free of charge from Intel (Literature Hotline: 1-800-548-4725).

## Using PLDasm

**Source file** *A list of instructions written in a PLD programming language, such as PLDasm, used to create a JEDEC file for a PLD application.*

**Compile** *Translate a source file into binary code that can be processed by a computer.*

**Disassemble** *Convert a JEDEC file back to a source file.*

**File extension** *A three-letter suffix following the name of a disk file that identifies the file's function.*

**Graphics shell** *A screen created by a computer program with a number of menu options for invoking various functions of the program.*

*Disclaimer:* The description of the PLDasm programming language in this text is not intended to replace the material in the Intel manual that accompanies the software. References to PLDshell and PLDasm assume that you have a copy of the Intel manual and will rely on that for the majority of your software support. The material on PLDasm in this book is primarily in the form of application-specific examples and some supplementary material on the PLDasm language.

The manual accompanying the PLDshell Plus/PLDasm program gives detailed instructions for using this software package. Two files supplied with PLDshell Plus, TEMPLATE.PDS and SUMMARY.PDS, contain sample uses of many of the standard PLDasm **source file** statements and can be used as a starting point for creating your own source files.

We will learn enough about PLDasm to write and **compile** the source files for a number of PLD examples. Any source files (with a .PDS **file extension**) and JEDEC files (with a .JED extension) generated by the PLD examples in this text are stored on a diskette, available to instructors. The JEDEC files can be used directly to program an Intel PLD or adapted to any programming equipment by use of translation software, if necessary.

---

If you are using software other than PLDasm, you should still be able to use the PLD examples.

Most development software can **disassemble** a JEDEC file developed using different software and produce a usable source file. Consult the manual with your PLD software for details on the disassembly procedure.

---

PLDasm runs under a **graphics shell** called PLDshell Plus, a screen with a number of menu options that will run the various programs needed to create a JEDEC file from a source file. Once you have installed PLDshell Plus/PLDasm on your computer, you can run the program by typing PLDSHELL at the command prompt, provided you are in the same directory as PLDshell is installed in. (For more information on running a personal computer, see your operator's manual or your local computer expert.)

### Source File

We can create a PLDasm source file (a .PDS file) with any text editor or word processor *provided we store the text as ASCII code.* The file *must not* include any of the formatting codes inserted by most word processors.

A .PDS file contains the following sections:

---

**Declaration:**

[⟨header⟩]
CHIP ⟨application name⟩ ⟨device⟩
⟨pin list⟩
[⟨string substitutions⟩]

**Design Section (One or More of the Following):**

EQUATIONS
    ⟨Boolean equations⟩
T_TAB
    ⟨truth table⟩
STATE
    ⟨state equations for synchronous sequential circuits (state machines)⟩

---

> **Simulation Section (Optional):**
>
> SIMULATION
>     ⟨simulation commands⟩

Text shown in [square brackets] represents an optional part of the file. Text in ⟨angle brackets⟩ gives a description of the sort of text required at this point in the file. UPPER-CASE TEXT must appear exactly as shown.

## Declaration Section

The header consists of six identifying lines of text (title, pattern number, revision number, author, company, and date) that enable the designer to keep track of design development. These lines are all optional.

The next line, starting with the keyword CHIP, specifies the design name and the target device (e.g., 85C220, 85C090, etc.).

Since a PLD has no defined function in its unprogrammed state, we must assign input and output variable names to the pins. The input and output names are then related by Boolean equations or truth tables.

Pins can be assigned explicitly in the declaration section or automatically by PLDasm. If assigned in the declaration section, each function can be assigned by typing the keyword PIN, followed by the pin number and the name of the input or output signal at that pin:

PIN 1 clock

PIN 2 ina

PIN 3 inb

PIN 15 out

If we are explicitly assigning pin numbers, we must consult the device data sheet to ensure that we assign input signals to input pins, output signals to output pins, and functions such as Clock and Output Enable to their respective inputs. For instance, the Clock input on an 85C220/224 is always at pin 1.

We can also write out the pin names, without pin numbers, in the same order as the pins to which they are assigned. If we choose this option, the number of pin names must equal the number of available pins, including any pins not connected (NC).

If assigned automatically, the keyword PIN is followed by the signal name only, with no pin number:

PIN   clock
PIN   ina
PIN   inb
PIN   out

**Design Section**

> **Syntax** *The "grammar" of a computer language (i.e., the rules of construction of language statements).*

> **String** *A sequence of ASCII characters.*
>
> **String substitution** *Assignment of a relatively simple string to represent a relatively complex string, thus simplifying any source file statements in which the string is used.*
>
> **Comment** *Explanatory text in a source file that is ignored by the computer when the source file is compiled.*

Circuit design can be described in three different ways in PLDasm: by Boolean equation, truth table, or state equations.

Of the three possible types of design sections, we will, for now, examine the Boolean equations section only. The word EQUATIONS begins this section. Boolean expressions must be entered using the following symbols:

| Function | Symbol | Example | |
|----------|--------|---------|---|
| NOT | / | / A | NOT A |
| AND | * | A * / B | A AND (NOT B) |
| OR | + | A + B | A OR B |
| XOR | :+: | A :+: B | A XOR B |

These symbols follow the usual Boolean rules of precedence (AND before OR unless otherwise indicated by parentheses). Combinational outputs are assigned by the equal sign (=). Registered outputs are assigned by the symbol :=. Some examples of permissible **syntax** are:

F1 := A * / B + / A * B
F2 := / (A * B + C)
F3 = / ((A + B) * (A + C) * (B + C))
/ F3 = (A + B) * (A + C) * (B + C)
F4 = (A :+: B) * C + A * B

Note that the third and fourth example lines are two ways of writing the same thing.

A **comment** is explanatory text that is ignored by PLDasm when creating a JEDEC file. We write comments into the source file to explain the program to someone who did not write it or to remind ourselves what we did when we look back at it six months from now. Comments begin with a leading semicolon and end at the end of a line.

; This is a comment.

; A comment goes to the end of a line,
so this new line is not part of it.

Let us create a simple PLDasm source file, using the elements we have examined so far.

**EXAMPLE 8.5**    Write a PLDasm source file to program an 85C220 EPLD as a 4-to-1 multiplexer, as shown in Figure 8.6a.

**Solution**
```
Title      4 to multiplexer
Pattern    pds
Revision   1.0
Author     R.Dueck
Company    Seneca College
Date       April 2, 1993

; Source file for a 4-to-1 multiplexer, based on an
; 85C220 device.

CHIP   4to1mux   85C220

PIN  2  D0      ; Data Inputs
PIN  3  D1
PIN  4  D2
PIN  5  D3

PIN  6  S0      ; Select Inputs
PIN  7  S1

PIN  8  EN      ; Tristate ENABLE

PIN 19  Y       ; Output


;    The Boolean equations are defined in the block below, which
; must begin with the keyword EQUATIONS.  There is only one
; equation since the PLD circuit has only one output.

EQUATIONS

Y = D0*/S1*/S0 + D1*/S1*S0 + D2*S1*/S0 + D3*S1*S0

; Output Y is enabled when the EN input is HIGH.  The
; tristate enable function is indicated by the signal
; extension .trst.

Y.trst = EN
```

If we wish, we can define one or more **string substitutions** in the declaration section. A **string** is a sequence of ASCII characters. If we are likely to use a long Boolean expression repeatedly, we can make the file more readable by making a string stand for the Boolean expression. The substitute string is just a typing shortcut or a way to understand a PLDasm statement more easily. It acts as a direct replacement of text in a PLDasm statement.

For example, we might want to replace the states of two inputs *A* and *B* by a string IN*x*, where *x* is the binary value of the input combination. The four required string substitutions are:

```
STRING    IN0    '(/ A * / B)'
STRING    IN1    '(/ A *   B)'
STRING    IN2    '( A * / B)'
STRING    IN3    '( A *   B)'
```

Anywhere in the source file that we want to write "/ A * B", we simply write "IN1". The same is true for other combinations of *A* and *B*.

---

**EXAMPLE 8.6**

Rewrite the EQUATIONS section of the PLDasm file of Example 8.5 to include a string substitution as follows:

| Original | Substitute |
|----------|------------|
| / S0 * / S1 | SEL0 |
| / S0 *  S1 | SEL1 |
| S0 * / S1 | SEL2 |
| S0 *  S1 | SEL3 |

**Solution**  The following string substitutions must be included in the declaration section of the source file:

STRING      SEL0    '(/ S1 * / S0)'

STRING      SEL1    '(/ S1 *  S0)'

STRING      SEL2    '( S1 * / S0)'

STRING      SEL3    '( S1 *  S0)'

The EQUATIONS section is modified as follows:

EQUATIONS

Y = D0 * SEL0 + D1 * SEL1 + D2 * SEL2 + D3 * SEL3

The rest of the source file remains unchanged.

---

**EXAMPLE 8.7**

Write a PLDasm source file to implement the logic functions of Example 8.1, using an 85C224 EPLD.

**Solution**

```
Title      Basic Logic Gates
Pattern    pds
Revision   1.0
Author     R. Dueck
Company    Seneca College
Date       November 12, 1992

;     This file generates the JEDEC file for seven basic logic
; functions (NOT, AND, OR , NAND, NOR, XOR, XNOR) to fit into an
; 85C224 EPLD.

CHIP gates 85C224

; pin list
;
; Since the pin numbers are not specified, no explicit pin
; assignments are made.  This is done automatically when the
; source file is compiled.

PIN      F1                ; Output Variables
PIN      F2
PIN      F3
PIN      F4
PIN      F5
```

```
PIN     F6
PIN     F7

PIN     A               ; Input Variables
PIN     B
PIN     C
PIN     D
PIN     E
PIN     F
PIN     G
PIN     H
PIN     J
PIN     K
PIN     L
PIN     M
PIN     N


EQUATIONS

F1 = /A                 ; NOT function
F2 = B*C                ; AND function
F3 = D+E                ; OR function
F4 = /(F*G)             ; NAND function
F5 = /(H+J)             ; NOR function
F6 = K :+: L            ; XOR function
F7 = /(M :+: N)         ; XNOR function
```

When the source file in Example 8.7 is compiled, it generates a report file (GATES. RPT) and the JEDEC file (GATES.JED).

The report file, listed below, shows the Boolean equations in their simplified SOP form, as well as a pin diagram for the programmed device. Since the pin assignment was done automatically, this diagram is crucial; without it you would not know how to connect the device inputs and outputs and could therefore inadvertently damage the device.

Since the EPLD is CMOS, *unused inputs* should be *grounded* and *unused outputs* should be left *open*. This is also indicated on the pin diagram.

The report file also includes information about the number of product terms used out of the possible total and the configuration of the device macrocells. The outputs are listed as "CONF," meaning COmbinational, No Feedback.

The JEDEC file for this application is included on the instructor's diskette as \INTRO\GATES.JED.

```
INTEL Logic Optimizing Compiler Utilization Report
FIT Release [ 2.1 ] SID [ 2.84 ]


***** Design implemented successfully

TITLE     BASIC LOGIC GATES
PATTERN   PDS
REVISION  1.0
AUTHOR    R. DUECK
COMPANY   SENECA COLLEGE
DATE      NOVEMBER 12, 1992
```

```
;     THIS FILE GENERATES THE JEDEC FILE FOR SEVEN BASIC LOGIC
; FUNCTIONS (NOT, AND, OR , NAND, NOR, XOR, XNOR) TO FIT INTO AN
; 85C224 EPLD.

OPTIONS
     TURBO = ON
     SECURITY = OFF

CHIP  GATES 85C224

PIN     23     A
PIN     14     B
PIN     13     C
PIN     11     D
PIN     10     E
PIN      9     F
PIN      8     G
PIN      7     H
PIN      6     J
PIN      5     K
PIN      4     L
PIN      3     M
PIN      2     N

PIN     20     F1
PIN     19     F2
PIN     18     F3
PIN     17     F4
PIN     16     F5
PIN     15     F6
PIN     21     F7


EQUATIONS

F1 = /A
F1.TRST = VCC

F2 = B * C
F2.TRST = VCC

/F3 = /D * /E
F3.TRST = VCC

/F4 = F * G
F4.TRST = VCC

F5 = /H * /J
F5.TRST = VCC

F6 = K * /L
   + /K * L
F6.TRST = VCC

F7 = M * N
   + /M * /N
F7.TRST = VCC


INTEL Logic Optimizing Compiler Utilization Report

    GATES.rpt

***** Design implemented successfully
```

```
            85C224
         - - - - -
  Gnd - | 1     24 | - Vcc
    N - | 2     23 | - A
    M - | 3     22 | - Gnd
    L - | 4     21 | - F7
    K - | 5     20 | - F1
    J - | 6     19 | - F2
    H - | 7     18 | - F3
    G - | 8     17 | - F4
    F - | 9     16 | - F5
    E - |10     15 | - F6
    D - |11     14 | - B
  GND - |12     13 | - C
         - - - - -
```

CMOS Device:  ground unused inputs and I/Os
RESERVED = Leave pins unconnected on board.
Gnd  = unused input or I/O pin.
N.C. = unconnected pins


**OUTPUTS**

| Name | Pin | Resource | MCell | PTerms |
|------|-----|----------|-------|--------|
| F7 | 21 | CONF | 2 | 2/ 8 |
| F1 | 20 | CONF | 3 | 1/ 8 |
| F2 | 19 | CONF | 4 | 1/ 8 |
| F3 | 18 | CONF | 5 | 1/ 8 |
| F4 | 17 | CONF | 6 | 1/ 8 |
| F5 | 16 | CONF | 7 | 1/ 8 |
| F6 | 15 | CONF | 8 | 2/ 8 |


**INPUTS**

| Name | Pin | Resource | MCell | PTerms |
|------|-----|----------|-------|--------|
| A | 23 | INP | - | - |
| B | 14 | INP | - | - |
| C | 13 | INP | - | - |
| D | 11 | INP | - | - |
| E | 10 | INP | - | - |
| F | 9 | INP | - | - |
| G | 8 | INP | - | - |
| H | 7 | INP | - | - |
| J | 6 | INP | - | - |
| K | 5 | INP | - | - |
| L | 4 | INP | - | - |
| M | 3 | INP | - | - |
| N | 2 | INP | - | - |

**UNUSED RESOURCES**

| Name | Pin | Resource | MCell | PTerms |
|------|-----|----------|-------|--------|
| - | 1 | INPUT | - | - |
| - | 22 | MCELL | 1 | 8 |

```
**PART UTILIZATION**

 7/ 8  MacroCells (87%), 16% of used Pterms Filled
13/14  Input Pins (92%)
       PTerms Used 14%

**RESOURCE MNEMONICS**
INP  = Pin Input to Logic Array
CONF = Comb.       pin Output, No        Feedback

Macrocell Interconnection Cross Reference

FEEDBACKS:          M M M M M M M
                    0 0 0 0 0 0 0
                    2 3 4 5 6 7 8

F7 ....... CONF @M2 -> . . . . . . .  @21
F1 ....... CONF @M3 -> . . . . . . .  @20
F2 ....... CONF @M4 -> . . . . . . .  @19
F3 ....... CONF @M5 -> . . . . . . .  @18
F4 ....... CONF @M6 -> . . . . . . .  @17
F5 ....... CONF @M7 -> . . . . . . .  @16
F6 ....... CONF @M8 -> . . . . . . .  @15


INPUTS:

N ........ INP @2  -> * . . . . . .
M ........ INP @3  -> * . . . . . .
L ........ INP @4  -> . . . . . . *
K ........ INP @5  -> . . . . . . *
J ........ INP @6  -> . . . . . * .
H ........ INP @7  -> . . . . . * .
G ........ INP @8  -> . . . . * . .
F ........ INP @9  -> . . . . * . .
E ........ INP @10 -> . . . * . . .
D ........ INP @11 -> . . . * . . .
C ........ INP @13 -> . . * . . . .
B ........ INP @14 -> . . * . . . .
A ........ INP @23 -> . * . . . . .
                      F F F F F F F
                      7 1 2 3 4 5 6
```

```
. = not connected      x = no connection possible
* = signal feeds cell  ? = error, unable to fit
```

## Simulation Section

> **Flow control** *Commands that control the order of statement execution in a computer program.*
>
> **Nested** *Flow control statements are nested when they are placed within the body of other flow control statements.*

The simulation section of a PLDasm source file allows us to check the operation of an EPLD design in two ways:

1. The simulation statements generate a timing diagram of the device that we can then view as waveforms. This allows us to confirm that the actual operation of the device is the same as its intended operation.

2. The simulator automatically creates a series of test vectors that are appended to the device JEDEC file. The vectors contain the same logic levels as the generated waveforms. Each time there is a signal change, the simulator creates a new test vector. These vectors allow device testing when the EPLD is programmed.

Several simulation keywords allow us to simulate most EPLD designs:

*VECTOR.* Allows us to group several Boolean variables so that they are treated as a single multibit variable. For example, suppose we want to group eight inputs, *A* through *H,* in one vector called IN. We can create the vector as follows (*H* is the most significant bit):

VECTOR IN := [ H, G, F, E, D, C, B, A ]

The value of this vector is generally a hexadecimal number—it could also be a decimal or binary value—consisting of the values of bits *HGFEDCBA*. For example, if *HGFEDCBA* = 01101111, the value of IN is 6F. This is used as a shorthand method of representing several logic levels, such as all circuit inputs, as a single value.

*SETF.* Sets a specified variable or variables to a known value. If the variable is listed with a NOT function (/), it is set LOW. Otherwise it is set HIGH.

Vectors can be set to numerical values that determine the Boolean states of the vector. Hexadecimal values are indicated by a leading 0x (e.g., 0xFF = $FF_{16}$). Binary values are prefixed by #b (e.g., #b10010011 = $10010011_2$).

The statement

SETF /QA QB QC /QD /CLOCK IN := 0xC8

results in the following variable states:

$QA = 0, QB = 1, QC = 1, QD = 0, CLOCK = 0, IN = 11001000$

(The vector *IN* is the same as defined above.) The 0x prefix indicates a hexadecimal value of *IN.*

*PRLDF.* Preloads the macrocell registers to a specific value. PRLDF works the same as SETF but is specifically designed for testing macrocell registers. This command allows us to define the first register state after power-up and to test the "next state" of the registers after any "present state." This is important in testing the operation of synchronous sequential circuits.

The following statement sets the registers *QA, QB, QC,* and *QD* to 0.

PRLDF /QA /QB /QC /QD

*CLOCKF.* Generates one clock pulse. If the clock input is initially LOW, the pulse is LOW-HIGH-LOW. Otherwise it is HIGH-LOW-HIGH. If you have named your clock input *CLK,* the following statements generate two LOW-HIGH-LOW pulses:

SETF /CLK

CLOCKF CLK

CLOCKF CLK

*FOR loop.* The PLDasm simulator includes a number of **flow control** commands that allow us to specify conditions under which simulation statements are executed. These include structures such as the FOR loop, the WHILE loop, and the IF/THEN/ELSE structure. These are all standard flow control forms in a number of different programming languages. We will examine only the FOR loop.

The FOR loop executes one or more simulation statements a fixed number of times before continuing through the program. The general form of this structure is:

```
FOR ⟨counter⟩ := ⟨first⟩ TO ⟨last⟩ DO
BEGIN
        ⟨simulation statements⟩
END
```

Suppose we wish to generate eight clock pulses. Rather than writing eight CLOCKF statements, we can write the following loop:

```
SETF /CLK
FOR i := 0 TO 7 DO
BEGIN
        CLOCKF CLK
END
```

The ⟨counter⟩ is a variable, often shown as a single letter such as *i, j,* or *k,* which keeps track of the number of times the loop is executed. It can be used to modify statements within the body of the loop.

The following statements initialize the vector *IN* to 0, then increment its value 255 times to show how the device responds to all 256 possible input states.

```
SETF IN := 0
FOR j := 1 TO 255 DO
BEGIN
        IN := j
END
```

FOR loops can be **nested.** That is, one FOR loop can be inside another. The following statements set the vector *IN* to 0, increment it 32 times, and apply 8 clock pulses after *each* increment.

```
SETF IN := 0 /CLK
FOR j := 0 TO 31 DO
BEGIN
        IN := j                ; Increment IN 32 times
        FOR k := 0 TO 7 DO
        BEGIN                  ; Clock 8 times
            CLOCKF CLK         ; after each increment
```

                    END
               END

---

**EXAMPLE 8.8**
Write a PLDasm source file to implement the 4-bit rotation circuit of Example 8.2 for an 85C220 EPLD as a target device. Include a simulation section that tests the programming equations by rotating test data 1000 and 1100 all possible ways.

**Solution**

```
Title          4-bit Data Rotation
Revision       1.0
Author         R.Dueck
Company        Seneca College
Date           November 16, 1992

; 4-bit Parallel Load circuit for Rotate Right function
;
;    This circuit rotates incoming data to the right by the
; number of bits specified by the binary value of S1 S0.  For
; input ABCD, the rotation is as follows.
;
;              S1 S0    QA QB QC QD   Rotation
;
;              0  0     A  B  C  D        0
;              0  1     D  A  B  C        1
;              1  0     C  D  A  B        2
;              1  1     B  C  D  A        3
;

CHIP    rotate4      85C220

        ; PINLIST

PIN     1    CLOCK

PIN     2    S1      ; Select Inputs
PIN     3    S0

PIN     4    A       ; Parallel Inputs
PIN     5    B
PIN     6    C
PIN     7    D

PIN     19   QA      ; Outputs
PIN     18   QB
PIN     17   QC
PIN     16   QD


        ; String Substitution

STRING R0 '(/S1*/S0)'
STRING R1 '(/S1* S0)'
STRING R2 '( S1*/S0)'
STRING R3 '( S1* S0)'


        ; Boolean equation section

EQUATIONS

QA := R0*A + R1*D + R2*C + R3*B
```

```
QB := R0*B + R1*A + R2*D + R3*C
QC := R0*C + R1*B + R2*A + R3*D
QD := R0*D + R1*C + R2*B + R3*A

; Simulation section

SIMULATION

    VECTOR SELECT := [ S1, S0 ]          ; Select rotation value
    VECTOR PARIN  := [ A, B, C, D ]      ; Parallel In
    VECTOR PAROUT := [ QA, QB, QC, QD ]  ; Parallel Out

    ; Set all registers to known values (power-up state)
    PRLDF /QA /QB /QC /QD

    ; Set all inputs to known values
    SETF /CLOCK SELECT := 0  PARIN := 8

FOR i := 0 TO 1 DO
;
; Rotate two sets of data (1000 and 1100)
; all four possible ways.
;
BEGIN
   FOR j := 0 TO 3 DO
;
; FOR loop to count up 4 clocks
; Increment SELECT before each clock.
;
   BEGIN
       SETF SELECT := j
       CLOCKF CLOCK
   END
   SETF PARIN := 0xC
END
```

Figure 8.28 shows the pin diagram generated by PLDasm when it compiles ROTATE4.PDS.

The simulation section of ROTATE4.PDS generates the waveforms illustrated in Figure 8.29. The simulator also generates the test vectors shown in Figure 8.30. Each vector represents one signal change in Figure 8.29. Compare a few vectors to the timing diagram to see that this is so.

**Figure 8.28**

**Pin Diagram of 4-Bit Rotation Circuit**

```
            85C220
         - - - - -
CLOCK -|  1     20 |- Vcc
   S1 -|  2     19 |- QA
   S0 -|  3     18 |- QB
    A -|  4     17 |- QC
    B -|  5     16 |- QD
    C -|  6     15 |- Gnd
    D -|  7     14 |- Gnd
  Gnd -|  8     13 |- Gnd
  Gnd -|  9     12 |- Gnd
  GND -| 10     11 |- Gnd
         - - - - -
```

**Figure 8.29**

**Simulation Waveforms for 4-Bit Rotation Circuit**

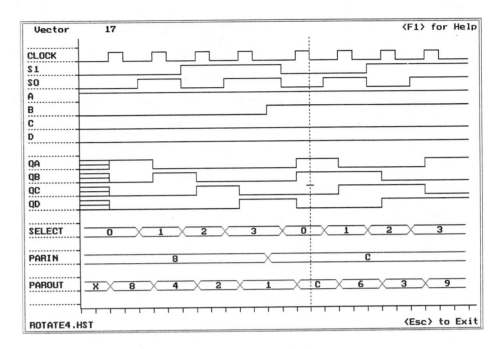

**Figure 8.30**

**Test Vectors Generated by the Simulation Section of ROTATE4.PDS**

```
V0001 XXXXXXXXXNXXXXXXXXN*
V0002 0001000XXNXXXXXXXXXN*
V0003 0001000XXNXXXXXXXXXN*
V0004 1001000XXNXXXXXLLLHN*
V0005 0001000XXNXXXXXLLLHN*
V0006 0011000XXNXXXXXLLLHN*
V0007 1011000XXNXXXXXLLHLN*
V0008 0011000XXNXXXXXLLHLN*
V0009 0101000XXNXXXXXLLHLN*
V0010 1101000XXNXXXXXLHLLN*
V0011 0101000XXNXXXXXLHLLN*
V0012 0111000XXNXXXXXLHLLN*
V0013 1111000XXNXXXXXHLLLN*
V0014 0111000XXNXXXXXHLLLN*
V0015 0111100XXNXXXXXHLLLN*
V0016 0001100XXNXXXXXHLLLN*
V0017 1001100XXNXXXXXLLHHN*
V0018 0001100XXNXXXXXLLHHN*
V0019 0011100XXNXXXXXLLHHN*
V0020 1011100XXNXXXXXLHHLN*
V0021 0011100XXNXXXXXLHHLN*
V0022 0101100XXNXXXXXLHHLN*
V0023 1101100XXNXXXXXHHLLN*
V0024 0101100XXNXXXXXHHLLN*
V0025 0111100XXNXXXXXHHLLN*
V0026 1111100XXNXXXXXHLLHN*
V0027 0111100XXNXXXXXHLLHN*
V0028 0111100XXNXXXXXHLLHN*
```

**EXAMPLE 8.9**

Write a PLDasm source file to implement the 8-bit parallel/serial shift register of Example 8.4. The target device is an 85C22V10 EPLD.

**Solution**

```
Title           8-bit parallel load shift register
Revision        1.0
Author          R.Dueck
Company         Seneca College
Date            November 13, 1992

; 8-bit Serial Shift Register with Parallel Load

CHIP    8BPARSR     85C22V10

        ; PINLIST

PIN     1    CLOCK

PIN     2    SI        ; Serial Input

PIN     3    A         ; Parallel Inputs
PIN     4    B
PIN     5    C
PIN     6    D
PIN     7    E
PIN     8    F
PIN     9    G
PIN     10   H

PIN     11   SL        ; Shift/Load (= 0 for Parallel Load
                       ;                = 1 for Serial Shift)

PIN     23   QA        ; Outputs
PIN     22   QB
PIN     21   QC
PIN     20   QD
PIN     19   QE
PIN     18   QF
PIN     17   QG
PIN     16   QH


        ; Boolean equation section

EQUATIONS

QA := SI*SL + A*/SL
QB := QA*SL + B*/SL
QC := QB*SL + C*/SL
QD := QC*SL + D*/SL
QE := QD*SL + E*/SL
QF := QE*SL + F*/SL
QG := QF*SL + G*/SL
QH := QG*SL + H*/SL


        ; Simulation section

SIMULATION

    VECTOR PARIN := [ A, B, C, D, E, F, G, H ]
    VECTOR PAROUT := [ QA, QB, QC, QD, QE, QF, QG, QH ]

    ; Set all inputs to known values

    SETF /CLOCK /SI  /SL  PARIN := 0x80

    ; Set all registers to known values (power-up state)
```

```
PRLDF /QA /QB /QC /QD /QE /QF /QG /QH

; Clock an input signal   0-->1-->0

CLOCKF CLOCK

; Set serial shift mode
SETF SL

; FOR loop to count up 8 clocks

FOR j := 0 TO 7 DO
BEGIN
     CLOCKF CLOCK
END

; Fill register with 1s
SETF SI

; FOR loop to count up 8 clocks

FOR j := 0 TO 7 DO
BEGIN
    CLOCKF CLOCK
END

; Parallel loading

SETF /SL PARIN := 0xCA
CLOCKF CLOCK

SETF PARIN := 0
CLOCKF CLOCK
```

Pin Diagram:

```
              85C22V10
             - - - - -
   CLOCK  -| 1      24 |- Vcc
      SI  -| 2      23 |- QA
       A  -| 3      22 |- QB
       B  -| 4      21 |- QC
       C  -| 5      20 |- QD
       D  -| 6      19 |- QE
       E  -| 7      18 |- QF
       F  -| 8      17 |- QG
       G  -| 9      16 |- QH
       H  -|10      15 |- Gnd
      SL  -|11      14 |- Gnd
     GND  -|12      13 |- Gnd
             - - - - -
```

**Architecture cell**  A programmable cell within a generic array logic device that, along with several other such cells, sets the configuration of an output logic macrocell.

**Cell**  A fuse location in a programmable logic device, specified by the intersection of an input line and a product line.

**Checksum** An error-checking code derived from the accumulated sum of the data being checked.

**Comment** Explanatory text in a source file that is ignored by the computer when the source file is compiled.

**Compile** Translate a source file into binary code that can be processed by a computer.

**Disassemble** Convert a JEDEC file back to a source file.

**Erasable programmable logic device (EPLD)** A PLD with programmable input/output macrocells whose AND matrix can be erased by exposure to ultraviolet light of a specified wavelength.

**File extension** A three-letter suffix following the name of a disk file that identifies the file's function.

**Flow control** Commands that control the order of statement execution in a computer program.

**FPLA** Field programmable logic array. Programmable logic whose AND and OR matrices are both programmable.

**Generic array logic (GAL)** A type of programmable logic whose outputs can be configured in a variety of different ways (e.g., registered, combinational, or tristate), depending on the design requirements.

**Global architecture cell** An architecture cell that affects all OLMCs within a GAL device.

**Graphics shell** A screen created by a computer program with a number of menu options for invoking various functions of the program.

**Input line** A line that applies the true or complement form of an input variable to the AND matrix of a PLD.

**Input line number** A number assigned to a true or complement input line in a PAL AND matrix.

**JEDEC** Joint Electron Device Engineering Council.

**JEDEC file** An industry-standard form of text file indicating which fuses are blown and which are intact in a programmable logic device.

**Local architecture cell** An architecture cell that affects only one OLMC within a GAL device.

**Multiplexer** A circuit that selects one of several signals to be directed to a single output.

**Nested** Flow control statements are nested when they are placed within the body of other flow control statements.

**OTP** One-time programmable. Refers to a programmable device, such as a PLD, that cannot be erased after being programmed.

**Output logic macrocell (OLMC)** A programmable output cell used as the output of a generic array logic device that can be configured as a registered, combinational, or tristate output of either polarity, with or without feedback, or as an input.

**PAL** Programmable array logic. Programmable logic with a fixed OR matrix and a programmable AND matrix.

**Product line** A single line on a logic diagram used to represent all inputs to an AND gate (i.e., one product term) in a PLD sum-of-products array.

**Product line first cell number** The lowest cell number on a particular product line in a PAL AND matrix where all cells are consecutively numbered.

**Programmable logic device (PLD)** A logic device whose function can be programmed by the user, usually in sum-of-products form.

**PROM** Programmable read-only memory. Programmable logic with a fixed AND matrix and a programmable OR matrix.

**Register** A digital circuit such as a flip-flop or array of flip-flops that stores one or more bits of digital information.

**Registered output** An output of a programmable array logic (PAL) device having a flip-flop (usually D-type) that stores the output state.

**Source file** A list of instructions written in a PLD programming language, such as PLDasm, used to create a JEDEC file for a PLD application.

**String** A sequence of ASCII characters.

**String substitution** Assignment of a relatively simple string to represent a relatively complex string, thus simplifying any source file statements in which the string is used.

**Superset** Containing all features of another device plus some additional features not contained in the original device.

**Syntax** The "grammar" of a computer language (i.e., the rules of construction of language statements).

**Test vector** A list of inputs and expected outputs used to test the function of a programmed PLD.

**Text file** An ASCII-coded document stored on a magnetic disk.

**Windowed CERDIP** A ceramic dual in-line package having a quartz window that passes ultraviolet light.

# PROBLEMS

### Section 8.1 Introduction to Programmable Logic
### Section 8.2 PAL Fuse Matrix and Combinational Outputs
### Section 8.3 PAL Outputs With Programmable Polarity

**8.1** Draw a diagram showing the basic configuration and symbology for a PLD sum-of-products array.

**8.2** What do PROM, FPLA, and PAL stand for? Briefly describe the differences between these types of PLD.

**8.3** Draw a basic PAL circuit having four inputs, eight product terms, and one active-LOW combinational output. Draw fuses on your diagram showing how to make the following Boolean expression:

$$\overline{F} = \overline{A} B \overline{C} + \overline{B} C D + \overline{A} C D + A \overline{C} D$$

**8.4** Modify the PAL circuit drawn in Problem 8.3 to make two outputs having eight product terms and programmable polarity. Draw fuses on the diagram for each of the following functions:

$$\overline{F1} = \overline{A} B \overline{C} + \overline{B} C D + \overline{A} C D + A \overline{C} D$$
$$F2 = \overline{A} B \overline{C} + \overline{B} C D + \overline{A} C D + A \overline{C} D$$

**\*8.5** Make a photocopy of Figure 8.5 (PAL16L8 logic diagram). Draw fuses on the logic diagram showing how to implement a transparent latch of the type illustrated in Figure 6.19.

**8.6** Make a photocopy of Figure 8.8 (PAL20P8 logic diagram). Draw fuses on the PAL20P8 logic diagram showing how to make a BCD-to-2421 code converter, as developed in Example 4.8.

Table 8.8 shows how the two codes relate to each other.

The Boolean equations for the BCD-to-2421 decoder are:

$$Y_4 = D_4 + D_3 D_2 + D_3 D_1$$
$$Y_3 = D_4 + D_3 D_2 + D_3 \overline{D_1}$$
$$Y_2 = D_4 + \overline{D_3} D_2 + D_3 \overline{D_2} D_1$$
$$Y_1 = D_1$$

**Table 8.8** BCD and 2421 Code

| Decimal Equivalent | BCD Code | | | | 2421 Code | | | |
|---|---|---|---|---|---|---|---|---|
| | $D_4$ | $D_3$ | $D_2$ | $D_1$ | $Y_4$ | $Y_3$ | $Y_2$ | $Y_1$ |
| 0 | 0 | 0 | 0 | 0 | 0 | 0 | 0 | 0 |
| 1 | 0 | 0 | 0 | 1 | 0 | 0 | 0 | 1 |
| 2 | 0 | 0 | 1 | 0 | 0 | 0 | 1 | 0 |
| 3 | 0 | 0 | 1 | 1 | 0 | 0 | 1 | 1 |
| 4 | 0 | 1 | 0 | 0 | 0 | 1 | 0 | 0 |
| 5 | 0 | 1 | 0 | 1 | 1 | 0 | 1 | 1 |
| 6 | 0 | 1 | 1 | 0 | 1 | 1 | 0 | 0 |
| 7 | 0 | 1 | 1 | 1 | 1 | 1 | 0 | 1 |
| 8 | 1 | 0 | 0 | 0 | 1 | 1 | 1 | 0 |
| 9 | 1 | 0 | 0 | 1 | 1 | 1 | 1 | 1 |

**8.7** Repeat Problem 8.6 for a 2421-to-BCD code converter.

### Section 8.4 PAL and GAL Devices With Registered Outputs
### Section 8.5 UV-Erasable Programmable Logic Devices (EPLD)

**8.8** The circuit in Figure 8.31 transfers one of two 4-bit numbers, $A$ or $B$, into the D flip-flops when a clock pulse is applied.

  **a.** Briefly explain how the circuit works.

  **b.** Refer to the logic diagram of the GAL16V8 in Figure 8.16. Assign pin numbers from the GAL16V8 to the circuit of Figure 8.31 and write the Boolean equations necessary to program the GAL for this function.

**8.9** Modify the circuit in Figure 8.31 so that it transfers *one of four* 4-bit numbers, *A, B, C,* or *D,* to a 4-bit output. Use an 85C224 EPLD as the target device. Assign pin numbers to the circuit inputs and outputs and write the necessary Boolean equations. (You will need to make the I/O macrocells combinational (=), not registered (:=). Why?)

**8.10** Describe the features of an 85C220 EPLD that form a superset of the features of a 20-pin PAL or GAL.

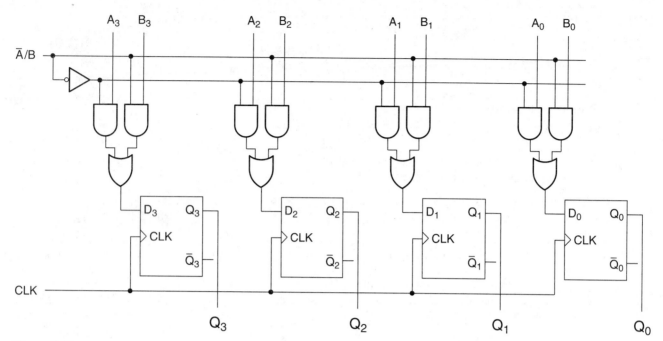

**Figure 8.31**

**Problem 8.8 and 8.9 Circuit**

### Section 8.6 Development Software

**8.11** Write a PLDasm source file to implement the logic function of Problem 8.5. Target device is an 85C220 EPLD.

**8.12** Write a PLDasm source file to implement the logic function described in Problem 8.9. Include a simulation section in the file to confirm the circuit operation. (Use the following test data: $A = 1000$, $B = 0101$, $C = 1001$, $D = 0110$.)

**8.13** **a.** Write the Boolean equations for a BCD-to-Excess-3 code converter, as developed in Problem 4.15. The truth table for the code converter is shown in Table 8.9 for reference.

   **b.** Write a source file in PLDasm to program the code converter into an 85C220 EPLD. Include a simulation section with two vectors, BCD and XS3, that confirms the operation of the converter.

**8.14** Repeat Problem 8.13 for an Excess-3-to-BCD code converter, as developed in Problem 4.16.

**Table 8.9** BCD and Excess-3 Code

| Decimal Equivalent | BCD Code | | | | Excess-3 | | | |
|---|---|---|---|---|---|---|---|---|
| | $D_4$ | $D_3$ | $D_2$ | $D_1$ | $E_4$ | $E_3$ | $E_2$ | $E_1$ |
| 0 | 0 | 0 | 0 | 0 | 0 | 0 | 1 | 1 |
| 1 | 0 | 0 | 0 | 1 | 0 | 1 | 0 | 0 |
| 2 | 0 | 0 | 1 | 0 | 0 | 1 | 0 | 1 |
| 3 | 0 | 0 | 1 | 1 | 0 | 1 | 1 | 0 |
| 4 | 0 | 1 | 0 | 0 | 0 | 1 | 1 | 1 |
| 5 | 0 | 1 | 0 | 1 | 1 | 0 | 0 | 0 |
| 6 | 0 | 1 | 1 | 0 | 1 | 0 | 0 | 1 |
| 7 | 0 | 1 | 1 | 1 | 1 | 0 | 1 | 0 |
| 8 | 1 | 0 | 0 | 0 | 1 | 0 | 1 | 1 |
| 9 | 1 | 0 | 0 | 1 | 1 | 1 | 0 | 0 |

**\*\*8.15** Write a PLDasm source file for a dual *switchable* BCD/Excess-3 or Excess-3/BCD decoder, using an 85C220 EPLD. Each of the two circuits has four code inputs, four code outputs, and a *CODE SELECT (CS)* input. When $CS = 0$, the circuit converts BCD inputs to Excess-3 outputs. When $CS = 1$, the conversion is from Excess-3 to BCD.

# Digital Arithmetic and Arithmetic Circuits

● CHAPTER OUTLINE

**9.1** Digital Arithmetic

**9.2** Representing Signed Binary Numbers

**9.3** Signed Binary Arithmetic

**9.4** Hexadecimal Arithmetic

**9.5** Binary Adders and Subtractors

**9.6** Register Arithmetic Circuits

**9.7** BCD Adders

**9.8** Programmable Logic Implementation of Arithmetic Circuits

● CHAPTER OBJECTIVES

Upon successful completion of this chapter, you will be able to:

- Add or subtract two unsigned binary numbers.

- Write a signed binary number in true-magnitude, 1's complement, or 2's complement form.

- Add or subtract two signed binary numbers.

- Explain the concept of overflow.

- Calculate the maximum sum or difference of two signed binary numbers that will not result in an overflow.

- Add or subtract two hexadecimal numbers.

- Derive the logic gate circuits for full and half adders, given their truth tables.

- Demonstrate the use of full and half adder circuits in arithmetic and other applications.

- Add and subtract $n$-bit binary numbers, using parallel binary adders and logic gates.

- Explain the difference between ripple carry and parallel carry.

- Design a circuit to detect sign-bit overflow in a parallel adder.

- Draw circuits to perform "register arithmetic" and BCD arithmetic, using various MSI devices, and explain their operation.

- Program EPLD devices to perform various arithmetic functions, such as parallel adders, overflow detectors, 1's complementers, and ALUs.

There are two ways of performing binary arithmetic: with unsigned binary numbers or with signed binary numbers. Signed binary numbers incorporate a bit defining the sign of a number; unsigned binary numbers do not. Several ways of writing signed binary numbers are true-magnitude form, which maintains the magnitude of the number in binary value, and 1's complement and 2's complement forms, which modify the magnitude but are more suited to digital circuitry.

Hexadecimal arithmetic is used for calculations that would be awkward in binary due to the large number of bits involved. Important applications of hexadecimal arithmetic are found in microcomputer systems.

There are a number of different digital circuits for performing digital arithmetic, most of which are based on the parallel binary adder, which in turn is based on the full adder and half adder circuits. The half adder adds two bits and produces a sum and a carry. The full adder also allows for an input carry from a previous adder stage. Parallel adders have many full adders in cascade, with carry bits connected between the stages.

Specialized adder circuits are used for adding and subtracting binary numbers, generating logic functions, and adding numbers in binary-coded decimal (BCD) form.

### 9.1

## Digital Arithmetic

> **Signed binary number** *A binary number of fixed length whose sign is represented by one bit, usually the most significant bit, and whose magnitude is represented by the remaining bits.*
>
> **Unsigned binary number** *A binary number whose sign is not specified by a sign bit. A positive sign is assumed unless explicitly stated otherwise.*

Digital arithmetic usually means binary arithmetic, or perhaps BCD arithmetic. Binary arithmetic can be performed using **signed binary numbers,** in which the MSB of each number indicates a positive or negative sign, or **unsigned binary numbers,** in which the sign is presumed to be positive.

The usual arithmetic operations of addition and subtraction can be performed using signed or unsigned binary numbers. Signed binary arithmetic is often used in digital circuits for two reasons:

1. Calculations involving real-world quantities require us to use both positive and negative numbers.
2. It is easier to build circuits to perform some arithmetic operations, such as subtraction, with certain types of signed numbers than with unsigned numbers.

## Unsigned Binary Arithmetic

> **Operand** *A number upon which an arithmetic function operates (e.g., in the expression $x + y = z$, $x$ and $y$ are the operands).*
>
> **Sum** *The result of an addition operation.*
>
> **Carry** *A digit that is "carried over" to the next most significant position when the sum of two single digits is too large to be expressed as a single digit.*
>
> **Sum bit (single-bit addition)** *The least significant bit of the sum of two 1-bit binary numbers.*
>
> **Carry bit** *A bit that holds the value of a carry (0 or 1) resulting from the sum of two binary numbers.*

### Addition

When we add two numbers, they combine to yield a result called the **sum.** If the sum is larger than can be contained in one digit, the operation generates a second digit, called the **carry.** The two numbers being added are called the **operands.**

For example, in the decimal addition $9 + 6 = 15$, 6 and 9 are the operands, 15 is the sum, and the "1" in 15 is the carry.

Four binary sums give us all of the possibilities for adding two $n$-bit binary numbers:

$$0 + 0 = 00$$
$$1 + 0 = 01$$
$$1 + 1 = 10 \qquad (1_{10} + 1_{10} = 2_{10})$$
$$1 + 1 + 1 = 11 \quad (1_{10} + 1_{10} + 1_{10} = 3_{10})$$

Each of these results consists of a **sum bit** and a **carry bit.** For the first two results above, the carry bit is 0. The final sum in the table is the result of adding a carry bit from a sum in a less significant position.

When we add two 1-bit binary numbers in a logic circuit, the result *always* consists of a sum bit and a carry bit, even when the carry is 0, since each bit corresponds to a measurable voltage at a specific circuit location. Just because the value of the carry is 0 does not mean it has ceased to exist.

---

**EXAMPLE 9.1**

Calculate the sum $10010 + 1010$.

**Solution**

```
              ┌──── (Carry from sum of 2nd LSBs)
          1
        10010
    +    1010
        ─────
        11100
```

**EXAMPLE 9.2**   Calculate the sum 10111 + 10010.

**Solution**

```
          ┌─┬┬───── (Carry bits)
          1  11
            10111
        +   10010
            101001
```

---

> **Section Review Problems for Section 9.1a**
>
> 9.1. Add 11111 + 1001.
>
> 9.2. Add 10011 + 1101.

## Subtraction

> **Difference** *The result of a subtraction operation.*
>
> **Minuend** *The first number in a subtraction operation.*
>
> **Subtrahend** *The second operand in a subtraction operation.*
>
> **Borrow** *A digit brought back from a more significant position when the subtrahend digit is larger than the minuend digit.*

In unsigned binary subtraction, two operands, called the **subtrahend** and the **minuend,** are subtracted to yield a result called the **difference.** In the operation $x = a - b$, $x$ is the difference, $a$ is the minuend, and $b$ is the subtrahend. To remember which comes first, think of the minuend as the number that is di*mini*shed (i.e., something is taken away from it).

Unsigned binary subtraction is based on the following four operations:

$$0 - 0 = 0$$
$$1 - 0 = 1$$
$$1 - 1 = 0$$
$$10 - 1 = 1 \quad (2_{10} - 1_{10} = 1_{10})$$

The last operation shows how to obtain a positive result when subtracting a 1 from a 0: **borrow** 1 from the next most significant bit.

---

**Borrowing Rules:**

1. If you are borrowing from a position that contains a 1, leave behind a 0 in the borrowed-from position.

2. If you are borrowing from a position that already contains a 0, you must borrow from a more significant digit that contains a 1. All 0s up to that point become 1s, and the last borrowed-from digit becomes a 0.

---

**EXAMPLE 9.3**    Subtract 1110 − 1001.

**Solution**

(New 2nd LSB) ⎤ ⎡ (Bit borrowed from 2nd LSB)
```
        01
      11 1 0
    −  1001
      ─────
      0101
```

---

**EXAMPLE 9.4**    Subtract 10000 − 101.

**Solution**

```
                              1
    10000   (Original      01110   (After borrowing
  −   101   problem)     −   101   from higher-order bits)
                          ──────
                            1011
```

---

**Section Review Problems for Section 9.1b**

9.3. Subtract 10101 − 10010.

9.4. Subtract 10000 − 1111.

---

# 9.2

# Representing Signed Binary Numbers

**Sign bit**  *A bit, usually the MSB, that indicates whether a signed binary number is positive or negative.*

**Magnitude bits**  *The bits of a signed binary number that tell us how large the number is (i.e., its magnitude).*

**True-magnitude form**  *A form of signed binary number whose magnitude is represented in true binary.*

**1's complement**  *A form of signed binary notation in which negative numbers are created by complementing all bits of a number, including the sign bit.*

> **2's complement** *A form of signed binary notation in which negative numbers are created by adding 1 to the 1's complement form of the number.*
>
> **Note:** *Positive numbers are the same in all three notations.*

Binary arithmetic operations are performed by digital circuits that are designed for a fixed number of bits, since each bit has a physical location within a circuit. It is useful to have a way of representing binary numbers within this framework that accounts not only for the magnitude of the number, but for the sign as well.

This can be accomplished by designating one bit of a binary number, usually the most significant bit, as the **sign bit** and the rest as **magnitude bits.** When the number is negative, the sign bit is 1, and when the number is positive, the sign bit is 0.

There are several ways of writing the magnitude bits, each having its particular advantages. **True-magnitude** form represents the magnitude in straight binary form, which is relatively easy for a human operator to read. Complement forms, such as **1's complement** and **2's complement,** modify the magnitude so that it is more suited to digital circuitry.

## True-Magnitude Form

In true-magnitude form, the magnitude of a number is translated into its true binary value. The sign is represented by the MSB, 0 for positive and 1 for negative.

---

**EXAMPLE 9.5**

Write the following numbers in 6-bit true-magnitude form:

   a. $25_{10}$     b. $-25_{10}$     c. $12_{10}$     d. $-8_{10}$

**Solution**   Translate the magnitudes of each number into 5-bit binary, padding with leading zeros as required, and set the sign bit to 0 for a positive number and 1 for a negative number.

   a. 011001     b. 111001     c. 001100     d. 101000

---

## 1's Complement Form

True-magnitude and 1's complement forms of binary numbers are the same for positive numbers—the magnitude is represented by the true binary value and the sign bit is 0. We can generate a negative number in one of two ways:

1. Write the positive number of the same magnitude as the desired negative number. Complement each bit, including the sign bit.
2. Subtract the $n$-bit positive number from a binary number consisting of $n$ 1s.

---

**EXAMPLE 9.6**

Convert the following numbers to 8-bit 1's complement form:

   a. $57_{10}$     b. $-57_{10}$     c. $72_{10}$     d. $-72_{10}$

**Solution**  Positive numbers are the same as numbers in true-magnitude form. Negative numbers are the bitwise complements of the corresponding positive number.

    **a.**   $57_{10} = 00111001$

    **b.**  $-57_{10} = 11000110$

    **c.**   $72_{10} = 01001000$

    **d.**  $-72_{10} = 10110111$

We can also generate an 8-bit 1's complement negative number by subtracting its positive magnitude from 11111111 (eight 1s). For example, for part b:

$$
\begin{array}{r}
11111111 \\
-\underline{00111001} \ \ (\ \ 57_{10}) \\
11000110 \ \ (-57_{10})
\end{array}
$$

## 2's Complement Form

Positive numbers in 2's complement form are the same as in true-magnitude and 1's complement forms. We create a negative number by adding 1 to the 1's complement form of the number.

**EXAMPLE 9.7**  Convert the following numbers to 8-bit 2's complement form:

    a. $57_{10}$      b. $-57_{10}$      c. $72_{10}$      d. $-72_{10}$

**Solution**     **a.**   $57 = 00111001$

    **b.**  $-57 = 11000110$   (1's complement)

$$
\begin{array}{r}
\underline{\phantom{1100011}1} \\
11000111
\end{array}
$$

          $11000111$   (2's complement)

    **c.**   $72 = 01001000$

    **d.**  $-72 = 10110111$   (1's complement)

$$
\begin{array}{r}
\underline{\phantom{1011011}1} \\
10111000
\end{array}
$$

          $10111000$   (2's complement)

A negative number in 2's complement form can be made positive by 2's complementing it again. Try it with the negative numbers in Example 9.7.

## 9.3

# Signed Binary Arithmetic

**Signed binary arithmetic**  *Arithmetic operations performed using signed binary numbers.*

## Signed Addition

Signed addition is done in the same way as unsigned addition. The only difference is that both operands *must* have the same number of magnitude bits, and each has a sign bit.

**EXAMPLE 9.8** Add $+30_{10}$ and $+75_{10}$. Write the operands and the sum as 8-bit signed binary numbers.

**Solution**

$$
\begin{array}{rl}
+30 & \quad 00011110 \\
+75 & +01001011 \\
\hline
+105 & \quad 01101001
\end{array}
$$

 (Magnitude bits)

(Sign bit)

## Subtraction

The real advantage of complement notation becomes evident when we subtract signed binary numbers. In complement notation, we add a negative number instead of subtracting a positive number. We thus have only one kind of operation—addition—and can use the same circuitry for both addition and subtraction.

This idea does not work for true-magnitude numbers. In the complement forms, the magnitude bits change depending on the sign of the number. In true-magnitude form, the magnitude bits are the same regardless of the sign of the number.

Let us subtract $80_{10} - 65_{10} = 15_{10}$ using 1's complement and 2's complement addition. We will also show that the method of adding a negative number to perform subtraction is not valid for true-magnitude signed numbers.

### 1's Complement Method

> **End-around carry** *An operation in 1's complement subtraction where the carry bit resulting from a sum of two 1's complement numbers is added to that sum.*

Add the 1's complement values of 80 and $-65$. If the sum results in a carry beyond the sign bit, perform an **end-around carry.** That is, add the carry to the sum.

$$80_{10} = 01010000$$

$$65_{10} = 01000001$$
$$-65_{10} = 10111110 \quad \text{(1's complement)}$$

$$
\begin{array}{rl}
80 & \quad 01010000 \\
-\ 65 & +\ 10111110 \\
\hline
 & 1\ 00001110 \\
 & \qquad \longrightarrow 1 \quad \text{(End-around carry)} \\
+\ 15 & \quad 00001111
\end{array}
$$

## 2's Complement Method

Add the 2's complement values of 80 and $-65$. If the sum results in a carry beyond the sign bit, discard it.

$$80_{10} = 01010000$$

$$
\begin{aligned}
65_{10} &= 01000001 \\
-65_{10} &= 10111110 \quad \text{(1's complement)} \\
+ &\quad\quad\quad 1 \\
\hline
&\phantom{=}10111111 \quad \text{(2's complement)}
\end{aligned}
$$

$$
\begin{array}{rl}
80 & 01010000 \\
-\ 65 & +\ 10111111 \\
\hline
+\ 15 & 1\ 00001111 \\
& \underline{\qquad\qquad}\ \text{(Discard carry)}
\end{array}
$$

## True-Magnitude Method

$$80_{10} = 01010000$$

$$
\begin{aligned}
65_{10} &= 01000001 \\
-65_{10} &= 11000001
\end{aligned}
$$

$$
\begin{array}{rl}
80 & 01010000 \\
-\ 65 & +\ 11000001 \\
\hline
? & 1\ 00010001
\end{array}
$$

If we perform an end-around carry, the result is $00010010 = 18_{10}$. If we discard the carry, the result is $00010001 = 17_{10}$. Neither answer is correct. Thus, adding a negative true-magnitude number is not equivalent to subtraction.

## Negative Sum or Difference

All examples to this point have given positive-valued results. When a 2's complement addition or subtraction yields a negative sum or difference, we can't just read the magnitude from the result, since a 2's complement operation modifies the bits of a negative number. We must calculate the 2's complement of the sum or difference, which will give us the positive number that has the same magnitude.

**EXAMPLE 9.9**    Subtract $65_{10} - 80_{10}$ in 2's complement form.

**Solution**
$$65_{10} = \quad 01000001$$

$$
\begin{aligned}
80_{10} &= \quad 01010000 \\
-80_{10} &= \quad 10101111 \quad \text{(1's complement)} \\
+ &\quad\quad\quad\quad 1 \\
\hline
&\phantom{=}\quad 10110000 \quad \text{(2's complement)}
\end{aligned}
$$

$$
\begin{array}{rl}
65 & 01000001 \\
-\underline{80} & +\ \underline{10110000} \\
& 11110001
\end{array}
$$

Take the 2's complement of the difference to find the positive number with the same magnitude.

$$
\begin{array}{rl}
& 11110001 \\
& 00001110 \quad \text{(1's complement)} \\
+ & \underline{\hspace{1.5em}1} \\
& 00001111 \quad \text{(2's complement)}
\end{array}
$$

$00001111 = +15_{10}$. We generated this number by complementing 11110001. Thus, $11110001 = -15_{10}$.

## Range of Signed Numbers

Table 9.1 4-bit 2's Complement Numbers

| Decimal | 2's Complement |
|---------|----------------|
| +7 | 0111 |
| +6 | 0110 |
| +5 | 0101 |
| +4 | 0100 |
| +3 | 0011 |
| +2 | 0010 |
| +1 | 0001 |
| 0 | 0000 |
| −1 | 1111 |
| −2 | 1110 |
| −3 | 1101 |
| −4 | 1100 |
| −5 | 1011 |
| −6 | 1010 |
| −7 | 1001 |
| −8 | 1000 |

The largest positive number in 2's complement notation is a 0 followed by $n$ 1s for a number with $n$ magnitude bits. For instance, the largest positive 4-bit number is 0111 = $+7_{10}$. The negative number with the largest magnitude is *not* the 2's complement of the largest positive number. We can find the largest negative number by extension of a sequence of 2's complement numbers.

The 2's complement form of $-7_{10}$ is 1000 + 1 = 1001. The positive and negative numbers with the next largest magnitudes are 0110 (= $+6_{10}$) and 1010 (= $-6_{10}$). If we continue this process, we will get the list of numbers in Table 9.1.

We have generated the 4-bit negative numbers from $-1_{10}$ (1111) through $-7_{10}$ (1001) by writing the 2's complement forms of the positive numbers 1 through 7. Notice that these numbers count down in binary sequence. The next 4-bit number in the sequence (which is the only binary number we have left) is 1000. By extension, 1000 = $-8_{10}$. This number is its own 2's complement. (Try it.) It exemplifies a general rule for the $n$-bit negative number with the largest magnitude.

> A 2's complement number consisting of a 1 followed by $n$ 0s is equal to $-2^n$. Therefore, the range of a signed number, $x$, is $-2^n \le x \le 2^n - 1$ for a number with $n$ magnitude bits.

**EXAMPLE 9.10**

Write the largest positive and negative numbers for an 8-bit signed number in decimal and 2's complement notation.

**Solution**

$01111111 = +127$ (7 magnitude bits: $2^7 - 1 = 127$)

$10000000 = -128$ (1 followed by seven 0s: $-2^7 = -128$)

**EXAMPLE 9.11**

Write $-16_{10}$

a. As an 8-bit 2's complement number

b. As a 5-bit 2's complement number

(8-bit numbers are more common than 5-bit numbers in digital systems, but it is useful to see how we must write the same number differently with different numbers of bits.)

**Solution**    **a.** An 8-bit number has 7 magnitude bits and 1 sign bit.

$$16_{10} = 00010000_2$$

$$-16_{10} = 11101111 \quad \text{(1's complement)}$$
$$+ \quad\quad\ 1$$
$$\overline{11110000} \quad \text{(2's complement)}$$

**b.** A 5-bit number has 4 magnitude bits and 1 sign bit. Four magnitude bits are not enough to represent $16_{10}$. However, a 1 followed by $n$ 0s is equal to $-2^n$. For a 1 and four 0s, $-2^n = -2^4 = -16$. Thus, $10000_2 = -16_{10}$.

---

The last five bits of the binary equivalent of $-16$ are the same in both the 5-bit and 8-bit numbers. The 8-bit number is padded with leading 1s. This same general pattern applies for any negative power of 2. ($-2^n = n$ 0s preceded by all 1s within the defined number size.)

---

**Section Review Problem for Section 9.3**

Write $-32_{10}$

9.5.  As an 8-bit 2's complement number

9.6  As a 6-bit 2's complement number

---

## Sign Bit Overflow

**Overflow** *An erroneous carry into the sign bit of a signed binary number that results from a sum larger than can be represented by the number of magnitude bits.*

Signed addition of positive numbers is performed in the same way as unsigned addition. The only problem occurs when the number of bits in the sum of two numbers exceeds the number of magnitude bits and **overflows** into the sign bit. This causes the number to appear to be negative when it is not. For example, the sum $75 + 96 = 171$ causes an overflow in 8-bit signed addition. In unsigned addition the binary equivalent is:

$$1001011$$
$$+ \ 1100000$$
$$\overline{10101011}$$

In signed addition, the sum is the same, but has a different meaning.

$$
\begin{array}{r}
0\ \ 1001011 \\
+\ 0\ \ 1100000 \\
\hline
1\ \ 0101011
\end{array}
$$

(Sign bit) ____⌐ ⌐____⌐____ (Magnitude bits)

The sign bit is 1, indicating a negative number, which cannot be true, since the sum of two positive numbers is always positive.

---

A sum of positive signed binary numbers must not exceed $2^n - 1$ for numbers having $n$ magnitude bits. Otherwise, there will be an overflow into the sign bit.

---

## Overflow in Negative Sums

Overflow can also occur with large negative numbers. For example, the addition of $-80_{10}$ and $-65_{10}$ should produce the result:

$$-80_{10} + (-65_{10}) = -145_{10}$$

In 2's complement notation, we get:

$$
\begin{array}{rll}
+80_{10} = & 01010000_2 & \\
-80_{10} = & 10101111 & \text{(1's complement)} \\
+ & \underline{\hspace{1.2cm}1} & \\
& 10110000 & \text{(2's complement)}
\end{array}
$$

$$
\begin{array}{rll}
+65_{10} = & 01000001_2 & \\
-65_{10} = & 10111110 & \text{(1's complement)} \\
+ & \underline{\hspace{1.2cm}1} & \\
& 10111111 & \text{(2's complement)}
\end{array}
$$

$$
\begin{array}{rll}
-80 & 10110000 & \\
+\ (-65) & +\ \underline{10111111} & \\
? & 1\ 01101111 &
\end{array}
$$

⌐____⌐____ (Incorrect magnitude = $111_{10}$)
⌐_____ (Erroneous sign bit = 0)
⌐_____ (Discard carry)

This result shows a positive sum of two negative numbers—clearly incorrect. We can extend the statement we made earlier about permissible magnitudes of sums to include negative as well as positive numbers.

---

A sum of signed binary numbers must be within the range of $-2^n \leq \text{sum} \leq 2^n - 1$ for numbers having $n$ magnitude bits. Otherwise, there will be an overflow into the sign bit.

---

For an 8-bit signed number in 2's complement form, the permissible range of sums is $10000000 \leq \text{sum} \leq 01111111$. In decimal, this range is $-128 \leq \text{sum} \leq +127$.

A sum of two positive numbers is always positive. A sum of two negative numbers is always negative. Any 2's complement addition or subtraction operation that appears to contradict these rules has produced an overflow into the sign bit.

**EXAMPLE 9.12**

Which of the following sums will produce a sign bit overflow in 8-bit 2's complement notation? How can you tell?

**a.** $67_{10} + 33_{10}$

**b.** $67_{10} + 63_{10}$

**c.** $-96_{10} - 22_{10}$

**d.** $-96_{10} - 42_{10}$

**Solution**    A sign bit overflow is generated if the sum of two positive numbers appears to produce a negative result or the sum of two negative numbers appears to produce a positive result. In other words, overflow occurs if the operand sign bits are both 1 and the sum sign bit is 0 or vice versa. We know this will happen if an 8-bit sum is outside the range $(-128 \leq \text{sum} \leq +127)$.

**a.**

$$
\begin{array}{rl}
+67_{10} & 01000011 \\
+33_{10} & 00100001 \\
\hline
100_{10} & 01100100
\end{array}
$$
(No overflow; sum of positive numbers is positive.)

**b.**

$$
\begin{array}{rl}
+67_{10} & 01000011 \\
+63_{10} & 00111111 \\
\hline
130 & 10000010
\end{array}
$$
(Overflow; sum of positive numbers is negative. Sum > +127; out of range.)

**c.**

$$
\begin{array}{rll}
+96 = & 01100000 & \\
-96 = & 10011111 & \text{(1's complement)} \\
+ & \underline{\qquad 1} & \\
 & 10100000 & \text{(2's complement)}
\end{array}
$$

$$
\begin{array}{rll}
+22 = & 00010110 & \\
-22 & 11101001 & \text{(1's complement)} \\
+ & \underline{\qquad 1} & \\
 & 11101010 & \text{(2's complement)}
\end{array}
$$

$$
\begin{array}{rl}
-96 & 10100000 \\
-22 & 11101010 \\
\hline
-118 & 1\ 10001010
\end{array}
$$

(Magnitude bits)
(Sign bit)
(Discard carry)

(No overflow; sum of two negative numbers is negative.)

**d.**

$$
\begin{array}{rll}
+96 = & 01100000 & \\
-96 = & 10011111 & \text{(1's complement)} \\
+ & \underline{\qquad 1} & \\
 & 10100000 & \text{(2's complement)}
\end{array}
$$

$$+42 = \quad 00101010$$
$$-42 \qquad 11010101 \quad \text{(1's complement)}$$
$$+\underline{\qquad 1}$$
$$\qquad 11010110 \quad \text{(2's complement)}$$

$$-96 \qquad\qquad 10100000$$
$$\underline{-42} \qquad\qquad \underline{11010110}$$
$$-138 \qquad\qquad 1 \ 01110110$$

(Magnitude bits)
(Sign bit)
(Discard carry)

(Overflow; sum of two negative numbers is positive. Sum $< -128$; out of range.)

---

> The carry bit generated in 1's and 2's complement operations is not the same as an overflow bit. (See Example 9.12, parts c and d.) An overflow is a change in the sign bit, which leads us to believe that the number is opposite in sign from its true value. A carry is the result of an operation carrying beyond the physical limits of an $n$-bit number. It is similar to the idea of an odometer rolling over from 999999.9 to 1 000000.0. There are not enough places to hold the new number, so it goes back to the beginning and starts over.

## 9.4

# Hexadecimal Arithmetic

(This section may be omitted without loss of continuity.)

The main reason to be familiar with addition and subtraction in the hexadecimal system is that it is useful for calculations related to microcomputer and memory systems. Microcomputer systems often use binary numbers of 8, 16, 20, or 32 bits. Rather than write out all these bits, we use hex numbers as shorthand. Binary numbers having 8, 16, 20, or 32 bits can be represented by 2, 4, 5, or 8 hex digits, respectively.

### Hex Addition

Hex addition is very much like decimal addition, except that we must remember how to deal with the hex digits A to F. A few sums are helpful:

$$F + 1 = 10$$

$$F + F = 1E$$

$$F + F + 1 = 1F$$

The positional multipliers for the hexadecimal system are powers of 16. Thus, the most significant bit of the first sum is in the 16's column. The equivalent sum in decimal is:

$$15_{10} + 1_{10} = 16_{10} = 10_{16}$$

The second sum is the largest possible sum of two hex digits; the carry to the next position is 1. This shows that the sum of two hex digits will never produce a carry larger than 1. The second sum can be calculated as follows:

$$F_{16} + F_{16} = 15_{10} + 15_{10}$$
$$= 30_{10}$$
$$= 16_{10} + 14_{10}$$
$$= 10_{16} + E_{16}$$
$$= 1E_{16}$$

The third sum shows that if there is a carry from a previous sum, the carry to the next bit will still be 1.

---

It is useful to think of any digits larger than 9 as their decimal equivalents. For any sum greater than $15_{10}$ ($F_{16}$), subtract $16_{10}$, convert the difference to its hex equivalent, and carry 1 to the next digit position.

---

**EXAMPLE 9.13**    Add $6B3_{16} + A9C_{16}$.

**Solution**

| Hex | Decimal Equivalents |
|-----|---------------------|
| 6B3 | ( 6) (11) ( 3) |
| + A9C | + (10) ( 9) (12) |
| | (16) (20) (15) |

For sums greater than 15, subtract 16 and carry 1 to the next position:

| | Hex | Decimal Equivalents |
|-----|-----|---------------------|
| (Carry) ——— | 11 | ( 1) ( 1) |
| | 6B3 | ( 6) (11) ( 3) |
| + | A9C | + (10) ( 9) (12) |
| | 114F | ( 1) ( 1) ( 4) (15) |

Sum: $6B3_{16} + A9C_{16} = 114F_{16}$.

## Hex Subtraction

There are two ways to subtract hex numbers. The first reverses the addition process in the previous section. The second is a complement form of subtraction.

---

**EXAMPLE 9.14**    Subtract $6B3_{16} - 49C_{16}$.

**Solution**

| Hex | Decimal Equivalent |
|-----|---------------------|
| 6B3 | (6) (11) ( 3) |
| − 49C | − (4) ( 9) (12) |

To subtract the least significant digits, we must borrow $10_{16}$ ($16_{10}$) from the previous position. This leaves the subtraction looking like this:

|   | **Hex** | **Decimal Equivalent** |
|---|---------|------------------------|
| (Borrow)_____1 | | |
| | 6A3 | (6) (10) (16 + 3) |
| − | 49C | − (4) ( 9) (12) |
| | 217 | (2) ( 1) ( 7) |

The second subtraction method is a complement method, where, as in 2's complement subtractions, we add a negative number to subtract a positive number.

Calculate the 15's complement of a hex number by subtracting it from a number having the same number of digits, all Fs. Calculate the 16's complement by adding 1 to this number. This is the negated value of the number.

**EXAMPLE 9.15**   Negate the hex number 15AC by calculating its 16's complement.

**Solution**

$$
\begin{array}{r}
\text{FFFF} \\
- \ \text{15AC} \\
\hline
\text{EA53} \quad \text{(15's complement)} \\
+ \quad\quad 1 \\
\hline
\text{EA54} \quad \text{(16's complement)}
\end{array}
$$

The original value, 15AC, can be restored by calculating the 16's complement of EA54. Try it.

**EXAMPLE 9.16**   Subtract 8B63 − 55D7 using the complement method.

**Solution**   Find the 16's complement of 55D7.

$$
\begin{array}{r}
\text{FFFF} \\
- \ \text{55D7} \\
\hline
\text{AA28} \quad \text{(15's complement)} \\
+ \quad\quad 1 \\
\hline
\text{AA29} \quad \text{(16's complement)}
\end{array}
$$

Therefore, −55D7 = AA29.

$$
\begin{array}{r}
\text{8B63} \\
+ \ \text{AA29} \\
\hline
1 \ \ \text{358C}
\end{array}
$$

(Discard_____⌋
carry)

Difference: 8B63 − 55D7 = 358C.

## 9.5

# Binary Adders and Subtractors

## *Half and Full Adders*

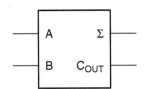

**Figure 9.1**
**Half Adder**

> **Half adder** *A circuit that will add two bits and produce a sum bit and a carry bit.*
>
> **Full adder** *A circuit that will add a carry bit from another full or half adder and two operand bits to produce a sum bit and a carry bit.*

There are only three possible sums of two 1-bit binary numbers:

$$0 + 0 = 00$$
$$0 + 1 = 01$$
$$1 + 1 = 10$$

**Table 9.2** Half Adder Truth Table

| *A* | *B* | *C_OUT* | Σ |
|-----|-----|---------|---|
| 0 | 0 | 0 | 0 |
| 0 | 1 | 0 | 1 |
| 1 | 0 | 0 | 1 |
| 1 | 1 | 1 | 0 |

We can build a simple combinational logic circuit to produce the above sums. Let us designate the bits on the left side of the above equalities as inputs to the circuit and the bits on the right side as outputs. Let us call the LSB of the output the sum bit, symbolized by $\Sigma$, and the MSB of the output the carry bit, designated $C_{OUT}$.

Figure 9.1 shows the logic symbol of the circuit, which is called a **half adder.** Its truth table is given in Table 9.2. Since addition is subject to the commutative property, $(A + B = B + A)$, the second and third lines of the truth table are the same.

The Boolean functions of the two outputs, derived from the truth table, are:

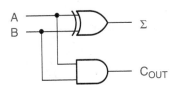

**Figure 9.2**
**Half Adder Circuit**

$$C_{OUT} = AB$$
$$\Sigma = \overline{A}B + A\overline{B} = A \oplus B$$

The corresponding logic circuit is shown in Figure 9.2.

The half adder circuit cannot account for an *input* carry, that is, a carry from a lower-order 1-bit addition. A **full adder,** shown in Figure 9.3, can add two 1-bit numbers *and* accept a carry bit from a previous adder stage. Operation of the full adder is based on the following sums:

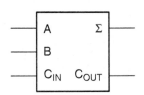

**Figure 9.3**
**Full Adder**

$$0 + 0 + 0 = 00$$
$$0 + 0 + 1 = 01$$
$$0 + 1 + 1 = 10$$
$$1 + 1 + 1 = 11$$

**Table 9.3** Full Adder Truth Table

| A | B | $C_{IN}$ | $C_{OUT}$ | $\Sigma$ |
|---|---|---|---|---|
| 0 | 0 | 0 | 0 | 0 |
| 0 | 0 | 1 | 0 | 1 |
| 0 | 1 | 0 | 0 | 1 |
| 0 | 1 | 1 | 1 | 0 |
| 1 | 0 | 0 | 0 | 1 |
| 1 | 0 | 1 | 1 | 0 |
| 1 | 1 | 0 | 1 | 0 |
| 1 | 1 | 1 | 1 | 1 |

Designating the left side of the above equalities as circuit inputs $A$, $B$, and $C_{IN}$ and the right side as outputs $C_{OUT}$ and $\Sigma$, we can make the truth table in Table 9.3. (The second and third of the above sums each account for three lines in the full adder truth table.)

The unsimplified Boolean expressions for the outputs are:

$$C_{OUT} = \overline{A}\,B\,C_{IN} + A\,\overline{B}\,C_{IN} + A\,B\,\overline{C}_{IN} + A\,B\,C_{IN}$$

$$\Sigma = \overline{A}\,\overline{B}\,C_{IN} + \overline{A}\,B\,\overline{C}_{IN} + A\,\overline{B}\,\overline{C}_{IN} + A\,B\,C_{IN}$$

There are a couple of ways to simplify these expressions.

## Karnaugh Map Method

Since we have expressions for $\Sigma$ and $C_{OUT}$ in sum-of-products form, let us try to use the Karnaugh maps in Figure 9.4 to simplify them. The expression for $\Sigma$ doesn't reduce at all. The simplified expression for $C_{OUT}$ is:

$$C_{OUT} = A\,B + A\,C_{IN} + B\,C_{IN}$$

The corresponding logic circuits for $\Sigma$ and $C_{OUT}$, shown in Figure 9.5, don't give us much of a simplification.

## Boolean Algebra Method

The simplest circuit for $C_{OUT}$ and $\Sigma$ involves the Exclusive OR function, which we cannot derive from K-map groupings. This can be shown by Boolean algebra, as follows:

$$C_{OUT} = \overline{A}\,B\,C_{IN} + A\,\overline{B}\,C_{IN} + A\,B\,\overline{C}_{IN} + A\,B\,C_{IN}$$
$$= (\overline{A}\,B + A\,\overline{B})\,C_{IN} + A\,B\,(\overline{C}_{IN} + C_{IN})$$
$$= (A \oplus B)\,C_{IN} + A\,B$$

$$\Sigma = (\overline{A}\,\overline{B} + A\,B)\,C_{IN} + (\overline{A}\,B + A\,\overline{B})\,\overline{C}_{IN}$$
$$= (\overline{A \oplus B})\,C_{IN} + (A \oplus B)\,\overline{C}_{IN} \qquad \text{Let } x = A \oplus B$$
$$= \overline{x}\,C_{IN} + x\,\overline{C}_{IN}$$

**Figure 9.4**

**K-Maps for a Full Adder**

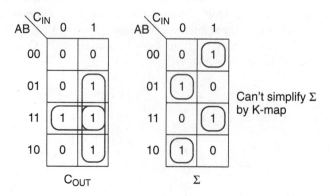

**Figure 9.5**

**Full Adder From K-Map Simplification**

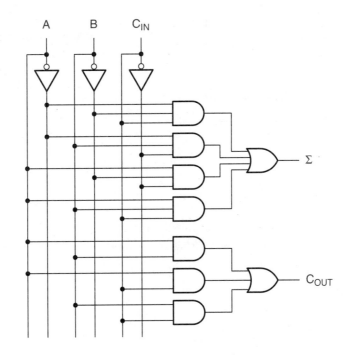

**Figure 9.6**

**Full Adder From Logic Gates**

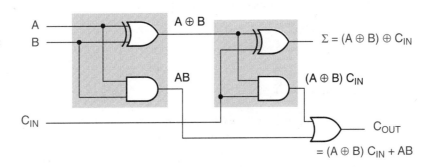

$$= x \oplus C_{IN}$$
$$= (A \oplus B) \oplus C_{IN}$$

The simplified expressions are as follows:

$$C_{OUT} = (A \oplus B)\, C_{IN} + A\, B$$
$$\Sigma = (A \oplus B) \oplus C_{IN}$$

Figure 9.6 shows the logic circuit derived from these equations. If you refer back to the half adder circuit in Figure 9.2, you will see that the full adder can be constructed from two half adders and an OR gate, as shown in Figure 9.7.

**Figure 9.7**

**Full Adder From Two Half Adders**

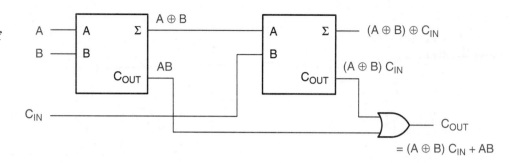

---

**EXAMPLE 9.17**

Evaluate the Boolean expression for $\Sigma$ and $C_{OUT}$ of the full adder in Figure 9.8 for the following input values. What is the binary value of the outputs in each case?

    **a.** $A = 0$, $B = 0$, $C_{IN} = 1$

    **b.** $A = 1$, $B = 0$, $C_{IN} = 0$

    **c.** $A = 1$, $B = 0$, $C_{IN} = 1$

    **d.** $A = 1$, $B = 1$, $C_{IN} = 0$

**Solution**

The output of a full adder for any set of inputs is simply given by $C_{OUT}\,\Sigma = A + B + C_{IN}$. For each of the stated sets of inputs:

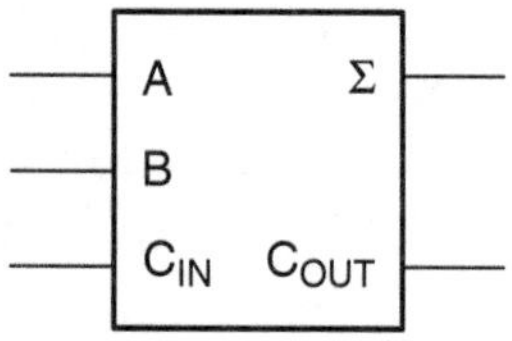

    **a.** $C_{OUT}\,\Sigma = A + B + C_{IN} = 0 + 0 + 1 = 01$

    **b.** $C_{OUT}\,\Sigma = A + B + C_{IN} = 1 + 0 + 0 = 01$

    **c.** $C_{OUT}\,\Sigma = A + B + C_{IN} = 1 + 0 + 1 = 10$

    **d.** $C_{OUT}\,\Sigma = A + B + C_{IN} = 1 + 1 + 0 = 10$

We can verify each of these sums algebraically by plugging the specified inputs into the full adder Boolean equations:

$$C_{OUT} = (A \oplus B)\, C_{IN} + A\,B$$

$$\Sigma = (A \oplus B) \oplus C_{IN}$$

    **a.** $C_{OUT} = (0 \oplus 0) \cdot 1 + 0 \cdot 0$

          $= 0 \cdot 1 + 0$

          $= 0 + 0 = 0$

      $\Sigma = (0 \oplus 0) \oplus 1$

          $= 0 \oplus 1 = 1$     (Binary equivalent: $C_{OUT}\,\Sigma = 01$)

    **b.** $C_{OUT} = (1 \oplus 0) \cdot 0 + 1 \cdot 0$

          $= 1 \cdot 0 + 0$

          $= 0 + 0 = 0$

      $\Sigma = (1 \oplus 0) \oplus 0$

          $= 1 \oplus 0 = 1$     (Binary equivalent: $C_{OUT}\,\Sigma = 01$)

**Figure 9.8**

**Example 9.17**

**Full Adder**

**c.** $C_{OUT} = (1 \oplus 0) \cdot 1 + 1 \cdot 0$

$= 1 \cdot 1 + 0$

$= 1 + 0 = 1$

$\Sigma = (1 \oplus 0) \oplus 1$

$= 1 \oplus 1 = 0$    (Binary equivalent: $C_{OUT} \Sigma = 10$)

**d.** $C_{OUT} = (1 \oplus 1) \cdot 0 + 1 \cdot 1$

$= 0 \cdot 0 + 1$

$= 0 + 1 = 1$

$\Sigma = (1 \oplus 1) \oplus 0$

$= 0 \oplus 0 = 0$    (Binary equivalent: $C_{OUT} \Sigma = 10$)

In each case, the binary equivalent is the same as the number of HIGH inputs, regardless of which inputs they are.

**EXAMPLE 9.18**    Combine a half adder and a full adder to make a circuit that will add two 2-bit numbers. Check that the circuit will work by adding the following numbers and writing the binary equivalents of the inputs and outputs:

**a.** $A_2 A_1 = 01, B_2 B_1 = 01$

**b.** $A_2 A_1 = 11, B_2 B_1 = 10$

**Solution**    The 2-bit adder is shown in Figure 9.9. The half adder combines $A_1$ and $B_1$; $A_2$, $B_2$, and $C_1$ are added in the full adder. The carry output, $C_1$, of the half adder is connected to the carry input of the full adder. (A half adder can be used only in the LSB of a multiple-bit addition.)

**Figure 9.9**

**Example 9.18**
**2-Bit Adder**

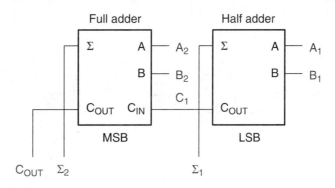

**Sums:**

**a.** $01 + 01 = 010$

$A_1 = 1, B_1 = 1$          $C_1 = 1, \Sigma_1 = 0$

$A_2 = 0, B_2 = 0, C_1 = 1$    $C_2 = 0, \Sigma_2 = 1$

(Binary equivalent: $A_2 A_1 + B_2 B_1 = C_2 \Sigma_2 \Sigma_1 = 010$)

**b.** $11 + 10 = 101$

$$A_1 = 1, B_1 = 0 \qquad\qquad C_1 = 0, \Sigma_1 = 1$$

$$A_2 = 1, B_2 = 1, C_1 = 0 \quad C_2 = 1, \Sigma_2 = 0$$

(Binary equivalent: $A_2 A_1 + B_2 B_1 = C_2 \Sigma_2\Sigma_1 = 101$)

## Parallel Binary Adder/Subtractor

**Parallel binary adder** *A circuit, consisting of* n *full adders, that will add two* n-*bit binary numbers. The output consists of* n *sum bits and a carry bit.*

**Ripple carry** *A method of passing carry bits from one stage of a parallel adder to the next by connecting $C_{OUT}$ of one full adder to $C_{IN}$ of the following stage.*

**Cascade** *To connect an output of one device to an input of another, often for the purpose of expanding the number of bits available for a particular function.*

As Example 9.18 implies, a binary adder can be expanded to any number of bits by using a full adder for each bit addition and connecting their carry inputs and outputs in **cascade**. Figure 9.10 shows four full adders connected as a 4-bit **parallel binary adder.**

The first stage (LSB) can be either a full adder with its carry input forced to logic 0 or a half adder, since there is no previous stage to provide a carry. The addition is done one bit at a time, with the carry from each adder propagating to the next stage.

**Figure 9.10**

**4-Bit Parallel Binary Adder**

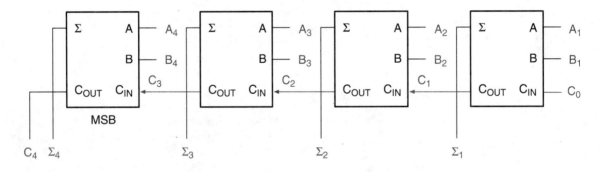

**EXAMPLE 9.19**

Verify the summing operation of the circuit in Figure 9.10 by calculating the output for the following sets of inputs:

**a.** $A_4 A_3 A_2 A_1 = 0101, B_4 B_3 B_2 B_1 = 1001$

**b.** $A_4 A_3 A_2 A_1 = 1111, B_4 B_3 B_2 B_1 = 0001$

**Solution** At each stage, $A + B + C_{IN} = C_{OUT} \Sigma$.

**a.** $0101 + 1001 = 1110$

$\qquad (5_{10} + 9_{10} = 14_{10})$

$$A_1 = 1, B_1 = 1, C_0 = 0; C_1 = 1, \Sigma_1 = 0$$

$$A_2 = 0, B_2 = 0, C_1 = 1; C_2 = 0, \Sigma_2 = 1$$

$$A_3 = 1, B_3 = 0, C_2 = 0; C_3 = 0, \Sigma_3 = 1$$

$$A_4 = 0, B_4 = 1, C_3 = 0; C_4 = 0, \Sigma_4 = 1$$

(Binary equivalent: $C_4 \Sigma_4 \Sigma_3 \Sigma_2 \Sigma_1 = 01110$)

**b.** $1111 + 0001 = 10000$
$(15_{10} + 1_{10} = 16_{10})$

$$A_1 = 1, B_1 = 1, C_0 = 0; C_1 = 1, \Sigma_1 = 0$$

$$A_2 = 1, B_2 = 0, C_1 = 1; C_2 = 1, \Sigma_2 = 0$$

$$A_3 = 1, B_3 = 0, C_2 = 1; C_3 = 1, \Sigma_3 = 0$$

$$A_4 = 1, B_4 = 0, C_3 = 1; C_4 = 1, \Sigma_4 = 0$$

(Binary equivalent: $C_4 \Sigma_4 \Sigma_3 \Sigma_2 \Sigma_1 = 10000$)

Two parallel binary adders available in integrated circuit form are the 74LS83A (TTL) and 4008B (CMOS) 4-bit parallel binary adders. The Motorola version of the 4008B adder is designated MC14008B. The logic circuit diagrams for these chips are shown in Figures 9.11 and 9.12.

The 74LS83A and MC14008B are functionally similar. Four full adders are implemented on each chip and connected internally to give four sum outputs and a carry output. Figure 9.13 shows one full adder section of the MC14008B device. The sum and carry functions of this circuit can be verified algebraically as follows:

$$\Sigma = \overline{A\,B + \overline{A + B}} \oplus C_{IN} \quad \text{(From logic diagram)}$$

$$= \overline{A\,B + \overline{A}\,\overline{B}} \oplus C_{IN}$$

$$= \overline{\overline{A \oplus B}} \oplus C_{IN}$$

$$= (A \oplus B) \oplus C_{IN}$$

$$C_{OUT} = (A + B)C_{IN} + A\,B \quad \text{(From logic diagram)}$$

$$= A\,C_{IN} + B\,C_{IN} + A\,B$$

Compare these with the full adder equations derived earlier.

The internal carries in the CMOS device are achieved by a system called **ripple carry.** This is the carry method used in the parallel binary adder in Figure 9.10. The carry output of one full adder cascades directly to the carry input of the next. Every time a carry bit changes, it "ripples" through some or all of the following stages. A sum is not complete until the carry from another stage has arrived.

A potential problem with this design is that the adder circuitry, being nonideal, does not switch instantaneously. A carry propagating through a ripple adder adds delays to the summation time and, more importantly, can introduce unwanted intermediate states.

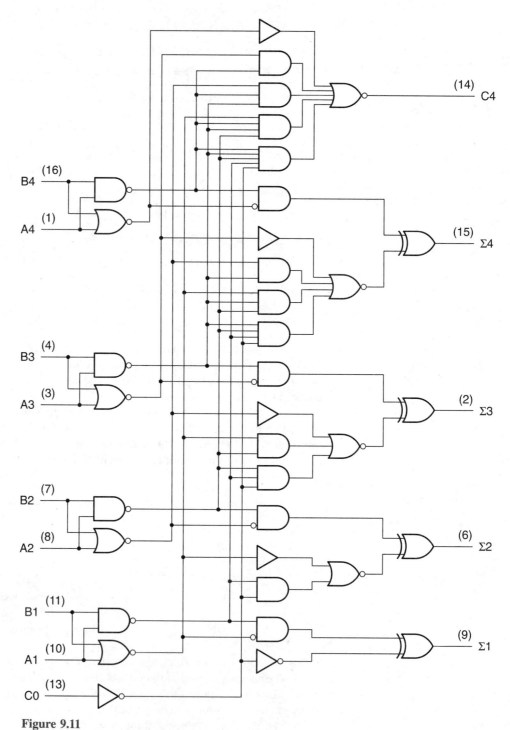

**Figure 9.11**

**74LS83A Parallel Binary Adder** (Reprinted by permission of Texas Instruments)

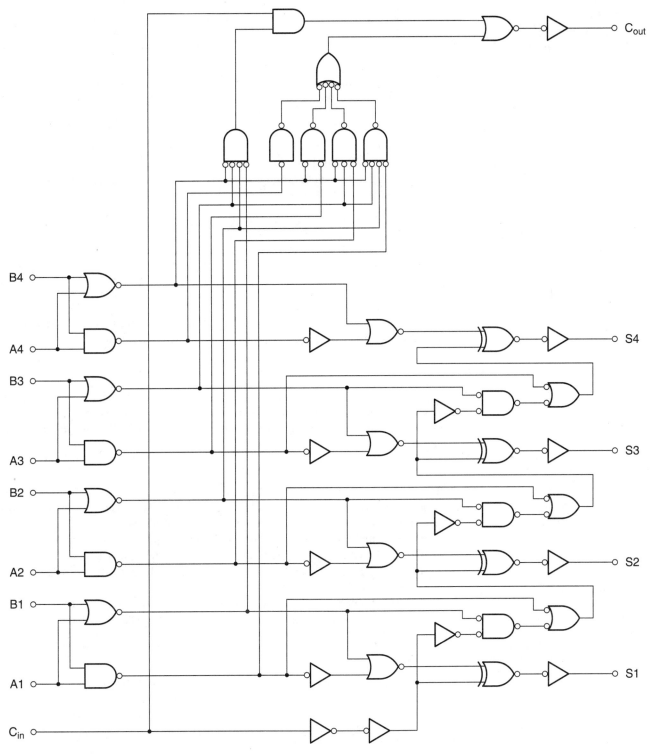

**Figure 9.12**
**MC14008B Parallel Binary Adder** (Reprinted with permission of Motorola)

**Figure 9.13**

**One Full Adder From MC14008B Parallel Adder**

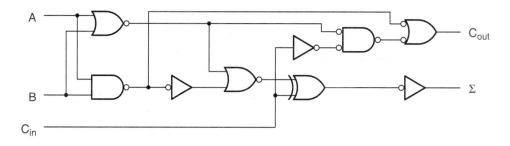

Examine the sum (1111 + 0001 = 10000). For a parallel adder having a ripple carry, the output goes through the following series of changes as the carry bit propagates through the circuit:

$$C_4 \; \Sigma_4 \; \Sigma_3 \; \Sigma_2 \; \Sigma_1 = 01111$$
$$01110$$
$$01100$$
$$01000$$
$$10000$$

If the output of the full adder is being used to drive another circuit, these unwanted intermediate states may cause erroneous operation of the load circuit.

### Fast Carry

**Fast carry (or parallel carry)** *A gate network that generates a carry bit directly from all incoming operand bits, independent of the operation of each full adder stage.*

Both the 74LS83A and MC14008B parallel binary adders have a feature called **fast carry** or **parallel carry,** which generates the adder's carry output ($C_4$). The 74LS83A also incorporates this feature into its internal carries between full adder stages.

The idea behind fast carry is that the circuit will examine all the *A* and *B* bits and produce a carry bit that uses fewer levels of gating than a ripple carry circuit. Also, since there is a carry bit gate network for each internal stage, the propagation delay is the same for each full adder, regardless of the input operands.

The algebraic relation between operand bits and fast carry output is presented below, without proof. It can be developed from the circuit of the MC14008B adder by tracing the logic of the multiple-input AND-shaped gates at the top of the diagram in Figure 9.12.

$$C_4 = A_4 \; B_4 + A_3 \; B_3 \; (A_4 + B_4) + A_2 \; B_2 \; (A_4 + B_4)(A_3 + B_3)$$
$$+ \; A_1 \; B_1 \; (A_4 + B_4)(A_3 + B_3)(A_2 + B_2)$$
$$+ \; C_0 \; (A_4 + B_4)(A_3 + B_3)(A_2 + B_2)(A_1 + B_1)$$

We can make some intuitive sense of the above expression by examining it a term at a time. The first term says if the MSBs of both operands are 1, there will be a carry (e.g., 1000 + 1000 = 10000; carry generated).

The second term says if both second bits are 1 AND at least one MSB is 1, there will be a carry (e.g., 0100 + 1100 = 10000, or 1100 + 1100 = 11000; carry generated in either case). This pattern can be followed logically through all the terms.

The internal carry bits are generated by similar circuits found at the carry input of each full adder stage in the 74LS83A parallel adder. In general, we can generate each internal carry by expanding the following expression:

$$C_n = A_n B_n + C_{n-1} (A_n + B_n)$$

The algebraic expressions for the remaining carry bits are:

$$C_1 = A_1 B_1 + C_0 (A_1 + B_1)$$

$$C_2 = A_2 B_2 + A_1 B_1 (A_2 + B_2) + C_0 (A_2 + B_2)(A_1 + B_1)$$

$$C_3 = A_3 B_3 + A_2 B_2 (A_3 + B_3) + A_1 B_1 (A_3 + B_3)(A_2 + B_2)$$
$$+ C_0 (A_3 + B_3)(A_2 + B_2)(A_1 + B_1)$$

---

**Section Review Problem for Section 9.5a**

9.7. Refer to the logic diagrams for the 74LS83A and MC14008B parallel binary adders (Figures 9.11 and 9.12). How many gates must a carry bit propagate through in each device if the effect of the carry input ripples through to the $\Sigma_4$ bit? (See Figure 9.14 on page 371 and Figure 9.15 on page 372.)

---

## 2's Complement Subtractor

Recall the technique for subtracting binary numbers in 2's complement notation. For example, to find the difference 0101 − 0011 by 2's complement subtraction:

1.  Find the 2's complement of 0011:

$$
\begin{array}{ll}
0011 & \\
1100 & \text{(1's complement)} \\
\underline{+1} & \\
1101 & \text{(2's complement)}
\end{array}
$$

2.  Add the 2's complement of the subtrahend to the minuend:

$$
\begin{array}{ll}
\phantom{+\ }0101 & (+5) \\
+\ \underline{1101} & (-3) \\
1\ 0010 & (\ \ 2)
\end{array}
$$

(Discard carry) ⎯⎯⏌

We can easily build a circuit to perform 2's complement subtraction, using a parallel binary adder and an inverter for each bit of one of the operands. The circuit shown in Figure 9.16 performs the operation $(A - B)$.

The four inverters generate the 1's complement of $B$. The parallel adder generates the 2's complement by adding the carry bit (held at logic 1) to the 1's complement at the

**Figure 9.16**

**Z's Complement Subtractor**

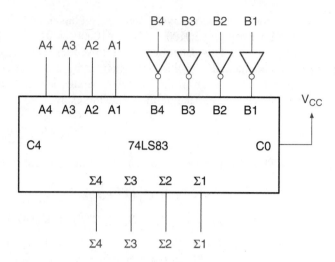

*B* inputs. Algebraically, this is expressed as:

$$A - B = A + (-B) = A + \overline{B} + 1$$

where $\overline{B}$ is the 1's complement of *B*, and $(\overline{B} + 1)$ is the 2's complement of *B*.

**EXAMPLE 9.20**    Verify the operation of the 2's complement subtractor in Figure 9.16 by subtracting:

**a.** $1001 - 0011$ (unsigned)

**b.** $0100 - 0111$ (signed)

**Solution**    Let $\overline{B}$ be the 1's complement of *B*.

**a.** Inverter inputs *(B):*        0011

Inverter outputs $(\overline{B})$:        1100

Sum $(A + \overline{B} + 1)$:
```
            1001      (  9)
            1100    + (−3)
        +      1     ───────
          1 0110      (  6)
```
(Discard carry) ⌐

**b.** Inverter inputs *(B):*        0111

Inverter outputs $(\overline{B})$:        1000

Sum $(A + \overline{B} + 1)$:
```
            0100      (  4)
            1000    + (−7)
        +      1     ───────
```
Negative result:        1101      (−3) ←┐

1's complement of 1101:        0010

```
                    +      1
```
2's complement of 1101:        0011      (  3) ┘

**Figure 9.17**

**Z's Complement Adder/Subtractor**

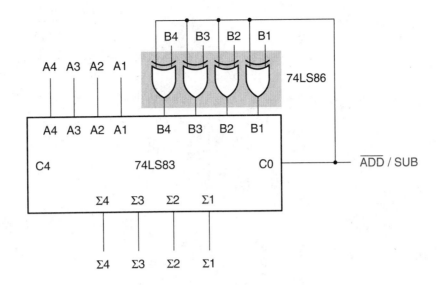

**Figure 9.18**

**XOR as a Programmable Inverter**

### Parallel Binary Adder/Subtractor

Figure 9.17 shows a 74LS83A parallel binary adder configured as a programmable adder/subtractor. The Exclusive OR gates work as programmable inverters to pass $B$ to the parallel adder in either true or complement form, as shown in Figure 9.18.

The $\overline{ADD}/SUB$ input is tied to the XOR inverter/buffers and to the carry input of the parallel adder. When $\overline{ADD}/SUB = 1$, $B$ is complemented and the 1 from the carry input is added to the complement sum. The effect is to subtract $(A - B)$. When $\overline{ADD}/SUB = 0$, the $B$ inputs are presented to the adder in true form and the carry input is 0. This produces an output equivalent to $(A + B)$.

This circuit can add or subtract 4-bit signed or unsigned binary numbers.

**EXAMPLE 9.21**     Combine two 4008B parallel binary adders and any other required logic to make an 8-bit parallel binary adder.

**Solution**     The 4008B parallel binary adder can be expanded by 4 bits by connecting another 4008B as shown in Figure 9.19. Two 4070B Quad Exclusive OR gate ICs are used to make the programmable inverters required for 2's complement addition and subtraction.

The circuit can be used to add or subtract 8-bit signed or unsigned binary numbers.

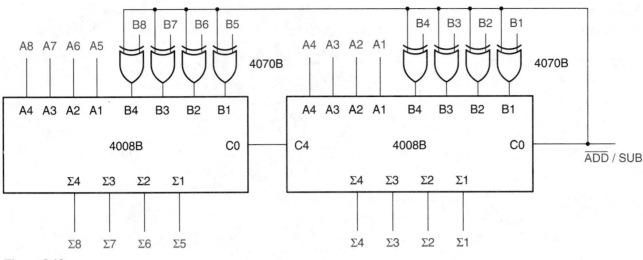

**Figure 9.19**

**Example 9.21**
**8-Bit Adder/Subtractor**

## Overflow Detection

Recall from Example 9.12 the condition for detecting a sign bit overflow in a sum of two binary numbers.

> If the sign bits of both operands are the same and the sign bit of the sum is different from the operand sign bits, an overflow has occurred.

This implies that overflow is not possible if the sign bits of the operands are different from each other. This is true because the sum of two opposite-sign numbers will always be smaller in magnitude than the larger of the two operands.

Here are two examples:

1. $(+15) + (-7) = (+8)$; $+8$ has a smaller magnitude than $+15$.
2. $(-13) + (+9) = (-4)$; $-4$ has a smaller magnitude than $-13$.

No carry into the sign bit will be generated in either case.

An 8-bit parallel binary adder will add two signed binary numbers as follows:

**Table 9.4** Overflow Detector Truth Table

| $S_A$ | $S_B$ | $S_\Sigma$ | $V$ |
|-------|-------|------------|-----|
| 0 | 0 | 0 | 0 |
| 0 | 0 | 1 | 1 |
| 0 | 1 | 0 | 0 |
| 0 | 1 | 1 | 0 |
| 1 | 0 | 0 | 0 |
| 1 | 0 | 1 | 0 |
| 1 | 1 | 0 | 1 |
| 1 | 1 | 1 | 0 |

$$S_A\ A_7\ A_6\ A_5\ A_4\ A_3\ A_2\ A_1 \quad (S_A = \text{Sign bit of } A)$$
$$\underline{S_B\ B_7\ B_6\ B_5\ B_4\ B_3\ B_2\ B_1} \quad (S_B = \text{Sign bit of } B)$$
$$S_\Sigma\ \Sigma_7\ \Sigma_6\ \Sigma_5\ \Sigma_4\ \Sigma_3\ \Sigma_2\ \Sigma_1 \quad (S_\Sigma = \text{Sign bit of sum})$$

From our condition for overflow detection, we can make a truth table for an overflow variable, $V$, in terms of $S_A$, $S_B$, and $S_\Sigma$. Let us specify that $V = 1$ when there is an overflow condition. This condition occurs when $(S_A = S_B) \neq S_\Sigma$. Table 9.4 shows the truth table for the overflow detector function.

The SOP Boolean expression for the overflow detector is:

$$V = S_A\ S_B\ \overline{S_\Sigma} + \overline{S_A}\ \overline{S_B}\ S_\Sigma$$

**Figure 9.20**

**Overflow Detector**

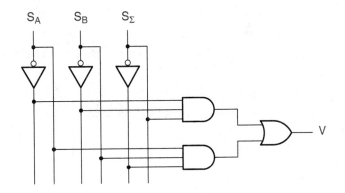

Figure 9.20 shows a logic circuit that will detect a sign bit overflow in a parallel binary adder. The inputs $S_A$, $S_B$, and $S_\Sigma$ are the MSBs of the adder $A$ and $B$ inputs and $\Sigma$ outputs, respectively.

**EXAMPLE 9.22**    Modify the 8-bit binary adder/subtractor circuit of Figure 9.19 to include a circuit to indicate sign bit overflow.

**Solution**    Figure 9.21 represents the 8-bit adder/subtractor with an overflow detector of the type shown in Figure 9.20.

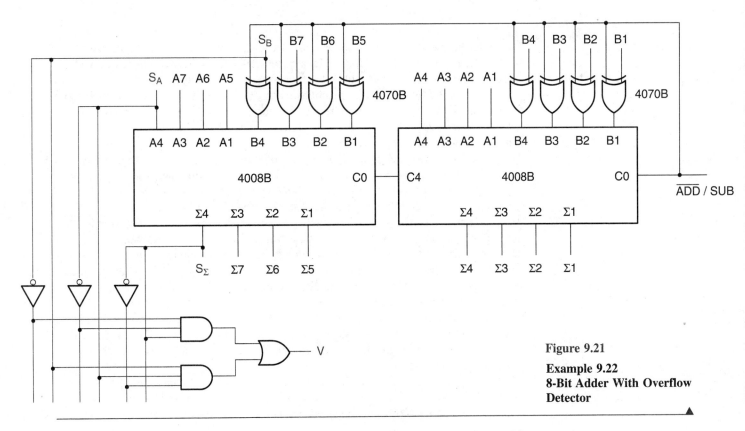

**Figure 9.21**

**Example 9.22**
**8-Bit Adder With Overflow Detector**

**Section Review Problem for Section 9.5b**

9.8. What is the permissible range of values of a sum or difference, $x$, in a 12-bit parallel binary adder if it is written as:

   a. A signed binary number?

   b. An unsigned binary number?

**9.6**

# Register Arithmetic Circuits

> **Register** *A digital circuit that can store a binary number until needed (e.g., a multibit flip-flop or latch).*
>
> **Accumulator** *A special register that stores the accumulating total of several binary additions and subtractions.*
>
> **Arithmetic logic unit (ALU)** *A combinational logic circuit that performs several selectable arithmetic and logic functions on two binary numbers.*
>
> **Data bus** *A group of parallel conductors carrying related logic signals, such as multibit data.*

One of the most important applications of digital arithmetic is the digital computer. Arithmetic in computers is performed by a circuit called the **arithmetic logic unit (ALU),** which, at the very least, will add and subtract two binary numbers. Often the ALU will execute several other arithmetic functions, such as increment (add 1 to a number) and decrement (subtract 1 from a number), and a number of logic functions, such as AND, OR, NOT, and XOR. (Strictly speaking, an adder/subtractor circuit is not really an ALU, since it does not perform logic functions.)

The ALU is frequently used in conjunction with a special **register** (a multibit latch or flip-flop that stores a binary number) called the **accumulator.** The accumulator stores an accumulating total of additions and subtractions. Figure 9.22 shows a block diagram of a basic ALU with an accumulator and another register.

To understand the register arithmetic circuit, we must introduce a few simple computer concepts. The first is that of the **data bus.** Most computers are "bus-oriented" or "bus-organized," meaning that data are simultaneously available to a number of devices connected in parallel, only a few of which are enabled at any time. The connecting data bus in Figure 9.22 is 8 bits wide. That is, it is made up of 8 parallel conductors. This is indicated in shorthand notation by the slash through the wire and the number 8.

Those devices not enabled are electrically disconnected from the data bus by tristate logic. Recall from the section on parallel data transfer in Chapter 6 that when a tristate output is enabled, its logic state is HIGH or LOW, as usual, and when it is disabled, it acts like an open circuit.

The control lines $\overline{L}_A$, $\overline{L}_B$, $\overline{E}_U$, and $\overline{E}_A$ control the flow of data to and from the bus by enabling and disabling the tristate logic in the various registers. The $L$ lines load data from the bus into its corresponding register. The $E$ lines enable the register outputs and send data from the register to the bus. The $S_U$ line selects the add and subtract functions: $S_U = 0$ to add and $S_U = 1$ to subtract.

Other devices are required to control the arithmetic functions and supply the operand data. These are not shown.

When two numbers are added or subtracted, the first number is loaded from the data bus into the accumulator. The second operand then goes from the bus to the $B$ register. The adder/subtractor combines the contents of the accumulator and $B$ register. This result is transferred back to the accumulator.

New operands for further additions or subtractions are loaded into the $B$ register only and added to or subtracted from the accumulator contents. The new results are again stored in the accumulator.

**EXAMPLE 9.23**

Describe the steps necessary for the register arithmetic circuit in Figure 9.22 to add the binary numbers 01011101 and 01010001, store the result, subtract 01110101, store the result, and send the final result back to the data bus.

$$(01011101 + 01010001) - 01110101 = 00111001$$

Assume that the contents of the accumulator and $B$ register are initially zero.

**Solution**    Figure 9.23 shows the steps for the required arithmetic operations.

**a.** The registers are initially cleared. All control lines are inactive initially and after each step.

**b.** The number 01011101 is placed on the data bus and $\overline{L}_A = 0$, storing the number in the accumulator. Since there is no tristate connection between the accumulator output and the adder/subtractor, the number immediately appears in the adder/subtractor ($A + 0 = A$).

**Figure 9.22**
**Simple ALU and Registers**

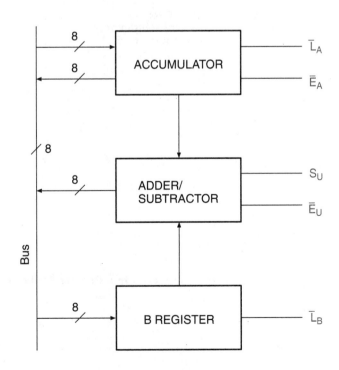

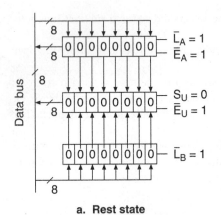

a.  Rest state

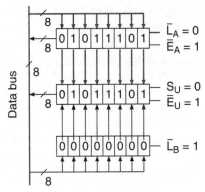

b.  Load accumulator

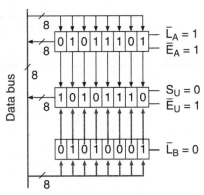

c.  Load B register.
Adder / subtractor
generates (A + B).

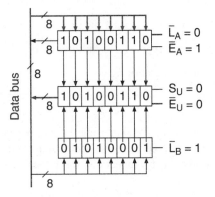

d.  Store sum in accumulator

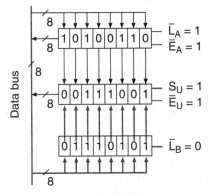

e.  Load B register.
Adder / subtractor
generates (A – B).

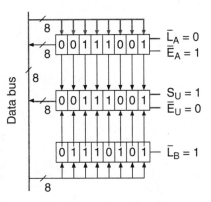

f.  Store accumulated
total

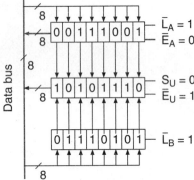

g.  Send accumulated
total to data bus

Figure 9.23

Example 9.23

**An Addition and Subtraction in a Register Arithmetic Circuit**

**c.** The number 01010001 is placed on the data bus, and the $B$ register input control line activates, storing the number in the $B$ register. Since $S_U = 0$, the adder/subtractor generates the sum of $A$ and $B$ ($= 10101110$).

**d.** The output control of the adder/subtractor and the input control of the accumulator activate simultaneously ($\overline{E}_U = 0$, $\overline{L}_A = 0$) to store the sum $A + B$ in the accumulator. (Shortly after the sum is transferred, the adder/subtractor generates the sum of the accumulated total and the $B$ register contents. Figure 9.23d shows the original adder/subtractor contents, just before the change occurs and after the contents have been transferred to the accumulator.)

**e.** The number 01110101 is placed on the data bus. $\overline{L}_B$ goes LOW, storing the number in the $B$ register. $S_U$ goes HIGH, which adds the previously accumulated total (10101110) to the 2's complement of 01110101 (2's complement: 10001011). This is equivalent to $A$ minus $B$.

**f.** The number at the adder/subtractor output (00111001) is transferred to the accumulator, as in step d. Figure 9.23f shows the adder/subtractor output *just before* it changes as a result of the new value in the accumulator.

**g.** $E_A$ goes LOW, sending the accumulator contents (00111001) to the data bus, where they would probably be transferred to an output register or a memory location. The adder/subtractor output is the sum of the accumulator and $B$ register data. Since $E_U$ is HIGH, this value does not affect the number on the bus.

Figure 9.24 represents a register arithmetic circuit assembled using standard TTL components. This circuit combines the concepts of the parallel binary adder/subtractor and parallel data transfer. For a review of parallel data transfer using 74LS/HC374 Octal Flip-Flops and 74LS/HC244 Octal Tristate Buffers, refer to Examples 6.10 and 6.11 in Chapter 6 (Latches and Flip-Flops).

The '374 Octal Flip-Flops act as the accumulator and $B$ register. They will store 8 bits of input data (*8D* to *1D*) when the logic level at the input marked *CLK* (clock) makes a LOW-to-HIGH transition. This occurs when $\overline{L}_A$ (for the accumulator) or $\overline{L}_B$ (for the $B$ register) is held LOW and a HIGH-to-LOW transition is applied to the input marked $\overline{CLK}$, as shown in Figure 9.24b.

The '244 Tristate Buffers allow the contents of the accumulator or the contents of the adder/subtractor to be sent to the data bus. If the buffers are disabled ($\overline{E}_A$ or $\overline{E}_U$ is in the HIGH state), the register outputs are electrically disconnected from the data bus.

The following steps add two 8-bit numbers, $A$ and $B$:

1. $A$ is placed on the bus and loaded into the accumulator ($\overline{L}_A = 0$, $CLK = $ LOW-to-HIGH).

2. $B$ appears on the bus, $\overline{L}_B = 0$, and $CLK = $ LOW-to-HIGH, storing the data in the $B$ register.

3. $S_U = 0$ to add the numbers.

4. The result is transferred to the accumulator ($\overline{L}_A = 0$, $\overline{E}_U = 0$, $CLK = $ LOW-to-HIGH).

5. To send the accumulated total to the data bus, $\overline{E}_A = 0$ to enable the accumulator output buffers.

The subtraction $A - B$ is the same, except that in steps 3 and 4 $S_U = 1$.

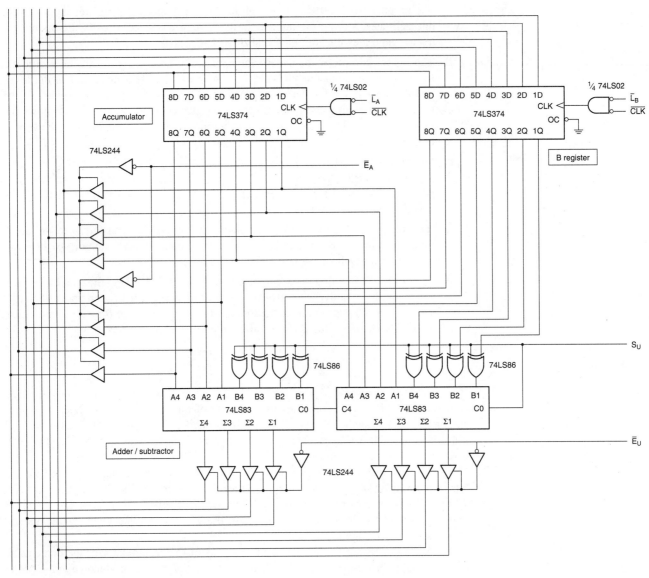

**a. Circuit**

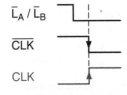

**b. Register clocking**

**Figure 9.24**

**Register Arithmetic Circuit
Built From TTL Devices**

## 9.7

# BCD Adders

(This section may be omitted without loss of continuity.)

> **BCD adder**  *A parallel adder whose output is in groups of 4 bits, each group representing a BCD digit.*

It is sometimes convenient to have the output of an adder circuit available as a BCD number, particularly if the result is to be displayed numerically. The problem is that most parallel adders have binary outputs, and 6 of the 16 possible 4-bit binary sums—1010 to 1111—are not within the range of the BCD code.

BCD numbers range from 0000 to 1001, or 0 to 9 in decimal. The unsigned sum of any two BCD numbers plus an input carry can range from 00000 ($= 0000 + 0000 + 0$) to 10011 ($= 1001 + 1001 + 1 = 19_{10}$).

For any sum up to 1001, the BCD and binary values are the same. Any sum greater than 1001 must be modified, since it requires a second BCD digit. For example, the binary value of $19_{10}$ is $10011_2$. The BCD value of $19_{10}$ is $0001\ 1001_{BCD}$. (The most significant digit of a sum of two BCD digits and a carry will never be larger than 1, since the largest such sum is $19_{10}$.)

Table 9.5 shows the complete list of possible binary sums of two BCD digits ($A$ and $B$) and a carry ($C$), their decimal equivalents, and their corrected BCD values. The MSD of the BCD sum is shown only as a carry bit, with leading zeros suppressed.

**Table 9.5**  Binary Sums of Two BCD Digits and a Carry Bit

| Binary Sum ($A + B + C$) | Decimal | Corrected BCD (Carry + BCD) |
|---|---|---|
| 00000 | 0 | 0 + 0000 |
| 00001 | 1 | 0 + 0001 |
| 00010 | 2 | 0 + 0010 |
| 00011 | 3 | 0 + 0011 |
| 00100 | 4 | 0 + 0100 |
| 00101 | 5 | 0 + 0101 |
| 00110 | 6 | 0 + 0110 |
| 00111 | 7 | 0 + 0111 |
| 01000 | 8 | 0 + 1000 |
| 01001 | 9 | 0 + 1001 |
| 01010 | 10 | 1 + 0000 |
| 01011 | 11 | 1 + 0001 |
| 01100 | 12 | 1 + 0010 |
| 01101 | 13 | 1 + 0011 |
| 01110 | 14 | 1 + 0100 |
| 01111 | 15 | 1 + 0101 |
| 10000 | 16 | 1 + 0110 |
| 10001 | 17 | 1 + 0111 |
| 10010 | 18 | 1 + 1000 |
| 10011 | 19 | 1 + 1001 |

**Figure 9.25**

**BCD Adder (1½ Digit Output)**

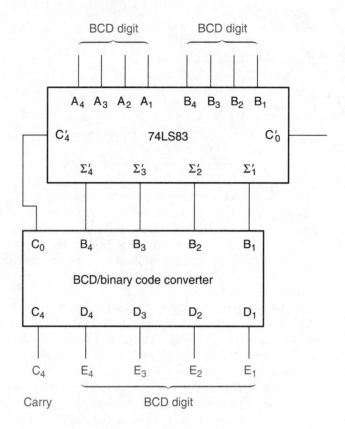

Figure 9.25 shows how we can add two BCD digits and get a corrected output. The **BCD adder** circuit consists of a standard 4-bit parallel adder to get the binary sum and a code converter to translate it into BCD.

The Binary-to-BCD code converter operates on the binary inputs as follows:

1. A carry output is generated if the binary sum is in the range $01010 \leq$ sum $\leq 10011$ (BCD equivalent: $1\ 0000 \leq$ sum $\leq 1\ 1001$).

2. If the binary sum is less than 01001, the output is the same as the input.

3. If the sum is in the range $01010 \leq$ sum $\leq 10011$, the four LSBs of the input must be corrected to a BCD value. This can be done by adding 0110 to the four LSBs of the input and discarding any resulting carry. We add $0110_2$ ($6_{10}$) because we must account for six unused codes.

Let's look at how each of these requirements can be implemented by a digital circuit.

## Carry Output

The carry output will be automatically 0 for any uncorrected sum from 00000 to 01001 and automatically 1 for any sum from 10000 to 10011. Thus, if the binary adder's carry output, which we will call $C_4'$, is 1, the BCD adder's carry output, $C_4$, will also be 1.

Any sum falling between these ranges, that is, between 01010 and 01111, must have its MSB modified. This modifying condition is a function, designated $C_4''$, of the

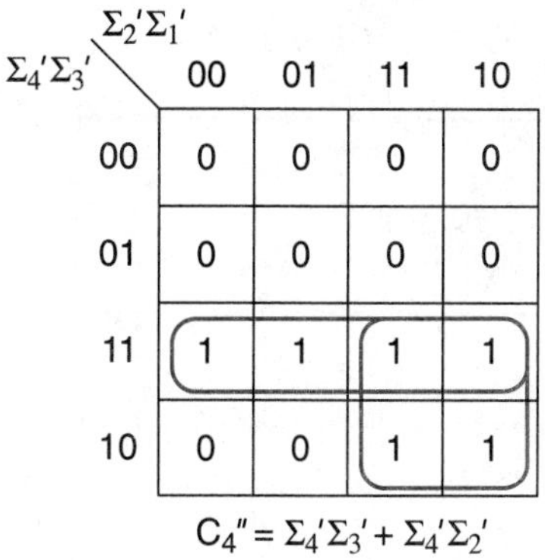

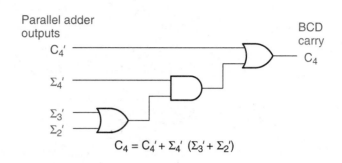

$$C_4'' = \Sigma_4'\Sigma_3' + \Sigma_4'\Sigma_2'$$

**Figure 9.26**

**Carry as a Function of Sum Bits When $C_4' = 0$**

**Figure 9.27**

**BCD Carry**

binary adder's sum outputs when its carry output is 0. This function can be simplified by a Karnaugh map, as shown in Figure 9.26, resulting in the following Boolean expression.

$$C_4'' = \Sigma_4' \, \Sigma_3' + \Sigma_4' \, \Sigma_2'$$

The BCD carry output $C_4$ is given by:

$$
\begin{aligned}
C_4 &= C_4' + C_4'' \\
&= C_4' + \Sigma_4' \, \Sigma_3' + \Sigma_4' \, \Sigma_2' \\
&= C_4' + \Sigma_4' \, (\Sigma_3' + \Sigma_2')
\end{aligned}
$$

The BCD carry circuit is shown in Figure 9.27.

### Sum Correction

The four LSBs of the binary adder output need to be corrected if the sum is 01010 or greater and need not be corrected if the binary sum is 01001 or less. This condition is indicated by the BCD carry. Let us designate the binary sum outputs as $\Sigma_4' \, \Sigma_3' \, \Sigma_2' \, \Sigma_1'$ and the BCD sum outputs as $\Sigma_4 \, \Sigma_3 \, \Sigma_2 \, \Sigma_1$.

If $C_4 = 0$, $\Sigma_4 \, \Sigma_3 \, \Sigma_2 \, \Sigma_1 = \Sigma_4' \, \Sigma_3' \, \Sigma_2' \, \Sigma_1' + 0000$;

If $C_4 = 1$, $\Sigma_4 \, \Sigma_3 \, \Sigma_2 \, \Sigma_1 = \Sigma_4' \, \Sigma_3' \, \Sigma_2' \, \Sigma_1' + 0110$.

Figure 9.28 shows a BCD adder, complete with a binary adder, BCD carry, and sum correction. A second parallel adder is used for sum correction. The $B$ inputs are the uncorrected binary sum inputs. The $A$ inputs are either 0000 or 0110, depending on the value of the BCD carry.

### Multiple-Digit BCD Adders

Several BCD adders can be cascaded to add multidigit BCD numbers. Figure 9.29 shows a $4\frac{1}{2}$-digit BCD adder. The carry output of the most significant digit is considered to be a

**Figure 9.28**
**BCD Adder**

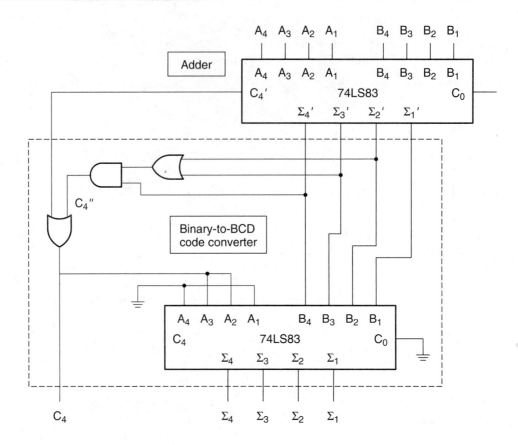

**Figure 9.29**
**4½-Digit BCD Adder**

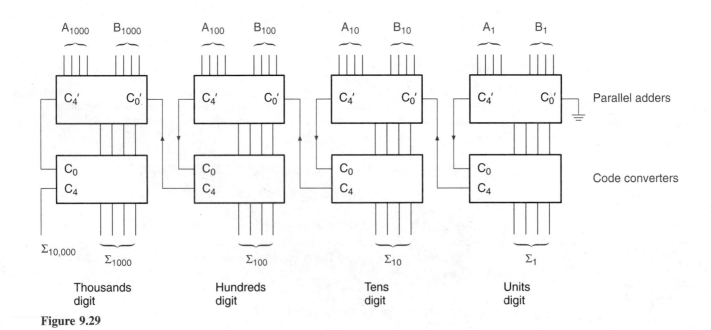

half-digit since it can only be 0 or 1. The output range of the $4\frac{1}{2}$-digit BCD adder is 00000 to 19999.

BCD adders are cascaded by connecting the code converter carry output of one stage to the binary adder carry input of the next most significant stage. Each BCD output digit represents a decade, designated as the units, tens, hundreds, thousands, and ten thousands digits.

---

**Section Review Problem for Section 9.7**

9.9. What is the maximum BCD sum of two 3-digit numbers with no carry input? How many digits are required to display this result on a numerical output?

---

## 9.8

# Programmable Logic Implementation of Arithmetic Circuits

n this section we will construct PLDasm source files to program an Intel or equivalent EPLD to perform one of several arithmetic functions. These include a 4-bit adder with ripple carry, an 8-bit adder/subtractor with overflow indication, and a 4-bit BCD adder.

### PLD Implementation of a Parallel Adder

It is fairly easy to construct a programmable logic device (PLD) version of a 4-bit parallel binary adder with ripple carry. We can program such a device using either the Boolean equations for the sum and carry outputs or a truth table. Since we have not yet used a truth table in a PLD example, we will use that method.

Intel's Programmable Logic Compiler/Simulator, PLDasm, can accept a truth table as the description of a PLD design. A full adder truth table, shown in Table 9.6, demonstrates the design syntax. The adder inputs are *A, B,* and *C0.* The outputs are *C1* and *SUM.*

The table begins with the PLDasm keyword T_TAB. Input and output variables follow in parentheses, separated by the symbol ">>" for a combinational circuit. The input and output values for each line of the truth table are separated by a colon (:).

**Table 9.6** Full Adder Truth Table Demonstrating PLDasm Syntax

| T_TAB | (A | B | C0 | >> | C1 | SUM) |
|-------|----|----|----|----|----|------|
| | 0 | 0 | 0 | : | 0 | 0 |
| | 0 | 0 | 1 | : | 0 | 1 |
| | 0 | 1 | 0 | : | 0 | 1 |
| | 0 | 1 | 1 | : | 1 | 0 |
| | 1 | 0 | 0 | : | 0 | 1 |
| | 1 | 0 | 1 | : | 1 | 0 |
| | 1 | 1 | 0 | : | 1 | 0 |
| | 1 | 1 | 1 | : | 1 | 1 |

**EXAMPLE 9.24**    Write a PLDasm source file to program an 85C220 EPLD as a 4-bit parallel binary adder with ripple carry.

**Solution**    File names: \ARITH\ADD4RIPL.PDS; \ARITH\ADD4RIPL.JED)

```
Title      4-bit Parallel Binary Adder with Ripple Carry
Pattern    pds
Revision   0
Author     R. Dueck
Company    Seneca College

CHIP add4ripl 85C220

; pin list

PIN  1     A1     ; A inputs
PIN  2     A2
PIN  3     A3
PIN  4     A4

PIN  5     B1     ; B inputs
PIN  6     B2
PIN  7     B3
PIN  8     B4

PIN  9     C0     ; Carry input

PIN  19    C4     ; Carry output
PIN  18    S4     ; Sum Outputs
PIN  17    S3
PIN  16    S2
PIN  15    S1

PIN  14    C3     ; Intermediate Carries
PIN  13    C2
PIN  12    C1

; One truth table for each full adder stage.
; Carry output of one adder is carry input of next adder.

T_TAB      (A1  B1  C0  >>  C1  S1)
            0   0   0   :   0   0
            0   0   1   :   0   1
            0   1   0   :   0   1
            0   1   1   :   1   0
            1   0   0   :   0   1
            1   0   1   :   1   0
            1   1   0   :   1   0
            1   1   1   :   1   1

T_TAB      (A2  B2  C1  >>  C2  S2)
            0   0   0   :   0   0
            0   0   1   :   0   1
            0   1   0   :   0   1
            0   1   1   :   1   0
            1   0   0   :   0   1
            1   0   1   :   1   0
            1   1   0   :   1   0
            1   1   1   :   1   1
```

```
T_TAB      (A3   B3   C2   >>   C3   S3)
            0    0    0    :    0    0
            0    0    1    :    0    1
            0    1    0    :    0    1
            0    1    1    :    1    0
            1    0    0    :    0    1
            1    0    1    :    1    0
            1    1    0    :    1    0
            1    1    1    :    1    1

T_TAB      (A4   B4   C3   >>   C4   S4)
            0    0    0    :    0    0
            0    0    1    :    0    1
            0    1    0    :    0    1
            0    1    1    :    1    0
            1    0    0    :    0    1
            1    0    1    :    1    0
            1    1    0    :    1    0
            1    1    1    :    1    1


SIMULATION

     VECTOR A := [A4 A3 A2 A1]
     VECTOR B := [B4 B3 B2 B1]
     VECTOR SUM := [C4 S4 S3 S2 S1]

     TRACE_ON  A B SUM

; Test following possibilities: A+0, B+0, A+F, B+F for all
; values of A or B.
FOR i:= 0 TO 3 DO
BEGIN
     IF (i<2) THEN
     BEGIN
          SETF A:=0 B:=0 /C0
     END

     IF (i=2) THEN
     BEGIN
          SETF A:=0 B:=15 /C0
     END

     IF (i=3) THEN
     BEGIN
          SETF A:=15 B:=0 /C0
     END

     FOR j:=0 TO 15 DO
          BEGIN
               IF ((i=0)+(i=2)) THEN
               BEGIN
                    SETF A:=j
               END

               ELSE
               BEGIN
                    SETF B:=j
               END
          END
     SETF C0
END
```

```
SETF A:=15  B:=15  /CO
SETF A:=15  B:=15  CO


TRACE_OFF
```

## Adder/Subtractor With Overflow Indication

Recall the 8-bit adder/subtractor with overflow detection, illustrated in Example 9.22. This circuit requires the following components:

| Quantity | Description |
|---|---|
| 2 | 4-bit parallel binary adder |
| 2 | Quad 2-input XOR gate |
| 1 | Hex inverter |
| 1 | Quad 2-input AND gate * |
| 1 | Quad 2-input OR gate * |

\* (The AND and OR can be replaced by a Triple 3-input NAND.)

We can implement this entire circuit on a single 85C090 EPLD.

**EXAMPLE 9.25**

Write the equations section of a PLDasm source file to implement an 8-bit parallel binary adder/subtractor with overflow detection. The target device is an 85C090 EPLD.

**Solution**

The complete file is included on the instructor's diskette.
(File names: \ARITH\ADDSUB8.PDS; \ARITH\ADDSUB8.JED)

```
;    The following string substitutions are used to modify the
; B inputs to a 8-bit Parallel Adder, depending on the function.
; ADDSUB=0 for ADD, ADDSUB=1 for SUBTRACT.

STRING B80 '(B8 :+: ADDSUB)'
STRING B70 '(B7 :+: ADDSUB)'
STRING B60 '(B6 :+: ADDSUB)'
STRING B50 '(B5 :+: ADDSUB)'
STRING B40 '(B4 :+: ADDSUB)'
STRING B30 '(B3 :+: ADDSUB)'
STRING B20 '(B2 :+: ADDSUB)'
STRING B10 '(B1 :+: ADDSUB)'


EQUATIONS

;    The following equations implement an 8-bit Parallel Binary
; Adder.


S1 = (A1 :+: B10) :+: ADDSUB
C1 = (A1 :+: B10)*ADDSUB + A1*B10

S2 = (A2 :+: B20) :+: C1
C2 = (A2 :+: B20)*C1 + A2*B20
```

```
S3 = (A3 :+: B3O) :+: C2
C3 = (A3 :+: B3O)*C2 + A3*B3O

S4 = (A4 :+: B4O) :+: C3
C4 = (A4 :+: B4O)*C3 + A4*B4O

S5 = (A5 :+: B5O) :+: C4
C5 = (A5 :+: B5O)*C4 + A5*B5O

S6 = (A6 :+: B6O) :+: C7
C6 = (A6 :+: B6O)*C5 + A6*B6O

S7 = (A7 :+: B7O) :+: C6
C7 = (A7 :+: B7O)*C6 + A7*B7O

S8 = (A8 :+: B8O) :+: C7

OVF = /S8*A8*B8 + S8*/A8*/B8


***** Design implemented successfully
```

```
                      85C090
                   - - - - -
           Gnd -|  1     40 |- Vcc
            A8 -|  2     39 |- A4
            A7 -|  3     38 |- A3
            A6 -|  4     37 |- A2
            A5 -|  5     36 |- A1
        ADDSUB -|  6     35 |- S8
            C1 -|  7     34 |- S7
            C2 -|  8     33 |- S6
            C3 -|  9     32 |- S5
            C4 -| 10     31 |- S4
            C5 -| 11     30 |- S3
            C6 -| 12     29 |- S2
            C7 -| 13     28 |- S1
           Gnd -| 14     27 |- OVF
           Gnd -| 15     26 |- Gnd
            B8 -| 16     25 |- B4
            B7 -| 17     24 |- B3
            B6 -| 18     23 |- B2
            B5 -| 19     22 |- B1
           GND -| 20     21 |- Gnd
                   - - - - -
```

## BCD Adder

The easiest way to implement a 4-bit BCD adder in an EPLD is in two separate portions: a parallel binary adder and a binary-to-BCD code converter. The binary adder can be implemented by truth table, as in Example 9.24, or by Boolean equations, as in Example 9.25. The code converter is done most straightforwardly by truth table.

**EXAMPLE 9.26**     Write the equations and truth table sections of a PLDasm source file to implement a 4-bit (i.e., single-digit-plus-carry) BCD adder. Out-of-range outputs (>19) should make all outputs go HIGH. The target device is an 85C090.

**Solution**   The complete source file is included on the instructor's diskette.
(File names: \ARITH\BCDADD.PDS; \ARITH\BCDADD.JED)

```
EQUATIONS

S1 = A1 :+: B1 :+: C0            ; Sum and carry equations for a
C1 = (A1 :+: B1)*C0 + A1*B1      ; parallel binary adder.

S2 = A2 :+: B2 :+: C1
C2 = (A2 :+: B2)*C1 + A2*B2

S3 = A3 :+: B3 :+: C2
C3 = (A3 :+: B3)*C2 + A3*B3

S4 = A4 :+: B4 :+: C3
C4B = (A4 :+: B4)*C3 + A4*B4

; Truth table for converting a binary sum to a BCD digit.
; Outputs are all HIGH for an out-of-range result (>19).

T_TAB  ( C4B S4 S3 S2 S1 >> C4D D4 D3 D2 D1)

              0  0  0  0  0  :  0  0  0  0  0
              0  0  0  0  1  :  0  0  0  0  1
              0  0  0  1  0  :  0  0  0  1  0
              0  0  0  1  1  :  0  0  0  1  1
              0  0  1  0  0  :  0  0  1  0  0
              0  0  1  0  1  :  0  0  1  0  1
              0  0  1  1  0  :  0  0  1  1  0
              0  0  1  1  1  :  0  0  1  1  1
              0  1  0  0  0  :  0  1  0  0  0
              0  1  0  0  1  :  0  1  0  0  1
              0  1  0  1  0  :  1  0  0  0  0
              0  1  0  1  1  :  1  0  0  0  1
              0  1  1  0  0  :  1  0  0  1  0
              0  1  1  0  1  :  1  0  0  1  1
              0  1  1  1  0  :  1  0  1  0  0
              0  1  1  1  1  :  1  0  1  0  1
              1  0  0  0  0  :  1  0  1  1  0
              1  0  0  0  1  :  1  0  1  1  1
              1  0  0  1  0  :  1  1  0  0  0
              1  0  0  1  1  :  1  1  0  0  1
              1  0  1  X  X  :  1  1  1  1  1
              1  1  X  X  X  :  1  1  1  1  1
```

## GLOSSARY

**1's complement**  A form of signed binary notation in which negative numbers are created by complementing all bits of a number, including the sign bit.

**2's complement**  A form of signed binary notation in which negative numbers are created by adding 1 to the 1's complement form of the number.

**Accumulator**  A special register that stores the accumulating total of several binary additions and subtractions.

**Arithmetic logic unit (ALU)**  A combinational logic circuit that performs several selectable arithmetic and logic functions on two binary numbers.

**BCD adder**  A parallel adder whose output is in groups of 4 bits, each group representing a BCD digit.

**Borrow**  A digit brought back from a more significant position when the subtrahend digit is larger than the minuend digit.

**Carry**   A digit that is "carried over" to the next most significant position when the sum of two single digits is too large to be expressed as a single digit.

**Carry bit**   A bit that holds the value of a carry (0 or 1) resulting from the sum of two binary numbers.

**Cascade**   To connect an output of one device to an input of another, often for the purpose of expanding the number of bits available for a particular function.

**Data bus**   A group of parallel conductors carrying related logic signals, such as multibit data.

**Difference**   The result of a subtraction operation.

**End-around carry**   An operation in 1's complement subtraction where the carry bit resulting from a sum of two 1's complement numbers is added to that sum.

**Fast carry (or parallel carry)**   A gate network that generates a carry bit directly from *all* incoming operand bits, independent of the operation of each full adder stage.

**Full adder**   A circuit that will add a carry bit from another full or half adder and two operand bits to produce a sum bit and a carry bit.

**Half adder**   A circuit that will add two bits and produce a sum bit and a carry bit.

**Magnitude bits**   The bits of a signed binary number that tell us how large the number is (i.e., its magnitude).

**Minuend**   The first number in a subtraction operation.

**Operand**   A number upon which an arithmetic function operates (e.g., in the expression $x + y = z$, $x$ and $y$ are the operands).

**Overflow**   An erroneous carry into the sign bit of a signed binary number that results from a sum larger than can be represented by the number of magnitude bits.

**Parallel binary adder**   A circuit, consisting of $n$ full adders, that will add two $n$-bit binary numbers. The output consists of $n$ sum bits and a carry bit.

**Register**   A digital circuit that can store a binary number until needed (e.g., a multibit flip-flop or latch).

**Ripple carry**   A method of passing carry bits from one stage of a parallel adder to the next by connecting $C_{OUT}$ of one full adder to $C_{IN}$ of the following stage.

**Sign bit**   A bit, usually the MSB, that indicates whether a signed binary number is positive or negative.

**Signed binary arithmetic**   Arithmetic operations performed using signed binary numbers.

**Signed binary number**   A binary number of fixed length whose sign is represented by one bit, usually the most significant bit, and whose magnitude is represented by the remaining bits.

**Subtrahend**   The second operand in a subtraction operation.

**Sum**   The result of an addition operation.

**Sum bit (single-bit addition)**   The least significant bit of the sum of two 1-bit binary numbers.

**True-magnitude form**   A form of signed binary number whose magnitude is represented in true binary.

**Unsigned binary number**   A binary number whose sign is not specified by a sign bit. A positive sign is assumed unless explicitly stated otherwise.

## ● P R O B L E M S

### Section 9.1 Digital Arithmetic

**9.1**   Add the following unsigned binary numbers.

  **a.**  10101 + 1010

  **b.**  10101 + 1011

  **c.**  1111 + 1111

  **d.**  11100 + 1110

  **e.**  11001 + 10011

  **f.**  111011 + 101001

**9.2**   Subtract the following unsigned binary numbers.

  **a.**  1100 − 100

  **b.**  10001 − 1001

  **c.**  10101 − 1100

  **d.**  10110 − 1010

  **e.**  10110 − 1001

  **f.**  10001 − 1111

  **g.**  100010 − 10111

  **h.**  1100011 − 100111

### Section 9.2 Representing Signed Binary Numbers
### Section 9.3 Signed Binary Arithmetic

**9.3**   Write the following decimal numbers in 8-bit true-magnitude, 1's complement, and 2's complement forms.

**a.** −110

**b.** 67

**c.** −54

**d.** −93

**e.** 0

**f.** −1

**g.** 127

**h.** −127

**9.4**  Perform the following arithmetic operations in the true-magnitude (addition only), 1's complement, and 2's complement systems. Use 8-bit numbers consisting of a sign bit and 7 magnitude bits. (The numbers shown are in the decimal system.)

Convert the results back to decimal to prove the correctness of each operation. Also demonstrate that the idea of adding a negative number to perform subtraction is not valid for the true-magnitude form.

**a.**  37 + 25

**b.**  85 + 40

**c.**  95 − 63

**d.**  63 − 95

**e.**  −23 − 50

**f.**  120 − 73

**g.**  73 − 120

**9.5**  What are the largest positive and negative numbers, expressed in 2's complement notation, that can be represented by an 8-bit signed binary number?

**9.6**  Perform the following *signed* binary operations, using 2's complement notation where required. State whether or not sign bit overflow occurs. Give the signed decimal equivalent values of the sums in which overflow does not occur.

**a.**  01101 + 00110

**b.**  01101 + 10110

**c.**  01110 − 01001

**d.**  11110 + 00010

**e.**  11110 − 00010

**9.7**  Without doing any binary complement arithmetic, indicate which of the following operations will result in 2's complement overflow. (Assume 8-bit representation consisting of a sign bit and

7 magnitude bits.) Explain the reasons for each choice.

**a.**  −109 + 36

**b.**   109 + 36

**c.**    65 + 72

**d.**  −110 − 29

**e.**   117 + 11

**f.**   117 − 11

**9.8**  Explain how you can know, by examining sign or magnitude bits of the numbers involved, when overflow has occurred in 2's complement addition or subtraction.

## Section 9.4 Hexadecimal Arithmetic

**9.9**  Add the following hexadecimal numbers.

**a.**  $27_{16} + 16_{16}$

**b.**  $87_{16} + 99_{16}$

**c.**  $A55_{16} + C5_{16}$

**d.**  $C7F_{16} + 380_{16}$

**e.**  $1FFF_{16} + A80_{16}$

**9.10**  Subtract the following hexadecimal numbers.

**a.**  $F86_{16} − 614_{16}$

**b.**  $E72_{16} − 229_{16}$

**c.**  $37FF_{16} − 137F_{16}$

**d.**  $5764_{16} − ACB_{16}$

**e.**  $7D30_{16} − 5D33_{16}$

**f.**  $5D33_{16} − 7D30_{16}$

**g.**  $813A_{16} − A318_{16}$

## Section 9.5 Binary Adders and Subtractors

**9.11**  Write the truth table for a half adder, and from the table derive the Boolean expressions for both $C_o$ (carry output) and $\Sigma$ (sum output) in terms of inputs $A$ and $B$. Draw the half adder circuit.

**9.12**  Write the truth table for a full adder, and from the table derive the simplest possible Boolean expressions for $C_o$ and $\Sigma$ in terms of $A$, $B$, and $C_I$ (carry input).

**9.13**  From the equations in Problems 9.11 and 9.12, draw a circuit showing a full adder constructed from two half adders.

**9.14** Evaluate the Boolean expression for $\Sigma$ and $C_{OUT}$ of the full adder in Figure 9.7 for the following inputs values. What is the binary value of the outputs in each case?

    **a.** $A = 0$, $B = 0$, $C_{IN} = 0$
    **b.** $A = 0$, $B = 1$, $C_{IN} = 0$
    **c.** $A = 0$, $B = 1$, $C_{IN} = 1$
    **d.** $A = 1$, $B = 1$, $C_{IN} = 1$

**9.15** Verify the summing operation of the circuit in Figure 9.10, as follows. Determine the output of each full adder based on the inputs shown below. Calculate each sum manually and compare it to the 5-bit output ($C_4\ \Sigma_4\ \Sigma_3\ \Sigma_2\ \Sigma_1$) of the parallel adder circuit.

    **a.** $A_4 A_3 A_2 A_1 = 0100$, $B_4 B_3 B_2 B_1 = 1001$
    **b.** $A_4 A_3 A_2 A_1 = 1010$, $B_4 B_3 B_2 B_1 = 0110$
    **c.** $A_4 A_3 A_2 A_1 = 0101$, $B_4 B_3 B_2 B_1 = 1101$
    **d.** $A_4 A_3 A_2 A_1 = 1111$, $B_4 B_3 B_2 B_1 = 0111$

**\*9.16** Write the algebraic expressions for the sum and carry functions of the LSB full adder of a 74LS83A Parallel Binary Adder, as shown in Figure 9.11. Use Boolean algebra to verify that these functions are equivalent to the sum and carry functions of the full adder developed in Section 9.5.

**\*9.17** Determine the algebraic function of the carry into the $\Sigma_4$ bit of the 74LS83A Parallel Binary Adder in Figure 9.11.

**\*9.18** The following equation describes the carry output function for a 4008B Parallel Binary Adder:

$$C_{OUT} = A_4 B_4 + A_3 B_3 (A_4 + B_4)$$
$$+ A_2 B_2\ (A_4 + B_4)(A_3 + B_3)$$
$$+ A_1 B_1\ (A_4 + B_4)(A_3 + B_3)(A_2 + B_2)$$
$$+ C_{IN}\ (A_4 + B_4)(A_3 + B_3)(A_2 + B_2)$$
$$(A_1 + B_1)$$

Briefly explain how to interpret the third term of this equation.

**9.19** Draw a 4-bit 2's complement subtractor based on a 7483 Parallel Binary Adder. Explain how the circuit produces the 2's complement of $B$ for the subtraction $A - B$.

**9.20** Draw a 4-bit 2's complement adder/subtractor based on a 7483 Parallel Binary Adder. Explain how the circuit is programmed to add or subtract and how it produces the 2's complement of $B$ for the subtraction $A - B$.

**9.21** Draw a circuit that will detect an overflow condition in a 4-bit 2's complement adder/subtractor. The detector output should go HIGH upon overflow detection. Draw the circuit truth table, explain what all input and output variables are, and show any Boolean equations you need to complete the circuit design.

**9.22** What is the permissible range of values that a sum or difference, $x$, can have in a 16-bit parallel binary adder if it is written as:

    **a.** A signed binary number
    **b.** An unsigned binary number

## Section 9.6 Register Arithmetic Circuits

**9.23** The register arithmetic circuit shown in Figure 9.23 is to perform the following binary operation:

    $00111001 + 10010010 - 01110110$

Complete the table at the top of page 370 for each step of the operation, showing the logic levels of the control inputs and the contents of each register for every step. Assume that the registers are all cleared at the beginning of the operation.

## Section 9.7 BCD Adders

**9.24** What is the maximum BCD sum of two 3-digit BCD numbers plus an input carry? How many digits are needed to display the result?

**9.25** What is the maximum BCD sum of two 4-digit BCD numbers plus an input carry? How many digits are needed to display the result?

**\*9.26** Based on the answers to Problems 9.24 and 9.25, formulate a general rule to calculate the maximum BCD sum of two $n$-digit BCD numbers plus a carry bit.

**\*9.27** Derive the Boolean expression for a BCD carry output as a function of the sum of two BCD digits.

**9.28** Draw the circuit for a binary-to-BCD code converter.

**9.29** Draw the block diagram of a circuit that will add two 3-digit BCD numbers and display the result as a series of decimal digits. How many digits will the output display?

| Accumulator | | B Register | | Adder/Subtractor | | $\overline{L_A}$ | $\overline{L_B}$ | $S_U$ | $\overline{E_U}$ |
|---|---|---|---|---|---|---|---|---|---|
| $A_7$ | $A_0$ | $B_7$ | $B_0$ | $\Sigma_7$ | $\Sigma_0$ | | | | |
| 00000000 | | 00000000 | | 00000000 | | 1 | 1 | 0 | 1 |

### Section 9.8 Programmable Logic Implementation of Arithmetic Circuits

**\*\*9.30** Refer to the fast-carry adder equations given in Section 9.5. Use PLDasm to implement a 4-bit fast-carry adder in an 85C22V10 EPLD. (The number of product terms generated by the equation for $C_4$ is too large for one macrocell. This equation must be partitioned into two macrocells.)

**9.31** Expand the 4-bit ripple-carry adder programmed in Example 9.24 to an 8-bit device. Use either Boolean equations or truth tables.

**\*9.32** A 9's complementer, used in BCD arithmetic, converts a 4-bit BCD number to its 9's complement value when an input *COMP* is HIGH. The outputs are forced to 0000 when an input *ZERO* goes HIGH.

The truth table in Table 9.7 shows the inputs ($B_4$ $B_3$ $B_2$ $B_1$) and outputs ($D_4$ $D_3$ $D_2$ $D_1$) of the 9's complementer when *COMP* = 1. (Recall from Chapter 5, Number Systems and Codes, that a 9's complement of a BCD digit is derived by subtracting the digit from 9.)

Write a PLDasm source file to program an 85C220 EPLD as a 4-bit (single digit) 9's complementer.

Table 9.7 9's Complement Inputs and Outputs

| $B_4$ | $B_3$ | $B_2$ | $B_1$ | $D_4$ | $D_3$ | $D_2$ | $D_1$ |
|---|---|---|---|---|---|---|---|
| 0 | 0 | 0 | 0 | 1 | 0 | 0 | 1 |
| 0 | 0 | 0 | 1 | 1 | 0 | 0 | 0 |
| 0 | 0 | 1 | 0 | 0 | 1 | 1 | 1 |
| 0 | 0 | 1 | 1 | 0 | 1 | 1 | 0 |
| 0 | 1 | 0 | 0 | 0 | 1 | 0 | 1 |
| 0 | 1 | 0 | 1 | 0 | 1 | 0 | 0 |
| 0 | 1 | 1 | 0 | 0 | 0 | 1 | 1 |
| 0 | 1 | 1 | 1 | 0 | 0 | 1 | 0 |
| 1 | 0 | 0 | 0 | 0 | 0 | 0 | 1 |
| 1 | 0 | 0 | 1 | 0 | 0 | 0 | 0 |

## ▼ *Answers to Section Review Questions*

### Section 9.1a

**9.1** 101000;  **9.2.** 100000.

### Section 9.1b

**9.3.** 11;  **9.4.** 1.

### Section 9.3

**9.5.** 11100000;  **9.6.** 100000.

### Section 9.5a

**9.7.** Figures 9.14 and 9.15 show the propagation paths for the carry bits.

74LS83A:   Parallel carry
           4 gates (inverter, 4-in AND, NOR, XOR)
MC14008B:   Ripple carry
           13 gates (2 inverters, 3 gates in each of 3 adder
                    stages, XNOR, inverter)

### Section 9.5b

**9.8a.** Signed: $-2048 \le x \le +2047$ (11 magnitude bits, 1 sign bit) **9.8b.** Unsigned: $0 \le x \le +4095$ (12 magnitude bits, no sign bit: positive implied)

### Section 9.7

**9.9.** Maximum BCD sum = 1001 1001 1001 + 1001 1001 1001 = 1 1001 1001 1000$_{BCD}$ = 1998$_{10}$. This sum requires a $3\frac{1}{2}$-digit numerical display.

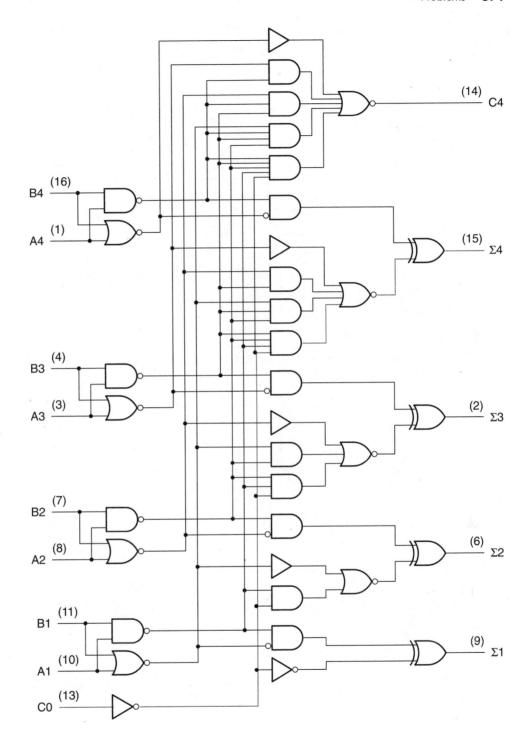

**Figure 9.14**

**74LS83A Parallel Binary Adder** (Reprinted by permission of Texas Instruments)

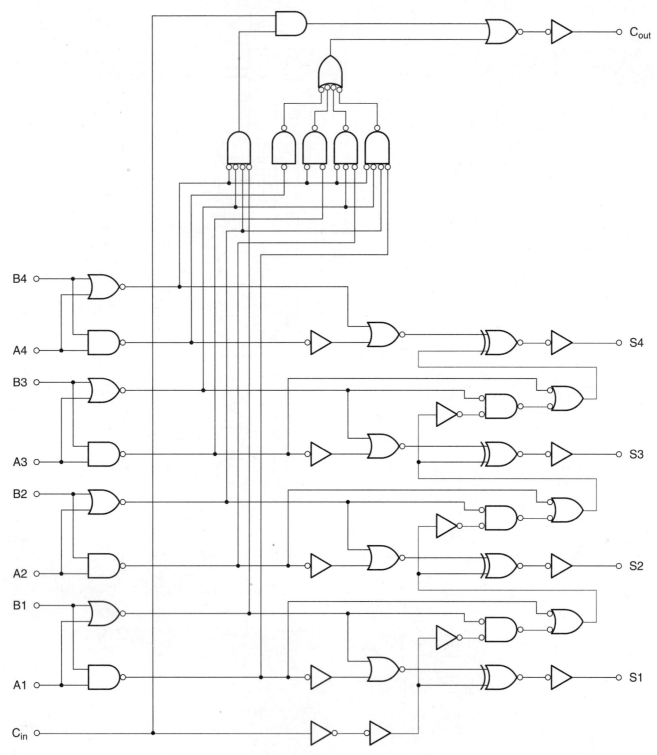

**Figure 9.15**

**MC14008B Parallel Binary Adder** (Reprinted with permission of Motorola)

# MSI Combinational Logic

10

● CHAPTER OUTLINE

**10.1** Decoders

**10.2** Encoders

**10.3** Magnitude Comparators

**10.4** Parity Generators and Checkers

**10.5** Multiplexers

**10.6** Demultiplexers

**10.7** Combinational Logic Applications of PLDs

● CHAPTER OBJECTIVES

Upon successful completion of this chapter, you will be able to:

- Define the functions of a wide variety of MSI combinational devices and use them individually and in combination with other devices.

- Design simple binary decoders using logic gates.

- Explain the operation and use of seven-segment displays and BCD-to-seven-segment decoders, including internal Boolean functions and special features, such as ripple blanking.

- Define the function of an MSI priority encoder and use it for various applications.

- Define the operation of MSI magnitude comparators and use them in various applications.

- Explain the use of parity as an error-checking system and use MSI parity generators and checkers in circuits.

- Describe the circuit and operation of a simple multiplexer.

- Draw logic circuits for multiplexer applications such as single-channel data selection, Boolean function generation, time-division multiplexing (TDM), waveform generation, parallel-to-serial conversion, and multibit data selection.

- Describe and use MSI demultiplexers for applications such as demultiplexing TDM signals.

- Define the operation of a CMOS transmission gate and its use in multiplexers and demultiplexers.

- Program PLDs to perform simple combinational logic functions.

●INTRODUCTION

Msi stands for medium-scale integration. Digital devices have historically been classified in terms of their circuit complexity and are generally divided into classes such as small, medium, large, and very large scale integration. Medium-scale integration circuits are devices having relatively complex functions on a single chip, which would require from 12 to 100 logic gates to accomplish the same function.

A variety of combinational logic functions are common enough to have MSI devices to perform them. These include arithmetic and numeric functions such as binary addition, parity generation, and comparison; code conversion devices; display driving; and switching functions such as multiplexing and demultiplexing.

Programmable logic devices (PLDs) are often used to perform functions similar to those of standard fixed MSI devices. PLD also has the flexibility to perform several functions in one device, resulting in a saving of IC packaging.

## 10.1

# Decoders

**Decoder**  *A digital circuit designed to detect the presence of a particular digital state.*

**MSI**  *Medium-scale integration. A single-chip logic device whose equivalent function could be performed by a logic gate circuit consisting of 12 to 100 gates.*

**Digital signal processor (DSP)**  *A specialized microprocessor used to modify digital signals, such as audio or video data, for various special purposes. For instance, a DSP can process an audio signal to introduce reverberation or a video signal to generate graphic images.*

The general function of a **decoder** is to activate one or more circuit outputs upon detection of a particular digital state. The simplest decoder is a single logic gate, such as a NAND or an AND, whose output activates when all its inputs are HIGH. When combined with one or more inverters, a NAND or AND can detect any unique combination of binary input values.

An extension of this type of decoder is an **MSI** device containing several such gates, each of which responds to a different input state. Usually, for an $n$-bit input, there are $2^n$ logic gates, each of which decodes a different combination of input variables. A variation is a BCD device with four input variables and ten outputs, each of which activates for a different BCD input.

Some types of decoders translate binary inputs to other forms, such as the decoders that drive numerical display outputs, those familiar figure-8 arrangements of LED or LCD outputs ("segments"). The decoder has one output for every segment in the display. These segments illuminate in unique combinations for each input code.

Figure 10.1

**Single-Gate Decoders**

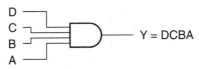

a. Active-HIGH indication

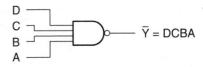

b. Active-LOW indication

## Single-Gate Decoders

The simplest decoder is a single gate, sometimes in combination with one or more inverters, used to detect the presence of one particular binary value. Figure 10.1 shows two such decoders, both of which detect an input $DCBA = 1111$.

The decoder in Figure 10.1a generates a logic HIGH when its input is 1111. The decoder in Figure 10.1b responds to the same input but makes the output LOW instead.

Now is a good time to talk about notation used for MSI circuitry. In Figure 10.1, we designated $D$ as the most significant bit of the input and $A$ the least significant bit. We will continue this convention for alphabetically labeled multibit inputs.

Sometimes, an input or output will have the same alphabetic designation as all other similar inputs or outputs and be distinguished from the others by a subscript, such as $D_5$ or $Y_7$. In such a case, the lowest number subscript is the least significant bit.

In Boolean expressions, we will indicate the active levels of inputs and outputs separately. For example, in Figure 10.1, the inputs to both gates are the same, so we write $DCBA$ for the inputs of both gates. The gates in Figures 10.1a and b have outputs with opposite active levels, so we write the output variables ($Y$ and $\overline{Y}$) oppositely.

**EXAMPLE 10.1**

Figure 10.2 shows three single-gate decoders. For each decoder, state the output active level and the input code that activates the decoder. Also write the Boolean expression of each output.

**Solution**    Each decoder is a NAND or an AND gate. These gates require all inputs HIGH to make the output active. Because of the inverters, each circuit has a different code that fulfils this requirement.

Figure 10.2a: Output: Active LOW

   Input code: $DCBA = 1001$

   $\overline{Y} = D\,\overline{C}\,\overline{B}\,A$

Figure 10.2c: Output: Active HIGH

   Input code: $DCBA = 1010$

   $Y = D\,\overline{C}\,B\,\overline{A}$

Figure 10.2b: Output: Active LOW

   Input code: $CBA = 001$

   $\overline{Y} = \overline{C}\,\overline{B}\,A$

Figure 10.2

**Example 10.1
Single-Gate Decoders**

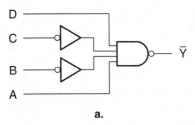

a.

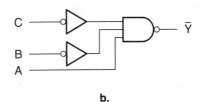

b.

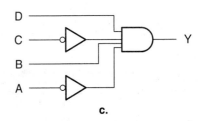

c.

Single-gate decoders are often used to activate other digital circuits under various operating conditions, particularly if there is a choice of circuits to activate. A decoder circuit with an $n$-bit input can activate up to $2^n$ load circuits.

**E X A M P L E   1 0 . 2**

A digital data channel in a television studio is used to send information to one of five different destinations, depending on the application, as shown in Figure 10.3.

The destinations are:

Channel 1: Digital audio tape (DAT)

Channel 2: Digital videotape

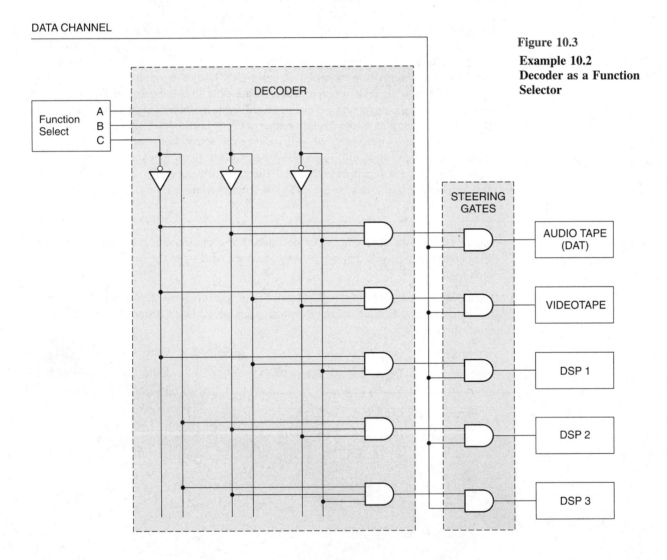

**Figure 10.3**

**Example 10.2**
**Decoder as a Function Selector**

Channel 3: **Digital signal processor** (DSP) 1

Channel 4: DSP 2

Channel 5: DSP 3

Briefly explain how the decoder selects one of the data destinations in Figure 10.3. What is the maximum number of channels this system could accommodate?

**Solution**    The channel selector consists of a decoder and five steering gates. Each steering gate is inhibited, thus blocking the data stream on that channel, as long as its input from the decoder is LOW. When the decoder receives a valid Function Select code, $CBA$, the corresponding decoder gate output goes HIGH, enabling the steering gate for that channel.

The decoder responds to the following input codes:

Channel 1: DAT recorder; $CBA = 001$

Channel 2: Videotape; $CBA = 010$

Channel 3: DSP 1; $CBA = 011$

Channel 4: DSP 2; $CBA = 100$

Channel 5: DSP 3; $CBA = 101$

Note that the code $CBA$ is the binary equivalent of the selected channel number.

Since the channel select code $CBA$ has 3 bits, there could be a maximum of 8 channels in this system.

(Note: A circuit that directs a digital signal from a single source to one of several destinations by means of a decoder is called a demultiplexer. We will examine this circuit in more detail later in the chapter.)

## MSI Decoders

The decoder portion of the circuit in Figure 10.3 is similar to the decoders available as medium-scale integration devices. Three of the most popular MSI decoders are the 74HC139A Dual 2-to-4 Line, the 74HC138A 3-to-8 Line, and the 74HC154 4-to-16 Line decoders, all of which are also available as TTL devices (74LS138A, 74LS139, and 74LS154, respectively). These circuits have sufficient internal AND or NAND gates to decode every possible binary input value. In addition, these devices have one or more *ENABLE* inputs, allowing us to inhibit or enable the entire device.

### 2-to-4 Line (74HC139A)

The 74HC139A Dual 2-to-4 Line Decoder is represented in Figure 10.4. This part contains two separate decoders in one IC package. The circuit is called 2-to-4 line because each decoder activates one of the four output lines for every possible 2-bit input combination, $BA$. The circuit could also be called a 1-of-4 decoder.

**Table 10.1** Truth Table for a 2-to-4 Line Decoder

| $\overline{G}$ | $B$ | $A$ | Active Output |
|:--:|:--:|:--:|:--:|
| 0 | 0 | 0 | $Y_0$ |
| 0 | 0 | 1 | $Y_1$ |
| 0 | 1 | 0 | $Y_2$ |
| 0 | 1 | 1 | $Y_3$ |
| 1 | X | X | None |

In general, a decoder having $n$ binary inputs has $m = 2^n$ outputs and is designated an "$n$-to-$m$ line" or "1-of-$m$ decoder."

Table 10.1 shows the truth table for a 74HC139A decoder.

**Figure 10.4**

**74HC139A
Dual 2-to-4 Line Decoder**

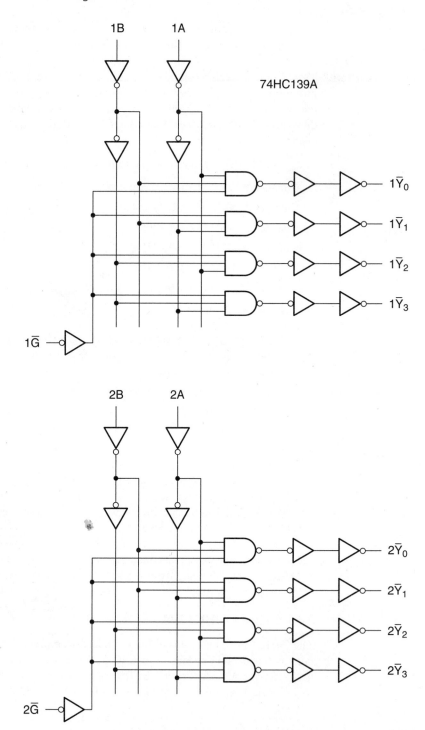

The subscript of each active output is the same number as the binary input. For example, when inputs $BA = 10$, output $\overline{Y}_2 = 0$ provided $\overline{G} = 0$; otherwise $\overline{Y}_2 = 1$. The other decoder outputs behave similarly. When $\overline{G} = 1$, *all* outputs are HIGH, regardless of the state of inputs $BA$.

EXAMPLE 10.3

A pharmaceutical company has an isolated room for growing experimental bacterial cultures. Temperature, humidity, and light level in the culture room are controlled from an associated control room. The culture room and control room are separated by an airlock, accessible by a sliding door from either room, where workers must observe strict decontamination procedures when passing in either direction. Figure 10.5 shows a plan of the three rooms and connecting doors.

Since it is extremely important that contamination in either direction be avoided, both doors into the airlock are monitored. Door status is displayed in the control room by the circuit shown in Figure 10.6. Each door switch is open when the corresponding sliding door is open. A key switch automatically activates the monitor circuit when the control room is unlocked. Briefly explain the operation of the monitor circuit.

**Figure 10.5**

**Example 10.3 Floor Plan for Isolated Culture Room and Control Room**

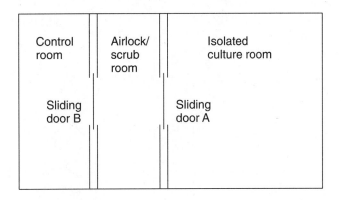

**Figure 10.6**

**Example 10.3 Door Monitors**

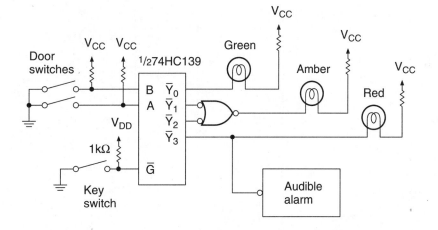

**Solution**  The monitor circuit uses one of two decoders in a 74HC139A to illuminate one of three lamps under the following conditions:

1. If both doors are closed, $BA = 00$, activating output $\overline{Y}_0$. This lights the green lamp connected to this output. (The other two lamps are off.)

2. If either door opens, either $BA = 01$ or $BA = 10$, activating output $\overline{Y}_1$ or $\overline{Y}_2$. The logic gate combines these two outputs so that an amber lamp illuminates in either

case, warning an operator that one door is open in the airlock. (The green light goes off.)

3. If both doors open, $BA = 11$, activating $\overline{Y}_3$. A red lamp lights and an alarm sounds. (Both green and amber lights are off.)

The key switch enables the monitor circuit by making $\overline{G} = 0$. If this key switch is open, $\overline{G} = 1$ and all outputs are HIGH. Under this condition, all lamps are off.

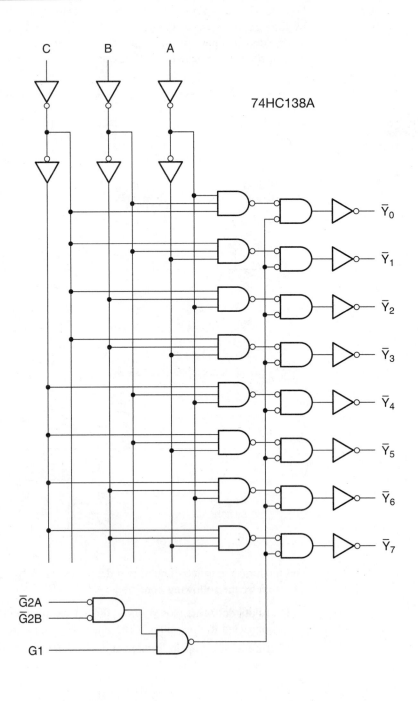

**Figure 10.7**

**74HC138A 3-to-8 Line Decoder**

## 3-to-8 Line (74HC138A)

Figure 10.7 shows the circuit of a 74HC138A 3-to-8 line decoder. This device has eight NAND gates that decode three binary inputs, *CBA*. Three inputs, *G1*, $\overline{G2A}$, and $\overline{G2B}$, enable or inhibit the active-LOW decoder outputs. Table 10.2 represents the truth table for this device. Note that (*G1* must be HIGH) AND ($\overline{G2A}$ AND *G2B* must be LOW) to activate any output.

**Table 10.2**  Truth Table for a 74HC138A Decoder

| *G1* | $\overline{G2A}$ | $\overline{G2B}$ | *C* | *B* | *A* | Active Output |
|------|------|------|-----|-----|-----|---------------|
| 1 | 0 | 0 | 0 | 0 | 0 | $Y_0$ |
| 1 | 0 | 0 | 0 | 0 | 1 | $Y_1$ |
| 1 | 0 | 0 | 0 | 1 | 0 | $Y_2$ |
| 1 | 0 | 0 | 0 | 1 | 1 | $Y_3$ |
| 1 | 0 | 0 | 1 | 0 | 0 | $Y_4$ |
| 1 | 0 | 0 | 1 | 0 | 1 | $Y_5$ |
| 1 | 0 | 0 | 1 | 1 | 0 | $Y_6$ |
| 1 | 0 | 0 | 1 | 1 | 1 | $Y_7$ |
| X | X | 1 | X | X | X | None |
| X | 1 | X | X | X | X | None |
| 0 | X | X | X | X | X | None |

**EXAMPLE 10.4**

Draw a circuit showing how to replace the decoder and steering gates in Figure 10.3 (Example 10.2) with a 74HC138A decoder.

**Solution**

Examination of the 74HC138A decoder in Figure 10.7 reveals that the logic function is very similar to that of the decoder/steering gate circuit of Figure 10.3. The primary difference is the active levels of the circuit outputs. We can rectify this problem by using one of the active-LOW *ENABLE* inputs ($\overline{G2A}$) as a data input, as shown in Figure 10.8 (see page 382). This input signal passes through the enable circuit and to all output gates. The decoder enables one of the outputs according to input code *CBA*, just as it did in Figure 10.3.

## 4-to-16 Line (74HC154)

Figure 10.9 represents the internal circuit of a 74HC154 4-to-16 line decoder. This device is similar to the 74HC139A and 74HC138A decoders in that it has $2^n$ active-LOW outputs for *n* binary inputs, each of which is activated by a different binary input code. Like the other devices, it has inputs, $\overline{G1}$ and $\overline{G2}$, that enable or inhibit the outputs. Both $\overline{G1}$ and $\overline{G2}$ must be LOW to enable any output; otherwise all outputs are HIGH. Table 10.3 shows the truth table for the 74HC154 decoder.

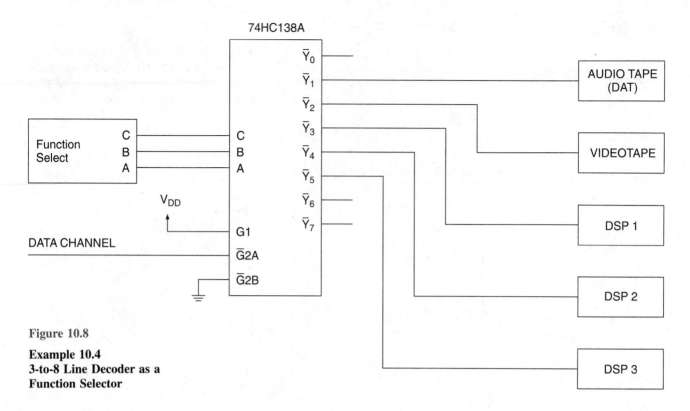

**Figure 10.8**

**Example 10.4**
**3-to-8 Line Decoder as a**
**Function Selector**

**Table 10.3** Truth Table for a 74HC154 Decoder

| $\overline{G1}$ | $\overline{G2}$ | $D$ | $C$ | $B$ | $A$ | Active Output |
|---|---|---|---|---|---|---|
| 0 | 0 | 0 | 0 | 0 | 0 | $Y_0$ |
| 0 | 0 | 0 | 0 | 0 | 1 | $Y_1$ |
| 0 | 0 | 0 | 0 | 1 | 0 | $Y_2$ |
| 0 | 0 | 0 | 0 | 1 | 1 | $Y_3$ |
| 0 | 0 | 0 | 1 | 0 | 0 | $Y_4$ |
| 0 | 0 | 0 | 1 | 0 | 1 | $Y_5$ |
| 0 | 0 | 0 | 1 | 1 | 0 | $Y_6$ |
| 0 | 0 | 0 | 1 | 1 | 1 | $Y_7$ |
| 0 | 0 | 1 | 0 | 0 | 0 | $Y_8$ |
| 0 | 0 | 1 | 0 | 0 | 1 | $Y_9$ |
| 0 | 0 | 1 | 0 | 1 | 0 | $Y_{10}$ |
| 0 | 0 | 1 | 0 | 1 | 1 | $Y_{11}$ |
| 0 | 0 | 1 | 1 | 0 | 0 | $Y_{12}$ |
| 0 | 0 | 1 | 1 | 0 | 1 | $Y_{13}$ |
| 0 | 0 | 1 | 1 | 1 | 0 | $Y_{14}$ |
| 0 | 0 | 1 | 1 | 1 | 1 | $Y_{15}$ |
| X | 1 | X | X | X | X | None |
| 1 | X | X | X | X | X | None |

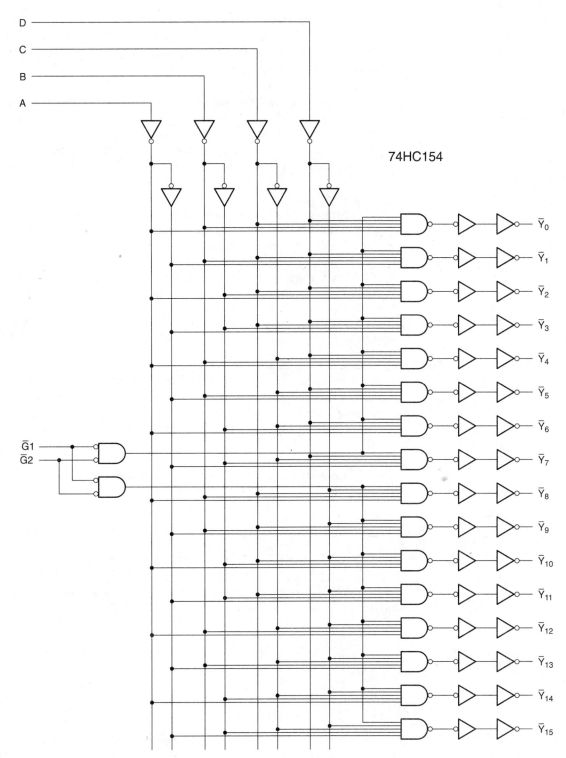

**Figure 10.9**
**74HC154 4-to-16 Line Decoder**

**EXAMPLE 10.5**

MSI decoders are sometimes used in microprocessor systems as "address decoders." A microprocessor has a parallel set of outputs called the address bus that selects a unique location in the system memory (an address) in which to store data or from which to retrieve it.

A computer with 1 megabyte of memory requires 20 address lines to address each byte individually. (This is because $2^{20} = 1,048,576 = 1M$; M is analogous, but not equal, to the metric prefix "mega-.") If the memory chips used in the system have a smaller capacity than 1M, the most significant bits on the address bus are decoded and used to select one memory device out of several available. The remaining address lines are connected to all memory devices in parallel to select the specific location in the enabled memory device.

Suppose a microcomputer system has 1 megabyte of random-access memory (RAM) divided into 16 devices, each having a capacity of 64 Kbytes ($1K = 2^{10} = 1024$; this is analogous, but not equal, to the metric prefix "kilo-"). A logic LOW at an enabling input, $\overline{G}$, selects a particular RAM device.

Draw a sketch showing how a 74HC154 decoder could be used to select 1 of 16 RAM chips. Briefly explain how the circuit works. (Other RAM inputs may be ignored for this example.)

Also, make a table listing the active decoder output for each selected device and the range of hexadecimal addresses on the entire address bus for each selected RAM.

**Solution**

Assume that the 20 address lines required to address 1M of memory are labeled $A_{19}$ to $A_0$. Each RAM device can store 64K bytes of data. Therefore, each RAM requires 16 address lines ($A_{15}$ to $A_0$) to address each byte uniquely ($64K = 2^6 \times 2^{10} = 2^{16}$). The remaining 4 lines ($A_{19}$ to $A_{16}$) are used to select a particular RAM chip in the system.

Figure 10.10 illustrates how a 74HC154 decoder can select and enable a particular RAM chip. Each decoder output connects to the *ENABLE* input, $\overline{G}$, of a different RAM device.

Table 10.4 shows the selected RAM and system address range for each decoder binary input.

**Table 10.4** Decoded Addresses

| D | C | B | A | Active Output | Selected RAM | Hex Address Range |
|---|---|---|---|---|---|---|
| 0 | 0 | 0 | 0 | $Y_0$ | 0 | 00000 to 0FFFF |
| 0 | 0 | 0 | 1 | $Y_1$ | 1 | 10000 to 1FFFF |
| 0 | 0 | 1 | 0 | $Y_2$ | 2 | 20000 to 2FFFF |
| 0 | 0 | 1 | 1 | $Y_3$ | 3 | 30000 to 3FFFF |
| 0 | 1 | 0 | 0 | $Y_4$ | 4 | 40000 to 4FFFF |
| 0 | 1 | 0 | 1 | $Y_5$ | 5 | 50000 to 5FFFF |
| 0 | 1 | 1 | 0 | $Y_6$ | 6 | 60000 to 6FFFF |
| 0 | 1 | 1 | 1 | $Y_7$ | 7 | 70000 to 7FFFF |
| 1 | 0 | 0 | 0 | $Y_8$ | 8 | 80000 to 8FFFF |
| 1 | 0 | 0 | 1 | $Y_9$ | 9 | 90000 to 9FFFF |
| 1 | 0 | 1 | 0 | $Y_{10}$ | A | A0000 to AFFFF |
| 1 | 0 | 1 | 1 | $Y_{11}$ | B | B0000 to BFFFF |
| 1 | 1 | 0 | 0 | $Y_{12}$ | C | C0000 to CFFFF |
| 1 | 1 | 0 | 1 | $Y_{13}$ | D | D0000 to DFFFF |
| 1 | 1 | 1 | 0 | $Y_{14}$ | E | E0000 to EFFFF |
| 1 | 1 | 1 | 1 | $Y_{15}$ | F | F0000 to FFFFF |

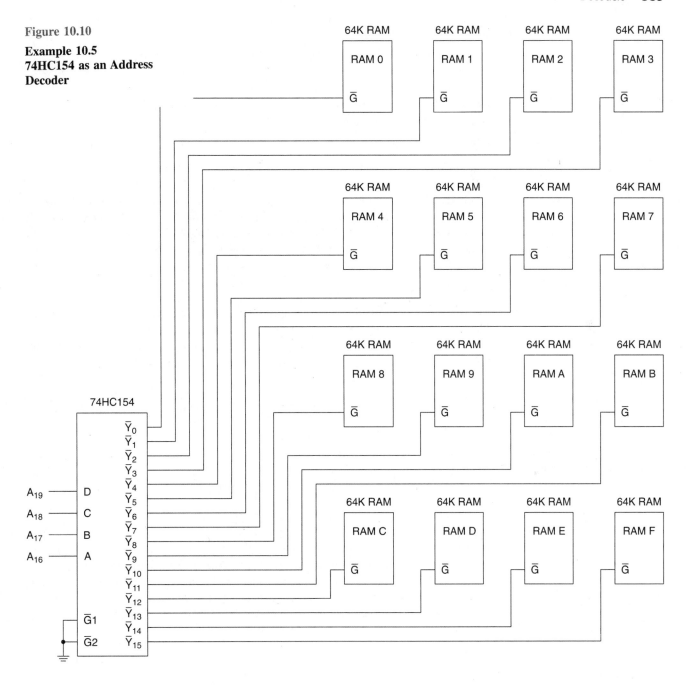

**Figure 10.10**

**Example 10.5**
**74HC154 as an Address Decoder**

One example address range: When RAM 5 is selected, $A_{19}A_{18}A_{17}A_{16}$ = 0101. The remaining 16 address lines, $A_{15}$ to $A_0$, can take on any binary value from all 0s to all 1s. Thus, if RAM 5 is selected, the minimum address in this range is

$$0101\ 0000\ 0000\ 0000\ 0000_2 = 50000_{16}$$

The maximum address is $0101\ 1111\ 1111\ 1111\ 1111_2 = 5FFFF_{16}$.

(Note that only the most significant digit changes from device to device. The lowest four hex digits are the same in all cases.)

## BCD-to-Seven-Segment Decoders

**Seven-segment display** *An array of seven independently controlled light-emitting diode (LED) or liquid crystal display (LCD) elements, shaped like a figure-8, which can be used to display decimal digits and other characters by turning on the appropriate elements.*

**Common-anode display** *A seven-segment LED display in which the anodes of all the LEDs are connected to the circuit supply voltage. Each segment is illuminated by a logic LOW at its cathode.*

**Common-cathode display** *A seven-segment LED display in which the cathodes of all LEDs are connected together and grounded. A logic HIGH illuminates a segment when applied to its anode.*

### Display

The **seven-segment display** is probably the most familiar digital device in general use. In fact, it is often what the layperson thinks you mean when you talk about digital circuits. This numerical display device, shown in Figure 10.11, is used to display decimal digits (and sometimes hexadecimal digits or other alphabetic characters). It is called a seven-segment display because it consists of seven luminous segments, usually LEDs or liquid crystals, arranged in a figure-8. We can display any decimal digit by turning on the appropriate elements. The elements are designated by the lowercase letters *a* through *g,* beginning with *a* at the top and progressing clockwise.

Figure 10.12 shows the usual convention for decimal digits. Some variation is possible with this type of display. For example, we could have drawn the digits 6 and 9 with "tails" (i.e., with segment *a* illuminated for 6 or segment *d* for 9). By convention, we display the digit 1 by illuminating segments *b* and *c,* although segments *e* and *f* would also work.

The electrical requirements for an LED circuit are simple. Since an LED is a diode, it conducts when its anode is positive with respect to its cathode, as shown in Figure 10.13a. A decoder/driver for an LED display will illuminate an element by completing this circuit, either by supplying $V_{CC}$ or ground. A series resistor limits the current to prevent the diode from burning out and to regulate its brightness. If the anode is +5 volts with respect to the cathode, the resistor value should be in the range of 220 $\Omega$ to 470 $\Omega$. Depending on the device, a seven-segment decoder/driver may or may not have internal series resistors.

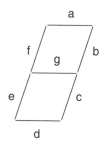

**Figure 10.11**

**Seven-Segment Display**

**Figure 10.12**

**Decimal Digits on a Seven-Segment Display**

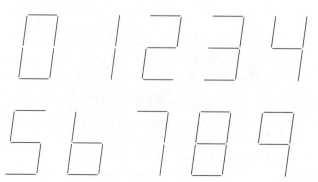

**Figure 10.13**

**Seven-Segment LED Connections**

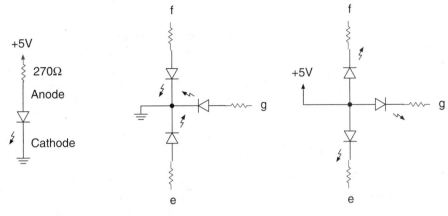

a. Circuit requirements for an illuminated LED

b. Common cathode (active HIGH)

c. Common anode (active LOW)

Seven-segment displays are configured as **common anode** or **common cathode.** Figure 10.13b and c shows the two different types of connection for segments *e, f,* and *g.* In a common-anode display, the anodes of all LEDs are connected together and brought out to one or more pin connections on the display package. The anode pins are wired externally to the circuit power supply. We illuminate the segments by applying logic LOWs to individual cathodes.

Similarly, the common-cathode display has the cathodes of the segments brought out to one or more common pins. These pins must be grounded. The segments illuminate when a decoder/driver makes their individual anodes HIGH. The two types of displays allow the use of either active-HIGH or active-LOW circuits to drive the LEDs, thus giving the designer some flexibility.

---

**EXAMPLE 10.6**

Sketch the segment patterns required to display all 16 hexadecimal digits on a seven-segment display. What changes from the patterns in Figure 10.12 need to be made?

**Solution**    The segment patterns are shown in Figure 10.14. Hex digits "B" and "D" must be displayed as lowercase letters, "b" and "d," to avoid confusion between "B" and "8" and between "D" and "0." To make "6" distinct from "b," "6" must be given a tail (segment *a*), and to make "6" and "9" symmetrical, "9" should also have a tail (segment *d*).

**Figure 10.14**

**Example 10.6 Hex Digits on Seven-Segment Display**

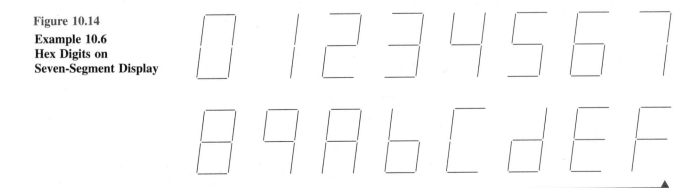

## Decoder

A BCD-to-seven-segment decoder is a circuit with a 4-bit input for a BCD digit and seven outputs for segment selection. To display a number, the decoder must translate the input bits to a combination of active outputs. For example, the input digit $DCBA = 0000$ must illuminate segments $a$, $b$, $c$, $d$, $e$, and $f$ to display the digit 0. We can make a truth table for each of the outputs, showing which must be active for every digit we wish to display.

The truth table for a common-cathode decoder (active-HIGH outputs) is given in Table 10.5.

The operation of each segment is determined by a Boolean function of the input variables, $DCBA$. From the truth table, the function for segment $a$ is:

$$a = \overline{D}\,\overline{C}\,\overline{B}\,\overline{A} + \overline{D}\,\overline{C}\,B\,\overline{A} + \overline{D}\,\overline{C}\,B\,A + \overline{D}\,C\,\overline{B}\,A + \overline{D}\,C\,B\,A$$
$$+ D\,\overline{C}\,\overline{B}\,\overline{A} + D\,\overline{C}\,\overline{B}\,A$$

If we assume that inputs 1010 to 1111 are never going to be used ("don't care states," symbolized by X), we can make any of these states produce HIGH or LOW decoder outputs, depending on which is most convenient for simplifying the segment functions. Figure 10.15a shows a Karnaugh map simplification for segment $a$. The resulting function is:

$$a = D + C\,A + \overline{C}\,\overline{A} + B\,A$$

The resulting partial decoder is shown in Figure 10.15b.

Each of the other LED segments is controlled by a similar Boolean function.

Table 10.5  Truth Table for Common-Cathode BCD-to-Seven-Segment Decoder

| Digit | DCBA | a | b | c | d | e | f | g |
|---|---|---|---|---|---|---|---|---|
| 0 | 0000 | 1 | 1 | 1 | 1 | 1 | 1 | 0 |
| 1 | 0001 | 0 | 1 | 1 | 0 | 0 | 0 | 0 |
| 2 | 0010 | 1 | 1 | 0 | 1 | 1 | 0 | 1 |
| 3 | 0011 | 1 | 1 | 1 | 1 | 0 | 0 | 1 |
| 4 | 0100 | 0 | 1 | 1 | 0 | 0 | 1 | 1 |
| 5 | 0101 | 1 | 0 | 1 | 1 | 0 | 1 | 1 |
| 6 | 0110 | 0 | 0 | 1 | 1 | 1 | 1 | 1 |
| 7 | 0111 | 1 | 1 | 1 | 0 | 0 | 0 | 0 |
| 8 | 1000 | 1 | 1 | 1 | 1 | 1 | 1 | 1 |
| 9 | 1001 | 1 | 1 | 1 | 0 | 0 | 1 | 1 |
| Invalid | 1010 | X | X | X | X | X | X | X |
| range | 1011 | X | X | X | X | X | X | X |
| | 1100 | X | X | X | X | X | X | X |
| | 1101 | X | X | X | X | X | X | X |
| | 1110 | X | X | X | X | X | X | X |
| | 1111 | X | X | X | X | X | X | X |

X = don't care

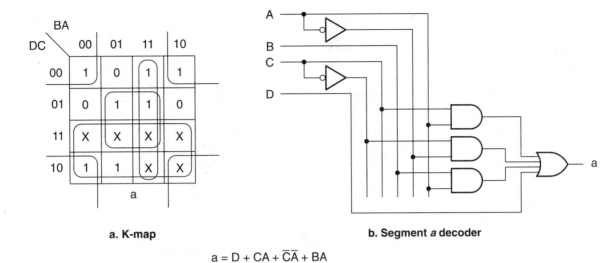

**a. K-map**

**b. Segment _a_ decoder**

$$a = D + CA + \overline{C}\overline{A} + BA$$

**Figure 10.15**
**Possible Implementation of Segment _a_ Decoder Function**

## MSI Seven-Segment Decoders

**Ripple blanking** _A technique used in a multiple-digit numerical display that suppresses leading or trailing 0s in the display, but leaves internal 0s._

Two TTL MSI seven-segment decoders are the 74LS48 common-cathode (active-HIGH) decoder, shown in Figure 10.16, and the 74LS47 common-anode (active-LOW) decoder, shown in Figure 10.17. The 74LS48 decoder is an obsolete device. It is included to show the operation of a common-cathode (active-HIGH) decoder.

The 74LS48 decoder requires no external resistors, as they are built into the circuit outputs. The 74LS47 does require series resistors.

In addition to the standard decimal digit outputs, these decoders are designed to have specific LED patterns for input codes 1010 to 1111, as shown in Figure 10.18. These output patterns are not just randomly chosen; they have been selected to simplify the decoding circuit as much as possible.

Table 10.6 shows the truth table for the 74LS48 decoder, including the 1010 to 1111 states.

When we derived the decoder function for segment _a_, shown in Figure 10.15, we simplified its Boolean function by SOP simplification. The segment _a_ output for the 74LS48 and 74LS47 decoders is derived from a POS simplification of the segment function. The K-map and POS solution for the 74LS48 segment _a_ function are shown in Figure 10.19.

The simplified expression for the active-HIGH segment _a_ function is:

$$a = (\overline{C} + A)(\overline{D} + \overline{B})(D + C + B + \overline{A})$$
$$= (\overline{\overline{C}\,\overline{A}})(\overline{D\,B})(\overline{\overline{D}\,\overline{C}\,\overline{B}\,A})$$

This function can be traced out on the 74LS48 logic diagram in Figure 10.16.

**Figure 10.16**

**74LS48 BCD-to-Seven-Segment Decoder, Obsolete**
(Reprinted by permission of Texas Instruments)

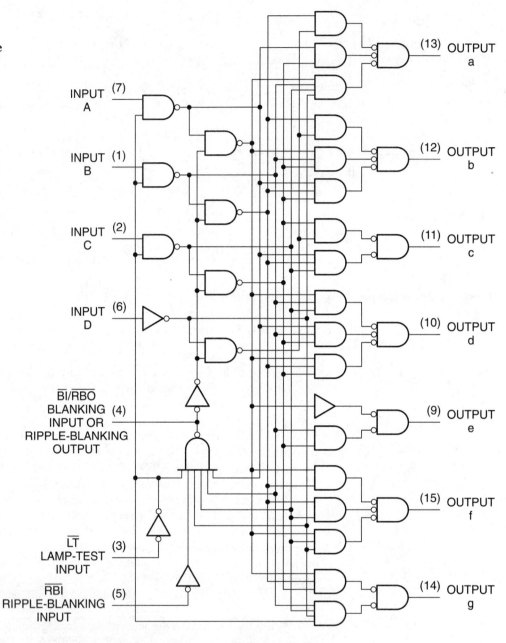

The segment *a* function for the 74LS47 decoder, which can be traced out on Figure 10.17, is the complement of the 74LS48 function. This is given by:

$$a = \overline{(\overline{C\,\overline{A}})(\overline{D\,B})(\overline{\overline{D}\,\overline{C}\,\overline{B}\,A})}$$

$$= C\,\overline{A} + D\,B + \overline{D}\,\overline{C}\,\overline{B}\,A$$

The complete list of segment functions for the 74LS47 decoder is given below:

**Figure 10.17**
**74LS47 BCD-to-Seven-Segment Decoder** (Reprinted by permission of Texas Instruments)

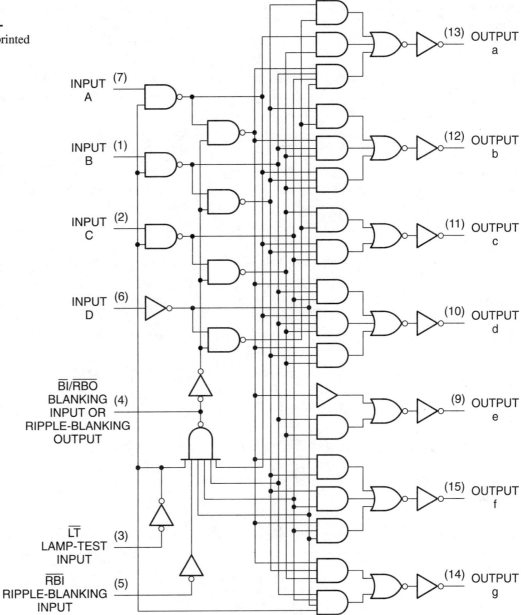

$$a = C\overline{A} + DB + \overline{D}\,\overline{C}\,\overline{B}A$$

$$b = C\overline{B}A + CB\overline{A} + DB$$

$$c = \overline{C}B\overline{A} + DC$$

$$d = \overline{C}\,\overline{B}A + CB\overline{A} + CBA$$

$$e = C\overline{B} + A$$

$$f = \overline{D}\,\overline{C}A + CB + BA$$

$$g = \overline{D}\,\overline{C}\,\overline{B} + CBA$$

**Figure 10.18**

**Outputs From 74LS47/48 Decoders**

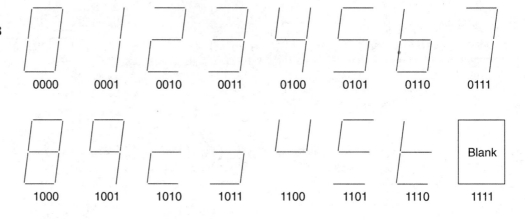

**Table 10.6** Truth Table for 74LS48 Decoder (Common Cathode)

| Digit | DCBA | a | b | c | d | e | f | g |
|-------|------|---|---|---|---|---|---|---|
| 0 | 0000 | 1 | 1 | 1 | 1 | 1 | 1 | 0 |
| 1 | 0001 | 0 | 1 | 1 | 0 | 0 | 0 | 0 |
| 2 | 0010 | 1 | 1 | 0 | 1 | 1 | 0 | 1 |
| 3 | 0011 | 1 | 1 | 1 | 1 | 0 | 0 | 1 |
| 4 | 0100 | 0 | 1 | 1 | 0 | 0 | 1 | 1 |
| 5 | 0101 | 1 | 0 | 1 | 1 | 0 | 1 | 1 |
| 6 | 0110 | 0 | 0 | 1 | 1 | 1 | 1 | 1 |
| 7 | 0111 | 1 | 1 | 1 | 0 | 0 | 0 | 0 |
| 8 | 1000 | 1 | 1 | 1 | 1 | 1 | 1 | 1 |
| 9 | 1001 | 1 | 1 | 1 | 0 | 0 | 1 | 1 |
|   | 1010 | 0 | 0 | 0 | 1 | 1 | 0 | 1 |
|   | 1011 | 0 | 0 | 1 | 1 | 0 | 0 | 1 |
|   | 1100 | 0 | 1 | 0 | 0 | 0 | 1 | 1 |
|   | 1101 | 1 | 0 | 0 | 1 | 0 | 1 | 1 |
|   | 1110 | 0 | 0 | 0 | 1 | 1 | 1 | 1 |
|   | 1111 | 0 | 0 | 0 | 0 | 0 | 0 | 0 |

**Figure 10.19**

**Segment *a* Function (74LS48 Decoder)**

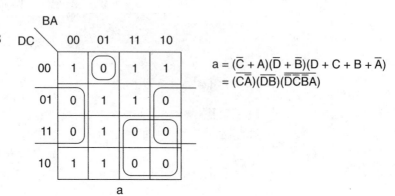

$$a = (\overline{C} + A)(\overline{D} + \overline{B})(D + C + B + \overline{A})$$
$$= (\overline{CA})(\overline{DB})(\overline{\overline{D}\,\overline{C}\,\overline{B}A})$$

The segment functions for the 74LS48 decoder are the complements of those above.

The 74LS47 and 74LS48 decoders include two additional features: lamp test ($\overline{LT}$) and **ripple blanking** ($\overline{RBI}/\overline{RBO}$).

The lamp test feature allows us to test whether all display segments are functioning properly. If the display is working, all segments light when the $\overline{LT}$ input is pulled LOW.

The ripple blanking feature allows for suppression of leading or trailing 0s in a multiple-digit display, while allowing 0s to be displayed in the middle of a number.

Each display decoder has a Ripple Blanking input ($\overline{RBI}$) and a Ripple Blanking output ($\overline{RBO}$), which are connected in cascade, as shown in Figure 10.20.

If the decoder input $DCBA$ is 0000, it displays digit 0 if $\overline{RBI} = 1$ and shows a blank if $\overline{RBI} = 0$.

If $\overline{RBI} = 1$ OR $DCBA$ is (NOT 0000), then $\overline{RBO} = 1$.

When we cascade two or more displays, these conditions suppress leading or trailing zeros (but not both) and still display internal zeros. To suppress leading zeros in a

**Figure 10.20**

**Zero Suppression in Seven-Segment Displays**

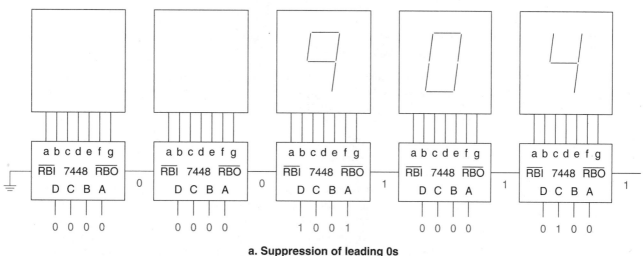

a. Suppression of leading 0s

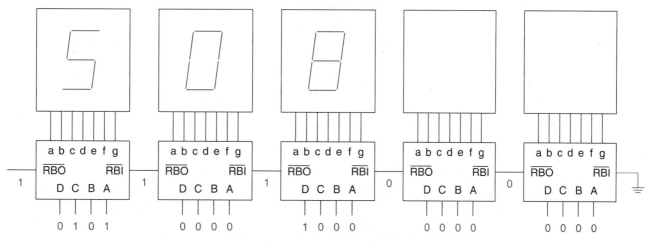

b. Suppression of trailing 0s

display, ground the $\overline{RBI}$ of the MSD display and connect the $\overline{RBO}$ of each decoder to the $\overline{RBI}$ of the next least significant digit, as shown in Figure 10.20a. Any zeros preceding the first nonzero digit (9 in this case) will be blanked, as $\overline{RBI} = 0$ AND $DCBA = 0000$ for each of these decoders. The 0 inside the number 904 is displayed, since its $\overline{RBI} = 1$.

Trailing 0s are suppressed by reversing the order of $\overline{RBI}$ and $\overline{RBO}$ from the above example. $\overline{RBI}$ is grounded for the least significant digit, and the $\overline{RBO}$ for each decoder cascades to the $\overline{RBI}$ of the next most significant digit. This is shown in Figure 10.20b.

---

**Section Review Problem for Section 10.1**

10.1. When would it be logical to suppress trailing zeros in a multiple-digit display, and when should trailing zeros be displayed?

---

## 10.2

# Encoders

**Encoder** *A circuit that produces a digital code at its outputs in response to one or more active input lines.*

**Priority encoder** *An encoder that will produce a binary output corresponding to the subscript of the highest-priority active input. This is usually defined as the input with the largest subscript.*

The function of a digital **encoder** is complementary to that of a digital decoder. A decoder activates a specified output for a unique digital input code. An encoder operates in the reverse direction, producing a particular digital code (e.g., a binary or BCD number) at its outputs when a specific input is activated.

Figure 10.21 shows an 8-bit binary encoder. The circuit generates a unique 3-bit binary output for every active input provided *only one input* is active at a time.

The encoder has only 8 permitted input states out of a possible 256. Table 10.7 shows the allowable input states, which yield the Boolean equations used to design the encoder. These Boolean equations are:

$$Q_2 = D_7 + D_6 + D_5 + D_4$$
$$Q_1 = D_7 + D_6 + D_3 + D_2$$
$$Q_0 = D_7 + D_5 + D_3 + D_1$$

The $D_0$ input is not connected to any of the encoding gates, since all outputs are in their LOW (inactive) state when the 000 code is selected.

### Priority Encoder

The shortcoming of the encoder circuit shown in Figure 10.21 is that it can generate wrong codes if more than one input is active at the same time. For example, if we make $D_3$ and $D_5$ HIGH at the same time, the output is neither 011 or 101, but 111; the output code does not correspond to either active input.

**Figure 10.21**
**8-bit Encoder (No Input Priority)**

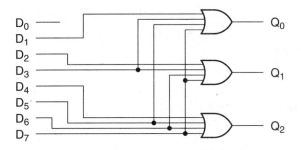

**Table 10.7** Partial Truth Table for an 8-Bit Encoder

| $D_7$ | $D_6$ | $D_5$ | $D_4$ | $D_3$ | $D_2$ | $D_1$ | $Q_2$ | $Q_1$ | $Q_0$ |
|---|---|---|---|---|---|---|---|---|---|
| 0 | 0 | 0 | 0 | 0 | 0 | 0 | 0 | 0 | 0 |
| 0 | 0 | 0 | 0 | 0 | 0 | 1 | 0 | 0 | 1 |
| 0 | 0 | 0 | 0 | 0 | 1 | 0 | 0 | 1 | 0 |
| 0 | 0 | 0 | 0 | 1 | 0 | 0 | 0 | 1 | 1 |
| 0 | 0 | 0 | 1 | 0 | 0 | 0 | 1 | 0 | 0 |
| 0 | 0 | 1 | 0 | 0 | 0 | 0 | 1 | 0 | 1 |
| 0 | 1 | 0 | 0 | 0 | 0 | 0 | 1 | 1 | 0 |
| 1 | 0 | 0 | 0 | 0 | 0 | 0 | 1 | 1 | 1 |

One solution to this problem is to assign a priority level to each input and, if two or more are active, make the output code correspond to the highest-priority input. As you might expect, this is called a **priority encoder.** Highest priority is assigned to the input whose subscript has the largest numerical value.

**EXAMPLE 10.7**

Figure 10.22a to c shows a priority encoder with three different combinations of inputs. Determine the resulting output code for each figure. Inputs and outputs are active HIGH.

**Solution**

Figure 10.22a: The highest-priority active input is $D_5$. $D_4$ and $D_1$ are ignored. $Q_2Q_1Q_0 = 101$.

**Figure 10.22**
**Example 10.7**
**Priority Encoder Inputs**

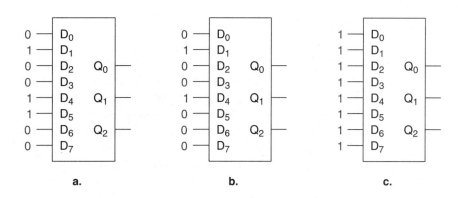

a.

b.

c.

Figure 10.22b: The highest-priority active input is $D_4$. $D_1$ is ignored. $Q_2Q_1Q_0 = 100$.

Figure 10.22c: The highest-priority active input is $D_7$. All other inputs are ignored. $Q_2Q_1Q_0 = 111$.

### Priority Encoder Circuit

> The encoding principle of a priority encoder is that a low-priority input must not change the code resulting from a high-priority input.

For example, if inputs $D_3$ and $D_5$ are both active, the correct output code is $Q_2Q_1Q_0 = 101$. The code for $D_3$ would be $Q_2Q_1Q_0 = 011$. $D_3$ must not make $Q_1 = 1$. Thus, the Boolean expressions for $Q_2$, $Q_1$, and $Q_0$ covering only these two codes are:

$$Q_2 = D_5 \qquad \text{(HIGH if } D_5 \text{ is active.)}$$

$$Q_1 = D_3\overline{D_5} \qquad \text{(HIGH if } D_3 \text{ is active AND } D_5 \text{ is not active.)}$$

$$Q_0 = D_3 + D_5 \qquad \text{(HIGH if } D_3 \text{ OR } D_5 \text{ is active.)}$$

Restating the encoding principle, a bit goes HIGH (1) if it is part of the code for an active input AND (2) it is *not* kept LOW by an input with a higher priority. We can use this principle to develop a mechanical method for generating the Boolean equations of the outputs.

1. Write the codes in order from highest to lowest priority.

| $Q_2$ | $Q_1$ | $Q_0$ | Code value |
|-------|-------|-------|------------|
| 1 | 1 | 1 | 7 |
| 1 | 1 | 0 | 6 |
| 1 | 0 | 1 | 5 |
| 1 | 0 | 0 | 4 |
| 0 | 1 | 1 | 3 |
| 0 | 1 | 0 | 2 |
| 0 | 0 | 1 | 1 |
| 0 | 0 | 0 | 0 |

2. Examine each code. For a code with value $n$, add a $D_n$ term to each $Q$ equation where there is a 1. For example, for code 111, add the term $D_7$ to the equations for $Q_2$, $Q_1$, and $Q_0$. For code 110, add the term $D_6$ to the equations for $Q_2$ and $Q_1$. (Steps 1 and 2 generate the nonpriority encoder equations listed earlier.)

3. Modify any $D_n$ terms to ensure correct priority. Every time you write a $D_n$ term, look up. For each previous code with a 0 in the same column, AND the $D_n$ term with a corresponding $\overline{D}$. For example, code 101 generates a $D_5$ term in the equations for $Q_2$ and $Q_0$. The term in the $Q_2$ equation need not be modified,

because there are no previous codes with a 0 in the same column. The term in the $Q_0$ equation must be modified since there is a 0 in the $Q_0$ column for code 110. This generates the term $\overline{D}_6 D_5$.

The equations from the 8-bit encoder of Figure 10.21 are modified by the priority encoding principle as follows:

$$Q_2 = D_7 + D_6 + D_5 + D_4$$
$$Q_1 = D_7 + D_6 + \overline{D}_5\overline{D}_4 D_3 + \overline{D}_5\overline{D}_4 D_2$$
$$Q_0 = D_7 + \overline{D}_6 D_5 + \overline{D}_6\overline{D}_4 D_3 + \overline{D}_6\overline{D}_4\overline{D}_2 D_1$$

## MSI Priority Encoders

Eight-bit priority encoders are available in both TTL and CMOS technologies. Figure 10.23 shows a CMOS device: the 4532B Priority Encoder. This device has active-HIGH inputs and outputs, as well as Enable Input and Output ($E_{in}$ and $E_{out}$) and Group Select *(GS)* pins for cascading to other encoders.

Table 10.8 shows the function of the 4532B encoder. The following information about the encoder can be extracted from the truth table:

1. The circuit generates nonzero output codes only if $E_{in} = 1$.
2. $E_{out} = 1$ only if $E_{in} = 1$ AND all $D$ inputs = 0. In cascade (when $E_{out}$ is tied to $E_{in}$ of a lower-priority device) this can be used to signify that no inputs are active on the high-priority encoder.
3. The $D_0$ input has no effect on the output codes, but does affect $E_{out}$ and *GS*.
4. *GS* = 1 if any $D$ (including $D_0$) = 1 AND $E_{in} = 1$. This feature can be used to encode an extra output bit if two or more encoders are cascaded. Note that in cascade, *GS* = 1 on only one device at a time: the encoder with the highest-priority active line. The same conditions that make *GS* = 1 ensure that $E_{out} = 0$. This disables any lower-priority devices, preventing their *GS* and code outputs from becoming active.

**Table 10.8** Truth Table of 4532B Priority Encoder

| $D_7$ | $D_6$ | $D_5$ | $D_4$ | $D_3$ | $D_2$ | $D_1$ | $D_0$ | $E_{in}$ | $E_{out}$ | GS | $O_2$ | $O_1$ | $O_0$ |
|---|---|---|---|---|---|---|---|---|---|---|---|---|---|
| 0 | 0 | 0 | 0 | 0 | 0 | 0 | 0 | 1 | 1 | 0 | 0 | 0 | 0 |
| 0 | 0 | 0 | 0 | 0 | 0 | 0 | 1 | 1 | 0 | 1 | 0 | 0 | 0 |
| 0 | 0 | 0 | 0 | 0 | 0 | 1 | X | 1 | 0 | 1 | 0 | 0 | 1 |
| 0 | 0 | 0 | 0 | 0 | 1 | X | X | 1 | 0 | 1 | 0 | 1 | 0 |
| 0 | 0 | 0 | 0 | 1 | X | X | X | 1 | 0 | 1 | 0 | 1 | 1 |
| 0 | 0 | 0 | 1 | X | X | X | X | 1 | 0 | 1 | 1 | 0 | 0 |
| 0 | 0 | 1 | X | X | X | X | X | 1 | 0 | 1 | 1 | 0 | 1 |
| 0 | 1 | X | X | X | X | X | X | 1 | 0 | 1 | 1 | 1 | 0 |
| 1 | X | X | X | X | X | X | X | 1 | 0 | 1 | 1 | 1 | 1 |
| X | X | X | X | X | X | X | X | 0 | 0 | 0 | 0 | 0 | 0 |

**LOGIC DIAGRAM**
**(Positive Logic)**

**LOGIC EQUATIONS**

$E_{out} = Ein \cdot \overline{D0} \cdot \overline{D1} \cdot \overline{D2} \cdot \overline{D3} \cdot \overline{D4} \cdot \overline{D5} \cdot \overline{D6} \cdot \overline{D7}$
$Q0 = Ein \cdot (D1 \cdot \overline{D2} \cdot \overline{D4} \cdot \overline{D6} + D3 \cdot \overline{D4} \cdot \overline{D6} + D5 \cdot \overline{D6} + D7)$
$Q1 = Ein \cdot (D2 \cdot \overline{D4} \cdot \overline{D5} + D3 \cdot \overline{D4} \cdot \overline{D5} + D6 + D7)$
$Q2 = Ein \cdot (D4 + D5 + D6 + D7)$
$GS = Ein \cdot (D0 + D1 + D2 + D3 + D4 + D5 + D6 + D7)$

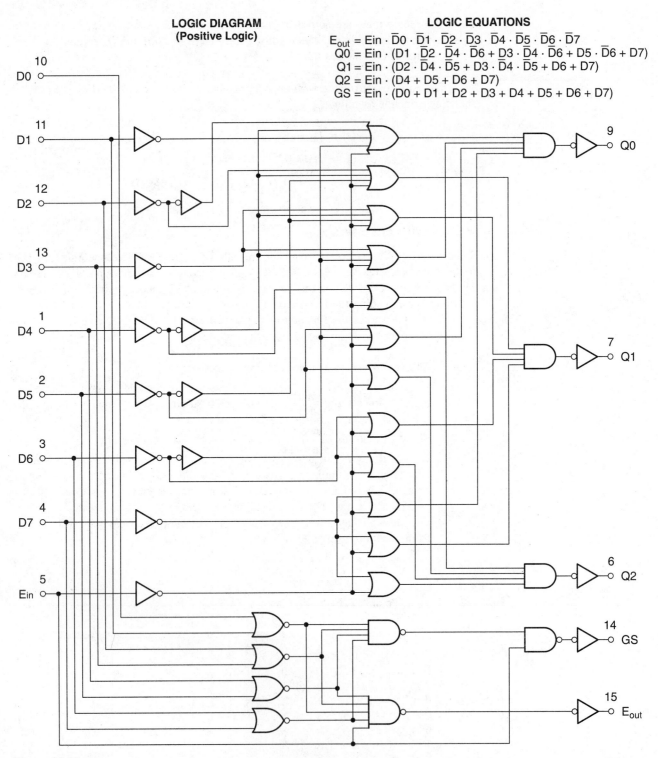

**Figure 10.23**
**MC14532B Priority Encoder** (Reprinted with permission of Motorola)

**EXAMPLE 10.8**

Connect two 4532B 8-bit priority encoders and any additional logic required to make a 16-bit priority encoder. Briefly explain the circuit operation.

**Solution**

The required circuit is shown in Figure 10.24. The output codes for $D_{15}$ and $D_7$ are the same in the lower 3 bits. (1111/0111). This is true for every other pair of matching inputs. Thus, each of the lower 3 bits (outputs $Q_2$, $Q_1$, and $Q_0$) can be paralleled via an OR gate.

The GS output, designated $Q_3$ in the circuit, is used to distinguish between upper and lower priority groups. If an input in the upper group is selected, $GS = 1$, producing a code in the range 1000 to 1111. Otherwise, $GS = 0$, indicating a code in the lower group (0000 to 0111).

The cascade connection ($E_{out}$ to $E_{in}$) disables the lower-priority encoder when an input is active in the upper group. If no upper group lines are active, $E_{out} = 1$, $GS = 0$, and codes may be selected from the lower group, which ranges from 0000 to 0111.

**Figure 10.24**

**Example 10.8**

**16-Bit Priority Encoder**

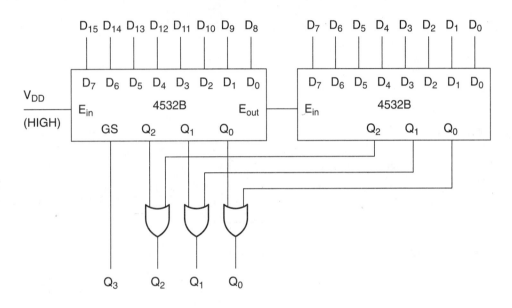

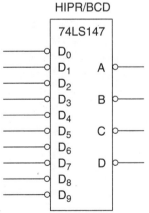

**Figure 10.25**

**BCD Priority Encoder**

## BCD Priority Encoders

The 74LS147 10-Line to 4-Line Priority Encoder is a TTL device that encodes one of ten active inputs into a 4-bit BCD code. It is functionally similar to the 4532B 8-bit priority encoder with two main differences:

1. Its inputs and outputs are active LOW, thus requiring us to invert the outputs to interpret them as BCD numbers.

2. There are no cascading inputs or outputs.

Figure 10.25 shows the logic symbol for this device.

**EXAMPLE 10.9**

A common application of a 10-line to 4-line encoder is the encoding of switch closures on a decimal keypad, shown in Figure 10.26a. The keypad is a rectangular array of push-button switches, such as the one shown in Figure 10.26b. (The switches are called "key switches," in the sense of "piano key," not "lock and key.") When a switch is pressed, it makes its associated output LOW. Normally the output is HIGH, due to its pull-up resistor.

Draw a circuit showing how to encode the output of each switch as its binary equivalent, using a 74LS147 priority encoder.

**Figure 10.26**

**Example 10.9**
**Decimal Keypad**

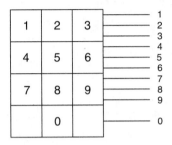   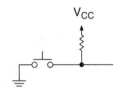

**a. Keypad layout**       **b. Individual pushbutton switch**

**Figure 10.27**

**Example 10.9**
**Priority Encoder for**
**Decimal Keypad**

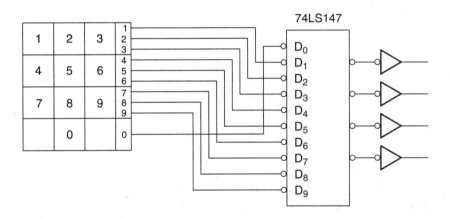

**Solution**

Figure 10.27 shows a decimal keypad connected to the encoder. The key switches are connected directly to the encoder inputs, as the inputs are active LOW and each switch produces a LOW when pressed. Since the outputs are active LOW, they must be inverted to interpret the output as the correct BCD digits.

---

**Section Review Problem for Section 10.2**

10.2. What is the Boolean expression for output $\overline{D}$, the most significant bit, of the 74LS147 BCD Priority Encoder? What is the expression for $\overline{C}$? (Keep in mind that the inputs and outputs are active LOW.)

## 10.3

# Magnitude Comparators

> **Magnitude comparator** *A circuit that compares two n-bit binary numbers, indicates whether or not the numbers are equal and/or which one is larger.*

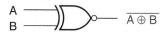

**Figure 10.28**

**Exclusive NOR Gate**

If we are interested in finding out whether or not two binary numbers are the same, we can use a **magnitude comparator.** The simplest comparison circuit is the Exclusive NOR gate, whose circuit symbol is shown in Figure 10.28 and whose truth table is given in Table 10.9.

The output of the XNOR gate is 1 if its inputs are the same ($A = B$) and 0 if they are different. For this reason, the XNOR gate is sometimes called a coincidence gate.

We can use several XNORs to compare each bit of two multibit binary numbers. Figure 10.29 shows a 2-bit comparator with one output that goes HIGH if all bits of $A$ and $B$ are identical.

If the MSB of $A$ equals the MSB of $B$, the output of the upper XNOR is HIGH. If the LSBs are the same, the output of the lower XNOR is HIGH. If both these conditions are satisfied, then $A = B$, which is indicated by a HIGH at the AND output. This general principle applies to any number of bits.

**Table 10.9** Exclusive NOR Function

| $A$ | $B$ | $\overline{A \oplus B}$ |
|:---:|:---:|:---:|
| 0 | 0 | 1 |
| 0 | 1 | 0 |
| 1 | 0 | 0 |
| 1 | 1 | 1 |

$$(A = B) = (\overline{A_{n-1} \oplus B_{n-1}})(\overline{A_{n-2} \oplus B_{n-2}}) \ldots (\overline{A_1 \oplus B_1})(\overline{A_0 \oplus B_0})$$

for two *n*-bit numbers

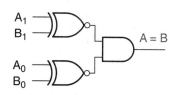

**Figure 10.29**

**2-Bit Magnitude Comparator**

Some magnitude comparators also include an output that activates if $A > B$ and another that is active when $A < B$. Figure 10.30 shows the comparator of Figure 10.29 expanded to include the "greater than" and "less than" functions.

Let us analyze the ($A > B$) circuit. The ($A > B$) function has two AND-shaped gates, which compare $A$ and $B$ bit by bit to see which is larger.

1. The 2-input gate examines the MSBs of $A$ and $B$. If $A_1 = 1$ AND $B_1 = 0$, then we know that $A > B$. (This implies one of the following inequalities: $10 > 00$; $10 > 01$; $11 > 00$; or $11 > 01$.)

2. If $A_1 = B_1$, then we don't know whether or not $A > B$. To find out, we must compare the next most significant bits, $A_0$ and $B_0$. The 3-input gate makes this comparison. Since this gate is enabled by the XNOR that compares the two MSBs, it is active only when $A_1 = B_1$. This yields the term $(\overline{A_1 \oplus B_1})\overline{A_0}B_0$ in the Boolean expression for the ($A > B$) function.

3. If $A_1 = B_1$ AND $A_0 = 1$ AND $B_0 = 0$, the 3-input gate has a HIGH output, telling us, via the OR gate, that $A > B$. (This is true. The only possibilities are ($01 > 00$) and ($11 > 10$).)

Similar logic works in the ($A < B$) circuit, except that inversion is on the $A$ bits rather than the $B$ bits. Alternatively, we can simplify either the ($A > B$) or the ($A < B$)

**Figure 10.30**

**2-Bit Comparator With A = B, A>B, and A<B Outputs**

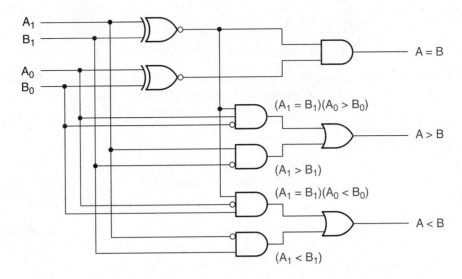

function by using a NOR function. For instance, if we have developed a circuit to indicate $(A = B)$ and $(A < B)$, we can make the $(A > B)$ function from the other two, as follows:

$$(A > B) = \overline{(A = B) + (A < B)}$$

This Boolean expression says that if $A$ is not equal to or less than $B$, it must be greater than $B$.

Figure 10.31 shows a 4-bit comparator with $(A = B)$, $(A < B)$, and $(A > B)$ outputs.

---

The Boolean expressions for the outputs are:

$$(A = B) = \overline{(A_3 \oplus B_3)}\,\overline{(A_2 \oplus B_2)}\,\overline{(A_1 \oplus B_1)}\,\overline{(A_0 \oplus B_0)}$$

$$(A < B) = \overline{A}_3 B_3 + \overline{(A_3 \oplus B_3)}\overline{A}_2 B_2 + \overline{(A_3 \oplus B_3)}\,\overline{(A_2 \oplus B_2)}\overline{A}_1 B_1$$

$$+ \,\overline{(A_3 \oplus B_3)}\,\overline{(A_2 \oplus B_2)}\,\overline{(A_1 \oplus B_1)}\overline{A}_0 B_0$$

$$(A > B) = \overline{(A = B) + (A < B)}$$

---

This comparison technique can be expanded to as many bits as necessary. A 4-bit comparator requires four AND-shaped gates for its $(A < B)$ function. We can interpret the Boolean expression for this function as follows.

$(A < B)$ if:

1. The MSB of $A$ is less than the MSB of $B$, OR

2. The MSBs are equal, but the second bit of $A$ is less than the second bit of $B$, OR

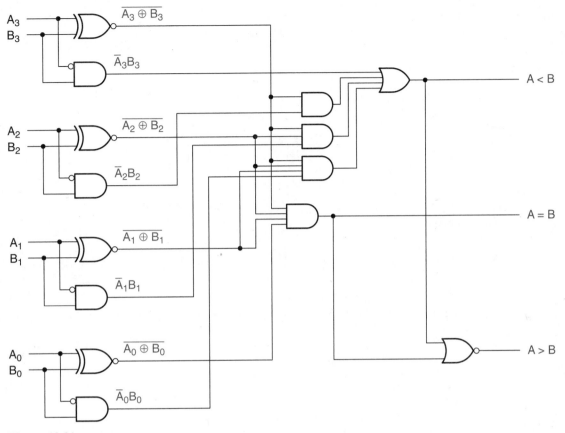

**Figure 10.31**

**4-Bit Magnitude Comparator**

3. The first two bits are equal, but the third bit of $A$ is less than the third bit of $B$, OR

4. The first three bits are equal, but the LSB of $A$ is less than the LSB of $B$.

Expansion to more bits would use the same principle of comparing bits one at a time, beginning with the MSBs.

**EXAMPLE 10.10**

A digital thermometer has two input probes. A circuit in the thermometer converts the measured temperature at each probe to a 4-bit number, as shown by the block in Figure 10.32.

In addition to measuring the temperature at each input, the thermometer has a comparison function that indicates whether the temperature at one input is greater than, equal to, or less than the temperature at the other input.

Draw a block diagram showing how a magnitude comparator could be connected to light a green LED for ($A > B$), an amber LED for ($A = B$), and a red LED for ($A < B$).

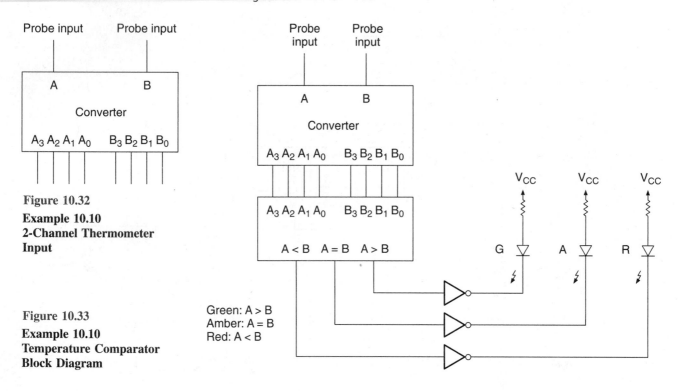

**Figure 10.32**

**Example 10.10 2-Channel Thermometer Input**

**Figure 10.33**

**Example 10.10 Temperature Comparator Block Diagram**

Green: A > B
Amber: A = B
Red: A < B

**Solution**   Figure 10.33 shows the block diagram of the magnitude comparator connected to the thermometer's digital output. When one of the comparator outputs goes HIGH, the output of the corresponding inverter goes LOW. This supplies a current path to ground for the related LED.

Two 4-bit magnitude comparators available as MSI devices are the 74LS85 Magnitude Comparator, fabricated using TTL technology, and its CMOS equivalent, the 4585B Magnitude Comparator. In addition to the $(A = B)$, $(A > B)$, and $(A < B)$ outputs, which indicate HIGH for a true condition, these devices also have inputs for $(A = B)$, $(A > B)$, and $(A < B)$. This enables several devices to be connected in cascade so that words larger than 4 bits can be compared.

**EXAMPLE 10.11**   Suppose the converter on the digital thermometer in Example 10.10 is changed so that it represents the temperature at each probe input as an 8-bit number. Modify the comparison circuit to accommodate 8 bits. (Use 74LS85 comparators.)

**Solution**   Figure 10.34 shows the new 8-bit comparison circuit. The two magnitude comparators are cascaded by connecting the $(A = B)$, $(A > B)$, and $(A < B)$ outputs of the first (least significant) comparator to the corresponding inputs of the second comparator. For reasons we shall see shortly, the $(A = B)$ input on the least significant stage must be tied HIGH for proper operation.

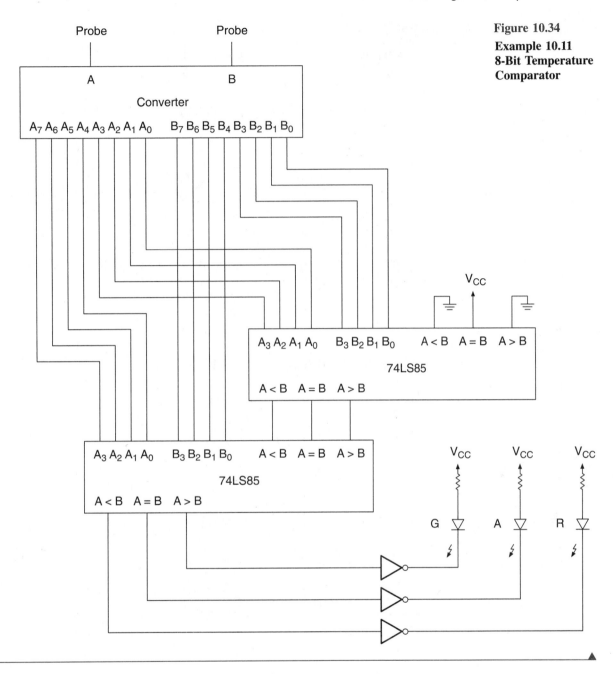

**Figure 10.34**
**Example 10.11**
**8-Bit Temperature**
**Comparator**

Table 10.10 shows the truth table for the 74LS85 comparator. This table is broken up into three sections:

1. The first eight lines show the response to unequal values at the comparing inputs.

2. The next three lines show the response to equal values at the comparing inputs but indication of $(A < B)$, $(A = B)$, or $(A > B)$ at a previous stage.

3. The last two lines are a "residue" of the cascading logic and have no meaning.

**Table 10.10** Truth Table for 74LS85 Magnitude Comparator

| Comparing Inputs | | | | Cascading Inputs | | | Outputs | | |
|---|---|---|---|---|---|---|---|---|---|
| $A_3, B_3$ | $A_2, B_2$ | $A_1, B_1$ | $A_0, B_0$ | $A > B$ | $A < B$ | $A = B$ | $A > B$ | $A < B$ | $A = B$ |
| $A_3 > B_3$ | X | X | X | X | X | X | 1 | 0 | 0 |
| $A_3 < B_3$ | X | X | X | X | X | X | 0 | 1 | 0 |
| $A_3 = B_3$ | $A_2 > B_2$ | X | X | X | X | X | 1 | 0 | 0 |
| $A_3 = B_3$ | $A_2 < B_2$ | X | X | X | X | X | 0 | 1 | 0 |
| $A_3 = B_3$ | $A_2 = B_2$ | $A_1 > B_1$ | X | X | X | X | 1 | 0 | 0 |
| $A_3 = B_3$ | $A_2 = B_2$ | $A_1 < B_1$ | X | X | X | X | 0 | 1 | 0 |
| $A_3 = B_3$ | $A_2 = B_2$ | $A_1 = B_1$ | $A_0 > B_0$ | X | X | X | 1 | 0 | 0 |
| $A_3 = B_3$ | $A_2 = B_2$ | $A_1 = B_1$ | $A_0 < B_0$ | X | X | X | 0 | 1 | 0 |
| $A_3 = B_3$ | $A_2 = B_2$ | $A_1 = B_1$ | $A_0 = B_0$ | 1 | 0 | 0 | 1 | 0 | 0 |
| $A_3 = B_3$ | $A_2 = B_2$ | $A_1 = B_1$ | $A_0 = B_0$ | 0 | 1 | 0 | 0 | 1 | 0 |
| $A_3 = B_3$ | $A_2 = B_2$ | $A_1 = B_1$ | $A_0 = B_0$ | X | X | 1 | 0 | 0 | 1 |
| $A_3 = B_3$ | $A_2 = B_2$ | $A_1 = B_1$ | $A_0 = B_0$ | 1 | 1 | 0 | 0 | 0 | 0 |
| $A_3 = B_3$ | $A_2 = B_2$ | $A_1 = B_1$ | $A_0 = B_0$ | 0 | 0 | 0 | 1 | 1 | 0 |

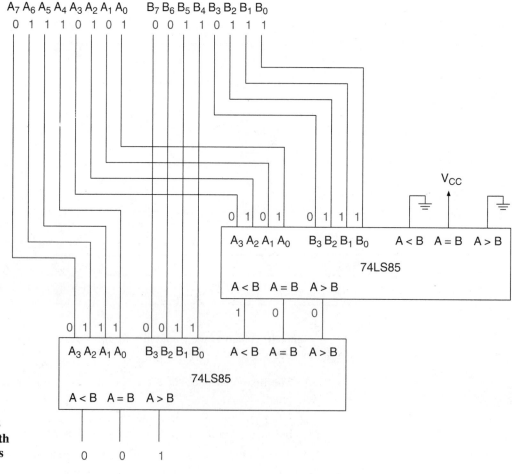

**Figure 10.35**

**Comparison of A and B With Inequalities in Both Upper and Lower 4 Bits**

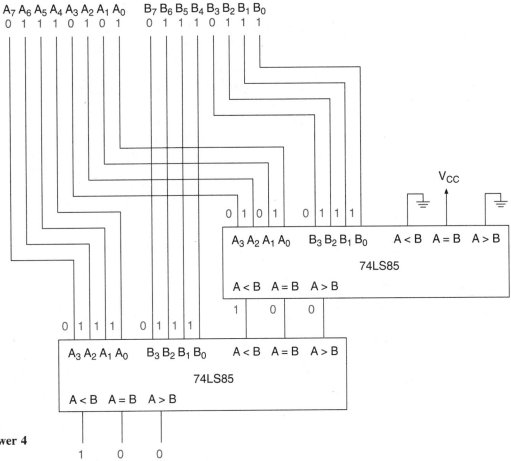

**Figure 10.36**

**Comparison of A and B With Inequalities in Lower 4 Bits Only**

The first eight lines of the 74LS85 truth table show that if $A$ and $B$ are unequal, the $(A > B)$ or $(A < B)$ outputs so indicate without regard to the states of the cascading inputs. This means that a comparison of inputs at a less significant stage does not over-rule a comparison at a more significant stage.

For example, suppose we compare two 8-bit numbers, $A = 0111\ 0101$ and $B = 0011$ $0111$, as illustrated in Figure 10.35.

The comparator looking at the lower 4 bits of $A$ and $B$ sees that 0101 is less than 0111 and passes that information to the second comparator by making the $(A < B)$ line HIGH. This does not affect the decision of the second comparator, which looks at the upper 4 bits of $A$ and $B$ and sees that 0111 is greater than 0011. The circuit determines that $A > B$, which is true.

The second group of input conditions in Table 10.10 (lines 9 through 11) indicates that if the values at the comparing inputs are equal, then comparison of any less significant bits will be passed straight to the outputs. Of the three cascading inputs, $(A = B)$ has priority, since it will activate the $(A = B)$ output regardless of the states of the other two cascading inputs.

Suppose we change the comparator inputs to $A = 0111\ 0101$ and $B = 0111\ 0111$, as illustrated in Figure 10.36. The upper 4 bits of $A$ and $B$ are identical to each other, and the lower 4 bits are the same as they were in the previous example. The $(A < B)$ line of the first comparator goes HIGH, as before. (See the third line of the truth table for this condition.) But now, since the upper 4 bits are equal, the second comparator transmits this $(A < B)$ condition straight to its $(A < B)$ output. (The tenth line of the truth table specifies this condition.)

The cascading conditions indicate that the $(A = B)$ input on the least significant stage should be HIGH. Conditions $(A < B)$ and $(A > B)$ are transmitted regardless of the states of the cascading inputs, but $(A = B)$ must have a HIGH value for transmission to the next stage.

The last two lines of the truth table have no useful function; they represent the "residue" of the cascading logic. The line showing $(A > B) = (A < B) = 1$ implies that in a less significant stage $A$ is both greater than *and* less than $B$—obviously nonsense. The line $(A > B) = (A < B) = (A = B) = 0$ seems to say that at a previous stage, $A$ is neither greater than nor less than nor equal to $B$; again, this makes no sense.

---

**Section Review Problem for Section 10.3**

10.3. Figure 10.37 shows a 74LS85 magnitude comparator with the input logic levels as shown. What information does this give us about any previous stages? What are the logic levels of the $A > B$, $A = B$, and $A < B$ outputs?

**Figure 10.37**

**Section Review Problem 74LS85 Magnitude Comparator**

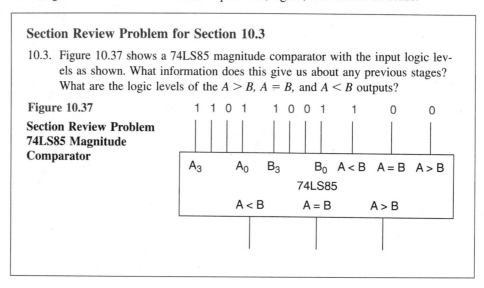

---

## 10.4

# Parity Generators and Checkers

**Parity**  *A system that checks for errors in a multibit binary number by counting the number of 1s.*

**EVEN parity**  *An error-checking system that requires a binary number to have an even number of 1s.*

**ODD parity**  *An error-checking system that requires a binary number to have an odd number of 1s.*

**Parity bit**  *A bit appended to a binary number to make the number of 1s even or odd, depending on the type of parity.*

When data are transmitted from one device to another, it is necessary to have a system of checking for errors in transmission. These errors, which appear as incorrect bits, occur as a result of electrical limitations such as line capacitance or induced noise.

**Figure 10.38**
**Parity Error Checking**

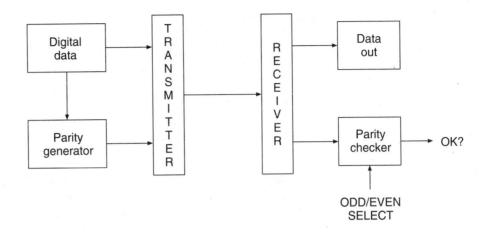

Parity error checking is a way of encoding information about the correctness of data before they are transmitted. The data can then be verified at the system's receiving end. Figure 10.38 shows a block diagram of a parity error-checking system.

The parity generator in Figure 10.38 examines the outgoing data and adds a bit called the **parity bit,** which makes the number of 1s in the transmitted data odd or even, depending on the type of parity. As you might expect, data with **EVEN parity** have an even number of 1s, including the parity bit, and data with **ODD parity** have an odd number of 1s.

(An even number is divisible by 2 without a remainder. An odd number leaves a remainder of 1 when divided by 2. Thus 0, 2, 4, 6 . . . are even, and 1, 3, 5, 7 . . . are odd.)

The data receiver "knows" whether to expect EVEN or ODD parity. If the incoming number of 1s matches the expected parity, the parity checker responds by indicating that correct data have been received. Otherwise, the parity checker indicates an error.

**EXAMPLE 10.12**

Data are transmitted from a personal computer to a video display terminal in groups of 7 data bits plus a parity bit. What should the parity bit, $P$, be for each of the following data if the parity is EVEN? If the parity is ODD?

   **a.** 0110110

   **b.** 1000000

   **c.** 0010101

**Solution**

a. 0110110    Four 1s in data. (Four is an even number.)

                EVEN parity: $P = 0$

                ODD parity: $P = 1$

b. 1000000    One 1 in data. (One is an odd number.)

                EVEN parity: $P = 1$

                ODD parity: $P = 0$

c. 0010101    Three 1s in data. (Three is an odd number.)

                EVEN parity: $P = 1$

                ODD parity: $P = 0$

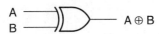

**Figure 10.39**
**Exclusive OR Gate**

**Table 10.11** Exclusive OR
Truth Table

| *A* | *B* | *A* ⊕ *B* |
|---|---|---|
| 0 | 0 | 0 |
| 0 | 1 | 1 |
| 1 | 0 | 1 |
| 1 | 1 | 0 |

An Exclusive OR gate can be used as a parity generator or as a parity checker. Figure 10.39 shows the gate, and Table 10.11 is the XOR truth table. Notice that each line of the XOR truth table has an even number of 1s if we include the output column.

Figure 10.40 shows the block diagram of a circuit that will generate an EVEN parity bit from 2 data bits, *A* and *B,* and transmit the 3 bits one after the other, that is, serially, to a data receiver.

Figure 10.41 shows a parity checker for the parity generator in Figure 10.40. Data are received serially, but read in parallel. The parity bit is re-created from the received values of *A* and *B,* and then compared to the received value of *P* to give an error indication, *P'*. If *P* and *A* ⊕ *B* are the same, then *P'* = 0 and the transmission is correct. If *P* and *A* ⊕ *B* are different, then *P'* = 1 and there has been an error in transmission.

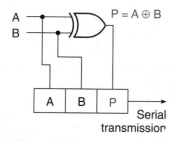

**Figure 10.40**
**Even Parity Generation**

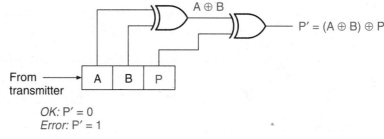

**Figure 10.41**
**Even Parity Checking**

**E X A M P L E   1 0 . 1 3**

The following data and parity bits are transmitted four times: *ABP* = 101.

**a.** State the type of parity used.

**b.** The transmission line over which the data are transmitted is particularly noisy and the data arrive differently each time as follows:

  **1.** *ABP* = 101

  **2.** *ABP* = 100

  **3.** *ABP* = 111

  **4.** *ABP* = 110

Indicate the output *P'* of the parity checker in Figure 10.41 for each case, and state what the output means.

**Solution**   **a.** The system is using EVEN parity.

**b.** The parity checker produces the following responses:

1. $ABP = 101$
$A \oplus B = 1 \oplus 0 = 1$
$P' = P \oplus (A \oplus B) = 1 \oplus 1 = 0$   (Data received correctly.)

2. $ABP = 100$
$A \oplus B = 1 \oplus 0 = 1$
$P' = P \oplus (A \oplus B) = 1 \oplus 0 = 1$   (Transmission error. Parity bit incorrect.)

3. $ABP = 111$
$A \oplus B = 1 \oplus 1 = 0$
$P' = P \oplus (A \oplus B) = 0 \oplus 1 = 1$   (Transmission error. Data bit $B$ incorrect.)

4. $ABP = 110$
$A \oplus B = 1 \oplus 1 = 0$
$P' = P \oplus (A \oplus B) = 0 \oplus 0 = 0$   (Transmission error undetected. $B$ and $P$ incorrectly received.)

The second and third cases in Example 10.13 show that parity error detection cannot tell which bit is incorrect.

The fourth case points out the major flaw of parity error detection: An even number of errors cannot be detected. This is true whether the parity is EVEN or ODD. If a group of bits has an even number of 1s, a single error will change that to an odd number of 1s, but a double error will change it back to even. (Try a few examples to convince yourself that this is true.)

An ODD parity generator and checker can be made using an Exclusive NOR, rather than an Exclusive OR, gate. If a set of transmitted data bits requires a 1 for EVEN parity, it follows that it requires a 0 for ODD parity. This implies that EVEN and ODD parity generators must have opposite-sense outputs.

**EXAMPLE 10.14**   Modify the circuits in Figures 10.40 and 10.41 to operate with ODD parity. Verify their operation with the data bits $AB = 11$ and $AB = 01$.

**Solution**   Figure 10.42a shows an ODD parity generator, and Figure 10.42b shows an ODD parity checker. The checker circuit still has an Exclusive OR output since it presents the same error codes as an EVEN parity checker. The parity bit is re-created at the receive end of the transmission path and compared with the received parity bit. If they are the same, $P' = 0$ (correct transmission). If they are different, $P' = 1$ (transmission error).

**Verification:**

*Generator:*

Data:   $AB = 11$   Parity:   $P = \overline{A \oplus B} = \overline{1 \oplus 1} = 1$

*Checker:*

$P' = (\overline{A \oplus B}) \oplus P = (\overline{1 \oplus 1}) \oplus 1 = 1 \oplus 1 = 0$   (Correct transmission.)

*Generator:*

$$\text{Data:} \quad AB = 01 \qquad \text{Parity:} \quad P = \overline{A \oplus B} = \overline{0 \oplus 1} = 0$$

*Checker:*

$$P' = (\overline{A \oplus B}) \oplus P = (\overline{0 \oplus 1}) \oplus 0 = 0 \oplus 0 = 0 \quad \text{(Correct transmission.)}$$

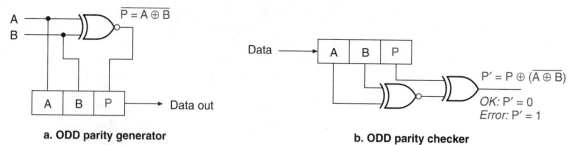

**a. ODD parity generator**        **b. ODD parity checker**

**Figure 10.42**

**Example 10.14**
**ODD Parity Generator and Checker**

Parity generators and checkers can be expanded to any number of bits by using an XOR gate for each pair of bits and combining the gate outputs in further stages of 2-input XOR gates. The true form of the generated parity bit is $P_E$, the EVEN parity bit. The complement form of the bit is $P_O$, the ODD parity bit.

Table 10.12 shows the XOR truth table for 4 data bits and the ODD and EVEN parity bits. The EVEN parity bit $P_E$ is given by $(A \oplus B) \oplus (C \oplus D)$. The ODD parity bit $P_O$ is given by $\overline{P_E} = \overline{(A \oplus B) \oplus (C \oplus D)}$. For every line in Table 10.12, the bit combination $ABCDP_E$ has an even number of 1s, and the group $ABCDP_O$ has an odd number of 1s.

**Table 10.12**  EVEN and ODD Parity Bits for 4 Data Bits

| $A$ | $B$ | $C$ | $D$ | $A \oplus B$ | $C \oplus D$ | $P_E$ | $P_O$ |
|---|---|---|---|---|---|---|---|
| 0 | 0 | 0 | 0 | 0 | 0 | 0 | 1 |
| 0 | 0 | 0 | 1 | 0 | 1 | 1 | 0 |
| 0 | 0 | 1 | 0 | 0 | 1 | 1 | 0 |
| 0 | 0 | 1 | 1 | 0 | 0 | 0 | 1 |
| 0 | 1 | 0 | 0 | 1 | 0 | 1 | 0 |
| 0 | 1 | 0 | 1 | 1 | 1 | 0 | 1 |
| 0 | 1 | 1 | 0 | 1 | 1 | 0 | 1 |
| 0 | 1 | 1 | 1 | 1 | 0 | 1 | 0 |
| 1 | 0 | 0 | 0 | 1 | 0 | 1 | 0 |
| 1 | 0 | 0 | 1 | 1 | 1 | 0 | 1 |
| 1 | 0 | 1 | 0 | 1 | 1 | 0 | 1 |
| 1 | 0 | 1 | 1 | 1 | 0 | 1 | 0 |
| 1 | 1 | 0 | 0 | 0 | 0 | 0 | 1 |
| 1 | 1 | 0 | 1 | 0 | 1 | 1 | 0 |
| 1 | 1 | 1 | 0 | 0 | 1 | 1 | 0 |
| 1 | 1 | 1 | 1 | 0 | 0 | 0 | 1 |

**EXAMPLE 10.15**    Use Table 10.12 to draw a 4-bit parity generator and a 4-bit parity checker that can generate and check either EVEN or ODD parity, depending on the state of one select input.

**Solution**    Figure 10.43 shows the circuit for a 4-bit parity generator. The XOR gate at the output is configured as a programmable inverter to give $P_E$ or $P_O$. When $\overline{EVEN}/ODD = 0$, the parity output is not inverted and the circuit generates $P_E$. When $\overline{EVEN}/ODD = 1$, the XOR inverts the parity bit, giving $P_O$.

The 4-bit parity checker, shown in Figure 10.44, is the same circuit, with an additional XOR gate to compare the parity bit re-created from data and the previously encoded parity bit.

**Figure 10.43**

**Example 10.15**
**4-Bit Parity Generator**

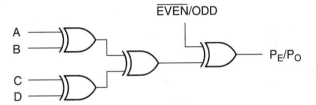

**Figure 10.44**

**Example 10.15**
**4-Bit Parity Checker**

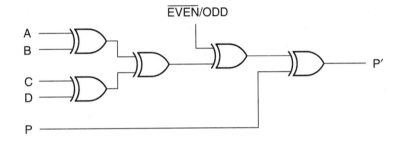

**EXAMPLE 10.16**    Draw the circuit for an 8-bit EVEN/ODD parity generator.

**Solution**    An 8-bit parity generator is an expanded version of the 4-bit generator in the previous example. The circuit is shown in Figure 10.45.

**Figure 10.45**

**Example 10.16**
**8-Bit Parity Generator**

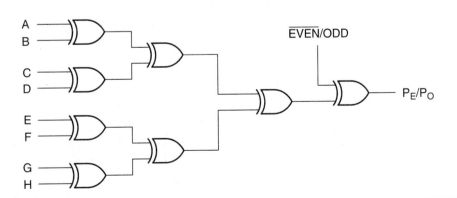

**Figure 10.46**

**74180 Parity Generator/Checker**
(Reprinted by permission of Texas Instruments)

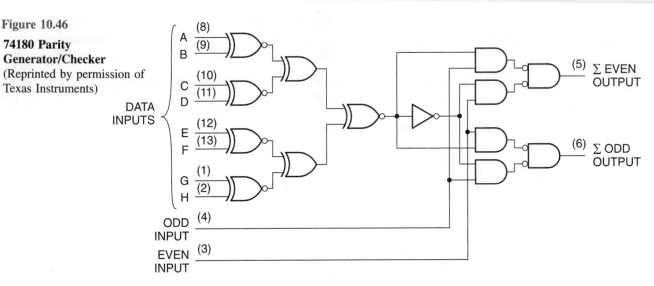

**Table 10.13**   Truth Table of a 74180 Parity Generator/Checker

| Inputs | | | Outputs | |
|---|---|---|---|---|
| No. of 1s at *A* Through *H* | EVEN | ODD | ΣEVEN | ΣODD |
| EVEN | 1 | 0 | 1 | 0 |
| ODD | 1 | 0 | 0 | 1 |
| EVEN | 0 | 1 | 0 | 1 |
| ODD | 0 | 1 | 1 | 0 |
| X | 1 | 1 | 0 | 0 |
| X | 0 | 0 | 1 | 1 |

A TTL device that can be used as a parity generator or checker is the 74180 9-bit Parity Generator/Checker. Nine bits are useful for systems having 8 data bits and 1 parity bit. Figure 10.46 shows the logic circuit of the 74180.

In addition to *EVEN* and *ODD* parity bit outputs, labeled Σ*EVEN* and Σ*ODD,* the 74180 also has *EVEN* and *ODD* inputs, one of which acts as the ninth bit, so that several devices can be cascaded. Table 10.13 shows the truth table for this device.

The 74180 can be configured to produce an active-HIGH or LOW indication of parity when used as a parity checker. Also, the Σ*EVEN* and Σ*ODD* outputs can be forced HIGH or LOW regardless of parity.

When two or more devices are cascaded, the least significant stage must have its *ODD* and *EVEN* inputs tied HIGH or LOW, depending on what the final Σ*EVEN* and Σ*ODD* outputs are supposed to do.

▼ ————————————————————————————————————————————

**EXAMPLE 10.17**    Draw a circuit that will produce EVEN and ODD parity bits from 16 data bits, using 74180 parity generator/checkers.

**Solution**   Use two 74180 chips, as shown in Figure 10.47. To make the Σ*EVEN* output of the first stage LOW when there are an even number of 1s on inputs *A* to *H,* tie the *EVEN* input

**Figure 10.47**

**Example 10.17**
**Parity Bit From 16 Data Bits**

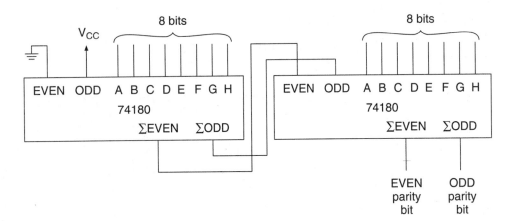

of the first device LOW and the *ODD* input HIGH. This configuration also ensures that the $\Sigma ODD$ output is HIGH when there is an even number of 1s on inputs $A$ to $H$.

The 16-bit-generated ODD and EVEN parity bits are available at the $\Sigma EVEN$ and $\Sigma ODD$ outputs of the second 74180 device.

---

**Section Review Problems for Section 10.4**

10.4.  Data (including a parity bit) are detected at a receiver configured for checking ODD parity. Which of the following data do we know are incorrect? Could there be errors in the remaining data? Explain.
   a. 010010
   b. 011010
   c. 1110111
   d. 1010111
   e. 1000101

# 10.5

# Multiplexers

**Multiplexer**  *A circuit that directs one of several digital signals to a single output, depending on the states of several select inputs.*

**Data inputs**  *Those inputs to a multiplexer that feed a digital signal to the output when selected.*

**Select inputs**  *The multiplexer inputs that select a digital input channel.*

Figure 10.48 shows a digital audio switching system. The system shown can select a signal from one of four sources and direct it to a digital signal processor (DSP) at its output. The digital audio sources include two compact disc players ($CD_1$ and $CD_2$) and two digital audio tape players ($DAT_1$ and $DAT_2$). (We assume we have direct access to

**Figure 10.48**

**Multiplexing 4 Digital Channels**

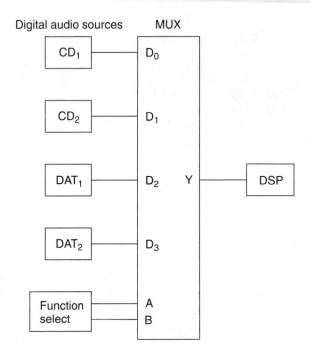

the audio signals in digital form.) The switching from source to output is done by a device called a **multiplexer,** shown as the block labeled MUX, the standard abbreviation for this device.

The MUX inputs that connect to the digital sources are called the **data inputs,** designated $D_0$ to $D_3$. One, but only one, of the sources is connected to the MUX output at any time. The particular source-to-output path is selected by the states of inputs $B$ and $A$, which are called the **select inputs.** The select inputs enable the data input whose subscript is the same number as the binary value $BA$.

**EXAMPLE 10.18**

Make a table listing which digital audio source in Figure 10.48 is routed to the DSP for each combination of the multiplexer select inputs, $B$ and $A$.

**Solution**

| $B$ | $A$ | Selected Input | Selected Source |
|-----|-----|----------------|-----------------|
| 0 | 0 | $D_0$ | $CD_1$ |
| 0 | 1 | $D_1$ | $CD_2$ |
| 1 | 0 | $D_2$ | $DAT_1$ |
| 1 | 1 | $D_3$ | $DAT_2$ |

The multiplexer in Figure 10.48 uses internal logic gates to pass or block a signal. Each channel has its own AND gate that is enabled or inhibited by the select inputs. These AND gate outputs are combined in an output OR gate, as shown in Figure 10.49.

Each binary combination of $B$ and $A$ will direct a different data channel to the output. For example, if $BA = 00$, the input inverters enable the top AND gate. ($D_0 \cdot \overline{B} \cdot \overline{A}$ $= D_0 \cdot \overline{0} \cdot \overline{0} = D_0 \cdot 1 \cdot 1 = D_0$.) All other ANDs are disabled. Thus, the signal at $D_0$ is routed to output $Y$, and signals $D_1$ to $D_3$ are blocked.

**Figure 10.49**

**Internal Circuitry of 4-to-1 MUX**

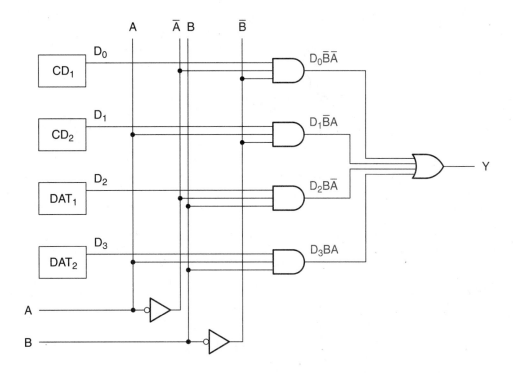

## MSI Multiplexers

MSI multiplexers are available in several configurations to suit a variety of design needs. Some commercially available TTL devices are the 74LS150 16-to-1, 74LS151 8-to-1, 74LS153 Dual 4-to-1, and 74LS157 Quadruple 2-to-1 multiplexers. High-Speed CMOS versions of some of these devices are also available (74HC151, 74HC153, and 74HC157A). Several of these devices are represented in Figure 10.50.

Standard (Metal-Gate) CMOS devices include the 4067B 16-channel and 4097B Dual 8-channel multiplexer/demultiplexers.

Chips having single multiplexers (74LS150, 74LS/HC151, and 4067B) are suited to data channel selection, as in Example 10.18, and to other applications where several serial streams of data must be sent to the same destination.

Chips with more than one device per package (74LS/HC153, 74LS/HC157, and 4097B) can be used to send parallel sets of data from two or more sources. For example, the 74LS/HC157 can send a 4-bit word to a 4-bit output from one of two different locations. These multiple-device chips use the same decoder for all multiplexers in the package.

The next several examples will examine some of the many possible applications of multiplexers.

## Single-Channel Data Selection

We have already seen, in Figure 10.48, how we can use a multiplexer to direct one of several streams of data to a single destination. We can use an MSI multiplexer to perform the same function.

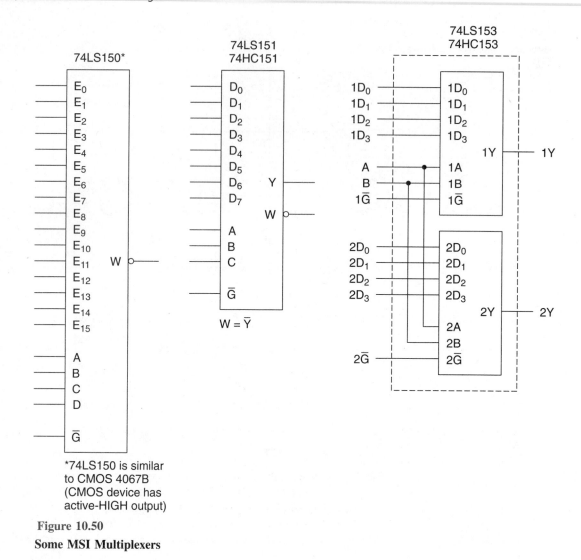

**Figure 10.50**

**Some MSI Multiplexers**

**EXAMPLE 10.19**    Show how to connect a 74HC151 8-to-1 Multiplexer to switch one of eight digitized audio signals to a single digital audio output, $Y$.

**Solution**    Figure 10.51 shows the required connection.

## Boolean Function Generator

Since a multiplexer is really just a sum-of-products combinational circuit, we can use it to generate a hard-to-simplify Boolean function. In such an application, the $Y$ output is a Boolean function of the select input variables. The data inputs are tied HIGH or LOW, depending on the required logic level for the corresponding select input combination.

**Figure 10.51**

**Example 10.19**
**Digital Channel Selection**
**(8-to-1 MUX)**

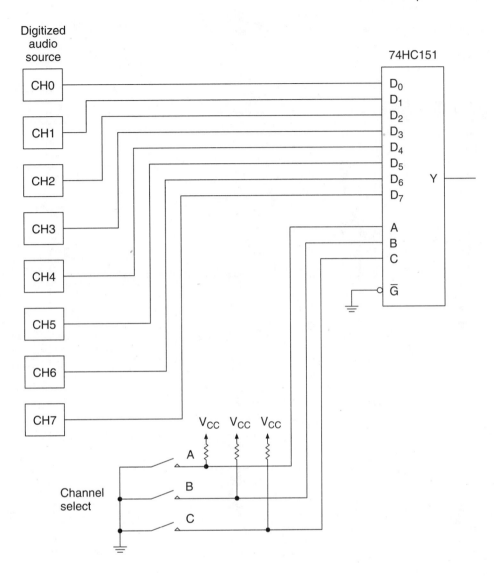

---

**EXAMPLE 10.20**    Simplify the Boolean expression represented by Table 10.14 as much as possible. Show both the simplest SOP logic gate circuit and the connections to a 74LS150 16-to-1 multiplexer needed to generate the same Boolean function.

**Solution**    Figure 10.52a shows the Karnaugh map for the function shown by Table 10.14. Since no cells containing a 1 are adjacent, no SOP simplification is possible. (Although a POS simplification eliminates a few variables in some of the terms, it yields seven terms, compared to six SOP terms. This is no improvement.)

The Boolean expression is given by:

$$Y = \overline{D}\,\overline{C}\,\overline{B}\,A + \overline{D}\,\overline{C}\,B\,\overline{A} + \overline{D}\,C\,B\,\overline{A}$$

$$+ \overline{D}\,C\,B\,A + D\,\overline{C}\,\overline{B}\,\overline{A} + D\,\overline{C}\,B\,A$$

**Table 10.14**  Truth Table for Example 10.20

| D | C | B | A | Y |
|---|---|---|---|---|
| 0 | 0 | 0 | 0 | 0 |
| 0 | 0 | 0 | 1 | 1 |
| 0 | 0 | 1 | 0 | 1 |
| 0 | 0 | 1 | 1 | 0 |
| 0 | 1 | 0 | 0 | 1 |
| 0 | 1 | 0 | 1 | 0 |
| 0 | 1 | 1 | 0 | 0 |
| 0 | 1 | 1 | 1 | 1 |
| 1 | 0 | 0 | 0 | 1 |
| 1 | 0 | 0 | 1 | 0 |
| 1 | 0 | 1 | 0 | 0 |
| 1 | 0 | 1 | 1 | 1 |
| 1 | 1 | 0 | 0 | 0 |
| 1 | 1 | 0 | 1 | 0 |
| 1 | 1 | 1 | 0 | 0 |
| 1 | 1 | 1 | 1 | 0 |

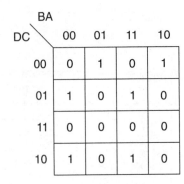

a. K-map

**Figure 10.52**

**Example 10.20 MUX as a Logic Function Generator**

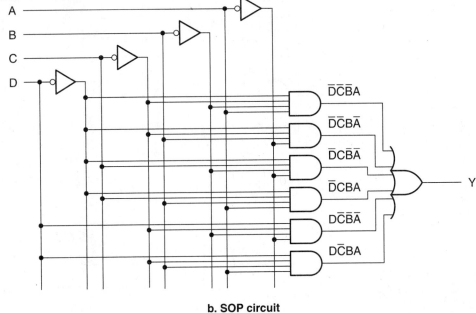

b. SOP circuit

Figure 10.52b shows this SOP circuit. The multiplexer connections are shown in Figure 10.53. According to the truth table, when *DCBA* has one of the decimal equivalent values 1, 2, 4, 7, 8, or 11, the circuit output is HIGH. Since the 74LS150 has an inverting output, inputs $E_1$, $E_2$, $E_4$, $E_7$, $E_8$, and $E_{11}$ are tied LOW. All other inputs are HIGH.

**Figure 10.53**

**Example 10.20 MUX Circuit**

**EXAMPLE 10.21**    Show how we can implement the Boolean function of Table 10.14, using a 74HC151 8-to-1 multiplexer.

**Solution**    We can also use an 8-to-1 multiplexer to implement a 4-bit Boolean function, such as the one from Table 10.14. In order to do so, we must break up the truth table into an upper and a lower half, as shown in Table 10.15.

 Note that the input on any line differs only by the value of $D$; the values of $CBA$ are identical in each column.

 Inputs $C$, $B$, and $A$ connect to the select inputs of the MUX, as before. There is no select input for $D$, so we must use the data inputs $D_0$ through $D_7$ to switch this variable to the output. The data in Table 10.15 yield the following possible cases.

**Table 10.15** Truth Table for Example 10.21

| $D = 0$ | | | | $D = 1$ | | | |
|---|---|---|---|---|---|---|---|
| $C$ | $B$ | $A$ | $Y$ | $C$ | $B$ | $A$ | $Y$ |
| 0 | 0 | 0 | 0 | 0 | 0 | 0 | 1 |
| 0 | 0 | 1 | 1 | 0 | 0 | 1 | 0 |
| 0 | 1 | 0 | 1 | 0 | 1 | 0 | 0 |
| 0 | 1 | 1 | 0 | 0 | 1 | 1 | 1 |
| 1 | 0 | 0 | 1 | 1 | 0 | 0 | 0 |
| 1 | 0 | 1 | 0 | 1 | 0 | 1 | 0 |
| 1 | 1 | 0 | 0 | 1 | 1 | 0 | 0 |
| 1 | 1 | 1 | 1 | 1 | 1 | 1 | 0 |

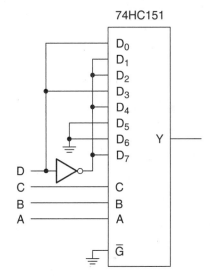

**Figure 10.54**

**Example 10.21**
**8-to-1 MUX Implementing a**
**4-Bit Boolean Function**

1. $Y = D$. This is true when $CBA = 000$ or $011$. To implement these lines on the truth table, connect $D_0$ and $D_3$ to input $D$.
2. $Y = \overline{D}$. This is the case when $CBA = 001$, $010$, $100$, or $111$. To generate the correct outputs for these inputs, invert input $D$ and connect the inverter output to $D_1$, $D_2$, $D_4$, and $D_7$.
3. $Y = 0$, regardless of the value of $D$. This is true when $CBA = 101$ or $110$. To satisfy these conditions, inputs $D_5$ and $D_6$ should be tied LOW.

Figure 10.54 shows the 8-to-1 MUX with the inputs specified above.

 One case that does not occur in Table 10.15 is when $Y = 1$, regardless of the value of $D$. In such a case, we would tie HIGH the data input corresponding to the value of $CBA$.

## Time-Dependent Multiplexer Applications

> **Counter** *A sequential digital circuit whose output produces a fixed sequence of binary states when the circuit is clocked (e.g., the output state of a 4-bit binary counter progresses in order from 0000 to 1111).*

Some of the most important applications of multiplexers require a digital circuit called a **counter** to be connected to the MUX select inputs.

We briefly examined a simple counter circuit in Example 6.13 in the chapter on flip-flops. We will study counters in detail in a later chapter, but for now just realize that a simple type of counter generates a repeating binary sequence from 0 to the maximum number of 1s that the outputs can accommodate.

For example, a 3-bit binary counter generates the binary sequence from 000 to 111 (8 states) and repeats indefinitely. The outputs of a 4-bit counter progress from 0000 to 1111 (16 states) and repeat. Figure 10.55 represents a timing diagram of a 4-bit binary counter.

**Figure 10.55**

**Timing Diagram of a 4-Bit Binary Counter**

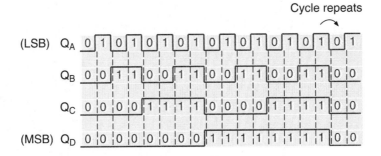

## Time-Division Multiplexing

> **Time-division multiplexing** *A technique of using one transmission line to send many signals simultaneously by making them share the line for equal fractions of time.*

**Time-division multiplexing** (TDM) is a technique used to send several streams of data down the same transmission line simultaneously. Telephone companies use TDM to maximize the use of their phone lines. Speech or data is digitally encoded for transmission. Each speech or data channel becomes a multiplexer data input that shares time with all other channels on a single phone line. A counter on the MUX select inputs selects the speech channels one after the other in a continuous sequence. The counter must switch the channels fast enough so that there is no apparent interruption of the transmitted conversation or data stream.

This is a highly simplified description of an extremely complex system. It doesn't account for many of the practical details involved, such as assignment of a phone line to a particular channel, what to do with idle lines, how to multiplex speech in two directions, frequency limitations of the phone lines, and so forth.

**EXAMPLE 10.22**

Draw a diagram of a circuit that uses a 74HC151 8-to-1 Multiplexer to share one telephone line among eight digitally encoded speech channels. Draw a timing diagram showing the select input and $Y$ output waveforms.

**Solution**

Figure 10.56 shows the required multiplexer circuit. Each speech channel is connected to a data input, and a 3-bit binary counter is connected to the select inputs. The output phone line connects to the MUX output. (This is identical to the circuit in Figure 10.51, except that the logic switches have been replaced by a counter.)

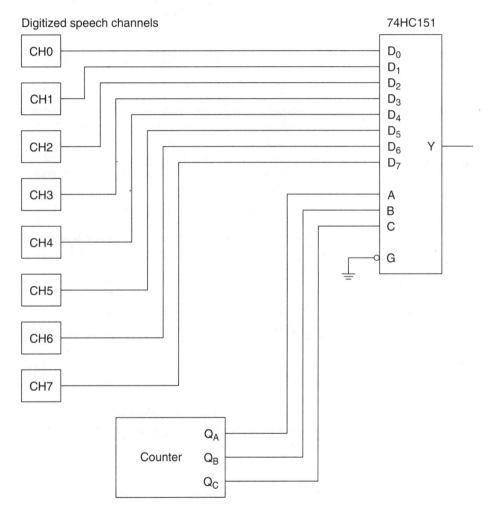

**Figure 10.56**
**Example 10.22**
**Time-Division Multiplexing of Telephone Channels**

The timing diagram is shown in Figure 10.57. The output is shown as a "bus waveform": two parallel lines, at logic HIGH and LOW levels, with a crossover at transition points. We use this notation when it is more important to show that new data are on the line than it is to give the particular value of those data.

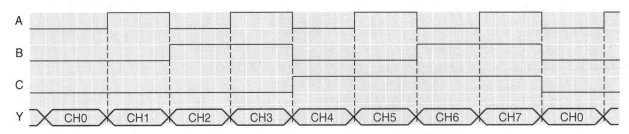

**Figure 10.57**
**Example 10.22**
**Timing Diagram**

## Waveform Generation

A multiplexer and counter can be used as a programmable waveform generator. This is similar to using the MUX as a logic function generator, except that we introduce a time element; a counter selects all data inputs in a continuous sequence. The output waveform can be programmed to any pattern by switching the logic levels on the data inputs.

This is an easy way to generate an asymmetrical waveform, a task that is more complicated when using a binary counter and output decoder.

**EXAMPLE 10.23**    Draw a circuit that uses a 74HC151 8-to-1 Multiplexer to generate a programmable 8-bit repeating pattern. Draw the timing diagram of the select inputs and the output waveform for the following pattern of data inputs.

$$D_0 \qquad D_7$$
$$01100101$$

**Solution**    Figure 10.58a shows the waveform generator circuit. The output waveform with respect to the counter inputs is shown in Figure 10.58b. This pattern is relatively difficult to generate by other means since it has several unequal HIGHs and LOWs in one period.

The circuit in Figure 10.58 can also generate symmetrical binary waveforms whose frequency can easily be changed by altering the switch settings.

**EXAMPLE 10.24**    The programmable waveform generator in Figure 10.58 generates a symmetrical pulse waveform having a frequency of 1 kHz when the data inputs are set as follows:

$$D_0 \qquad D_7$$
$$00001111$$

How should the switches be set to generate a symmetrical 2-kHz waveform? A symmetrical 4-kHz waveform?

**Solution**
$$D_0 \qquad D_7$$
2 kHz:    00110011

$$D_0 \qquad D_7$$
4 kHz:    01010101

**Figure 10.58**

**Example 10.23
Programmable Waveform
Generator**

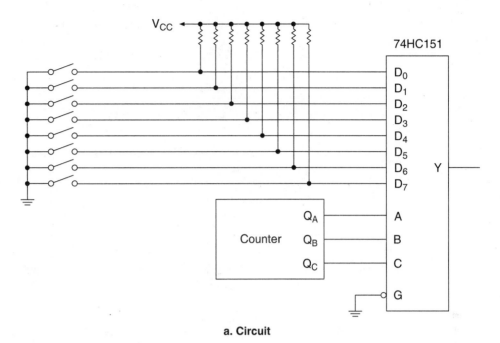

a. Circuit

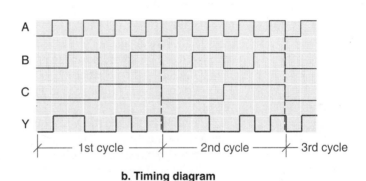

b. Timing diagram

## Parallel-to-Serial Conversion

Figure 10.59 shows a common format for serially transmitting ASCII data. Between characters, the transmission line is in an idle state, represented by a logic HIGH. The transmitted character begins with a start bit (a logic LOW), followed in sequence by the data bits, sent LSB first. The character ends with an optional parity bit and at least one stop bit (HIGH).

Figure 10.59a shows the general format of a transmitted character. Figure 10.59b shows the waveform when ASCII "A" ($= 1000001_2$) is transmitted with EVEN parity (we don't count the stop bit).

A multiplexer used for parallel-to-serial conversion is connected in much the same way as the waveform generator. Parallel data are applied to the MUX data inputs, and the output is converted to a serial output waveform by a counter connected to the MUX select inputs. The serial data to be transmitted can be changed by changing the parallel data at the MUX inputs.

**Figure 10.59**
**Serial Transmission**

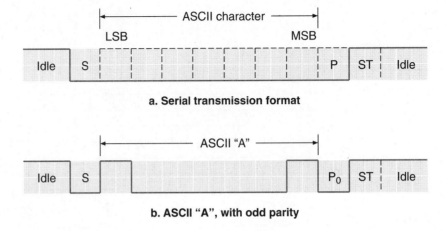

a. Serial transmission format

b. ASCII "A", with odd parity

**EXAMPLE 10.25**

Draw a block diagram showing how to connect a computer having an 8-bit parallel output to a video display terminal with a serial input using a 74HC151 8-to-1 MUX. The circuit should transmit 8 data bits and remain idle for the period of 8 counter states to give some idle time, which allows the parallel inputs to change between characters. Show logic levels at the MUX data inputs for transmission of ASCII "M" (= $1001101_2$).

Also draw the timing diagram, showing inputs $A$, $B$, $C$, and $\overline{G}$ and output $W$. Assume that there is no parity bit.

**Solution**

Figure 10.60a shows the MUX connections for an 8-bit parallel-to-serial converter. The select inputs are connected to the 3 LSBs of a 4-bit binary counter. The counter's MSB is connected to the multiplexer's Strobe ("enable" or "gating") input, $\overline{G}$. The outputs $Y$ and $W$ are enabled only when $\overline{G} = 0$. This occurs only in the first half of the counter cycle. When $\overline{G} = 1$, in the second half of the counter cycle, output $Y = 0$ and output $W = 1$.

Since $W = 1$ when the MUX output is disabled, it can be used to generate the idle state and, by implication, the stop bit. Since $W$ is an inverting output, all data inputs must be the complement of the intended output.

The ASCII code for "M" is $4D_{16} = 100\ 1101_2$. The complement of this code is 011 0010. These bits are transmitted after the start bit, LSB first, MSB last. Thus, the logic levels of the data inputs, from $D_7$ to $D_0$, are 011 0010 1. These are shown in Figure 10.60a. Figure 10.60b shows the timing diagram.

## Multichannel Data Selection

The multiplexer applications examined so far assume that the output is a single bit or stream of bits. Some applications require several bits to be selected in parallel.

Figure 10.61 shows a circuit, based on a 74LS157 Quad 2-to-1 multiplexer, that will direct one of two BCD digits to a seven-segment display. The bits $A_4A_3A_2A_1$ act as a single input, since when the MUX select input ($\overline{A}/B$) is LOW, these inputs are all connected to the outputs $Y_4Y_3Y_2Y_1$. Similarly, when the select input is HIGH, inputs $B_4B_3B_2B_1$ are connected to the outputs.

Thus, one of two 4-bit inputs is directed to a 4-bit output. The seven-segment display in Figure 10.61 will display "2" if $\overline{A}/B = 0$ ($A$ inputs selected) and "9" if $\overline{A}/B = 1$ ($B$ inputs selected).

**Figure 10.60**

**Example 10.25**
**MUX as a Parallel-to-Serial Converter**

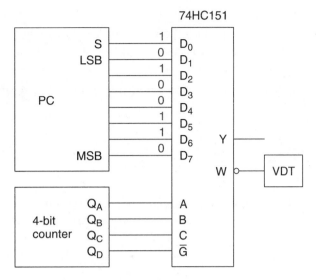

a. Block diagram

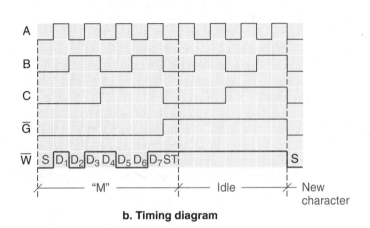

b. Timing diagram

**Figure 10.61**

**Quad 2-to-1 MUX as a BCD Digit Selector**

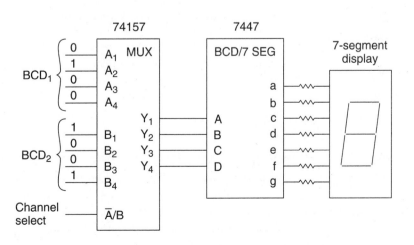

---

**Section Review Problem for Section 10.5**

10.5.  What is the most important difference between a Boolean function generator and a programmable waveform generator, each based on a multiplexer?

---

## 10.6

## Demultiplexers

---

**Demultiplexer**  *A circuit that uses a binary decoder to direct a digital signal from a single source to one of several destinations.*

---

A **demultiplexer** performs the reverse function of a multiplexer. A multiplexer (MUX) directs one of several input signals to a single output; a demultiplexer (DMUX) directs a single input signal to one of several outputs. In both cases, the selected input or output is chosen by the state of an internal decoder.

In Example 10.2, we examined the use of a decoder in a demultiplexer application. Figure 10.3 showed a circuit that directs a stream of data to a digital audio tape recorder, a videotape recorder, or one of three digital signal processors, depending on the states of a decoder.

Each AND gate in the demultiplexer enables or inhibits the signal output according to the state of the select inputs, thus directing the data to one of the output lines. For instance, $CBA = 010$ directs incoming digital data to channel 2, the video channel.

Example 10.2 implies that any decoder with an *ENABLE* input can also be used as a demultiplexer. When such a decoder is used as a DMUX, we feed the incoming data to what would be the decoder *ENABLE*.

Three TTL devices are commonly used as both demultiplexers and decoders. These are the 74LS138 3-Line to 8-Line, 74LS139A Dual 2-Line to 4-Line, and 74154 4-Line to 16-Line decoder/demultiplexers. These demultiplexers are all available in High-Speed CMOS versions: 74HC138, 74HC139A, and 74HC154. (When used as demultiplexers, it is clearer to refer to these devices as 1-to-4, 1-to-8, or 1-to-16 line devices, since they each take a single data input and distribute it to several outputs.)

Figure 10.62 illustrates the use of one half of a 74LS139A device as both decoder and demultiplexer. In Figure 10.62a, $\overline{G}$ is tied LOW. When an output is selected by $B$ and $A$, it goes LOW, acting as a decoder with active-LOW outputs. In Figure 10.62b, $\overline{G}$ acts as a demultiplexer data input. The data are directed to the output selected by $B$ and $A$. Since both $\overline{G}$ and the outputs are active LOW, the device transmits the data in true form.

### Demultiplexing a TDM Signal

In Example 10.22, we looked at a multiplexer used to send eight digital speech channels across a single telephone line. Obviously, such a system is not of much value if the signals cannot be sorted out at the receiving end. The received speech data must be demultiplexed and sent to their appropriate destinations.

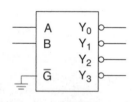

**a. Decoder**

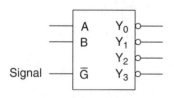

**b. Demultiplexer**

**Figure 10.62**

**74HC139A as Decoder or Demultiplexer.**

The process is just the reverse of multiplexing; data are sent to an output selected by a counter at the DMUX select inputs. (We assume that the counters at the MUX and DMUX select inputs are somehow synchronized or possibly, if they are located close together, are the same counter.)

---

**EXAMPLE 10.26**

Draw a demultiplexing circuit that will take the multiplexed output of the circuit in Figure 10.56 and distribute it to eight different local telephone circuits.

**Solution**

Since we have eight different channels, we should use a 74LS138 1-to-8 demultiplexer.

There are three gating (*ENABLE*) inputs, designated *G1, $\overline{G2A}$,* and *$\overline{G2B}$*. One of these can be used as the data input; the other two must be tied to their active levels to enable the circuit.

Since the outputs of the 74LS138 are active LOW, either $\overline{G2A}$ or $\overline{G2B}$ should be used as the data input, since the inversions at input and output would then cancel. The select inputs are driven by a 3-bit counter synchronized with the MUX counter in Figure 10.56. The required circuit is shown in Figure 10.63.

**Figure 10.63**

**Example 10.26
Time-Division Demultiplexer**

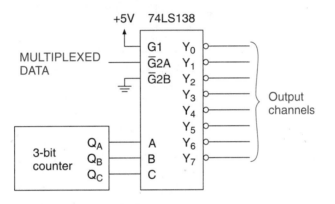

---

## CMOS Analog Multiplexer/Demultiplexer

> **CMOS transmission gate** *A CMOS device that will pass an analog or digital signal in either direction, when enabled. Also called an analog switch. There is no TTL equivalent.*

An interesting device used in some CMOS MSI multiplexers and demultiplexers, as well as in other applications, is the **CMOS transmission gate** or analog switch. This device has the property, unknown to TTL, of allowing signals to pass in two directions instead of only one. There is no "forward" direction for this device.

Figure 10.64 shows several symbols indicating the development of the transmission gate concept. Figure 10.64a and b show amplifiers whose output and input are clearly defined by the direction of the triangular amplifier symbol. A signal has one possible direction of flow. Figure 10.64b includes an active-LOW gating input, which can turn the signal on and off.

**Figure 10.64**
**Line Drivers**

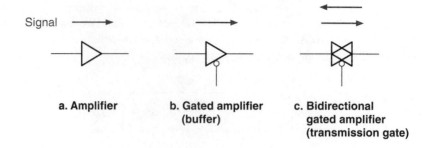

a. Amplifier

b. Gated amplifier
(buffer)

c. Bidirectional
gated amplifier
(transmission gate)

**Figure 10.65**

**4-Channel CMOS**
**MUX/DMUX**

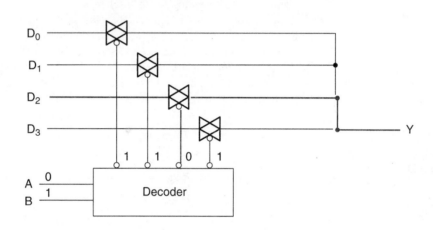

Figure 10.64c shows two opposite-direction overlapping amplifier symbols, with a gating input to enable or inhibit the bidirectional signal flow. The signal through the transmission gate can be either analog or digital.

Several CMOS MUX/DMUX chips use these transmission gates to send signals in either direction. Figure 10.65 illustrates the design principle as applied to a 4-channel MUX/DMUX.

If four signals are to be multiplexed, they are connected to inputs $D_0$ to $D_3$. The decoder, activated by $B$ and $A$, selects which one of the four transmission gates is enabled. Figure 10.65 shows channel 2 active ($BA = 10$).

Since all transmission gate outputs are connected together, any selected channel connects to $Y$, resulting in a multiplexed output. To use the circuit in Figure 10.65 as a demultiplexer, the inputs and outputs are merely reversed.

▼

**EXAMPLE 10.27**
A CMOS 4097B Dual 8-channel MUX/DMUX can be used simultaneously as a multiplexer on one half of the device and as a demultiplexer on the other side.

A circuit in a recording studio uses one side of a 4097B MUX/DMUX to multiplex eight digital audio channels into a digital signal processor, using time-division multiplexing. The other half of the 4097B takes the processed signals from a DSP output and distributes them to eight channels on a digital audio tape unit. Draw the circuit.

(We assume that a DSP is fast enough to keep up with eight audio channels in sequence without losing fidelity. This is a reasonable assumption, given the operating speed of today's processors.)

**Figure 10.66**

**Example 10.27**
**4097B MUX/DMUX as a**
**Time-Division MUX/DMUX**

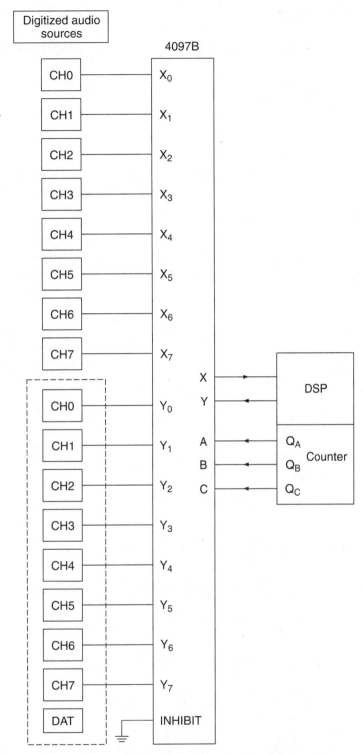

**Solution**    Figure 10.66 shows a possible circuit. The counter can be part of the DSP. An audio source channel is selected by the counter inputs, and data are sent to the DSP, where they are processed and sent to the same channel of the DAT. The counter advances by 1, selecting a new channel and repeating the process.

## 10.7

# Combinational Logic Applications of PLDs

Programmable logic devices (PLDs) lend themselves easily to the implementation of combinational logic functions. As they are in the same range of package size as many MSI devices, we can program a PLD to replace the function of one or more standard MSI devices.

Plain economic facts tell us that it is not worthwhile to replace a standard MSI device with a PLD. Most MSI chips cost between 50 cents and a dollar. The least expensive PLDs are about $2; the more complex devices in windowed CERDIP packages can be quite expensive ($15 and up). PLDs become economically viable when they replace many chips and perform custom functions, particularly in high-end computer-based systems.

The PLD examples shown in this section and later chapters are intended to give you some idea of the possibilities of PLDs. It is up to you to apply some creative thinking to your own design problems to which PLDs might be a solution.

---

**EXAMPLE 10.28**

Program the function of an 8-bit priority encoder into an 85C220 EPLD. Use the truth table of the 4532B encoder (Table 10.8) as a basis for this design.

**Solution**    File name: (\MSI\HIPRI8.PDS) The JEDEC file for this device is found in \MSI\HIPRI8.JED on the instructor's diskette.

```
Title       8-bit Priority Encoder
Pattern     pds
Revision    1.0
Author      R.Dueck
Company     Seneca College
Date        November 27, 1992

CHIP hipri8 85C220

;     This design duplicates the CMOS 4532B 8-bit priority
;  encoder.
;
; Pin List:
;
PIN 1 D0    ; Priority Inputs
PIN 2 D1
PIN 3 D2
PIN 4 D3
PIN 5 D4
PIN 6 D5
PIN 7 D6
PIN 8 D7

PIN 19 Q2   ; Output Code = Q2 Q1 Q0
PIN 18 Q1   ; Output Code is the binary equivalent of the
PIN 17 Q0   ; highest-priority active input.

PIN 11 Ein  ; Enable Input (for cascading)
PIN 12 Eout ; Enable Output (for cascading)
PIN 15 GS   ; Group Select (for cascading)
            ;
            ;     Cascade function:  A valid nonzero output code is
            ; produced only when Ein=1.  If no priority input is
```

```
;    active, Eout=1, enabling a lower priority device
;    if Eout(MSB) is connected to Ein(LSB).
;       Group Select activates on a valid nonzero
;    priority if Ein=1.
;
```

| T_TAB ( | D7 | D6 | D5 | D4 | D3 | D2 | D1 | D0 | Ein | >> | Eout | GS | Q2 | Q1 | Q0 ) |
|---------|----|----|----|----|----|----|----|----|-----|----|------|----|----|----|-------|
|         | 0  | 0  | 0  | 0  | 0  | 0  | 0  | 0  | 1   | :  | 1    | 0  | 0  | 0  | 0     |
|         | 0  | 0  | 0  | 0  | 0  | 0  | 0  | 1  | 1   | :  | 0    | 1  | 0  | 0  | 0     |
|         | 0  | 0  | 0  | 0  | 0  | 0  | 1  | X  | 1   | :  | 0    | 1  | 0  | 0  | 1     |
|         | 0  | 0  | 0  | 0  | 0  | 1  | X  | X  | 1   | :  | 0    | 1  | 0  | 1  | 0     |
|         | 0  | 0  | 0  | 0  | 1  | X  | X  | X  | 1   | :  | 0    | 1  | 0  | 1  | 1     |
|         | 0  | 0  | 0  | 1  | X  | X  | X  | X  | 1   | :  | 0    | 1  | 1  | 0  | 0     |
|         | 0  | 0  | 1  | X  | X  | X  | X  | X  | 1   | :  | 0    | 1  | 1  | 0  | 1     |
|         | 0  | 1  | X  | X  | X  | X  | X  | X  | 1   | :  | 0    | 1  | 1  | 1  | 0     |
|         | 1  | X  | X  | X  | X  | X  | X  | X  | 1   | :  | 0    | 1  | 1  | 1  | 1     |
|         | X  | X  | X  | X  | X  | X  | X  | X  | 0   | :  | 0    | 0  | 0  | 0  | 0     |

SIMULATION

```
VECTOR CODE := [Q2 Q1 Q0]
VECTOR HIPRI := [D7 D6 D5 D4 D3 D2 D1 D0]

SETF /Ein

FOR i := 0 TO 1 DO
BEGIN
     SETF HIPRI:= 0
     SETF HIPRI:= 1
     SETF HIPRI:= 3
     SETF HIPRI:= 7
     SETF HIPRI:= 0xF
     SETF HIPRI:= 0x1F
     SETF HIPRI:= 0x3F
     SETF HIPRI:= 0x7F
     SETF HIPRI:= 0xFF
     SETF Ein
END
```

Pin Diagram:

```
              85C220
         - - - - -
    D0 -| 1    20|- Vcc
    D1 -| 2    19|- Q2
    D2 -| 3    18|- Q1
    D3 -| 4    17|- Q0
    D4 -| 5    16|- Gnd
    D5 -| 6    15|- GS
    D6 -| 7    14|- Gnd
    D7 -| 8    13|- Gnd
   Gnd -| 9    12|- EOUT
   GND -|10    11|- EIN
         - - - - -
```

A PLD can be programmed to implement a function that would normally take several SSI/MSI chips. The next few examples will show how we can combine in one circuit a number of the MSI devices we have studied and how this circuit can be replaced by a single PLD.

**EXAMPLE 10.29**    A 74HC151 8-to-1 multiplexer is used in a recording studio to switch one of seven digital audio sources to a digital audio tape (DAT) recorder. Figure 10.67 shows a circuit to control the channel selection. Pushing one of seven interlocked pushbutton switches sets a latch, making its output $QS_n$ HIGH. The latched output stays HIGH until a different

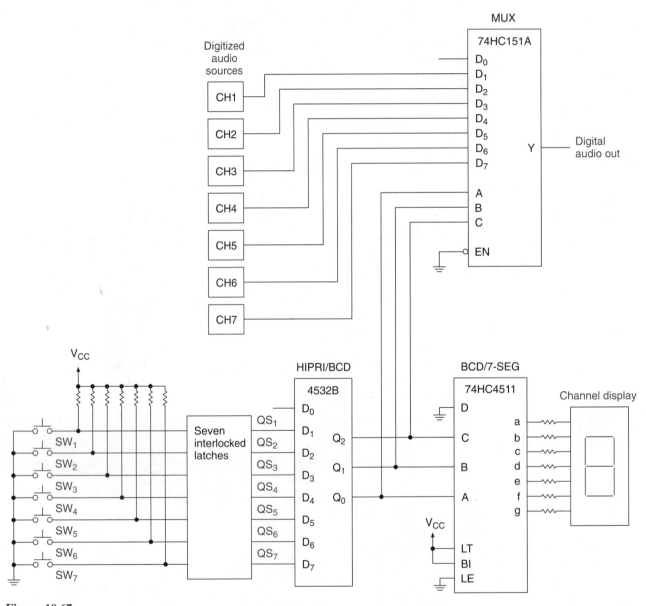

**Figure 10.67**

**Example 10.29**
**7-Channel Digital Audio Switch**

switch is pressed. Pressing a pushbutton also makes all other $QS$ outputs LOW by resetting all latches other than $QS_n$.

Briefly explain the circuit operation.

**Solution**    The interlocked latches produce one and only one HIGH output at a time. The 4532B priority encoder converts this HIGH output to a 3-bit code corresponding to the subscript of the active encoder input. This selects a MUX channel and also displays a single-digit channel number. The 74HC4511 BCD-to-Seven-Segment Decoder converts the 4532B output code to a numerical pattern and drives the channel display. Its lamp test, blanking input, and latch enable functions are all disabled.

Figure 10.68 shows four interlocked latches. Each latch is set by one of the four switches and reset by any one of the three remaining ones. The circuit allows us to make any one line HIGH by pressing a momentary-contact pushbutton.

**EXAMPLE 10.30**    Draw a circuit showing one of seven interlocked latches, using commercially available High-Speed CMOS gates. Calculate the number of IC packages needed to make the seven latches. (You may wish to refer to Table 4.9, Available SSI Gates, to select the required parts.)

**Solution**    Figure 10.69 (see page 437) illustrates the output latched by switch $SW_1$. The latch is set when $SW_1$ is pressed and reset when any other switch is pressed.

The latch itself is constructed from two NAND gates of a 74HC00 Quad 2-input NAND IC. The *RESET* circuit is made from two gates of a 74HC10 Triple 3-input NAND and one gate of a 74HC02 Quad 2-input NOR.

Table 10.16 (see page 437) lists the number of gates used for one latch, for seven latches, and the total number of IC packages required.

**EXAMPLE 10.31**    List the total number of IC packages required to make the seven-channel digital audio switch of Example 10.29.

**Solution**

| Device | Description | Packages |
|---|---|---|
| 74HC00 | Quad 2-input NAND | 4 |
| 74HC02 | Quad 2-input NOR | 2 |
| 74HC10 | Triple 3-input NAND | 5 |
| 74HC151A | 8-to-1 MUX | 1 |
| 74HC4511 | BCD/seven-segment decoder | 1 |
| 4532B | 8-bit priority encoder | 1 |
| | Total | 14 |

**Figure 10.68**
**Four Interlocked Latches**

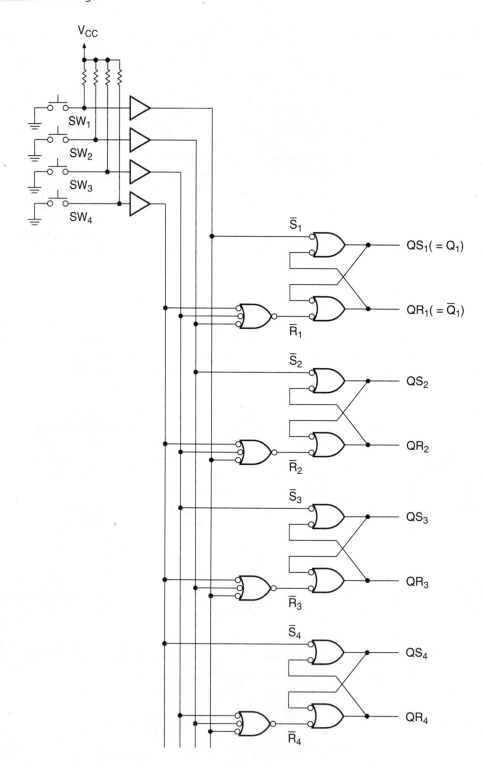

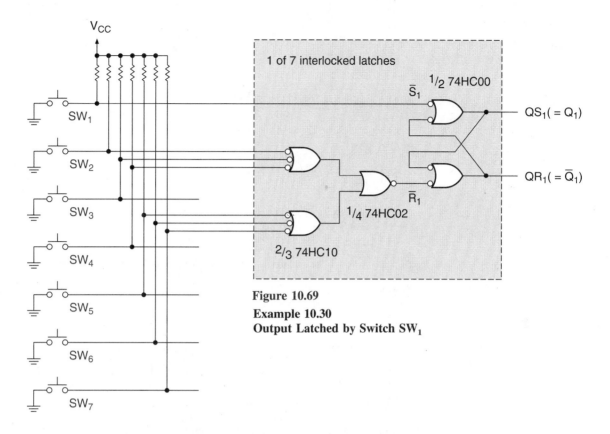

**Figure 10.69**

**Example 10.30**
**Output Latched by Switch SW$_1$**

**Table 10.16** Parts Required for Seven Interlocked Latches

|          | One Latch | Seven Latches | Gates per Package | Number of Packages |
|----------|-----------|---------------|-------------------|--------------------|
| 74HC00   | 2         | 14            | 4                 | 4                  |
| 74HC02   | 1         | 7             | 4                 | 2                  |
| 74HC10   | 2         | 14            | 3                 | 5                  |

**EXAMPLE 10.32**

Write a PLDasm source file that incorporates the function of the seven-channel digital audio switch of Example 10.29 in a single PLD. Use an 85C090 EPLD.

**Solution**

Before programming the EPLD, we must define a pair of Boolean equations to describe an SR latch. Rather than $Q$ and $\overline{Q}$, let us call the latch output $QS$ and $QR$.

As shown in Figure 10.70, we can develop two equations to represent the latch:

$$QS = \overline{S} + R \cdot QS$$

$$QR = \overline{R} + S \cdot QR$$

We do not need to program a priority encoder into the circuit; the latch outputs supply unique states that can be decoded directly to select a data channel and display the channel number.

The PLDasm source file (\MSI\AUDIO_7.PDS) is listed below. It is included, along with the device JEDEC file (\MSI\AUDIO_7.PDS), on the instructor's diskette.

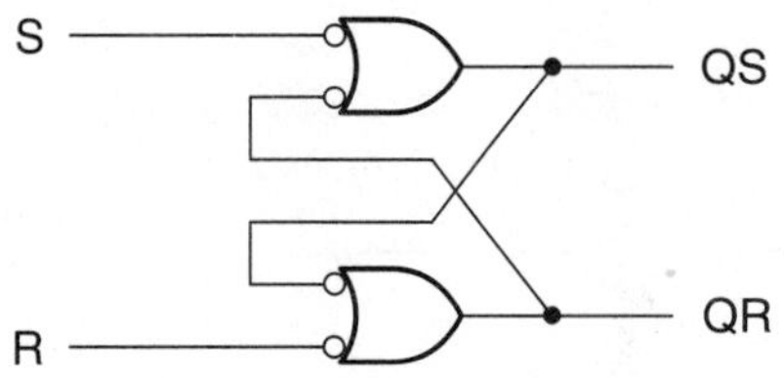

**Figure 10.70**

**Example 10.32**
**SR Latch Equations**

$$QS = \overline{S} + \overline{QR}$$
$$QR = \overline{R} + \overline{QS}$$

$$QS = \overline{S} + (\overline{\overline{R} + \overline{QS}})$$
$$= \overline{S} + R \cdot QS$$

$$QR = \overline{R} + (\overline{\overline{S} + \overline{QR}})$$
$$= \overline{R} + S \cdot QR$$

```
Title      7-channel digital audio switch and display driver
Pattern    pds
Revision   1.0
Author     R.Dueck
Company    Seneca College
Date       November 30, 1992

;     This circuit has seven momentary pushbutton inputs which
; make an input LOW to select one of seven digital data
; channels.  Each switch is latched asynchronously. The switch
; resets all other latches when pressed so that only one channel
; is active at any time.
;     There is also a BCD-to-seven-segment decoder to display
; the selected channel.  The decoder outputs are disabled when
; any switch is pressed to suppress an asynchronous "race
; condition" that otherwise occurs.
;     The circuit is 7-channel since the 85C090 EPLD has 12 inputs
; and 24 I/O macrocells.  We need:
;
;          7 momentary switch inputs
;          7 data inputs
;
;          14 latch outputs (Set and Reset for each switch)
;          1 data output
;          7 display decoder outputs
;
;          Total: 14 inputs and 22 outputs
;

CHIP audio_7 85C090


PIN  2 SW1        ; Momentary switch inputs
PIN  3 SW2
PIN  4 SW3
PIN 16 SW4
PIN 17 SW5
PIN 18 SW6
PIN 19 SW7

PIN  5 QS1        ; Latching outputs
PIN 36 QR1
PIN  6 QS2
PIN 35 QR2
PIN  7 QS3
```

```
PIN 34 QR3
PIN  8 QS4
PIN 33 QR4
PIN  9 QS5
PIN 32 QR5
PIN 10 QS6
PIN 31 QR6
PIN 11 QS7
PIN 30 QR7


PIN 39 D1              ; Digital Audio Inputs
PIN 38 D2
PIN 37 D3
PIN 25 D4
PIN 24 D5
PIN 23 D6
PIN 22 D7

PIN 26 Y              ; Digital Audio Out

PIN 12 a              ; Seven-segment Display Driver outputs
PIN 13 b
PIN 14 c
PIN 15 d
PIN 29 e
PIN 28 f
PIN 27 g

     ; Any switch pressed makes all other latches Reset.

STRING R1 '( SW2*SW3*SW4*SW5*SW6*SW7 )'
STRING R2 '( SW1*SW3*SW4*SW5*SW6*SW7 )'
STRING R3 '( SW1*SW2*SW4*SW5*SW6*SW7 )'
STRING R4 '( SW1*SW2*SW3*SW5*SW6*SW7 )'
STRING R5 '( SW1*SW2*SW3*SW4*SW6*SW7 )'
STRING R6 '( SW1*SW2*SW3*SW4*SW5*SW7 )'
STRING R7 '( SW1*SW2*SW3*SW4*SW5*SW6 )'

STRING RALL '(SW1*SW2*SW3*SW4*SW5*SW6*SW7 )'


EQUATIONS

QS1 = /SW1 + R1*QS1        ; Latch Set and Reset equations
QR1 = /R1 + SW1*QR1

QS2 = /SW2 + R2*QS2
QR2 = /R2 + SW2*QR2

QS3 = /SW3 + R3*QS3
QR3 = /R3 + SW3*QR3

QS4 = /SW4 + R4*QS4
QR4 = /R4 + SW4*QR4

QS5 = /SW5 + R5*QS5
QR5 = /R5 + SW5*QR5

QS6 = /SW6 + R6*QS6
QR6 = /R6 + SW6*QR6

QS7 = /SW7 + R7*QS7
QR7 = /R7 + SW7*QR7
```

```
; Digital Data Selection

Y = D1*QS1 + D2*QS2 + D3*QS3 + D4*QS4 + D5*QS5 + D6*QS6 + D7*QS7


; Output Decoder

a.trst = RALL          ;      Disable decoder outputs when
b.trst = RALL          ; any channel select switch is pressed.
c.trst = RALL          ; (ie. Reset ALL)
d.trst = RALL          ;      These equations MUST come before
e.trst = RALL          ; the truth table to keep them within
f.trst = RALL          ; the EQUATIONS section of the file.
g.trst = RALL


T_TAB ( QS1 QS2 QS3 QS4 QS5 QS6 QS7 >> a b c d e f g )

         1   0   0   0   0   0   0  : 0 1 1 0 0 0 0
         0   1   0   0   0   0   0  : 1 1 0 1 1 0 1
         0   0   1   0   0   0   0  : 1 1 1 1 0 0 1
         0   0   0   1   0   0   0  : 0 1 1 0 0 1 1
         0   0   0   0   1   0   0  : 1 0 1 1 0 1 1
         0   0   0   0   0   1   0  : 1 0 1 1 1 1 1
         0   0   0   0   0   0   1  : 1 1 1 0 0 0 0



SIMULATION

VECTOR DATA_IN := [ D7, D6, D5, D4, D3, D2, D1]
VECTOR SEV_SEG := [ a, b, c, d, e, f, g]
VECTOR SWITCHES := [SW7, SW6, SW5, SW4, SW3, SW2, SW1]

; All waveforms are included in the file AUDIO_7.HST
; The waveforms for SW1 to SW7, the decoder hex value,
; and the MUX output Y only are included in the file
; AUDIO_7.TRF.

TRACE_ON  SW1 SW2 SW3 SW4 SW5 SW6 SW7 SEV_SEG Y

SETF SWITCHES := 0x7F     ; Set all switch inputs HIGH.

FOR i:= 0 TO 0x40 DO
;
;     This loop sets one and only one data input HIGH each time
; the "IF" condition is true.  Each time this occurs, it
; activates each channel in turn to see which MUX input is HIGH.
; On the first pass, Channel 1 has the HIGH data.  On the second
; pass, Channel 2 is HIGH, and so on until all MUX inputs have
; been tested.
;
BEGIN
    IF ((i=1)+(i=2)+(i=4)+(i=8)+(i=0x10)+(i=0x20)+(i=0x40)) THEN
      BEGIN
            SETF DATA_IN := i
            SETF SWITCHES := 0x7E
            SETF SWITCHES := 0x7F
            SETF SWITCHES := 0x7D
            SETF SWITCHES := 0x7F
            SETF SWITCHES := 0x7B
            SETF SWITCHES := 0x7F
            SETF SWITCHES := 0x77
            SETF SWITCHES := 0x7F
            SETF SWITCHES := 0x6F
            SETF SWITCHES := 0x7F
```

```
        SETF SWITCHES := 0x5F
        SETF SWITCHES := 0x7F
        SETF SWITCHES := 0x3F
        SETF SWITCHES := 0x7F
    END
END

TRACE_OFF


Pin Diagram:

             85C090
          - - - - -
   Gnd -| 1      40|- Vcc
   SW1 -| 2      39|- D1
   SW2 -| 3      38|- D2
   SW3 -| 4      37|- D3
   QS1 -| 5      36|- QR1
   QS2 -| 6      35|- QR2
   QS3 -| 7      34|- QR3
   QS4 -| 8      33|- QR4
   QS5 -| 9      32|- QR5
   QS6 -|10      31|- QR6
   QS7 -|11      30|- QR7
     A -|12      29|- E
     B -|13      28|- F
     C -|14      27|- G
     D -|15      26|- Y
   SW4 -|16      25|- D4
   SW5 -|17      24|- D5
   SW6 -|18      23|- D6
   SW7 -|19      22|- D7
   GND -|20      21|- Gnd
          - - - - -
```

## GLOSSARY

**CMOS transmission gate**   A CMOS device that will pass an analog or digital signal in either direction, when enabled. Also called an analog switch. There is no TTL equivalent.

**Common-anode display**   A seven-segment LED display in which the anodes of all the LEDs are connected to the circuit supply voltage. Each segment is illuminated by a logic LOW at its cathode.

**Common-cathode display**   A seven-segment LED display in which the cathodes of all LEDs are connected together and grounded. A logic HIGH illuminates a segment when applied to its anode.

**Counter**   A sequential digital circuit whose output produces a fixed sequence of binary states when the circuit is clocked (e.g., the output state of a 4-bit binary counter progresses in order from 0000 to 1111).

**Data inputs**   Those inputs to a multiplexer that feed a digital signal to the output when selected.

**Decoder**   A digital circuit designed to detect the presence of a particular digital state.

**Demultiplexer**   A circuit that uses a binary decoder to direct a digital signal from a single source to one of several destinations.

**Digital signal processor (DSP)**   A specialized microprocessor used to modify digital signals, such as audio or video data, for various special purposes. For instance, a DSP can process an audio signal to introduce reverberation or a video signal to generate graphic images.

**Encoder**   A circuit that produces a digital code at its outputs in response to one or more active input lines.

**EVEN parity**   An error-checking system that requires a binary number to have an even number of 1s.

**Magnitude comparator** A circuit that compares two $n$-bit binary numbers, indicates whether or not the numbers are equal or which one is larger.

**MSI** Medium-scale integration. A single-chip logic device whose equivalent function could be performed by a logic gate circuit consisting of 12 to 100 gates.

**Multiplexer** A circuit that directs one of several digital signals to a single output, depending on the states of several select inputs.

**ODD parity** An error-checking system that requires a binary number to have an odd number of 1s.

**Parity** A system that checks for errors in a multibit binary number by counting the number of 1s.

**Parity bit** A bit appended to a binary number to make the number of 1s in the number even or odd, depending on the type of parity.

**Priority encoder** An encoder that will produce a binary output corresponding to the subscript of the highest-

priority active input. This is usually defined as the input with the largest subscript.

**Ripple blanking** A technique used in a multiple-digit numerical display that suppresses leading or trailing 0s in the display, but leaves internal 0s.

**Select inputs** The multiplexer inputs that select a digital input channel.

**Seven-segment display** An array of seven independently controlled light-emitting diode (LED) or liquid crystal display (LCD) elements, shaped like a figure-8, which can be used to display decimal digits and other characters by turning on the appropriate elements.

**Time-division multiplexing** A technique of using one transmission line to send many signals simultaneously by making them share the line for equal fractions of time.

## ● PROBLEMS

### Section 10.1 Decoders

**10.1** When a HIGH is on the outputs of each of the decoding circuits shown in Figure 10.71, what is the binary code appearing at the inputs? Write the Boolean expression for each decoder output. $D$ is the most significant bit.

**10.2** Draw a decoder circuit for each of the following Boolean expressions:

   **a.** $\overline{Y} = \overline{D}\,C\,\overline{B}\,A$

   **b.** $\overline{Y} = \overline{D}\,\overline{C}\,B\,A$

   **c.** $Y = \overline{D}\,\overline{C}\,B\,A$

   **d.** $\overline{Y} = D\,C\,\overline{B}\,\overline{A}$

   **e.** $Y = D\,C\,\overline{B}\,A$

**10.3** Sketch the logic diagram of a 2-to-4 line decoder with active-HIGH outputs and an active-LOW *ENABLE* input.

**10.4** Sketch the logic diagram of a 3-to-8 line decoder with active-HIGH outputs and two *ENABLE* inputs, one active HIGH and the other active LOW.

**10.5** For a generalized $n$-to-$m$ line decoder, state the value of $m$ if $n$ is:

   **a.** 5

   **b.** 6

   **c.** 8

Write the equation giving the general relation between $n$ and $m$.

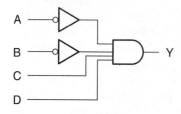
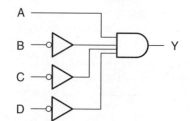
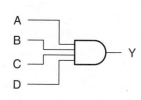

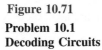

**Figure 10.71**

**Problem 10.1**
**Decoding Circuits**

**Figure 10.72**

**Problem 10.6**
**Input Waveforms**

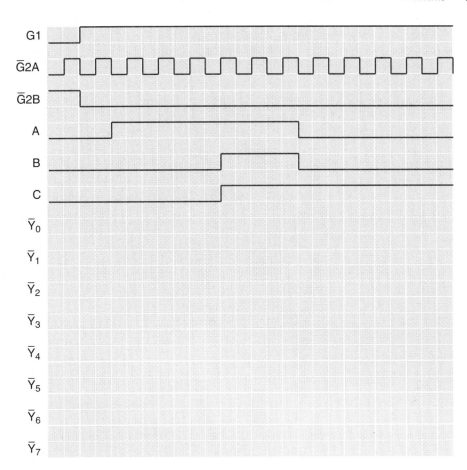

10.6     A 74HC138A 3-to-8 Line Decoder has the waveforms of Figure 10.72 applied to its inputs. Complete the timing diagram by sketching the output waveforms.

10.7     Repeat Problem 10.6 for the waveforms shown in Figure 10.73 (see page 444).

*10.8    Sketch a logic diagram showing how to connect the two 2-to-4 line decoders in a 74HC139 device and any other required logic to make a 3-to-8 line decoder. (Hint: Use $\overline{G}$ as the third input.)

*10.9    A microcomputer system has a RAM capacity of 2 megabytes, split into 256-KB portions. Each RAM device is enabled by a LOW at a $\overline{G}$ input. Draw a logic diagram that shows how a 74HC138 3-to-8 Line Decoder can select one particular RAM device.

*10.10   Repeat Problem 10.9, using a 74HC139A Dual 2-to-4 Line Decoder instead of the 74HC138 decoder.

10.11    Write the truth table for a BCD-to-seven-segment decoder for a common-anode display.

10.12    Use the truth table derived in Problem 10.11 to derive Boolean equations for each segment driver. Simplify the expressions as much as possible, using any convenient method.

*10.13   Write the complete list of segment driver functions for a 74LS48 decoder.

10.14    Draw a circuit consisting of four common-cathode seven-segment displays, each driven by a 74LS47 decoder/driver. The circuit should be configured to suppress all leading zeros. Show the displayed digits and $\overline{RBO}/\overline{RBI}$ logic levels for each of the following inputs: 100, 217, 1024.

**Figure 10.73**

**Problem 10.7**
**Input Waveforms**

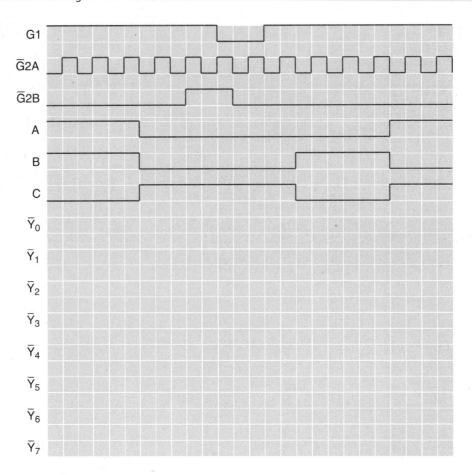

## Section 10.2 Encoders

**10.15** Figure 10.74 shows a BCD priority encoder with three different sets of inputs. Determine the resulting output code for each input combination. Inputs and outputs are active HIGH.

**\*\*10.16** Derive the Boolean equations for the output of a BCD priority encoder, based on the priority encoding principle stated in Section 10.2.

**\*10.17** Use the 16-bit priority encoder derived in Example 10.8 to design a hexadecimal keypad encoder. The hex keypad is similar to the decimal keypad in Example 10.9.

**10.18** Refer to Figure 10.24, the 16-bit priority encoder built from two 4532B 8-bit encoders. Table 10.17 shows the output codes produced by the circuit when each input is at the highest priority (e.g., when $D_0$ is the highest priority,

$Q_3Q_2Q_1Q_0 = 1000$; this, of course, is an error). State why the codes are not correct and list the most likely sources of the errors.

**10.19** Refer to the 16-bit priority encoder circuit in Figure 10.24. The circuit is tested as follows: All $D$ inputs are connected to logic switches, and all switches are made HIGH. $D_{15}$ is switched LOW, then $D_{14}$, and so on in sequence until all inputs are LOW. Table 10.18 shows the resulting codes. What is the most likely source of the errors?

**\*\*10.20** Draw a circuit showing how to connect four 4532B priority encoders to make a 32-bit priority encoder. Prove that the circuit works properly by briefly explaining the input and output behavior of the circuit for each of the following inputs:

**Figure 10.74**

**Problem 10.15**
**BCD Priority Encoder**

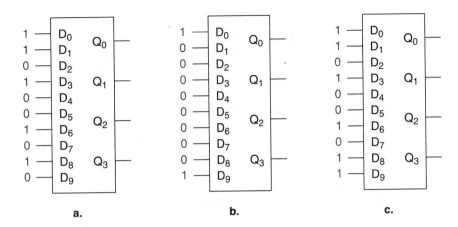

**a.**      **b.**      **c.**

**Table 10.17** Truth Table for Problem 10.18

| Highest-Priority Input | Output Code | | | |
|---|---|---|---|---|
| | $Q_3$ | $Q_2$ | $Q_1$ | $Q_0$ |
| $D_0$ | 1 | 0 | 0 | 0 |
| $D_1$ | 1 | 0 | 0 | 1 |
| $D_2$ | 1 | 0 | 1 | 0 |
| $D_3$ | 1 | 0 | 1 | 1 |
| $D_4$ | 1 | 1 | 0 | 0 |
| $D_5$ | 1 | 1 | 0 | 1 |
| $D_6$ | 1 | 1 | 1 | 0 |
| $D_7$ | 1 | 1 | 1 | 1 |
| $D_8$ | 1 | 0 | 0 | 0 |
| $D_9$ | 1 | 0 | 0 | 1 |
| $D_{10}$ | 1 | 0 | 1 | 0 |
| $D_{11}$ | 1 | 0 | 1 | 1 |
| $D_{12}$ | 1 | 1 | 0 | 0 |
| $D_{13}$ | 1 | 1 | 0 | 1 |
| $D_{14}$ | 1 | 1 | 1 | 0 |
| $D_{15}$ | 1 | 1 | 1 | 1 |

**Table 10.18** Truth Table for Problem 10.19

| Highest-Priority Input | Output Code | | | |
|---|---|---|---|---|
| | $Q_3$ | $Q_2$ | $Q_1$ | $Q_0$ |
| $D_{15}$ | 1 | 1 | 1 | 1 |
| $D_{14}$ | 1 | 1 | 1 | 0 |
| $D_{13}$ | 1 | 1 | 0 | 1 |
| $D_{12}$ | 1 | 1 | 0 | 0 |
| $D_{11}$ | 1 | 1 | 0 | 0 |
| $D_{10}$ | 1 | 1 | 0 | 0 |
| $D_9$ | 1 | 1 | 0 | 0 |
| $D_8$ | 1 | 1 | 0 | 0 |
| $D_7$ | 1 | 1 | 0 | 0 |
| $D_6$ | 1 | 1 | 0 | 0 |
| $D_5$ | 1 | 1 | 0 | 0 |
| $D_4$ | 1 | 1 | 0 | 0 |
| $D_3$ | 0 | 0 | 1 | 1 |
| $D_2$ | 0 | 0 | 1 | 0 |
| $D_1$ | 0 | 0 | 0 | 1 |
| $D_0$ | 0 | 0 | 0 | 0 |

**a.** $D_{24}$ active

**b.** $D_{23}$ active

**c.** $D_{16}$ active

**d.** $D_{15}$ active

**e.** $D_8$ active

**f.** $D_7$ active

**\*\*10.21** Repeat Problem 10.20 for a 24-bit priority encoder, using three 4532B priority encoders. Explain the operation for cases c through f in Problem 10.20.

### Section 10.3 Magnitude Comparators

**10.22** Briefly explain the operation of the $A < B$ portion of the 2-bit magnitude comparator shown in Figure 10.30.

**10.23** Draw the $A < B$ portion of a 4-bit magnitude comparator and briefly explain its operation.

**10.24** Draw the circuit for a 3-bit magnitude comparator that has outputs for $A = B$, $A > B$, and $A < B$ functions.

**10.25**   Write the Boolean expressions for the $(A = B)$, $(A < B)$, and $(A > B)$ outputs of a 6-bit magnitude comparator.

**10.26**   Figure 10.75 shows an 8-bit magnitude comparator with logic levels specified for all inputs. State the logic levels for the $(A = B)$, $(A < B)$, and $(A > B)$ outputs.

**10.27**   Repeat Problem 10.26 for the comparator circuit in Figure 10.76.

**10.28**   Figure 10.77 shows a 74LS85 4-bit magnitude comparator with its input logic levels specified. State the logic levels for the $(A = B)$, $(A < B)$, and $(A > B)$ outputs. What information does the diagram tell us about any previous stages of a cascaded comparator?

**10.29**   Repeat Problem 10.28 for the comparator shown in Figure 10.78 (see page 448).

**10.30**   Repeat Problem 10.28 for the comparator shown in Figure 10.79 (see page 448).

**\*10.31**   The digital thermometer in Figure 10.34 is modified to give a 6-bit output. Show how to connect two 7485 magnitude comparators to compare the two 6-bit outputs.

## Section 10.4 Parity Generators and Checkers

**10.32**   What parity bit, $P$, should be added to the following data if the parity is EVEN? If the parity is ODD?

   **a.**  1111100

   **b.**  1010110

   **c.**  0001101

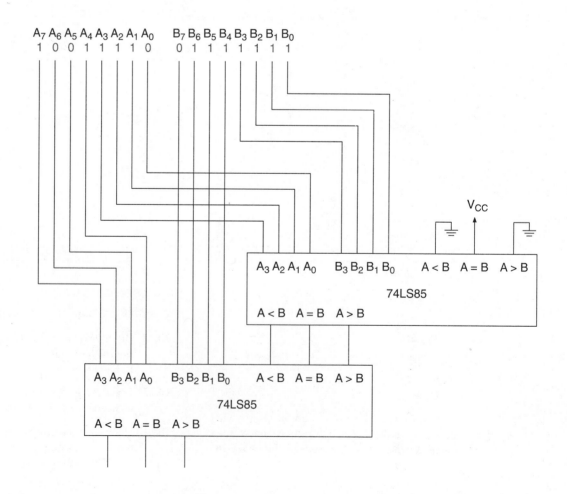

**Figure 10.75**

**Problem 10.26**
**8-Bit Magnitude**
**Comparator**

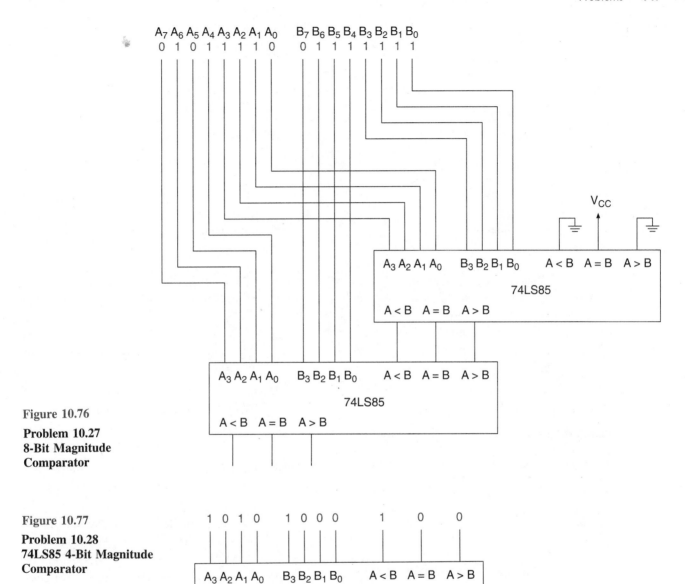

**Figure 10.76**

**Problem 10.27**
**8-Bit Magnitude**
**Comparator**

**Figure 10.77**

**Problem 10.28**
**74LS85 4-Bit Magnitude**
**Comparator**

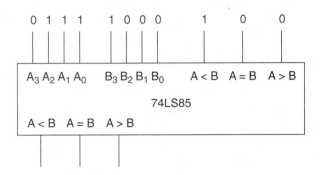

**Figure 10.78**
**Problem 10.29**
**74LS85 4-Bit Magnitude Comparator**

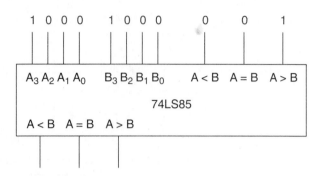

**Figure 10.79**
**Problem 10.30**
**74LS85 4-Bit Magnitude Comparator**

**10.33** The following data are transmitted in a serial communication system ($P$ is the parity bit). What parity is being used in each case?

**a.** $ABCDEFGHP = 010000101$

**b.** $ABCDEFGHP = 011000101$

**c.** $ABCDP = 01101$

**d.** $ABCDEP = 101011$

**e.** $ABCDEP = 111011$

**10.34** The data $ABCDEFGHP = 110001100$ are transmitted in a serial communication system. Give the output $P'$ of a receiver parity checker for the following received data. State the meaning of the output $P'$ for each case.

**a.** $ABCDEFGHP = 110101100$

**b.** $ABCDEFGHP = 110001101$

**c.** $ABCDEFGHP = 110001100$

**d.** $ABCDEFGHP = 110010100$

**10.35** Draw the logic circuit of a 5-bit parity generator with a switchable *EVEN/ODD* output.

**10.36** Draw the 5-bit parity checker corresponding to the parity generator in Problem 10.35.

**\*10.37** Draw a circuit that will produce EVEN and ODD parity bits from 12 data bits, using 74180 parity generator/checkers.

## Section 10.5 Multiplexers

**10.38** Make a table listing which digital audio source in Figure 10.80 is routed to output $Y$ for each combination of the multiplexer select inputs, $C$, $B$, and $A$. (CD = compact disc player; DAT = digital audio tape deck.)

**10.39** Draw the internal logic circuit for the 74LS151A 8-to-1 line multiplexer in Figure 10.80.

**10.40** Implement the logic function represented by Table 10.19, using a 74LS150 16-to-1 line multiplexer.

**10.41** Implement the logic function of Table 10.19 with a 74HC151 8-to-1 multiplexer.

**10.42** Implement the logic function represented by Table 10.20, using a 74150 16-to-1 multiplexer.

**10.43** Implement the logic function of Table 10.20 with a 74HC151 8-to-1 multiplexer.

**\*10.44** Modify the circuit in Figure 10.56 so that it can transmit 16 digitized speech channels over the same phone line. Use a second multiplexer, a 4-bit counter, and any other required logic.

**10.45** Draw the circuit of a programmable waveform generator based on a 74HC151 8-to-1 multiplexer.

**10.46** Draw the timing diagram of the select inputs and the output waveform of a MUX-based waveform generator with the following data inputs.

$$D_0 \qquad D_7$$
$$01100101$$

**10.47** A MUX-based waveform generator has the following data inputs.

$$D_0 \qquad D_7$$
$$01010101$$

These settings yield a symmetrical waveform with a frequency of 12 kHz. List the data input settings for the same circuit if the output is to have a symmetrical 6-kHz output.

**Figure 10.80**

**Problem 10.38**
**Digital Audio Multiplexer**

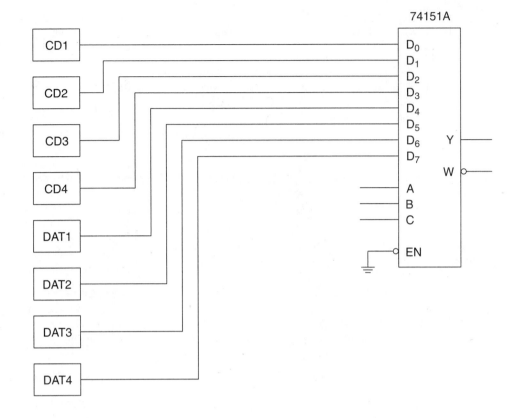

**Table 10.19** Truth Table for
Problem 10.40

| D | C | B | A | Y |
|---|---|---|---|---|
| 0 | 0 | 0 | 0 | 0 |
| 0 | 0 | 0 | 1 | 1 |
| 0 | 0 | 1 | 0 | 1 |
| 0 | 0 | 1 | 1 | 0 |
| 0 | 1 | 0 | 0 | 0 |
| 0 | 1 | 0 | 1 | 1 |
| 0 | 1 | 1 | 0 | 1 |
| 0 | 1 | 1 | 1 | 1 |
| 1 | 0 | 0 | 0 | 1 |
| 1 | 0 | 0 | 1 | 0 |
| 1 | 0 | 1 | 0 | 1 |
| 1 | 0 | 1 | 1 | 0 |
| 1 | 1 | 0 | 0 | 1 |
| 1 | 1 | 0 | 1 | 1 |
| 1 | 1 | 1 | 0 | 0 |
| 1 | 1 | 1 | 1 | 1 |

**Table 10.20** Truth Table for
Problem 10.42

| D | C | B | A | Y |
|---|---|---|---|---|
| 0 | 0 | 0 | 0 | 1 |
| 0 | 0 | 0 | 1 | 1 |
| 0 | 0 | 1 | 0 | 0 |
| 0 | 0 | 1 | 1 | 0 |
| 0 | 1 | 0 | 0 | 1 |
| 0 | 1 | 0 | 1 | 1 |
| 0 | 1 | 1 | 0 | 1 |
| 0 | 1 | 1 | 1 | 1 |
| 1 | 0 | 0 | 0 | 1 |
| 1 | 0 | 0 | 1 | 0 |
| 1 | 0 | 1 | 0 | 0 |
| 1 | 0 | 1 | 1 | 0 |
| 1 | 1 | 0 | 0 | 1 |
| 1 | 1 | 0 | 1 | 0 |
| 1 | 1 | 1 | 0 | 0 |
| 1 | 1 | 1 | 1 | 1 |

**10.48** Repeat Problem 10.47 if the waveform generator is to have a symmetrical 3-kHz output.

**10.49** Modify the serial-to-parallel circuit of Figure 10.60 to use a 74LS150 16-to-1 multiplexer. Show all input logic levels required to transmit an ASCII "m" and an 8-bit HIGH (idle state) output. Draw a timing diagram.

---

Problems 10.50 and 10.51 use a 74LS157 Quadruple 2-to-1 Multiplexer as well as one or more 74LS173A Quadruple D Flip-Flops. The '173A consists of four D flip-flops with common clock and clear functions as well as input and output enable functions. The data at the $D$ inputs are latched into all four flip-flops on the rising edge of the clock if $M$ and $N$ are both LOW. These data are presented to the $Q$ outputs if $G1$ and $G2$ are both LOW.

---

**10.50** Figure 10.81 shows a 74LS157 multiplexer that switches one of two 4-bit numbers to the inputs of a 74LS173A Quad Flip-Flop. The waveforms

shown in Figure 10.82 are applied to the circuit. Complete the timing diagram in Figure 10.82 by drawing the waveforms for the MUX outputs and the flip-flop outputs.

**\*10.51** Figure 10.83 (see page 452) shows a system that transfers a 4-bit number to the output of a 74LS157 multiplexer manually, from four data switches, or automatically, from a data bus via the output of a '173A quad flip-flop. This MUX output is latched into another '173A flip-flop and decoded to display the number as a decimal digit on a seven-segment display.

The waveforms of Figure 10.84 are applied to the circuit inputs. Complete the timing diagram by drawing the waveforms for the MUX $B$ inputs, MUX outputs, and '173A flip-flop (latch) outputs.

**10.52** List in order the decimal digits displayed on the seven-segment display of Figure 10.83 when the waveforms of Figure 10.84 (see page 453) are applied.

### Section 10.6 Demultiplexers

**10.53** Draw a logic gate circuit that will perform the same function as a 74HC139A 1-to-4 Line Demultiplexer. Briefly describe the function of the circuit.

**Figure 10.81**

**Problem 10.50**

**74LS157 Multiplexer**

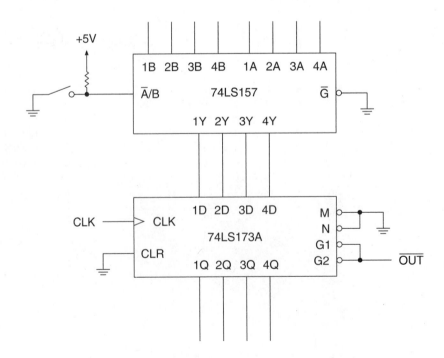

**Figure 10.82**

**Problem 10.50**
**Input Waveforms**

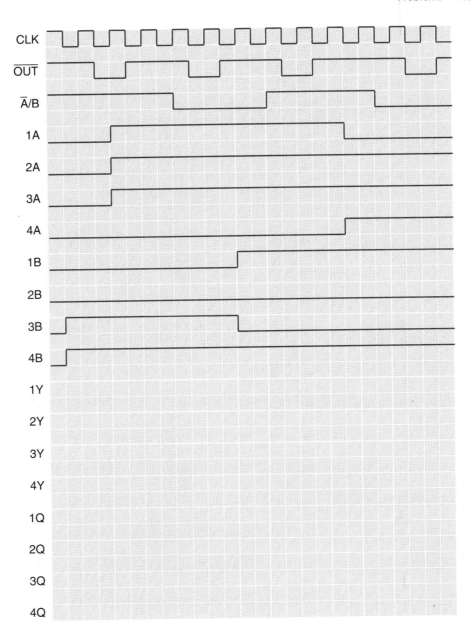

**10.54** Repeat Problem 10.53 for a 74HC138 1-to-8 Line Demultiplexer.

**\*10.55** Draw a circuit with a 74154 1-to-16 Line Demultiplexer and a 4-bit binary counter that will demultiplex the 16 speech channels multiplexed by the circuit in Problem 10.44.

**10.56** Draw a circuit showing how a 4097B Dual 8-channel MUX/DMUX can be used to multiplex eight transmitted digital audio channels onto a phone line and demultiplex eight received audio channels from another phone line.

**\*\*10.57** Draw a timing diagram of the audio outputs of each transmitted channel in the MUX/DMUX system described in Problem 10.56 and how they appear in sequence on the phone line. Also draw a timing diagram of the received audio channels showing how they come into the circuit in sequence and how they are distributed to their destination lines in sequence.

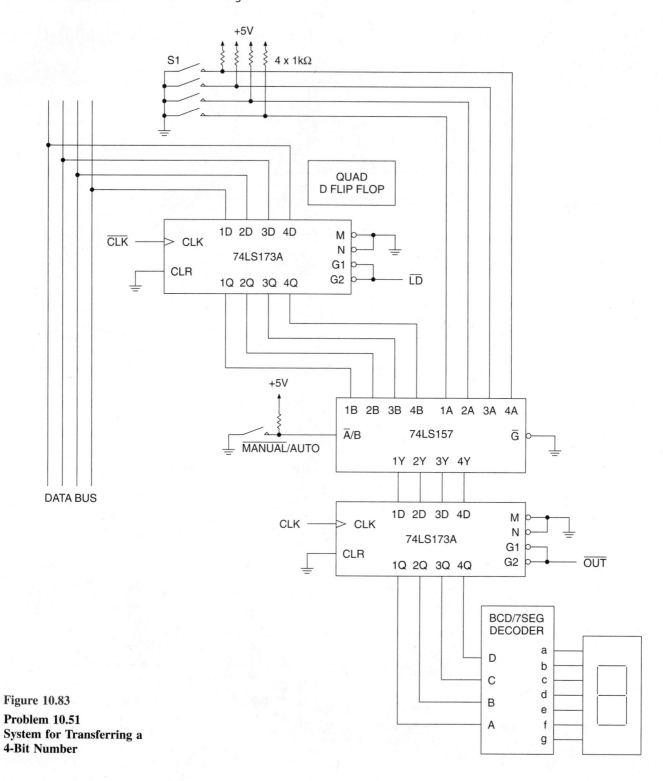

**Figure 10.83**

**Problem 10.51**
**System for Transferring a**
**4-Bit Number**

**Figure 10.84**

**Problem 10.51**
**Input Waveforms**

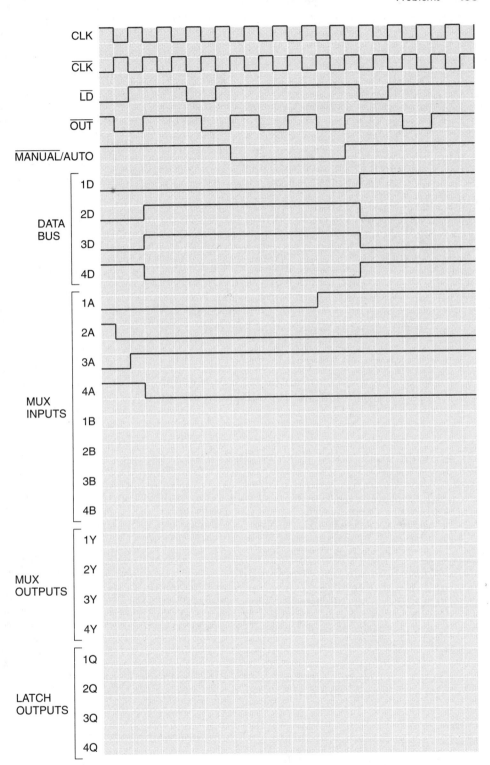

## Section 10.7 Combinational Logic Applications of PLDs

**10.58**  Write a PLDasm source file that programs an 85C220 EPLD as a hexadecimal-to-seven-segment decoder with an output for a decimal point segment. Label the hex inputs *DCBA* and the seven segment outputs *aout, bout, cout, dout, eout, fout,* and *gout*. The decimal point input is *DP* and the decimal point output is *dpout*. All inputs and outputs are active HIGH.

**10.59**  Write a PLDasm source file to program an 85CO90 EPLD as a 3½-digit hexadecimal-to-seven-segment decoder. The half digit is off when its corresponding input is LOW and displays a numeral "1" when its input is HIGH.

Label the hex inputs as *D1, C1, B1,* and *A1* and the segment outputs as *ao1, bo1, co1, do1, eo1, fo1,* and *go1* for the first digit and similarly for the second and third digits. Label the input for the half digit as *A4* and its outputs as *bo4* and *co4*.

**10.60**  Modify the PLDasm source file developed in Example 10.28 to make a BCD priority encoder

with active-HIGH outputs. The cascading outputs are the same as in Example 10.28.

**10.61**  Write a PLDasm source file to program an 85C22V10 EPLD as an 8-bit equality comparator. (No provision for $(A < B)$ or $(A > B)$ functions is required.)

**\*\*10.62**  Write a PLDasm source file that programs an 85C22V10 EPLD as a 7-bit comparator with $(A = B)$, $(A < B)$, $(A > B)$, $(A \leq B)$, and $(A \geq B)$ functions. (Hint: All other functions can be derived from $(A = B)$ and $(A < B)$ functions. However, $(A < B)$ must be partitioned into several smaller functions because of the large number of product terms.)

**10.63**  Write a PLDasm source file to program an 85C22V10 EPLD as an 8-channel digital MUX/DMUX. There should be eight MUX inputs and a MUX output, a DMUX input and eight DMUX outputs, three select inputs that select the same channel of the MUX and DMUX, and an active-HIGH tristate *ENABLE* input.

## ▼ Answers to Section Review Questions

### Section 10.1

**10.1.** Trailing zeros could logically be suppressed after a decimal point or if there are digits displaying a power-of-10 exponent (e.g., 455. or 4.55 02), that is, if they are nonsignificant. The zeros should be displayed if they set the location of the decimal point (e.g., 450).

### Section 10.2

**10.2.** $\overline{D} = \overline{D}_9 + \overline{D}_8$; $\overline{C} = D_9 D_8 \overline{D}_7 + D_9 D_8 \overline{D}_6 + D_9 D_8 \overline{D}_5 + D_9 D_8 \overline{D}_4$ (Factor out $D_9 D_8$ from the expression for $\overline{C}$ and compare the rest of the expression to that for $C$ in the 4532B encoder. Recall that the inputs and outputs for the 4532B encoder are active HIGH.)

### Section 10.3

**10.3.** $A < B$ in the previous stages (i.e., in the least significant bits). However, since $A_2 > B_2$ in the stage shown (i.e.,

in the most significant stage) and all other bits in this stage are equal, $A > B$ overall.

Outputs: $(A > B) = 1$, $(A = B) = 0$, $(A < B) = 0$.

### Section 10.4

**10.4.** Parts a and c are incorrect for sure because each has an even number of 1s. Parts b, d, and e could have an even number of errors, which is undetectable by parity checking.

### Section 10.5

**10.5.** The output of the waveform generator is a function of time as well as of the logic states at the MUX data inputs. The output of the Boolean function generator is a function of the data inputs only. The time element is introduced into the waveform generator by a binary counter at its select inputs.

# Introduction to Counters

● CHAPTER CONTENTS

**11.1**  Digital Counters
**11.2**  Asynchronous Counters

**11.3**  Decoding a Counter
**11.4**  Synchronous Counters

● CHAPTER OBJECTIVES

Upon successful completion of this chapter, you will be able to:

- Define and calculate the modulus of a counter circuit.

- Determine, from its modulus, the number of outputs required by a counter.

- Draw the count-sequence table, state diagram, and timing diagram of a counter.

- Determine the recycle point of a counter's sequence.

- Calculate the frequencies of each counter output, given the input clock frequency.

- Draw the circuit of a binary asynchronous counter, having either an UP or a DOWN sequence, using positive or negative edge-triggered JK flip-flops.

- Draw the circuit of a truncated-sequence asynchronous counter.

- Decode the output of a counter and use a decoder for various applications, such as event sequencing.

- Explain the origin of glitches in asynchronous counter and decoder outputs, determine where they occur in a timing diagram, and make circuit modifications to eliminate them.

- Explain the differences between asynchronous and synchronous counters.

- Draw a circuit for any full-sequence synchronous UP or DOWN counter.

- Draw the circuit for a bidirectional full-sequence synchronous counter.

- Determine the count sequence, state diagram, timing diagram, and modulus of any synchronous counter.

- Complete the state diagram of a synchronous counter to account for unused states.

Counters are digital sequential circuits that generate a fixed or programmable sequence of binary states at their outputs. We saw some applications of counters in conjunction with various MSI devices in Chapter 10, but we did not examine the circuits of the counters themselves.

Digital counters are based on clocked flip-flops, usually JK flip-flops. They are configured either as synchronous counters or as asynchronous counters, meaning that the flip-flops they consist of are clocked either from the same source or independently. Counters can count forward or backward or, in some cases, in a nonlinear sequence.

## 11.1

# Digital Counters

**Counter** *A sequential digital circuit whose output produces a fixed sequence of binary states when the circuit is clocked.*

**Count sequence** *The specific series of output states through which a counter progresses.*

**Recycle** *Make a transition from the last state of a count sequence to the first state.*

The simplest definition of a **counter** is "a circuit that counts pulses." This is not the most general or most accurate definition, but it does give a reasonably good understanding of the function of a counter. Knowing no more than this very simple definition, let us look at two examples of how we might use a counter.

**EXAMPLE 11.1**

The manager of a supermarket wants to know how many people visit her store during each hour of the business day. She assumes that it won't be much more than about 1000 per hour. She has a friend who is pretty good with electronics, who installs a monitoring system that counts the number of people entering the store. Once an hour the system updates a computer disk file and starts counting again from 0. (The store has only one entrance and a separate exit, so it's easy to monitor the traffic passing through one door.) Explain how the monitoring system could use a digital counter to keep track of the number of customers entering the supermarket.

**Solution**  Figure 11.1 shows a possible solution to the problem. An optical sensor generates a pulse for each person passing through the store entrance. The binary value of the counter output increases by 1 for every pulse at its *CLOCK* input. Once every hour, a computer stores the counter value and then clears the counter by pulsing an active-LOW Clear input. The counter then starts over from 0.

The counter represents the number of people entering the supermarket as a binary number. Since no more than about 1000 people are expected to visit the store in any

**Figure 11.1**

**Example 11.1**
**Customer Traffic Counter**

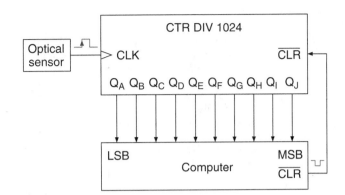

given hour, the counter should have 10 bits. ($2^{10}$ = 1024; this is indicated by the designation CTR DIV 1024.)

If this estimate is too low, the counter will "roll over" or **recycle** after the output exceeds 1023 (= 11 1111 1111$_2$), much as the odometer in an automobile will roll over after the mileage exceeds the number of digits it can display.

By convention, $Q_A$ is the least significant bit of a counter, for reasons that will become apparent later in the chapter. If a counter circuit has both $Q$ and $\overline{Q}$ outputs, the output value is the combined state of the $Q$ outputs, unless otherwise specified.

Example 11.1 showed how a counter can be used to count events or objects, such as a person passing an optical sensor. We can also use a counter to initiate a circuit action every time it counts a fixed number of pulses.

**EXAMPLE 11.2**

Example 6.9 showed how eight D flip-flops can be used to receive a serially transmitted ASCII character. Suppose this circuit receives a steady stream of ASCII characters. Show how we can use a counter to count the number of incoming serial bits and thus transfer one complete ASCII character in parallel to an octal D flip-flop and repeat this action for each new incoming character. (You might want to briefly review Example 6.9.)

**Solution**    Since there are 8 bits in an ASCII character (a start bit and a 7-bit ASCII character), we should count eight clock pulses before transferring a character to the octal flip-flop. (We are not counting the stop bit in this simplified case.)

Figure 11.2 shows a circuit that will accomplish this task. The counter has 3 bits and will count progressively from $Q_2\,Q_1\,Q_0$ = 000 to 111. This is a total of eight states, as shown by the designation CTR DIV 8.

The ASCII character is available in the D flip-flops after eight positive edges on the *CLK* input. After the eighth positive edge, the output is $Q_2\,Q_1\,Q_0$ = 000. This state enables the 4-input NOR gate, shown in its DeMorgan equivalent form. The NOR gate passes the *CLK* waveform in complement form to the *LOAD* terminal during this state only; otherwise *LOAD* = 0. This action produces one positive-going pulse *LOAD* waveform for every eight pulses on the *CLK* waveform. The positive-going *LOAD* pulse clocks the ASCII character into the octal flip-flops at a point halfway between positive edges of the *CLK* waveform.

**Figure 11.2**

**Example 11.2**
**Counter Used to Load an ASCII Character into an Octal Flip-Flop**

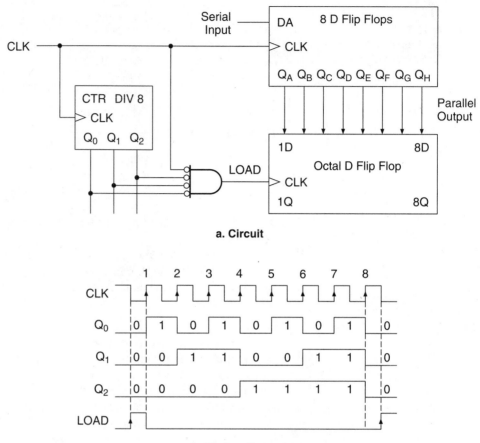

a. Circuit

b. Timing diagram

Since the counter is running continuously, it loads a new ASCII character into the octal flip-flop once every eight clock pulses.

There are numerous other applications of counter circuits, such as frequency division, frequency counting, and timekeeping. We will see some of these and other applications throughout the chapter. As a beginning, let us look at the underlying basic theory of digital counters.

## Basic Concepts of Digital Counters

**State diagram** *A diagram showing the progression of states of a sequential circuit.*

**Modulus** *The number of states through which a counter sequences before repeating.*

**Modulo-*n* (or mod-*n*) counter** *A counter with a modulus of* n.

**UP counter** *A counter with an ascending sequence.*

**DOWN counter** *A counter with a descending sequence*

A counter is a digital circuit that has a number of binary outputs whose states progress through a fixed sequence. This **count sequence** can be ascending, descending, or nonlinear.

The count sequence of a counter is usually defined by its **modulus,** that is, the number of states through which the counter progresses. An **UP counter** with a modulus of 12 counts through 12 states from 0000 up to 1011 (0 to 11 in decimal), recycles to 0000, and continues. A **DOWN counter** with a modulus of 12 counts from 1011 down to 0000, recycles to 1011, and continues downward. Both types of counter are called **modulo-12,** or just **mod-12** counters, since they both have sequences of 12 states.

### State Diagram

The states of a counter can be represented by a **state diagram.** Figure 11.3 compares the state diagram of a mod-12 UP counter to an analog clock face. Each counter state is illustrated in the state diagram by a circle containing its binary value. The progression is shown by a series of directional arrows.

Both the clock face and the state diagram represent a closed system of counting. In each case, when we reach the end of the count sequence, we start over from the beginning of the cycle.

For instance, if it is 10:00 a.m. and we want to meet a friend in four hours, we know we should turn up for the appointment at 2:00 p.m. We arrive at this figure by starting at 10 on the clock face and counting 4 digits forward in a "clockwise" circle. This takes us 2 digits past 12, the "recycle point" of the clock face.

Similarly, if we want to know the eighth state after 0111 in a mod-12 UP counter, we start at state 0111 and count eight positions in the direction of the arrows. This brings us to state 0000 (the recycle point) in five counts and then on to state 0011 in another three counts.

**Figure 11.3**

**Mod-12 State Diagram and Analog Clock Face**

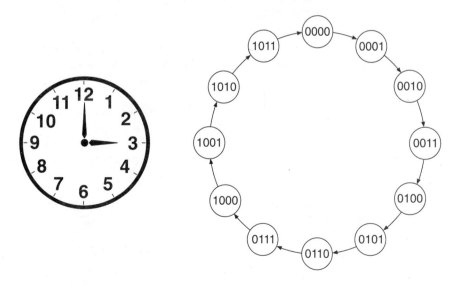

**Figure 11.4**
**State Diagram of a Mod-16 Counter**

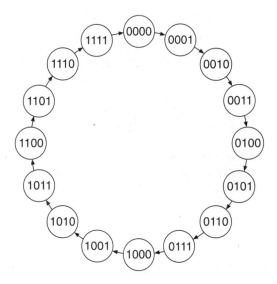

## Number of Bits and Maximum Modulus

**Maximum modulus ($m_{max}$)** *The largest number of counter states that can be represented by* n *bits.* ($m_{max} = 2^n$.)

**Full-sequence counter** *A counter whose modulus is the same as its maximum modulus.* (m = $2^n$ *for an* n-*bit counter.*)

**Binary counter** *A counter with a binary count sequence whose modulus is $2^n$ for an* n-*bit counter.*

**Truncated-sequence counter** *A counter whose modulus is less than its maximum modulus.* (m < $2^n$ *for an* n-*bit counter.*)

You have probably noticed that the states of the mod-12 count sequence each consist of 4 bits. Counter states are always written with a fixed number of bits, since each bit represents the logic level of a physical location in the counter circuit. We use 4 bits for a simple reason: we need that many to count from 0000 to 1011.

The **maximum modulus** of a 4-bit counter is 16 (= $2^4$). The count sequence of a mod-16 UP counter is from 0000 to 1111 (0 to 15 in decimal), as illustrated in the state diagram of Figure 11.4.

In general, an *n*-bit UP counter has a maximum modulus of $2^n$ and a count sequence from 0 to $2^n - 1$ (i.e., all 0s to all 1s). Since a mod-16 counter has a modulus of $2^n$ (= $m_{max}$), we say that it is a **full-sequence counter.** We can also call this a **binary counter** since it generates the complete *n*-bit binary sequence. A counter, such as a mod-12, with a modulus of less than $2^n$ is called a **truncated-sequence counter.**

## Count-Sequence Table and Timing Diagram

**Count-sequence table** *A list of counter states in the order of the count sequence.*

Two ways to represent a count sequence other than with a state diagram are by a **count-sequence table** and by a timing diagram. The count-sequence table is simply a list of counter states in the same order as the count sequence. Table 11.1 shows the count-sequence tables of a mod-16 UP counter and a mod-12 UP counter.

We can derive timing diagrams from each of these tables. We know that each counter advances by one state with each applied clock pulse. The mod-16 count sequence shows us that the $Q_A$ waveform changes state with each clock pulse. $Q_B$ changes with every two clock pulses, $Q_C$ with every four, and $Q_D$ with every eight. Figure 11.5 shows

**Table 11.1** Count-Sequence Tables for Mod-16 and Mod-12 UP Counters

| $Q_D$ | $Q_C$ | $Q_B$ | $Q_A$ | | $Q_D$ | $Q_C$ | $Q_B$ | $Q_A$ | |
|---|---|---|---|---|---|---|---|---|---|
| 0 | 0 | 0 | 0 | ← | 0 | 0 | 0 | 0 | ← |
| 0 | 0 | 0 | 1 | | 0 | 0 | 0 | 1 | |
| 0 | 0 | 1 | 0 | | 0 | 0 | 1 | 0 | |
| 0 | 0 | 1 | 1 | | 0 | 0 | 1 | 1 | |
| 0 | 1 | 0 | 0 | | 0 | 1 | 0 | 0 | |
| 0 | 1 | 0 | 1 | | 0 | 1 | 0 | 1 | |
| 0 | 1 | 1 | 0 | | 0 | 1 | 1 | 0 | |
| 0 | 1 | 1 | 1 | | 0 | 1 | 1 | 1 | |
| 1 | 0 | 0 | 0 | | 1 | 0 | 0 | 0 | |
| 1 | 0 | 0 | 1 | | 1 | 0 | 0 | 1 | |
| 1 | 0 | 1 | 0 | | 1 | 0 | 1 | 0 | |
| 1 | 0 | 1 | 1 | | 1 | 0 | 1 | 1 | |
| 1 | 1 | 0 | 0 | | | | | | |
| 1 | 1 | 0 | 1 | | | | | | |
| | | | | | **Mod-12** | | | | |
| 1 | 1 | 1 | 0 | | | | | | |
| 1 | 1 | 1 | 1 | | | | | | |
| | | **Mod-16** | | | | | | | |

**Figure 11.5**

**Timing Diagram of a Mod-16 UP Counter**

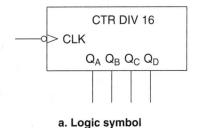

a. Logic symbol

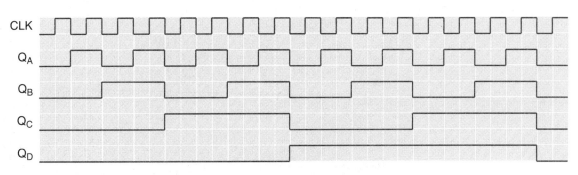

b. Timing diagram

**Figure 11.6**

**Timing Diagram of a Mod-12 UP Counter**

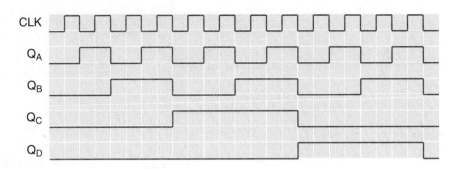

this pattern for the mod-16 UP counter, assuming that the counter is a negative edge-triggered device.

We can construct a similar timing diagram, illustrated in Figure 11.6, for a mod-12 UP counter. The changes of state can be monitored by noting where $Q_A$ (the least significant bit) changes. This occurs on each negative edge of the $CLK$ waveform. The sequence progresses by 1 with each $CLK$ pulse until the outputs all go to 0 on the first $CLK$ pulse after state $Q_D Q_C Q_B Q_A = 1011$.

**EXAMPLE 11.3**

Draw the state diagram, count-sequence table, and timing diagram for a mod-12 DOWN counter.

**Solution**

Figure 11.7 shows the state diagram for the mod-12 DOWN counter. The states are identical to those of a mod-12 UP counter, but they progress in the opposite direction. Table 11.2 shows the count-sequence table of this circuit.

The timing diagram of this counter is illustrated in Figure 11.8. The output starts in state $Q_D Q_C Q_B Q_A = 1011$ and counts DOWN until it reaches 0000. On the next pulse, it recycles to 1011 and starts over.

**Figure 11.7**

**Example 11.3 Timing Diagram of a Mod-12 DOWN Counter**

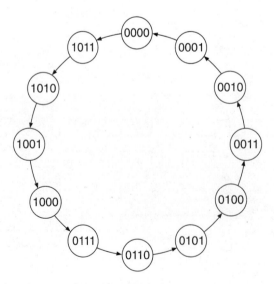

Table 11.2 Count Sequence of a Mod-12 DOWN Counter

| $Q_D$ | $Q_C$ | $Q_B$ | $Q_A$ |
|-------|-------|-------|-------|
| 1 | 0 | 1 | 1 |
| 1 | 0 | 1 | 0 |
| 1 | 0 | 0 | 1 |
| 1 | 0 | 0 | 0 |
| 0 | 1 | 1 | 1 |
| 0 | 1 | 1 | 0 |
| 0 | 1 | 0 | 1 |
| 0 | 1 | 0 | 0 |
| 0 | 0 | 1 | 1 |
| 0 | 0 | 1 | 0 |
| 0 | 0 | 0 | 1 |
| 0 | 0 | 0 | 0 |

**Figure 11.8**

**Example 11.3
Timing Diagram of a
Mod-12 DOWN Counter**

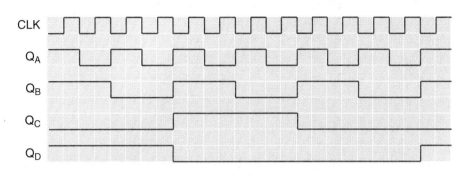

---

**Section Review Problem for Section 11.1**

11.1. How many outputs does a mod-24 counter require? Is this a full-sequence or a truncated-sequence counter? Explain your answer.

---

## 11.2

## Asynchronous Counters

**Asynchronous counter**  *A counter whose bits are not clocked by the same source and thus do not change in synchronization with one another.*

**Ripple counter**  *An asynchronous counter. More specifically, a counter where one bit is clocked by transitions on the next least significant bit.*

In Chapter 6 (Latches and Flip-Flops), we briefly examined a simple circuit called a mod-8 **asynchronous counter,** shown again in Figure 11.9. The circuit consists of three negative edge-triggered JK flip-flops, permanently configured for toggle mode, and a clock input. When a series of clock pulses is applied, the output $Q_C Q_B Q_A$ will advance in binary order from 000 to 111 and repeat, as shown in the timing diagram of Figure 11.9b and the state diagram of Figure 11.9c.

Recall that a JK flip-flop will toggle (change its output to the opposite state) with every applied *CLK* pulse if $J = K = 1$.

The *CLK* waveform in Figure 11.9b is the clock input to the first flip-flop. Every time *CLK* makes a HIGH-to-LOW transition, $Q_A$ toggles. The first flip-flop output completes one cycle for every two *CLK* cycles.

The $Q_A$ output is the clock input to the second flip-flop. Since the second flip-flop's *J* and *K* inputs are always HIGH, $Q_B$ will toggle every time $Q_A$ makes a HIGH-to-LOW transition. The second flip-flop completes one cycle for every two cycles of $Q_A$, or for every four *CLK* cycles. Similarly, $Q_C$ completes one cycle for every eight *CLK* cycles.

Theoretically, this type of counter can be expanded to as many bits as are required by adding a flip-flop for every bit and configuring it for toggle mode. The *CLK* input of

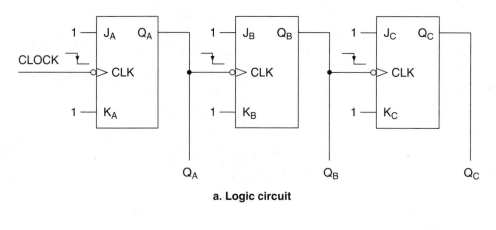

**a. Logic circuit**

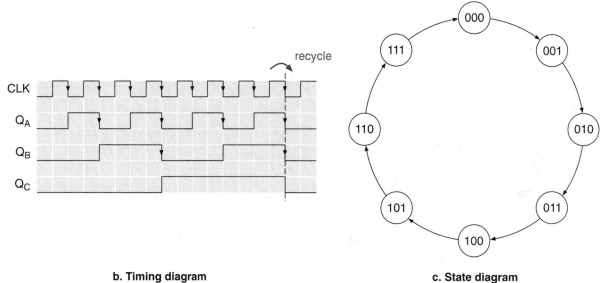

**b. Timing diagram**

**c. State diagram**

**Figure 11.9**

**Asynchronous Binary Counter**

each new flip-flop must also be connected to the output of the previous stage. In practice, propagation delays of the flip-flops limit the number of bits in the counter output.

This type of counter is sometimes called a **ripple counter** because output changes are not simultaneous but are delayed by a few nanoseconds through each flip-flop and appear to "ripple" through the counter.

The counter shown in Figure 11.9 has a modulus of 8. Any such counter without additional circuitry always has a modulus given by $m = 2^n$ for $n$ flip-flops. It is therefore a full-sequence counter, as the output will count to the maximum modulus.

**EXAMPLE 11.4**     Draw the logic circuit of a mod-32 asynchronous counter built from 74LS76A negative edge-triggered JK flip-flops.

**Solution**     Number of flip-flops required: 5 (since $2^5 = 32$).

The important features of the circuit, shown in Figure 11.10, are:

**Figure 11.10**

**Example 11.4**
**Mod-32 Asynchronous**
**Counter Circuit**

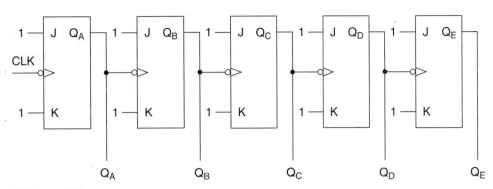

PRE and CLR inputs of all flip-flops (not shown) should be tied HIGH.

- Five flip-flops
- Flip-flops are negative edge-triggered
- Outputs are taken from $Q$ of each flip-flop
- Each $J$ and $K$ input is tied HIGH (toggle mode)
- Clock pulse is connected to the first flip-flop $CLK$ input
- $CLK$ inputs of the remaining flip-flops are connected to the $Q$ output of the previous stage.

(Not shown: $\overline{PRE}$ and $\overline{CLR}$ inputs, which must be disabled by tying them HIGH.)

---

> **Section Review Problem for Section 11.2a**
>
> 11.2. If a 10-bit asynchronous counter is clocked by a waveform with a frequency of 1024 kHz and a duty cycle of 25%, what are the frequency and duty cycle of the waveform at the most significant bit output?

## Asynchronous UP Counters and DOWN Counters

The asynchronous counters in Figures 11.9 and 11.10 are both UP counters due to three features of their construction:

1. The flip-flops are negative edge-triggered (as opposed to positive edge-triggered).
2. The counter outputs are taken from the flip-flop $Q$ outputs (as opposed to the $\overline{Q}$s).
3. The $Q$ (as opposed to the $\overline{Q}$) output of each flip-flop feeds the $CLK$ input of the next one.

If we change any one of these features, we will make the circuit into a DOWN counter. Change any two and we again have an UP counter. Change all three and we have a DOWN counter.

**EXAMPLE 11.5**

The circuit in Figure 11.11 is constructed from CMOS 4027B flip-flops (positive edge-triggered). Analyze the operation of the circuit by drawing its timing diagram. The initial state is $Q_C Q_B Q_A = 111$. Use the timing diagram to generate the count-sequence table. What is the pattern of the outputs?

**Figure 11.11**

**Example 11.5
Counter Circuit Using
Positive Edge-Triggered
Flip-Flops**

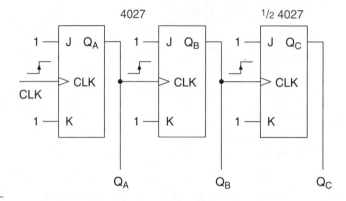

**Table 11.3** Count Sequence of Circuit in Figure 11.11

| $Q_C$ | $Q_B$ | $Q_A$ |
|-------|-------|-------|
| 1 | 1 | 1 |
| 1 | 1 | 0 |
| 1 | 0 | 1 |
| 1 | 0 | 0 |
| 0 | 1 | 1 |
| 0 | 1 | 0 |
| 0 | 0 | 1 |
| 0 | 0 | 0 |

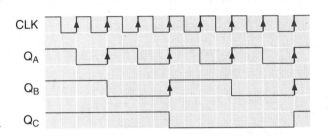

**Figure 11.12**

**Example 11.5
Timing Diagram of
Counter in Figure 11.11**

**Solution**

Figure 11.12 shows the timing diagram of the circuit, with the outputs initially at $Q_C Q_B Q_A = 111$. Since the flip-flops are positive edge-triggered, $Q_A$ changes on each positive *CLK* edge. Every positive edge at $Q_A$ makes $Q_B$ toggle. Similarly, $Q_C$ toggles on the positive edges of $Q_B$.

Table 11.3 shows the count sequence of the circuit as derived from its timing diagram. We can see from this table that the circuit in Figure 11.11 is a DOWN counter.

**EXAMPLE 11.6**

Analyze the circuit of Figure 11.13 by drawing its timing diagram. What pattern does the circuit output $\overline{Q}_C \overline{Q}_B \overline{Q}_A$ follow? All flip-flops are initially cleared.

**Solution**

The timing diagram of Figure 11.14 shows both the $Q$ and $\overline{Q}$ outputs of the JK flip-flops. The first state of the $Q$ outputs is $Q_C Q_B Q_A = 000$.

A negative edge on each $Q$ triggers the next flip-flop in the sequence, as it did with the asynchronous counters of Figures 11.9 and 11.10. These outputs count UP in binary sequence.

The $\overline{Q}$ outputs are shown on the last three lines of the diagram. We derive these waveforms by inverting those for the $Q$ outputs. The $\overline{Q}$ outputs start in the state 111 and count DOWN to 000, at which point they recycle to 111.

**Figure 11.13**

**Example 11.6**
**Counter With Output State Taken From $\overline{Q}$ Outputs**

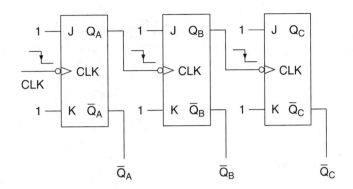

**Figure 11.14**

**Example 11.6**
**Timing Diagram of Counter in Figure 11.13**

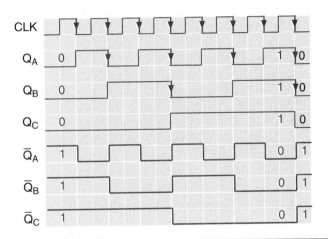

The principle for designing asynchronous UP or DOWN counters is based on the number of inversions between flip-flop stages. We can think of three places where a clock signal can be inverted:

1.  The output of the flip-flop that drives the *CLK* of the next stage
2.  The *CLK* input of the next flip-flop
3.  The flip-flop output that acts as the circuit output

> If there is an even number of inversions in the *Q-CLK-Q* path, the circuit is a DOWN counter. If there is an odd number of inversions, the circuit counts UP.

This principle is summarized by the diagrams in Figure 11.15. For example, the first asynchronous counter we examined (Figure 11.9) is like the top left diagram in Figure 11.15. It has one inversion in the *Q-CLK-Q* path: at the *CLK* input. Thus, the circuit is an UP counter (it has an odd number of inversions).

A circuit like that in the bottom left diagram in Figure 11.15 has two inversions between stages: the output of the driving flip-flop and the *CLK* input. This circuit counts DOWN.

**Figure 11.15**

**Q-CLK-Q Connections for UP or DOWN Counters**

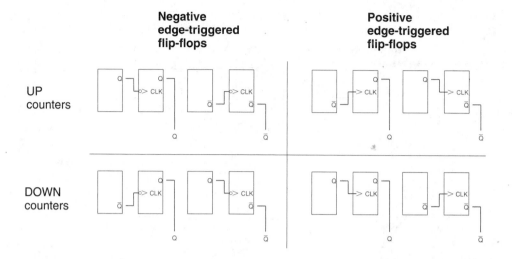

> **Section Review Problem for Section 11.2b**
>
> 11.3. Can the direction of a counter's sequence be reversed by inverting the *CLK* input to the first flip-flop? Why or why not? What effect will this inversion have on the circuit?

## Truncated-Sequence Counters

> **Recycle state** *A temporary state that is decoded by external logic to truncate a count sequence. The recycle state is not counted as part of the count sequence.*
>
> **Glitch** *An unwanted voltage spike that can cause erroneous switching in a digital circuit.*

Each of the asynchronous counter circuits examined so far has had a power-of-2 modulus; that is, each was a full-sequence counter. Many applications, such as BCD counting (mod-10) or digital clocks (mod-12), require a modulus other than a power of 2. To make such a circuit, we must find a way to truncate, or cut short, the count sequence before it reaches the maximum modulus, as specified by the number of flip-flops in the circuit.

A simple way to do this is to use the *CLR* inputs of the counter's JK flip-flops to set the $Q$ outputs to 0 at any desired ending point of the count sequence. A logic gate, such as a NAND or an AND gate, can be used to decode the last state in the count sequence and activate the flip-flops' clear function.

Figure 11.16 shows an asynchronous decade (mod-10) counter. A decade counter, which must count 0000 to 1001, requires four flip-flops. For a 4-bit counter, $m_{max} = 2^4 = 16$. The circuit in Figure 11.16 is a standard 4-bit asynchronous counter with an additional decoding gate to truncate the mod-16 sequence.

The outputs of the counter progress through the states $0000_2$ to $1001_2$ (= $0_{10}$ to $9_{10}$) normally. The 11th output state ($1010_2 = 10_{10}$) is the first time in the count sequence that $Q_B$ and $Q_D$ outputs are both HIGH at the same time. They make the NAND output,

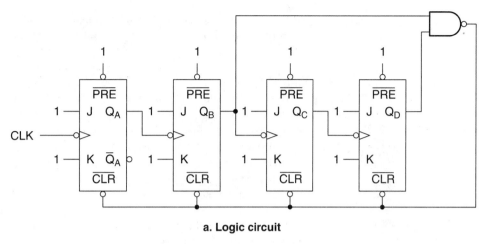

**a. Logic circuit**

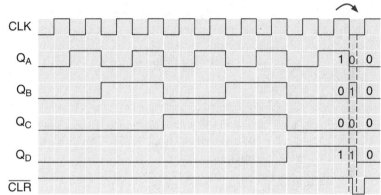

**b. Timing diagram (recycle point in exaggerated scale)**

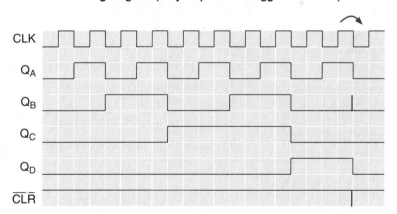

**c. Timing diagram (recycle shown as spike of $Q_B$ and $\overline{CLR}$)**

**Figure 11.16**

**Asynchronous Decade Counter**

and thus the $\overline{CLR}$ inputs, LOW, and all flip-flop outputs immediately go to 0. As soon as the flip-flops reset, the NAND output goes HIGH again (since the conditions making it LOW have disappeared) and the count resumes from $Q_D\,Q_C\,Q_B\,Q_A = 0000$.

The 11th state in the mod-10 sequence is called the **recycle state.** The counter output is in this state for a very short time, about 35 ns for LSTTL devices. We can calculate this time by adding the propagation delay times of the decoding gate ($t_{pHL} = 15$ ns) and the flip-flop ($t_{pHL}$, from $CLR$ to $Q = 20$ ns). Thus, the outputs are in the 11th state for so short a time that we don't count it. This state appears on an oscilloscope as a spike; it doesn't last long enough to show as a square pulse.

The timing diagrams in Figure 11.16b and c show the CLR pulse in two scales: in an exaggerated scale to show the operation of the Clear logic and in a more realistic scale that you would see on an oscilloscope. For many applications the **glitches** (voltage spikes) introduced on some of the outputs (e.g., on $Q_B$) are not a problem, although if they are intolerable, there are ways of eliminating them. We will look at a way to eliminate glitches in a later section.

We can set an UP counter to any modulus by decoding the binary value of its modulus. In Figure 11.16, the counter has a modulus of 10. The binary value of the desired modulus is $Q_D\,Q_C\,Q_B\,Q_A = 1010$. This number is decoded by an AND or a NAND gate with inputs fed from $Q_B$ and $Q_D$, the HIGH bits of the desired modulus.

**EXAMPLE 11.7**

Draw the logic circuit, count sequence, and timing diagram for a mod-11 asynchronous UP counter built from 4027B CMOS JK flip-flops (positive edge-triggered, active-HIGH Preset and Clear).

**Solution**

Count range: 0000 to 1010.

Number of flip-flops: 4.

$J = K = 1$ for all flip-flops.

Connect $\overline{Q}$ to $CLK$ (one inversion in $Q$-$CLK$-$Q$ path.)

Disable all Presets by grounding them (active-HIGH Preset).

Use a 4073B 3-input AND gate to reset counter.

AND gate inputs: $Q_D$, $Q_B$, and $Q_A$ (since the recycle state for a mod-11 counter is $Q_D\,Q_C\,Q_B\,Q_A = 1011$).

AND gate output: Connect to $CLR$ inputs of flip-flops.

All these design criteria are combined in the logic circuit shown in Figure 11.17a. The timing diagram for this counter is shown in Figure 11.17b. Note the following:

1. The AND gate output is normally LOW, but goes HIGH in the recycle state because the $CLR$ is active HIGH with this type of flip-flop.
2. There is a glitch at the recycle point on the $CLR$ line and also on the $Q_A$ line. The glitch is present on $Q_A$ because the logic level normally goes HIGH at this point, but is reset as soon as the recycle state is decoded and the flip-flop is cleared.
3. The other lines have no glitches, either because they are HIGH before the recycle state ($Q_B$ and $Q_D$) or because, although LOW, they don't normally change at this point in the timing diagram ($Q_C$ would normally change at pulse 12).

**Table 11.4** Count Sequence for a Mod-11 UP Counter

| $Q_D$ | $Q_C$ | $Q_B$ | $Q_A$ |
|---|---|---|---|
| 0 | 0 | 0 | 0 |
| 0 | 0 | 0 | 1 |
| 0 | 0 | 1 | 0 |
| 0 | 0 | 1 | 1 |
| 0 | 1 | 0 | 0 |
| 0 | 1 | 0 | 1 |
| 0 | 1 | 1 | 0 |
| 0 | 1 | 1 | 1 |
| 1 | 0 | 0 | 0 |
| 1 | 0 | 0 | 1 |
| 1 | 0 | 1 | 0 |
| ( 1 | 0 | 1 | 1 ) |

**Figure 11.17**

**Example 11.7
Mod-11 UP Counter**

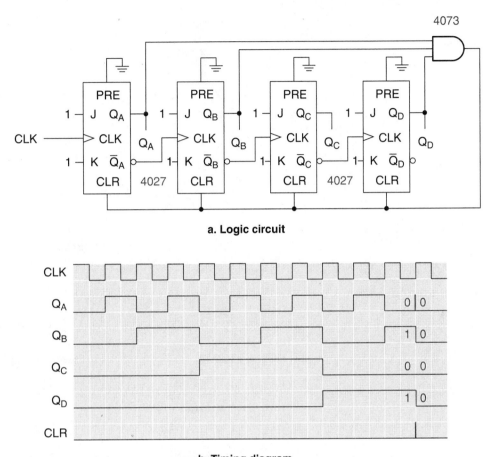

a. Logic circuit

b. Timing diagram

The count sequence for this circuit is shown in Table 11.4, with the recycle state in parentheses.

---

**EXAMPLE 11.8**

Draw the logic circuit and timing diagram of a DOWN counter that counts from 1010 to 0000, then recycles. Use 4027B CMOS JK flip-flops.

**Solution**

Figure 11.18a shows the logic circuit of the counter, and the timing diagram is in Figure 11.18b. The flip-flops are connected as a 4-bit asynchronous DOWN counter, with the $Q$ outputs of each flip-flop connected to the *CLK* input of the next one (there are no inversions in the *Q-CLK-Q* path). The flip-flops are permanently configured for toggle mode.

The normal sequence for this counter is from 1111 to 0000 in descending binary order. However, we want to truncate the sequence to eliminate the portion from 1111 to 1011, leaving only the sequence from 1010 to 0000.

If we use the asynchronous *CLR* inputs to truncate the sequence, we must detect when the counter state is $Q_D\, Q_C\, Q_B\, Q_A = 1111$, then clear $Q_C$ and $Q_A$. This is done by the 4-input AND gate, whose inputs are connected to all flip-flop outputs and whose output is connected to the *CLR* inputs of flip-flops *A* and *C*.

**Figure 11.18**

**Example 11.8**
**Mod-11 DOWN Counter**

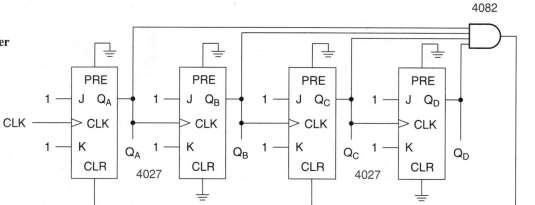

a. Logic circuit

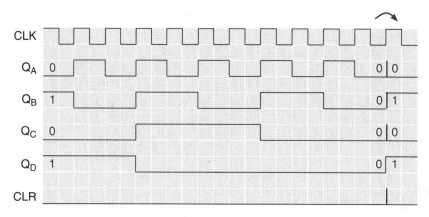

b. Timing diagram

---

**Section Review Problem for Section 11.2c**

11.4. A 4-bit asynchronous counter has a modulus of 12. On which $Q$ outputs, if any, do glitches appear at the recycle point?

## 11.3

# Decoding a Counter

**Decoder** *A digital circuit designed to detect the presence of a particular digital state or states.*

Sometimes an application requires a circuit that will detect a particular output state of a counter and initiate some circuit action. The simplest example of this is the NAND or

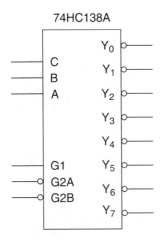

74HC138A

AND **decoder** gate in a truncated-sequence counter, which clears the counter's flip-flops upon detection of the recycle state.

A more generalized decoder, such as the 74HC138A 3-to-8 line decoder, described in Chapter 10, can also be used with a counter to sequence a series of circuit actions. The logic symbol of this device is shown in Figure 11.19.

The 74HC138A decoder has three data inputs and eight active-LOW outputs. Each binary input code *CBA* activates a different output.

When a decoder output becomes active, any circuit connected to it "knows" that a particular point in the count sequence has been reached and that it can take appropriate action.

**Figure 11.19**

**74HC138A 3-to-8 Line Decoder**

**EXAMPLE 11.9**

A common process that repeats sequentially is the firing of spark plugs in an internal combustion engine.

Each spark plug is assigned a number from 0 to 7. For the sake of the example, assume the spark plugs fire in sequence from 0 to 7. In an actual engine, the firing order would not be from lowest to highest number, but rather in some staggered order.

Assume that a LOW input to a spark plug interface will trigger a high-voltage circuit that fires the corresponding spark plug. This interface can be shown as a black box.

Design a digital circuit that could control the firing of spark plugs in an eight-cylinder engine, using a mod-8 counter and a 3-to-8 line decoder with active-LOW outputs.

Show the timing diagram of the counter and decoder outputs.

**Solution**

Figure 11.20 (see page 474) shows a mod-8 counter decoded by a 74HC138A and the counter/decoder timing diagram. The decoder has three enable inputs, *G1* (active HIGH), $\overline{G2A}$, and $\overline{G2B}$ (both active LOW), all of which must be active to allow any of the outputs to go LOW.

As shown in the timing diagram in Figure 11.20b, each decoder output goes LOW only when the binary value of the counter state is the same number as the subscript of the output. For example, when $Q_C Q_B Q_A = 101$, output $Y_5$ goes LOW. Each time an output goes LOW, its corresponding spark plug fires.

**Section Review Problem for Section 11.3a**

11.5. A design for a mod-16 counter specifies that a decoder must indicate the state just before the counter's recycle point by producing a HIGH output. What is the simplest circuit that will perform this function?

## Propagation Delay, Glitches, and Strobing

**Strobing** *A technique used to eliminate decoder glitches by activating the decoder outputs only when there are no glitch states present.*

**Figure 11.20**

**Example 11.9**
**Spark Plug Sequencer**

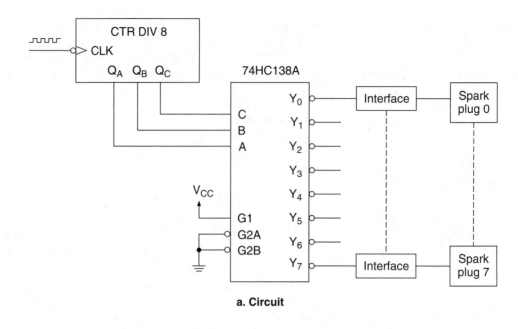

a. Circuit

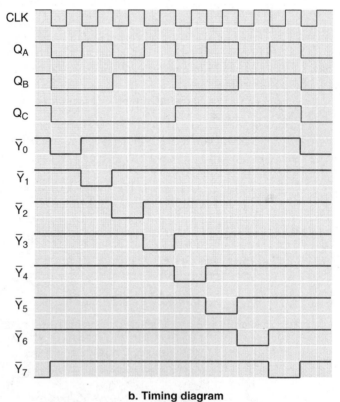

b. Timing diagram

The main disadvantage of an asynchronous counter is that it is subject to glitches introduced by the propagation delay of the flip-flops. Figure 11.21 shows the output transition of a flip-flop when the *CLK* input is pulsed. If the *J* and *K* inputs are both HIGH, the output will toggle, after a short delay, whenever *CLK* goes from HIGH to LOW.

Propagation delays for a 74LS76A JK flip-flop range from 15 ns (typical) to 20 ns (maximum) for both $t_{pLH}$ and $t_{pHL}$. For a 74HC76 flip-flop, the propagation delay is about 25 ns.

Figure 11.22 shows the effect of propagation delay on a mod-8 asynchronous counter. A negative-going transition at one stage initiates a transition at the next. There is a delay of several nanoseconds at each transition and, as a result, the counter generates unwanted intermediate output states.

Let us examine the transition from $Q_C\,Q_B\,Q_A = 001$ to $Q_C\,Q_B\,Q_A = 010$. To make this transition, $Q_A$ goes LOW, causing flip-flop $B$ to toggle, but not instantaneously. There is a delay between the HIGH-to-LOW transition on $Q_A$ and the resulting transition on $Q_B$. Since $Q_A$ is LOW for 15 to 20 ns before $Q_B$ goes HIGH, the output during this period is $Q_C\,Q_B\,Q_A = 000$. The actual count sequence at this point is:

| $Q_C$ | $Q_B$ | $Q_A$ |
|-------|-------|-------|
| 0 | 0 | 1 |
| (0 | 0 | 0) |
| 0 | 1 | 0 |

Next look at the 111–000 transition. $Q_A$ goes LOW when its *CLK* input goes LOW. After a delay, this makes $Q_B$ go LOW, which in turn makes $Q_C$ go LOW after yet another delay.

**Figure 11.21**

**Propagation Delay in a JK Flip-Flop**

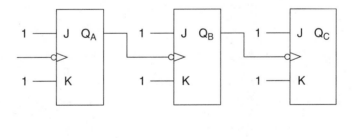

**Figure 11.22**

**Propagation Delay in an Asynchronous Counter**

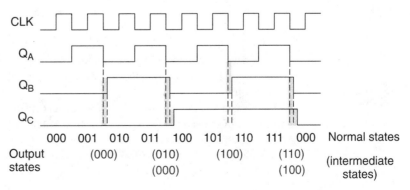

If you look carefully, you will see that the actual count sequence at this point is:

| $Q_C$ | $Q_B$ | $Q_A$ |
|-------|-------|-------|
| 1 | 1 | 1 |
| (1 | 1 | 0) |
| (1 | 0 | 0) |
| 0 | 0 | 0 |

The changes in the outputs do not occur simultaneously, but begin with the least significant bit and propagate through the higher-order bits.

Since the circuit is negative edge-triggered, the intermediate states occur every time there is a negative-going transition on any of the $Q$-to-$CLK$ connections. For example, there is an intermediate state on the 011–100 transition (both $Q_A$ and $Q_B$ make HIGH-to-LOW transitions), but not on the 100–101 transition ($Q_A$ makes a LOW-to-HIGH transition and nothing else switches).

By examining the timing diagram for the mod-8 counter, we can derive the actual count sequence as shown below. Each of the states in parentheses is an intermediate state lasting a few nanoseconds, depending on the propagation delay of each flip-flop.

| $Q_C$ | $Q_B$ | $Q_A$ | Decimal Equivalent |
|-------|-------|-------|--------------------|
| 0 | 0 | 0 | 0 |
| 0 | 0 | 1 | 1 |
| (0 | 0 | 0) | (0) |
| 0 | 1 | 0 | 2 |
| 0 | 1 | 1 | 3 |
| (0 | 1 | 0) | (2) |
| (0 | 0 | 0) | (0) |
| 1 | 0 | 0 | 4 |
| 1 | 0 | 1 | 5 |
| (1 | 0 | 0) | (4) |
| 1 | 1 | 0 | 6 |
| 1 | 1 | 1 | 7 |
| (1 | 1 | 0) | (6) |
| (1 | 0 | 0) | (4) |
| 0 | 0 | 0 | 0 | Recycle |

The problem with these intermediate states is that they produce glitches on the outputs of a decoder connected to the counter outputs. If you are decoding the counter output and using it to sequence other operations, the glitches might shuffle the required order of operation.

For instance, the engine ignition sequencer in Example 11.9 is supposed to fire the spark plugs in the sequence 0-1-2-3-4-5-6-7. If the interface circuit is fast enough to respond to all the glitch states in the counter analyzed above, the spark plugs will fire in the sequence 0-1-(0)-2-3-(2)-(0)-4-5-(4)-6-7-(6)-(4), which is clearly unacceptable.

**EXAMPLE 11.10**    Draw the timing diagram, including propagation delays and decoder glitches, of a negative edge-triggered asynchronous mod-8 counter and a decoder with active-LOW outputs.

**Solution**    Figure 11.23 shows the required timing diagram. The glitch states shown are the same as those we worked out for the counter in Figure 11.22. At the scale shown, the $CLK$ period is about 400 ns and the propagation delays are about 20 ns. Thus, the $CLK$ frequency is about 2.5 MHz.

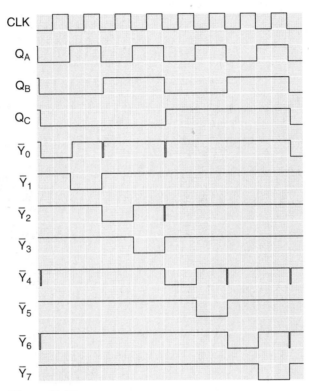

**Figure 11.23**
**Example 11.10**
**Glitches on a Decoder**

The easiest way to eliminate the glitch states is a technique called **strobing,** which enables decoder outputs only when there are no glitch states.

We can see from Figure 11.23 that decoder glitches occur just after the negative edges of the $CLK$ waveform, since this is when the flip-flops toggle. If the decoder outputs are switched off just after this time (i.e., during the LOW half-cycle of the $CLK$ waveform), the output glitches are eliminated.

For a circuit using positive edge-triggered flip-flops, the glitches would occur immediately after the positive $CLK$ edges and could be eliminated by enabling the decoder outputs on the LOW half-cycle of the $CLK$ waveform.

**EXAMPLE 11.11**     A 74HC138A decoder is used to decode a negative edge-triggered mod-8 asynchronous counter. Show how to connect the decoder to eliminate glitches on its outputs.

**Solution**     Since the 74HC138A decoder has three *ENABLE* inputs, we can use one of them as an output strobe. We want to activate the outputs only when the clock is HIGH, since all glitches occur in the LOW half-cycle of the clock waveform. We can do this by connecting *G1* to *CLK,* as illustrated in Figure 11.24a. The timing diagram for this circuit is shown in Figure 11.24b.

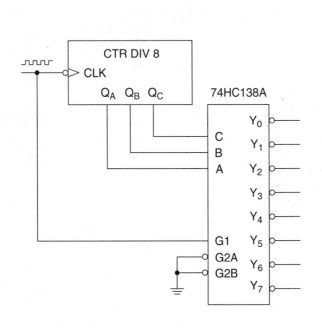

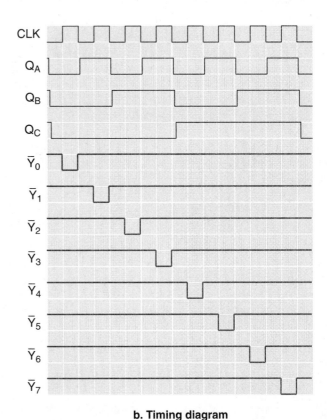

a. Circuit

b. Timing diagram

**Figure 11.24**
**Example 11.11**
**Use of Strobing to Eliminate Decoder Glitches**

---

**Section Review Problem for Section 11.3b**

11.6. A 4-bit negative edge-triggered asynchronous counter is fully decoded, with decoder outputs designated $Y_0$ to $Y_{15}$. If no steps are taken to suppress decoder glitches, on how many outputs will glitches appear during the transition from $Q_D\,Q_C\,Q_B\,Q_A = 1111$ to 0000? On which decoder outputs do the glitches appear?

## Maximum Frequency of an Asynchronous Counter

The propagation delay of internal flip-flops determines the maximum frequency at which an asynchronous counter can operate. If the clock speed is too fast, the least significant bit of the counter will change before the previous transition has finished propagating through the circuit. As a result, some states in the count sequence never occur.

Figure 11.25 illustrates how propagation delay limits the clock frequency of an asynchronous counter. Each of the three timing diagrams shows the transition of a mod-16

**Figure 11.25**

**Maximum CLK Frequency of an Asynchronous Counter**

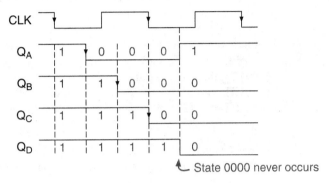

a. $t_{pLH} = t_{pHL} = 20$ ns, $T_{CLK} = 60$ ns

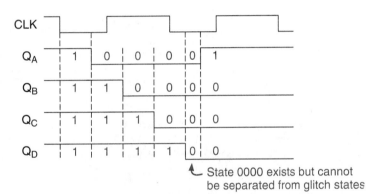

b. $t_{pLH} = t_{pHL} = 20$ ns, $T_{CLK} = 70$ ns

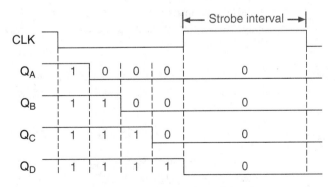

c. $t_{pLH} = t_{pHL} = 20$ ns, $T_{CLK} = 160$ ns
**Glitches confined to LOW half-cycle of CLK waveform**

counter from state 1111 to 0000. This represents the worst-case transition, since all four flip-flops change, each adding a delay. Propagation delays are 20 ns in all cases, but the clock frequencies are different for each diagram.

Figure 11.25a shows a *CLK* waveform with a period shorter than the total propagation delay of the flip-flops. The *CLK* period is 60 ns ($f = 1/T = 16.7$ MHz), and the total delay from state 1111 to 0000 is 80 ns. We can see that $Q_A$ changes to the *next* state just as $Q_D$ changes for the original state transition; state 0000 never occurs.

Figure 11.25b shows the situation with a slightly slower clock frequency. The *CLK* period is 70 ns ($f = 14.3$ MHz). In this case, the state 0000 appears for 10 ns. If we are decoding this counter, we cannot use the strobing technique previously described to eliminate the glitch states, since there is never a glitch-free *CLK* state; the transitions on the $Q$ outputs extend beyond the HIGH part of the *CLK* cycle.

Figure 11.25c shows the timing diagram when the *CLK* period is 160 ns ($f = 6.25$ MHz). In this situation, the state 0000 is present for the entire time that the *CLK* signal is HIGH. This allows us to strobe the output of a decoder and eliminate glitch states.

Figure 11.25c illustrates the following principle:

---

For a negative edge-triggered asynchronous counter, the LOW portion of the *CLK* waveform must be at least as long as the sum of the flip-flop propagation delays.

$$t_{l(CLK)} \geq (t_{pA} + t_{pB} + \ldots + t_{pn}) \text{ for an } n\text{-bit counter}$$

---

For a *CLK* waveform with a 50% duty cycle, this implies:

---

$$T_{CLK} \geq 2(t_{pA} + t_{pB} + \ldots + t_{pn}) \text{ for an } n\text{-bit counter; and } f_{max} = 1/T_{CLK}$$

---

We can run the counter at a higher frequency if we use a duty cycle on the *CLK* input that is less than 50%. As long as we allow time during the LOW part of the *CLK* cycle for the propagation delays to elapse, we can eliminate glitches by using a strobe input on the decoder.

---

**EXAMPLE 11.12**

Calculate the maximum frequency of a fully decoded mod-32 counter whose flip-flops have propagation delays of 25 ns if the *CLK* waveform has a duty cycle of 50%. How long are the decoder output pulses?

**Solution**  A mod-32 counter has five flip-flops. The minimum LOW clock half-cycle is the sum of propagation delays ($= 5 \times 25$ ns $= 125$ ns). Minimum clock period is given by ($2 \times 5 \times 25$ ns $= 250$ ns).

$$f_{max} = 1/T_{CLK} = 1/250 \text{ ns} = 4 \text{ MHz}$$

Since the decoder is strobed by the HIGH part of the *CLK* waveform, the decoder pulses are the same length as the LOW part of the *CLK* cycle:

$$t_{pulse} = t_{h(CLK)} = t_{l(CLK)} = 125 \text{ ns}$$

**EXAMPLE 11.13**    Repeat the problem of Example 11.12 for a *CLK* duty cycle of 10%.

**Solution**    If we want to eliminate decoder glitches by strobing the decoder outputs, the value of $t_{l(CLK)}$ must be the same as in Example 11.12 (125 ns).

A duty cycle of 0.1 (10%) implies that $t_{h(CLK)} = 0.1T_{CLK}$ and $t_{l(CLK)} = 0.9T_{CLK}$. Thus,

$$t_{h(CLK)} = t_{l(CLK)}/9 = 125 \text{ ns}/9 = 13.9 \text{ ns}$$

$$f_{max} = 1/T_{CLK} = 1/(125 \text{ ns} + 13.9 \text{ ns}) = 1/158.9 \text{ ns} = 7.2 \text{ MHz}$$

The decoder pulses are the same as $t_{h(CLK)}$ (= 13.9 ns).

---

**Section Review Problem for Section 11.3c**

11.7.  Calculate the maximum frequency of a fully decoded asynchronous mod-8 counter whose flip-flops have propagation delays of 20 ns each. The decoder must have no glitch states on its outputs. Assume a 50% duty cycle on the *CLK* input.

---

## 11.4

# Synchronous Counters

**Synchronous counter**  *A counter whose flip-flops are all clocked by the same source and thus change in synchronization with one another.*

The main characteristics that distinguish a **synchronous counter** from an asynchronous counter are the connections to the clock lines and to the synchronous inputs of the flip-flops.

In an asynchronous counter (Figure 11.26b), the *J* and *K* inputs of the flip-flops are tied HIGH, placing them permanently in toggle mode. The flip-flop states are changed by clocking each one from the output of the previous stage. Each time a flip-flop is clocked, it toggles.

In a synchronous counter (Figure 11.26a), the flip-flops are all clocked from a single source. Thus, we cannot use the same *J* and *K* connections as we did for an asynchronous counter. If we did, the outputs would all toggle together between two states (e.g., 0000 to 1111 and back, indefinitely).

To make the circuit act as a counter, we must use at least two operating modes of the JK flip-flops, not just the toggle mode. Table 11.5 lists the operating modes of a JK flip-flop.

In a synchronous counter, we use combinational logic to read the present state of the counter outputs and set the operating mode of each flip-flop to make it go to the desired next state upon receiving a clock pulse.

**Table 11.5** Operating Modes of a JK Flip-Flop

| J | K | Q | $\overline{Q}$ | |
|---|---|---|---|---|
| 0 | 0 | $Q_0$ | $\overline{Q}_0$ | No change |
| 0 | 1 | 0 | 1 | Reset |
| 1 | 0 | 1 | 0 | Set |
| 1 | 1 | $\overline{Q}_0$ | $Q_0$ | Toggle |

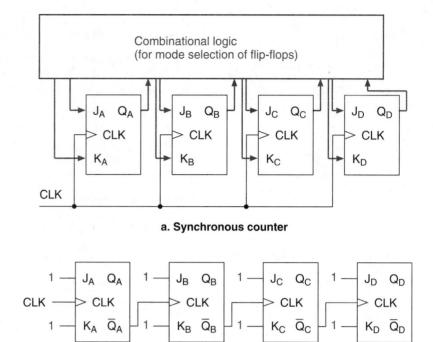

a. Synchronous counter

b. Asynchronous counter

**Figure 11.26**

**Clock Connections and Mode Selection for Synchronous and Asynchronous Counters**

Since the flip-flops of a synchronous counter are all clocked simultaneously, we do not encounter the same glitch problems inherent in asynchronous counters. If the flip-flops do not have equal propagation delays, or if not all flip-flops are changing on a given transition, there will be a short time when the counter has some intermediate output states. However, in a synchronous counter, the propagation delays are not additive; there is a maximum of one flip-flop delay for the whole counter.

The combinational logic in a synchronous circuit adds its own delays, but since these occur at the J and K inputs after each transition, they do not affect the output states directly. The combinational logic delays will affect the minimum time between clock pulses and therefore the maximum frequency of the counter.

## Full-Sequence Counters

A 3-bit synchronous binary counter is shown in Figure 11.27. Let us analyze its count sequence in detail so that we can see how the J and K inputs are affected by the Q outputs and how transitions between states are made.

Assume that the counter output is initially $Q_C Q_B Q_A = 000$. Before any clock pulses are applied, the J and K inputs are at the following states:

$$J_C = K_C = Q_A \cdot Q_B = 0 \cdot 0 = 0 \quad \text{(No change)}$$

$$J_B = K_B = Q_A = 0 \quad \text{(No change)}$$

$$J_A = K_A = 1 \text{ (Constant)} \quad \text{(Toggle)}$$

**Figure 11.27**

**Synchronous Binary Counter**

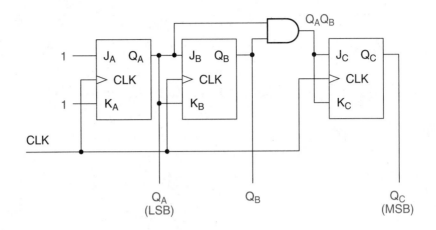

The transitions of the outputs after the clock pulse are:

$$Q_C: 0 \rightarrow 0 \quad \text{(No change)}$$
$$Q_B: 0 \rightarrow 0 \quad \text{(No change)}$$
$$Q_A: 0 \rightarrow 1 \quad \text{(Toggle)}$$

The output goes from $Q_C\, Q_B\, Q_A = 000$ to $Q_C\, Q_B\, Q_A = 001$. (See Figure 11.28.) The transition is defined by the values of $J$ and $K$ *before* the clock pulse, since the propagation delays of the flip-flops prevent the new output conditions from changing the $J$ and $K$ values until after the transition.

The new conditions of the $J$ and $K$ inputs are:

$$J_C = K_C = Q_A \cdot Q_B = 1 \cdot 0 = 0 \quad \text{(No change)}$$
$$J_B = K_B = Q_A = 1 \quad\quad\quad\quad\quad \text{(Toggle)}$$
$$J_A = K_A = 1 \text{ (Constant)} \quad\quad\quad \text{(Toggle)}$$

The transitions of the outputs generated by the second clock pulse are:

$$Q_C: 0 \rightarrow 0 \quad \text{(No change)}$$
$$Q_B: 0 \rightarrow 1 \quad \text{(Toggle)}$$
$$Q_A: 1 \rightarrow 0 \quad \text{(Toggle)}$$

**Figure 11.28**

**Timing Diagram for a Synchronous Mod-8 Counter**

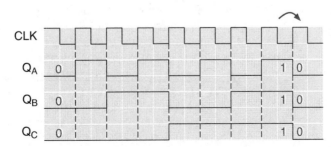

The new output is $Q_C Q_B Q_A = 010$, since both $Q_A$ and $Q_B$ change and $Q_C$ stays the same. The $J$ and $K$ conditions are now:

$$J_C = K_C = Q_A \cdot Q_B = 0 \cdot 1 = 0 \quad \text{(No change)}$$

$$J_B = K_B = Q_A = 0 \quad\quad\quad\quad \text{(No change)}$$

$$J_A = K_A = 1 \text{ (Constant)} \quad\quad \text{(Toggle)}$$

The output transitions are:

$$Q_C\text{: } 0 \rightarrow 0 \quad \text{(No change)}$$

$$Q_B\text{: } 1 \rightarrow 1 \quad \text{(No change)}$$

$$Q_A\text{: } 0 \rightarrow 1 \quad \text{(Toggle)}$$

The output is now $Q_C Q_B Q_A = 011$, which results in the $JK$ conditions:

$$J_C = K_C = Q_A \cdot Q_B = 1 \cdot 1 = 1 \quad \text{(Toggle)}$$

$$J_B = K_B = Q_A = 1 \quad\quad\quad\quad \text{(Toggle)}$$

$$J_A = K_A = 1 \text{ (Constant)} \quad\quad \text{(Toggle)}$$

The above conditions result in output transitions:

$$Q_C\text{: } 0 \rightarrow 1 \quad \text{(Toggle)}$$

$$Q_B\text{: } 1 \rightarrow 0 \quad \text{(Toggle)}$$

$$Q_A\text{: } 1 \rightarrow 0 \quad \text{(Toggle)}$$

All the outputs toggle and the new output state is $Q_C Q_B Q_A = 100$. The $J$ and $K$ values repeat the above pattern in the second half of the counter cycle (states 100 to 111). Go through the exercise of calculating the $J$, $K$, and $Q$ values for the rest of the cycle. Compare the result with the timing diagram in Figure 11.28.

In the counter we have just analyzed, the combinational circuit generates either a toggle ($JK = 11$) or a no change ($JK = 00$) state at each point through the count sequence. We could use any combination of $JK$ modes (no change, reset, set, or toggle) to make the transitions from one state to the next. For instance, instead of using only the no change and toggle modes, the 000–001 transition could also be done by making $Q_A$ set ($J_A = 1$, $K_A = 0$) and $Q_B$ and $Q_C$ reset ($J_B = 0$, $K_B = 1$ and $J_C = 0$, $K_C = 1$). To do so we would need a different set of combinational logic in the circuit.

The simplest synchronous counter design uses only the no change ($JK = 00$) or toggle ($JK = 11$) modes, since the $J$ and $K$ inputs of each flip-flop can be connected together. The no change and toggle modes allow us to make any transition (i.e., not just in a linear sequence), even though for truncated sequence and nonbinary counters this is not usually the most efficient design. We will examine alternate design methods in Chapter 12.

There is a simple progression of algebraic expressions for the $J$ and $K$ inputs of a synchronous binary (full-sequence) counter, which uses only the no change and toggle states:

$$J_A = K_A = 1$$

$$J_B = K_B = Q_A$$

$$J_C = K_C = Q_A \cdot Q_B$$

$$J_D = K_D = (Q_A \cdot Q_B) \cdot Q_C$$

$$J_E = K_E = ((Q_A \cdot Q_B) \cdot Q_C) \cdot Q_D$$

etc.

The $J$ and $K$ inputs of each stage are the ANDed outputs of all previous stages. This implies that a flip-flop toggles only when the outputs of *all* previous stages are HIGH. For example, $Q_C$ doesn't change unless *both* $Q_A$ AND $Q_B$ are HIGH (and therefore $J_C = K_C = 1$) before the clock pulse. Look at the timing diagram of Figure 11.28 to confirm this.

**EXAMPLE 11.14**    Draw the logic circuit of a mod-16 synchronous counter, using negative edge-triggered JK flip-flops. Verify the count sequence by analyzing the $J$, $K$, and $Q$ values for all transitions. Draw the timing diagram of the counter.

**Solution**    The $J$ and $K$ equations for a mod-16 counter follow the algebraic progression previously outlined.

$$J_A = K_A = 1$$

$$J_B = K_B = Q_A$$

$$J_C = K_C = Q_A \cdot Q_B$$

$$J_D = K_D = (Q_A \cdot Q_B) \cdot Q_C$$

These equations can be used to design the counter shown in Figure 11.29. The counter is the same whether the flip-flops are positive or negative edge-triggered. The only difference is the edges of the *CLK* waveform on which the $Q$ waveforms toggle.

Table 11.6 shows the values of $J$ and $K$ for each flip-flop, both before and after a clock pulse is applied, for the complete count sequence. (The sequence is finished when it repeats.) Each line of the table shows one transition. Since $J$ and $K$ of each flip-flop

**Figure 11.29**

**Example 11.14 Synchronous Mod-16 Counter**

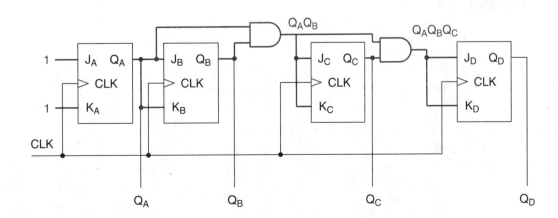

Table 11.6  Count Sequence and Flip-Flop Transitions for a Mod-16 Synchronous Counter

| Q Outputs Before CLK Pulse | | | | J and K Before CLK Pulse | | | | | | | | Q Outputs After CLK Pulse | | | |
|---|---|---|---|---|---|---|---|---|---|---|---|---|---|---|---|
| $Q_D$ | $Q_C$ | $Q_B$ | $Q_A$ | $J_D$ | $K_D$ | $J_C$ | $K_C$ | $J_B$ | $K_B$ | $J_A$ | $K_A$ | $Q_D$ | $Q_C$ | $Q_B$ | $Q_A$ |
| 0 | 0 | 0 | 0 | 0 | 0 | 0 | 0 | 0 | 0 | 1 | 1 | 0 | 0 | 0 | 1 |
| 0 | 0 | 0 | 1 | 0 | 0 | 0 | 0 | 1 | 1 | 1 | 1 | 0 | 0 | 1 | 0 |
| 0 | 0 | 1 | 0 | 0 | 0 | 0 | 0 | 0 | 0 | 1 | 1 | 0 | 0 | 1 | 1 |
| 0 | 0 | 1 | 1 | 0 | 0 | 1 | 1 | 1 | 1 | 1 | 1 | 0 | 1 | 0 | 0 |
| 0 | 1 | 0 | 0 | 0 | 0 | 0 | 0 | 0 | 0 | 1 | 1 | 0 | 1 | 0 | 1 |
| 0 | 1 | 0 | 1 | 0 | 0 | 0 | 0 | 1 | 1 | 1 | 1 | 0 | 1 | 1 | 0 |
| 0 | 1 | 1 | 0 | 0 | 0 | 0 | 0 | 0 | 0 | 1 | 1 | 0 | 1 | 1 | 1 |
| 0 | 1 | 1 | 1 | 1 | 1 | 1 | 1 | 1 | 1 | 1 | 1 | 1 | 0 | 0 | 0 |
| 1 | 0 | 0 | 0 | 0 | 0 | 0 | 0 | 0 | 0 | 1 | 1 | 1 | 0 | 0 | 1 |
| 1 | 0 | 0 | 1 | 0 | 0 | 0 | 0 | 1 | 1 | 1 | 1 | 1 | 0 | 1 | 0 |
| 1 | 0 | 1 | 0 | 0 | 0 | 0 | 0 | 0 | 0 | 1 | 1 | 1 | 0 | 1 | 1 |
| 1 | 0 | 1 | 1 | 0 | 0 | 1 | 1 | 1 | 1 | 1 | 1 | 1 | 1 | 0 | 0 |
| 1 | 1 | 0 | 0 | 0 | 0 | 0 | 0 | 0 | 0 | 1 | 1 | 1 | 1 | 0 | 1 |
| 1 | 1 | 0 | 1 | 0 | 0 | 0 | 0 | 1 | 1 | 1 | 1 | 1 | 1 | 1 | 0 |
| 1 | 1 | 1 | 0 | 0 | 0 | 0 | 0 | 0 | 0 | 1 | 1 | 1 | 1 | 1 | 1 |
| 1 | 1 | 1 | 1 | 1 | 1 | 1 | 1 | 1 | 1 | 1 | 1 | 0 | 0 | 0 | 0 |
| | (Recycle) | | | | | | | | | | | | | | |

are connected together, the only possible modes are no change ($JK = 00$) and toggle ($JK = 11$).

For example, the transition from 0011 to 0100 is shown on the fourth line of the table. Before the clock pulse, the synchronous inputs are:

$$J_D = K_D = Q_C\,Q_B\,Q_A = 0 \cdot 1 \cdot 1 = 0 \quad \text{(No change)}$$

$$J_C = K_C = Q_B\,Q_A \qquad = 1 \cdot 1 \quad = 1 \quad \text{(Toggle)}$$

$$J_B = K_B = Q_A \qquad\qquad = 1 \qquad\quad \text{(Toggle)}$$

$$J_A = K_A = 1 \qquad\qquad\qquad\qquad \text{(Toggle)}$$

The resulting transitions are:

$$Q_D\colon 0 \to 0 \quad \text{(No change)}$$

$$Q_C\colon 0 \to 1 \quad \text{(Toggle)}$$

$$Q_B\colon 1 \to 0 \quad \text{(Toggle)}$$

$$Q_A\colon 1 \to 0 \quad \text{(Toggle)}$$

**Figure 11.30**

**Example 11.14**
**Timing Diagram of a Mod-16 Synchronous Counter**

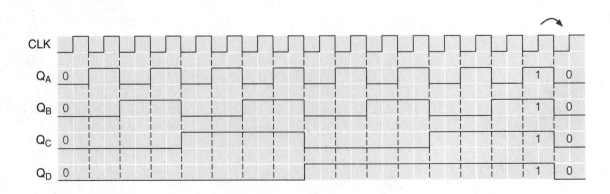

Similar analysis applies to all other transitions.

Figure 11.30 shows the timing diagram of the counter. Since the flip-flops are negative edge-triggered, the transitions between states occur only on the negative edges of the *CLK* waveform. The synchronous inputs (*J* and *K*) required for each transition are those present *just before* the negative edge of the *CLK* pulse.

We could have used a 3-input AND gate, rather than the 2-input, for the function $Q_C Q_B Q_A$ in the counter of Figure 11.29. Further stages would have ANDs with four, five, and six inputs and so on for each stage. There are advantages both to using 2-input gates and to using gates with more inputs at each stage.

By using 2-input gates, we can expand the counter simply by adding a flip-flop and a 2-input AND for each new stage. Since these gates come in packages of four, we reduce the number of IC packages we need.

By expanding the number of inputs to each new AND gate, we reduce the propagation delays of the combinational part of the circuit. The inputs of each AND gate connect directly to the flip-flop outputs without passing through several previous ANDs. The reduced propagation delay increases the maximum frequency of the counter. However, if we expand beyond the point where we need to combine four flip-flop outputs, we run into problems, since 5-input ANDs are not generally available.

---

**Section Review Problem for Section 11.4a**

11.8.  List the two main differences between an asynchronous and a synchronous counter circuit.

---

## Synchronous DOWN Counters

Full-sequence synchronous DOWN counters, which count from their maximum modulus to zero, can be constructed in almost the same way as full-sequence synchronous UP counters. The only difference is that the flip-flop *JK* inputs are driven by AND gates connected to the $\overline{Q}$ outputs of the flip-flops, rather than the *Q* outputs. Figure 11.31 shows a 4-bit synchronous DOWN counter.

**Figure 11.31**
**Synchronous DOWN Counter**

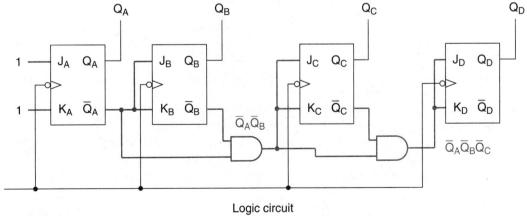

Logic circuit

**Table 11.7** Partial Count Sequence and Flip-Flop Transitions for a Mod-16 Synchronous DOWN Counter

| $Q$ Outputs Before $CLK$ Pulse | | | | $J$ and $K$ Before $CLK$ Pulse | | | | | | | | $Q$ Outputs After $CLK$ Pulse | | | |
|---|---|---|---|---|---|---|---|---|---|---|---|---|---|---|---|
| $Q_D$ | $Q_C$ | $Q_B$ | $Q_A$ | $J_D$ | $K_D$ | $J_C$ | $K_C$ | $J_B$ | $K_B$ | $J_A$ | $K_A$ | $Q_D$ | $Q_C$ | $Q_B$ | $Q_A$ |
| 1 | 1 | 1 | 1 | 0 | 0 | 0 | 0 | 0 | 0 | 1 | 1 | 1 | 1 | 1 | 0 |
| 1 | 1 | 1 | 0 | 0 | 0 | 0 | 0 | 1 | 1 | 1 | 1 | 1 | 1 | 0 | 1 |
| 1 | 1 | 0 | 1 | 0 | 0 | 0 | 0 | 0 | 0 | 1 | 1 | 1 | 1 | 0 | 0 |
| 1 | 1 | 0 | 0 | 0 | 0 | 1 | 1 | 1 | 1 | 1 | 1 | 1 | 0 | 1 | 1 |
| . | . | . | | | | | | | | | | | . | | |
| . | . | . | | | | | | | | | | | . | | |
| . | . | . | | | | | | | | | | | . | | |
| 0 | 0 | 1 | 0 | 0 | 0 | 0 | 0 | 1 | 1 | 1 | 1 | 0 | 0 | 0 | 1 |
| 0 | 0 | 0 | 1 | 0 | 0 | 0 | 0 | 0 | 0 | 1 | 1 | 0 | 0 | 0 | 0 |
| 0 | 0 | 0 | 0 | 1 | 1 | 1 | 1 | 1 | 1 | 1 | 1 | 1 | 1 | 1 | 1 |
| | (Recycle) | | | | | | | | | | | | | | |

From the diagram in Figure 11.31, we can derive the following equations for the $J$ and $K$ inputs of the flip-flops:

$$J_A = K_A = 1$$
$$J_B = K_B = \overline{Q}_A$$
$$J_C = K_C = \overline{Q}_A \cdot \overline{Q}_B$$
$$J_D = K_D = \overline{Q}_A \cdot \overline{Q}_B \cdot \overline{Q}_C$$

As can be done with a synchronous UP counter, this pattern theoretically can be extended for as many flip-flops as you care to add. The practical limits are the propagation delays of the AND gates and the current-driving capability of the clock generator.

We can use these equations to generate the count sequence of the DOWN counter circuit, much as we did in Example 11.14. Table 11.7 shows the first and last few entries in the table. As an exercise, you can complete the table.

The timing diagram of a mod-16 synchronous DOWN counter is shown in Figure 11.32.

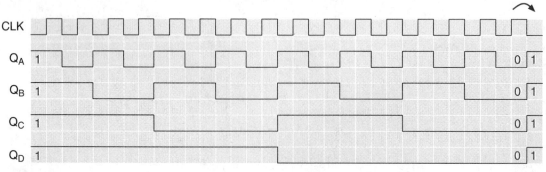

**Figure 11.32**
**Timing Diagram**

> **Section Review Problem for Section 11.4b**
>
> 11.9.  Write the *J* and *K* equations for the fifth and sixth flip-flops of a 6-bit synchronous DOWN counter.

### Synchronous UP/DOWN Counters

Figure 11.33a shows a circuit that combines a synchronous UP counter and a synchronous DOWN counter. The direction of counting is selected by an additional external control input *(UP/$\overline{DOWN}$)*. The circuit counts UP when *UP/$\overline{DOWN}$* = 1 and DOWN when *UP/$\overline{DOWN}$* = 0. The state diagram of the counter appears in Figure 11.33b.

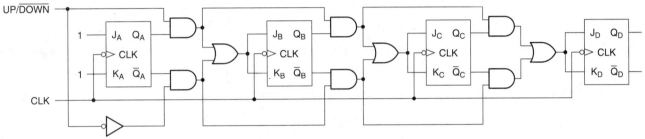

Mod-16 bidirectional counter

**a. Logic circuit**

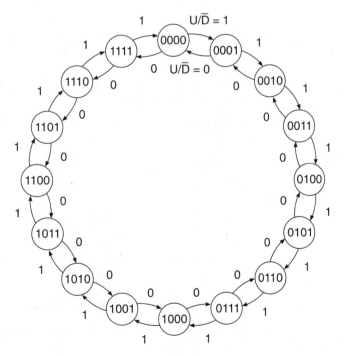

**Figure 11.33**
**4-Bit UP/DOWN Counter**

**b. State diagram**

Let us examine the operation of the counter in Figure 11.33a. $J_A$ and $K_A$ are permanently wired for toggle mode, since this connection is normally present in both an UP counter and a DOWN counter. The other flip-flops will be controlled by the $Q$ or the $\overline{Q}$ inputs, depending on the selected count direction.

If $UP/\overline{DOWN} = 1$, the first AND gate in the top row will enable $Q_A$ and direct it to $J_B$, $K_B$, and the remaining UP-count logic. Meanwhile, $\overline{Q}_A$ is inhibited, since the input to its AND gate is 0. The entire chain of UP-count logic is enabled in this mode and the DOWN-count logic is inhibited. Thus, for $UP/\overline{DOWN} = 1$:

$$J_A = K_A = 1$$
$$J_B = K_B = Q_A$$
$$J_C = K_C = Q_A \cdot Q_B$$
$$J_D = K_D = (Q_A \cdot Q_B) \cdot Q_C$$

By similar reasoning, we can see that $\overline{Q}_A$ and the rest of the DOWN-count logic are enabled when $UP/\overline{DOWN} = 0$ and that $Q_A$ and the UP-count logic are inhibited. In this mode:

$$J_A = K_A = 1$$
$$J_B = K_B = \overline{Q}_A$$
$$J_C = K_C = \overline{Q}_A \cdot \overline{Q}_B$$
$$J_D = K_D = \overline{Q}_A \cdot \overline{Q}_B \cdot \overline{Q}_C$$

**EXAMPLE 11.15**

The $UP/\overline{DOWN}$ counter in Figure 11.33 is initially at state $Q_D \, Q_C \, Q_B \, Q_A = 0000$ and $UP/\overline{DOWN} = 1$. After five $CLK$ pulses, $UP/\overline{DOWN}$ switches to 0. Eight $CLK$ pulses are applied and then $UP/\overline{DOWN} = 1$ again. After two more $CLK$ pulses, the clock stops.

Give the count-sequence table and draw the corresponding timing diagram, showing the $UP/\overline{DOWN}$ and $CLK$ inputs and all $Q$ outputs.

**Solution**   Table 11.8 shows the count-sequence table. The timing diagram of this sequence is shown in Figure 11.34.

**Figure 11.34**

**Example 11.15 Timing Diagram**

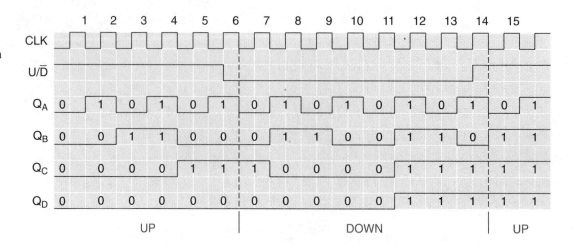

**Table 11-8** Possible UP/DOWN Count Sequence

| UP/$\overline{DOWN}$ | $Q_D$ | $Q_C$ | $Q_B$ | $Q_A$ | |
|:---:|:---:|:---:|:---:|:---:|:---|
| 1 | 0 | 0 | 0 | 0 | |
| 1 | 0 | 0 | 0 | 1 | |
| 1 | 0 | 0 | 1 | 0 | UP |
| 1 | 0 | 0 | 1 | 1 | |
| 1 | 0 | 1 | 0 | 0 | |
| 1 | 0 | 1 | 0 | 1 | |
| 0 | 0 | 1 | 0 | 0 | |
| 0 | 0 | 0 | 1 | 1 | |
| 0 | 0 | 0 | 1 | 0 | |
| 0 | 0 | 0 | 0 | 1 | DOWN |
| 0 | 0 | 0 | 0 | 0 | |
| 0 | 1 | 1 | 1 | 1 | |
| 0 | 1 | 1 | 1 | 0 | |
| 0 | 1 | 1 | 0 | 1 | |
| 1 | 1 | 1 | 1 | 0 | UP |
| 1 | 1 | 1 | 1 | 1 | |

## Truncated-Sequence Synchronous Counters

As is the case with asynchronous counters, it may be necessary to design a synchronous counter with a modulus other than a power of 2. Truncating a synchronous counter can be done asynchronously or synchronously.

Asynchronous truncation uses the flip-flops' Preset and Clear inputs and some external decoding logic to detect a desired recycle point. Synchronous truncation uses logic gates to select a desired count sequence by controlling the flip-flop operating modes, as determined by the $J$ and $K$ inputs.

An asynchronously truncated synchronous counter is easier to design than a synchronously truncated circuit but is subject to glitching at the recycle point. The design of a synchronously truncated counter is much more complicated but produces smooth transitions throughout its count sequence. We will come back to the design of synchronous counters in the next chapter and will concentrate here on analysis.

### Determining the Modulus of a Synchronous Counter

We can use the technique described in Example 11.14 to analyze any synchronous counter, as follows.

1. Determine the equations for the $J$ and $K$ logic functions in terms of the $Q$ outputs for all flip-flops. (For counters other than straight binary full-sequence types, the equations will *not* be the same as the algebraic progressions previously listed.)

2. Lay out a table with headings for the present state of the counter ($Q$ outputs before clock pulse), each $J$ and $K$ (before clock pulse), and next state of the counter ($Q$ outputs after the clock pulse).

3. Choose a starting point for the count sequence, usually 0000 or sometimes 1111 (for a 4-bit counter), and enter the starting point in the Present State column.

4. Substitute the $Q$ values of the initial Present State into the $J$ and $K$ equations and enter the results under the columns for the $J$ and $K$ states.

5. For each $JK$ pair, determine if the corresponding flip-flop will not change ($JK = 00$), reset ($JK = 01$), set ($JK = 10$), or toggle ($JK = 11$) on the next clock pulse.

6. Look at the $Q$ values for every flip-flop. Change them according to the function determined in step 5, and enter them in the column for the counter's Next State.

7. Enter the result from step 6 on the next line of the column for the counter's Present State (i.e., this line's Next State is the next line's Present State).

8. Repeat the above process until the result in the Next State column is the same as the initial state.

Alternatively, at step 3 you can list all the possible Present States of the counter in binary order. The procedure remains the same until step 6. Steps 7 and 8 become unnecessary. You should convert the resultant state table to a state diagram to illustrate the counter's operation.

**EXAMPLE 11.16**

Find the count sequence of the synchronous counter shown in Figure 11.35 and, from the count-sequence table, draw the timing diagram and state diagram. What is the modulus of the counter?

**Figure 11.35**

**Example 11.16
Counter of Unknown
Modulus**

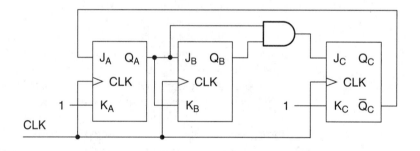

**Solution**    For reference, the JK flip-flop function table is:

| $J$ | $K$ | $Q$ | $\overline{Q}$ | |
|---|---|---|---|---|
| 0 | 0 | $Q_0$ | $\overline{Q}_0$ | No change |
| 0 | 1 | 0 | 1 | Reset |
| 1 | 0 | 1 | 0 | Set |
| 1 | 1 | $\overline{Q}_0$ | $Q_0$ | Toggle |

The $J$ and $K$ equations are:

$$J_C = Q_A \cdot Q_B \quad J_B = Q_A \quad J_A = \overline{Q}_C$$
$$K_C = 1 \qquad\qquad K_B = Q_A \quad K_A = 1$$

Table 11.9  Count Sequence

| Present State | | | $J_C$ | $K_C$ | | $J_B$ | $K_B$ | | $J_A$ | $K_A$ | | Next State | | |
|---|---|---|---|---|---|---|---|---|---|---|---|---|---|---|
| $Q_C$ | $Q_B$ | $Q_A$ | | | | | | | | | | $Q_C$ | $Q_B$ | $Q_A$ |
| 0 | 0 | 0 | 0 | 1 | (R) | 0 | 0 | (NC) | 1 | 1 | (T) | 0 | 0 | 1 | |
| 0 | 0 | 1 | 0 | 1 | (R) | 1 | 1 | (T) | 1 | 1 | (T) | 0 | 1 | 0 | |
| 0 | 1 | 0 | 0 | 1 | (R) | 0 | 0 | (NC) | 1 | 1 | (T) | 0 | 1 | 1 | |
| 0 | 1 | 1 | 1 | 1 | (T) | 1 | 1 | (T) | 1 | 1 | (T) | 1 | 0 | 0 | |
| 1 | 0 | 0 | 0 | 1 | (R) | 0 | 0 | (NC) | 0 | 1 | (R) | 0 | 0 | 0 | (Recycle) |

**Figure 11.36**

**Example 11.16**
**Timing Diagram and State Diagram**

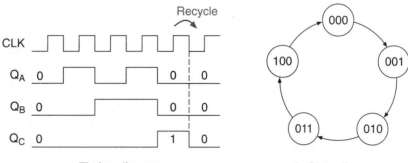

a. Timing diagram                    b. State diagram

The output transitions can be determined from the values of the $J$ and $K$ functions before each clock pulse, as shown in Table 11.9.

Since there are five unique output states, the counter's modulus is 5.

The timing diagram and state diagram are shown in Figure 11.36. Since this circuit produces one pulse on $Q_C$ for every five clock pulses, we can use it as a divide-by-5 circuit.

▲

The analysis in Example 11.16 did not account for the fact that the counter uses only five of a possible eight output states. In any truncated-sequence counter, it is a good idea to determine the next state for each unused state to ensure that if the counter powers up in one of these unused states, it will eventually enter the main sequence.

▼

**EXAMPLE 11.17**

Extend the analysis of the counter in Example 11.16 to include its unused states. Redraw the counter's state diagram to show how these unused states enter the main sequence (if they do).

**Solution**

$$J_C = Q_A \cdot Q_B \quad J_B = Q_A \quad J_A = \overline{Q}_C$$
$$K_C = 1 \quad\quad\quad K_B = Q_A \quad K_A = 1$$

Table 11.10 shows the resulting count sequence. Figure 11.37 shows the complete state diagram, including unused states.

**Table 11.10** Count Sequence Including Unused States

| Present State | | | | | | | | | | | | | Next State | | | |
|---|---|---|---|---|---|---|---|---|---|---|---|---|---|---|---|---|
| $Q_C$ | $Q_B$ | $Q_A$ | $J_C$ | $K_C$ | | $J_B$ | $K_B$ | | $J_A$ | $K_A$ | | $Q_C$ | $Q_B$ | $Q_A$ | |
| 0 | 0 | 0 | 0 | 1 | (R) | 0 | 0 | (NC) | 1 | 1 | (T) | 0 | 0 | 1 | |
| 0 | 0 | 1 | 0 | 1 | (R) | 1 | 1 | (T) | 1 | 1 | (T) | 0 | 1 | 0 | |
| 0 | 1 | 0 | 0 | 1 | (R) | 0 | 0 | (NC) | 1 | 1 | (T) | 0 | 1 | 1 | |
| 0 | 1 | 1 | 1 | 1 | (T) | 1 | 1 | (T) | 1 | 1 | (T) | 1 | 0 | 0 | |
| 1 | 0 | 0 | 0 | 1 | (R) | 0 | 0 | (NC) | 0 | 1 | (R) | 0 | 0 | 0 | (Recycle) |
| 1 | 0 | 1 | 0 | 1 | (R) | 1 | 1 | (T) | 0 | 1 | (R) | 0 | 1 | 0 | (Unused |
| 1 | 1 | 0 | 0 | 1 | (R) | 0 | 0 | (NC) | 0 | 1 | (R) | 0 | 1 | 0 | states) |
| 1 | 1 | 1 | 1 | 1 | (T) | 1 | 1 | (T) | 0 | 1 | (R) | 0 | 0 | 0 | |

**Figure 11.37**
**Example 11.17**
**Complete State Diagram**

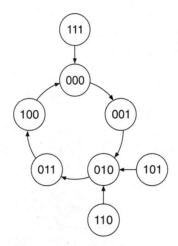

## ● G L O S S A R Y

**Asynchronous counter** A counter whose bits are not clocked by the same source and thus do not change in synchronization with one another.

**Binary counter** A counter with a binary count sequence whose modulus is $2^n$ for an $n$-bit counter.

**Count sequence** The specific series of output states through which a counter progresses.

**Count-sequence table** A list of counter states in the order of the count sequence.

**Counter** A sequential digital circuit whose output produces a fixed sequence of binary states when the circuit is clocked.

**Decoder** A digital circuit designed to detect the presence of a particular digital state or states.

**DOWN counter** A counter with a descending sequence.

**Full-sequence counter** A counter whose modulus is the same as its maximum modulus. ($m = 2^n$ for an $n$-bit counter.)

**Glitch** An unwanted voltage spike that can cause erroneous switching in a digital circuit.

**Maximum modulus** ($m_{max}$) The largest number of counter states that can be represented by $n$ bits ($m_{max} = 2^n$).

**Modulo-$n$ (or mod-$n$) counter** A counter with a modulus of $n$.

**Modulus** The number of states through which a counter sequences before repeating.

**Recycle** Make a transition from the last state of a count sequence to the first state.

**Recycle state**  A temporary state that is decoded by external logic to truncate a counter sequence. The recycle state is not counted as part of the count sequence.

**Ripple counter**  An asynchronous counter. More specifically, a counter where one bit is clocked by transitions on the next least significant bit.

**State diagram**  A diagram showing the progression of states of a sequential circuit.

**Strobing**  A technique used to eliminate decoder glitches by activating the decoder outputs only when there are no glitch states present.

**Synchronous counter**  A counter whose flip-flops are all clocked by the same source and thus change in synchronization with one another.

**Truncated-sequence counter**  A counter whose modulus is less than its maximum modulus. ($m < 2^n$ for an $n$-bit counter.)

**UP counter**  A counter with an ascending sequence.

## ● PROBLEMS

### Section 11.1 Digital Counters

**\*11.1**  A parking lot at a football stadium is monitored before a game to determine whether or not there is available space for more cars. When a car enters the lot, the driver takes a ticket from a dispenser, which also produces a pulse for each ticket taken.

  The parking lot has space for 4096 cars. Draw a block diagram that shows how you can use a digital counter to light a LOT FULL sign after 4096 cars have entered. (Assume that no cars leave the lot until after the game, so you don't need to keep track of cars leaving the lot.) How many bits should the counter have?

**\*11.2**  How should the circuit in Example 11.2 be modified so that one ASCII character is transferred to the octal flip-flop every 16 pulses?

**\*11.3**  Figure 11.38 shows a mod-16 counter that controls the operation of two digital sequential circuits, labeled Cct.1 and Cct.2. Cct.1 is positive edge-triggered and clocked by counter output $Q_B$. Cct.2 is negative edge-triggered and clocked by $Q_D$.

  **a.** Draw the timing diagram for one complete cycle of the circuit operation. Draw arrows on the active edges of the waveforms that activate Cct.1 and Cct.2.

  **b.** State how many times Cct.1 is clocked for each time that Cct.2 is clocked.

**11.4**  Draw the timing diagram for one complete cycle of a mod-8 counter, including waveforms for $CLK$, $Q_A$, $Q_B$, and $Q_C$, where $Q_A$ is the LSB.

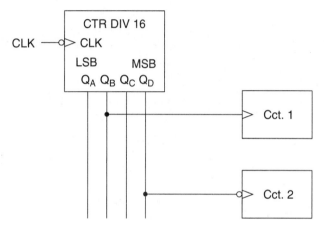

**Figure 11.38**
**Problem 11.3**
**Mod-16 Counter**

**11.5**  How many bits are required to make a counter with a modulus of 64? Why? What is the maximum count of such a counter?

**11.6**  **a.** Draw the state diagram of a mod-10 UP counter.

  **b.** Use the state diagram drawn in part a to answer the following questions.
    i. The counter is at state 0111. What is the count after seven clock pulses are applied?
    ii. After five clock pulses, the counter output is at 0001. What was the counter state prior to the clock pulses?
    \*iii. The counter output is at 1000 after 15 clock pulses. What was the original output state?

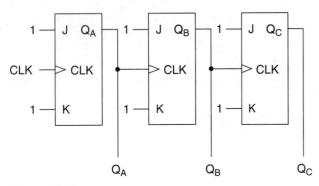

**Figure 11.39**

**Problem 11.14**
**Counter**

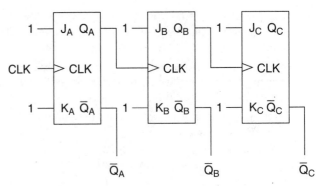

**Figure 11.40**

**Problem 11.15**
**Counter**

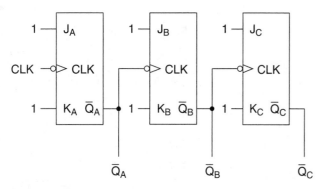

**Figure 11.41**

**Problem 11.16**
**Counter**

**11.7** What is the maximum modulus of a 6-bit counter? A 7-bit counter? 8-bit?

**11.8** Draw the count-sequence table and timing diagram of a mod-10 UP counter.

**11.9** Draw the state diagram, count-sequence table, and timing diagram of a mod-10 DOWN counter.

**11.10** A mod-16 counter is clocked by a waveform having a frequency of 48 kHz. What is the frequency of each of the waveforms at $Q_D$, $Q_C$, $Q_B$, and $Q_A$?

**\*\*11.11** A mod-10 counter is clocked by a waveform having a frequency of 48 kHz. What is the frequency of the $Q_D$ output waveform? The $Q_A$ waveform? Why is it difficult to determine the frequencies of $Q_B$ and $Q_C$?

## Section 11.2 Asynchronous Counters

**11.12** Draw the logic circuit of a 3-bit asynchronous counter, using negative edge-triggered JK flip-flops. How high can the counter count? What is its modulus?

**11.13** Repeat Problem 11.12 for a 4-bit asynchronous counter.

**11.14** Draw the timing diagram for one complete cycle of the counter shown in Figure 11.39.

**11.15** Repeat Problem 11.14 for the circuit in Figure 11.40.

**11.16** Repeat Problem 11.14 for the circuit in Figure 11.41.

**11.17** Draw the circuit for a mod-16 asynchronous DOWN counter made from negative edge-triggered JK flip-flops. Draw the timing diagram and state diagram of the counter.

**11.18** Draw the timing diagram of a 4-bit asynchronous counter as monitored at $Q_D$ $Q_C$ $Q_B$ $Q_A$. Next draw the timing diagram of the same circuit at $\overline{Q}_D$ $\overline{Q}_C$ $\overline{Q}_B$ $\overline{Q}_A$. How are the resulting count sequences different?

**\*11.19** The input *CLK* waveform of the counter in Figure 11.9 is inverted. How does this affect the timing diagram?

**11.20** Draw the count-sequence table and timing diagram of the circuit in Figure 11.42, and including glitches, for *CLK*, $Q_A$, $Q_B$, $Q_C$, $Q_D$, and *CLR*. Draw one full cycle of the circuit output. Assume that all $Q$s are initially LOW.

**Figure 11.42**

**Problem 11.20
Counter**

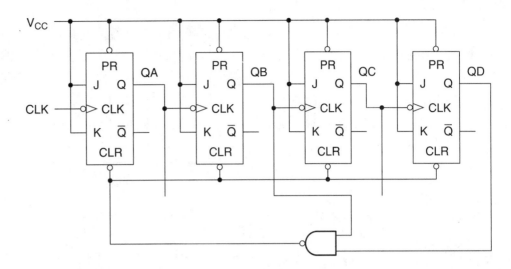

**Figure 11.43**

**Problem 11.21
Counter**

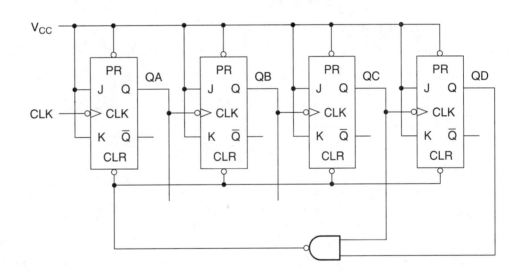

**11.21** Repeat Problem 11.20 for the logic diagram shown in Figure 11.43.

**11.22** Draw the circuit of an asynchronous mod-10 DOWN counter made from negative edge-triggered JK flip-flops.

**\*\*11.23** **a.** Draw the circuit of a mod-10 UP counter whose *starting point* is $Q_D$ $Q_C$ $Q_B$ $Q_A = 0011$. The counter should be constructed from 74HC76 flip-flops (negative edge-triggered, active-LOW Preset and Clear).

**b.** Draw the state diagram, count-sequence table, and timing diagram of the counter in part a.

### Section 11.3 Decoding a Counter

**\*11.24** A design for a mod-16 DOWN counter specifies that a decoder must indicate the state just before the counter's recycle point by producing a HIGH output. Draw the simplest circuit that will perform this function.

**\*11.25** Modify the circuit you drew for Problem 11.24 to give a LOW indication of the last state before the recycle point.

**11.26** **a.** Figure 11.44 shows a mod-8 asynchronous counter with a single decoding gate. Draw the timing diagram of the counter and decoder for one complete cycle, showing the

**Figure 11.44**

**Problem 11.27**
**Mod-8 Asynchronous**
**Counter**

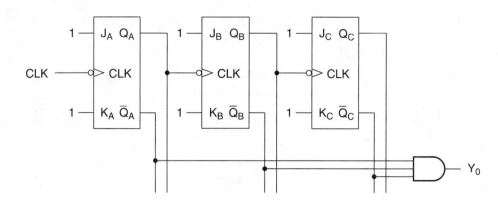

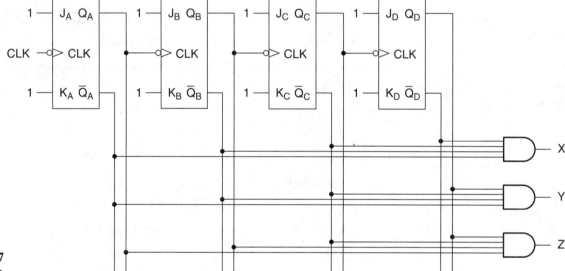

**Figure 11.45**

**Problem 11.27**
**4-Bit Counter**

effect of all propagation delays. Show waveforms for *CLK*, $Q_A$, $Q_B$, $Q_C$, $Q_D$, and $Y_0$.

**b.** Modify the circuit to eliminate any glitches on the decoder output.

**\*\*11.27** A 4-bit counter is partially decoded as shown in Figure 11.45.

    **a.** Draw the timing diagram of the circuit, showing the clock waveform, the *Q* outputs (including propagation delay), and the decoder outputs (including glitches).

    **b.** Modify the circuit in Figure 11.45 to eliminate any decoding glitches.

**11.28** A negative edge-triggered mod-16 asynchronous counter is fully decoded, with decoder outputs labeled $Y_0$ to $Y_{15}$. If no steps are taken to suppress decoder glitches, on which output(s) will glitches appear for each of the following transitions on $Q_D Q_C Q_B Q_A$?

    **a.** 0011–0100

    **b.** 0101–0110

    **c.** 0111–1000

    **d.** 1000–1001

**\*11.29** A mod-16 counter is constructed from flip-flops having a propagation delay of 15 ns. The input clock pulse has a period of 90 ns and a duty cycle of 50%. Explain, using a timing diagram, whether or not the maximum frequency of the counter has been exceeded.

**11.30** Calculate the maximum frequency of a mod-16 counter constructed of negative edge-triggered flip-flops having a propagation delay of 15 ns. Assume that the counter output is fully decoded and that decoder glitches must be suppressed. The clock input has a duty cycle of 50%.

**11.31** Repeat Problem 11.30 for a clock duty cycle of 20%.

**\*\*11.32** Repeat Problem 11.30 for a clock duty cycle of 20% and positive edge-triggered flip-flops.

## Section 11.4 Synchronous Counters

**11.33** Draw the circuit for a synchronous mod-16 UP counter made from negative edge-triggered JK flip-flops.

**11.34** Explain how to change the circuit drawn in Problem 11.33 to make it count DOWN.

**11.35** Explain how to extend the counter drawn in Problem 11.33 to make it into a mod-64 counter.

**11.36** Write the $J$ and $K$ equations for the MSB of a synchronous mod-256 (8-bit) UP counter.

**11.37** Repeat Problem 11.36 for a mod-256 DOWN counter.

**Table 11.11** Problem 11.39 Count-Sequence Table

| CLK | UP/$\overline{DOWN}$ | $Q_D$ | $Q_C$ | $Q_B$ | $Q_A$ |
|-----|----------|-------|-------|-------|-------|
| 0 | 1 | 0 | 0 | 0 | 0 |
| 1 | 1 | | | | |
| 2 | 1 | | | | |
| 3 | 1 | | | | |
| 4 | 0 | | | | |
| 5 | 0 | | | | |
| 6 | 0 | | | | |
| 7 | 0 | | | | |
| 8 | 0 | | | | |
| 9 | 0 | | | | |
| 10 | 1 | | | | |
| 11 | 1 | | | | |

**11.38 a.** The mod-16 bidirectional counter shown in Figure 11.33 is initially in the state shown in Table 11.11. Complete this count-sequence table for the number of clock pulses shown.

**b.** Draw the timing diagram corresponding to the sequence in Table 11.11.

**11.39** Analyze the operation of the synchronous counter in Figure 11.46 by drawing a state table showing all transitions, including unused states. Use this state table to draw a state diagram and a timing diagram. What is the counter's modulus?

**Figure 11.46**

**Problem 11.39
Synchronous Counter**

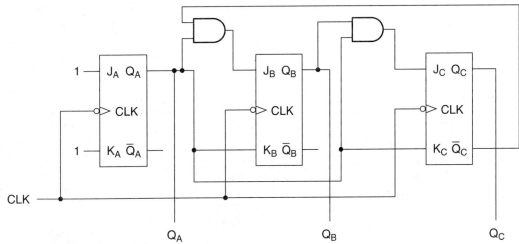

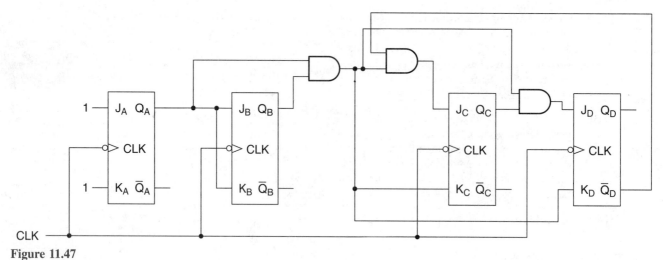

**Figure 11.47**

**Problem 11.40**
**Synchronous Counter**

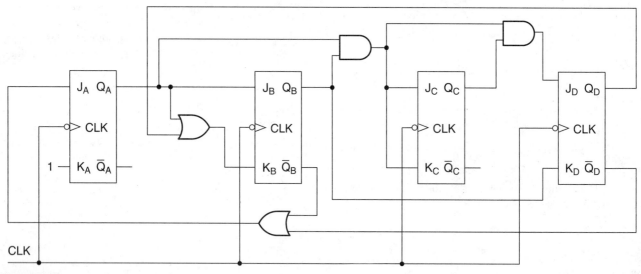

**Figure 11.48**

**Problem 11.41**
**Counter**

**11.40** **a.** Write the equations for the $J$ and $K$ inputs of each flip-flop of the synchronous counter represented in Figure 11.47.

   **b.** $Q_D \, Q_C \, Q_B \, Q_A = 1010$. Use the equations from part a to predict the circuit outputs after the circuit is clocked three times.

**11.41** Analyze the operation of the counter shown in Figure 11.48. Predict the count sequence by determining the $J$ and $K$ inputs and resulting transitions for each counter output state. Draw the state diagram and the timing diagram. Assume that all flip-flop outputs are initially 0.

# ▼ Answers to Section Review Questions

### Section 11.1

**11.1.** A mod-24 counter requires five outputs. It is a truncated-sequence counter, since the maximum modulus for a 5-bit counter is 32. Twenty-four is less than $m_{max}$.

### Section 11.2a

**11.2.** $f = 1$ kHz, $DC = 50\%$.

### Section 11.2b

**11.3.** The count sequence will not be reversed, because the inverter affects only the incoming *CLK* waveform, not any of the following *Q-CLK-Q* connections. The inversion will change the *circuit* from positive to negative edge-triggered or vice versa.

### Section 11.2c

**11.4.** $Q_C$.

### Section 11.3a

**11.5.** A 4-input AND gate, connected to all four *Q* outputs.

### Section 11.3b

**11.6.** 3 outputs. $Y_{14}$, $Y_{12}$, and $Y_8$. Transition: 1111–(1110)–(1100)–(1000)–0000

### Section 11.3c

**11.7.** 8.33 MHz.

### Section 11.4a

**11.8.**
1. Clock: asynchronous: flip-flops clock separately.
   synchronous: flip-flops clock together.
2. Mode: asynchronous: flip-flops permanently in toggle mode.
   synchronous: flip-flop modes change throughout cycle.

### Section 11.4b

**11.9.**
$$J_E = K_E = \overline{Q}_A \cdot \overline{Q}_B \cdot \overline{Q}_C \cdot \overline{Q}_D$$
$$J_F = K_F = \overline{Q}_A \cdot \overline{Q}_B \cdot \overline{Q}_C \cdot \overline{Q}_D \cdot \overline{Q}_E$$

12

# MSI Counters and Synchronous Counter Design

## ● CHAPTER OUTLINE

**12.1** Asynchronous MSI Counters

**12.2** Synchronous Presettable MSI Counters

**12.3** IEEE/ANSI Notation

**12.4** Design of Synchronous Sequential Circuits

**12.5** Sequential Logic Applications of PLDs

## ● CHAPTER OBJECTIVES

Upon successful completion of this chapter, you will be able to:

- Use several types of asynchronous and synchronous MSI counters for various applications.

- Set an MSI counter for any modulus up to its maximum.

- Use a bidirectional MSI counter for various applications.

- Extend the maximum modulus of a counter by cascading it with one or more counters.

- Determine the overall modulus of a counter circuit consisting of two or more MSI devices.

- Design a synchronous sequential circuit having any specified output sequence.

- Program a PLD as a synchronous sequential circuit.

## ● INTRODUCTION

A wide variety of medium-scale integration (MSI) counters is commercially available in both CMOS and TTL technologies. As a result, modern digital applications seldom make use of counters constructed directly from flip-flops.

MSI counters are available as full-sequence binary or BCD (mod-10) counters, and nearly all have some provision for truncation, either synchronously or asynchronously. Many also have presettable inputs that allow the counter state to be set to any value when a Control input, usually called *LOAD,* is made active.

Any MSI counter can be cascaded, or extended, asynchronously by connecting the most significant bit output to the *CLK* input of another counter. Some synchronous MSI counters have provision for synchronous cascading as well.

In addition to the analysis of MSI counters, we will study the design of synchronous sequential circuits. This is a large class of circuits, of which synchronous counters form only a small part.

We will also study programmable logic devices used in counter and other synchronous applications.

**Figure 12.1**

**Asynchronous MSI Counters**

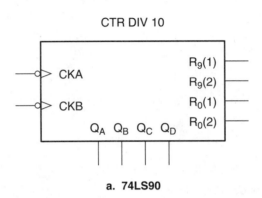

a. 74LS90

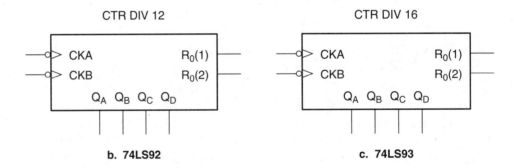

b. 74LS92          c. 74LS93

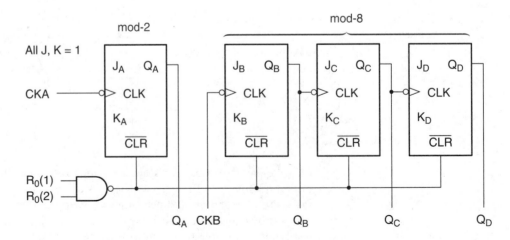

d. Internal configuration of 74LS93

## 12.1

# Asynchronous MSI Counters

**Biquinary sequence**  *A mod-10 count sequence consisting of two groups of five states. The sequence has a 50% duty cycle on its most significant bit.*

A family of simple and versatile asynchronous counters is the 74LS90/92/93 series of asynchronous MSI counters. Logic symbols for each of these are shown in Figure 12.1.

Each version has a different modulus: the 74LS90 is a decade, or mod-10, counter; the 74LS92 has a modulus of 12, which is useful as a divide-by-12 circuit for digital timekeeping; the 74LS93 is a 4-bit binary (mod-16) device.

These counters are internally configured so that the least significant bit is clocked separately from the rest of the counter. Thus, each device consists of a mod-2 counter (a flip-flop) and a mod-$(n/2)$ counter. For example, the 74LS93 device is made up of a mod-2 and a mod-8 counter, as shown in Figure 12.1d. Each counter has its own clock input: $CKA$ clocks the LSB and $CKB$ the remaining bits. To extend the modulus to 16, the external $CLK$ pulse should be connected to $CKA$ and $Q_A$ to $CKB$, as illustrated in Figure 12.2. That is, the two parts of the counter should be connected in cascade.

The 74LS92 counter actually consists of a mod-2 counter, clocked by $CKA$, and a pair of cascaded counters, mod-3 for $Q_B$ and $Q_C$, mod-2 for $Q_D$, with the mod-3 counter clocked by $CKB$. This internal arrangement is shown in Figure 12.3.

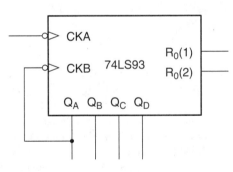

**Figure 12.2**

**Configuration of a 74LS93 Counter as a Mod-16 Binary Counter**

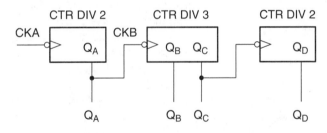

**a.  Block diagram of a 74LS92 counter**

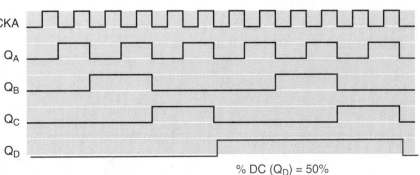

% DC ($Q_D$) = 50%
$f_D$ = f/12 for clock frequency f

**b.  7492 Timing diagram**

**Figure 12.3**

**Internal Block Diagram and Timing Diagram of a 74LS92 Mod-12 Counter**

**Table 12.1** Count Sequence of
a 74LS92 Mod-12 Counter

| $Q_D$ | $Q_C$ | $Q_B$ | $Q_A$ |
|-------|-------|-------|-------|
| 0 | 0 | 0 | 0 |
| 0 | 0 | 0 | 1 |
| 0 | 0 | 1 | 0 |
| 0 | 0 | 1 | 1 |
| 0 | 1 | 0 | 0 |
| 0 | 1 | 0 | 1 |
| 1 | 0 | 0 | 0 |
| 1 | 0 | 0 | 1 |
| 1 | 0 | 1 | 0 |
| 1 | 0 | 1 | 1 |
| 1 | 1 | 0 | 0 |
| 1 | 1 | 0 | 1 |

The circuit of the 74LS92 counter gives rise to the count sequence represented in Table 12.1.

Note that the mod-3 sequence for $Q_C Q_B$ is 00, 01, 10, with each change occurring on the 1-to-0 transition (negative edge) of $Q_A$. $Q_D$ changes on the negative edge of $Q_C$. The resultant sequence of $Q_D Q_C Q_B Q_A$ does not count directly from 0000 to 1011. Rather, it counts from 0000 to 0101, then from 1000 to 1101. This is still a total of 12 states, thus making the device a mod-12 counter.

The major advantage of this sequence is that it generates a symmetrical waveform when the counter is used as a divide-by-12 circuit. The $Q_D$ waveform has a 50% duty cycle and a frequency given by $f/12$ for a clock frequency of $f$. You can see this by examining the timing diagram in Figure 12.3b. This is not the standard output waveform of a mod-12 binary-sequence counter.

The internal circuit of the 74LS92 uses the synchronous inputs of the flip-flops to truncate the count sequence. Therefore, the output waveforms have no glitches. The counter is still considered asynchronous because not all internal clocks are common.

The 74LS90 counter consists of a mod-2 counter clocked by *CKA* and a mod-5 counter clocked by *CKB*. This device can be configured to yield a standard binary count or a **biquinary** count (two groups of five with a 50% duty cycle on the MSB output).

---

**EXAMPLE 12.1**

Figure 12.4 shows a 74LS90 decade counter connected two ways: with the counter clocked by *CKA* and with it clocked by *CKB*. In Figure 12.4a, the sequence of outputs from most to least significant bit is $Q_D Q_C Q_B Q_A$. In Figure 12.4b, the sequence is $Q_A Q_D Q_C Q_B$. All *R*, or *RESET*, inputs are disabled by grounding them.

Draw the timing diagram and write the count sequence for each circuit. Briefly describe the difference between the two sequences.

**Solution**    The 74LS90 counter consists of a mod-2 counter, $Q_A$, clocked by *CKA*, and a mod-5 counter, $Q_D Q_C Q_B$, clocked by *CKB*. The $Q_D Q_C Q_B$ waveforms count from 000 to 100 and recycle synchronously.

**Figure 12.4**

**Example 12.1
Configuring a 74LS90
Counter**

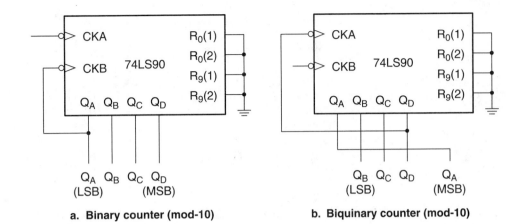

a. **Binary counter (mod-10)**

b. **Biquinary counter (mod-10)**

**Figure 12.5**

**Example 12.1
74LS90 Timing Diagrams**

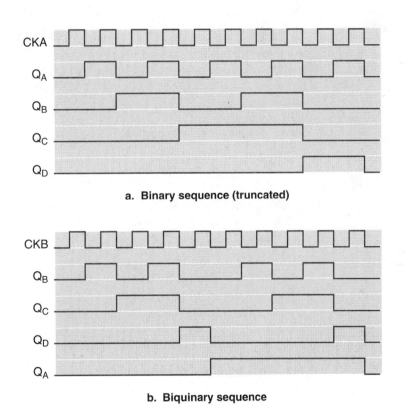

a. **Binary sequence (truncated)**

b. **Biquinary sequence**

If $CKB$ is driven by $Q_A$, the counter generates a mod-10 binary count sequence, as shown in Figure 12.5a.

If the circuit clock is $CKB$ and if $CKA$ is driven by $Q_D$, the 74LS90 generates a biquinary sequence, illustrated in Figure 12.5b. Note that this sequence produces a symmetrical divide-by-10 waveform at $Q_A$.

The count sequences are given in Table 12.2.

Table 12.2 74LS90 Count Sequences

| $Q_D$ | $Q_C$ | $Q_B$ | $Q_A$ | | $Q_A$ | $Q_D$ | $Q_C$ | $Q_B$ | |
|---|---|---|---|---|---|---|---|---|---|
| 0 | 0 | 0 | 0 | <— | 0 | 0 | 0 | 0 | <— |
| 0 | 0 | 0 | 1 | | 0 | 0 | 0 | 1 | |
| 0 | 0 | 1 | 0 | | 0 | 0 | 1 | 0 | |
| 0 | 0 | 1 | 1 | | 0 | 0 | 1 | 1 | |
| 0 | 1 | 0 | 0 | | 0 | 1 | 0 | 0 | |
| 0 | 1 | 0 | 1 | | 1 | 0 | 0 | 0 | |
| 0 | 1 | 1 | 0 | | 1 | 0 | 0 | 1 | |
| 0 | 1 | 1 | 1 | | 1 | 0 | 1 | 0 | |
| 1 | 0 | 0 | 0 | | 1 | 0 | 1 | 1 | |
| 1 | 0 | 0 | 1 | | 1 | 1 | 0 | 0 | |
| **Binary Sequence** | | | | | **Biquinary Sequence** | | | | |

Figure 12.6 shows the *RESET* circuitry for the 74LS90/92/93 counters. Each counter has a reset-to-0 function, and the 74LS90 also has a reset-to-9 function. These inputs can be used to clear the counter at the beginning or end of the count sequence.

Since the internal *RESET* gates are NANDs, both $R_0(1)$ and $R_0(2)$ must be HIGH to make $Q_D Q_C Q_B Q_A = 0000$. For the 74LS90 counter, when $R_9(1)$ and $R_9(2)$ are both HIGH, $Q_D Q_C Q_B Q_A = 1001$. The *RESET* inputs can be inhibited by grounding one or both of the inputs for each function. If no *RESET* inputs are used, it is good practice to tie all of them LOW.

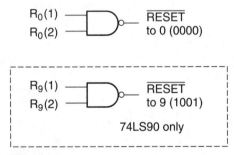

Figure 12.6
**Reset Inputs for 74LS90/92/93 Counters**

**EXAMPLE  1 2 . 2**     Draw the circuit of a 74LS93 binary counter configured as a mod-10 counter. No external logic is required.

**Solution**     A mod-10 counter has a count sequence from 0000 to 1001. Since the counter resets asynchronously, its output goes to a temporary recycle state, $Q_D Q_C Q_B Q_A = 1010$. Connect $R_0(1)$ and $R_0(2)$ to $Q_D$ and $Q_B$, as in Figure 12.7, to configure the 74LS93 as a mod-10 counter.

**Figure 12.7**

**Example 12.2
Mod-10 Counter From
74LS93**

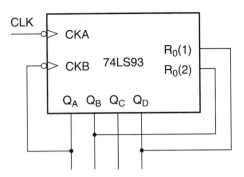

## 12.2

# Synchronous Presettable MSI Counters

> **Presettable counter** *A counter whose contents can be set to any value by loading a binary number directly into the internal flip-flops.*

Synchronous MSI counters are available in a number of variations. For a particular variation, there is usually a choice of modulus, mod-16 or mod-10, and most synchronous counters are **presettable.** That is, the counter output can be set to any value by loading a binary value directly into the flip-flops. Presets can be synchronous, in which case they load the flip-flops on an active *CLK* edge, or asynchronous, in which case they do not wait for the *CLK*.

Most MSI synchronous counters are easily expandable since they have outputs that can connect directly to a *CLK* or *ENABLE* input of another, higher-order counter. The cascading can be asynchronous or synchronous.

When counters are cascaded asynchronously, a special cascading output of one counter (usually called Ripple Clock Out) connects to the *CLK* of the next. In a synchronous cascade connection, the *CLK*s on all counter chips are driven by a common source. A higher-order counter is generally enabled by a cascading output of the previous device at the appropriate time.

Some MSI counters are bidirectional. That is, they can count UP or DOWN, depending on the state of one or more control inputs.

Table 12.3 shows variations in counter features for several TTL and High-Speed CMOS devices.

## Bidirectional Synchronous Counters

Both the 74LS191 and 74LS193 counters are bidirectional 4-bit binary counters, but the method of selecting the count direction is quite different in each device. Figure 12.8 shows the logic symbols for each device, as well as the symbol for the 74LS163A counter, which is unidirectional.

**Table 12.3** MSI Synchronous Counter Features

| Device | | | | |
| BCD (mod-10) | Binary (mod-16) | DOWN/UP Method | Presets | Cascading |
| --- | --- | --- | --- | --- |
| 74LS160A | 74LS161A | None | Async | Sync/Async |
| 74HC160 | 74HC161 | None | Async | Sync/Async |
| 74LS162A | 74LS163A | None | Sync | Sync/Async |
| 74HC162 | 74HC163 | None | Sync | Sync/Async |
| 74LS190 | 74LS191 | $D/U$ control input | Async | Sync/Async |
| 74LS192 | 74LS193 | Two clocks | Async | Async |

**Figure 12.8**

**MSI Synchronous Binary Counter**

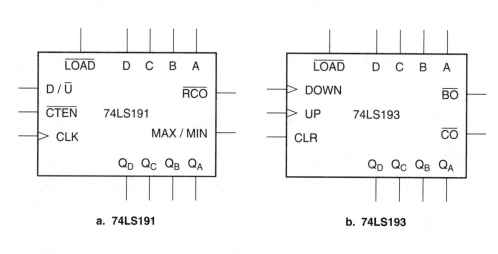

a. 74LS191

b. 74LS193

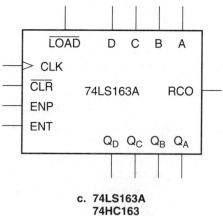

c. 74LS163A
74HC163

The 74LS191 counter has a $DOWN/\overline{UP}$ control input and a single clock. To count UP, $D/\overline{U}$ and $\overline{CTEN}$ (Count Enable) must be LOW. To count DOWN, $D/\overline{U} = 1$ and $\overline{CTEN} = 0$. The 74LS193 device has two clock inputs: one for counting UP, the other for counting DOWN. To enable either clock input, the other one must be HIGH.

The logic circuits of these devices are shown in Figures 12.9 and 12.10.

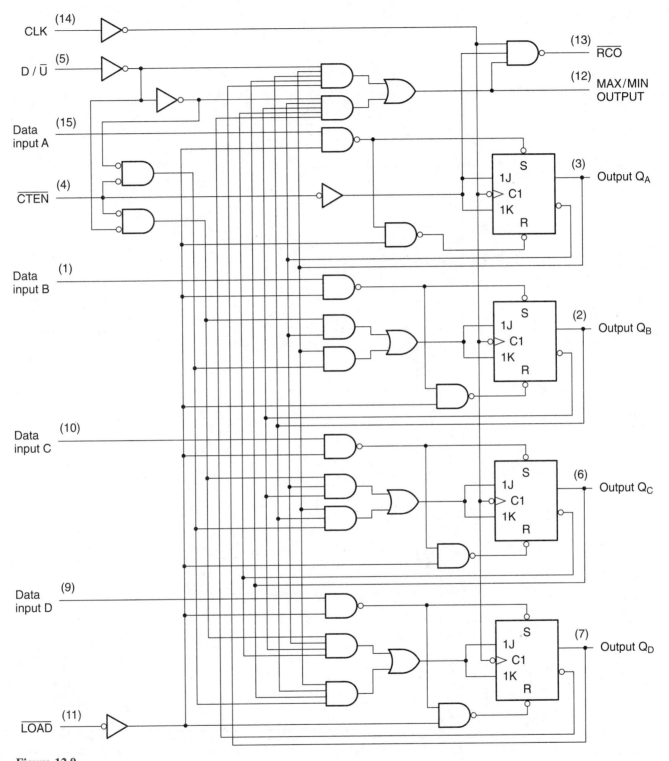

**Figure 12.9**
**74LS191 Counter** (Reprinted by permission of Texas Instruments)

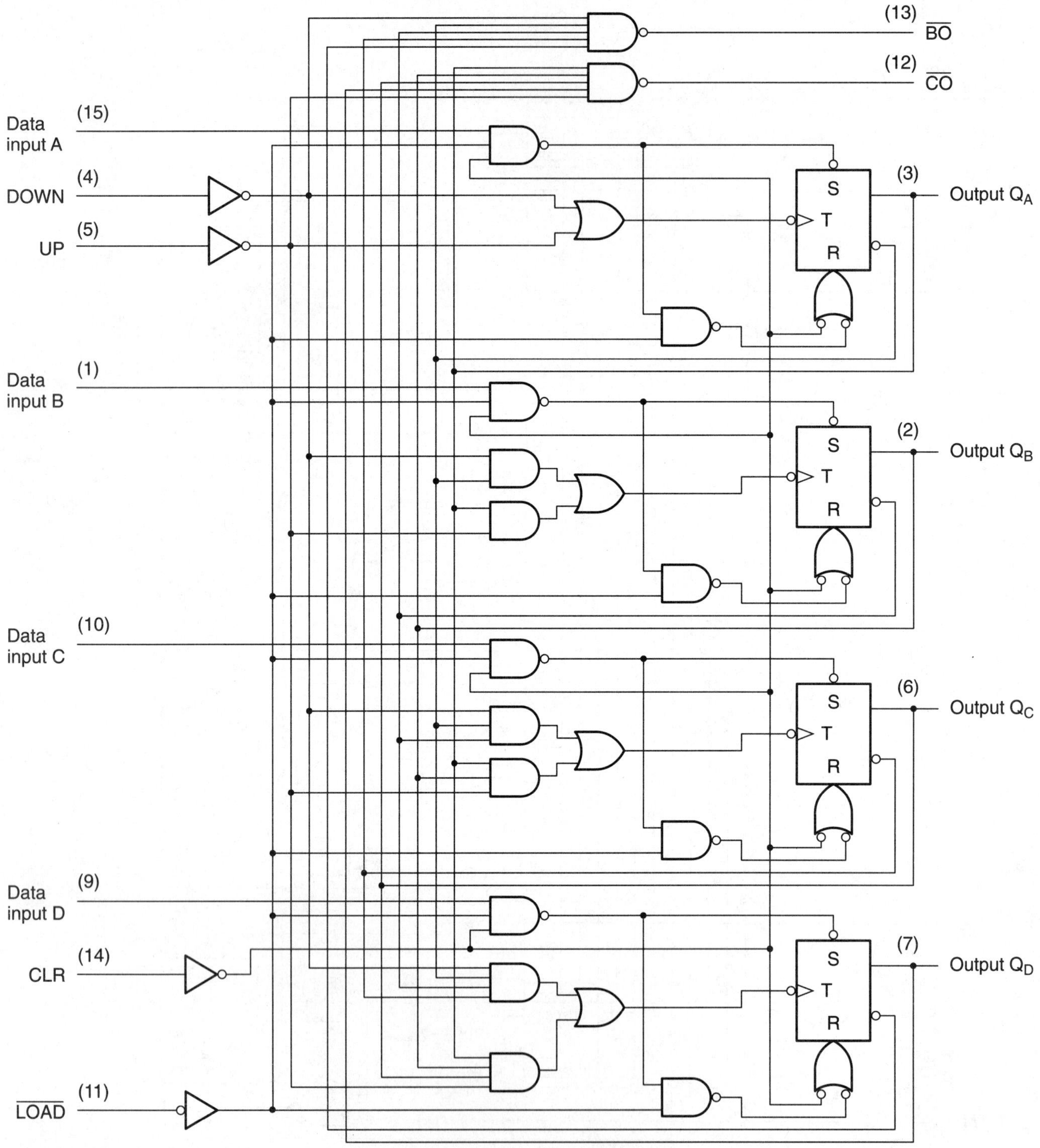

**Figure 12.10**

**74LS193 Counter** (Reprinted by permission of Texas Instruments)

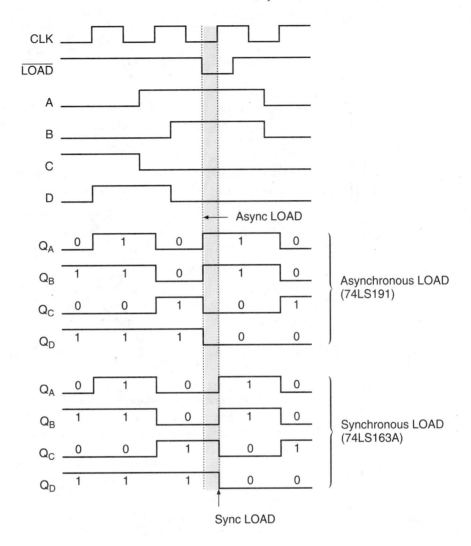

**Figure 12.11**

**Synchronous and Asynchronous LOAD Function**

## Presettable MSI Synchronous Counters

All three counters shown in Figure 12.8 have Preset inputs, labeled *D, C, B,* and *A.* When the $\overline{LOAD}$ input of any of these devices is pulled LOW, the bits *DCBA* are loaded directly into the internal flip-flops of the counter. In the case of the 74LS191 and '193 counters, this load operation is asynchronous; the bits are loaded without waiting for an active edge on the *CLK* input. The 74LS163A counter is loaded synchronously, that is, by the *CLK.*

Figure 12.11 shows partial timing diagrams of a 74LS191 and a 74LS163A counter, each with the same set of inputs. The counters progress from 1010 to 1100, then are loaded with 0011 and continue to count up. The outputs of the '191 counter go from 1100 to 0011 as soon as $\overline{LOAD}$ goes LOW. The '163A outputs do not change until the next positive edge on the *CLK* input.

Even though the 74LS191 counter is synchronous, one or more outputs could glitch at the recycle point if the $\overline{LOAD}$ input were used to truncate the '191 count sequence.

**EXAMPLE 12.3**

An automatic packing system in a food processing plant is used to pack cans of fruit juice in shrink-wrapped packs of 12. Figure 12.12 illustrates a part of the system used for keeping track of the number of cans before they are packed.

When a can is filled, it passes optical sensor $F$, which generates a pulse. The can passes along the packing line to a scale, where it is weighed to see if it is properly filled. If the can is underweight, the scale output $S$ goes HIGH and the can is removed from the line. If a can is rejected, the scale output remains HIGH briefly and another optical sensor $R$ produces a pulse.

When the $\overline{WRAP}$ output goes LOW, it activates the next part of the packing line, which shrink-wraps the cans and starts a new cycle.

**a.** Briefly explain how the circuit keeps track of the number of cans that have been properly filled.

**b.** Complete the timing diagram in Figure 12.13 and briefly explain how we should interpret the diagram.

**Solution**

**a.** The scale output $S$ is normally LOW, meaning that the 74LS191 counter normally counts UP. Each filled can makes sensor $F$ produce a pulse, incrementing the counter by 1. If the can is underweight, $S$ changes the count direction by going HIGH. The pulse from sensor $R$ decrements the counter by 1. When 12 cans pass the line without being rejected, the counter rolls over to 1101, activating the $\overline{WRAP}$ output. The $\overline{WRAP}$ output also loads 0001 into the counter asynchronously so that the cycle can start again.

To work properly, the circuit must have a chance to reject any can. Thus, the machine cannot wrap the pack until after the 12th can could be rejected. We must wait until the $F$ pulse *after* the 12th can to wrap and reset.

This means that one can has already been counted for the next cycle. Therefore, the new cycle must begin at 0001. It's OK for the *first* cycle to begin at 0000, since it has no previous history of pulses from $F$; this shows how we can use the Preset inputs to treat the first counter cycle differently from the rest.

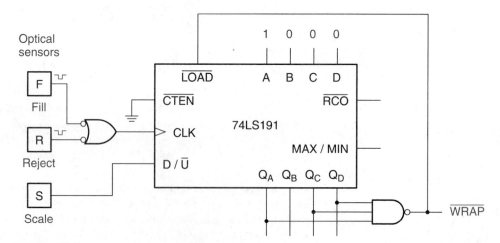

**Figure 12.12**

**Example 12.3**
**Can Counter in a Packing Plant**

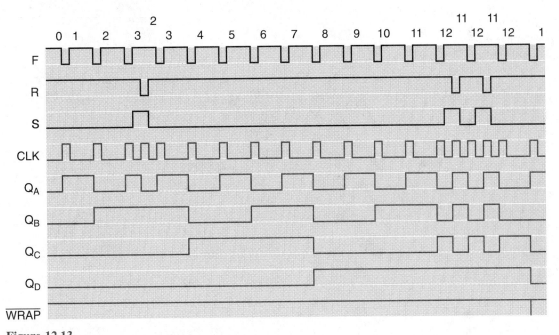

**Figure 12.13**

**Example 12.3**
**Timing Diagram for Circuit in Figure 12.12**

**b.** The completed timing diagram is illustrated in Figure 12.13. The following action occurs:

- Two cans are accepted.
- The third can is rejected.
- All cans are accepted until number 12.
- Two cans are rejected.
- The 12th can is accepted. The pack is wrapped and the cycle starts over.

Even though the output reaches 1100 three times, the pack is not wrapped until the 12th can is accepted.

**EXAMPLE 12.4**    Show how a 74LS193 counter can be used instead of the 74LS191 in the packing system of Example 12.3.

**Solution**    Figure 12.14 shows the required circuit.

The 74LS193 counter requires a HIGH at one clock input and a pulse at the other to count either UP or DOWN. We can use a NAND gate at each clock input, since the inhibit state of a NAND output is HIGH.

The $S$ output will enable the UP gate if LOW and the DOWN gate if HIGH. The 74LS193 is clocked UP or DOWN through the enabled gate by pulses from the $F$ or $R$ sensor.

The rest of the circuit operates the same as in Figure 12.12.

**Figure 12.14**
**Example 12.4**
**Can Counter Based on a**
**74LS193 Counter**

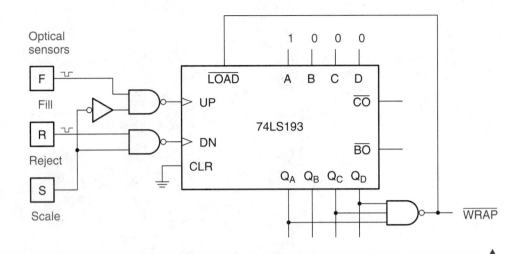

## Cascading MSI Counters

> **Terminal count** *The state of a counter's output in the last state before the recycle point.*

As we have seen in Chapter 11, a counter made up of individual flip-flops can be extended indefinitely, either synchronously or asynchronously. It should be possible to increase the modulus of an MSI counter asynchronously by using its MSB waveform to clock another MSI counter and thus add another set of bits to the counter output.

One difficulty is that the synchronous MSI counters examined so far are all positive edge-triggered devices and thus require a $\overline{Q}$ output to clock another counter. However, these counters have only $Q$ outputs. Fortunately, the MSI counters listed in Table 12.3 all have internal cascading logic so that direct connections can be made from one counter to another.

**Figure 12.15**
**Cascading Circuitry of**
**74LS191 Counter**

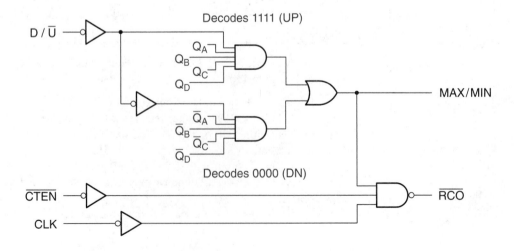

The 74LS191 counter has two outputs used for cascading:

1. *MAX/MIN,* which goes HIGH for one clock cycle during the **terminal count** (i.e., when $Q_DQ_CQ_BQ_A = 1111$ during an UP count and when $Q_DQ_CQ_BQ_A = 0000$ during a DOWN count)
2. $\overline{RCO}$ (Ripple Clock Out), which reproduces one *CLK* cycle when the *MAX/MIN* output is HIGH AND $\overline{CTEN}$ (Count Enable) is LOW

Figure 12.15 shows the cascading circuitry for a 74LS191 4-bit binary counter.

**EXAMPLE 12.5**

The timing diagram of a 74LS191 counter is shown in Figure 12.16. Complete the timing diagram by drawing the waveforms for the *MAX/MIN* and $\overline{RCO}$ outputs.

**Solution**

The *MAX/MIN* and $\overline{RCO}$ waveforms are shown in Figure 12.16. The $D/\overline{U}$ input enables one of two internal decoding gates, depending on the count direction. For the UP count, the 1111 decoder is enabled, and for the DOWN count, the 0000 decoder is enabled. Thus *MAX/MIN* is HIGH for one clock cycle if (the direction is UP AND $Q_DQ_CQ_BQ_A = 1111$) OR (the direction is DOWN AND $Q_DQ_CQ_BQ_A = 0000$). The corresponding Boolean expression is:

$$MAX/MIN = (\text{UP} \cdot Q_DQ_CQ_BQ_A) + (\text{DOWN} \cdot \overline{Q_D}\,\overline{Q_C}\,\overline{Q_B}\,\overline{Q_A})$$

This gives an active-HIGH indication of the terminal count.

The $\overline{RCO}$ output reproduces the *CLK* input waveform as long as $\overline{CTEN}$ and *MAX/MIN* are both active. The double inversion generated by the *CLK* inverter and the NAND output ensure that the $\overline{RCO}$ output is in phase with the *CLK* input. Since *MAX/MIN* is active for only one clock cycle in the entire count sequence, the $\overline{RCO}$ output produces one *CLK* pulse at the end of the sequence.

**Figure 12.16**

**Example 12.5
MAX/MIN and RCO
Waveforms (74LS191)**

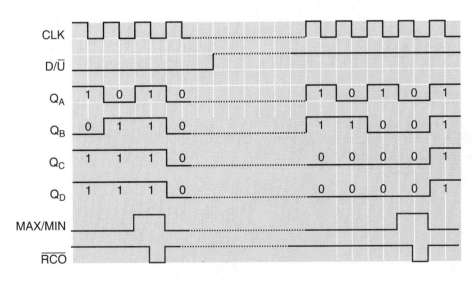

The $\overline{RCO}$ output of a 74LS191 counter can be used to cascade the counter either synchronously or asynchronously.

To cascade two or more counters asynchronously, connect the $\overline{RCO}$ output of one counter to the *CLK* of the next one, as shown in Figure 12.17. The $\overline{RCO}$ output generates the required positive edge exactly at the end of the lower-order device's count sequence. This ensures that the count sequences of both counters are properly synchronized.

Figure 12.18 shows two 74LS191 counters cascaded synchronously. The *CLK* inputs are common, as are the $D/\overline{U}$ inputs, ensuring that both devices are clocked together and in the same direction. The $\overline{RCO}$ output of the lower-order counter connects to the $\overline{CTEN}$ input of the higher-order device. The cascading logic of the first counter feeds the logical product of the outputs ($Q_D Q_C Q_B Q_A$ or $\overline{Q_D}\,\overline{Q_C}\,\overline{Q_B}\,\overline{Q_A}$) to the flip-flops of the second counter. Recall that these are the Boolean conditions for synchronous UP or DOWN counting.

The overall modulus of several full-sequence counters in cascade is the product of the individual counter moduli. (For example, a mod-16 counter in cascade with a mod-16 counter, as in Figure 12.18, has a total modulus of $16 \times 16 = 256$. A mod-16 and a mod-10 counter in cascade have a total modulus of $16 \times 10 = 160$.)

The lower-order counter in Figure 12.18 produces an $\overline{RCO}$ pulse once for every 16 *CLK* pulses. The higher-order counter goes through its count sequence once every 16 $\overline{RCO}$ pulses. The total number of input *CLK* pulses needed to make the second counter recycle is $16 \times 16 = 256$.

**Figure 12.17**

**Asynchronous Cascading of 74LS191 Counters**

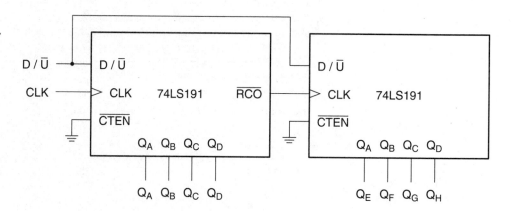

**Figure 12.18**

**Synchronous Cascading of 74LS191 Counters**

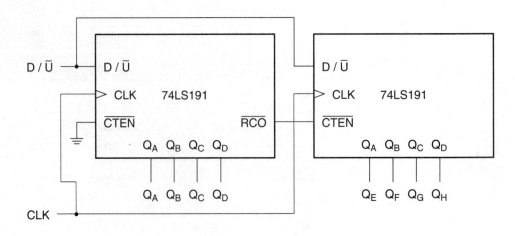

> **Section Review Problem for Section 12.2a**
>
> 12.1.  What is the modulus of a counter circuit consisting of two 74LS190 decade counters connected in cascade? What is the total modulus of three cascaded 74LS190 counters?

### Application: Frequency Counter

A frequency counter is a circuit that measures the frequency of a time-varying input waveform. The waveform could be analog or digital, but we will concern ourselves only with digital waveforms. (An analog frequency counter requires a wave-shaping circuit that will "square up" a waveform and thus make it act like a digital pulse.)

The fundamental design concept of this circuit is that a counter is enabled for a fixed period of time known as the "sampling interval." During the sampling interval, the circuit counts pulse transitions on the input waveform. When the sampling interval ends, the circuit output displays the counter state, usually in decimal numerical form. Since the input frequency is proportional to the number of input pulse transitions, we can adjust the sampling interval to display the output in Hz, kHz, or any other convenient unit.

---

**EXAMPLE 12.6**

Figure 12.19 shows the circuit of a four-digit frequency counter based on four synchronously cascaded 74HC160 decade counters. The Count function is enabled when $ENT = ENP = 1$. The counter has an asynchronous $RESET$ function. $RCO$ is active HIGH.

The 74HC4511 devices are BCD-to-seven-segment decoder/drivers with $D$ latch inputs. The latches in each device are transparent when $\overline{LE} = 0$.

A 74HC4538 Dual Monostable Multivibrator provides timing pulses required for circuit operation. One of the two monostables is set for a pulse width of 1 μs and the other for 1 ms.

Briefly explain the operation of the circuit. What unit of frequency should we use for the circuit output value?

**Solution**

The frequency counter in Figure 12.19 counts the negative edges of a pulse waveform at the $INPUT$ terminal during the 1-ms sampling interval set by the lower of the two monostables in the diagram. Each of the four counters represents a BCD digit that is decoded by a 74HC4511 latching BCD-to-seven-segment decoder/driver.

The cascaded counters have a total modulus of 10,000 (0000–9999). When one counter recycles, its $RCO$ output goes HIGH, enabling the $ENT$ input of the following counter. On the next pulse, both counters increment and $ENT$ goes LOW. The latter counter is disabled until the next time $ENT$ goes HIGH.

The sampling interval begins when $\overline{COUNT}$ is made momentarily LOW. This can be done with a pushbutton switch for a single sample or by a counter or astable circuit that repetitively pulses the $\overline{COUNT}$ input for continuous sampling. (If a counter or astable is used, it should have a period greater than the sampling interval, 1 ms in this case.)

A LOW at the $\overline{COUNT}$ input triggers both monostables. The upper one pulses the $\overline{R}$ inputs of the counters for 1 μs, setting each one to 0000. The lower monostable enables the counters by making each $ENP$ input HIGH for 1 ms. This allows them to increment with each negative edge of the waveform at $INPUT$. (The $CLK$ inputs respond to positive edges, but the gate at the input inverts the incoming waveform.)

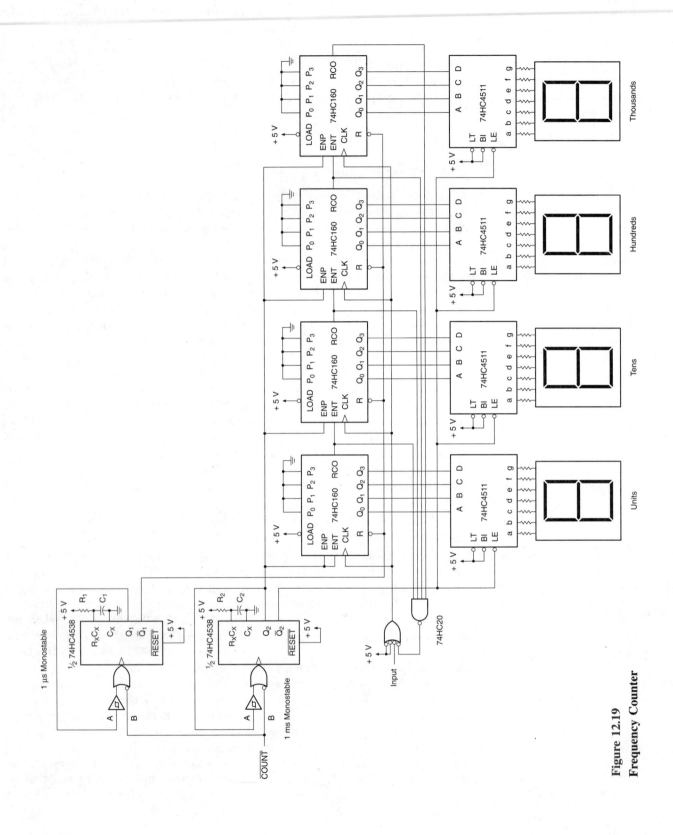

**Figure 12.19**
**Frequency Counter**

The 74HC4511 decoder/drivers have latching inputs that are enabled when $\overline{LE}$ is LOW. When $\overline{LE}$ goes HIGH, the last input value is retained. The 1-ms monostable enables the latch inputs during the sampling interval. Thus, the output display counts during this interval and displays the final count value after the monostable has timed out.

The two NAND gates prevent the frequency counter from overflowing past 9999. When the output reaches 9999, all *RCO* outputs are HIGH. This condition disables the *CLK* lines; no more input pulses will pass to the counters until they are reset.

The output is displayed in kHz. For example, suppose the circuit counts 100 input pulses. The period of each pulse is 1 ms/100 = 0.01 ms. This yields a frequency of $f = 1/T = 1/(0.01$ ms$) = 100$ kHz.

## Mod-n Counters

We can use the Preset inputs and cascading outputs of an MSI counter to construct a counter of any modulus. Figure 12.20 shows a 74LS191 counter configured in two ways: as a mod-9 UP counter and as a mod-9 DOWN counter. Each of these circuits has the same function: it will produce a pulse at its *MAX/MIN* output for every nine pulses at its *CLK* input.

**Figure 12.20**

**74LS191 as a Mod-9 Counter**

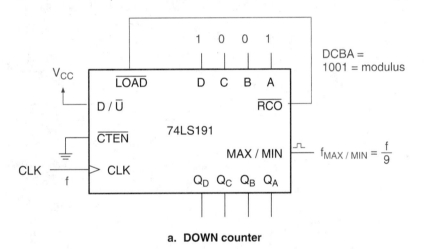

a. DOWN counter

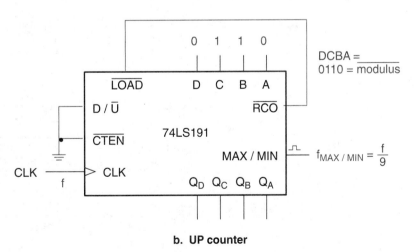

b. UP counter

Feedback from $\overline{RCO}$ asynchronously truncates the count sequence by loading a new value into the counter at the terminal count. For the DOWN counter, the loaded value is 1001, the same as the modulus, which makes the circuit count from 9 to 0. The value loaded into the UP counter is 0110, the 1's complement of the modulus. This value makes the circuit count from 6 to 15, which takes nine clock pulses.

Each of these sequences actually contains 10 output states, not 9, but as we shall see in the next example, the last clock pulse in the sequence generates 2 states, not 1. Thus, the sequence takes 9 clock pulses, as required.

---

**E X A M P L E   1 2 . 7**

Draw the timing diagrams for each of the circuits in Figure 12.20 to show that the *MAX/ MIN* output of each counter produces a pulse for every nine CLK pulses. Include waveforms for *CLK*, *Q* outputs, $\overline{RCO/LOAD}$, and *MAX/MIN*.

**Figure 12.21**

**Example 12.7
Mod-9 Timing Diagrams
(74LS191)**

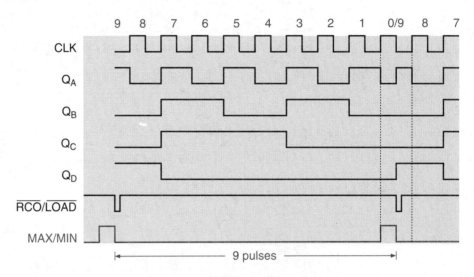

a. **DOWN counter**

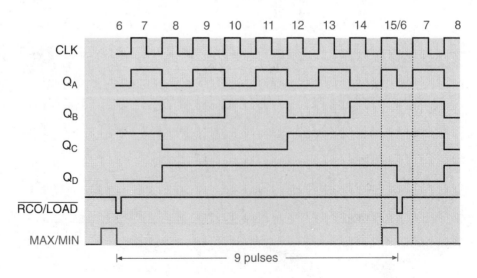

b. **UP counter**

**Solution**   Figure 12.21 shows the timing diagrams of the two circuits. The DOWN counter diagram begins with the $\overline{LOAD}$ pulse loading the binary value 1001 into the counter. With each positive edge of the *CLK* waveform, the counter decrements until it reaches 0000. This value makes the *MAX/MIN* output HIGH.

Halfway through the next *CLK* cycle, $\overline{RCO}$ goes LOW, following the *CLK*. This loads the value 1001 into the counter, thus removing the condition that enables both $\overline{RCO}$ and *MAX/MIN*. The count resumes from 1001 on the next pulse.

Since there are eight clock pulses from 1000 to 0001 and one pulse for the 1001/0000 recycle point, there are nine *CLK* pulses between *MAX/MIN* pulses.

The UP counter operates in a similar way. Complete analysis of the timing diagram for this circuit is left as an exercise.

---

**Section Review Problem for Section 12.2b**

12.2.  A 74LS191 counter is configured to count DOWN, with $\overline{RCO}$ and $\overline{LOAD}$ connected. What binary value should be at the Preset inputs *DCBA* if the counter is to produce a pulse at the *MAX/MIN* output every 12 *CLK* pulses? What value should *DCBA* be if the 74LS191 is configured as an UP counter with all other conditions the same?

## Multichip Mod-n Counters

In an earlier section, we saw how the modulus of a cascaded counter circuit is the product of individual counter moduli, *provided the counters were all full-sequence types*. For a circuit with presettable MSI counters that do not count to full sequence, we must use a different method to calculate the total modulus.

Figure 12.22 shows two 74LS191 counters cascaded so that both will load a preset value when the higher-order counter rolls over from 1111 to 0000. The bits are shown least to most significant for the whole counter. The lower-order counter loads 1001 and the higher-order counter loads 1100 when the $\overline{LOAD}$ pulse is applied. The binary value of the presets is the same as the overall modulus. That is, since $11001001_2 = 201_{10}$, the counter has a modulus of 201.

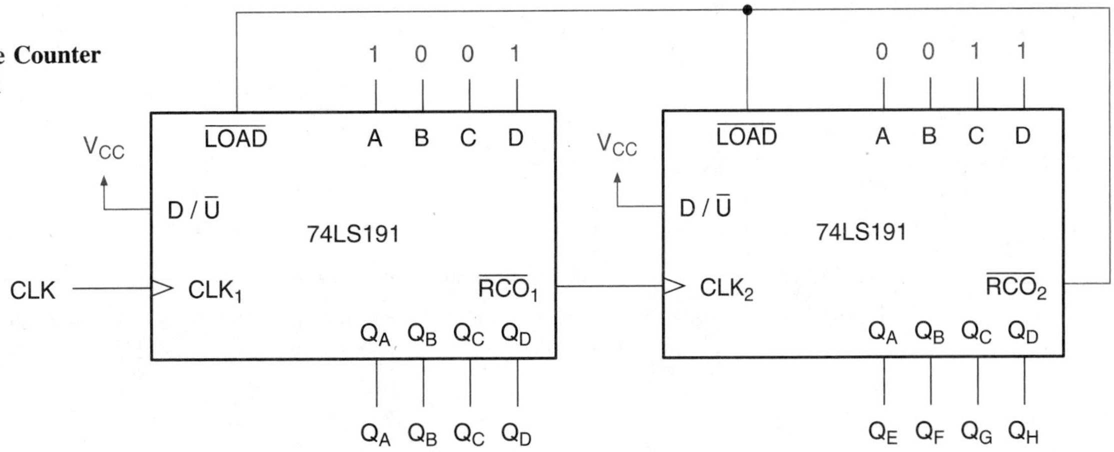

**Figure 12.22**
**Truncated-Sequence Counter Using Two 74LS191 Counters**

**EXAMPLE 12.8**

Use timing diagrams and other analytical techniques to prove that the circuit of Figure 12.22 has a modulus of 201.

**Solution**

Both counters in Figure 12.22 are configured to count DOWN. The timing diagram of Figure 12.23 shows the first 9 pulses after $\overline{LOAD}$ and the last 16 pulses before the next $\overline{LOAD}$.

The first 9 pulses recycle the lower-order counter and decrement the higher-order counter to 1011. The lower-order counter will count DOWN from 1111 to 0000 and recycle, decrementing the higher-order counter once at the end of every cycle (i.e., after every 16 *CLK* pulses) until the higher-order counter reaches 0000.

This, however, is not the end of the cycle. Figure 12.23b shows that at this point $MAX/MIN_2$ goes HIGH, but $\overline{RCO_2}$ does not activate until the second counter receives a *CLK* pulse from $\overline{RCO_1}$. This does not happen until the first counter receives 16 more pulses at its *CLK* input. Only then will $\overline{RCO_2}$ go LOW, loading the preset values into the two counters.

The total number of clock pulses for a complete cycle is $9 + (12 \times 16) = 9 + 192 = 201$. This is the decimal equivalent of the binary Preset inputs if we choose bit $H$ as most significant bit. $HGFEDCBA = 11001001_2 = 128_{10} + 64_{10} + 8_{10} + 1_{10} = 201$. Thus, the modulus of the counter is given by the binary value of the presets.

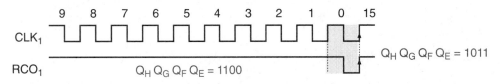

a. 9 → 0 on lower-order counter

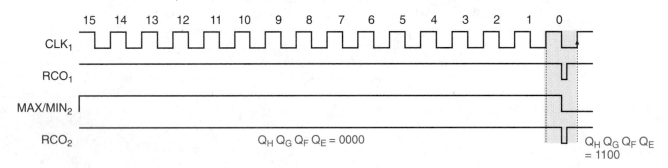

**Figure 12.23**

**Example 12.8 Timing Diagram for Mod-201 Counter**

b. Last 16 pulses

---

**Section Review Problem for Section 12.2c**

12.3. What is the modulus of the circuit in Figure 12.22 if we change the presets to $HGFEDCBA = 10011100$? (The presets are still 9 and 12, but on opposite counters.)

## 12.3

# IEEE/ANSI Notation

**Common control block** *The portion of an IEEE/ANSI symbol that shows the function of those Control inputs (e.g., a common clock or clear input) that act on several outputs simultaneously.*

**Dependency notation** *A way of writing variables in an IEEE/ANSI symbol to indicate which outputs depend on the action of which inputs and how.*

**Interconnection dependency** *A dependency that directly transfers a 1 or 0 state from one function to another. (Symbol: Z)*

**Mode dependency** *A dependency that selects one of several possible operating modes (e.g., LOAD or COUNT in a presettable counter). (Symbol: M)*

**AND dependency** *An enable/inhibit function with the same properties as an AND gate. (Symbol: G)*

**Control dependency** *An enable function such as an edge-triggered clock or a level-sensitive latch enable. (Symbol: C)*

IEEE/ANSI Standard 91-1984 specifies standard ways of drawing complex devices such as MSI counters. We will learn about the application of this standard by examining the symbols for the 74LS93, 74LS90, 74LS161A/163A, and 74LS191 counters.

## 74LS93 and 74LS90 Asynchronous Counters

Figure 12.24 shows the IEEE/ANSI symbols for the 74LS93 and 74LS90 asynchronous counters. Each of these symbols contains two main parts, as do the symbols of the other counters we will examine. The bottom rectangular portion of the symbol represents the flip-flops of the counter. The top indented portion of the symbol is the **common control block.** This shows how the Control inputs $R_0(1)$, $R_0(2)$, $R_9(1)$, and $R_9(2)$ affect the states of outputs $Q_D$, $Q_C$, $Q_B$, and $Q_A$.

The elements of the 74LS93 counter are divided into a mod-2 and mod-8 counter, indicated by the symbols DIV2 and DIV8. The outputs of the mod-8 portion are grouped by a bracket that shows the order of the most and least significant bits. Each of these elements is incremented (shown by the "+") by the negative edge of an input clock pulse, *CKA* or *CKB*, indicated by the triangular "edge-triggering" symbol and the "arrow" (negation symbol) at each clock input.

The common control block has an embedded AND gate that combines the two *RESET* inputs $R_0(1)$ and $R_0(2)$. The symbol CT = 0 at the output of the embedded AND function indicates that when both $R_0(1)$ and $R_0(2)$ go HIGH, the count is set to 0.

The symbol for the 74LS90 counter is similar, except that the lower counter element is mod-5, indicated by the symbol DIV5. In addition to the reset-to-0 function, there are two reset-to-9 inputs. The relation between the $R_9(1)$ and $R_9(2)$ Control inputs and the flip-flop outputs is specified by the **dependency notation** shown in the common control block and the flip-flop blocks.

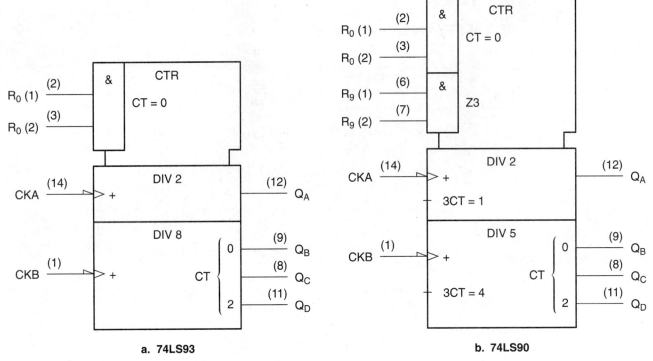

**Figure 12.24**
**IEEE/ANSI Symbols for Asynchronous MSI Counters**

Two functions are linked by a dependency when the numerical suffix of one function symbol is the same as the numerical prefix of the other. For instance, the lower embedded AND in the common control block of the 74LS90 counter shows an **interconnection dependency** (symbol = Z) between the $R_9(1)$ and $R_9(2)$ inputs and the count outputs. This is symbolized by Z3 at the embedded AND output and the notation $3CT = 1$ and $3CT = 4$ at the counter element inputs.

This notation tells us that when $R_9(1)$ and $R_9(2)$ are both HIGH, then $Q_A$ is set to 1 and $Q_D Q_C Q_B$ is set to 4 (= $100_2$). Thus, $Q_D Q_C Q_B Q_A$ is set to 1001.

## 74LS161A/163A Synchronous Presettable Counters

The 74LS161A and 74LS163A counters are identical except that the '161A has an asynchronous clear, whereas the '163A has a synchronous clear. The IEEE/ANSI symbol for the '161A counter, shown in Figure 12.25, indicates that this counter has several dependencies not used in the 74LS90/93 counters.

Two inputs indicate a **mode dependency,** symbolized by *M1* and *M2*. These modes are opposite states of the $\overline{LOAD}$ input. When *M1* is active ($\overline{LOAD}$ function; $\overline{LOAD}$ = 0), the flip-flops are loaded with the data at *DCBA* upon receiving a positive edge on the *CLK* input *(C5)*. This is shown in the counter data elements by the notation *1,5D*. The dependency chains are *M1–1D* and *C5–5D*.

The count sequence has four dependencies: mode dependency *M2* (*COUNT* function; $\overline{LOAD}$ = 1); **AND dependencies** *G3* and *G4;* and **control dependency** *C5.*

**Figure 12.25**

**IEEE/ANSI Symbol for 74LS161A Counter (Asynchronous Clear)**

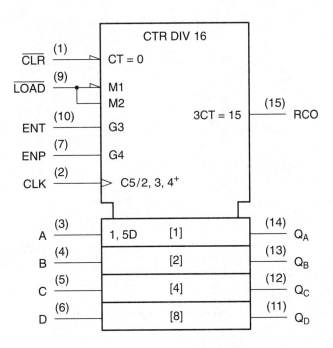

$G3$ and $G4$ are Enable functions with the enable and inhibit properties of an AND gate. When the associated control input is HIGH, it enables the count; when LOW it inhibits. The control dependency, $C5$, is on the positive edge of the $CLK$ input.

The counter increments if there is a positive edge on the $CLK$ ($C5$), the device is in mode 2, and both $ENT$ ($G3$) and $ENP$ ($G4$) are active. The order of most to least significant bits is shown by the numbers in square brackets: [8], [4], [2], [1].

The $RCO$ (Ripple Clock Output) activates on the terminal count ($CT = 15$), provided $ENT$ ($G3$) is active. This is shown by the notation $3CT = 15$ at the $RCO$ output.

The counter is cleared when $\overline{CLR} = 0$, indicated by the notation $CT = 0$ in the common control block. This is an asynchronous clear, since there is no dependency on the $CLK$ input. The symbol for the '163A counter, which has a synchronous Clear function, uses the notation $5CT = 0$ to indicate the required control dependency. Otherwise, the symbols for the '161A and '163A are identical.

### 74LS191 Bidirectional Synchronous Presettable Counter

The symbol for the 74LS191 counter, illustrated in Figure 12.26, has no new notation, but the dependencies are arranged differently than those of the other counters we have examined. The $\overline{LOAD}$ input of the '191 counter asynchronously loads the preset values when active, as indicated by the dependency chain $C5$–$5D$. The UP count progresses with each clock pulse when mode 3 *(UP)* and $G1$ (Count Enable) are active, indicated by the notation $1,3^+$ at the $CLK$ input. The DOWN count is selected by mode 2 *(DOWN)* and $G1$, shown as $1,2^-$.

*MAX/MIN* activates upon the terminal count, 0000 in DOWN mode *(M2)* and 1111 in UP mode *(M3)*. The notation for this is $2(CT = 0)$ or $3(CT = 15)$. Each *MAX/MIN* active state imposes an interconnection dependency *(Z6)* on $\overline{RCO}$. $\overline{RCO}$ depends on $G1$ (Count Enable), $G4$ (inverted $CLK$), and $Z6$ *(MAX/MIN)*.

**Figure 12.26**
**IEEE/ANSI Symbol for 74LS191 Bidirectional Synchronous Presettable Counter**

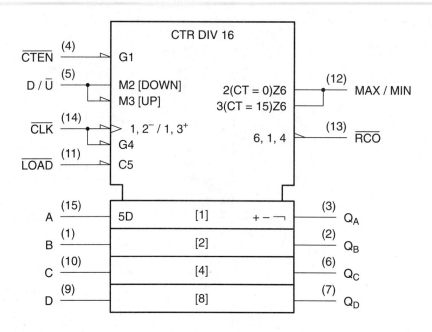

## 12.4

# Design of Synchronous Sequential Circuits

**Synchronous sequential circuit** *A circuit whose outputs progress through a sequence of states in response to a synchronous clock signal and possibly other input variables (e.g., a counter or shift register).*

**State machine** *A synchronous sequential circuit.*

**Moore-type state machine** *A state machine whose outputs depend only on the present state of the internal flip-flops.*

**Mealy-type state machine** *A state machine whose outputs depend on the present state of the internal flip-flops as well as the next state of the flip-flops, as defined by the inputs to the combinational part of the circuit.*

**Excitation table** *A table showing the required input conditions for every possible transition of a flip-flop output.*

$S$ynchronous counters are examples of a larger class of circuits called **synchronous sequential circuits** or **state machines.** Figure 12.27 shows a general block diagram of a type of synchronous sequential circuit called a **Moore-type state machine.**

The circuit consists of several main sections:

1. One or more flip-flops that act as memory storage to remember the present state of the circuit

2. Combinational logic, such as gates (e.g., $J_C = Q_B Q_A$) or hardwired flip-flop connections (e.g., $J_B = Q_A$) that direct the circuit to its next state

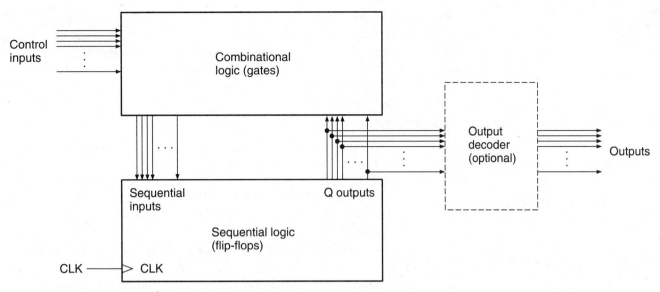

**Figure 12.27**
**Moore-Type State Machine**

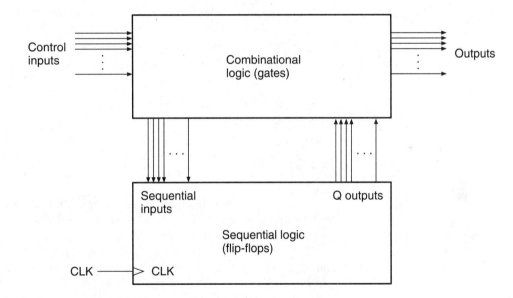

**Figure 12.28**
**Mealy-Type State Machine**

3. Optional control inputs that select one of several state sequences (e.g., an *UP/DOWN* input)

4. Outputs that may be decoded by combinational logic or wired directly to the flip-flop *Q* outputs

5. A clock input that is common to all flip-flops

Another class of state machine, called a **Mealy-type state machine,** is illustrated in Figure 12.28. The major difference between Moore and Mealy machines is that the combinational logic of a Mealy machine acts as a next-state decoder and an output decoder; the Mealy machine's outputs are functions of the flip-flop states and the input variable states.

Since the combinational inputs are partially responsible for determining the next state of the machine, we can also say that the present state of a Mealy machine is a function of the machine's present state and its next state. That is, where you are has something to do with where you are going. If a particular state can go in two directions, depending on a combinational input, the output potentially will be different in each case.

## Design Procedure

There are several steps involved in the design of a synchronous sequential circuit:

1. Define the problem. Before you can begin design of a circuit, you have to know what its purpose is and what it should do under all possible conditions.
2. Draw a state diagram showing the progression of states under various input conditions and what outputs the circuit should produce, if any.
3. Make a state table that lists all possible present states and the next state for each one.
4. Use flip-flop **excitation tables** to determine at what states the flip-flop synchronous inputs must be to make the circuit go from each present state to its next state.
5. The logic levels of the synchronous inputs are Boolean functions of the flip-flop outputs and the control inputs. Simplify the expression for each input and write the simplified Boolean expression.
6. Use the Boolean expressions found in step 5 to draw the required logic circuit.

### Flip-Flop Excitation Tables

In the synchronous counter circuits we examined in Chapter 11, we used JK flip-flops configured to operate only in toggle or no change mode. We can use any type of flip-flop for a synchronous sequential circuit. If we choose to use JK flip-flops, we can use any of the modes (no change, reset, set, or toggle) to make transitions from one state to another.

A flip-flop excitation table shows all possible transitions of a flip-flop output and the synchronous input levels needed to effect these transitions. Table 12.4 is the excitation table of a JK flip-flop.

If we want a flip-flop to make a transition from 0 to 1, we can use either the toggle function ($JK = 11$) or the set function ($JK = 10$). *It doesn't matter what K is as long as J = 1.* This is reflected by the variable pair (1 X) beside the $0 \rightarrow 1$ entry in Table 12.4. The X is a don't care state, a 0 or 1 depending on which is more convenient for the simplification of the Boolean function of the *J* or *K* input affected.

Table 12.5 shows a condensed version of the JK flip-flop excitation table.

## Design of a Synchronous Mod-12 Counter

We will follow the procedure outlined earlier to design a synchronous mod-12 counter circuit, using JK flip-flops. The aim is to derive the Boolean equations of all *J* and *K* inputs and to draw the counter circuit.

**1.** *Define the problem.* The circuit must count in binary sequence from 0000 to 1011 and repeat. The output progresses by 1 for each applied clock pulse. Since the outputs

**Table 12.4** Excitation Table of a JK Flip-Flop

| Transition | Function | J | K | |
|---|---|---|---|---|
| 0 → 0 | No change or | 0 | 0 | |
| | Reset | 0 | 1 | (0 X) |
| 0 → 1 | Toggle or | 1 | 1 | |
| | Set | 1 | 0 | (1 X) |
| 1 → 0 | Toggle or | 1 | 1 | |
| | Reset | 0 | 1 | (X 1) |
| 1 → 1 | No change or | 0 | 0 | |
| | Set | 1 | 0 | (X 0) |

X = Don't care

**Table 12.5** Excitation Table of a JK Flip-Flop

| Transition | J | K |
|---|---|---|
| 0 → 0 | 0 | X |
| 0 → 1 | 1 | X |
| 1 → 0 | X | 1 |
| 1 → 1 | X | 0 |

**Figure 12.29**

**State Diagram for a Mod-12 Counter**

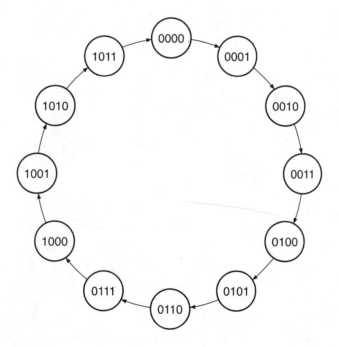

are 4-bit numbers, we require four flip-flops. The circuit output is the same as the undecoded flip-flop states, and there are no control inputs. Therefore, it is a Moore machine.

**2.** *Draw a state diagram.* The state diagram for this problem is shown in Figure 12.29.

**3.** *Make a state table showing each present state and the corresponding next state.*

**4.** *Use flip-flop excitation tables to fill in the J and K entries in the state table.* Table 12.6 shows the combined result of steps 3 and 4.

Table 12.6  State Table for a Mod-12 Counter

| Present State | | | | Next State | | | | | | | | | | | | |
|---|---|---|---|---|---|---|---|---|---|---|---|---|---|---|---|---|
| $Q_D$ | $Q_C$ | $Q_B$ | $Q_A$ | $Q_D$ | $Q_C$ | $Q_B$ | $Q_A$ | $J_D$ | $K_D$ | $J_C$ | $K_C$ | $J_B$ | $K_B$ | $J_A$ | $K_A$ | |
| 0 | 0 | 0 | 0 | 0 | 0 | 0 | 1 | 0 | X | 0 | X | 0 | X | 1 | X | |
| 0 | 0 | 0 | 1 | 0 | 0 | 1 | 0 | 0 | X | 0 | X | 1 | X | X | 1 | |
| 0 | 0 | 1 | 0 | 0 | 0 | 1 | 1 | 0 | X | 0 | X | X | 0 | 1 | X | |
| 0 | 0 | 1 | 1 | 0 | 1 | 0 | 0 | 0 | X | 1 | X | X | 1 | X | 1 | |
| 0 | 1 | 0 | 0 | 0 | 1 | 0 | 1 | 0 | X | X | 0 | 0 | X | 1 | X | |
| 0 | 1 | 0 | 1 | 0 | 1 | 1 | 0 | 0 | X | X | 0 | 1 | X | X | 1 | |
| 0 | 1 | 1 | 0 | 0 | 1 | 1 | 1 | 0 | X | X | 0 | X | 0 | 1 | X | |
| 0 | 1 | 1 | 1 | 1 | 0 | 0 | 0 | 1 | X | X | 1 | X | 1 | X | 1 | |
| 1 | 0 | 0 | 0 | 1 | 0 | 0 | 1 | X | 0 | 0 | X | 0 | X | 1 | X | |
| 1 | 0 | 0 | 1 | 1 | 0 | 1 | 0 | X | 0 | 0 | X | 1 | X | X | 1 | |
| 1 | 0 | 1 | 0 | 1 | 0 | 1 | 1 | X | 0 | 0 | X | X | 0 | 1 | X | |
| 1 | 0 | 1 | 1 | 0 | 0 | 0 | 0 | X | 1 | 0 | X | X | 1 | X | 1 | |
| 1 | 1 | 0 | 0 | X | X | X | X | X | X | X | X | X | X | X | X | Unused |
| 1 | 1 | 0 | 1 | X | X | X | X | X | X | X | X | X | X | X | X | States |
| 1 | 1 | 1 | 0 | X | X | X | X | X | X | X | X | X | X | X | X | |
| 1 | 1 | 1 | 1 | X | X | X | X | X | X | X | X | X | X | X | X | |

We assume for now that states 1100 to 1111 never occur. If we assign them to be don't care states, they can be used to simplify the $J$ and $K$ expressions we derive from the state table.

Let us examine one transition to show how the table is completed. The transition from $Q_D Q_C Q_B Q_A = 0101$ to $Q_D Q_C Q_B Q_A = 0110$ consists of the following individual flip-flop transitions:

$$Q_D: 0 \rightarrow 0 \quad \text{(No change or reset; } J_D K_D = 0X\text{)}$$
$$Q_C: 1 \rightarrow 1 \quad \text{(No change or set; } J_C K_C = X0\text{)}$$
$$Q_B: 0 \rightarrow 1 \quad \text{(Toggle or set; } J_B K_B = 1X\text{)}$$
$$Q_A: 1 \rightarrow 0 \quad \text{(Toggle or reset; } J_A K_A = X1\text{)}$$

The other lines of the table are completed similarly.

**5.** *Simplify the Boolean expression for each input.* Table 12.6 can be treated as eight truth tables, one for each $J$ or $K$ input. We can simplify each function by using Boolean algebra or a Karnaugh map.

Figure 12.30 shows K-map simplification for all eight synchronous inputs. These maps yield the following simplified Boolean expressions:

$$J_A = 1$$
$$K_A = 1$$
$$J_B = Q_A$$
$$K_B = Q_A$$

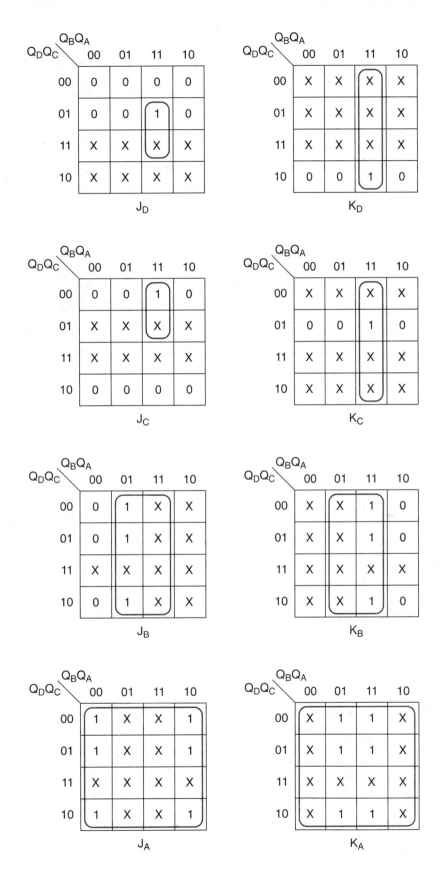

**Figure 12.30**

**K-Map Simplification of Table 12.6**

$$J_C = \overline{Q}_D Q_B Q_A$$

$$K_C = Q_B Q_A$$

$$J_D = Q_C Q_B Q_A$$

$$K_D = Q_B Q_A$$

**6.** *Draw the required logic circuit.* Figure 12.31 shows the circuit corresponding to the above Boolean expressions. We have assumed that states 1100 to 1111 will never occur in the operation of the mod-12 counter. This is normally the case, but when the circuit is powered up, there is no guarantee that the flip-flops will be in any particular state.

If a counter powers up in an unused state, the circuit should enter the main sequence after one or more clock pulses. To test whether or not this happens, let us make a state table, applying each unused state to the $J$ and $K$ equations as implemented, to see what the next state is for each case. This analysis is shown in Table 12.7.

Figure 12.32 shows the complete state diagram for the designed mod-12 counter. If the counter powers up in an unused state, it will enter the main sequence in no more than four clock pulses.

If we want an unused state to make a transition directly to 0000 in one clock pulse, we have a couple of options:

1. We could reset the counter asynchronously and otherwise leave the design as is.

**Figure 12.31**

**Synchronous Mod-12 Counter**

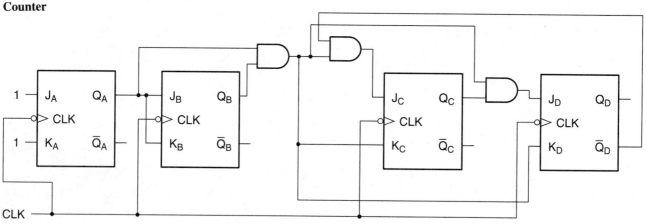

Table 12.7 Resolution of Unused States

| Present State | | | | $J_D$ | $K_D$ | $J_C$ | $K_C$ | $J_B$ | $K_B$ | $J_A$ | $K_A$ | Next State | | | |
|---|---|---|---|---|---|---|---|---|---|---|---|---|---|---|---|
| $Q_D$ | $Q_C$ | $Q_B$ | $Q_A$ | | | | | | | | | $Q_D$ | $Q_C$ | $Q_B$ | $Q_A$ |
| 1 | 1 | 0 | 0 | 0 | 0 | 0 | 0 | 0 | 0 | 1 | 1 | 1 | 1 | 0 | 1 |
| 1 | 1 | 0 | 1 | 0 | 0 | 0 | 0 | 1 | 1 | 1 | 1 | 1 | 1 | 1 | 0 |
| 1 | 1 | 1 | 0 | 0 | 0 | 0 | 0 | 0 | 0 | 1 | 1 | 1 | 1 | 1 | 1 |
| 1 | 1 | 1 | 1 | 1 | 1 | 0 | 1 | 1 | 1 | 1 | 1 | 0 | 0 | 0 | 0 |

**Figure 12.32**

**Complete State Diagram of Mod-12 Counter in Figure 12.31**

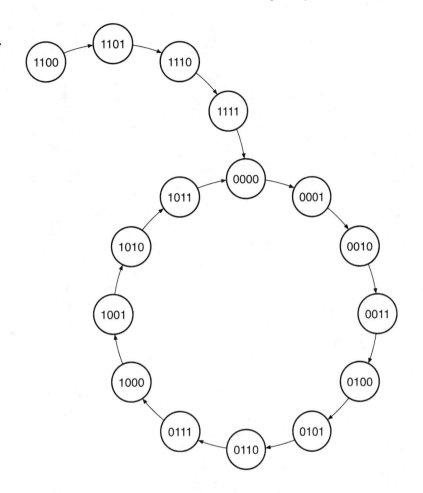

2. We could rewrite the state table to specify these transitions, rather than making the unused states don't care states.

Figure 12.33 shows the mod-12 circuit with an *RC* circuit that automatically resets the flip-flops on power-up. When the power is off, the capacitor is completely discharged. Upon power-up, the capacitor charges to $V_{CC}$ at a rate determined by the values of *R* and *C*. The charging curve is illustrated in Figure 12.34.

The flip-flop $\overline{CLR}$ inputs are all held LOW by the *RC* circuits when the power is turned on, thus making the counter output 0000. After the capacitors charge, the $\overline{CLR}$ inputs are held HIGH and the circuit can count normally.

We can design a counter that synchronously goes to 0000 by making that the next state for each unused state. This design is left as an exercise.

## Counters With Nonstandard Sequences

Our design procedure can also be used to design sequential circuits with nonstandard count sequences. One convention that makes the design process easier is to list the present states, including unused states, in binary order, even if the circuit does not count in binary order. Since the synchronous input functions *(J, K)* are Boolean functions of the present-state variables, keeping them in truth table order allows us easily to use familiar techniques of simplification, such as K-maps.

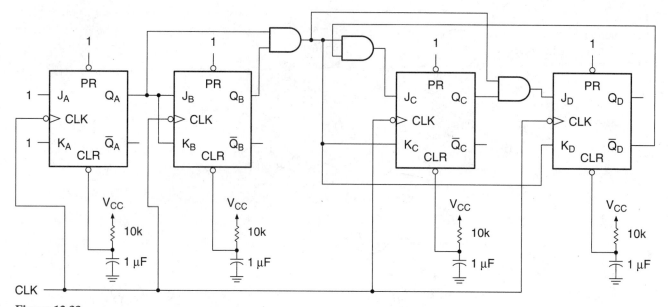

**Figure 12.33**

**Mod-12 Counter That Resets to 0000 on Power-Up**

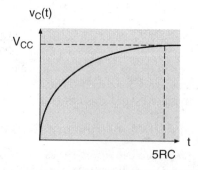

**a. RC circuit**

**b. Capacitor voltage after power-up**

**Figure 12.34**

**RC Power-Up Reset Circuit**

**EXAMPLE 12.9**

Design a synchronous sequential circuit that will produce the biquinary sequence shown in the timing diagram of Figure 12.35. Draw the state diagram of the circuit you design, including any unused states. Test the unused states to ensure that they enter the main sequence in one or more clock pulses.

**Figure 12.35**

**Example 12.9
Biquinary Waveforms**

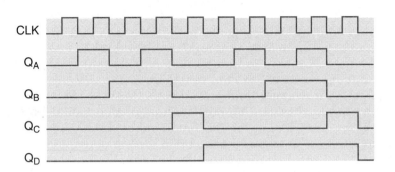

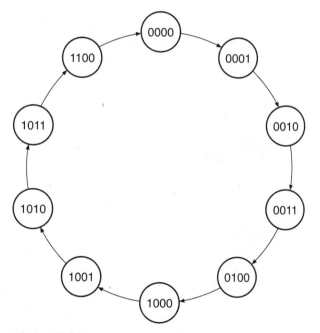

**Figure 12.36**

**Example 12.9
State Diagram of a Biquinary Sequence**

**Solution**

Figure 12.36 shows the state diagram of the biquinary sequence. Table 12.8 is the state table for the sequence specified in Figures 12.35 and 12.36.

The sequence of present states is shown in binary order, even though the circuit output does not progress in this sequence. This makes it easier to load the resulting $J$ and $K$ values into a K-map, which is ordered in terms of the present states.

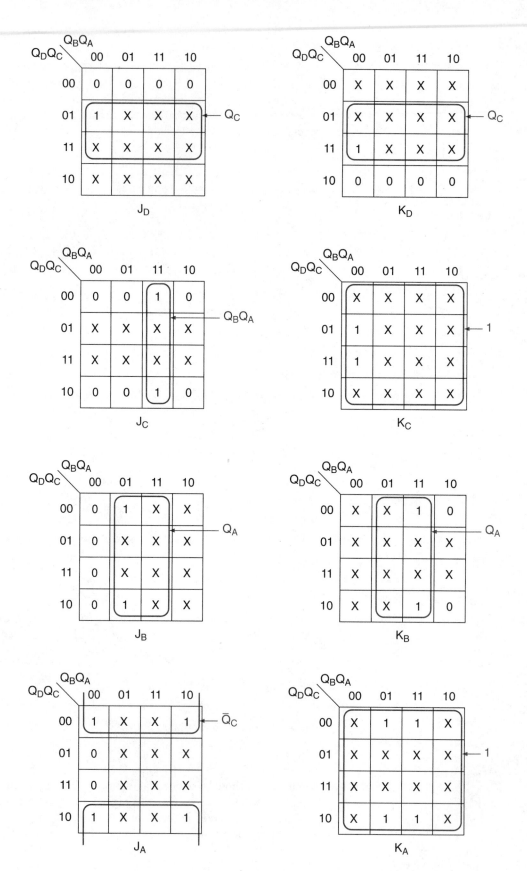

**Figure 12.37**

**Example 12.9
K-Map Simplification for
the Biquinary Sequence
Generator**

**Table 12.8** State Table for a Biquinary Sequence Counter

| Present State | | | | Next State | | | | $J_D$ | $K_D$ | $J_C$ | $K_C$ | $J_B$ | $K_B$ | $J_A$ | $K_A$ |
|---|---|---|---|---|---|---|---|---|---|---|---|---|---|---|---|
| $Q_D$ | $Q_C$ | $Q_B$ | $Q_A$ | $Q_D$ | $Q_C$ | $Q_B$ | $Q_A$ | | | | | | | | |
| 0 | 0 | 0 | 0 | 0 | 0 | 0 | 1 | 0 | X | 0 | X | 0 | X | 1 | X |
| 0 | 0 | 0 | 1 | 0 | 0 | 1 | 0 | 0 | X | 0 | X | 1 | X | X | 1 |
| 0 | 0 | 1 | 0 | 0 | 0 | 1 | 1 | 0 | X | 0 | X | X | 0 | 1 | X |
| 0 | 0 | 1 | 1 | 0 | 1 | 0 | 0 | 0 | X | 1 | X | X | 1 | X | 1 |
| 0 | 1 | 0 | 0 | 1 | 0 | 0 | 0 | 1 | X | X | 1 | 0 | X | 0 | X |
| 0 | 1 | 0 | 1 | X | X | X | X | X | X | X | X | X | X | X | X |
| 0 | 1 | 1 | 0 | X | X | X | X | X | X | X | X | X | X | X | X |
| 0 | 1 | 1 | 1 | X | X | X | X | X | X | X | X | X | X | X | X |
| 1 | 0 | 0 | 0 | 1 | 0 | 0 | 1 | X | 0 | 0 | X | 0 | X | 1 | X |
| 1 | 0 | 0 | 1 | 1 | 0 | 1 | 0 | X | 0 | 0 | X | 1 | X | X | 1 |
| 1 | 0 | 1 | 0 | 1 | 0 | 1 | 1 | X | 0 | 0 | X | X | 0 | 1 | X |
| 1 | 0 | 1 | 1 | 1 | 1 | 0 | 0 | X | 0 | 1 | X | X | 1 | X | 1 |
| 1 | 1 | 0 | 0 | 0 | 0 | 0 | 0 | X | 1 | X | 1 | 0 | X | 0 | X |
| 1 | 1 | 0 | 1 | X | X | X | X | X | X | X | X | X | X | X | X |
| 1 | 1 | 1 | 0 | X | X | X | X | X | X | X | X | X | X | X | X |
| 1 | 1 | 1 | 1 | X | X | X | X | X | X | X | X | X | X | X | X |

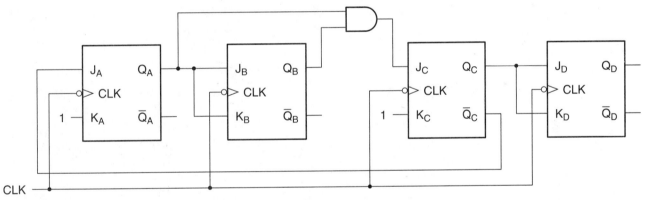

**Figure 12.38**

**Example 12.9**
**Biquinary Sequence Counter**

The $J$ and $K$ inputs indicated in Table 12.8 are loaded into Karnaugh maps, as shown in Figure 12.37. This yields the following Boolean equations for the synchronous inputs:

$$J_D = Q_C \qquad\qquad J_B = Q_A$$

$$K_D = Q_C \qquad\qquad K_B = Q_A$$

$$J_C = Q_B Q_A \qquad\qquad J_A = \overline{Q}_C$$

$$K_C = 1 \qquad\qquad K_A = 1$$

Figure 12.38 represents the counter derived from these Boolean equations.

**Table 12.9**  Resolution of Unused States

| Present State | | | | $J_D$ | $K_D$ | $J_C$ | $K_C$ | $J_B$ | $K_B$ | $J_A$ | $K_A$ | Next State | | | |
|---|---|---|---|---|---|---|---|---|---|---|---|---|---|---|---|
| $Q_D$ | $Q_C$ | $Q_B$ | $Q_A$ | | | | | | | | | $Q_D$ | $Q_C$ | $Q_B$ | $Q_A$ |
| 0 | 1 | 0 | 1 | 1 | 1 | 0 | 1 | 1 | 1 | 0 | 1 | 1 | 0 | 1 | 0 |
| 0 | 1 | 1 | 0 | 1 | 1 | 0 | 1 | 0 | 0 | 0 | 1 | 1 | 0 | 1 | 0 |
| 0 | 1 | 1 | 1 | 1 | 1 | 1 | 1 | 1 | 1 | 0 | 1 | 1 | 0 | 0 | 0 |
| 1 | 1 | 0 | 1 | 1 | 1 | 0 | 1 | 1 | 1 | 0 | 1 | 0 | 0 | 1 | 0 |
| 1 | 1 | 1 | 0 | 1 | 1 | 0 | 1 | 0 | 0 | 0 | 1 | 0 | 0 | 1 | 0 |
| 1 | 1 | 1 | 1 | 1 | 1 | 1 | 1 | 1 | 1 | 0 | 1 | 0 | 0 | 0 | 0 |

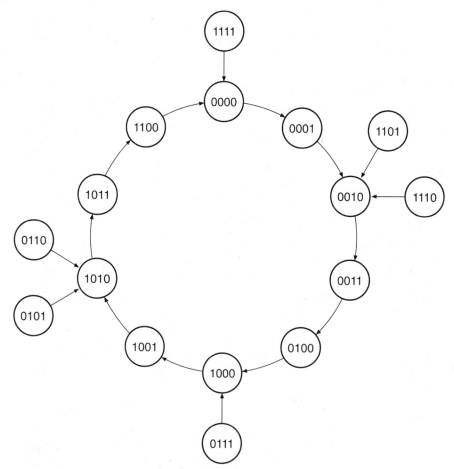

**Figure 12.39**

**Example 12.9**
**State Diagram for Biquinary Sequence Counter, Including Unused States**

There are six unused states in the count sequence. Each of them is tested as shown in Table 12.9. Figure 12.39 shows the revised state diagram, including the six unused states.

## Use of Control Inputs

Synchronous sequential circuits can be controlled by external inputs as well as by the clock. For example, if we want to make a bidirectional counter, we need a control input that will set the count direction. This control input becomes one of the present-state variables in the state table and will ultimately be part of the synchronous input equations defining the counter function.

Such a circuit is a Moore-type machine, as are all other synchronous circuits in this section. Even though the control input helps to define the next state of the counter, it does not define the output directly through the combinational logic of the counter circuit, as a Mealy-type machine would.

---

**EXAMPLE 12.10**    Design a bidirectional mod-8 synchronous counter whose direction is controlled by an input labeled $U$. For $U = 1$, the circuit counts UP. The state diagram is shown in Figure 12.40.

**Figure 12.40**

**Example 12.10 State Diagram of a Bidirectional Mod-8 Counter**

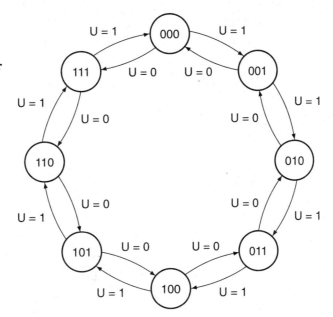

**Solution**    Table 12.10 (see page 543) shows the state table for the bidirectional counter. The input variable $U$ is shown as the most significant bit of the present state. This allows the entire UP and DOWN sequences to be grouped together. All present states, including those of the DOWN sequence, are shown in ascending binary order, in agreement with our convention.

Figure 12.41 (see page 542) shows the K-map simplification of the synchronous input expressions. The K-maps yield the following simplified expressions:

$$J_A = K_A = 1$$

$$J_B = K_B = U \cdot Q_A + \overline{U} \cdot \overline{Q}_A$$

$$J_C = K_C = U \cdot Q_A \cdot Q_B + \overline{U} \cdot \overline{Q}_A \cdot \overline{Q}_B$$

The logic circuit for these equations is shown in Figure 12.42 (see page 543).

**Figure 12.41**

**Example 12.10 K-Maps**

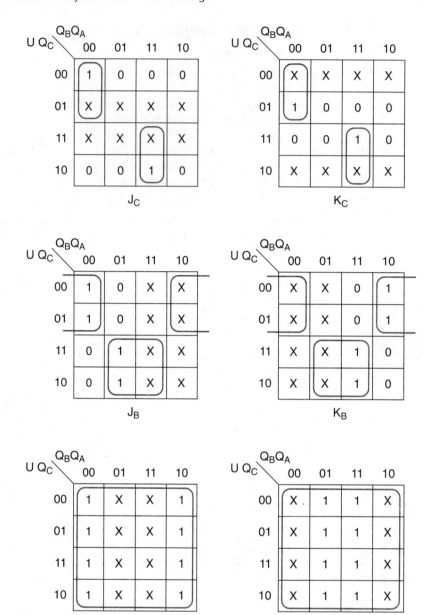

## 12.5

# Sequential Logic Applications of PLDs

*Disclaimer:* The description of the PLDasm programming language in this text is not intended to replace the material in the Intel manual that accompanies the software. References to PLDshell and PLDasm assume that you have a copy of the

Table 12.10  State Table for a Bidirectional Mod-8 Counter

| U | $Q_C$ | $Q_B$ | $Q_A$ | $Q_C$ | $Q_B$ | $Q_A$ | $J_C$ | $K_C$ | $J_B$ | $K_B$ | $J_A$ | $K_A$ | |
|---|---|---|---|---|---|---|---|---|---|---|---|---|---|
| | **Present State** | | | **Next State** | | | | | | | | | |
| 0 | 0 | 0 | 0 | 1 | 1 | 1 | 1 | X | 1 | X | 1 | X | |
| 0 | 0 | 0 | 1 | 0 | 0 | 0 | 0 | X | 0 | X | X | 1 | |
| 0 | 0 | 1 | 0 | 0 | 0 | 1 | 0 | X | X | 1 | 1 | X | |
| 0 | 0 | 1 | 1 | 0 | 1 | 0 | 0 | X | X | 0 | X | 1 | |
| 0 | 1 | 0 | 0 | 0 | 1 | 1 | X | 1 | 1 | X | 1 | X | **DOWN sequence** |
| 0 | 1 | 0 | 1 | 1 | 0 | 0 | X | 0 | 0 | X | X | 1 | |
| 0 | 1 | 1 | 0 | 1 | 0 | 1 | X | 0 | X | 1 | 1 | X | |
| 0 | 1 | 1 | 1 | 1 | 1 | 0 | X | 0 | X | 0 | X | 1 | |
| 1 | 0 | 0 | 0 | 0 | 0 | 1 | 0 | X | 0 | X | 1 | X | |
| 1 | 0 | 0 | 1 | 0 | 1 | 0 | 0 | X | 1 | X | X | 1 | |
| 1 | 0 | 1 | 0 | 0 | 1 | 1 | 0 | X | X | 0 | 1 | X | |
| 1 | 0 | 1 | 1 | 1 | 0 | 0 | 1 | X | X | 1 | X | 1 | |
| 1 | 1 | 0 | 0 | 1 | 0 | 1 | X | 0 | 0 | X | 1 | X | **UP sequence** |
| 1 | 1 | 0 | 1 | 1 | 1 | 0 | X | 0 | 1 | X | X | 1 | |
| 1 | 1 | 1 | 0 | 1 | 1 | 1 | X | 0 | X | 0 | 1 | X | |
| 1 | 1 | 1 | 1 | 0 | 0 | 0 | X | 1 | X | 1 | X | 1 | |

UP / $\overline{\text{DOWN}}$

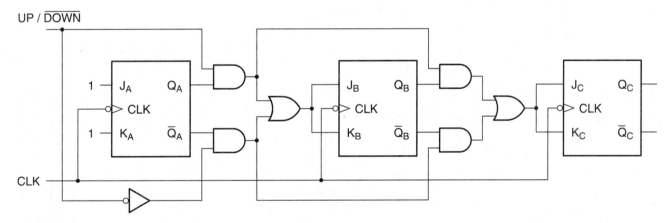

**Figure 12.42**

**Example 12.10**
**Mod-8 Bidirectional Counter**

Intel manual and will rely on that for the majority of your software support. The material on PLDasm in this book is primarily in the form of application-specific examples and some supplementary material on the PLDasm language.

If you are using software other than PLDasm, you should still be able to use the PLD examples by disassembling the JEDEC file for each example. These files are available to instructors who use this book.

We can use the Intel PLDasm programming language to program erasable programmable logic devices (EPLDs) as sequential circuits in one of two ways: with Boolean equations or with state tables. For circuits having many states but straightforward tran-

**Table 12.11** Excitation Table of a D Flip-Flop

| Transition | $D$ |
|---|---|
| $0 \to 0$ | 0 |
| $0 \to 1$ | 1 |
| $1 \to 0$ | 0 |
| $1 \to 1$ | 1 |

**Table 12.12** State Table for a 4-bit Binary Counter

| Present State | | | | Next State | | | | | | | |
|---|---|---|---|---|---|---|---|---|---|---|---|
| $Q_D$ | $Q_C$ | $Q_B$ | $Q_A$ | $Q_D$ | $Q_C$ | $Q_B$ | $Q_A$ | $D_D$ | $D_C$ | $D_B$ | $D_A$ |
| 0 | 0 | 0 | 0 | 0 | 0 | 0 | 1 | 0 | 0 | 0 | 1 |
| 0 | 0 | 0 | 1 | 0 | 0 | 1 | 0 | 0 | 0 | 1 | 0 |
| 0 | 0 | 1 | 0 | 0 | 0 | 1 | 1 | 0 | 0 | 1 | 1 |
| 0 | 0 | 1 | 1 | 0 | 1 | 0 | 0 | 0 | 1 | 0 | 0 |
| 0 | 1 | 0 | 0 | 0 | 1 | 0 | 1 | 0 | 1 | 0 | 1 |
| 0 | 1 | 0 | 1 | 0 | 1 | 1 | 0 | 0 | 1 | 1 | 0 |
| 0 | 1 | 1 | 0 | 0 | 1 | 1 | 1 | 0 | 1 | 1 | 1 |
| 0 | 1 | 1 | 1 | 1 | 0 | 0 | 0 | 1 | 0 | 0 | 0 |
| 1 | 0 | 0 | 0 | 1 | 0 | 0 | 1 | 1 | 0 | 0 | 1 |
| 1 | 0 | 0 | 1 | 1 | 0 | 1 | 0 | 1 | 0 | 1 | 0 |
| 1 | 0 | 1 | 0 | 1 | 0 | 1 | 1 | 1 | 0 | 1 | 1 |
| 1 | 0 | 1 | 1 | 1 | 1 | 0 | 0 | 1 | 1 | 0 | 0 |
| 1 | 1 | 0 | 0 | 1 | 1 | 0 | 1 | 1 | 1 | 0 | 1 |
| 1 | 1 | 0 | 1 | 1 | 1 | 1 | 0 | 1 | 1 | 1 | 0 |
| 1 | 1 | 1 | 0 | 1 | 1 | 1 | 1 | 1 | 1 | 1 | 1 |
| 1 | 1 | 1 | 1 | 0 | 0 | 0 | 0 | 0 | 0 | 0 | 0 |

**Figure 12.43**

**Example 12.11**
**K-Maps for a 4-Bit Counter Based on D Flip-Flops**

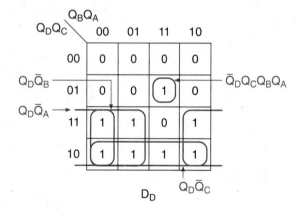

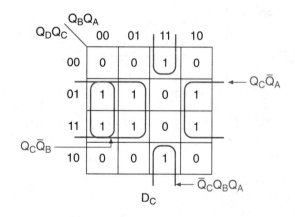

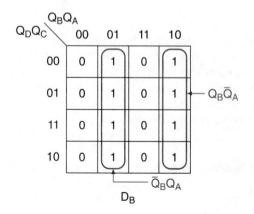

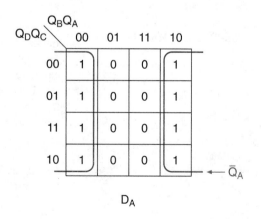

sitions, such as counters, it is easiest to use Boolean equations. For applications having fewer states but more-complex conditions for state transitions, the state table is probably easier. There is no hard and fast rule about this; after you gain some experience, your judgment will tell you the best way.

Let us look at three examples using various PLD programming techniques:

1. An 8-bit counter based on an 85C220 EPLD, programmed using Boolean equations
2. An 8-bit presettable counter for an 85C060 target device, using Boolean equations
3. A 4-bit binary counter that uses an 85C220 EPLD, programmed by a state table

## Two 8-Bit Counters (Boolean Equation Method)

Recall that the synchronous state machines used in previous examples have all used JK flip-flops. The registered outputs of an 85C220/224 device, however, are D flip-flops. We can still use these devices to make counters, but the excitation equations will be different from those used for JK flip-flops. Before we proceed with our counter design, we should briefly analyze the excitation equations required for a counter based on D flip-flops.

---

**EXAMPLE 12.11**    Derive the synchronous input equations of a 4-bit synchronous binary counter based on D flip-flops.

**Solution**    The first step in the counter design is to derive the excitation table of a D flip-flop. Recall that $Q$ follows $D$ when the flip-flop is clocked. Therefore, the destination state of $Q$ is the same as the input $D$ for any transition. This is illustrated in Table 12.11.

Next, we must construct a state table, Table 12.12, showing present and next states for all possible transitions. Notice that the binary value of $D_D D_C D_B D_A$ is the same as the next state of the counter.

This state table yields four Boolean equations, for $D_D$ through $D_A$, in terms of the present-state outputs. Figure 12.43 shows four Karnaugh maps used to simplify these functions.

The simplified equations are:

$$D_D = \overline{Q}_D Q_C Q_B Q_A + Q_D \overline{Q}_C + Q_D \overline{Q}_B + Q_D \overline{Q}_A$$

$$D_C = \overline{Q}_C Q_B Q_A + Q_C \overline{Q}_B + Q_C \overline{Q}_A$$

$$D_B = \overline{Q}_B Q_A + Q_B \overline{Q}_A$$

$$D_A = \overline{Q}_A$$

These equations represent the maximum SOP simplifications of the input functions. However, we can rewrite them to make them more compact. For example the equation for $D_D$ can be rewritten, using DeMorgan's Theorem and our knowledge of exclusive OR functions:

$$D_D = \overline{Q}_D Q_C Q_B Q_A + Q_D \overline{Q}_C + Q_D \overline{Q}_B + Q_D \overline{Q}_A$$
$$= \overline{Q}_D Q_C Q_B Q_A + Q_D(\overline{Q}_C + \overline{Q}_B + \overline{Q}_A)$$

$$= \overline{Q_D}(Q_C Q_B Q_A) + Q_D(\overline{Q_C Q_B Q_A})$$

$$= Q_D \oplus (Q_C Q_B Q_A)$$

We can write similar equations for the other $D$ inputs as follows:

$$D_C = Q_C \oplus (Q_B Q_A)$$

$$D_B = Q_A \oplus (Q_A)$$

$$D_A = Q_A \oplus 1$$

These equations follow a predictable pattern of expansion. Each equation for an input $D_n$ is simply $Q_n$ XORed with the logical product (AND) of all previous $Q$s.

---

**EXAMPLE 12.12**

Use the pattern of Boolean equations derived in Example 12.11 to program an 85C220 EPLD as an 8-bit binary counter.

**Solution**   A PLDasm source file for this application is found on the instructor's diskette as \MSICTR\CTR256.PDS. The compiled JEDEC file is \MSICTR\CTR256.JED.

The EQUATIONS section of the source file is as follows.

```
EQUATIONS

Q0 := /Q0
Q1 := Q1 :+: Q0
Q2 := Q2 :+: Q0*Q1
Q3 := Q3 :+: Q0*Q1*Q2
Q4 := Q4 :+: Q0*Q1*Q2*Q3
Q5 := Q5 :+: Q0*Q1*Q2*Q3*Q4
Q6 := Q6 :+: Q0*Q1*Q2*Q3*Q4*Q5
Q7 := Q7 :+: Q0*Q1*Q2*Q3*Q4*Q5*Q6
```

---

When compiled, the source file CTR256.PDS generates a report file, CTR256.RPT, which contains the simplified Boolean equations in SOP form. Compare these equations, listed below, to those derived from the Karnaugh maps in Example 12.11.

Each variable $Q_n$ in the report file incorporates a signal extension (.D) that indicates that the equation is for the $D$ input of the flip-flop. We could have written these signal extensions explicitly in the source file, but in this case the PLDasm compiler took care of it automatically.

```
Q0.D := /Q0

Q1.D := Q1 * /Q0
      + /Q1 * Q0

Q2.D := Q0 * Q1 * /Q2
      + /Q0 * Q2
      + /Q1 * Q2

Q3.D := Q0 * Q1 * Q2 * /Q3
      + /Q0 * Q3
      + /Q1 * Q3
      + /Q2 * Q3
```

```
Q4.D := Q0 * Q1 * Q2 * Q3 * /Q4
     + /Q0 * Q4
     + /Q1 * Q4
     + /Q2 * Q4
     + /Q3 * Q4

Q5.D := Q0 * Q1 * Q2 * Q3 * Q4 * /Q5
     + /Q0 * Q5
     + /Q1 * Q5
     + /Q2 * Q5
     + /Q3 * Q5
     + /Q4 * Q5

Q6.D := Q0 * Q1 * Q2 * Q3 * Q4 * Q5 * /Q6
     + /Q0 * Q6
     + /Q1 * Q6
     + /Q2 * Q6
     + /Q3 * Q6
     + /Q4 * Q6
     + /Q5 * Q6

Q7.D := Q0 * Q1 * Q2 * Q3 * Q4 * Q5 * Q6 * /Q7
     + /Q0 * Q7
     + /Q1 * Q7
     + /Q2 * Q7
     + /Q3 * Q7
     + /Q4 * Q7
     + /Q5 * Q7
     + /Q6 * Q7
```

The counter in Example 12.12 requires one additional product term for each additional bit. Thus, on the eighth bit ($Q_7$) we have reached the limit of product terms for the output macrocell. Special functions, such as Parallel Load, require more product terms, which are not available. To add such features, we need a device such as the 85C060, whose macrocells can be configured as JK flip-flops, which requires fewer product terms.

Recall the pattern of Boolean equations for an 8-bit counter:

$$J_0 = K_0 = 1$$

$$J_1 = K_1 = Q_0$$

$$J_2 = K_2 = Q_1 Q_0$$

$$J_3 = K_3 = Q_2 Q_1 Q_0$$

$$J_4 = K_4 = Q_3 Q_2 Q_1 Q_0$$

$$J_5 = K_5 = Q_4 Q_3 Q_2 Q_1 Q_0$$

$$J_6 = K_6 = Q_5 Q_4 Q_3 Q_2 Q_1 Q_0$$

$$J_7 = K_7 = Q_6 Q_5 Q_4 Q_3 Q_2 Q_1 Q_0$$

A macrocell using one of the above equations requires two product terms, one for $J$ and one for $K$. (The 85C060 EPLD splits the available product terms in a macrocell between the $J$ and $K$ equations.) This leaves six product terms for other functions.

**EXAMPLE 12.13**     Write the EQUATIONS section of a PLDasm source file to program an 85C060 EPLD as an 8-bit presettable binary counter having active-HIGH asynchronous *CLEAR*, active-LOW synchronous *LOAD*, and active-LOW Ripple Clock Output *(RCO)* functions.

**Solution**     The PLDasm file and its compiled JEDEC file are included on the instructor's diskette in files \MSI\CTR256.PDS and \MSI\CTR256PR.JED.

```
EQUATIONS

Q0.RSTF = CLEAR        ; Signal extension .RSTF shows
Q1.RSTF = CLEAR        ; that equation is for Asynchronous Reset.
Q2.RSTF = CLEAR
Q3.RSTF = CLEAR
Q4.RSTF = CLEAR
Q5.RSTF = CLEAR
Q6.RSTF = CLEAR
Q7.RSTF = CLEAR

Q0.J := CTEN*LOAD +  P0*/LOAD  ; J and K equations.  Count only
Q0.K := CTEN*LOAD + /P0*/LOAD  ; if CTEN=1 and LOAD=1.  Parallel
                               ; load if LOAD=0.  J and K are
                               ; complementary during LOAD.

Q1.J := Q0*CTEN*LOAD +  P1*/LOAD
Q1.K := Q0*CTEN*LOAD + /P1*/LOAD

Q2.J := Q1*Q0*CTEN*LOAD +  P2*/LOAD
Q2.K := Q1*Q0*CTEN*LOAD + /P2*/LOAD

Q3.J := Q2*Q1*Q0*CTEN*LOAD +  P3*/LOAD
Q3.K := Q2*Q1*Q0*CTEN*LOAD + /P3*/LOAD

Q4.J := Q3*Q2*Q1*Q0*CTEN*LOAD +  P4*/LOAD
Q4.K := Q3*Q2*Q1*Q0*CTEN*LOAD + /P4*/LOAD

Q5.J := Q4*Q3*Q2*Q1*Q0*CTEN*LOAD +  P5*/LOAD
Q5.K := Q4*Q3*Q2*Q1*Q0*CTEN*LOAD + /P5*/LOAD

Q6.J := Q5*Q4*Q3*Q2*Q1*Q0*CTEN*LOAD +  P6*/LOAD
Q6.K := Q5*Q4*Q3*Q2*Q1*Q0*CTEN*LOAD + /P6*/LOAD

Q7.J := Q6*Q5*Q4*Q3*Q2*Q1*Q0*CTEN*LOAD +  P7*/LOAD
Q7.K := Q6*Q5*Q4*Q3*Q2*Q1*Q0*CTEN*LOAD + /P7*/LOAD

/RCO = Q7*Q6*Q5*Q4*Q3*Q2*Q1*Q0
```

## 4-Bit Counter (State Table Method)

Programming an EPLD by the state table method in PLDasm is relatively simple except for one potential problem: the program must explicitly define all used states. (Unused states can be defined implicitly by a number of default options.) This is quite straightforward for a 4-bit counter, where there are only 16 states. It is much more difficult for an 8-bit counter, where we would need to list 256 unique states.

The state section of a PLDasm source file begins with the keyword STATE, followed by one of the keywords MOORE_MACHINE or MEALY_MACHINE. The remainder of the state section can be relatively simple or more complex, depending on the required application. The simplest state section consists of a set of state assignments and a set of equations that specify the transitions from one state to the next. The following is a source file for a 3-bit counter that uses this technique.

```
Title        3-bit Counter (Moore machine)
Pattern      pds
Revision     1.0
Author       R.Dueck
Company      Seneca College
Date         January 23, 1993

;    A 3-bit counter programmed by a state table.

CHIP   template        85C220

; PINLIST

PIN     1   CLOCK

PIN     19  Q0         ; state variables
PIN     18  Q1
PIN     17  Q2

STATE  MOORE_MACHINE

; State assignments, value of the machine variables for
; each state.

    S0 = /Q2 * /Q1 * /Q0
    S1 = /Q2 * /Q1 *  Q0
    S2 = /Q2 *  Q1 * /Q0
    S3 = /Q2 *  Q1 *  Q0
    S4 =  Q2 * /Q1 * /Q0
    S5 =  Q2 * /Q1 *  Q0
    S6 =  Q2 *  Q1 * /Q0
    S7 =  Q2 *  Q1 *  Q0


; state transitions

    S0 := VCC        -> S1   ; (VCC indicates that the transition
                             ; is unconditional.  A Boolean
    S1 := VCC        -> S2   ; expression would indicate an input
                             ; condition required for transition.)
    S2 := VCC        -> S3

    S3 := VCC        -> S4

    S4 := VCC        -> S5

    S5 := VCC        -> S6

    S6 := VCC        -> S7

    S7 := VCC        -> S0

SIMULATION

    VECTOR count := [ Q2, Q1, Q0 ]
```

```
        SETF CLOCK

        ;-- Preload registers to known a state

        PRLDF /Q2 /Q1 /Q0

        ;-- FOR loop to count 8 clocks

FOR i:= 0 TO 7 DO
     BEGIN
          CLOCKF CLOCK
     END
```

A more complex state section can account for several possible transition choices based on the condition of one or more circuit inputs, such as a count direction control. Such input states must be defined in a CONDITIONS subsection of the PLDasm source file.

A state machine can be programmed without specifying unused states if we use the machine default specifications. If it is not possible to resolve the conditions for a state transition, the machine goes to the specified default condition. This can be the present state (DEFAULT_BRANCH HOLD_STATE), the next state in the state assignment list (DEFAULT_BRANCH NEXT_STATE) or a specified state (DEFAULT_BRANCH ⟨state⟩).

If the outputs are not the same as the state variables and it is important to specify their status for an unresolved transition, they can be defined by the (OUTPUT_HOLD) or (DEFAULT_OUTPUT ⟨Boolean output conditions⟩) statements.

Table 12.13 summarizes the syntax for the state section of a PLDasm source file.

**Table 12.13**  State Commands for a PLDasm Source File

STATE             **One of:**
[MEALY_MACHINE]
or
[MOORE_MACHINE]

**Machine Defaults**
[OUTPUT_HOLD]
[DEFAULT_OUTPUT]
[DEFAULT_BRANCH ⟨specified state⟩]
or
[DEFAULT_BRANCH HOLD_STATE]
or
[DEFAULT_BRANCH NEXT_STATE]

**State Assignments**
Each state:
⟨state⟩ = ⟨Boolean combination of state variables⟩

**State Transitions**
⟨present state⟩ := ⟨condition 1⟩ → ⟨next state 1⟩
       + ⟨condition 2⟩ → ⟨next state 2⟩
       + ⟨condition 3⟩ → ⟨next state 3⟩

**Output Transitions**
⟨state⟩.OUTF = ⟨condition 1⟩ → ⟨output state 1⟩
       + ⟨condition 2⟩ → ⟨output state 2⟩
       + ⟨condition 3⟩ → ⟨output state 3⟩
(For all Moore machines, ⟨condition⟩ = VCC; i.e., output transitions are unconditional.)

**Input Conditions for State and Output Transitions**
CONDITIONS

⟨condition 1⟩ = ⟨Boolean combination 1 of input variables⟩
⟨condition 2⟩ = ⟨Boolean combination 2 of input variables⟩
⟨condition 3⟩ = ⟨Boolean combination 3 of input variables⟩

**EXAMPLE 12.14**    Write a PLDasm source file to program an 85C220 EPLD as a 4-bit bidirectional counter. Include active-LOW synchronous Clear and Ripple Clock Out Functions.

**Solution**

```
Title           4-bit Bidirectional Counter (Moore machine)
Pattern         pds
Revision        1.0
Author          R.Dueck
Company         Seneca College
Date            January 16, 1993

;     A 4-bit bidirectional counter programmed by a state table.
; The counter also has an active LOW synchronous clear input and
; an active LOW Ripple Clock Out (RCO).

CHIP    4ctmoore            85C220

; PINLIST

PIN     1     CLOCK

PIN     2     UPDOWN       ; Directional Input
PIN     3     CLEAR
PIN     14    RCO

PIN     19    Q0           ; state variables
PIN     18    Q1
PIN     17    Q2
PIN     16    Q3


; -- State Machine Format --
STATE   MOORE_MACHINE

    DEFAULT_BRANCH  S0 ; if state cannot be resolved, then CLEAR


; State assignments, value of the machine variables for
; each state.

        S0 = /Q3 * /Q2 * /Q1 * /Q0  ; 0x0
        S1 = /Q3 * /Q2 * /Q1 *  Q0  ; 0x1
        S2 = /Q3 * /Q2 *  Q1 * /Q0  ; 0x2
        S3 = /Q3 * /Q2 *  Q1 *  Q0  ; 0x3
        S4 = /Q3 *  Q2 * /Q1 * /Q0  ; 0x4
        S5 = /Q3 *  Q2 * /Q1 *  Q0  ; 0x5
        S6 = /Q3 *  Q2 *  Q1 * /Q0  ; 0x6
        S7 = /Q3 *  Q2 *  Q1 *  Q0  ; 0x7
        S8 =  Q3 * /Q2 * /Q1 * /Q0  ; 0x8
        S9 =  Q3 * /Q2 * /Q1 *  Q0  ; 0x9
        SA =  Q3 * /Q2 *  Q1 * /Q0  ; 0xA
        SB =  Q3 * /Q2 *  Q1 *  Q0  ; 0xB
        SC =  Q3 *  Q2 * /Q1 * /Q0  ; 0xC
        SD =  Q3 *  Q2 * /Q1 *  Q0  ; 0xD
        SE =  Q3 *  Q2 *  Q1 * /Q0  ; 0xE
        SF =  Q3 *  Q2 *  Q1 *  Q0  ; 0xF

; state transitions

        S0 := UP          -> S1
             + DOWN       -> SF

        S1 := UP          -> S2
             + DOWN       -> S0
```

```
S2 := UP          -> S3
    + DOWN        -> S1

S3 := UP          -> S4
    + DOWN        -> S2

S4 := UP          -> S5
    + DOWN        -> S3

S5 := UP          -> S6
    + DOWN        -> S4

S6 := UP          -> S7
    + DOWN        -> S5

S7 := UP          -> S8
    + DOWN        -> S6

S8 := UP          -> S9
    + DOWN        -> S7

S9 := UP          -> SA
    + DOWN        -> S8

SA := UP          -> SB
    + DOWN        -> S9

SB := UP          -> SC
    + DOWN        -> SA

SC := UP          -> SD
    + DOWN        -> SB

SD := UP          -> SE
    + DOWN        -> SC

SE := UP          -> SF
    + DOWN        -> SD

SF := UP          -> S0
    + DOWN        -> SE


CONDITIONS                ; Boolean combinations of input variables.

   UP =  UPDOWN * CLEAR
 DOWN = /UPDOWN * CLEAR

EQUATIONS
   /RCO = Q3*Q2*Q1*Q0

SIMULATION

   VECTOR count := [ Q3, Q2, Q1, Q0 ]

   ;-- Set all inputs to known values
   ;      1      1     1
   SETF CLOCK   CLEAR UPDOWN

   ;-- Preload registers to a known state

   PRLDF /Q3 /Q2 /Q1 /Q0

   ;-- Clock an input signal  1-->0-->1

   CLOCKF CLOCK
```

```
                    ;-- FOR loop to count 25 clocks
FOR i:= 0 TO 1 DO
BEGIN
    IF (i=1) THEN
    BEGIN
      SETF /UPDOWN
    END
    FOR j := 0 TO 24 DO
    BEGIN
        CLOCKF CLOCK
        IF ( j = 4 ) THEN       ; after 5 clocks
        BEGIN
            SETF /CLEAR         ; test the clear function
        END
        IF ( j = 5 ) THEN       ; then on next clock
        BEGIN
            SETF CLEAR          ; resume counting.
        END
    END
END
```

Each state in the counter of Example 12.14 has two possible destinations, based on the Boolean equations in the CONDITIONS section of the source file.

For example, the statement:

S1 := UP   → S2
+ DOWN   → S0

means, "If the counter is in state *S1,* go to state *S2* if *UP* is true and to state *S0* if *DOWN* is true."

*UP* and *DOWN* are Boolean conditions defined under CONDITIONS as:

UP = UPDOWN * CLEAR
DOWN = /UPDOWN * CLEAR

Thus, if UPDOWN = 1 AND CLEAR = 1, the circuit counts up; if UPDOWN = 0 AND CLEAR = 1, the circuit counts down.

Notice that there is no Boolean condition that includes /CLEAR. We could have included a CLEAR condition with each of the 16 state transition equations. Instead, we use the statement

DEFAULT_BRANCH S0

to implement the synchronous Clear function. If CLEAR = 0, neither of the state transition conditions *(UP* or *DOWN)* are met, and the circuit makes a transition to state *S0* $(Q_3 Q_2 Q_1 Q_0 = 0000)$ by default.

## GLOSSARY

**AND dependency**   An enable/inhibit function with the same properties as an AND gate. (Symbol: $G$)

**Biquinary sequence**   A mod-10 count sequence consisting of two groups of five states. The sequence has a 50% duty cycle on its most significant bit.

**Common control block**   The portion of an IEEE/ANSI symbol that shows the function of those control inputs (e.g., a common clock or clear input) that act on several outputs simultaneously.

**Control dependency**   An enable function such as an edge-triggered clock or a level-sensitive latch enable. (Symbol: $C$)

**Dependency notation**   A way of writing variables in an IEEE/ANSI symbol to indicate which outputs depend on the action of which inputs and how.

**Excitation table**   A table showing the required input conditions for every possible transition of a flip-flop output.

**Interconnection dependency**   A dependency that directly transfers a 1 or 0 state from one function to another. (Symbol: Z)

**Mealy-type state machine**   A state machine whose outputs depend on the present state of the internal flip-flops as well as the next state of the flip-flops, as defined by the inputs to the combinational part of the circuit.

**Mode dependency**   A dependency that selects one of several possible operating modes (e.g., *LOAD* or *COUNT* in a presettable counter). (Symbol: $M$)

**Moore-type state machine**   A state machine whose outputs depend only on the present state of the internal flip-flops.

**Presettable counter**   A counter whose contents can be set to any value by loading a binary number directly into the internal flip-flops.

**State machine**   A synchronous sequential circuit.

**Synchronous sequential circuit**   A circuit whose outputs progress through a sequence of states in response to a synchronous clock signal and possibly other input variables (e.g., a counter or shift register).

**Terminal count**   The state of a counter's output in the last state before the recycle point.

## PROBLEMS

### Section 12.1 Asynchronous MSI Counters

**12.1**   State the modulus for each of the following asynchronous MSI counters:

  **a.** 74LS90

  **b.** 74LS92

  **c.** 74LS93

**12.2**   Each of the three counters listed in Problem 12.1 has two separate portions, each clocked by a separate clock input.

  **a.** State the names of the two different clock inputs.

  **b.** Describe how the total modulus of each counter is divided between the two portions of the counter.

**12.3**   Draw a diagram showing how to connect the two clock inputs of a 74LS93 counter to make a counter with a total modulus of 16.

**12.4**   Draw a diagram of a 74LS93 counter configured as a mod-12 counter.

**12.5**   Repeat Problem 12.4 for a 74LS90 as a mod-8 counter.

**12.6**   Table 12.14 shows a count sequence known as a biquinary sequence. Sketch a diagram showing how to connect a 74LS90 counter to produce such a sequence.

**Table 12.14**  Biquinary Sequence

| $Q_A$ | $Q_D$ | $Q_C$ | $Q_B$ |
|-------|-------|-------|-------|
| 0 | 0 | 0 | 0 |
| 0 | 0 | 0 | 1 |
| 0 | 0 | 1 | 0 |
| 0 | 0 | 1 | 1 |
| 0 | 1 | 0 | 0 |
| 1 | 0 | 0 | 0 |
| 1 | 0 | 0 | 1 |
| 1 | 0 | 1 | 0 |
| 1 | 0 | 1 | 1 |
| 1 | 1 | 0 | 0 |

**12.7** Figure 12.44 represents two 74LS93 mod-16 counters connected in cascade. What is the total modulus of this counter circuit?

**12.8** Repeat Problem 12.7 if the two counters are replaced with 74LS90 decade counters.

**\*12.9** A digital pulse waveform of 1 Hz is applied to the *CKA* input of the 74LS90 counter shown in Figure 12.45. What is the period of the $Q_D$ waveform of the 74LS92 counter in Figure 12.45?

## Section 12.2 Synchronous Presettable MSI Counters

**12.10** Complete the timing diagram of the 74LS193 counter illustrated in Figure 12.46 (see page 556).

**12.11** Complete the timing diagram of the 74LS191 counter illustrated in Figure 12.47 (see page 557).

**12.12** Draw a circuit showing how to cascade two 74LS193 counters asynchronously.

**12.13** Draw a circuit illustrating how to connect two 74LS191 counters for synchronous cascading.

**12.14** Draw a circuit showing how to cascade two 74LS191 counters asynchronously.

**12.15** Show how to make a bidirectional counter that can count up or down between 0 and 255. Use 74LS191 counters.

**12.16** Draw the circuit of a synchronous DOWN counter whose outputs count from 383 to 0. Use 74LS191 counters.

**12.17** Briefly describe the difference between asynchronous and synchronous *LOAD* functions on a counter. Use 74LS193 and 74LS163 counters as examples.

**12.18** Refer to the frequency counter in Figure 12.19. What modification is required to make the circuit display the input frequency in Hz, rather than kHz?

**\*12.19** Refer to the frequency counter circuit represented in Figure 12.19. What difficulty would we encounter if we wanted to modify the circuit to display the input frequency in MHz? How could this difficulty be overcome?

**Figure 12.44**
**Problem 12.7**
**Two Mod-16 Counters Connected in Cascade**

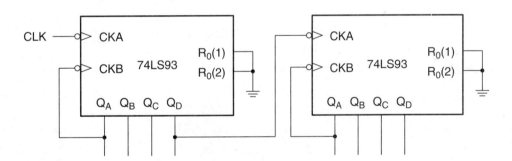

**Figure 12.45**
**Problem 12.9**
**74LS90 Counter**

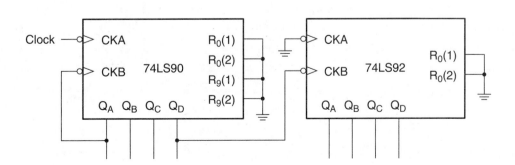

**Figure 12.46**

**Problem 12.10**
**74LS193 Counter**

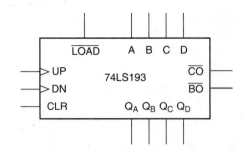

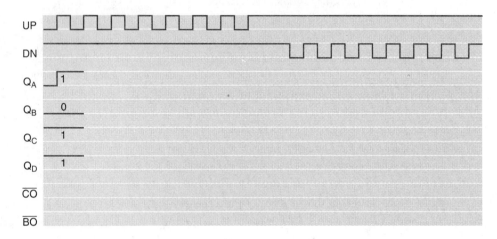

**Figure 12.47**

**Problem 12.11**
**74LS191 Counter**

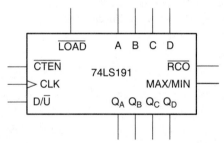

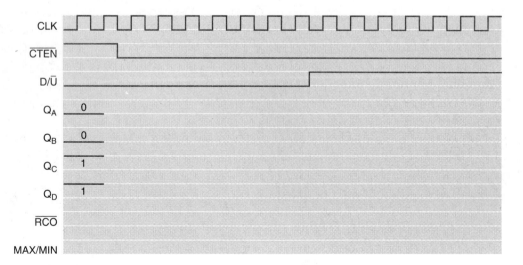

**Figure 12.48**

**Problem 12.20
Timing Diagram for
Frequency Counter Circuit**

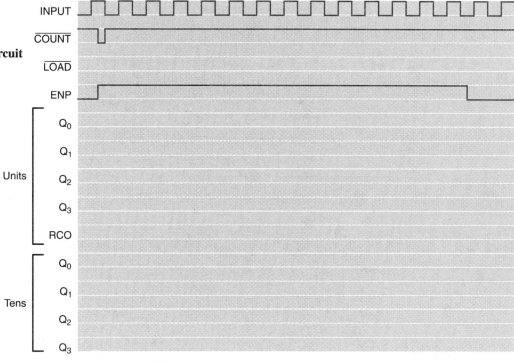

**12.20** The frequency counter shown in Figure 12.19 has a stored value from the previous measurement of 0083 kHz.

    **a.** Complete the timing diagram of the circuit, shown in Figure 12.48, for the next frequency measurement.

    **b.** What value is displayed after the sampling interval ends?

**12.21**  **a.** Draw a circuit showing how to connect a 74LS191 counter as an UP counter that produces an output pulse at its *MAX/MIN* output once every 12 clock pulses.

    **b.** Draw a timing diagram showing the operation of this circuit for one complete cycle. Include waveforms for *CLK*, $Q_D$, $Q_C$, $Q_B$, $Q_A$, $\overline{RCO/LOAD}$, and *MAX/MIN*.

**12.22** Repeat Problem 12.21 for a DOWN counter of the same modulus.

**12.23**  **a.** State the modulus of the counter circuit in Figure 12.49 (see page 558).

    **b.** Use timing diagrams and other analytical techniques, as in Example 12.8, to prove your answer to part a.

**12.24** Repeat Problem 12.23 for the case where both counters are configured to count UP, rather than DOWN.

**12.25** Draw a circuit showing how to configure two 74LS191 counters as a mod-175 counter.

## Section 12.3 IEEE/ANSI Notation

**12.26** Figure 12.50 shows the IEEE/ANSI symbol of a 74LS92 counter. Briefly describe the operation of the counter as symbolized by the notation in Figure 12.50 (see page 558).

**12.27** Repeat Problem 12.26 for the IEEE/ANSI symbol of the 74LS160 counter shown in Figure 12.51 (see page 558).

## Section 12.4 Design of Synchronous Sequential Circuits

**12.28** List the main sections of a Moore-type state machine.

**12.29** State the main difference between a Moore-type and a Mealy-type state machine.

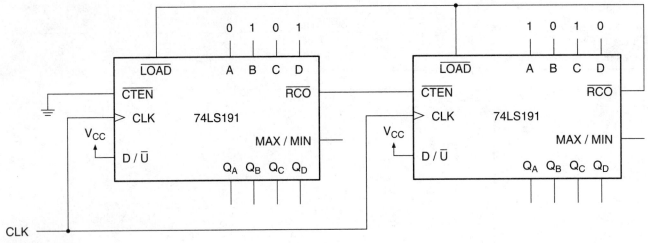

**Figure 12.49**

**Problem 12.23**
**Counter Circuit**

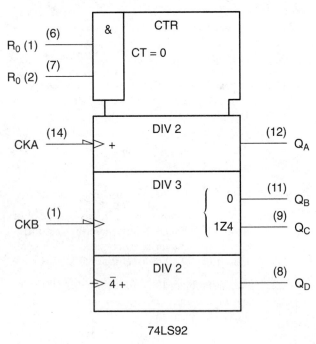

74LS92

**Figure 12.50**

**Problem 12.26**
**IEEE/ANSI Symbol of a 74LS92 Counter**

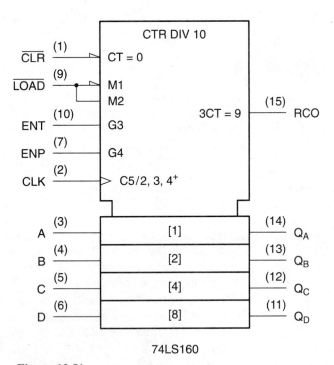

74LS160

**Figure 12.51**

**Problem 12.27**
**IEEE/ANSI Symbol of a 74LS160 Counter**

**12.30** **a.** Draw the timing diagram and state diagram of a synchronous mod-10 counter.

**b.** Show the work required to derive the equations for the $J$ and $K$ inputs for each flip-flop.

**c.** Draw the counter circuit, using negative edge-triggered JK flip-flops.

**\*12.31** Modify the mod-12 counter in Figure 12.31 to make it bidirectional. The count direction is controlled by an input called $U$. When $U = 1$ the circuit counts UP (0000 to 1101 and repeat). When $U = 0$, the circuit counts DOWN (1101 to 0000 and repeat). Unused states may be treated as don't care states.

**\*12.32** Draw the state diagram resulting from the circuit designed in Problem 12.31, including the progression of unused states.

**12.33** Design a synchronous mod-6 counter with a 50% duty cycle on its most significant bit. Test that all unused states enter the main sequence automatically.

## Section 12.5 Sequential Logic Applications of PLDs

**12.34** Write a PLDasm source file to program an 85C220 EPLD as a bidirectional mod-10 counter. Include an active-HIGH *ENABLE* input and an active-LOW *CLEAR* input.

**\*\*12.35** Write a PLDasm source file to implement a two-digit BCD counter using an 85C22V10 EPLD. (Outputs count from $00_{10}$ ($0000\ 0000_{BCD}$) to $99_{10}$ ($1001\ 1001_{BCD}$) and recycle.)

## ▼ Answers to Section Review Questions

### Section 12.2a
**12.1.** Two counters: mod-100; three counters: mod-1000.

### Section 12.2b
**12.2.** DOWN: $DCBA = 1100$ (= modulus)
UP: $DCBA = 0011$ (= $\overline{\text{modulus}}$)

### Section 12.2c
**12.3.** New modulus = $10011100_2 = 156_{10}$.

# Shift Registers

## ● CHAPTER OUTLINE

**13.1**  Basic Shift Register Configurations

**13.2**  Bidirectional Shift Registers

**13.3**  Shift Register Counters

**13.4**  IEEE/ANSI Notation

## ● CHAPTER OBJECTIVES

Upon successful completion of this chapter, you will be able to:

- Draw a logic circuit of a serial shift register and determine its contents over time, given any input data.

- Draw a timing diagram showing the operation of a serial shift register.

- Explain the operation of several MSI serial shift registers and use them in simple applications.

- Draw the logic circuit of a general parallel-load shift register.

- Draw a timing diagram showing the operation of a parallel-load shift register.

- Explain the operation of MSI parallel-load shift registers and use them in simple applications.

- Draw the general logic circuit of a bidirectional shift register and explain the concepts of right-shift and left-shift.

- Explain the operation of a bidirectional shift register, using timing diagrams and sequence tables.

- Configure a parallel-load shift register as a bidirectional shift register.

- Explain the operation of a universal shift register and how it can be used for simple applications.

- Construct ring counters and Johnson counters from flip-flops or from MSI shift registers.

- Design a decoder for a Johnson counter.

- Use a ring counter or a Johnson counter as an event sequencer.

- Compare binary, ring, and Johnson counters in terms of the modulus and the required decoding for each circuit.

- Read the IEEE/ANSI logic symbols for several common MSI shift registers.

## ● INTRODUCTION

The shift register is an important class of synchronous sequential circuit. Shift registers are widely used in various computer and communications applications, such as storing multibit binary numbers and translating them from serial to parallel form and vice versa.

We have already seen some of these applications in previous chapters. For example, we used eight D flip-flops to convert a serially transmitted ASCII character to parallel form.

The storage capacity of a shift register is the same as the numbers of flip-flops in the circuit. A circuit having $n$ flip-flops can store $n$ bits. There are three basic types of data movement in a shift register: serial, parallel, and rotation.

Serial data movement implies that all data in the register will move over by one flip-flop every time the clock is pulsed. Parallel movement occurs when all flip-flops are loaded at the same time. Rotation of data is similar to serial shifting, except that there is feedback from output to input, so that data are circulated continuously.

Rotation is used in some shift register circuits to make special types of counters—ring counters and Johnson counters. These counters are useful for event sequencing and have special advantages regarding the decoding of their outputs.

## 13.1

## Basic Shift Register Configurations

**Shift register** *A synchronous sequential circuit that will store and move* n-*bit data, either serially or in parallel, in* n *flip-flops.*

**Serial shifting** *Movement of data from one end of a shift register to the other at a rate of 1 bit per clock pulse.*

**Parallel shifting** *Movement of data into or out of all flip-flops of a shift register at the same time.*

**Parallel loading** *Parallel shifting of data into a shift register.*

**Rotation** *Serial shifting of data with the output(s) of the last flip-flop connected to the synchronous input(s) of the first flip-flop. The result is continuous circulation of the same data.*

A **shift register** is a synchronous sequential circuit used to store or move data. It consists of several flip-flops, connected so that data are transferred into and out of the flip-flops in a standard pattern.

Figure 13.1 represents three types of data movement in three 6-bit shift registers. The circuits each contain six flip-flops, configured to move data in one of the ways shown.

Figure 13.1a shows the operation of **serial shifting.** The stored data are taken in one at a time from the input and move one position toward the output with each applied clock pulse.

**Parallel shifting** is illustrated in Figure 13.1b. Data move simultaneously into all flip-flops when a clock pulse is applied. The data are available in parallel at the register outputs. **Parallel loading** refers to the parallel input operation only.

**Rotation,** depicted in Figure 13.1c, is similar to serial shifting in that data are shifted one place to the right with each clock pulse. In this operation, however, data are continuously circulated in the shift register by moving the rightmost bit back to the leftmost flip-flop with each clock pulse.

**Figure 13.1**

**Data Movements in a Shift Register**

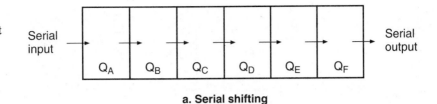

a. Serial shifting

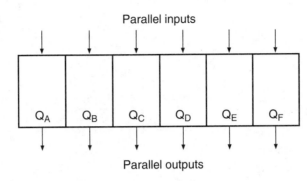

b. Parallel shifting

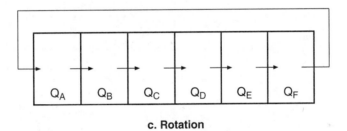

c. Rotation

**Figure 13.2**

**Serial Shift Register**

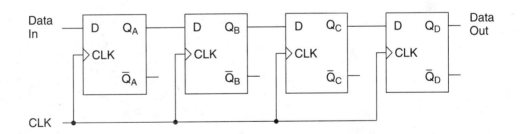

**Table 13.1** Excitation Table for a D Flip-Flop

| Transition | *D* |
|---|---|
| $0 \rightarrow 0$ | 0 |
| $0 \rightarrow 1$ | 1 |
| $1 \rightarrow 0$ | 0 |
| $1 \rightarrow 1$ | 1 |

## *Serial Shift Registers*

Figure 13.2 shows the most basic shift register circuit: the serial shift register, so called because data are shifted through the circuit in a linear or serial fashion. The circuit shown consists of four D flip-flops connected in cascade and clocked synchronously.

For a D flip-flop, $Q$ follows $D$, as indicated by the excitation table in Table 13.1. In the shift register of Figure 13.2, this implies that $Q$ of any flip-flop follows the $Q$ of the previous flip-flop, that is, the one to its left.

**Figure 13.3**

**Shifting a 1 Through the Shift Register**

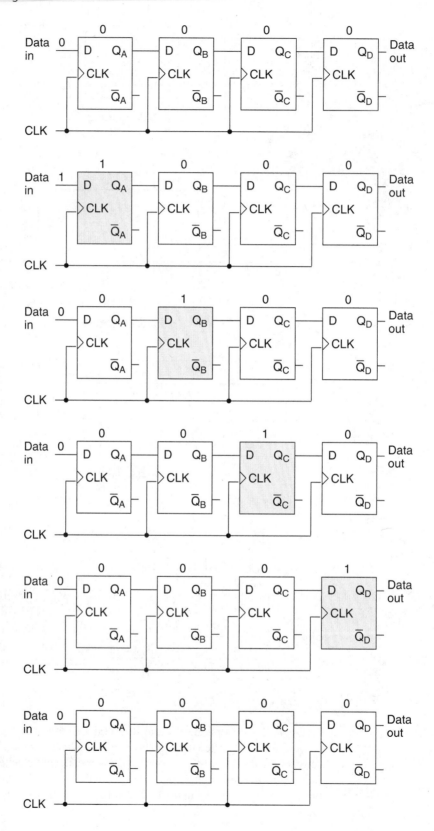

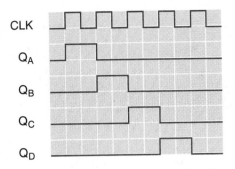

**Figure 13.4**

**Timing Diagram for Data Shown in Figure 13.3**

The Boolean equations for this circuit are:

$D_A$ = Data in

$D_B = Q_A$

$D_C = Q_B$

$D_D = Q_C$

The value of a bit stored in any flip-flop *after* a *CLK* pulse is the same as the bit in the flip-flop to its left *before* the pulse. The result is that when a *CLK* pulse is applied to the circuit, the contents of the flip-flops move one position to the right and the bit at the circuit input is shifted into $Q_A$. The bit stored in $Q_D$ is overwritten by the former value of $Q_C$ and is lost.

Let us track the progress of data through the circuit in two cases. All flip-flops are initially cleared in both cases.

*Case 1:* A 1 is clocked into the shift register, followed by a string of 0s, as shown in Figure 13.3. The flip-flop containing the 1 is shaded.

The 1 moves one position right with each clock pulse, the register filling up with 0s behind it. After four clock pulses, the 1 reaches the Data Out flip-flop. On the fifth pulse, the 0 coming behind overwrites the 1 at $Q_D$, leaving the register filled with 0s. Figure 13.4 represents the register timing diagram for Case 1.

*Case 2:* Figure 13.5 shows a shift register, initially cleared, being filled with 1s. As before, the initial 1 is clocked into the shift register and reaches the Data Out line on the fourth clock pulse. This time, the register fills up with 1s, not 0s, because the data input remains HIGH. The timing diagram for this case is shown in Figure 13.6.

---

**EXAMPLE 13.1**

The 4-bit serial shift register in Figure 13.2 has a sequence of input data applied, as shown in Table 13.2. Each new line represents the state of the input before a new *CLK* pulse.

Complete the table, showing the states of all flip-flop outputs after each *CLK* pulse. Also draw the timing diagram. Assume that all flip-flops are initially cleared.

**Solution**    Table 13.3 shows the movement of data in the 4-bit serial shift register. For every line, the leftmost bit $(Q_A)$ of the next state is the same as the input data. Bits $Q_B$, $Q_C$, and $Q_D$ of the next state are bits $Q_A$, $Q_B$, and $Q_C$ of the present state, shifted one place to the right. Notice that the $D$ values for each present state are the same as the next state $Q$ values.

**Figure 13.5**

**Filling a Shift Register With 1s**

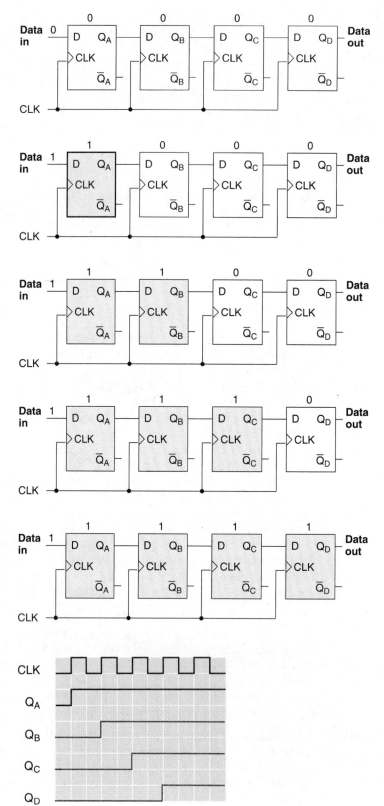

**Figure 13.6**

**Timing Diagram for Data Shown in Figure 13.5**

Table 13.2  Serial Data Movement in a Shift Register (Initial)

| Data In | Present State | | | | $D_A$ | $D_B$ | $D_C$ | $D_D$ | Next State | | | |
|---|---|---|---|---|---|---|---|---|---|---|---|---|
| | $Q_A$ | $Q_B$ | $Q_C$ | $Q_D$ | | | | | $Q_A$ | $Q_B$ | $Q_C$ | $Q_D$ |
| 1 | 0 | 0 | 0 | 0 | 1 | 0 | 0 | 0 | 1 | 0 | 0 | 0 |
| 0 | | | | | | | | | | | | |
| 1 | | | | | | | | | | | | |
| 0 | | | | | | | | | | | | |
| 1 | | | | | | | | | | | | |
| 1 | | | | | | | | | | | | |
| 1 | | | | | | | | | | | | |
| 1 | | | | | | | | | | | | |

Table 13.3  Serial Data Movement in a Shift Register (Completed)

| Data In | Present State | | | | $D_A$ | $D_B$ | $D_C$ | $D_D$ | Next State | | | |
|---|---|---|---|---|---|---|---|---|---|---|---|---|
| | $Q_A$ | $Q_B$ | $Q_C$ | $Q_D$ | | | | | $Q_A$ | $Q_B$ | $Q_C$ | $Q_D$ |
| 1 | 0 | 0 | 0 | 0 | 1 | 0 | 0 | 0 | 1 | 0 | 0 | 0 |
| 0 | 1 | 0 | 0 | 0 | 0 | 1 | 0 | 0 | 0 | 1 | 0 | 0 |
| 1 | 0 | 1 | 0 | 0 | 1 | 0 | 1 | 0 | 1 | 0 | 1 | 0 |
| 0 | 1 | 0 | 1 | 0 | 0 | 1 | 0 | 1 | 0 | 1 | 0 | 1 |
| 1 | 0 | 1 | 0 | 1 | 1 | 0 | 1 | 0 | 1 | 0 | 1 | 0 |
| 1 | 1 | 0 | 1 | 0 | 1 | 1 | 0 | 1 | 1 | 1 | 0 | 1 |
| 1 | 1 | 1 | 0 | 1 | 1 | 1 | 1 | 0 | 1 | 1 | 1 | 0 |
| 1 | 1 | 1 | 1 | 0 | 1 | 1 | 1 | 1 | 1 | 1 | 1 | 1 |

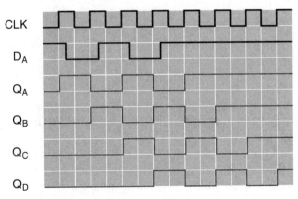

Figure 13.7

Example 13.1
Shift Register Timing Diagram

Figure 13.7 shows the timing diagram of the circuit with these inputs. The data at $D_A$ arrive asynchronously. The data at the $Q$ outputs are synchronized to the positive edge of the clock waveform.

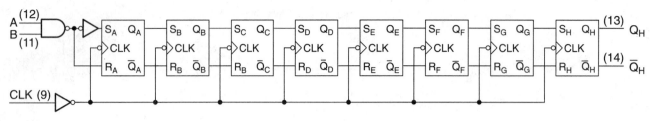

**Figure 13.8**

**74LS91 Serial-In-Serial-Out Shift Register**

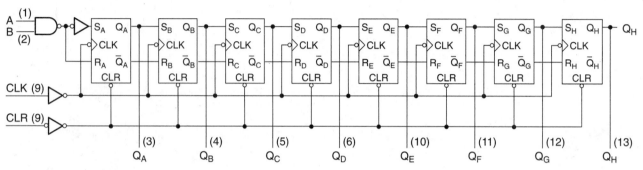

**Figure 13.9**

**74LS164 Serial-In-Parallel-Out Shift Register**

## MSI Serial Shift Registers

Two TTL serial shift registers are the 74LS91 Serial-In-Serial-Out Shift Register, shown in Figure 13.8, and the 74LS164 Serial-In-Parallel-Out Shift Register, shown in Figure 13.9. You can see from their logic diagrams that these devices are really the same circuit, except that the 74LS164 has a Clear input and all its flip-flop outputs are brought out to pin connections.

Both shift registers are constructed from SR flip-flops, whose excitation table is given in Table 13.4.

In the shift register circuit, the inverter at $S_A$ ensures that $S_A$ and $R_A$ are always in opposite states. Thus, the excitation table of the flip-flop should be modified, as in Table 13.5, to eliminate conditions that never occur. The SR flip-flops thus act the same as the D flip-flops in Figure 13.2. This can be verified by comparing $D$ in Table 13.1 and $S$ in Table 13.5.

Data enter the 74LS91 and 74LS164 shift registers at inputs $A$ and $B$. We can use one of these inputs as a data enable/inhibit line and the other as a data input. In the case of the 74LS91 shift register, data are available only at $Q_H$, eight clock pulses after they are entered. Data in the 74LS164 device can be shifted through the register in exactly the same way as through the 74LS91, but they are also available at every flip-flop output, if so desired.

The 74LS164 shift register can be used as a serial-to-parallel converter in data communications applications. In Chapter 6 (Example 6.9), we used eight D flip-flops as a serial shift register to convert a serially transmitted ASCII character to parallel form. (This is useful since computer systems store and move data in parallel, usually in multiples of 8, but data can be transmitted more economically between computers and peripheral devices in serial form, since it requires only one pair of wires for signal and ground.)

**Table 13.4** Excitation Table of an SR Flip-Flop

| Transition | S | R |
|---|---|---|
| $0 \rightarrow 0$ | 0 | X |
| $0 \rightarrow 1$ | 1 | 0 |
| $1 \rightarrow 0$ | 0 | 1 |
| $1 \rightarrow 1$ | X | 0 |

**Table 13.5** Excitation Table of an SR Flip-Flop (as Implemented)

| Transition | S | R |
|---|---|---|
| $0 \rightarrow 0$ | 0 | 1 |
| $0 \rightarrow 1$ | 1 | 0 |
| $1 \rightarrow 0$ | 0 | 1 |
| $1 \rightarrow 1$ | 1 | 0 |

The 74LS164 can replace the D flip-flops of Figure 6.36. The $A$ input of the '164 is used as a data input for the serially transmitted character. The $B$ input is tied HIGH to enable the data input. The $CLR$ input is tied HIGH to disable it. Eight clock pulses after the initial data are applied to the shift register, they are available in parallel form at outputs $Q_H$ through $Q_A$.

---

**Section Review Problem for Section 13.1a**

13.1. Can the D flip-flops in Figure 13.2 be replaced by SR flip-flops? By JK flip-flops? If so, what modifications to the existing circuit are required?

---

## Parallel-Load Shift Registers

**Parallel-load shift register**  *A shift register that can be preset to any value by directly loading a binary number into its internal flip-flops.*

**Parallel-access shift register**  *A shift register whose flip-flop outputs are individually accessible.*

**Right-shift**  *In an 8-bit shift register, serial shifting in the direction from $Q_A$ to $Q_H$.*

**Left-shift**  *In an 8-bit shift register, serial shifting in the direction from $Q_H$ to $Q_A$.*

In Chapter 12, we looked at a class of synchronous counter circuits that could be preset to any value. There are shift register circuits that have this feature as well. Figure 13.10 shows the general circuit of a **parallel-load shift register.** The operation of the parallel-load shift register is controlled by an input labeled $SHIFT/\overline{LOAD}$. When LOW, $SHIFT/\overline{LOAD}$ directs parallel data $A$ through $D$ directly into the flip-flops by enabling the righthand AND gates of each AND-OR circuit.

When $SHIFT/\overline{LOAD}$ is HIGH, it enables the lefthand AND gates, passing the $Q$ output of each flip-flop to the $D$ input of the next one. In other words, it enables the serial shift function, including the serial input at $D_A$.

Two representative parallel-load devices are the 74LS165A Parallel-Load Shift Register (serial out only) and the 74LS95B **Parallel-Access Shift Register** (serial and parallel out), shown in Figures 13.11 and 13.12.

### 74LS165A

The 74LS165A shift register consists of eight D flip-flops in cascade, as in a serial shift register. The parallel inputs are gated by NANDs into asynchronous $S$ and $R$ inputs on the individual flip-flops when the $SHIFT/\overline{LOAD}$ input is LOW. When $SHIFT/\overline{LOAD}$ is HIGH, the parallel inputs are inhibited and data shift serially from one flip-flop to the next via the $Q$-to-$D$ connections.

The 74LS165A has one serial input and two $CLK$ inputs. Since the two $CLK$s are ORed, either one can be used as an active-HIGH clock inhibit function. When HIGH, the clock inhibit function makes the OR gate output permanently HIGH, disabling the clock function.

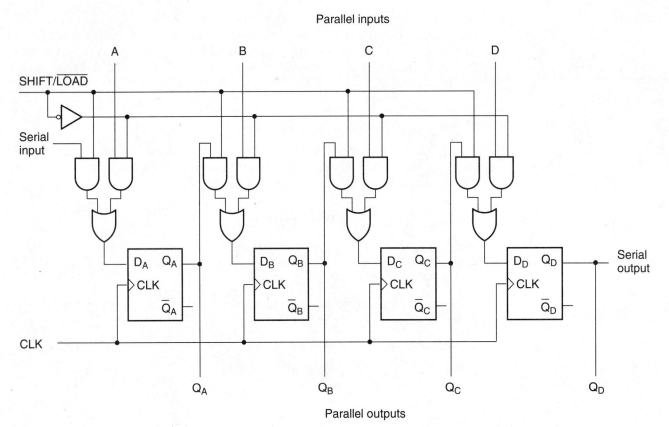

**Figure 13.10**

**Parallel-Load Shift Register**

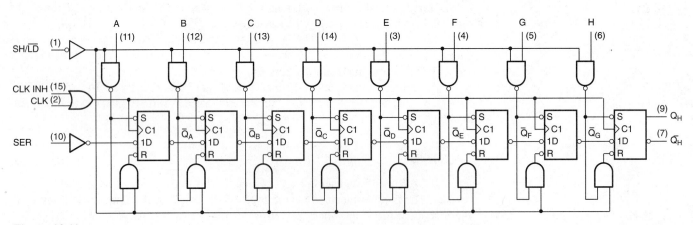

**Figure 13.11**

**74LS165A Parallel-Load Shift Register** (Reprinted by permission of Texas Instruments)

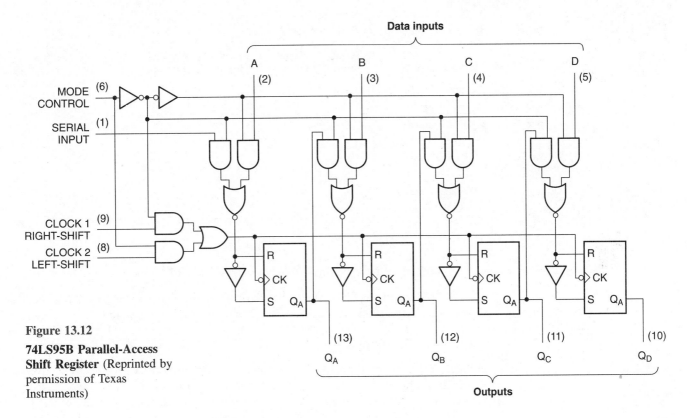

**Figure 13.12**

**74LS95B Parallel-Access Shift Register** (Reprinted by permission of Texas Instruments)

### 74LS95B

The 74LS95B shift register is very much like the general parallel-load shift register of Figure 13.10. Parallel data are loaded synchronously via the righthand AND gate of each AND-OR circuit and serially shifted via the lefthand AND gate. One or the other of these AND gates is always enabled by the *MODE CONTROL* input.

The 74LS95B also allows for bidirectional data shifting. By convention, all shift register circuits we have seen have had serial data entering $Q_A$ from the left and moving one flip-flop to the right with each clock pulse. We call this **right-shift.** We can also design a shift register to move data in the opposite direction. This is called **left-shift.** These terms have no physical meaning inside the shift register circuit, but relate only to our left-to-right reading conventions.

The 74LS95B shift register has two clocks, labeled *RIGHT-SHIFT* and *LEFT-SHIFT.* The *RIGHT-SHIFT* clock is enabled when *MODE CONTROL* = 0 (serial shift function). The *LEFT-SHIFT* clock is enabled when *MODE CONTROL* = 1 (parallel-load function).

It is possible to configure this shift register with external wiring so that it shifts to the left, that is, from $Q_D$ to $Q_A$. We will examine the left-shift function in a later section of this chapter. For now, assume that the *LEFT-SHIFT* input should be disabled by being tied LOW.

### Summary

With each new MSI shift register we have examined, one or more new features have been added. Table 13.6 shows a summary of features of the various devices studied to date. All devices are capable of serial-only shifting, should that be required.

Table 13.6 Summary of Shift Register Features

| Shift Register Type | Serial Input | Serial Output | Parallel Input | Parallel Output |
|---|---|---|---|---|
| Serial-In-Serial-Out (74LS91) | X | X | | |
| Serial-In-Parallel-Out (74LS164) | X | X | | X |
| Parallel Load (74LS165A) | X | X | X | |
| Parallel Access (74LS95B) | X | X | X | X |

## Applications of Parallel-Load Shift Registers

We will examine two applications of parallel-load shift registers: serial-to-parallel conversion for data transmission and power-of-2 multiplication and division.

### Serial Data Transmission

In Example 10.25 (Section 10.5, Multiplexers), we used an 8-to-1 multiplexer to convert an ASCII character from parallel to serial form for transmission between two devices. We can use a 74LS165A shift register for the same purpose, using a different technique.

The MUX circuit used a counter to direct the bits of an ASCII character to a common output in a continuous sequence. In the circuit using the '165A, the shift register loads data through its parallel inputs, then shifts them out serially.

---

**EXAMPLE 13.2**

Figure 13.13 shows a simple parallel-to-serial converter made from several chips we have already studied. The shift register accepts ASCII characters set by switches at its parallel-load inputs and transmits them serially. The circuit can be configured to transmit characters singly (jumper in position X2) or continuously (jumper in position X1).

Briefly describe the operation of the circuit and draw timing diagrams for the cases where it is set to transmit ASCII "M" continuously and as a single character. (Refer to Table 5.6 in Chapter 5 for the ASCII table.)

**Solution** Two pieces of information relate to the use of this circuit in a practical application:

1. The 555 timer is configured for astable operation to provide the circuit clock. The potentiometer is set for about 7 kΩ, giving a frequency of 9600 Hz, a standard clock speed for asynchronous transmission.

2. The output is converted from TTL voltage levels to ±12 V by a 1488 line driver chip. These voltages conform to a serial transmission specification called RS232. More information on RS232 is available in any good textbook on data communications.

When the jumper is in position X1, the 74LS191 counter synchrononizes the beginning of a character to the $\overline{RCO}$ pulse, which occurs once every 16 clock pulses. When $\overline{RCO}$ goes LOW, the ASCII data are loaded into the shift register, then shifted out when

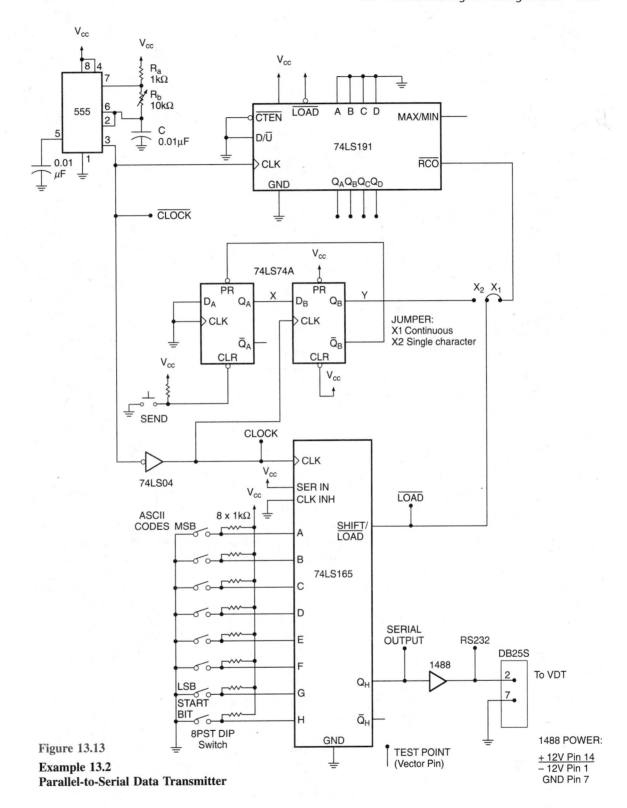

**Figure 13.13**

**Example 13.2**
**Parallel-to-Serial Data Transmitter**

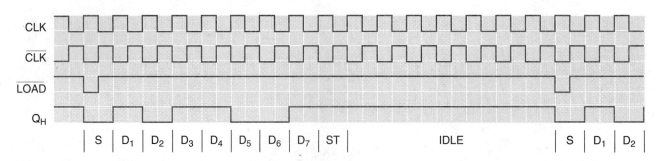

CLK

$\overline{\text{CLK}}$

$\overline{\text{LOAD}}$

$Q_H$

| S | $D_1$ | $D_2$ | $D_3$ | $D_4$ | $D_5$ | $D_6$ | $D_7$ | ST | | IDLE | | S | $D_1$ | $D_2$ |

**Figure 13.14**

**Example 13.2**
**Continuous Transmission of ASCII "M"**

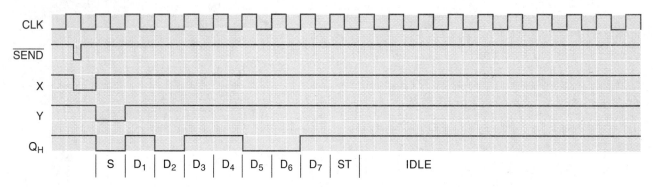

CLK

$\overline{\text{SEND}}$

X

Y

$Q_H$

| S | $D_1$ | $D_2$ | $D_3$ | $D_4$ | $D_5$ | $D_6$ | $D_7$ | ST | | IDLE |

**Figure 13.15**

**Example 13.2**
**Transmission of a Single ASCII "M"**

$\overline{RCO}$ goes HIGH. To maintain proper timing of the shift and load functions, the clocks for the counter and shift register are out of phase.

Eight clock pulses clear data out of the shift register. Since the serial input is HIGH, the register will fill with 1s and transmit them until an $\overline{RCO}$ pulse loads new ASCII data into the shift register. The stream of 1s acts as the idle state between successive characters.

Figure 13.14 shows the timing diagram of ASCII "M" being transmitted. ("M" = $4D_{16}$ = $1001101_2$. $S$ = start bit = 0. $D_7$ is the MSB.)

When the jumper is in position X2 and the *SEND* switch is pressed, the circuit transmits a single ASCII character. Since the *SEND* switch can be pressed any time, the chance of it activating exactly on a clock edge is very small. If *SEND* is not synchronized to the system clock, the start bit of the transmitted character will be of incorrect length, causing a transmission error.

The two D flip-flops synchronize the *SEND* pulse to the system clock. Figure 13.15 illustrates the sequence of events. *SEND* pulls the *CLR* input of the first-flip flop LOW, asynchronously setting $X$ LOW. On the next positive edge of *CLK*, $Y$ follows $X$ and also goes LOW. This loads the shift register with parallel ASCII data.

The complementary output of the second flip-flop asynchronously sets $X$ HIGH. On the next positive edge of *CLK*, $Y$ returns to the HIGH state and the data shifts out serially, as indicated in Figure 13.15.

## Arithmetic Operations

A shift register can be used to perform arithmetic operations, such as multiplying or dividing by a power of 2. For example, each number in the sequence of binary numbers $Q_D Q_C Q_B Q_A$ = 0011, 0110, and 1100 (decimal numbers 3, 6, and 12) is double the previous number. Assuming that the MSB is the farthest flip-flop to the right, a multiply-by-2 operation is the same as a shift-right, and a divide-by-2 operation is a shift-left.

> Circuit convention: MSB is rightmost flip-flop. Text convention: MSB is leftmost digit.

▼

**EXAMPLE 13.3**

Show how to connect a 74LS95B shift register so that it can multiply binary numbers by powers of 2. Draw the timing diagram showing the result of $2_{10} \times 4_{10}$. Assume that the shift register is initially cleared.

**Solution**

Figure 13.16a shows the shift register multiplier. The *MODE* pushbutton loads the multiplicand into the shift register on the first negative edge of the clock by selecting the parallel load function.

Two pulses on the *RIGHT-SHIFT* input multiply the binary input by 4, providing the serial input is LOW so that the register fills with 0s. The *LEFT-SHIFT* input is disabled by grounding it.

Figure 13.16 shows the load and shift functions in a timing diagram.

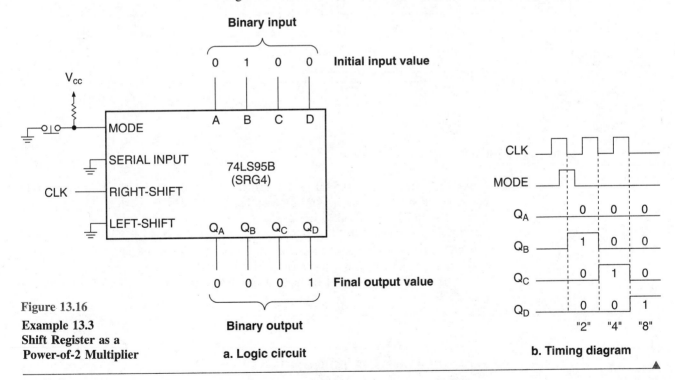

**Figure 13.16**

**Example 13.3
Shift Register as a
Power-of-2 Multiplier**

a. Logic circuit

b. Timing diagram

The shift register multiplier/divider in Example 13.3 is actually a very limited circuit. For example, suppose we want to multiply $1100 \times 100$. We load the shift register with 1100 and apply two clock pulses to shift the multiplicand two places. The product is 110000, but since we have only four places, the shift register holds the value 0000. The two MSBs have dropped into the "bit bucket."

In order to expand the usefulness of this application, we should cascade two or more shift registers.

**EXAMPLE 13.4**    Draw the circuit of two 74LS95B shift registers connected so that they can multiply a number by a power of 2 and obtain an 8-bit product.

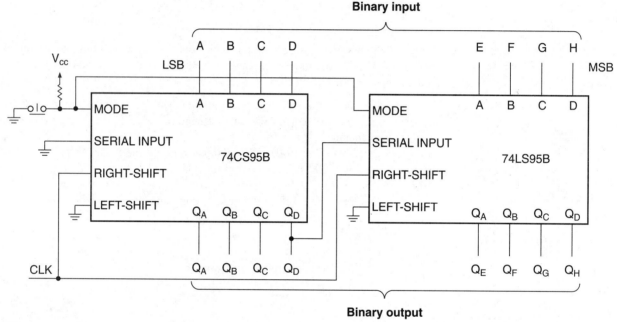

**Figure 13.17**

**Example 13.4**
**Two 74LS95Bs as an 8-Bit Shift Register**

**Solution**    Figure 13.17 shows two 74LS95B shift registers combined to make an 8-bit shift register multiplier. Since the circuit is synchronous, the *RIGHT-SHIFT* clocks are connected together. The *LEFT-SHIFT* clocks are disabled.

The serial input of the least significant shift register is tied LOW so that the circuit will fill with 0s as it clocks. The serial input of the most significant shift register is wired to $Q_D$ of the first device so that the chain of serial shifting is continuous.

The *MODE* inputs are common so that both chips are parallel-loaded simultaneously.

---

**Section Review Problem for Section 13.1b**

13.2. How many places and in what direction must you shift the input of an 8-bit shift register to obtain the result of $10101_2 \times 1000_2$?

## 13.2

## Bidirectional Shift Registers

**Bidirectional shift register** *A shift register that can serially shift bits left or right according to the state of a direction control input.*

Figures 13.18 and 13.19 show the general concept of right-shift and left-shift in a **bidirectional shift register.** The circuit is similar to that of a parallel-load shift register, except that instead of parallel and serial data paths, we have right and left data paths. Each path is enabled by its own set of AND gates when selected by the $RIGHT/\overline{LEFT}$ input.

When $RIGHT/\overline{LEFT} = 1$, the right-shift mode is selected, as shown in Figure 13.18. Serial data enter at $D_A$ and move toward $Q_D$, following the path through the lefthand AND gates of the AND-OR circuits. When $RIGHT/\overline{LEFT} = 0$, as shown in Figure 13.19, the righthand AND gates direct left-shifting data through the register from $D_D$ to $Q_A$.

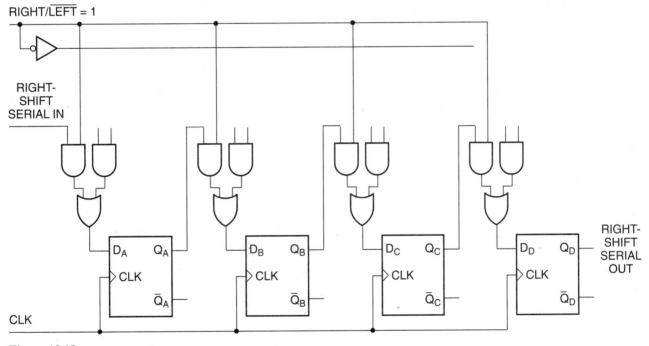

**Figure 13.18**

**Bidirectional Shift Register RIGHT-SHIFT**

**EXAMPLE 13.5**    Draw a circuit showing how to configure a 74LS95B shift register as a bidirectional shift register. Which logic level at the *MODE CONTROL* input corresponds to right-shift and which to left-shift?

RIGHT/$\overline{\text{LEFT}}$ = 0

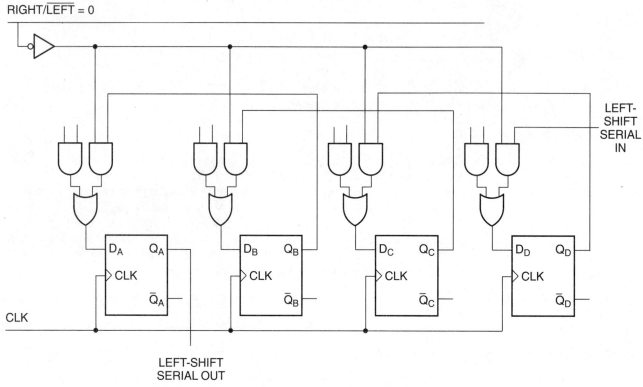

**Figure 13.19**

**Bidirectional Shift Register LEFT-SHIFT**

**Figure 13.20**

**Example 13.5**
**74LS95B as a Bidirectional**
**Serial Shift Register**

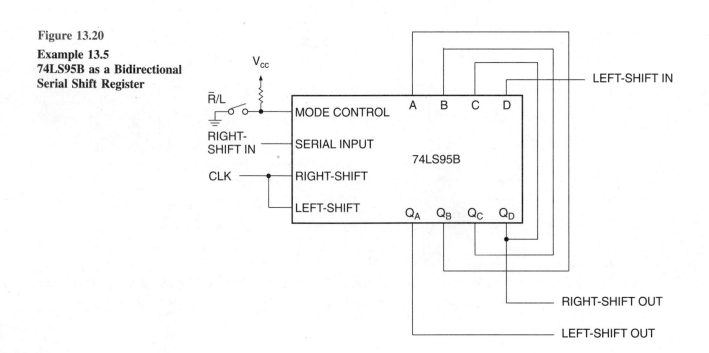

**Solution**    Figure 13.20 shows the connections required to make the 74LS95B into a bidirectional shift register. The 74LS95B requires no external wiring for right-shift mode other than a HIGH at the *MODE CONTROL* input and a *CLK* at the *RIGHT-SHIFT* input.

Left-shift mode requires a LOW at the *MODE CONTROL* input, a *CLK* at the *LEFT-SHIFT* input, and the wiring of the *Q* outputs to the parallel inputs, as shown. The *LEFT-SHIFT* data input is at parallel input *D*, $Q_D$ connects to parallel input *C* (the next least significant bit), and so on. The serial data output is at $Q_A$.

The *RIGHT-SHIFT* and *LEFT-SHIFT* inputs can be connected to the same clock, since only one of these inputs is enabled with either state of the *MODE CONTROL* input.

---

**EXAMPLE 13.6**    The bidirectional shift register shown in Figure 13.20 has input waveforms as shown in Figure 13.21. Complete the timing diagram by drawing the waveforms for outputs $Q_A$ through $Q_D$, and make a table showing the states of the outputs throughout the circuit operation. (Assume that the register is initially cleared.)

**Solution**    The *Q* waveforms are shown in Figure 13.21. In the right-shift mode (*MODE* = 0), the *SERIAL INPUT* data are loaded into $Q_A$ and shifted one place right with each negative *CLK* edge. The progress of the initial 0 and the first 1 are shown as arrows slanting diagonally down to the right.

When *MODE* = 1, the circuit is in left-shift mode. The 0s at parallel input *D* are loaded into $Q_D$ with each *CLK* pulse. The remaining data are shifted one place left, as shown by the upward-slanting diagonal arrows.

Table 13.7 shows the shift register operation.

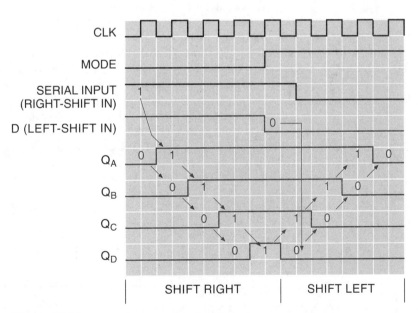

**Figure 13.21**
**Example 13.6**
**Timing Diagram of 74LS95B as Bidirectional Shift Register**

**Table 13.7**  Operation of a 74LS95B Shift Register (Bidirectional)

| MODE | RSI* | LSI** | $Q_A$ | $Q_B$ | $Q_C$ | $Q_D$ | |
|------|------|-------|-------|-------|-------|-------|-------------|
| 0 | 1 | 1 | 1 | 0 | 0 | 0 | Right shift |
| 0 | 1 | 1 | 1 | 1 | 0 | 0 | |
| 0 | 1 | 1 | 1 | 1 | 1 | 0 | |
| 0 | 1 | 1 | 1 | 1 | 1 | 1 | |
| 1 | 1 | 0 | 1 | 1 | 1 | 0 | Left shift |
| 1 | 0 | 0 | 1 | 1 | 0 | 0 | |
| 1 | 0 | 0 | 1 | 0 | 0 | 0 | |
| 1 | 0 | 0 | 0 | 0 | 0 | 0 | |

*RSI: *RIGHT-SHIFT* input
**LSI: *LEFT-SHIFT* input

## 74LS194A Universal Shift Register

Figure 13.22 shows the logic circuit of the 74LS194A Universal Shift Register. This device can serially shift data left or right, load parallel data, or hold data, depending on the states of inputs $S_1$ and $S_0$.

Each AND-NOR circuit and the $S_1$ and $S_0$ inverters act as a multiplexer to direct one of several possible data sources to the synchronous inputs of each flip-flop. For instance, if we trace the paths through the corresponding AND-NOR circuit, we find that the possible sources of data at $S_B$ and $R_B$, the synchronous inputs of the second flip-flop, are $Q_A$ ($S_1S_0 = 01$), $B$ ($S_1S_0 = 11$), $Q_C$ ($S_1S_0 = 10$), and $Q_B$ ($S_1S_0 = 00$). These are the inputs required for the right-shift, parallel-load, left-shift, and hold functions, respectively. All functions are synchronous, including the parallel-load and hold functions.

We have not seen the hold function in any other circuit we have examined. It is a synchronous no change function, implemented by feeding back the output of a flip-flop to its synchronous inputs.

In previous versions of the 74194 series, the hold function was not actively implemented with its own AND gate, but was a default when no other function was selected. Because of the way the clock was inhibited in these previous versions, mode inputs $S_1$ and $S_0$ could change without introducing possible glitches only when the *CLOCK* input was HIGH. The synchronous hold function makes this unnecessary in the 74LS194A version.

Table 13.8 summarizes the various possible inputs to each flip-flop as a function of $S_1$ and $S_0$.

**Table 13.8**  Flip-Flop Inputs as a Function of $S_1S_0$ (74LS194A)

| $S_1$ | $S_0$ | (Function) | $S_A\overline{R}_A$ | $S_B\overline{R}_B$ | $S_C\overline{R}_C$ | $S_D\overline{R}_D$ |
|-------|-------|----------------|---------|---------|---------|---------|
| 0 | 0 | (Hold) | $Q_A$ | $Q_B$ | $Q_C$ | $Q_D$ |
| 0 | 1 | (Shift right) | RSI* | $Q_A$ | $Q_B$ | $Q_C$ |
| 1 | 0 | (Shift left) | $Q_B$ | $Q_C$ | $Q_D$ | LSI** |
| 1 | 1 | (Parallel load) | $A$ | $B$ | $C$ | $D$ |

*RSI: *RIGHT-SHIFT* input
**LSI: *LEFT-SHIFT* input

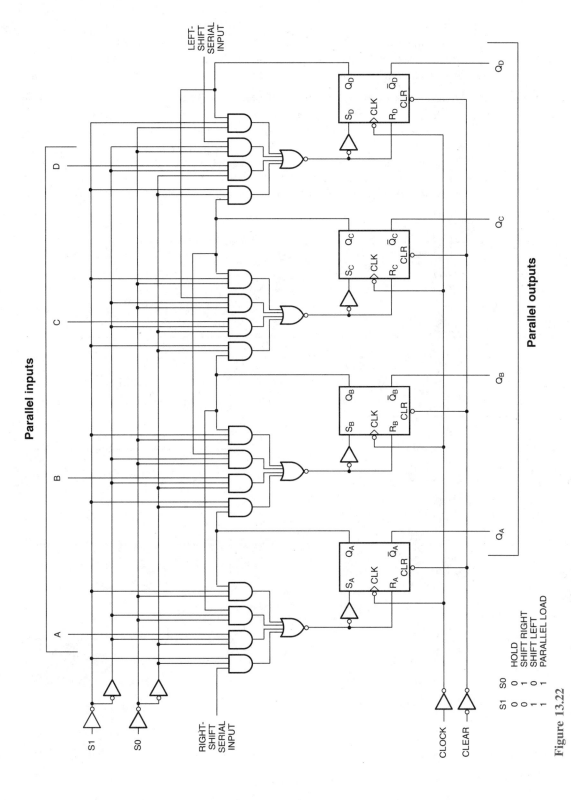

**Figure 13.22**

**74LS194A Universal Shift Register**

**EXAMPLE 13.7**    The waveforms given in Figure 13.23 are applied to the inputs of a 74LS194A Universal Shift Register. Complete the timing diagram by drawing the waveforms for the $Q$ outputs. Also make a table showing the output states for each clock pulse.

**Solution**    The $Q$ waveforms of the shift register are shown in Figure 13.23. In a couple of locations on the timing diagram, diagonal arrows show the progress of bits through the register. Downward-slanting arrows show a right-shift, and upward-slanting arrows show a left-shift.

   Since the timing diagram is very complex, it may be helpful to look at a table showing the same information, as in Table 13.9. The columns showing the serial and parallel inputs are filled in only at the points where their contents are being loaded into the register.

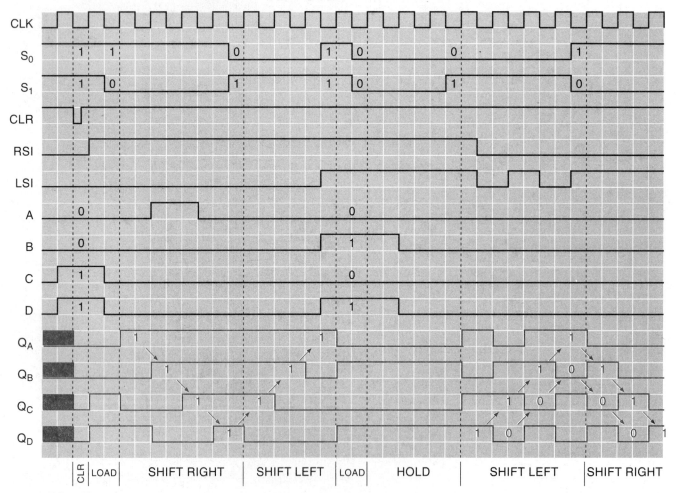

**Figure 13.23**

**Example 13.7**
**Timing Diagram**

**Table 13.9** Operation of 74LS194A Universal Shift Register

| $S_1$ | $S_0$ | RSI | LSI | A | B | C | D | $Q_A$ | $Q_B$ | $Q_C$ | $Q_D$ | Function |
|---|---|---|---|---|---|---|---|---|---|---|---|---|
| 1 | 1 | | | 0 | 0 | 1 | 1 | 0 | 0 | 1 | 1 | Parallel load |
| 0 | 1 | 1 | | | | | | 1 | 0 | 0 | 1 | Shift right |
| 0 | 1 | 1 | | | | | | 1 | 1 | 0 | 0 | |
| 0 | 1 | 1 | | | | | | 1 | 1 | 1 | 0 | |
| 0 | 1 | 1 | | | | | | 1 | 1 | 1 | 1 | |
| 1 | 0 | | 0 | | | | | 1 | 1 | 1 | 0 | Shift left |
| 1 | 0 | | 0 | | | | | 1 | 1 | 0 | 0 | |
| 1 | 0 | | 0 | | | | | 1 | 0 | 0 | 0 | |
| 1 | 1 | | | 0 | 1 | 0 | 1 | 0 | 1 | 0 | 1 | Parallel load |
| 0 | 0 | | | | | | | 0 | 1 | 0 | 1 | Hold |
| 0 | 0 | | | | | | | 0 | 1 | 0 | 1 | |
| 0 | 0 | | | | | | | 0 | 1 | 0 | 1 | |
| 1 | 0 | | 1 | | | | | 1 | 0 | 1 | 1 | Shift left |
| 1 | 0 | | 0 | | | | | 0 | 1 | 1 | 0 | |
| 1 | 0 | | ·1 | | | | | 1 | 1 | 0 | 1 | |
| 1 | 0 | | 0 | | | | | 1 | 0 | 1 | 0 | |
| 0 | 1 | 0 | | | | | | 0 | 1 | 0 | 1 | Shift right |
| 0 | 1 | 0 | | | | | | 0 | 0 | 1 | 0 | |
| 0 | 1 | 0 | | | | | | 0 | 0 | 0 | 1 | |

---

**Section Review Problem for Section 13.2**

13.3. Write the Boolean expression for flip-flop inputs $S_C$ AND $R_C$ in the 74LS194A Universal Shift Register. (Refer to Figure 13.22 and Table 13.9.)

---

## 13.3

# Shift Register Counters

> **Ring counter** *A serial shift register with feedback from the noninverting output of the last flip-flop to the input of the first.*
>
> **Johnson counter** *A serial shift register with feedback from the inverting output of the last flip-flop to the input of the first. Also called a twisted ring counter.*

By introducing feedback into a serial shift register, we can create a class of synchronous counters based on continuous circulation, or rotation, of data.

If we feed back the output of a serial shift register to its input without inversion, we create a circuit called a **ring counter.** If we introduce inversion into the feedback loop, we have a circuit called a **Johnson counter.** These circuits can be decoded more easily than binary counters of similar size and are particularly useful for event sequencing.

**Figure 13.24**

**Ring Counter**

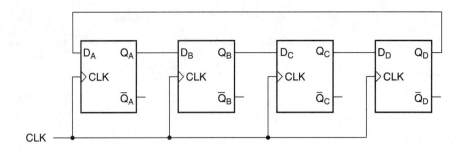

**Figure 13.25**

**Circulating a 1 in a Ring Counter**

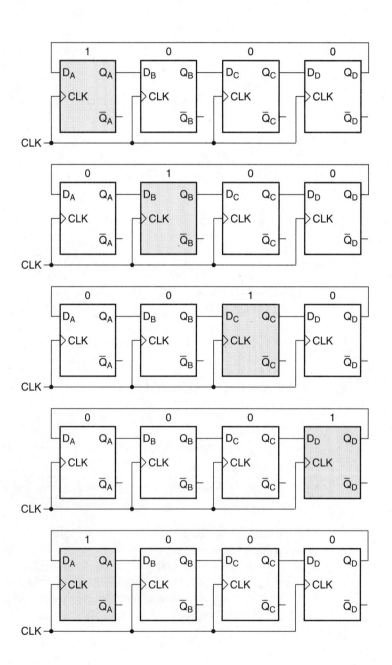

## Ring Counters

Figure 13.24 shows a 4-bit ring counter made from D flip-flops. This circuit could also be constructed from SR or JK flip-flops, as can any serial shift register.

A ring counter circulates the same data in a continuous loop. This assumes that the data have somehow been placed into the circuit upon initialization, usually by the asynchronous Preset and Clear inputs, which are not shown.

Figure 13.25 shows the circulation of a logic 1 through a 4-bit ring counter. If we assume that the circuit is initialized to the state $Q_A Q_B Q_C Q_D = 1000$, it is easy to see that the 1 is shifted one place right with each clock pulse. The feedback connection from $Q_D$ to $D_A$ ensures that the input of flip-flop $A$ will be filled by the contents of $Q_D$, thus recirculating the initial data. The final transition in the sequence shows the 1 recirculated to $Q_A$.

A ring counter is not restricted to circulating a logic 1. We can program the counter to circulate any data pattern we happen to find convenient.

Figure 13.26 shows a ring counter circulating a 0 by starting with an initial state of $Q_A Q_B Q_C Q_D = 0111$. The circuit is the same as before; only the initial state has changed. Figure 13.27 shows the timing diagrams for the circuit in Figures 13.25 and 13.26.

---

**EXAMPLE 13.8**

Example 11.9 (Section 11.3, Decoding a Counter) showed how a binary counter and decoder could be used to sequence the firing of eight spark plugs in an internal combustion engine. The application required eight output lines each to go LOW for one *CLK* cycle in eight, in sequence. The LOW at each decoder output fired a high-voltage circuit for each spark plug.

The 74198 shift register, shown in Figure 13.28 (see page 587), is an 8-bit version of the 74194 Universal Shift Register. $S_1 S_0 = 11$ selects the parallel-load function, and $S_1 S_0 = 01$ selects right-shift mode. Use this device to make a ring counter that performs the same function as the counter and decoder in Example 11.19. Draw the timing diagram and compare it to the one in Figure 11.24b in that example.

**Solution**  Figure 13.29a (see page 588) shows the connections required to make the 74198 shift register into a ring counter sequencer. The counter must be initialized by opening the switch on $S_0$ and clocking once (synchronous parallel load). After the initial parallel load, the switch must be closed to enable the right-shift function. The circuit produces the output waveforms shown in Figure 13.26b. The $Q_A$ through $Q_H$ waveforms are the same as the $\overline{Y}_0$ through $\overline{Y}_7$ waveforms in Figure 11.24. When any $Q$ output goes LOW, it fires its corresponding spark plug.

---

**EXAMPLE 13.9**

What changes are required to the circuit of Figure 13.29a to make it produce a firing sequence for six spark plugs?

**Solution**  Connect the *RIGHT-SHIFT* serial input of the 74198 shift register to $Q_F$ instead of $Q_H$. Data will still shift internally from $Q_F$ to $Q_G$, but will have no external effect. The ring counter effectively has been shortened by 2 bits.

**Figure 13.26**

**Circulating a 0 in a Ring Counter**

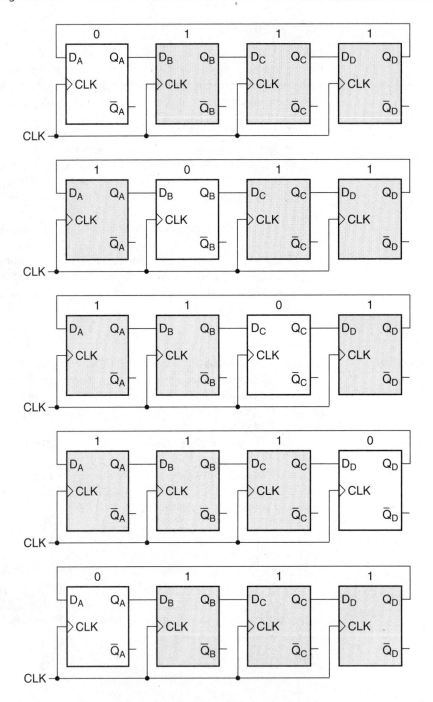

## Ring Counter Modulus and Decoding

The maximum modulus of a ring counter is the maximum number of unique states in its count sequence. In Examples 13.8 and 13.9, the ring counters had maximum moduli of 8 and 6, respectively.

We say that 8 and 6 are the maximum moduli of the ring counters shown, since we can change the modulus of a ring counter by loading different data at initialization.

**Figure 13.27**

**Timing Diagrams for Figures 13.25 and 13.26**

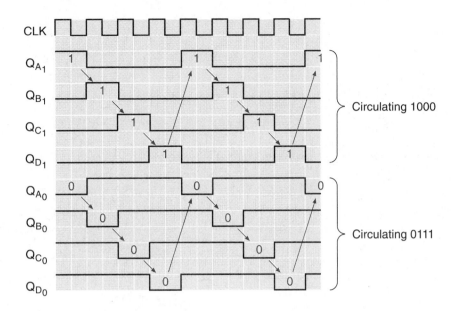

**Figure 13.28**

**Example 13.8**
**8-Bit Universal Shift Register**

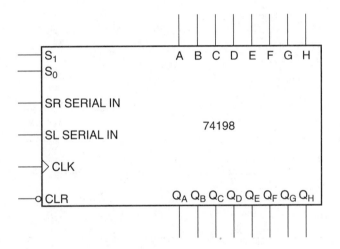

For example, if we load a 4-bit ring counter with the data $Q_A Q_B Q_C Q_D = 1000$, the following unique states are possible: 1000, 0100, 0010, and 0001. If we load the same circuit with the data $Q_A Q_B Q_C Q_D = 1010$, there are only two unique states: 1010 and 0101. Depending on which data are loaded, the modulus is 4 or 2.

Most input data in this circuit will yield a modulus of 4. Try a few combinations. Can you find input data that result in a modulus of 1?

> The maximum modulus of a ring counter is the same as the number of bits in its output.

A ring counter requires more flip-flops than a binary counter to produce the same number of unique states. Specifically, for $n$ flip-flops, a binary counter has $2^n$ unique states and a ring counter has $n$.

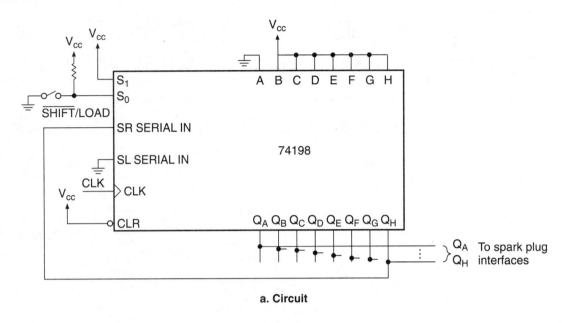

**a. Circuit**

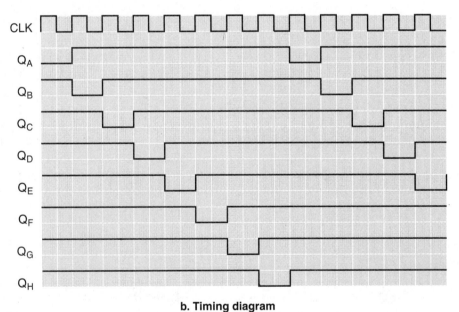

**Figure 13.29**

**Example 13.8**
**8-bit Ring Counter**
**Circulating 0111 1111**

**b. Timing diagram**

This is offset by the fact that a ring counter requires no decoding. A binary counter used to sequence eight events requires three flip-flops and eight 3-input decoding gates. To perform the same task, a ring counter requires eight flip-flops and no decoding gates. If we use an MSI counter and decoder for the binary counter and an 8-bit MSI shift register for the ring counter, we save one IC package by using the ring counter.

As the number of output states of an event sequencer increases, the complexity of the decoder for the binary counter also increases. A circuit requiring 16 output states can be implemented with a 4-bit binary counter and sixteen 4-input decoding gates. If you need 18 output states, you must have a 5-bit counter ($2^4 \leq 18 \leq 2^5$) and eighteen 5-input decoding gates.

The only required modification to the ring counter is one more flip-flop for each additional state. A 16-state ring counter needs 16 flip-flops, and an 18-state ring counter must have 18 flip-flops. No decoding is required for either circuit.

---

**Section Review Problem for Section 13.3a**

13.4.  Suppose the inputs to the spark plug interfaces required in Example 13.8 were active HIGH instead of active LOW. What changes must be made to the circuit in Figure 13.29a?

---

## Johnson Counters

**Table 13.10** Count Sequence of a 4-Bit Johnson Counter

| $Q_A$ | $Q_B$ | $Q_C$ | $Q_D$ |
|---|---|---|---|
| 0 | 0 | 0 | 0 |
| 1 | 0 | 0 | 0 |
| 1 | 1 | 0 | 0 |
| 1 | 1 | 1 | 0 |
| 1 | 1 | 1 | 1 |
| 0 | 1 | 1 | 1 |
| 0 | 0 | 1 | 1 |
| 0 | 0 | 0 | 1 |

Figure 13.30 shows a 4-bit Johnson counter constructed from D flip-flops. It is the same as a ring counter except for the inversion in the feedback loop where $\overline{Q}_D$ is connected to $D_A$. The circuit output is taken from flip-flop outputs $Q_A$ through $Q_D$. Since the feedback introduces a "twist" into the recirculating data, a Johnson counter is also called a twisted ring counter.

Figure 13.31 shows the progress of data through a Johnson counter that starts cleared ($Q_A Q_B Q_C Q_D = 0000$). The shaded flip-flops represents 1s, and the unshaded flip-flops are 0s.

Every 0 at $Q_D$ is fed back to $D_A$ as a 1 and vice versa. The count sequence for this circuit is given in Table 13.10. There are eight unique states in the count-sequence table.

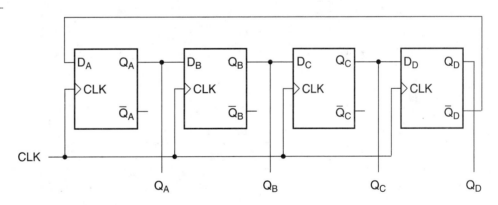

**Figure 13.30**
**Johnson Counter**

---

**EXAMPLE 13.10**

Draw the circuit of a 74LS164 Serial-In-Parallel-Out Shift Register connected as a Johnson counter. List the sequence of states in a table, assuming the counter is initially cleared, and draw the timing diagram.

**Solution**    The 74LS164 shift register has no output for $\overline{Q}_H$, so a 74LS04 inverter must be added to the circuit, as shown in Figure 13.32. Since the serial inputs, *A* and *B,* are internally ANDed, one of them must be tied HIGH to enable the feedback loop. The $\overline{CLR}$ input can be tied to a pushbutton to initialize the counter.

Table 13.11 shows the count sequence for the 8-bit Johnson counter.

The timing diagram for the Johnson counter is shown in Figure 13.33 (page 592).

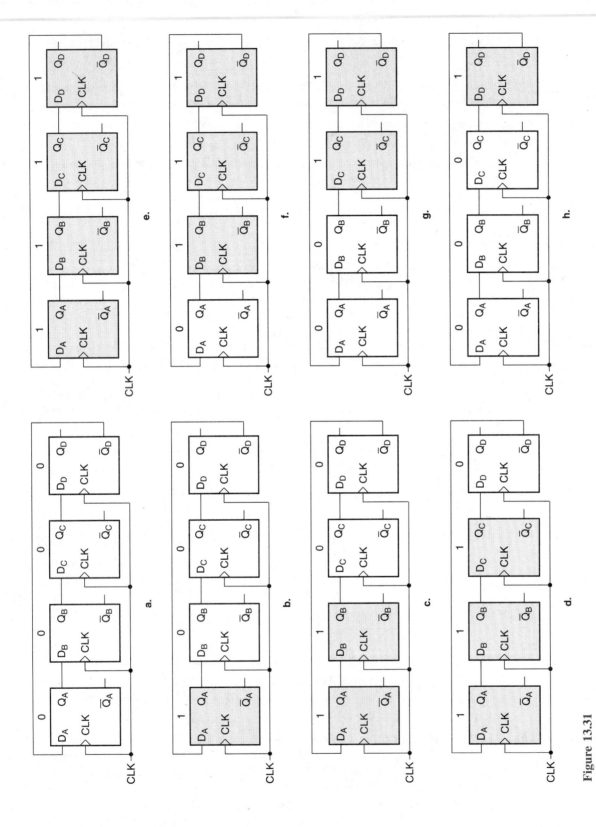

**Figure 13.31**

**Data Circulation in a 4-Bit Johnson Counter**

**Figure 13.32**

**Example 13.10
74LS164 as a Johnson
Counter**

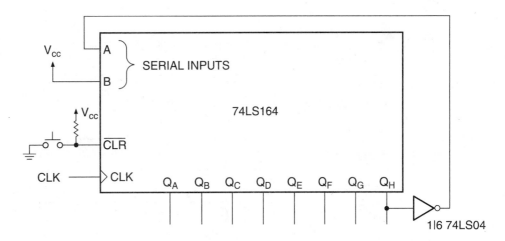

**Table 13.11**  Count Sequence for an 8-Bit Johnson Counter

| $Q_A$ | $Q_B$ | $Q_C$ | $Q_D$ | $Q_E$ | $Q_F$ | $Q_G$ | $Q_H$ |
|---|---|---|---|---|---|---|---|
| 0 | 0 | 0 | 0 | 0 | 0 | 0 | 0 |
| 1 | 0 | 0 | 0 | 0 | 0 | 0 | 0 |
| 1 | 1 | 0 | 0 | 0 | 0 | 0 | 0 |
| 1 | 1 | 1 | 0 | 0 | 0 | 0 | 0 |
| 1 | 1 | 1 | 1 | 0 | 0 | 0 | 0 |
| 1 | 1 | 1 | 1 | 1 | 0 | 0 | 0 |
| 1 | 1 | 1 | 1 | 1 | 1 | 0 | 0 |
| 1 | 1 | 1 | 1 | 1 | 1 | 1 | 0 |
| 1 | 1 | 1 | 1 | 1 | 1 | 1 | 1 |
| 0 | 1 | 1 | 1 | 1 | 1 | 1 | 1 |
| 0 | 0 | 1 | 1 | 1 | 1 | 1 | 1 |
| 0 | 0 | 0 | 1 | 1 | 1 | 1 | 1 |
| 0 | 0 | 0 | 0 | 1 | 1 | 1 | 1 |
| 0 | 0 | 0 | 0 | 0 | 1 | 1 | 1 |
| 0 | 0 | 0 | 0 | 0 | 0 | 1 | 1 |
| 0 | 0 | 0 | 0 | 0 | 0 | 0 | 1 |

## Johnson Counter Modulus and Decoding

> The maximum modulus of a Johnson counter is $2n$ for a circuit with $n$ flip-flops.

The two Johnson counters examined above illustrate this rule. The Johnson counter represents a compromise between binary and ring counters, whose maximum moduli are, respectively, $2^n$ and $n$ for an $n$-bit counter.

If it is used for event sequencing, a Johnson counter must be decoded, unlike a ring counter. Its output states are such that each state can be decoded uniquely by a 2-input AND or NAND gate, depending on whether you need active-HIGH or active-LOW indication. This yields a simpler decoder than is required for a binary counter.

Table 13.12 shows the decoding of a 4-bit Johnson counter.

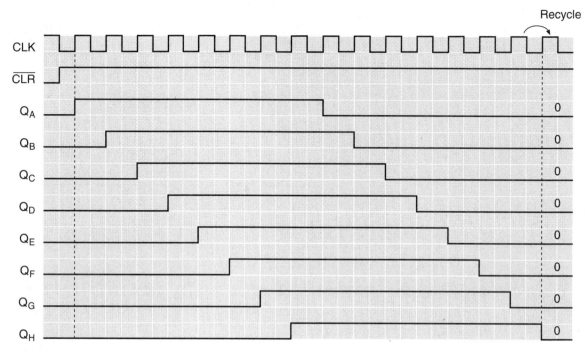

**Figure 13.33**

**Example 13.10**
**Johnson Counter Timing**

Table 13.12  Decoding a 4-Bit Johnson Counter

| $Q_A$ | $Q_B$ | $Q_C$ | $Q_D$ | Decoder Outputs | Comment |
|-------|-------|-------|-------|-----------------|---------|
| 0 | 0 | 0 | 0 | $\overline{Q}_A\overline{Q}_D$ | MSB = LSB = 0 |
| 1 | 0 | 0 | 0 | $Q_A\overline{Q}_B$ | |
| 1 | 1 | 0 | 0 | $Q_B\overline{Q}_C$ | Descending 10 pairs |
| 1 | 1 | 1 | 0 | $Q_C\overline{Q}_D$ | |
| 1 | 1 | 1 | 1 | $Q_AQ_D$ | MSB = LSB = 1 |
| 0 | 1 | 1 | 1 | $\overline{Q}_AQ_B$ | |
| 0 | 0 | 1 | 1 | $\overline{Q}_BQ_C$ | Descending 01 pairs |
| 0 | 0 | 0 | 1 | $\overline{Q}_CQ_D$ | |

Decoding a sequential circuit depends on the decoder responding uniquely to every possible state of the circuit outputs. If we want to use only 2-input gates in our decoder, it must recognize two variables for every state that are *both* active *only* in that state (AND/NAND function).

A Johnson counter decoder exploits what might be called the "1/0 interface" of the count-sequence table. Careful examination of Tables 13.10 and 13.11 reveals that for every state, except where the outputs are all 1s or all 0s, there is a side-by-side 10 or 01 pair that exists only in that state.

Each of these pairs can be decoded to give unique indication of a particular state. For example, the pair $Q_A\overline{Q}_B$ uniquely indicates the second state, since $Q_A = 1$ AND $Q_B = 0$ *only* in the second line of the count-sequence table. (This is true for any size of Johnson counter; compare the second lines of Tables 13.10 and 13.11.)

For the states where the outputs are all 1s or all 0s, the most significant AND least significant bits can be decoded uniquely, these being the only states where MSB = LSB.

Figure 13.34 shows the decoder circuit for a 4-bit Johnson counter. The output decoder of a Johnson counter does not increase in complexity as the modulus of the counter increases. The decoder will always consist of $2n$ 2-input AND (or NAND) gates for an $n$-bit counter (e.g., for an 8-bit Johnson counter, the decoder will consist of sixteen 2-input AND or NAND gates).

---

**EXAMPLE 13.11**    Draw the timing diagram of the Johnson counter decoder of Figure 13.34, assuming the counter is initially cleared. How does this compare to the ring counter output shown in Figure 13.29b?

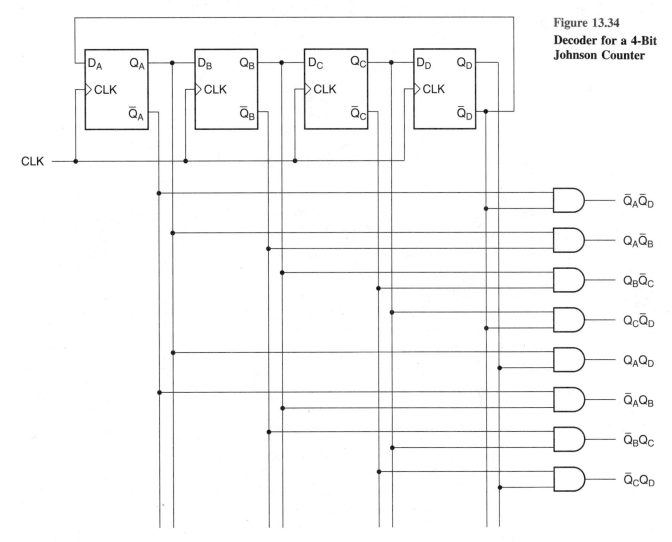

**Figure 13.34**

**Decoder for a 4-Bit Johnson Counter**

**Figure 13.35**

**Example 13.11**
**Johnson Counter Decoder Outputs**

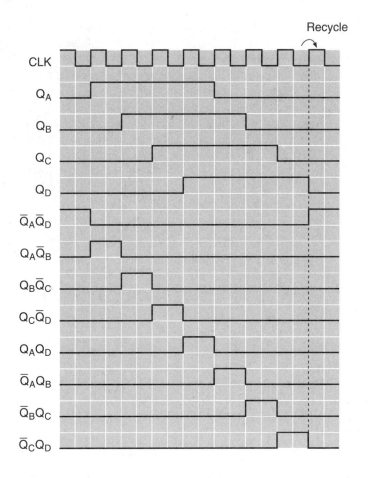

**Solution**  Figure 13.35 shows the timing diagram of the Johnson counter and its decoder outputs. The decoder output waveforms of Figure 13.35 differ from those of Figure 13.29b only by their active sense. The waveforms in Figure 13.35 indicate a decode by a HIGH output, whereas the waveforms in Figure 13.29b indicate an active-LOW decode.

**EXAMPLE 13.12**  Construct the count-sequence table of a 5-bit Johnson counter, assuming the counter is initially cleared. What changes must be made to the decoder part of the circuit in Figure 13.34 if it is to decode the 5-bit Johnson counter?

**Solution**  Table 13.13 shows the count sequence.

All gates used in the decoder of Figure 13.34 remain unchanged except those decoding the MSB/LSB pairs ($\overline{Q_A}\,\overline{Q_D}$ and $Q_A\overline{Q_D}$). Change these to decode $\overline{Q_A}\,\overline{Q_E}$ and $Q_A Q_E$. Add two new gates to decode $Q_D\overline{Q_E}$ (fifth state) and $\overline{Q_D}Q_E$ (tenth state).

---

**Section Review Problem for Section 13.3b**

13.5. How many flip-flops are required to produce 12 unique states in each of the following types of counters: binary counter, ring counter, and Johnson counter? How many and what types of decoding gates are required to produce an active-LOW decode for each type of counter?

**Table 13.13** Count Sequence of a 5-Bit Johnson Counter

| $Q_A$ | $Q_B$ | $Q_C$ | $Q_D$ | $Q_E$ |
|---|---|---|---|---|
| 0 | 0 | 0 | 0 | 0 |
| 1 | 0 | 0 | 0 | 0 |
| 1 | 1 | 0 | 0 | 0 |
| 1 | 1 | 1 | 0 | 0 |
| 1 | 1 | 1 | 1 | 0 |
| 1 | 1 | 1 | 1 | 1 |
| 0 | 1 | 1 | 1 | 1 |
| 0 | 0 | 1 | 1 | 1 |
| 0 | 0 | 0 | 1 | 1 |
| 0 | 0 | 0 | 0 | 1 |

## 13.4

# IEEE/ANSI Notation

(This section may be omitted without loss of continuity)

There is a standard way of drawing shift register symbols to conform with IEEE/ANSI Standard 91-1984. Let us look at the IEEE symbols for four of the shift registers examined in this chapter: 74LS91, 74LS164, 74LS95B, and 74LS194A.

### 74LS91 Serial-In-Serial-Out Shift Register

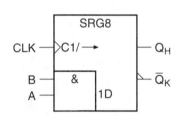

**Figure 13.36**

**74LS91 Serial-In-Serial-Out Shift Register**

Figure 13.36 shows the IEEE logic symbol for a 74LS91 shift register. The designation SRG8 tells us that the device is an 8-bit shift register.

Since the 74LS91 allows no external access to the individual flip-flops, only the true and complement forms of $Q_H$ are shown as outputs, and none of the flip-flops are shown individually. Inputs $A$ and $B$ are internally ANDed, as shown by the imbedded AND symbol.

Two functions in an IEEE/ANSI symbol are linked when the numeral prefix of one and the numeral suffix of the other are the same. The *1D* at the AND output indicates that the ANDed data depends on the clock input, symbolized by *C1*, and will be latched into flip-flop $A$ on the rising edge of the clock. (The *C* stands for Control Dependency.)

The right arrow at the clock input indicates that the clock controls a right-shift operation.

### 74LS164 Serial-In-Parallel-Out Shift Register

Figure 13.37 shows the IEEE symbol for the 74LS164 shift register. The 74LS164 is similar to the 74LS91 device, except that there is a common clear function and there is access to all internal flip-flop outputs. Each flip-flop is shown as an individual block, with the ANDed serial inputs $A$ and $B$ feeding the first flip-flop.

There is a common control block (the indented portion at the top of the diagram), which controls the shift and clear functions for all flip-flops. All flip-flops clear when $\overline{CLR} = 0$. This function is indicated inside the symbol by $R$, for Reset.

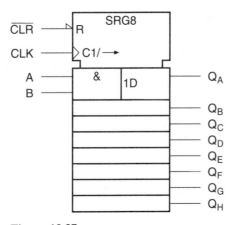

**Figure 13.37**

**74LS164 Serial-In-Parallel-Out Shift Register**

**Figure 13.38**

**74LS95B Parallel-Access Shift Register**

## 74LS95B Parallel-Access Shift Register

The IEEE symbol for the 74LS95B shift register is shown in Figure 13.38. This symbol has more complex dependency notation than the 74LS91 or 74LS164 symbols. The internal flip-flops are shown as four individual blocks at the bottom of the symbol, and the common control block at the top indicates the effect of the *MODE* and *CLK* inputs on the flip-flops. There is also dependency within the common control block; the *CLK* inputs are enabled and inhibited by the *MODE* input.

When *MODE* is HIGH, it enables the *CLK2* input (indicated by the dependency *M2–2C*). *CLK2* loads the flip-flops through the parallel inputs *A* through *D* (shown by dependency *C4–4D*). The whole chain of dependencies for the parallel-load function is *M2–2C4–4D* (*M* = Mode Dependency, *C* = Control Dependency, *D* = Data).

When *MODE* is LOW, *CLK1* is enabled (*M1–1C;* negation symbol on *M1* input). *CLK1* loads serial data into flip-flop *A* (*C3–3D*). The complete dependency chain is *M1–1C3–3D*. The right arrow on *CLK1* indicates a serial shift right with each clock pulse.

## 74LS194A Universal Shift Register

The 74LS194A shift register, shown in Figure 13.39, can serially shift data left and right and load parallel data. The common control block of the logic symbol shows a common clear input, a four-function mode control, and clock.

Inputs $S_1$ and $S_0$ have four possible binary values, 00, 01, 10, or 11, which select the shift register operating mode. This is indicated by the Mode notation, $M \frac{0}{3}$, in the common control block.

Mode 1 ($S_1 S_0 = 01$) is the right-shift mode. The right arrow at the *CLK* input depends on mode 1, as does the Shift-Right Serial input at flip-flop *A*.

When mode 2 is selected ($S_1 S_0 = 10$), the device operates in left-shift mode. The left arrow at the *CLK* and the Shift-Left Serial input at flip-flop *D* both depend on mode 2.

Mode 3 ($S_1 S_0 = 11$) selects the parallel-load function. Inputs *A* through *D* depend on mode 3.

Mode 0 ($S_1 S_0 = 00$) has no dependencies shown, since it is a hold function and there is no action taken that makes any external difference to the shift register.

**Figure 13.39**
**74LS194A Universal Shift Register**

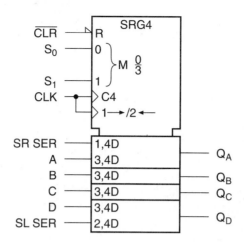

All flip-flop inputs depend on the clock *(C4–4D),* as well as the mode inputs. This shows that operation is synchronous in all modes.

## GLOSSARY

**Bidirectional shift register**   A shift register that can serially shift bits left or right according to the state of a direction control input.

**Johnson counter**   A serial shift register with feedback from the inverting output of the last flip-flop to the input of the first. Also called a twisted ring counter.

**Left-shift**   In an 8-bit shift register, serial shifting in the direction from $Q_H$ to $Q_A$.

**Parallel loading**   Parallel shifting of data into a shift register.

**Parallel shifting**   Movement of data into or out of all flip-flops of a shift register at the same time.

**Parallel-access shift register**   A shift register whose flip-flop outputs are individually accessible.

**Parallel-load shift register**   A shift register that can be preset to any value by directly loading a binary number into its internal flip-flops.

**Right-shift**   In an 8-bit shift register, serial shifting in the direction from $Q_A$ to $Q_H$.

**Ring counter**   A serial shift register with feedback from the noninverting output of the last flip-flop to the input of the first.

**Rotation**   Serial shifting of data with the output(s) of the last flip-flop connected to the synchronous input(s) of the first flip-flop. The result is continuous circulation of the same data.

**Serial shifting**   Movement of data from one end of a shift register to the other at a rate of 1 bit per clock pulse.

**Shift register**   A synchronous sequential circuit that will store and move *n*-bit data, either serially or in parallel, in *n* flip-flops.

## PROBLEMS

### Section 13.1 Basic Shift Register Configurations

**13.1**   Draw the circuit of a serial shift register constructed from negative edge-triggered JK flip-flops.

**13.2**   Write the Boolean expression for the *J* and *K* inputs of the shift register shown in Problem 13.1. Make a state table, similar to that in Table 13.2. Complete the table to show the circuit operation.

**13.3**   The following bits are applied to the input of a 6-bit serial shift register: 0111111. (0 is applied first.) Draw the timing diagram.

**13.4**   After the data in Problem 13.3 are applied to the 6-bit shift register, the serial input goes to 0 for the next eight clock pulses and then returns to 1. Write the internal states, $Q_A$ through $Q_F$, of the shift register flip-flops two clock pulses after the serial input goes to 0. Write the states after six, eight, and ten clock pulses.

**13.5**   Complete the timing diagram of Figure 13.40, which is for a serial shift register. Assume that the shift register is initially cleared. What happens to the state of the circuit if $D_A$ stays HIGH beyond the end of the diagram and the *CLK* input continues to pulse?

**13.6**   A 74LS91 Serial-In-Serial-Out Shift Register is initially cleared and has the following data clocked into its serial input: 1011001110. Draw a timing diagram of the circuit showing the *CLK,* serial input, and serial output.

**13.7**   Figure 13.41 shows the standard serial transmission format for a 7-bit ASCII character. Figure 13.42 shows three serially transmitted ASCII characters applied to the serial input of a 74LS164 shift register. Complete the timing diagram.

**\*13.8**   The ASCII characters transmitted to the shift register in Problem 13.7 can be read in parallel form at bits $Q_A$ (MSB) through $Q_G$ at the three points indicated by the circled letters *A, B,* and *C* in Figure 13.42. Use the ASCII table given

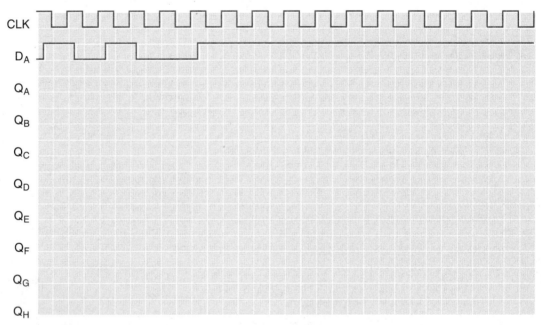

**Figure 13.40**

**Problem 13.5**
**Timing Diagram**

**Figure 13.41**

**Problems 13.7 and 13.8**
**Asynchronous ASCII**
**Character Format**

| | Start bit | | Data bits | | | | | Stop bit | Idle state |
|---|---|---|---|---|---|---|---|---|---|
| | 1 LSB | 2 | 3 | 4 | 5 | 6 | 7 MSB | | |

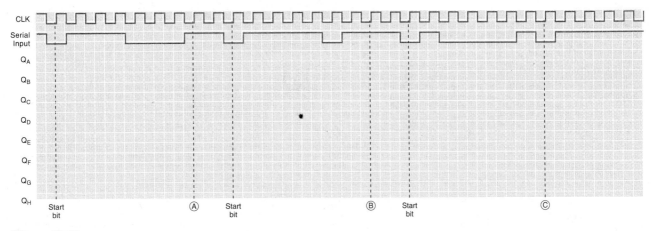

**Figure 13.42**
**Problems 13.7 and 13.8**
**Three ASCII Characters**

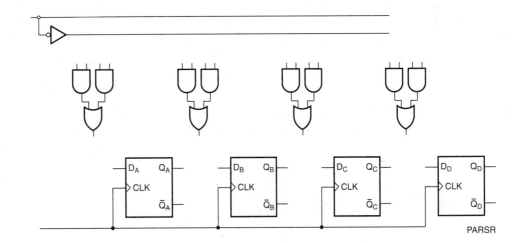

**Figure 13.43**
**Problem 13.9**
**Logic Circuit**

PARSR

in Table 5.6 to decipher the three-character message shown in Figure 13.42.

**13.9**  Complete the logic circuit shown in Figure 13.43 to make a parallel-in-serial-out shift register.

**\*13.10**  Draw a circuit using a 74LS165A and 74LS164 shift register that can transmit and receive ASCII characters. Draw the timing diagram of the circuit as it sends ASCII "a".

**13.11**  If a shift register is used as a multiplier or divider, what restriction is there on the multiplication or division?

**13.12**  Draw the timing diagram of an 8-bit shift register showing how it can multiply $14_{10}$ by $16_{10}$.

**\*13.13**  At what logic level must the serial input of a shift register be for a multiply-by-$2^n$ operation to work properly?

**\*13.14**  What happens to the result of a multiplication if bits are shifted out of the register? (e.g., when 0101 is shifted two places toward the MSB in a 4-bit register; equivalent to $5_{10} \times 4_{10}$).

## Section 13.2 Bidirectional Shift Registers

**\*13.15**  Complete the logic circuit shown in Figure 13.44 to make a bidirectional shift register.

**\*13.16**  Draw the circuit of two 74LS95B shift registers connected so that they can load an 8-bit number and divide it by a power of 2.

**Figure 13.44**

**Problem 13.15**
**Logic Circuit**

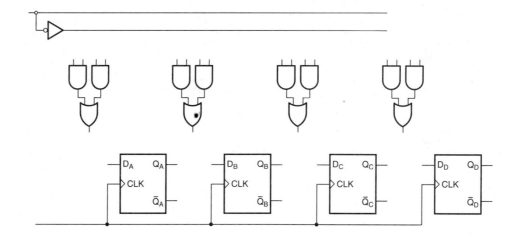

**13.17** Draw the timing diagram of an 8-bit shift register showing how it can divide $48_{10}$ by $8_{10}$.

## Section 13.3 Shift Register Counters

**13.18** What change is required to make the Johnson counter decoder of Figure 13.34 indicate a decode by a LOW output?

**13.19** Draw the circuit for a 6-bit Johnson counter made from positive edge-triggered D flip-flops.

**13.20** Construct the count-sequence table for the counter in Problem 13.19 if the initial state is 000000.

**13.21** Draw a circuit that will decode uniquely each state of a 6-bit Johnson counter, using only 2-input AND gates.

**\*13.22** A control sequence has ten steps, each activated by a logic HIGH. Use a counter and decoder in each of the following configurations to produce the required sequence: binary counter, ring counter, and Johnson counter. (You may use flip-flops or MSI shift registers for the latter two counters.) Make a timing diagram for each counter and decoder.

**\*\*13.23** Repeat Problem 13.22 for a sequence of nine steps, with an active-LOW output. (Hint: The Johnson counter does not have an even count, $2n$. You will have to account for this.)

**\*\*13.24** Design a 4-bit ring counter that can be asynchronously initialized to $Q_A Q_B Q_C Q_D = 1000$ by using only the *CLR* inputs of its flip-flops. No *PRESET*s allowed. (Hint: Use a circuit with a "double twist" in the data path.)

## ▼ Answers to Section Review Questions

### Section 13.1a

**13.1.** Both SR and JK flip-flops can replace the D flip-flops in the serial shift register of Figure 13.2. The $Q$ output of any stage connects to the $S$ or $J$ input of the next stage, and the $\overline{Q}$ output must connect to the $R$ or $K$ input of the next stage, ensuring that the synchronous inputs are always in opposite states. There must be an inverter from the data input to $R_A$ or $K_A$. $S_A$ or $J_A$ goes directly to the data input.

### Section 13.1b

**13.2.** Result: $10101000_2$. Shift *right* three places. (MSB is rightmost in a circuit, leftmost in a written number.)

### Section 13.2

**13.3.** $S_C = \overline{S}_1\overline{S}_0Q_C + \overline{S}_1S_0Q_B + S_1\overline{S}_0Q_D + S_1S_0C$

$R_C = \overline{\overline{S}_1\overline{S}_0Q_C + \overline{S}_1S_0Q_B + S_1\overline{S}_0Q_D + S_1S_0C}$

### Section 13.3a

**13.4.** The shift register would need to circulate the data $Q_A Q_B Q_C Q_D Q_E Q_F Q_G Q_H = 10000000$. Input $A$ of the 74198 shift register must be connected to a logic HIGH. Inputs $B$ through $H$ must be tied LOW.

### Section 13.3b

**13.5.**

| Counter | Flip-Flops | Decoding Gates |
|---------|-----------|----------------|
| Binary | 4 | 12 4-input NANDs |
| Ring | 12 | None |
| Johnson | 6 | 12 2-input NANDs |

# Electrical Characteristics of Logic Gates

## ● CHAPTER OUTLINE

**14.1** Electrical Characteristics of Logic Gates

**14.2** Propagation Delay

**14.3** Fanout

**14.4** Power Dissipation

**14.5** Noise Margin

**14.6** Interfacing TTL and CMOS Gates

## ● CHAPTER OBJECTIVES

Upon successful completion of this chapter, you will be able to:

- Name the various logic families most commonly in use today and state several advantages and disadvantages of each.

- Define propagation delay.

- Calculate propagation delay of simple circuits, using data sheets.

- Define fanout and calculate its value, using data sheets.

- Calculate power dissipation of TTL and CMOS circuits.

- Calculate noise margin of a logic gate from data sheets.

- Draw circuits that will interface various CMOS and TTL gates.

## ● INTRODUCTION

Our study of logic gates, flip-flops, and MSI devices in previous chapters has concentrated on digital logic and has largely ignored digital electronics. Digital logic devices are electronic circuits with their own characteristic voltages and currents. No serious study of digital circuitry is complete without some examination of this topic.

It is particularly important to understand the inputs and outputs of logic devices as electronic circuits. Knowing the input and output voltages and currents of these circuits is essential, since gate loading, power dissipation, noise voltages, and interfacing between logic families depend on them. The switching speed of device outputs is fundamental and may be a consideration when choosing the logic family for a circuit design.

Input and output voltages of logic devices are specified in manufacturers' data sheets, which allows us to take a "black box" approach initially. In Chapter 15 we will analyze the circuits of logic devices to understand them more fully.

## 14.1

# Electrical Characteristics of Logic Gates

**TTL**  *Transistor-transistor logic. A logic family based on bipolar transistors.*

**CMOS**  *Complementary metal-oxide semiconductor. A logic family based on metal-oxide-semiconductor field effect transistors (MOSFETs).*

**ECL**  *Emitter coupled logic. A high-speed logic family based on bipolar transistors.*

When we examine the electrical characteristics of logic circuits, we see them as practical, rather than ideal devices. We look at properties such as switching speed, power dissipation, noise immunity, and current-driving capability. There are several commonly available logic families in use today, each having a unique set of electrical characteristics that differentiates it from all the others. Each logic family gives superior performance in one or more of its electrical properties.

**CMOS** consumes very little power, has excellent noise immunity, and can be used with a wide range of power supply voltages. Its main drawback is its slow switching speed. In the latest high-speed versions of this technology, even this is ceasing to be a problem.

**TTL** is faster and has a larger current-driving capability. Its power consumption is higher than that of CMOS, and its power supply requirements are more rigid.

**ECL** is fast, making it the choice for high-speed applications. It is inferior to CMOS and TTL in terms of noise immunity and power consumption.

TTL and CMOS gates come in a wide range of subfamilies. Table 14.1 lists the TTL and CMOS variations of the Quadruple 2-input NAND gate. All gates listed have the same logic function but different electrical characteristics. Other gates would be similarly designated, with the last two or three digits indicating the gate function (e.g., a NOR gate would be designated 7402, 74S02, 74LS02, etc.).

We will examine four electrical characteristics of TTL and CMOS circuits: propagation delay, fanout, noise margin, and power dissipation. The first of these has to do with speed of output response to a change of input. The last three have to do with input and output voltages and currents. All four properties can be read directly from specifications given in a manufacturer's data sheet or derived from these specifications.

**Table 14.1** Designations for a Quad 2-Input NAND Gate in Various TTL and CMOS Families

|      | Numerical Designation | Logic Family |
|------|-----------------------|--------------|
| **TTL** | 7400 | Standard TTL |
|      | 74S00 | Schottky TTL |
|      | 74LS00 | Low Power Schottky TTL |
|      | 74AS00 | Advanced Schottky TTL |
|      | 74ALS00 | Advanced Low Power Schottky TTL |
|      | 74F00 | FAST Schottky TTL |
| **CMOS** | 4011B | B-series CMOS (metal gate) |
|      | 4011UB | Unbuffered Metal-Gate CMOS |
|      | 74C00 | Metal-Gate CMOS (TTL pinout) |
|      | 74HC00 | High-Speed CMOS (silicon gate) |
|      | 74HCT00 | High-Speed CMOS (TTL-compatible inputs) |

## 14.2

# Propagation Delay

$t_{pHL}$  *Propagation delay when the device output is changing from HIGH to LOW.*

$t_{pLH}$  *Propagation delay when the device output is changing from LOW to HIGH.*

Propagation delay occurs because the output of a logic gate or flip-flop cannot respond instantaneously to changes at its input. There is a short delay, on the order of several nanoseconds, between input change and output response. This is largely due to the charging and discharging of capacitances inherent in the switching transistors of the gate or flip-flop.

Figure 14.1 shows propagation delay in two gates: a 7400 TTL NAND gate and a 7408 TTL AND gate. Each gate has an identical input waveform, a LOW-HIGH-LOW pulse. After each input transition, the output changes after a short delay, $t_p$.

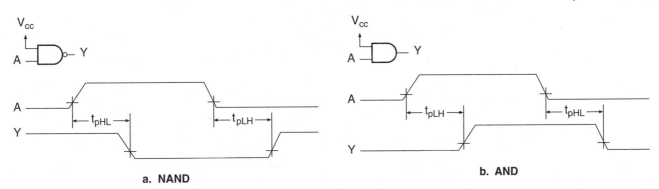

a. NAND          b. AND

**Figure 14.1**

**Propagation Delay in NAND and AND Gates**

Two delays are shown for each gate: $t_{pLH}$ and $t_{pHL}$. The *LH* and *HL* subscripts show the direction of change at the gate *output; LH* indicates that the output goes from LOW to HIGH, and *HL* shows the output changing from HIGH to LOW.

Propagation delay is the time between input and output voltages passing through a standard reference value. The reference voltage for standard TTL is 1.5 V. LSTTL and CMOS have different reference voltages, as follows.

---

**Propagation Delay for Various Logic Families:**

LSTTL: Time from 1.3 V at input to 1.3 V at output.
Other TTL: Time from 1.5 V at input to 1.5 V at output.
CMOS: Time from 50% of maximum input to 50% of maximum output.

---

Figures 14.2 and 14.3 show the data sheets for the 7400 and 7408 logic gates. The information on propagation delay is given in the block labeled "switching characteristics." Typical and maximum values are shown for $t_{pLH}$ and $t_{pHL}$. The manufacturer guarantees that the switching times of the gate outputs will not exceed the maximum values. Typical values indicate the speed that most gates can achieve.

---

Typical values of $t_{pLH}$ and $t_{pHL}$ should be regarded as "information only." Any design in which propagation delay is important should always be based on the maximum values of these parameters.

---

**EXAMPLE 14.1**   Use the data sheets in Figures 14.2 and 14.3, as well as a TTL data book, to find the maximum and typical propagation delays for each of the following logic gates: 7400 (Quad 2-input NAND), 7402 (Quad 2-input NOR), 7408 (Quad 2-input AND), 7410 (Triple 3-input NAND), 7432 (Quad 2-input OR).

| **Solution** | | **7400** | **7402** | **7408** | **7410** | **7432** |
|---|---|---|---|---|---|---|
| $t_{pLH}$ (typ) | | 11 ns | 12 ns | 17.5 ns | 11 ns | 10 ns |
| (max) | | 22 ns | 22 ns | 27 ns | 22 ns | 15 ns |
| $t_{pHL}$ (typ) | | 7 ns | 8 ns | 12 ns | 7 ns | 14 ns |
| (max) | | 15 ns | 15 ns | 19 ns | 15 ns | 22 ns |

Example 14.1 shows the variation of propagation delay among logic gates of the same family (74XX standard TTL). Since each logic function has a different circuit, its propagation delay will differ from those of gates with different functions. Gates with the same internal circuits, such as the 7400 and 7410 2- and 3-input NAND gates, have the same delays.

---

**EXAMPLE 14.2**   Use a TTL data book to find the maximum and typical propagation delays for each of the following logic gates: 7400, 74LS00, 74S00.

## SN5400, SN7400
## QUADRUPLE 2-INPUT POSITIVE-NAND GATES

### recommended operating conditions

| | | SN5400 | | | SN7400 | | | UNIT |
|---|---|---|---|---|---|---|---|---|
| | | MIN | NOM | MAX | MIN | NOM | MAX | |
| $V_{CC}$ | Supply voltage | 4.5 | 5 | 5.5 | 4.75 | 5 | 5.25 | V |
| $V_{IH}$ | High-level input voltage | 2 | | | 2 | | | V |
| $V_{IL}$ | Low-level input voltage | | | 0.8 | | | 0.8 | V |
| $I_{OH}$ | High-level output current | | | − 0.4 | | | − 0.4 | mA |
| $I_{OL}$ | Low-level output current | | | 16 | | | 16 | mA |
| $T_A$ | Operating free-air temperature | − 55 | | 125 | 0 | | 70 | °C |

### electrical characteristics over recommended operating free-air temperature range (unless otherwise noted)

| PARAMETER | TEST CONDITIONS † | | | SN5400 | | | SN7400 | | | UNIT |
|---|---|---|---|---|---|---|---|---|---|---|
| | | | | MIN | TYP‡ | MAX | MIN | TYP‡ | MAX | |
| $V_{IK}$ | $V_{CC}$ = MIN, | $I_I$ = − 12 mA | | | | − 1.5 | | | − 1.5 | V |
| $V_{OH}$ | $V_{CC}$ = MIN, | $V_{IL}$ = 0.8 V, | $I_{OH}$ = − 0.4 mA | 2.4 | 3.4 | | 2.4 | 3.4 | | V |
| $V_{OL}$ | $V_{CC}$ = MIN, | $V_{IH}$ = 2 V, | $I_{OL}$ = 16 mA | | 0.2 | 0.4 | | 0.2 | 0.4 | V |
| $I_I$ | $V_{CC}$ = MAX, | $V_I$ = 5.5 V | | | | 1 | | | 1 | mA |
| $I_{IH}$ | $V_{CC}$ = MAX, | $V_I$ = 2.4 V | | | | 40 | | | 40 | μA |
| $I_{IL}$ | $V_{CC}$ = MAX, | $V_I$ = 0.4 V | | | | − 1.6 | | | − 1.6 | mA |
| $I_{OS}$§ | $V_{CC}$ = MAX | | | − 20 | | − 55 | − 18 | | − 55 | mA |
| $I_{CCH}$ | $V_{CC}$ = MAX, | $V_I$ = 0 V | | | 4 | 8 | | 4 | 8 | mA |
| $I_{CCL}$ | $V_{CC}$ = MAX, | $V_I$ = 4.5 V | | | 12 | 22 | | 12 | 22 | mA |

† For conditions shown as MIN or MAX, use the appropriate value specified under recommended operating conditions.
‡ All typical values are at $V_{CC}$ = 5 V, $T_A$ = 25°C.
§ Not more than one output should be shorted at a time.

### switching characteristics, $V_{CC}$ = 5 V, $T_A$ = 25°C (see note 2)

| PARAMETER | FROM (INPUT) | TO (OUTPUT) | TEST CONDITIONS | | MIN | TYP | MAX | UNIT |
|---|---|---|---|---|---|---|---|---|
| $t_{PLH}$ | A or B | Y | $R_L$ = 400 Ω, | $C_L$ = 15 pF | | 11 | 22 | ns |
| $t_{PHL}$ | | | | | | 7 | 15 | ns |

NOTE 2: Load circuits and voltage waveforms are shown in Section 1.

Figure 14.2

**7400 Data** (Reprinted by permission of Texas Instruments)

**SN5408, SN7408**
**QUADRUPLE 2-INPUT POSITIVE-AND GATES**

### recommended operating conditions

| | | SN5408 | | | SN7408 | | | UNIT |
|---|---|---|---|---|---|---|---|---|
| | | MIN | NOM | MAX | MIN | NOM | MAX | |
| $V_{CC}$ | Supply voltage | 4.5 | 5 | 5.5 | 4.75 | 5 | 5.25 | V |
| $V_{IH}$ | High-level input voltage | 2 | | | 2 | | | V |
| $V_{IL}$ | Low-level input voltage | | | 0.8 | | | 0.8 | V |
| $I_{OH}$ | High-level output current | | | − 0.8 | | | − 0.8 | mA |
| $I_{OL}$ | Low-level output current | | | 16 | | | 16 | mA |
| $T_A$ | Operating free-air temperature | − 55 | | 125 | 0 | | 70 | °C |

### electrical characteristics over recommended operating free-air temperature range (unless otherwise noted)

| PARAMETER | TEST CONDITIONS† | | | SN5408 | | | SN7408 | | | UNIT |
|---|---|---|---|---|---|---|---|---|---|---|
| | | | MIN | TYP‡ | MAX | MIN | TYP‡ | MAX | |
| $V_{IK}$ | $V_{CC}$ = MIN, | $I_I$ = − 12 mA | | | − 1.5 | | | − 1.5 | V |
| $V_{OH}$ | $V_{CC}$ = MIN, | $V_{IH}$ = 2 V, $I_{OH}$ = − 0.8 mA | 2.4 | 3.4 | | 2.4 | 3.4 | | V |
| $V_{OL}$ | $V_{CC}$ = MIN, | $V_{IL}$ = 0.8 V, $I_{OL}$ = 16 mA | | 0.2 | 0.4 | | 0.2 | 0.4 | V |
| $I_I$ | $V_{CC}$ = MAX, | $V_I$ = 5.5 V | | | 1 | | | 1 | mA |
| $I_{IH}$ | $V_{CC}$ = MAX, | $V_I$ = 2.4 V | | | 40 | | | 40 | μA |
| $I_{IL}$ | $V_{CC}$ = MAX, | $V_I$ = 0.4 V | | | − 1.6 | | | − 1.6 | mA |
| $I_{OS}$§ | $V_{CC}$ = MAX | | − 20 | | − 55 | − 18 | | − 55 | mA |
| $I_{CCH}$ | $V_{CC}$ = MAX, | $V_I$ = 4.5 V | | 11 | 21 | | 11 | 21 | mA |
| $I_{CCL}$ | $V_{CC}$ = MAX, | $V_I$ = 0 V | | 20 | 33 | | 20 | 33 | mA |

† For conditions shown as MIN or MAX, use the appropriate value specified under recommended operating conditions.
‡ All typical values are at $V_{CC}$ = 5 V, $T_A$ = 25°C.
§ Not more than one output should be shorted at a time.

### switching characteristics, $V_{CC}$ = 5 V, $T_A$ = 25°C (see note 2)

| PARAMETER | FROM (INPUT) | TO (OUTPUT) | TEST CONDITIONS | | MIN | TYP | MAX | UNIT |
|---|---|---|---|---|---|---|---|---|
| $t_{PLH}$ | A or B | Y | $R_L$ = 400 Ω, | $C_L$ = 15 pF | | 17.5 | 27 | ns |
| $t_{PHL}$ | | | | | | 12 | 19 | ns |

NOTE 2: Load circuits and voltage waveforms are shown in Section 1.

**Figure 14.3**
**7408 Data** (Reprinted by permission of Texas Instruments)

| Solution | 7400 | 74LS00 | 74S00 |
|---|---|---|---|
| $t_{pLH}$ (typ) | 11 ns | 9 ns | 3 ns |
| (max) | 22 ns | 15 ns | 4.5 ns |
| $t_{pHL}$ (typ) | 7 ns | 10 ns | 3 ns |
| (max) | 15 ns | 15 ns | 5 ns |

All three gates in Example 14.2 have the same logic function (2-input NAND), but different propagation delay times. We might ask, "Why not always use the Schottky TTL gate (74S00), since it is the fastest?" The main reason is that it has the highest power dissipation of the gates shown. We wouldn't know this without looking up other specs on the data sheet. (We will learn how to do this later in the chapter.) Thus, it is important to make design decisions based on complete information, not just one parameter.

## Propagation Delay in Logic Circuits

A circuit consisting of two or more gates or flip-flops has a propagation delay that is the sum of delays *in the input-to-output path.* Delays in gates that do not affect the circuit output are disregarded. Figure 14.4 shows how propagation delay works in a simple logic circuit consisting of a 7408 AND gate and a 7432 OR gate. Changes at inputs $A$ and $B$ must propagate through both gates to affect the output. The total delay in such a case is the sum of $t_{p1}$ and $t_{p2}$. A change at input $C$ must pass only through gate 2. The circuit delay resulting from this change is only $t_{p2}$.

The timing diagram in Figure 14.4 shows the changes at inputs $A$, $B$, and $C$ and the resulting transitions at all gate outputs.

**Figure 14.4**

**Propagation Delays in a Logic Gate Circuit**

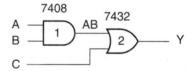

1. Both 1 and 2 change; $t_p = t_{pHL1} + t_{pHL2}$

2. No change from this transition

3. Only gate 2 changes; $t_p = t_{pLH2}$

1. When $A$ goes LOW, $AB$, the output of gate 1, also goes LOW after a maximum delay of $t_{pHL} = 19$ ns. This makes $Y$ go LOW after a further delay of up to $t_{pHL} = 22$ ns. Total delay: $t_p = t_{pHL1} + t_{pHL2} = 19$ ns + 22 ns = 41 ns, max.

2. The HIGH-to-LOW transition at input $B$ has no effect, since there is no difference between $0 \cdot 1$ and $0 \cdot 0$. $AB$ is already LOW.

3. The LOW-to-HIGH transition at input $C$ makes $Y$ go HIGH after a maximum delay of $t_{pLH2} = 15$ ns.

---

**EXAMPLE 14.3**

Figure 14.5 shows a simple logic circuit and the input waveforms at $A$, $B$, and $C$. Complete the timing diagram, showing propagation delays at all gate outputs. For the entire circuit, calculate the maximum propagation delays that result from each input change. The logic gates are as follows:

| Gate | Function | Designation |
|------|----------|-------------|
| 1 | NOR | 74LS02 |
| 2 | NAND | 74LS00 |
| 3 | NOR | 74LS02 |
| 4 | OR | 74LS32 |
| 5 | NAND | 74LS00 |

**Solution**

The completed timing diagram is shown in Figure 14.5. The delays are not to scale.

The first three transitions affect the output only via gates 3, 4, and 5. Even though the output of gate 1 changes during these transitions, its propagation delay is not added to the circuit total. This is because the condition $C = 0$ inhibits gate 2, preventing it from affecting the output of gate 5.

A change on input $C$ generates transition 4. The circuit output changes via gates 4 and 5 only.

The delay path for transitions 5 and 6 is through gates 1, 2, and 5. The condition $C = 1$ inhibits the path through gate 4. The propagation delay of gate 3 is not added to the circuit total.

Maximum propagation delays for the gates used are:

| | 74LS00 | 74LS02 | 74LS32 |
|-----------|--------|--------|--------|
| $t_{pLH}$ | 15 ns | 15 ns | 22 ns |
| $t_{pHL}$ | 15 ns | 15 ns | 22 ns |

Delays for each of the six transitions are:

1. $t_{p1} = t_{pHL3} + t_{pHL4} + t_{pLH5}$

    $= 15$ ns + 22 ns + 15 ns = 52 ns

2. $t_{p2} = t_{pLH3} + t_{pLH4} + t_{pHL5}$

    $= 15$ ns + 22 ns + 15 ns = 52 ns

3. $t_{p3} = t_{pHL3} + t_{pHL4} + t_{pLH5}$

    $= 15$ ns + 22 ns + 15 ns = 52 ns

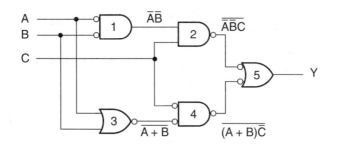

**Figure 14.5**

**Example 14.3
Logic Circuit and Timing
Diagram**

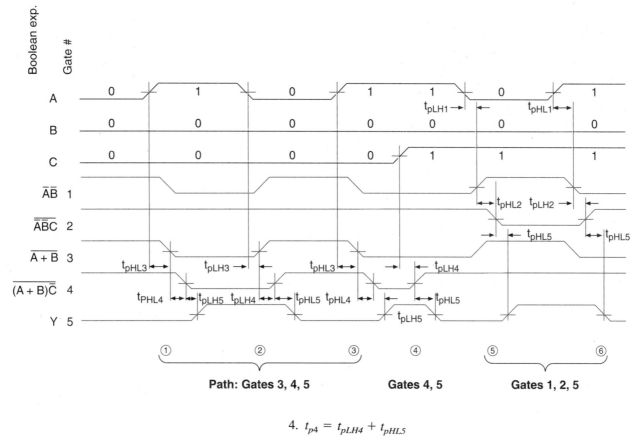

$$4.\ t_{p4} = t_{pLH4} + t_{pHL5}$$
$$= 22\ ns + 15\ ns = 37\ ns$$

$$5.\ t_{p5} = t_{pLH1} + t_{pHL2} + t_{pLH5}$$
$$= 15\ ns + 15\ ns + 15\ ns = 45\ ns$$

$$6.\ t_{p6} = t_{pHL1} + t_{pLH2} + t_{pHL5}$$
$$= 15\ ns + 15\ ns + 15\ ns = 45\ ns$$

**EXAMPLE 14.4**    Assuming that the input waveforms are the same as in Figure 14.5, calculate the propagation delays of the circuit in Example 14.3 if it is constructed from the following CMOS gates:

**Figure 14.6**

**CMOS Switching Characteristics** (Reprinted with permission of Motorola)

# CMOS B-SERIES GATES

## B-SERIES GATE SWITCHING TIMES

**SWITCHING CHARACTERISTICS*** ($C_L$ = 50 pF, $T_A$ = 25°C)

| Characteristic | Symbol | $V_{DD}$ Vdc | Min | Typ # | Max | Unit |
|---|---|---|---|---|---|---|
| Output Rise Time, All B-Series Gates | $^tTLH$ | | | | | ns |
| $^tTLH$ = (1.35 ns/pF) $C_L$ + 33 ns | | 5.0 | — | 100 | 200 | |
| $^tTLH$ = (0.60 ns/pF) $C_L$ + 20 ns | | 10 | — | 50 | 100 | |
| $^tTLH$ = (0.40 ns/pF) $C_L$ + 20 ns | | 15 | — | 40 | 80 | |
| Output Fall Time, All B-Series Gates | $^tTHL$ | | | | | ns |
| $^tTHL$ = (1.35 ns/pF) $C_L$ + 33 ns | | 5.0 | — | 100 | 200 | |
| $^tTHL$ = (0.60 ns/pF) $C_L$ + 20 ns | | 10 | — | 50 | 100 | |
| $^tTHL$ = (0.40 ns/pF) $C_L$ + 20 ns | | 15 | — | 40 | 80 | |
| Propagation Delay Time | $^tPLH, ^tPHL$ | | | | | ns |
| MC14001B, MC14011B only | | | | | | |
| $^tPLH, ^tPHL$ = (0.90 ns/pF) $C_L$ + 80 ns | | 5.0 | — | 125 | 250 | |
| $^tPLH, ^tPHL$ = (0.36 ns/pF) $C_L$ + 32 ns | | 10 | — | 50 | 100 | |
| $^tPLH, ^tPHL$ = (0.26 ns/pF) $C_L$ + 27 ns | | 15 | — | 40 | 80 | |
| All Other 2, 3, and 4 Input Gates | | | | | | |
| $^tPLH, ^tPHL$ = (0.90 ns/pF) $C_L$ + 115 ns | | 5.0 | — | 160 | 300 | |
| $^tPLH, ^tPHL$ = (0.36 ns/pF) $C_L$ + 47 ns | | 10 | — | 65 | 130 | |
| $^tPLH, ^tPHL$ = (0.26 ns/pF) $C_L$ + 37 ns | | 15 | — | 50 | 100 | |
| 8-Input Gates (MC14068B, MC14078B) | | | | | | |
| $^tPLH, ^tPHL$ = (0.90 ns/pF) $C_L$ + 155 ns | | 5.0 | — | 200 | 350 | |
| $^tPLH, ^tPHL$ = (0.36 ns/pF) $C_L$ + 62 ns | | 10 | — | 80 | 150 | |
| $^tPLH, ^tPHL$ = (0.26 ns/pF) $C_L$ + 47 ns | | 15 | — | 60 | 110 | |

*The formulas given are for the typical characteristics only at 25°C.

#Data labelled "Typ" is not to be used for design purposes but is intended as an indication of the IC's potential performance.

**FIGURE 1 – SWITCHING TIME TEST CIRCUIT AND WAVEFORMS**

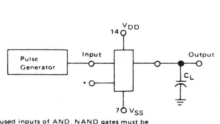

*All unused inputs of AND, NAND gates must be connected to $V_{DD}$.
All unused inputs of OR, NOR gates must be connected to $V_{SS}$.

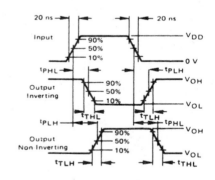

| Gate | Function | Designation |
|---|---|---|
| 1 | NOR | 4001B |
| 2 | NAND | 4011B |
| 3 | NOR | 4001B |
| 4 | OR | 4071B |
| 5 | NAND | 4011B |

CMOS switching characteristics are found in Figure 14.6. Assume that the supply voltage $(V_{DD})$ is 5 volts.

**Solution**

1. $t_{p1} = t_{pHL3} + t_{pHL4} + t_{pLH5}$

   $= 250 \text{ ns} + 300 \text{ ns} + 250 \text{ ns} = 800 \text{ ns}$

2. $t_{p2} = t_{pLH3} + t_{pLH4} + t_{pHL5}$

   $= 250 \text{ ns} + 300 \text{ ns} + 250 \text{ ns} = 800 \text{ ns}$

3. $t_{p3} = t_{pHL3} + t_{pHL4} + t_{pLH5}$

   $= 250 \text{ ns} + 300 \text{ ns} + 250 \text{ ns} = 800 \text{ ns}$

4. $t_{p4} = t_{pLH4} + t_{pHL5}$

   $= 300 \text{ ns} + 250 \text{ ns} = 550 \text{ ns}$

5. $t_{p5} = t_{pLH1} + t_{pHL2} + t_{pLH5}$

   $= 250 \text{ ns} + 250 \text{ ns} + 250 \text{ ns} = 750 \text{ ns}$

6. $t_{p6} = t_{pHL1} + t_{pLH2} + t_{pHL5}$

   $= 250 \text{ ns} + 250 \text{ ns} + 250 \text{ ns} = 750 \text{ ns}$

---

**Section Review Problem for Section 14.2**

14.1. Refer to the data sheets for the following High-Speed CMOS devices, found in Appendix A: 74HC00A, 74HC02A, 74HC32A. Use these data sheets to calculate the maximum propagation delays in the circuit of Figure 14.5 if the same input waveforms A, B, and C are applied as in Figure 14.5. Assume that $V_{CC}$ = 4.5 V and the operating temperature range is 25° C to −55° C.

# 14.3

# Fanout

**Fanout**  *The number of load gates that a logic gate output is capable of driving without possible logic errors.*

**Driving gate**  *A gate whose output supplies current to the inputs of other gates.*

**Load gate**  *A gate whose input current is supplied by the output of another gate.*

**Sourcing**  *A terminal on a gate or flip-flop is sourcing current when the current flows out of the terminal.*

**Sinking**  *A terminal on a gate or flip-flop is sinking current when the current flows into the terminal.*

$I_{OL}$  *Current measured at a device output when the output is LOW.*

$I_{OH}$  *Current measured at a device output when the output is HIGH.*

$I_{IL}$  *Current measured at a device input when the input is LOW.*

$I_{IH}$  *Current measured at a device input when the input is HIGH.*

$W$e have assumed that logic gates are able to drive any number of other logic gates. Since gates are electrical devices with finite current-driving capabilities, this is obviously not the case. The number of gates ("loads") a logic gate can drive is referred to as its **fanout.**

---

Fanout is simply an application of Kirchhoff's Current Law: The algebraic sum of currents at a node must be zero. Thus, the fanout of a logic gate is limited by:

a. The maximum current its output can supply safely in a given logic state ($I_{OH}$ or $I_{OL}$), and

b. The current requirements of the load to which it is connected ($I_{IH}$ or $I_{IL}$).

---

Figure 14.7 shows the fanout of an AND gate when its output is in the HIGH and LOW states. The AND gate, or **driving gate,** supplies current to the inputs of the other four gates, which are called the **load gates.**

Each load gate requires a fixed amount of input current, depending on which state it is in. The sum of these input currents equals the current supplied by the driving gate. The fanout is determined by the amount of current the driving gate can supply without damaging its output circuit.

The input and output currents of a gate are established by its internal circuitry. These values are usually the same for two gates in the same family, since the input and output circuitry of a gate is common to all members of the family. Exceptions may occur when the output of a particular type of gate, such as a 7406 high-current inverter, has additional output buffering.

Assuming that the gates in Figure 14.7 are of the same logic family, the sum of load gate input currents in Figure 14.7a is:

**Figure 14.7**
**Fanout**

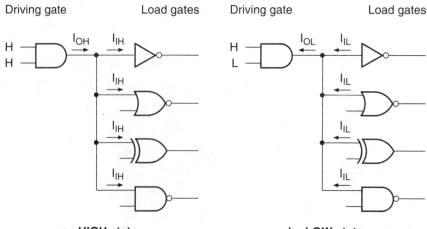

a.  **HIGH state**                    b.  **LOW state**

$$I_{IH1} + I_{IH2} + I_{IH3} + I_{IH4} = 4I_{IH}$$

By Kirchhoff's Current Law (KCL),

$$I_{OH} = 4I_{IH}$$

If $I_{OH}$ is at its maximum value (i.e., we can add no more loads without increasing $I_{OH}$ beyond its safe limit), then the fanout of the AND gate in its HIGH state is:

$$n_H = I_{OH}/I_{IH} = 4$$

Similarly,

$$I_{IL} + I_{IL} + I_{IL} + I_{IL} = 4I_{IL}$$
$$I_{OL} = 4I_{IL}$$

If $I_{OL}$ is at its maximum value, then the LOW-state fanout of the AND gate is:

$$n_L = I_{OL}/I_{IL} = 4$$

The required current specifications can be found on a data sheet as follows:

Parameters for driving gates:
$I_{OL}$: Output current when output is LOW
$I_{OH}$: Output current when output is HIGH

Parameters for load gates:
$I_{IL}$: Input current when input is LOW
$I_{IH}$: Input current when input is HIGH

**EXAMPLE 14.5**

Use a TTL data book to look up the maximum values of $I_{OH}$, $I_{OL}$, $I_{IH}$, and $I_{IL}$ for the following logic gates: 7400, 74S00, 74LS00, 7432, 74S32, 74LS32.

**Solution**

|  | 7400 | 74S00 | 74LS00 | 7432 | 74S32 | 74LS32 |
|---|---|---|---|---|---|---|
| $I_{OH}$ (mA) | −0.4 | −1 | −0.4 | −0.8 | −1 | −0.4 |
| $I_{OL}$ (mA) | 16 | 20 | 8 | 16 | 20 | 8 |
| $I_{IH}$ (μA) | 40 | 50 | 20 | 40 | 50 | 20 |
| $I_{IL}$ (mA) | −1.6 | −2 | −0.4 | −1.6 | −2 | −0.4 |

By convention, current entering a gate ($I_{IH}$, $I_{OL}$) is denoted as positive, and current leaving a gate ($I_{IL}$, $I_{OH}$) is denoted as negative. When current is leaving a gate, we say the gate is **sourcing** current. When current is entering a gate, we say the gate is **sinking** current.

Note that the output of a gate does not always source current, nor does an input always sink current. The current direction changes for the HIGH and LOW states at the same terminal. The reason for this will become apparent when we study the circuitry of logic gate inputs and outputs.

---

**EXAMPLE 14.6**

How many 74LS00 inputs can a 74LS00 NAND gate drive? (i.e., what is the fanout of a 74LS00 NAND gate?)

**Solution**    We must consider the following cases:

     **a.** When the output of the driving gate is LOW

     **b.** When the output of the driving gate is HIGH

**Output LOW:**

$$I_{OL} = 8 \text{ mA (sinking)}$$

$$I_{IL} = -0.4 \text{ mA (sourcing)}$$

$$n_L = 8 \text{ mA}/0.4 \text{ mA} = 20$$

**Output HIGH:**

$$I_{OH} = -0.4 \text{ mA (sourcing)}$$

$$I_{IH} = 20 \text{ μA (sinking)}$$

$$n_H = 0.4 \text{ mA}/20 \text{ μA} = 20$$

Since $n_L = n_H$, fanout is 20.

We disregard the negative sign in our calculations, since the input current of the load gate and output current of the driving gate are actually in the same direction. For example, even though $I_{OH}$ is leaving the driving gate (negative), $I_{IH}$ is entering the load gates (positive). These currents flow in the same direction. If we include the minus sign in our calculation, we get a negative value of fanout, which is meaningless.

---

The fanout in both HIGH and LOW states is the same in this case, but that is not always so. If the values of HIGH- and LOW-state fanout are different, the smallest value must be used. For example, if a gate can drive four loads in the HIGH state or eight in the LOW state, the fanout of the driving gate is four loads. If we attempt to drive eight loads, we can't guarantee enough driving current to supply all loads in both states.

If a gate from one logic family is used to drive gates from another logic family, we must use the output parameters ($I_{OL}$, $I_{OH}$) for the driving gate and the input parameters ($I_{IL}$, $I_{IH}$) for the load gates.

---

**EXAMPLE 14.7**

Calculate the maximum number of Schottky TTL loads (74SXX series) that a 74LS86 XOR gate can drive.

**Solution**    Driving gate:    74LS86    $I_{OH} = -0.4$ mA,
                                           $I_{OL} = 8$ mA

           Load gates:      74SXX      $I_{IH} = 50$ μA,
                                             $I_{IL} = -2$ mA

**Output LOW:**

$$I_{OL} = 8 \text{ mA (sinking)}$$

$$I_{IL} = -2 \text{ mA (sourcing)}$$

$$n_L = 8 \text{ mA/2 mA} = 4$$

**Output HIGH:**

$$I_{OH} = -0.4 \text{ mA (sourcing)}$$

$$I_{IH} = 50 \text{ }\mu\text{A (sinking)}$$

$$n_H = 0.4 \text{ mA/50 }\mu\text{A} = 8$$

Since $n_L < n_H$, fanout $= n_L = 4$.

We will examine the fanout of CMOS devices in a later section on interfacing between CMOS and TTL.

---

**Section Review Problem for Section 14.3**

14.2. The input and output currents $I_{OH}$, $I_{OL}$, $I_{IH}$, and $I_{IL}$ of a TTL device may be classified as source currents or sink currents. List each input or output current as a source or sink current.

---

## 14.4

# Power Dissipation

**Power dissipation** *The electrical energy used by a logic circuit in a specified period of time. Abbreviation:* $P_d$

$V_{CC}$ *TTL supply voltage.*

$I_{CC}$ *Total TTL supply current.*

$I_{CCH}$ *TTL supply current with all outputs HIGH.*

$I_{CCL}$ *TTL supply current with all outputs LOW.*

$V_{DD}$ *CMOS supply voltage.*

$I_{DD}$ *CMOS supply current under static (nonswitching) conditions.*

$I_T$ *When referring to CMOS supply current, the sum of static and dynamic supply currents.*

$C_{PD}$ *Internal capacitance of a High-Speed CMOS device used to calculate its power dissipation.*

Electronic logic gates require a certain amount of electrical energy to operate. The measure of the energy used over time is called **power dissipation.** Each of the different families of logic has a characteristic range of values for the power it consumes.

For TTL and CMOS, the power dissipation is calculated as follows:

TTL: $\qquad P_d = V_{CC}\, I_{CC}$

Metal-Gate CMOS: $\qquad P_d = V_{DD}\, I_T$

High-Speed CMOS: $\qquad P_d = V_{CC}\, I_T \qquad$ ($I_T$ = quiescent + dynamic supply current)

The main difference between the two families is the calculation of supply current.

The supply current in a TTL device is different when its outputs are HIGH than when they are LOW. Thus, supply current, $I_{CC}$, and therefore power dissipation, depends on the states of the device outputs. If the outputs are switching, $I_{CC}$ is proportional to output duty cycle.

In a CMOS device, very little power is consumed when the device outputs are static. Much more current is drawn from the supply when the outputs switch from one state to another. Thus, the power dissipation of a device depends on the switching frequency of its outputs.

## Power Dissipation in TTL Devices

Figures 14.8 and 14.9 show manufacturer's data for a 74LS00 and a 74S00 Quadruple 2-input NAND gate.

Two values are given for supply current in each data sheet. $I_{CCL}$ is the current drawn from the power supply when all four gate outputs are LOW. $I_{CCH}$ is the current drawn from the supply when all outputs are HIGH. If the gate outputs are not all at the same level, the supply current is the sum of currents given by:

$$I_{CC} = \frac{N_H}{N}\, I_{CCH} + \frac{N_L}{N}\, I_{CCL}$$

where $N$ is the total number of gates in the package

$N_H$ is the number of gates whose output is HIGH

$N_L$ is the number of gates whose output is LOW

The power dissipation of a TTL chip also depends on the duty cycle of the gate outputs. That is, it depends on the fraction of time that the chip's outputs are HIGH.

If we assume that, on average, the outputs of a chip are switching with a duty cycle of 50%, the supply current can be calculated as follows:

$$I_{CC} = (I_{CCH} + I_{CCL})/2$$

If the output duty cycle is other than 50%, the supply current is given by:

$$I_{CC} = DC\, I_{CCH} + (1 - DC)\, I_{CCL}$$

where $DC$ = duty cycle.

## SN54LS00, SN74LS00
## QUADRUPLE 2-INPUT POSITIVE-NAND GATES

### recommended operating conditions

| | | SN54LS00 | | | SN74LS00 | | | UNIT |
|---|---|---|---|---|---|---|---|---|
| | | MIN | NOM | MAX | MIN | NOM | MAX | |
| $V_{CC}$ | Supply voltage | 4.5 | 5 | 5.5 | 4.75 | 5 | 5.25 | V |
| $V_{IH}$ | High-level input voltage | 2 | | | 2 | | | V |
| $V_{IL}$ | Low-level input voltage | | | 0.7 | | | 0.8 | V |
| $I_{OH}$ | High-level output current | | | − 0.4 | | | − 0.4 | mA |
| $I_{OL}$ | Low-level output current | | | 4 | | | 8 | mA |
| $T_A$ | Operating free-air temperature | − 55 | | 125 | 0 | | 70 | °C |

### electrical characteristics over recommended operating free-air temperature range (unless otherwise noted)

| PARAMETER | TEST CONDITIONS † | | | SN54LS00 | | | SN74LS00 | | | UNIT |
|---|---|---|---|---|---|---|---|---|---|---|
| | | | | MIN | TYP‡ | MAX | MIN | TYP‡ | MAX | |
| $V_{IK}$ | $V_{CC}$ = MIN, | $I_I$ = − 18 mA | | | | − 1.5 | | | − 1.5 | V |
| $V_{OH}$ | $V_{CC}$ = MIN, | $V_{IL}$ = MAX, | $I_{OH}$ = − 0.4 mA | 2.5 | 3.4 | | 2.7 | 3.4 | | V |
| $V_{OL}$ | $V_{CC}$ = MIN, | $V_{IH}$ = 2 V, | $I_{OL}$ = 4 mA | | 0.25 | 0.4 | | 0.25 | 0.4 | V |
| | $V_{CC}$ = MIN, | $V_{IH}$ = 2 V, | $I_{OL}$ = 8 mA | | | | | 0.35 | 0.5 | |
| $I_I$ | $V_{CC}$ = MAX, | $V_I$ = 7 V | | | | 0.1 | | | 0.1 | mA |
| $I_{IH}$ | $V_{CC}$ = MAX, | $V_I$ = 2.7 V | | | | 20 | | | 20 | μA |
| $I_{IL}$ | $V_{CC}$ = MAX, | $V_I$ = 0.4 V | | | | − 0.4 | | | − 0.4 | mA |
| $I_{OS}$ § | $V_{CC}$ = MAX | | | − 20 | | − 100 | − 20 | | − 100 | mA |
| $I_{CCH}$ | $V_{CC}$ = MAX, | $V_I$ = 0 V | | | 0.8 | 1.6 | | 0.8 | 1.6 | mA |
| $I_{CCL}$ | $V_{CC}$ = MAX, | $V_I$ = 4.5 V | | | 2.4 | 4.4 | | 2.4 | 4.4 | mA |

† For conditions shown as MIN or MAX, use the appropriate value specified under recommended operating conditions.
‡ All typical values are at $V_{CC}$ = 5 V, $T_A$ = 25°C
§ Not more than one output should be shorted at a time, and the duration of the short-circuit should not exceed one second.

### switching characteristics, $V_{CC}$ = 5 V, $T_A$ = 25°C (see note 2)

| PARAMETER | FROM (INPUT) | TO (OUTPUT) | TEST CONDITIONS | | MIN | TYP | MAX | UNIT |
|---|---|---|---|---|---|---|---|---|
| $t_{PLH}$ | A or B | Y | $R_L$ = 2 kΩ, | $C_L$ = 15 pF | | 9 | 15 | ns |
| $t_{PHL}$ | | | | | | 10 | 15 | ns |

NOTE 2: Load circuits and voltage waveforms are shown in Section 1.

Figure 14.8

**74LS00 Data** (Reprinted by permission of Texas Instruments)

## SN54S00, SN74S00
## QUADRUPLE 2-INPUT POSITIVE-NAND GATES

### recommended operating conditions

| | | SN54S00 | | | SN74S00 | | | UNIT |
|---|---|---|---|---|---|---|---|---|
| | | MIN | NOM | MAX | MIN | NOM | MAX | |
| $V_{CC}$ | Supply voltage | 4.5 | 5 | 5.5 | 4.75 | 5 | 5.25 | V |
| $V_{IH}$ | High-level input voltage | 2 | | | 2 | | | V |
| $V_{IL}$ | Low-level input voltage | | | 0.8 | | | 0.8 | V |
| $I_{OH}$ | High-level output current | | | −1 | | | −1 | mA |
| $I_{OL}$ | Low-level output current | | | 20 | | | 20 | mA |
| $T_A$ | Operating free-air temperature | −55 | | 125 | 0 | | 70 | °C |

### electrical characteristics over recommended operating free-air temperature range (unless otherwise noted)

| PARAMETER | TEST CONDITIONS † | | | SN54S00 | | | SN74S00 | | | UNIT |
|---|---|---|---|---|---|---|---|---|---|---|
| | | | MIN | TYP‡ | MAX | MIN | TYP‡ | MAX | | |
| $V_{IK}$ | $V_{CC}$ = MIN, | $I_I$ = −18 mA | | | | −1.2 | | | −1.2 | V |
| $V_{OH}$ | $V_{CC}$ = MIN, | $V_{IL}$ = 0.8 V, $I_{OH}$ = −1 mA | 2.5 | 3.4 | | 2.7 | 3.4 | | V |
| $V_{OL}$ | $V_{CC}$ = MIN, | $V_{IH}$ = 2 V, $I_{OL}$ = 20 mA | | | 0.5 | | | 0.5 | V |
| $I_I$ | $V_{CC}$ = MAX, | $V_I$ = 5.5 V | | | 1 | | | 1 | mA |
| $I_{IH}$ | $V_{CC}$ = MAX, | $V_I$ = 2.7 V | | | 50 | | | 50 | µA |
| $I_{IL}$ | $V_{CC}$ = MAX, | $V_I$ = 0.5 V | | | −2 | | | −2 | mA |
| $I_{OS}$§ | $V_{CC}$ = MAX | | −40 | | −100 | −40 | | −100 | mA |
| $I_{CCH}$ | $V_{CC}$ = MAX, | $V_I$ = 0 V | | 10 | 16 | | 10 | 16 | mA |
| $I_{CCL}$ | $V_{CC}$ = MAX, | $V_I$ = 4.5 V | | 20 | 36 | | 20 | 36 | mA |

† For conditions shown as MIN or MAX, use the appropriate value specified under recommended operating conditions.
‡ All typical values are at $V_{CC}$ = 5 V, $T_A$ = 25°C.
§ Not more than one output should be shorted at a time, and the duration of the short-circuit should not exceed one second.

### switching characteristics, $V_{CC}$ = 5 V, $T_A$ = 25°C (see note 2)

| PARAMETER | FROM (INPUT) | TO (OUTPUT) | TEST CONDITIONS | MIN | TYP | MAX | UNIT |
|---|---|---|---|---|---|---|---|
| $t_{PLH}$ | A or B | Y | $R_L$ = 280 Ω, $C_L$ = 15 pF | | 3 | 4.5 | ns |
| $t_{PHL}$ | | | | | 3 | 5 | ns |
| $t_{PLH}$ | | | $R_L$ = 280 Ω, $C_L$ = 50 pF | | 4.5 | | ns |
| $t_{PHL}$ | | | | | 5 | | ns |

NOTE 2: Load circuits and voltage waveforms are shown in Section 1.

**Figure 14.9**
**74S00 Data** (Reprinted by permission of Texas Instruments)

**EXAMPLE 14.8**

Figure 14.10 shows a circuit constructed from the gates in a 74XX00 Quadruple 2-input NAND gate package. Use the data sheet shown in Figure 14.2 to determine the maximum power dissipation of the circuit if the input is $DCBA = 1001$ and the gates are 7400 NANDs. Refer to the data sheets in Figures 14.8 and 14.9 and repeat the calculation for 74LS00 and 74S00 gates.

**Figure 14.10**

**Power Dissipation of 74XX00 NAND**

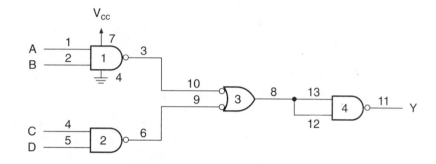

**Solution**

Gate 1: $\overline{AB} = 1$

Gate 2: $\overline{CD} = 1$

Gate 3: $AB + CD = 0$

Gate 4: $\overline{AB + CD} = 1$

Since three outputs are HIGH and one is LOW, the supply current is given by:

$$I_{CC} = \frac{N_H}{N} I_{CCH} + \frac{N_L}{N} I_{CCL}$$

$$= \frac{3}{4} I_{CCH} + \frac{1}{4} I_{CCL}$$

Maximum supply current for each device is:

$7400$:    $I_{CC} = 0.75(8 \text{ mA}) + 0.25(22 \text{ mA})$    $= 11.5 \text{ mA}$

$74LS00$: $I_{CC} = 0.75(1.6 \text{ mA}) + 0.25(4.4 \text{ mA}) =$    $2.3 \text{ mA}$

$74S00$:  $I_{CC} = 0.75(16 \text{ mA}) + 0.25(36 \text{ mA})$    $=$    $21 \text{ mA}$

Maximum power dissipation for each device is:

$7400$:    $P_d = V_{CC} I_{CC} = (5 \text{ V})(11.5 \text{ mA}) = 57.5 \text{ mW}$

$74LS00$: $P_d = V_{CC} I_{CC} = (5 \text{ V})(2.3 \text{ mA}) =$    $11.5 \text{ mW}$

$74S00$:  $P_d = V_{CC} I_{CC} = (5 \text{ V})(21 \text{ mA}) =$    $105 \text{ mW}$

(1 mW = 1 milliwatt = $10^{-3}$ W.)

**EXAMPLE 14.9**

Find the maximum power dissipation of the circuit in Figure 14.10 if the gates are 74LS00 and the gate outputs are switching with an average duty cycle of 30%.

**Solution**

$$I_{CC} = 0.3\ I_{CCH} + 0.7\ I_{CCL}$$

$$= 0.3(1.6\ \text{mA}) + 0.7(4.4\ \text{mA})$$

$$= 3.56\ \text{mA}$$

$$P_d = V_{CC}\ I_{CC} = (5\ \text{V})(3.56\ \text{mA}) = 17.8\ \text{mW}$$

## Power Dissipation in 4000B CMOS Devices

Figure 14.11 shows a data sheet for CMOS logic gates in the 4000B series. Until recently, this has been the standard CMOS family, although the newer High-Speed CMOS (74HCXX, 74HCTXX) devices are now being used for many applications.

CMOS gates draw the greatest power when the outputs are switching from one logic state to the other. This is because the large internal impedances of the logic gate limit supply current. A change of state requires the charging and discharging of internal capacitances, which draws a relatively large current from the power supply.

Total supply current has two components: a quiescent current, $I_{DD}$, that flows when the outputs are static (not switching) and a dynamic component that depends on frequency.

The values of quiescent and total current are found in the data sheet in Figure 14.11. The dynamic component is given as $(0.3\ \mu\text{A/kHz})f$ for a +5-V supply voltage. The combined current under this condition is given by $I_T = (0.3\ \mu\text{A/kHz})f + I_{DD}/N$ for a package having $N$ gates. We divide $I_{DD}$ by the number of gates because total supply current is specified *per gate* and quiescent current is specified *per package*.

The CMOS data sheet shows specifications for minimum and maximum temperatures, as well as room temperature ratings. Separate specifications are shown for AL and CL/CP devices. AL is the manufacturer's designation for military temperature rating. CL/CP signifies a commercial temperature rating for a device in a ceramic (CL) or plastic (CP) package. The AL devices are rated for operation over a larger range of temperatures and must meet other requirements more stringently than the CL/CP devices. For our purposes, we can take information from the 25°C column for the CL/CP device.

**EXAMPLE 14.10**

The circuit in Figure 14.10 is constructed from 4011B CMOS NAND gates. Calculate the maximum power dissipation of the circuit:

**a.** When the gate inputs are steady at the state $DCBA = 1010$

**b.** When the outputs are switching at an average frequency of 10 kHz

**c.** When the outputs are switching at an average frequency of 1 MHz

Supply voltage is 5 V.

**Solution**

**a.** The quiescent current is the same for a gate with a HIGH or a LOW output. From the data sheet, $I_{DDmax} = 1.0\ \mu\text{A}$ for $V_{DD} = 5$ V. $P_d = (1.0\ \mu\text{A})(5\ \text{V}) = 5.0\ \mu\text{W}$.

**b.** For $V_{DD} = 5$ V,

$$I_T \text{ (per gate)} = (0.3\ \mu\text{A/kHz})f + I_{DD}/N$$

$$I_T \text{ (per package)} = N(0.3\ \mu\text{A/kHz})f + I_{DD}$$

$$= 4(0.3\ \mu\text{A/kHz})(10\ \text{kHz}) + 1.0\ \mu\text{A}$$

$$= 12\ \mu\text{A} + 1.0\ \mu\text{A} = 13.0\ \mu\text{A}$$

$$P_d = (13.0\ \mu\text{A})(5\ \text{V}) = 65.0\ \mu\text{W}$$

$$P_d = 3(22 \text{ pF} + 10 \text{ pF})(5 \text{ V})^2 (1 \text{ MHz})$$
$$+ (22 \text{ pF})(5 \text{ V})^2 (1 \text{ MHz}) + 5 \text{ μW}$$
$$= 3(800 \text{ μW}) + 550 \text{ μW} + 5 \text{ μW}$$
$$= 2955 \text{ μW} = 2.95 \text{ mW}$$

**EXAMPLE 14.13**

Repeat the calculation of Example 14.11 to determine at what frequency the power dissipation of a High-Speed CMOS circuit exceeds that of an LSTTL circuit if both are configured as shown in Figure 14.12. Assume that $V_{CC} = 5$ V and the temperature is between 25°C and −55°C.

**Solution**  We have already calculated the LSTTL power dissipation as 36.5 mW. Assume we can neglect the quiescent power dissipation of the High-Speed CMOS circuit and that the load capacitance of all gates is the same (10 pF). The power dissipation is therefore given by:

$$P_d = 8(C_L + C_{PD}) V_{CC}^2 f$$

Solving for frequency:

$$f = P_d/8(C_L + C_{PD}) V_{CC}^2$$
$$= (36.5 \text{ mW})/(8(10 \text{ pF} + 22 \text{ pF})(25 \text{ V}^2))$$
$$= 5.7 \text{ MHz}$$

The power dissipation of a High-Speed CMOS implementation of the circuit in Figure 14.12 exceeds that of a similar LSTTL circuit at frequencies greater than 5.7 MHz. ▲

## 14.5
## Noise Margin

**Noise**  *Unwanted electrical signal, often resulting from electromagnetic radiation.*

**Noise margin**  *A measure of the ability of a logic circuit to tolerate noise.*

$V_{IH}$  *Voltage level required to make the input of a logic circuit HIGH.*

$V_{IL}$  *Voltage level required to make the input of a logic circuit LOW.*

$V_{OH}$  *Voltage measured at a device output when the output is HIGH.*

$V_{OL}$  *Voltage measured at a device output when the output is LOW.*

Electrical circuits are susceptible to **noise,** or unwanted electrical signals. Such signals are often induced by electromagnetic fields of motors, fluorescent lighting, high-frequency electronic circuits, and cosmic rays. They can cause erroneous operation of a digital circuit. Since it is impossible to eliminate all noise from a circuit, it is desirable to build a certain amount of tolerance, or **noise margin,** into digital devices used in the circuit.

## CMOS B-SERIES GATES

**ELECTRICAL CHARACTERISTICS** (Voltages Referenced to $V_{SS}$)

| Characteristic | | Symbol | $V_{DD}$ Vdc | $T_{low}$* Min | $T_{low}$* Max | 25°C Min | 25°C Typ # | 25°C Max | $T_{high}$* Min | $T_{high}$* Max | Unit |
|---|---|---|---|---|---|---|---|---|---|---|---|
| Output Voltage "0" Level | | $V_{OL}$ | 5.0 | — | 0.05 | — | 0 | 0.05 | — | 0.05 | Vdc |
| $V_{in} = V_{DD}$ or 0 | | | 10 | — | 0.05 | — | 0 | 0.05 | — | 0.05 | |
| | | | 15 | — | 0.05 | — | 0 | 0.05 | — | 0.05 | |
| "1" Level | | $V_{OH}$ | 5.0 | 4.95 | — | 4.95 | 5.0 | — | 4.95 | — | Vdc |
| $V_{in} = 0$ or $V_{DD}$ | | | 10 | 9.95 | — | 9.95 | 10 | — | 9.95 | — | |
| | | | 15 | 14.95 | — | 14.95 | 15 | — | 14.95 | — | |
| Input Voltage "0" Level | | $V_{IL}$ | | | | | | | | | Vdc |
| ($V_O$ = 4.5 or 0.5 Vdc) | | | 5.0 | — | 1.5 | — | 2.25 | 1.5 | — | 1.5 | |
| ($V_O$ = 9.0 or 1.0 Vdc) | | | 10 | — | 3.0 | — | 4.50 | 3.0 | — | 3.0 | |
| ($V_O$ = 13.5 or 1.5 Vdc) | | | 15 | — | 4.0 | — | 6.75 | 4.0 | — | 4.0 | |
| "1" Level | | $V_{IH}$ | | | | | | | | | |
| ($V_O$ = 0.5 or 4.5 Vdc) | | | 5.0 | 3.5 | — | 3.5 | 2.75 | — | 3.5 | — | Vdc |
| ($V_O$ = 1.0 or 9.0 Vdc) | | | 10 | 7.0 | — | 7.0 | 5.50 | — | 7.0 | — | |
| ($V_O$ = 1.5 or 13.5 Vdc) | | | 15 | 11.0 | — | 11.0 | 8.25 | — | 11.0 | — | |
| Output Drive Current (AL Device) | | $I_{OH}$ | | | | | | | | | mAdc |
| ($V_{OH}$ = 2.5 Vdc) Source | | | 5.0 | −3.0 | — | −2.4 | −4.2 | — | −1.7 | — | |
| ($V_{OH}$ = 4.6 Vdc) | | | 5.0 | −0.64 | — | −0.51 | −0.88 | — | −0.36 | — | |
| ($V_{OH}$ = 9.5 Vdc) | | | 10 | −1.6 | — | −1.3 | −2.25 | — | −0.9 | — | |
| ($V_{OH}$ = 13.5 Vdc) | | | 15 | −4.2 | — | −3.4 | −8.8 | — | −2.4 | — | |
| ($V_{OL}$ = 0.4 Vdc) Sink | | $I_{OL}$ | 5.0 | 0.64 | — | 0.51 | 0.88 | — | 0.36 | — | mAdc |
| ($V_{OL}$ = 0.5 Vdc) | | | 10 | 1.6 | — | 1.3 | 2.25 | — | 0.9 | — | |
| ($V_{OL}$ = 1.5 Vdc) | | | 15 | 4.2 | — | 3.4 | 8.8 | — | 2.4 | — | |
| Output Drive Current (CL/CP Device) | | $I_{OH}$ | | | | | | | | | mAdc |
| ($V_{OH}$ = 2.5 Vdc) Source | | | 5.0 | −2.5 | — | −2.1 | −4.2 | — | −1.7 | — | |
| ($V_{OH}$ = 4.6 Vdc) | | | 5.0 | −0.52 | — | −0.44 | −0.88 | — | −0.36 | — | |
| ($V_{OH}$ = 9.5 Vdc) | | | 10 | −1.3 | — | −1.1 | −2.25 | — | −0.9 | — | |
| ($V_{OH}$ = 13.5 Vdc) | | | 15 | −3.6 | — | −3.0 | −8.8 | — | −2.4 | — | |
| ($V_{OL}$ = 0.4 Vdc) Sink | | $I_{OL}$ | 5.0 | 0.52 | — | 0.44 | 0.88 | — | 0.36 | — | mAdc |
| ($V_{OL}$ = 0.5 Vdc) | | | 10 | 1.3 | — | 1.1 | 2.25 | — | 0.9 | — | |
| ($V_{OL}$ = 1.5 Vdc) | | | 15 | 3.6 | — | 3.0 | 8.8 | — | 2.4 | — | |
| Input Current (AL Device) | | $I_{in}$ | 15 | — | ±0.1 | — | ±0.00001 | ±0.1 | — | ±1.0 | μAdc |
| Input Current (CL/CP Device) | | $I_{in}$ | 15 | — | ±0.3 | — | ±0.00001 | ±0.3 | — | ±1.0 | μAdc |
| Input Capacitance ($V_{in}$ = 0) | | $C_{in}$ | — | — | — | — | 5.0 | 7.5 | — | — | pF |
| Quiescent Current (AL Device) (Per Package) | | $I_{DD}$ | 5.0 | — | 0.25 | — | 0.0005 | 0.25 | — | 7.5 | μAdc |
| | | | 10 | — | 0.50 | — | 0.0010 | 0.50 | — | 15.0 | |
| | | | 15 | — | 1.00 | — | 0.0015 | 1.00 | — | 30.0 | |
| Quiescent Current (CL/CP Device) (Per Package) | | $I_{DD}$ | 5.0 | — | 1.0 | — | 0.0005 | 1.0 | — | 7.5 | μAdc |
| | | | 10 | — | 2.0 | — | 0.0010 | 2.0 | — | 15.0 | |
| | | | 15 | — | 4.0 | — | 0.0015 | 4.0 | — | 30.0 | |
| Total Supply Current**† (Dynamic plus Quiescent, Per Gate, $C_L$ = 50 pF) | | $I_T$ | 5.0 | $I_T$ = (0.3 μA/kHz) f + $I_{DD}$/N | | | | | | | μAdc |
| | | | 10 | $I_T$ = (0.6 μA/kHz) f + $I_{DD}$/N | | | | | | | |
| | | | 15 | $I_T$ = (0.9 μA/kHz) f + $I_{DD}$/N | | | | | | | |

*$T_{low}$ = −55°C for AL Device, −40°C for CL/CP Device.
$T_{high}$ = +125°C for AL Device, 85°C for CL/CP Device.

#Data labelled "Typ" is not to be used for design purposes but is intended as an indication of the IC's potential performance.

**The formulas given are for the typical characteristics only at 25°C.

†To calculate total supply current at loads other than 50 pF:

$$I_T(C_L) = I_T(50 \text{ pF}) + (C_L - 50) \text{ Vfk}$$

where: $I_T$ is in μA (per package), $C_L$ in pF, V = ($V_{DD} - V_{SS}$) in volts, f in kHz is input frequency, and k = 0.001 × the number of exercised gates per package.

**Figure 14.11**

**B-Series CMOS Data** (Reprinted by permission of Motorola)

**c.**  $I_T$ (per package) = $N(0.3 \text{ μA/kHz})f + I_{DD}$
$$= 4(0.3 \text{ μA/kHz})(1 \text{ MHz}) + 1.0 \text{ μA}$$
$$= 1.2 \text{ mA} + 1.0 \text{ μA} = 1.201 \text{ mA}$$
$$P_d = (1.20025 \text{ mA})(5 \text{ V}) = 6.005 \text{ mW}$$

**EXAMPLE 14.11**

A circuit designer is trying to decide whether to build the circuit in Figure 14.12 with LSTTL or CMOS gates. One of the considerations is the power dissipation of the circuit over a range of operating frequencies.

She has assumed that the average duty cycle of the circuit outputs is 50%. At what operating frequency will the maximum power dissipation of a CMOS circuit exceed that of an LSTTL circuit, if both use a 5-volt supply?

**Figure 14.12**

**Example 14.11 Circuit**

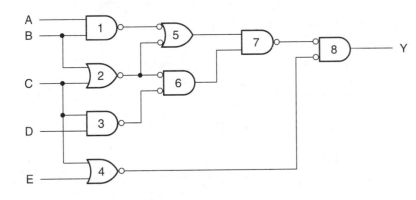

**Solution**   The circuit consists of one Quadruple 2-input NAND (74LS00 or 4011B) and one Quadruple 2-input NOR (74LS02 or 4001B).

$$\text{LSTTL: } I_{CC00} = (I_{CCH00} + I_{CCL00})/2$$

$$= (1.6 \text{ mA} + 4.4 \text{ mA})/2$$

$$= 3.0 \text{ mA}$$

$$P_{d00} = V_{CC} I_{CC} = (5 \text{ V})(3.0 \text{ mA}) = 15 \text{ mW}$$

$$I_{CC02} = (I_{CCH02} + I_{CCL02})/2$$

$$= (3.2 \text{ mA} + 5.4 \text{ mA})/2$$

$$= 4.3 \text{ mA}$$

$$P_{d02} = V_{CC} I_{CC} = (5 \text{ V})(4.3 \text{ mA}) = 21.5 \text{ mW}$$

$$P_d = P_{d00} + P_{d02} = 15 \text{ mW} + 21.5 \text{ mW} = 36.5 \text{ mW}$$

CMOS: Since supply current is specified per gate and each gate draws the same current, the total power dissipation is given by:

$$P_d = 8V_{DD} I_T$$

$$I_T = P_d/8V_{DD} = 36.5 \text{ mW}/(8 \text{ gates} \times 5 \text{ V}) = 912.5 \ \mu\text{A}$$

$$912.5 \ \mu\text{A} = (0.3 \ \mu\text{A/kHz})f + 1.0 \ \mu\text{A}$$

Neglect the quiescent current, since it is small compared to the total current.

$$912.5 \ \mu\text{A} = (0.3 \ \mu\text{A/kHz})f$$

$$f = 912.5 \ \mu\text{A}/(0.3 \ \mu\text{A/kHz}) = 3041 \text{ kHz} = 3.04 \text{ MHz}$$

The power dissipation of the CMOS circuit exceeds that of the LSTTL frequencies greater than about 3 MHz.

## Power Dissipation in High-Speed CMOS Devices

Calculation of power dissipation in High-Speed CMOS devices is similar to that (Metal-Gate) CMOS in that there is a static and dynamic component to calcul static component calculation is the same as Metal-Gate CMOS and is given by per package. However, the dynamic component calculation accounts for internal a capacitance and is given, per gate, by:

$$(C_L + C_{PD}) V_{CC}^2 f$$

where   $C_L$   is the gate load capacitance

$C_{PD}$   is the gate internal capacitance

$V_{CC}$   is the supply voltage

$f$   is the switching frequency of the gate output

Quiescent supply current $I_{CC}$ is specified only for $V_{CC} = 6.0$ V, regardless of tual supply voltage.

**EXAMPLE 14.12**

The circuit in Figure 14.10 is constructed from 74HC00A High-Speed CMOS NAN gates. Calculate the maximum power dissipation of the circuit:

**a.** When the gate inputs are steady at the state $DCBA = 1010$

**b.** When the outputs are switching at an average frequency of 10 kHz

**c.** When the outputs are switching at an average frequency of 1 MHz

Supply voltage is 5 V. Temperature range is 25°C to −55°C.

**Solution**   Refer to the 74HC00A data sheet in Appendix A.

**a.** $P_d = V_{CC} I_{CC} = (5 \text{ V})(1 \ \mu\text{A}) = 5 \ \mu\text{W}$. This is the quiescent power dissipation of the circuit.

**b.** The 74HC00A data sheet indicates that each gate has a maximum input capacitance, $C_{in}$ of 10 pF. Assume that this value represents the load capacitance of gates 1, 2, and 3 of the circuit in Figure 14.10. Further assume that gate 4 has a load capacitance of 0. The total power dissipation of the circuit is given by:

$$P_d = 3(22 \text{ pF} + 10 \text{ pF})(5 \text{ V})^2 (0.01 \text{ MHz})$$

$$+ (22 \text{ pF})(5 \text{ V})^2 (0.01 \text{ MHz}) + 5 \ \mu\text{W}$$

$$= 3(8 \ \mu\text{W}) + 5.5 \ \mu\text{W} + 5 \ \mu\text{W}$$

$$= 34.5 \ \mu\text{W}$$

**c.** For $f = 1$ MHz, total power dissipation is given by:

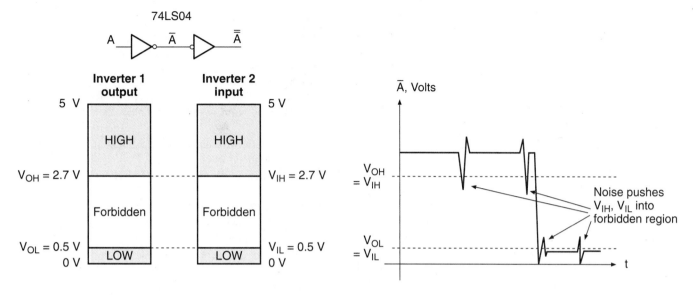

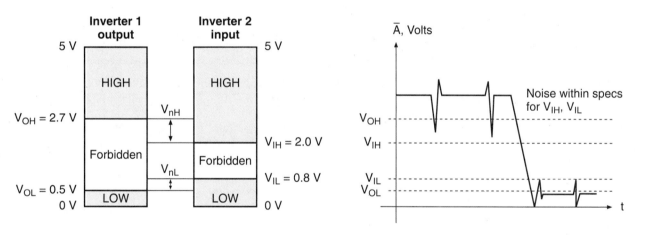

**Figure 14.13**

**Noise Margin**

b. **Nonzero noise margin**

In all circuits studied so far, we have assumed that logic HIGH is +5 volts and logic LOW is 0 volts. This is true as far as it goes, but ignores the nonideal nature of actual circuits.

In practice, there is a certain amount of tolerance on both the logic HIGH and LOW voltages; for TTL devices, a HIGH at a device input is anything above about +2 volts, and a LOW is any voltage below about +0.8 volts. Due to internal voltage drops, the HIGH output of a TTL gate is typically about +3.5 volts.

We can increase the noise margin of a device by adjusting the permissible logic HIGH and LOW levels at the device inputs and outputs.

Figure 14.13 shows one 74LS04 inverter driving another. In Figure 14.13a, the output of the first inverter and the input of the second have the same logic threshold. That is, the input of the second gate recognizes any voltage above 2.7 volts as HIGH

($V_{IH}$ = 2.7 V) and any voltage below 0.5 volts as LOW ($V_{IL}$ = 0.5 V). The output of the first inverter produces at least 2.7 volts when HIGH ($V_{OH}$ = 2.7 V) and no more than 0.5 volts as LOW ($V_{OL}$ = 0.5 V).

If there is noise on the line connecting the two gates, it will likely cause the voltage of the second gate input to penetrate into the forbidden region between logic HIGH and LOW levels. This is shown on the graph of the waveform in Figure 14.13a. When the voltage enters the forbidden region, the gate will not operate reliably. Its output may switch states when it is not supposed to.

Figure 14.13b shows the same circuit with different logic thresholds at input and output. The output of the first inverter is guaranteed to be *at least 2.7 volts* when HIGH ($V_{OH}$ = 2.7 V) and *no more than 0.5 volts* when LOW ($V_{OL}$ = 0.5 V). The second gate recognizes any input voltage *greater than 2 volts* as a HIGH ($V_{IH}$ = 2 V) and any input voltage *less than 0.8 volts* ($V_{IL}$ = 0.8 V) as a LOW.

The difference between logic thresholds allows for a small noise voltage, equal to or less than the difference, to be superimposed on the desired signal. It will not cause the input voltage of the second inverter to penetrate the forbidden region. This ensures reliable operation even in the presence of some noise.

For the 74LS04 inverter, the HIGH-state and LOW-state noise margins, $V_{nH}$ and $V_{nL}$, are:

$$V_{nH} = V_{OH} - V_{IH} = 2.7 \text{ V} - 2.0 \text{ V} = 0.7 \text{ V}$$

$$V_{nL} = V_{IL} - V_{OL} = 0.8 \text{ V} - 0.5 \text{ V} = 0.3 \text{ V}$$

---

**EXAMPLE 14.14**

Use the CMOS data sheet in Figure 14.11 to calculate the typical value of noise margin for a 4011B NAND gate for $V_{DD}$ = 5 V, 10 V, and 15 V. What percentage of the supply voltage is this noise margin?

**Solution**

$V_{DD}$ = 5 V

$$V_{nH} = V_{OH} - V_{IH}$$
$$= 5 \text{ V} - 2.75 \text{ V} = 2.25 \text{ V}$$
$$V_{nL} = V_{OL} - V_{IL}$$
$$= 2.25 \text{ V} - 0 \text{ V} = 2.25 \text{ V}$$
$$V_n/V_{DD} \times 100\% = 2.25 \text{ V/5 V} \times 100\% = 45\%$$

$V_{DD}$ = 10 V

$$V_{nH} = V_{OH} - V_{IH}$$
$$= 10 \text{ V} - 5.5 \text{ V} = 4.5 \text{ V}$$
$$V_{nL} = V_{OL} - V_{IL}$$
$$= 4.5 \text{ V} - 0 \text{ V} = 4.5 \text{ V}$$
$$V_n/V_{DD} \times 100\% = 4.5 \text{ V/10 V} \times 100\% = 45\%$$

$V_{DD}$ = 15 V

$$V_{nH} = V_{OH} - V_{IH}$$
$$= 15 \text{ V} - 8.25 \text{ V} = 6.75 \text{ V}$$
$$V_{nL} = V_{OL} - V_{IL}$$
$$= 6.75 \text{ V} - 0 \text{ V} = 6.75 \text{ V}$$
$$V_n/V_{DD} \times 100\% = 6.75 \text{ V/15 V} \times 100\% = 45\%$$

Note that the noise margins of LSTTL are in the range of several tenths of a volt. 4000B CMOS noise margins for the same supply voltage are 2.25 volts. This increased immunity to noise is one advantage that 4000B CMOS has over TTL.

---

**Section Review Problems for Section 14.5**

14.3. Given the choice between B-series CMOS and LSTTL, which should you choose for:

a. Maximum switching speed?

b. Lowest power dissipation at low speed?

c. Lowest power dissipation at high speed?

d. Highest noise immunity?

---

## 14.6

# Interfacing TTL and CMOS Gates

Interfacing different logic families is just an extension of the fanout and noise margin problems; you have to know what the load gates of a circuit require and what the driving gates can supply. In practice, this means you must know the specified values of input and output voltages and currents for the gates in question. Table 14.2 shows these specifications under worst-case conditions for a variety of TTL and CMOS families.

Table 14.2, which is derived from manufacturers' data books, can be used as a design tool whenever it is necessary to interface dissimilar logic families. Such an interfacing problem arises when an LSTTL gate drives CMOS loads or vice versa.

TTL families other than LSTTL are not generally used in conjunction with CMOS drivers, because CMOS gates do not have enough output current to drive these loads directly.

The most common interfacing problems are:

a. CMOS driving LSTTL

b. LSTTL driving a 74HCT device

c. LSTTL driving a 4000B or 74HC device

**Table 14.2** TTL and CMOS Input and Output Parameters

|  | 74 | 74S | 74LS | 74F | 74AS | 74ALS | 4000B | 74HC | 74HCT |
|---|---|---|---|---|---|---|---|---|---|
| $V_{OH}$ (V, min) | 2.4 | 2.7 | 2.7 | 2.7 | 3 | 3 | 4.95 | 3.7 | 3.7 |
| $V_{OL}$ (V, max) | 0.4 | 0.5 | 0.5 | 0.5 | 0.5 | 0.5 | 0.05 | 0.4 | 0.4 |
| $V_{IH}$ (V, min) | 2 | 2 | 2 | 2 | 2 | 2 | 3.5 | 3.15 | 2 |
| $V_{IL}$ (V, max) | 0.8 | 0.8 | 0.8 | 0.8 | 0.8 | 0.8 | 1.5 | 1.35 | 0.8 |
| $I_{OH}$ (mA, min) | −0.4 | −1 | −0.4 | −1 | −2 | −0.4 | −0.44 | * | * |
| $I_{OL}$ (mA, min) | 16 | 20 | 8 | 20 | 20 | 8 | 0.44 | * | * |
| $I_{IH}$ (mA, max) | 0.04 | 0.05 | 0.02 | 0.02 | 0.02 | 0.02 | 0.0003 | 0.0001 | 0.0001 |
| $I_{IL}$ (mA, max) | −1.6 | −2 | −0.4 | −0.6 | −0.5 | −0.1 | −0.0003 | 0.0001 | 0.0001 |

*High-Speed CMOS guaranteed to drive 10 LSTTL loads.

## CMOS Driving LSTTL

To design a CMOS-to-LSTTL interface, we examine the output parameters of the CMOS gates and the input parameters of LSTTL.

---

**EXAMPLE 14.15**

Use the information in Table 14.2 to determine the correct way to interface the output of any CMOS gate to one or more LSTTL loads. Draw the resultant circuit.

**Solution** We can extract the following data from Table 14.2.

| | **Driving Gates** | | | | **Load Gates** |
|---|---|---|---|---|---|
| | **4000B** | **74HC** | **74HCT** | | **74LS** |
| $V_{OH}$ | 4.95V | 3.7V | 3.7V | $V_{IH}$ | 2V |
| $V_{OL}$ | 0.05V | 0.4V | 0.4V | $V_{IL}$ | 0.8V |
| $I_{OH}$ | −0.44mA | * | * | $I_{IH}$ | 0.02mA |
| $I_{OL}$ | 0.44mA | * | * | $I_{IL}$ | −0.4mA |

*High-Speed CMOS guaranteed to drive 10 LSTTL loads.

1. LSTTL needs at least 2 V for a HIGH input; any CMOS output supplies more than 2 V in the HIGH state.
2. LSTTL recognizes any voltage below 0.8 V as a logic LOW; CMOS provides a LOW output of 0.05 V to 0.4 V, depending on the family.

These specifications tell us there is no voltage problem in a direct connection between a CMOS output and an LSTTL input.

We next examine the input and output currents to calculate the fanout of the configuration. Manufacturers' data sheets tell us that either High-Speed CMOS family is designed to drive 10 LSTTL loads, so no calculation is required.

For a 4000B CMOS driving gate, we calculate the fanout as follows:

**LOW State:**

$$I_{OL} = 0.44 \text{ mA (sinking)}$$

$$I_{IL} = -0.4 \text{ mA (sourcing)}$$

$$n_L = 0.44 \text{ mA/0.4 mA} = 1.1$$

**HIGH State:**

$$I_{OH} = -0.44 \text{ mA (sourcing)}$$

$$I_{IH} = 0.020 \text{ mA (sinking)}$$

$$n_H = 0.44 \text{ mA/0.020 mA} = 22$$

Fanout: $n = n_L = 1$

Any CMOS gate can drive at least one LSTTL load directly. The interfaces are shown in Figure 14.14.

**Figure 14.14**

**Example 14.15**
**CMOS Driving LSTTL**

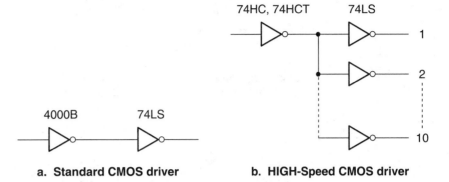

a. Standard CMOS driver          b. HIGH-Speed CMOS driver

## LSTTL Driving 74HCT CMOS

74HCT CMOS is designed to be driven directly by LSTTL. Its inputs have the same voltage specifications as LSTTL inputs and it has a very low level of input current in either the HIGH or LOW state. 74HCT logic has a large fanout when driven by an LSTTL gate. The connection is the same as in Figure 14.14b, except that the load and driving gates are reversed.

## LSTTL Driving 74HC or 4000B CMOS

Table 14.2 shows us that the input current of CMOS is so small that an LSTTL gate has no problem driving many CMOS loads. The difficulty with this interface is the required HIGH-state input voltage of the CMOS loads.

An LSTTL output is guaranteed to be at 2.7 volts in the HIGH state; a 4000B CMOS input requires 3.5 V, and a 74HC input needs 3.15 V. The required CMOS input level cannot be guaranteed in the HIGH state.

This problem can be overcome by using a pull-up resistor between the two gates, as shown in Figure 14.15.

**Figure 14.15**

**LSTTL Driving 4500B or 74HC CMOS**

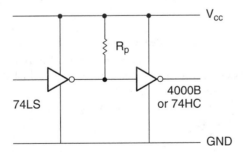

## Interfacing Devices With Different Power Supplies

The previous interfacing examples assume that both load and driving gates have the same power supply. This is not necessarily the case in practical designs.

For example, automotive electronic circuits are commonly powered by a +12-V supply, since this voltage is readily available from the car's electrical system. Metal-

Gate CMOS, with its high noise immunity and low power consumption, is a natural choice for such a logic circuit.

It might be necessary to include a few LSTTL devices when a relatively high output current or a faster switching speed is required for some part of the circuit. A TTL circuit needs a supply voltage of 5 V $\pm$ 0.5 V. If High-Speed CMOS is part of the design, it must have a supply voltage between +3 V and +6 V.

**EXAMPLE 14.16**

Give one or more reasons why you would choose to operate a mixed CMOS/LSTTL circuit from two power supplies (+12 V and +5 V) instead of a single +5-V supply.

**Solution** Examine the following data extracted from the B-series CMOS data sheet (Figure 14.11):

| T = 25°C | $V_{OL}$ (max) | $V_{OH}$ (min) | $V_{IL}$ (max) | $V_{IH}$ (min) |
|---|---|---|---|---|
| $V_{DD} = 5V$ | 0.05V | 4.95V | 1.5V | 3.5V |
| $V_{DD} = 10V$ | 0.05V | 9.95V | 3.0V | 7.0V |
| $V_{DD} = 15V$ | 0.05V | 14.95V | 4.0V | 11.0V |

From these data, we can calculate the HIGH- and LOW-state noise margins of a CMOS device at various power supply levels. (Refer to Example 14.14.)

| | $V_{nL}$ (min) | $V_{nH}$ (min) |
|---|---|---|
| $V_{DD} = 5V$ | 1.45V | 1.45V |
| $V_{DD} = 10V$ | 2.95V | 2.95V |
| $V_{DD} = 15V$ | 3.95V | 3.95V |

Noise immunity of a CMOS circuit increases with supply voltage. Thus, it may be preferable to run the CMOS part of the logic circuit from a +12-V supply.

If we use devices requiring different power supplies in the same circuit, we need some sort of device to translate between power supply levels. We can use a transistor, a buffer, or a special level translation device. Figure 14.16 shows interfacing circuits for various logic families.

Figure 14.16a shows an interfacing circuit between an LSTTL driver (+5-V supply) and a Metal-Gate CMOS load (+12-V supply). The transistor in Figure 14.16a isolates the two supplies and acts as a logic inverter.

Figure 14.16b and Figure 14.16c show an LSTTL and a CMOS gate driving a High-Speed CMOS load. The 4049/4050 buffer can accept input voltages ranging from −0.5 V to +15 V. An LSTTL or Metal-Gate CMOS device can drive it directly. The CMOS buffer can, in turn, drive several High-Speed CMOS loads.

Figure 14.16d shows how a special interfacing device can translate logic levels from CMOS to TTL or vice versa. The 4504B Hex Level Shifter has two supply volt-

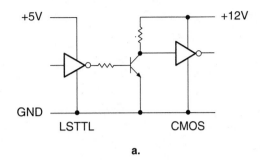

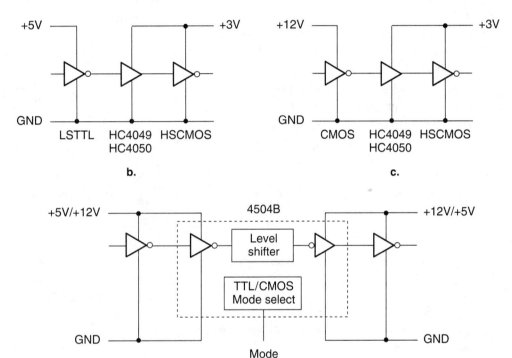

**Figure 14.16**

**Interfacing Devices With Different Power Supplies**

d.

ages and a CMOS ground. The state of the *MODE* input selects input and output logic levels. This device can be used to interface LSTTL and CMOS chips or Metal-Gate and High-Speed CMOS.

---

**Section Review Problem for Section 14.6**

14.4. A 4001B CMOS NAND driver is to be interfaced to a 74HC00A High-Speed CMOS NAND load, using a 4050 Noninverting Buffer. The Metal-Gate CMOS device has a supply voltage of +15 V. The 74HC00A gate has a power supply voltage of +4.5 V. What supply voltage should the 4050 buffer have? Why?

## GLOSSARY

**CMOS** Complementary metal-oxide semiconductor. A logic family based on metal-oxide-semiconductor field effect transistors (MOSFETs).

$C_{PD}$ Internal capacitance of a High-Speed CMOS device used to calculate its power dissipation.

**Driving gate** A gate whose output supplies current to the inputs of other gates.

**ECL** Emitter coupled logic. A high-speed logic family based on bipolar transistors.

**Fanout** The number of load gates that a logic gate output is capable of driving without possible logic errors.

$I_{CC}$ Total TTL supply current.

$I_{CCH}$ TTL supply current with all outputs HIGH.

$I_{CCL}$ TTL supply current with all outputs LOW.

$I_{DD}$ CMOS supply current under static (nonswitching) conditions.

$I_{IH}$ Current measured at a device input when the input is HIGH.

$I_{IL}$ Current measured at a device input when the input is LOW.

$I_{OH}$ Current measured at a device output when the output is HIGH.

$I_{OL}$ Current measured at a device output when the output is LOW.

$I_T$ When referring to CMOS supply current, the sum of static and dynamic supply currents.

**Load gate** A gate whose input current is supplied by the output of another gate.

**Noise** Unwanted electrical signal, often resulting from electromagnetic radiation.

**Noise margin** A measure of the ability of a logic circuit to tolerate noise.

**Power dissipation** The electrical energy used by a logic circuit in a specified period of time. Abbreviation: $P_d$

**Sinking** A terminal on a gate or flip-flop is sinking current when the current flows into the terminal.

**Sourcing** A terminal on a gate or flip-flop is sourcing current when the current flows out of the terminal.

$t_{pHL}$ Propagation delay when the device output is changing from HIGH to LOW.

$t_{pLH}$ Propagation delay when the device output is changing from LOW to HIGH.

**TTL** Transistor-transistor logic. A logic family based on bipolar transistors.

$V_{CC}$ TTL supply voltage.

$V_{DD}$ CMOS supply voltage.

$V_{IH}$ Voltage level required to make the input of a logic circuit HIGH.

$V_{IL}$ Voltage level required to make the input of a logic circuit LOW.

$V_{OH}$ Voltage measured at a device output when the output is HIGH.

$V_{OL}$ Voltage measured at a device output when the output is LOW.

## PROBLEMS

### Section 14.1 Electrical Characteristics of Logic Gates

**14.1** Briefly list the advantages and disadvantages of TTL, CMOS, and ECL logic gates.

### Section 14.2 Propagation Delay

**14.2** Explain how propagation delay is measured in TTL devices and CMOS devices. How do these measurements differ?

**14.3** Figure 14.17 shows the input and output waveforms of a logic gate. The graph shows two time scales: one for a TTL device and one for CMOS. Use the graph to calculate $t_{pHL}$ and $t_{pLH}$ for the TTL and the CMOS device.

**14.4** The inputs of the logic circuit in Figure 14.18 are in state 1 in the following table. The inputs change to state 2, then to state 3.

|         | A | B | C |
|---------|---|---|---|
| State 1 | 1 | 0 | 1 |
| State 2 | 0 | 0 | 1 |
| State 3 | 0 | 0 | 0 |

**a.** Draw a timing diagram that uses the above changes of input state to illustrate the effect of propagation delay in the circuit.

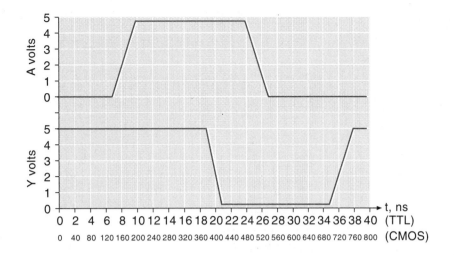

Figure 14.17

**Problem 14.3 Waveforms**

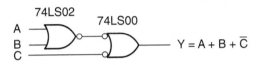

Figure 14.18

**Problems 14.4 to 14.6 Logic Circuit**

$Y = A + B + \overline{C}$

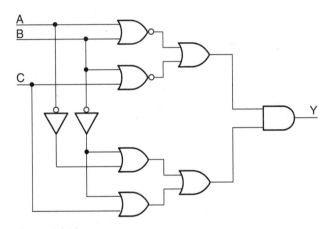

Figure 14.19

**Problem 14.7 Logic Circuit**

**b.** Calculate the maximum time it takes for the output to change when the inputs change from state 1 to state 2.

**c.** Calculate the maximum time it takes for the output to change when the inputs change from state 2 to state 3.

**14.5** Repeat the calculations of Problem 14.4, parts b and c, if the gates in Figure 14.18 are a 4001B NOR gate and a 4011B NAND gate.

**14.6** Repeat Problem 14.4, parts b and c, for a 74HC00 NAND and a 74HC02 NOR gate.

**\*14.7** Repeat Problems 14.4, 14.5, and 14.6 for the circuit shown in Figure 14.19. (Assume $V_{CC} = V_{DD} = 5$ V, T = 25°C, CL/CP device.)

|  | A | B | C |
|---|---|---|---|
| State 1 | 0 | 1 | 1 |
| State 2 | 0 | 0 | 1 |
| State 3 | 1 | 1 | 0 |

|  | LSTTL | CMOS | HSCMOS |
|---|---|---|---|
| NOR | 74LS02 | MC14001B | 74HC02A |
| INV. | 74LS04 | MC14069UB* | 74HC04A |
| AND | 74LS08 | MC14081B | 74HC08A |
| OR | 74LS32 | MC14071B | 74HC32A |

*MC14069UB:

$t_{pHL} = t_{pLH} = 125$ ns

**Figure 14.20**

**Problems 14.12 to 14.15**
**Logic Circuit**

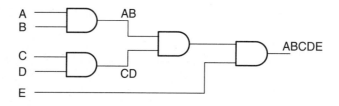

**Figure 14.21**

**Problem 14.16**
**Logic Circuit**

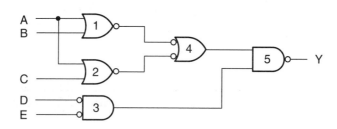

## Section 14.3 Fanout

**14.8** Calculate the maximum number of Low-Power Schottky TTL loads (74LSXX series) that a 74S86 XOR gate can drive.

**14.9** What is the maximum number of 74S32 OR gates that a 74LS00 NAND gate can drive?

**14.10** What is the maximum number of 7400 NAND gates that a 74S32 OR gate can drive?

**14.11** Calculate the maximum number of 74S32 OR gates that a 7400 NAND gate can drive.

## Section 14.4 Power Dissipation

**14.12** The circuit in Figure 14.20 is constructed from the gates of a 74LS08 AND device. Calculate the power dissipation of the circuit for the following input logic levels:

|    | $A$ | $B$ | $C$ | $D$ | $E$ |
|----|-----|-----|-----|-----|-----|
| a. | 0   | 0   | 0   | 0   | 0   |
| b. | 1   | 1   | 0   | 1   | 1   |
| c. | 1   | 1   | 1   | 1   | 0   |
| d. | 1   | 1   | 1   | 1   | 1   |

**14.13** The gate outputs in Figure 14.20 are switching at an average frequency of 100 kHz, with an average duty cycle of 60%. Calculate the power dissipation if the gates are all 74S08 AND gates.

**14.14** The gates in Figure 14.20 are 4081B CMOS gates.

**a.** Calculate the power dissipation of the circuit if the input state is $ABCDE = 010101$.

**b.** Calculate the circuit power dissipation if the outputs are switching at a frequency of 10 kHz, 50% duty cycle.

**c.** Repeat part b for a frequency of 2 MHz.

**14.15** Repeat Problem 14.14 for a circuit consisting of 74HC08A High-Speed CMOS AND gates.

**\*14.16** The circuit in Figure 14.21 consists of two 74LS00 NAND gates (gates 4 and 5) and three 74LS02 NOR gates (gates 1, 2, and 3). When this circuit is actually built, there will be two unused NAND gates and one unused NOR gate in the device packages.

　　Calculate the maximum total power dissipation of the circuit when its input state is $ABCDE = 01100$. Include all unused gates. (Connect unused gate inputs so that they will dissipate the least amount of power.)

## Section 14.5 Noise Margin

**14.17** Calculate the maximum noise margins, in both HIGH and LOW states, of:

**a.** A 4000B CMOS series gate ($V_{DD} = 5$ V)

**b.** A 4000B CMOS series gate ($V_{DD} = 15$ V)

**c.** A 74S00 NAND gate

**d.** A 74LS00 NAND gate

**e.** A 74AS00 NAND gate

**f.** A 74ALS00 NAND gate

**g.** A 74HC00 NAND gate ($V_{CC} = 5$ V)

**h.** A 74HCT00 NAND gate ($V_{CC} = 5$ V)

### Section 14.6 Interfacing TTL and CMOS Gates

Refer to Table 14.2 to answer the problems in this section.

**14.18** Briefly explain why a 4000B CMOS gate cannot drive a 74S load. Show calculations to support your argument.

**14.19** Calculate the fanout of a 4000B series CMOS gate when it drives an LSTTL circuit.

**\*14.20** Draw a circuit that would enable a 4000B CMOS

gate to drive 20 LSTTL gates without inverting the original CMOS logic level. Prove that the fanout is 20 in the new circuit.

**14.21** Why can an LSTTL gate drive a 74HCT gate directly, but not a 4000B or 74HC? Show calculations.

**14.22** Draw a circuit that allows an LSTTL gate to drive a 4000B or 74HC gate. Explain briefly how it works.

**14.23** Sketch a circuit showing how to interface a 4000B driver (+10-V supply) to a 74HC load (+6-V supply), using a 4504B level shifter.

**14.24** Repeat Problem 14.23 for a 74HC driver and a 4000B load (same supply voltages for each gate).

## ▼ *Answers to Section Review Questions*

### Section 14.2

**14.1.** $t_{p1} = t_{p2} = t_{p3} = t_{p5} = t_{p6} = 46$ ns, $t_{p4} = 30$ ns. (Note that some of these are faster than the LSTTL values calculated in Example 14.3.)

### Section 14.3

**14.2.** Source currents: $I_{OH}$, $I_{IL}$
Sink currents: $I_{OL}$, $I_{IH}$

### Section 14.5

**14.3.** a. TTL; b. CMOS; c. TTL; d. CMOS.

### Section 14.6

**14.4.** +4.5 V. The interface buffer should have the same supply as the load gates so that the buffer output and load input voltages are compatible.

# Logic Gate Circuitry

● CHAPTER OUTLINE

**15.1** Internal Circuitry of TTL Gates
**15.2** Internal Circuitry of NMOS and CMOS Gates

**15.3** TTL and CMOS Variations
**15.4** Other Logic Families

● CHAPTER OBJECTIVES

Upon successful completion of this chapter, you will be able to:

- Explain how a bipolar junction transistor can be used as a logic inverter.

- Illustrate the operation of TTL open-collector inverter and NAND gates.

- Describe the function of a TTL input transistor in all possible input states: HIGH, LOW, and open-circuit.

- Write the Boolean expression of a wired-AND circuit.

- Design a circuit that uses an open-collector gate to drive a high-current load.

- Calculate the value of a pull-up resistor at the output of an open-collector gate.

- Explain the difference between open-collector and totem pole outputs of a TTL gate.

- Explain the operation of a totem pole output.

- Illustrate how a totem pole output generates power line noise and describe how to remedy this problem.

- Illustrate why totem pole outputs cannot be tied together.

- Explain the operation of a tristate gate and name several of its advantages.

- Briefly explain the circuits of TTL NOR, AND, and OR gates.

- Describe the basic structure of a MOSFET and state its bias voltage requirements.

- Draw the circuit of an NMOS inverter and show how it works.

- Draw the circuit of a CMOS inverter and show how it works.

- Compare CMOS and NMOS and state the advantages of each.

- Draw the circuits of CMOS NAND, NOR, AND, and OR gates and explain the operation of each.

- Design a circuit using a CMOS transmission gate to enable and inhibit digital and analog signals.

- Interpret TTL data sheets to distinguish between the various TTL families.

- Describe the use of the Schottky barrier diode in TTL gates.

- Calculate speed-power product from data sheets.

- Compare Metal-Gate and High-Speed CMOS.

- State some basic facts about emitter coupled logic and describe some of its advantages.

- Describe BiCMOS logic.

Digital logic is based on transistor switching. Two major types of transistors, the bipolar junction transistor and the metal-oxide-semiconductor field effect transistor (MOSFET), form the basis of the major logic families in use today. Transistor-transistor logic (TTL) and emitter coupled logic (ECL) are based on the bipolar junction transistor. Complementary metal-oxide-semiconductor (CMOS) logic is based on the MOSFET.

In Chapter 14, we examined logic gates purely from a black box point of view. In this chapter, we will develop a fuller understanding of these devices as electronic circuits. We will begin with a brief study of bipolar transistor and MOSFET operating conditions. We will then see how these devices give rise to the electrical characteristics of simple logic gates.

We will compare the circuits and operating characteristics of various TTL and CMOS families. Finally, we will examine several specialized devices and logic families other than TTL and CMOS.

## 15.1

## Internal Circuitry of TTL Gates

> **Cutoff mode** *The operating mode of a bipolar transistor when there is no collector current flowing and the path from collector to emitter is effectively an open circuit. In a digital application, a transistor in cutoff mode is considered OFF.*
>
> **Saturation mode** *The operating mode of a bipolar transistor when an increase in base current will not cause a further increase in the collector current and the path from collector to emitter is very nearly (but not quite) a short circuit. This is the ON state of a transistor in a digital circuit.*

TTL has been around for a long time. The first transistor-transistor logic ICs were developed by Texas Instruments around 1965. Since then, there have been many improvements in the speed and power consumption of these devices, but the basic logic principles remain largely unchanged. Even though they are seldom used in modern designs, it makes sense to examine the internal circuitry of standard TTL gates such as the 7400 NAND, 7402 NOR, and 7404 inverter because the internal logic concepts are similar to the more advanced types of TTL.

The most important parts of the circuit, as far as a designer or technician is concerned, are the input and output circuits, because they are the only parts of the chip to which we have access. It is to these points that we interface other circuits and where we make diagnostic measurements. A basic understanding of the inputs and outputs of logic gate circuitry is helpful when we design or troubleshoot a digital circuit.

### *Bipolar Transistors as Logic Devices*

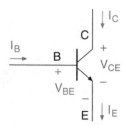

**Figure 15.1**

**Currents and Voltages in an NPN Bipolar Transistor**

The basic element of a TTL device is the bipolar junction transistor, illustrated in Figure 15.1. This is not the place to give a detailed analysis of the operation of a bipolar transistor, but a simplified summary of operating modes will be useful.

**Figure 15.2**

**NPN Bipolar Transistor as a Switch**

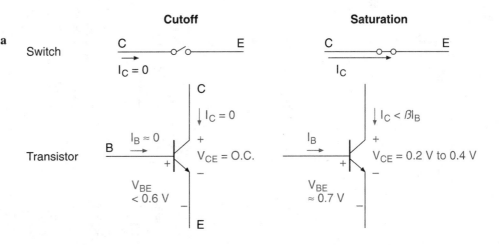

**Table 15.1**  Bipolar Transistor Characteristics

|          | Cutoff     | Active       | Saturation   |
| -------- | ---------- | ------------ | ------------ |
| $I_C$    | 0          | $= \beta I_B$ | $< \beta I_B$ |
| $V_{CE}$ | Open cct.  | $>0.8$ V     | 0.2 V–0.4 V  |
| $V_{BE}$ | $<0.6$ V   | 0.6 V–0.7 V  | $\approx 0.7$ V |

The bipolar transistor is a current amplifier having three terminals called the collector, emitter, and base. Current flowing into the base controls the amount of current flowing from the collector to the emitter. If base current is below a certain threshold, the transistor is in **cutoff mode** and no current flows in the collector. In this state, the base-emitter voltage is less than 0.6 V and the collector-emitter path acts like an open circuit. We can treat the collector-emitter path as an open switch, as shown in the lefthand diagram in Figure 15.2.

If the base current increases, the transistor enters the "active region," where the collector current is proportional to the base current by a current gain factor, $\beta$. This is the linear, or amplification, region of operation, used by analog amplifiers.

If the base current increases still further, collector current reaches a maximum value and will no longer increase with base current. This is called the **saturation mode** of the transistor. The saturated value of collector current, $I_{CS}$, is determined by (1) the resistance in the collector-emitter current path, (2) the voltage drop across the collector and emitter, $V_{CE}$, and (3) the collector supply voltage, $V_{CC}$. Base-emitter voltage is about 0.7 V and will not increase significantly with increasing base current. The voltage between collector and emitter is in the range from 0.2 V to 0.4 V. In this mode, we can treat the transistor as a closed switch, as shown in the righthand diagram of Figure 15.2.

Table 15.1 summarizes the voltages and currents in the cutoff, active, and saturation regions.

**EXAMPLE 15.1**

Figure 15.3 shows an NPN bipolar transistor connected in a common-emitter configuration. With the right choice of input voltages, this circuit acts as a digital inverter.

Analyze the circuit to show that it acts as an inverter if a logic HIGH is defined as $\geq 3$ V and a logic LOW is defined as $\leq 0.5$ V. Assume that $\beta = 100$, and assume that $V_{BE} = 0.7$ V and $V_{CE} = 0.2$ V in saturation.

**Figure 15.3**

**Example 15.1**
**Transistor as Inverter**

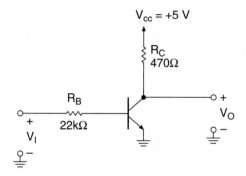

**Solution**   We will analyze the circuit with two input voltages: 3 V (logic HIGH) and 0.5 V (logic LOW). These two conditions are shown in Figure 15.4.

*HIGH input.*   We must prove that $V_I = 3$ V is sufficient to saturate the transistor. Let us assume that this is true and find out if calculations confirm our assumption.

Figure 15.4a shows the circuit with $V_I = 3$ V. By Kirchhoff's Voltage Law (KVL):

$$V_I = I_B R_B + V_{BE}, \text{ or}$$

$$I_B = (V_I - V_{BE})/R_B$$

If we assume that $I_B$ is sufficient to saturate the transistor, then:

$$I_B = (3 \text{ V} - 0.7 \text{ V})/22 \text{ k}\Omega$$

$$= 105 \text{ }\mu\text{A}$$

$$\beta I_B = (100)(105 \text{ }\mu\text{A}) = 10.5 \text{ mA}$$

Collector current won't increase beyond its saturated value, even if base current increases. Therefore, if the transistor is saturated, $\beta I_B$ will be larger than the current actually flowing in the collector-emitter path.

In saturation, the collector current can be calculated by KVL:

$$V_{CC} = I_C R_C + V_{CE}, \text{ or}$$

$$I_C = (V_{CC} - V_{CE})/R_C$$

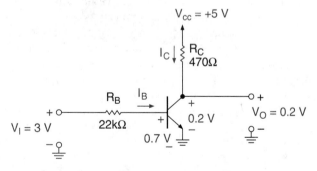

a. **HIGH input (saturation)**

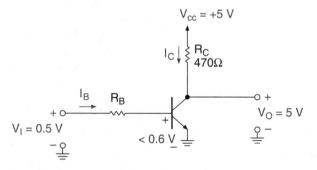

b. **LOW input (cutoff)**

**Figure 15.4**

**Example 15.1**
**Voltage and Current Analysis of Inverter**

$$I_C = (5 \text{ V} - 0.2 \text{ V})/470 \text{ }\Omega$$

$$= 10.2 \text{ mA}$$

Since $\beta I_B > I_C$, the transistor is saturated. Thus, an input voltage of 3 V will produce sufficient base current to saturate the transistor. The output is given by $V_O = V_{CE} = 0.2$ V, which is within the defined range of a logic LOW.

*LOW input.* Figure 15.4b shows the circuit with $V_I = 0.5$ V. By KVL:

$$V_I = I_B R_B + V_{BE}$$

$$V_{BE} = 0.5 \text{ V} - I_B R_B$$

Since $V_{BE}$ must be $<0.6$ V, the transistor is in cutoff mode. Thus, in the collector circuit:

$$V_{CC} = I_C R_C + V_{CE}$$

$$5 \text{ V} = (0)(470 \text{ }\Omega) + V_{CE}$$

$$V_O = V_{CE} = 5 \text{ V (logic HIGH)}$$

Table 15.2 summarizes the operation of the circuit as an inverter.

**Table 15.2**  Input and Output of Single-Transistor Inverter

| Input | | Output | |
|---|---|---|---|
| $V_I$ | Logic Level | $V_O$ | Logic Level |
| 0.5 V | LOW | 5 V | HIGH |
| 3 V | HIGH | 0.2 V | LOW |

## TTL Open-Collector Inverter and NAND Gate

**Open-collector output** *A TTL output where the collector of the LOW-state output transistor is brought out directly to the output pin. There is no built-in HIGH-state output circuitry, which allows two or more open-collector outputs to be connected without possible damage.*

Figure 15.5 shows the circuit of the simplest TTL gate: a 7405 inverter with **open-collector outputs.** This circuit performs the same function as the single-transistor inverter we examined in Example 15.1. These circuits differ most obviously in their input circuitry. The inverter circuit in Example 15.1 has a resistor as its input; the 7405 inverter has a transistor, $Q_1$, as its input. The input transistor allows faster switching of input states. This configuration is common to all standard TTL gates and will be examined in detail later in this section.

The logic function of the 7405 is performed by transistors $Q_2$ and $Q_3$. Output transistor $Q_3$ is switched ON and OFF by current flowing in the collector-emitter path of $Q_2$. When $Q_3$ is ON, $Y$ is LOW.

**Figure 15.5**

**Open-Collector Inverter (7405)**

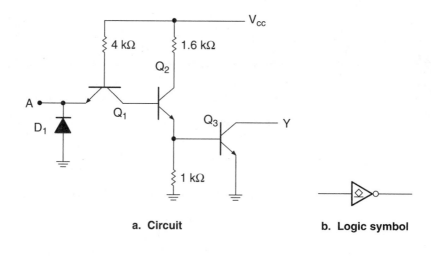

a. Circuit                    b. Logic symbol

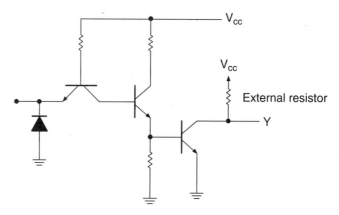

c. 7405 with HIGH-state pull-up

However, when $Q_3$ is OFF, $Y$ is floating. There is a high impedance between $Y$ and ground, so the output is not LOW. But there is no connection to $V_{CC}$ to make the output HIGH. In this condition, $Y$ is neither HIGH nor LOW.

To enable the output to produce a HIGH state, we need to add an external pull-up resistor. The value of this resistor depends on the current sinking capability of $Q_3$, specified in the data sheet as $I_{OL}$. We will do such calculations in a later example.

## TTL Inputs

Transistor $Q_1$ and diode $D_1$ make up the input circuit of the TTL inverter of Figure 15.5. The diode protects the input against small negative voltages. If the input goes more negative than about $-0.7$ V, the diode will conduct, effectively short-circuiting the input to ground plus one diode drop. This clamps the input to $-0.7$ V. $D_1$ has no logic function.

$Q_1$ can be treated as two back-to-back diodes, as shown in Figure 15.6. Figure 15.7 shows how the input responds to logic HIGH and LOW voltages.

**LOW Input.** When a TTL input is made LOW, the base-emitter junction of $Q_1$ acts as a forward-biased diode, creating a current path from $V_{CC}$ to ground via the input pin. This current makes up the majority of current $I_{IL}$, which has a maximum value of 1.6 mA in standard TTL.

**Figure 15.6**

**Diode Equivalent of TTL Input Transistor**

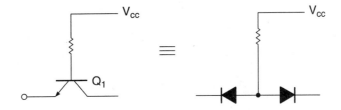

**Figure 15.7**

**HIGH and LOW Inputs at a TTL Gate**

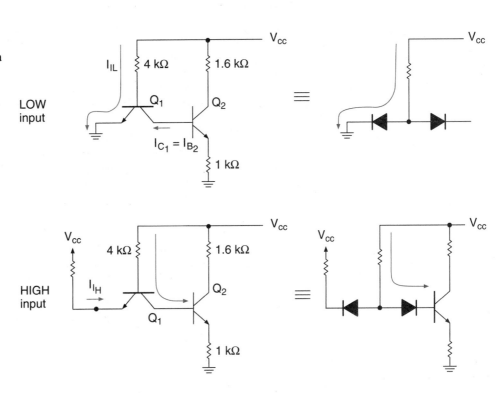

At the moment the input is made LOW, the transistor action of $Q_1$ transports charge away from the base of $Q_2$, pulling it LOW and keeping it in cutoff mode. This current dies out when the base charge of $Q_2$ has been depleted, shortly after the LOW is applied to the input pin. The diode formed by the base-collector junction of $Q_1$ does not carry sufficient current to turn on $Q_2$, since the base-emitter path is of much lower impedance.

**HIGH Input.** A HIGH at a TTL input reverse-biases the base-emitter junction of $Q_1$. Only a small leakage current, $I_{IH}$, flows. The maximum value of $I_{IH}$ is 40 μA.

Since the low-impedance current path to the input pin has not been established, current flows to the base of $Q_2$ via the forward-biased base-collector junction of $Q_1$. This current is sufficient to saturate $Q_2$.

**Open (Floating) TTL Input.** An open-circuit TTL input acts as a logic HIGH, as illustrated by Figure 15.8. A TTL input relies on a logic LOW to establish a low-impedance current path from $V_{CC}$ to the input pin. If the input is open, this LOW is not present and current flows in the base-collector junction of the transistor, by default. This is the same current that flows under the HIGH-input condition.

**Figure 15.8**

**LOW, HIGH, and Open TTL Inputs**

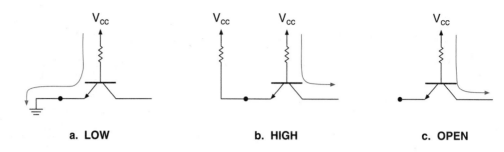

a. LOW        b. HIGH        c. OPEN

This HIGH is not stable; it can be converted to logic LOW by induced noise at the input pin. To avoid this uncertainty, an unused input should always be wired to a logic HIGH or LOW state.

## TTL Open-Collector Inverter

Figure 15.9 shows the operation of the 7405 open-collector inverter.

**LOW Input.** As was described above, a LOW input establishes a low-impedance path to ground, which draws current through the base-emitter junction of $Q_1$. This action also prevents base current from flowing in transistor $Q_2$, causing it to be in cutoff mode and making $I_{C2} = 0$.

Since $I_{B3}$ is derived from $I_{C2}$, $I_{B3} = 0$ and $Q_3$ is cut off, making a high-impedance path between the collector and emitter of $Q_3$. As was the case with the single-transistor inverter in Example 15.1, when $I_{C3} = 0$, then $V_O = V_{CE} = V_{CC}$. (Since no current flows through the pull-up resistor, the voltage must be the same at both ends.) *Output Y is HIGH.*

**HIGH Input.** When input $A$ is HIGH, the base-emitter junction of $Q_1$ does not have sufficient voltage across it to be forward-biased. Current flows through the base-collector junction of $Q_1$, saturating $Q_2$.

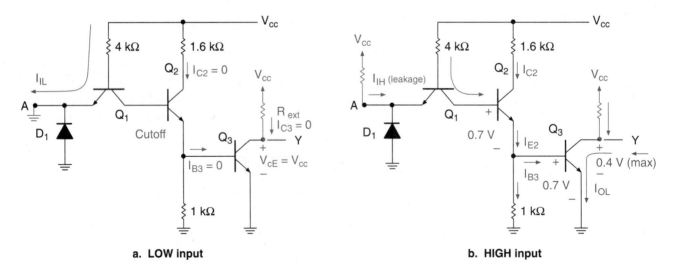

a. LOW input        b. HIGH input

**Figure 15.9**

**7405 Operation**

Since $Q_2$ is ON, current flows to the $Q_2$ emitter and splits through the 1-k$\Omega$ resistor and the base of $Q_3$. The output transistor, $Q_3$, turns ON, establishing a low-impedance current path from output $Y$ to ground. Current is limited by the external pull-up resistor, which must be chosen to keep $I_{OL}$ at or under its rated value of 16 mA. $V_{CE3}$ is about 0.2 V to 0.4 V. *Output Y is LOW.*

### TTL Open-Collector NAND

Figure 15.10 shows one gate of a 7401 Quadruple 2-input NAND gate with open-collector outputs. The circuit is the same as that of the 7405 inverter, except that the input transistor has a second emitter. Multiple-emitter transistors of this type are common in TTL circuits and can be modeled by the diode equivalent in Figure 15.10b. Figure 15.11 shows the response of the multiple-emitter input transistor to various combinations of logic levels.

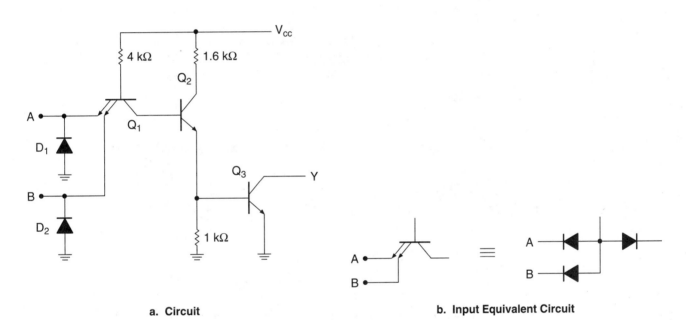

a. Circuit                                   b. Input Equivalent Circuit

**Figure 15.10**

**TTL NAND With Open Collector Output**

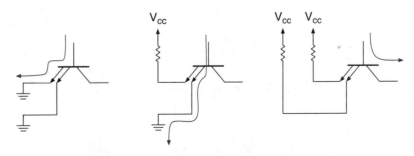

**Figure 15.11**

**Input Response of Multiple-Emitter Transistor**

a. **Both inputs LOW (LOW equivalent)**    b. **One input LOW (LOW equivalent)**    c. **Both inputs HIGH (HIGH equivalent)**

*If both inputs are LOW,* the NAND acts exactly the same as the 7405 inverter with a LOW input. (A low-impedance path is created through a base-emitter junction.) *Output Y is HIGH,* provided an external pull-up resistor is connected to output. A partial truth table for this condition is:

| A | B | Y |
|---|---|---|
| 0 | 0 | 1 |

*If one input is LOW,* the input acts the same as the inverter with a LOW input. The low-impedance current path through the one grounded emitter prevents sufficient base-collector current from flowing to forward-bias that junction. *Output Y is HIGH* if a pull-up resistor is connected to the output. A partial truth table is as follows:

| A | B | Y |
|---|---|---|
| 0 | 1 | 1 |
| 1 | 0 | 1 |

*If both inputs are HIGH,* the NAND circuit acts like the 7405 when its input is HIGH. (There is no base-emitter current path. A collector-emitter path is established by default.) *Output Y is LOW.* This condition can be represented by:

| A | B | Y |
|---|---|---|
| 1 | 1 | 0 |

Combining all these conditions, we get the standard NAND truth table:

| A | B | Y |
|---|---|---|
| 0 | 0 | 1 |
| 0 | 1 | 1 |
| 1 | 0 | 1 |
| 1 | 1 | 0 |

---

If one or more emitters of a TTL multiple-emitter input transistor is LOW, the input is a LOW equivalent. All emitters must be HIGH to make the transistor input a HIGH equivalent.

---

These statements lead to the familiar NAND-gate descriptive sentences, illustrated by the gate symbols in Figure 15.12:

a. At least one input LOW makes the output HIGH.

b. Both inputs HIGH make the output LOW.

---

**Section Review Problem for Section 15.1a**

15.1. What are the two main functions of the pull-up resistor on the output of an open-collector gate?

**Figure 15.12**

**DeMorgan Equivalent Forms of a NAND Gate**

| A | B | Y |
|---|---|---|
| 0 | 0 | 1 |
| 0 | 1 | 1 |
| 1 | 0 | 1 |

**a. At least one input LOW makes output HIGH**

| A | B | Y |
|---|---|---|
| 1 | 1 | 0 |

**b. Both inputs HIGH make output LOW**

## Open-Collector Applications

**Wired-AND** *A connection where open-collector outputs of logic gates are wired together. The logical effect is the ANDing of connected functions.*

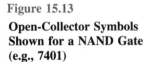

**Figure 15.13**

**Open-Collector Symbols Shown for a NAND Gate (e.g., 7401)**

A more common TTL output than the open collector is the totem pole output, which we will study later in this chapter. The totem pole output has its own internal pull-up circuit for HIGH outputs.

Gates with totem pole outputs cannot be used in all digital circuits. For example, open-collector gates are required when several outputs must be tied together, a connection called **wired-AND.** Totem pole outputs would be damaged by such a connection, since there is the possibility of conflict between an output HIGH and LOW state.

Open-collector outputs can also be used for applications requiring high current drive and for interfacing to circuits having supply voltages other than TTL levels.

A special symbol defined by IEEE/ANSI Standard 91-1984, an underlined square diamond, is shown in Figure 15.13. This symbol is added to a logic gate symbol to indicate that it has an open-collector output. Other symbols, such as a star (*), a dot (•), or the initials OC are also used.

### Wired-AND

A wired-AND connection combines the *outputs* of the connected gates in an AND function.

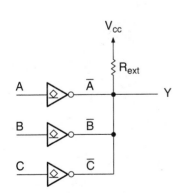

**Figure 15.14**

**Three Inverters in a Wired-AND Connection**

Figure 15.14 shows three open-collector inverters connected in a wired-AND configuration. The output transistors of the inverters are shown in Figure 15.15, with different possible ON and OFF states. The only way output $Y$ can remain HIGH is if all the transistors are in their OFF states, as in Figure 15.15c. This can happen only if the outputs of the inverters are all HIGH. This is the same as saying the outputs are ANDed together at $Y$.

The Boolean expression for $Y$ is:

$$Y = \overline{A} \cdot \overline{B} \cdot \overline{C}$$
$$= \overline{A + B + C}$$

By DeMorgan's Theorem, the wired-AND connection of inverter outputs is equivalent to a NOR function. Because of this DeMorgan equivalence, the connection is sometimes, not very accurately, called "wired-OR."

**Figure 15.15**
**Output Transistors of**
**Open-Collector Inverters in**
**a Wired-AND Connection**

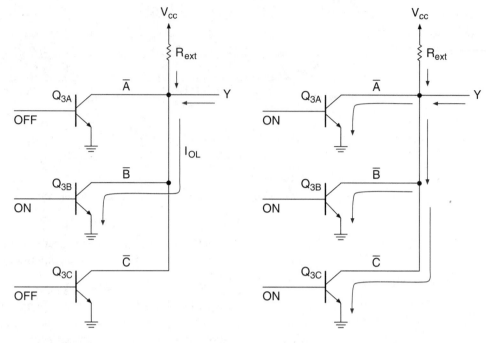

a.  **One inverter with LOW output**         b.  **All inverter outputs LOW**

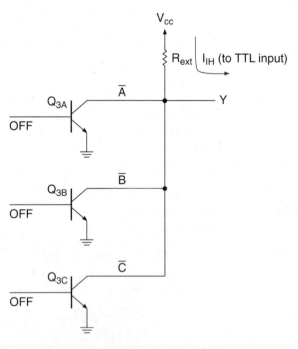

c.  **All inverter outputs HIGH**

**Figure 15.16**

**NAND Gates in Wired-AND Connection**

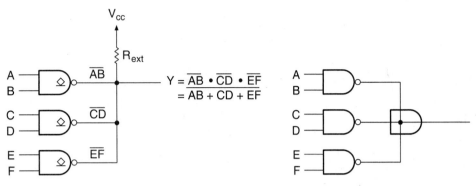

a. **Pull-up resistor and open-collector gates**          b. **Wired-AND symbol**

Figure 15.16 shows three NAND gates in a wired-AND connection. Since the output functions are ANDed, the Boolean expression for $Y$ is:

$$Y = \overline{AB} \cdot \overline{CD} \cdot \overline{EF}$$

$$= \overline{AB + CD + EF}$$

The resulting function is called AND-OR-INVERT. Normally this requires at least two types of logic gate—AND and NOR. The wired-AND configuration can synthesize any size of AND-OR-INVERT network using only NAND gates.

The wired-AND function is sometimes shown as an AND symbol around a soldered connection, as shown in Figure 15.16b.

---

**EXAMPLE 15.2**

Recall from Section 10.3 (Magnitude Comparators) that two $n$-bit numbers can be compared for equality by using an Exclusive NOR gate for each pair of bits and ANDing the output results. Thus, for two 4-bit numbers, the $A = B$ comparison function is given by:

$$(A = B) = (\overline{A_1 \oplus B_1})(\overline{A_2 \oplus B_2})(\overline{A_3 \oplus B_3})(\overline{A_4 \oplus B_4})$$

The 74LS266 Quadruple 2-input Exclusive NOR gate with open-collector outputs can be used to synthesize this logic function. Show how you would connect the gates of a 74LS266 chip to make the comparison function specified above. Verify the operation of the circuit with the following pairs of inputs:

    **a.** $A_4 A_3 A_2 A_1 = 1010$     $B_4 B_3 B_2 B_1 = 1010$     $(A = B)$

    **b.** $A_4 A_3 A_2 A_1 = 1110$     $B_4 B_3 B_2 B_1 = 1011$     $(A \neq B)$

**Solution**  Since the equality function is the AND product of four XNOR functions, the outputs of the gates in the 74LS266 should be connected as a wired-AND circuit. As we will see in a later section, this is possible only because the '266 outputs are open-collector. Figure 15.17 shows the required circuit. Figure 15.18 shows the input and output levels of the XNOR gates with the inputs specified.

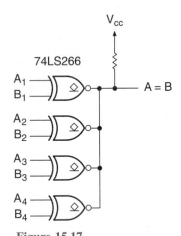

**Figure 15.17**

**Example 15.2**
**4-Bit Comparator Using
XNOR Gates in a
Wired-AND Connection**

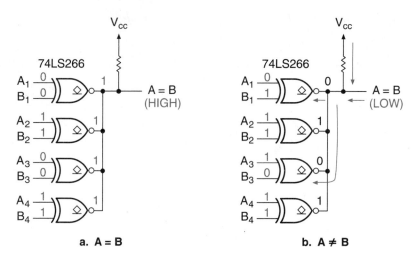

a. A = B

b. A ≠ B

**Figure 15.18**

**Example 15.2**
**Input and Output Levels of XNOR Gates**

For reference, the XNOR truth table is:

| A | B | $\overline{A \oplus B}$ |
|---|---|---|
| 0 | 0 | 1 |
| 0 | 1 | 0 |
| 1 | 0 | 0 |
| 1 | 1 | 1 |

Figure 15.18a shows that when all bits of *A* and *B* are the same, the gates all have HIGH outputs, giving a HIGH indication for equality. Figure 15.18b shows that if some bits of *A* and *B* are not equal, the corresponding gate output goes LOW, indicating inequality.

## High-Current Driver

Standard TTL outputs have higher current ratings in the LOW state than in the HIGH state. Thus, open-collector outputs are useful for driving loads that need more current than a standard TTL output can provide in the HIGH state. There are special TTL gates with higher ratings of $I_{OL}$ to allow even larger loads to be driven. Typical loads would be LEDs, incandescent lamps, and relay coils, all of which require currents in the tens of milliamperes.

**EXAMPLE 15.3**

A 7406 Hex Inverter Buffer/Driver contains six inverters, much like the 7404-type TTL inverters, except that the outputs are open-collector, rated for $I_{OLmax}$ of 40 mA and $V_{OHmax}$ of 30 V.

Draw a circuit showing how four of the six inverters in a 7406 chip can be used to turn on an interior light in an automobile when one or more of the doors is open.

**Solution**

Figure 15.19 shows a possible circuit for this application. The door switches and inverters run off the standard TTL voltage of +5 V. The lamp operates on the car's +12-V battery. (+12 V is a nominal value; the actual voltage is about +13.8 V.)

**Figure 15.19**

**Example 15.3**
**7406 as Lamp Driver**

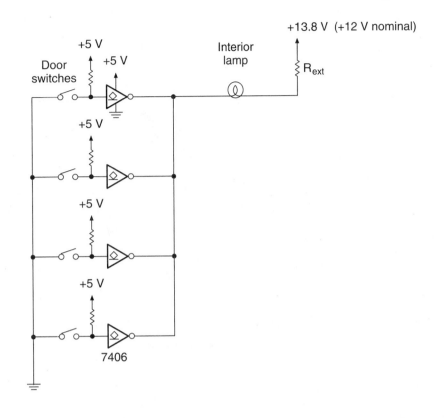

When a door opens, it pulls the input of the corresponding inverter HIGH, turning on its LOW output transistor. This completes a current path from +13.8 V to ground through the lamp, resistor, and output transistor, turning on the lamp. The external limiting resistor may not be necessary, depending on the resistance of the lamp filament. $I_{OL}$ must be limited to 40 mA or the current rating of the lamp, whichever is less.

Note that the lamp voltage need not be the same value as the TTL supply voltage, as long as it is within the specification for $V_{OH}$.

## Value of External Pull-up Resistor

The value of the pull-up resistor required by an open-collector circuit is calculated using manufacturer's specifications and the basic principles of circuit theory: Kirchhoff's Voltage and Current Laws (KVL and KCL) and Ohm's Law.

Figure 15.20 shows the circuit model for calculating the value of $R_{ext}$. It accounts for the current requirements of the loads, the LOW-state output voltage, and current-sinking capacity of the open-collector gate.

> The main rule in resistor selection is to keep the sum of currents into the open-collector output to less than $I_{OL}$.
>
> $$I_{OL} = I_R + nI_{IL}$$
> $$I_R = (V_{CC} - V_{OL})/R_{ext}$$

**Figure 15.20**

**Circuit Model for Pull-up
Resistor Calculation**

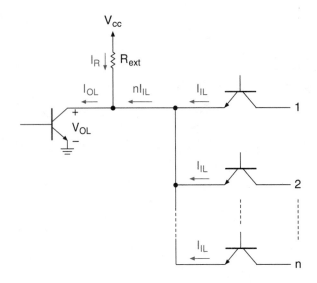

**EXAMPLE 15.4**

Calculate the minimum value of the pull-up resistor for the 4-bit comparator in Figure 15.17 if the circuit drives ten 74LS00 NAND gate inputs.

**Solution**

From 74LS00 specs:         $I_{IL} = 0.4$ mA

For 10 gates:               $nI_{IL} = 10I_{IL} = 4$ mA

From 74LS266 specs:        $I_{OL} = 8$ mA

$$I_R = I_{OL} - nI_{IL}$$

$$= 8 \text{ mA} - 4 \text{ mA}$$

$$= 4 \text{ mA}$$

For $I_{OL} = 8$ mA, $V_{OL} = 0.35$ V (typ.).

$$R_{ext} = (V_{CC} - V_{OL})/I_R$$

$$= (5 \text{ V} - 0.35 \text{ V})/4 \text{ mA}$$

$$= 4.65 \text{ V}/4 \text{ mA} = 1.16 \text{ k}\Omega$$

Use a 1.2-k$\Omega$ or 1.5-k$\Omega$ standard value resistor.

---

**Section Review Problem for Section 15.1b**

15.2. Calculate the minimum value of pull-up resistor required for the 4-bit comparator circuit of Figure 15.17 if it drives one input of a 74LS00 NAND gate. What is the minimum standard value of this resistor?

## Totem Pole Outputs

> **Totem pole output** *A type of TTL output with a HIGH and a LOW output transistor, only one of which is active at any time.*
>
> **Phase splitter** *A transistor in a TTL circuit that ensures that the LOW- and HIGH-state output transistors of a totem pole output are always in opposite phase (i.e., one ON, one OFF).*

Figure 15.21 shows one gate of a 7400 Quadruple 2-input NAND with **totem pole outputs.** The circuit is the same as that for a 7401 open-collector NAND except for a transistor, resistor, and diode, which make up the HIGH-state output circuitry of the NAND gate.

The totem pole output, shown in Figure 15.21b, has separate transistors to switch the output to the HIGH state ($Q_4$) and the LOW state ($Q_3$). These transistors are switched by $Q_2$, the **phase splitter.** Only one of them is ON at a time; the currents $I_{OH}$ and $I_{OL}$ never flow simultaneously.

The portion of the circuit consisting of $Q_4$, $D_3$, and the 130-$\Omega$ resistor replaces the external pull-up resistor required by the open-collector TTL output. Since the HIGH state is switched by its own transistor, we say that the circuit has an active pull-up.

The main advantage of the totem pole output over the open collector is that it can change states faster. The external pull-up resistance needed in an open-collector circuit slows down the output switching by contributing to the *RC* time constant of the output.

**Figure 15.21**

**NAND Gate With Totem Pole Output**

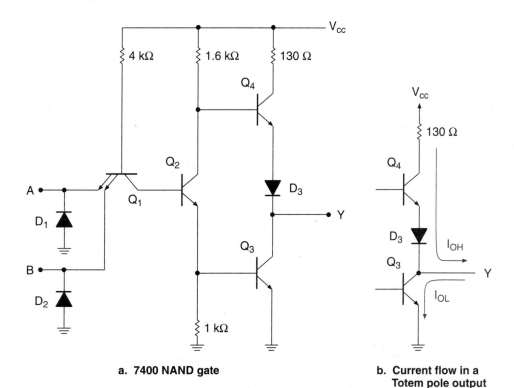

a. 7400 NAND gate

b. Current flow in a
   Totem pole output

**Figure 15.22**

**NAND Gate Operation**

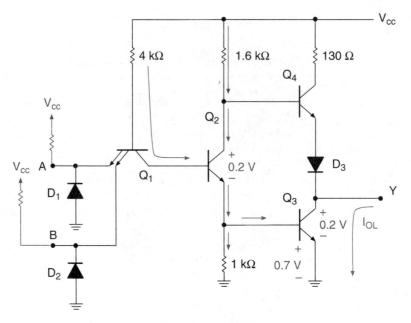

a. Both inputs HIGH

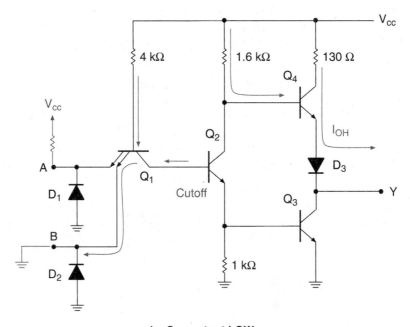

b. One output LOW

The HIGH-state transistor circuit, with its relatively low output impedance, reduces this time constant and thus improves switching speed.

Figure 15.22 shows the operation of the 7400 NAND gate for HIGH and LOW input conditions.

**HIGH Input.** When both inputs are HIGH, there is no low-impedance base-emitter current path in $Q_1$. The base-collector junction of $Q_1$ acts as a forward-biased diode. Base current flows in $Q_2$, saturating the transistor. Sufficient current flows to $Q_3$ to saturate it. $Y$ is connected to ground, via the collector-emitter path of $Q_3$. *The output is LOW.*

**LOW Input.** Figure 15.22b shows input $B$ of a 7400 NAND gate pulled LOW. The circuit operates the same way if $A$ or both $A$ and $B$ are LOW.

In this condition, a low-impedance path to ground is established through one of the base-emitter junctions of $Q_1$. This pulls the base of $Q_2$ LOW, causing it to be in cutoff mode. No current flows through the collector-emitter path of $Q_2$, so no base current flows in $Q_3$; it is also cut off.

Current flows through the 1.6-k$\Omega$ resistor to the base of $Q_4$, turning it ON. This connects the output, via $Q_4$, $D_3$, and the 130-$\Omega$ resistor, to $V_{CC}$. *The output is HIGH.*

$Q_4$ will not turn ON when $Q_3$ is ON. We can find out why by calculating $V_{BE4} + V_{D3}$. For $Q_4$ to conduct, two pn junctions ($D_3$ and the base-emitter junction of $Q_4$) must be forward-biased. Thus, $(V_{BE4} + V_{D3})$ must be greater than 0.6 V + 0.6 V = 1.2 V.

$$V_{BE4} + V_{D3} = V_{B4} - V_{CE3}$$

We can calculate $V_{B4}$ by adding up voltage drops, as follows:

$$V_{B4} = V_{CE2} + V_{BE3} = 0.2 \text{ V} + 0.7 \text{ V} = 0.9 \text{ V}$$

$Q_3$ is saturated, thus:

$$V_{CE3} = 0.2 \text{ V}$$

The difference between these voltages is:

$$V_{B4} - V_{CE3} = 0.9 \text{ V} - 0.2 \text{ V} = 0.7 \text{ V}$$

This is insufficient to forward-bias $BE_4$ and $D_3$. $Q_4$ stays OFF.

---

Without $D_3$ in the circuit,

$$V_{BE4} = (V_{CE2} + V_{BE3}) - V_{CE3}$$
$$= (0.2 \text{ V} + 0.7 \text{ V}) - 0.2 \text{ V}$$
$$= 0.7 \text{ V}$$

This is sufficient to saturate $Q_4$, even when $Q_3$ is ON. The diode is therefore necessary to keep $Q_4$ OFF when $Q_3$ is ON.

---

## Switching Noise

**Storage time** *Time required to transport stored charge away from the base region of a bipolar transistor before it can turn off.*

A totem pole output is an inherently noisy circuit. Noise is generated on the supply voltage line when the output switches from LOW to HIGH.

**Figure 15.23**

**Spikes on Power Line
During LOW-to-HIGH
Transition of Totem Pole
Output**

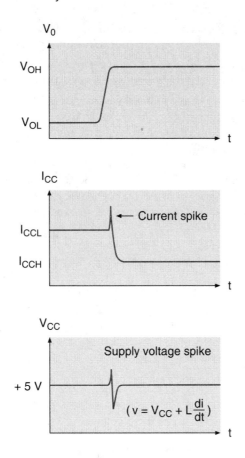

When the output is in a steady HIGH or LOW state, $Q_3$ and $Q_4$ are always in opposite phase. The design of the totem pole output is such that when $Q_3$ is ON, it is saturated, but when $Q_4$ is ON, it operates in the transistor's active, or linear, region.

A saturated transistor takes longer to shut off than an unsaturated one due to **storage time,** the time required to transport stored charge away from the base region of the transistor. Thus, $Q_3$ takes longer to turn off than $Q_4$.

When a totem pole output is LOW, $Q_3$ is ON and $Q_4$ is OFF. When the output changes state, $Q_4$ turns ON before $Q_3$ can turn OFF, due to the storage time of $Q_3$. For a few nanoseconds, both transistors are ON. This condition momentarily shorts $V_{CC}$ to ground, causing a surge of supply current, as shown in Figure 15.23.

The inductance of the power line produces a corresponding spike proportional to the instantaneous rate of change of the supply current ($v = L\,di/dt$, where $L$ is the power line inductance and $di/dt$ is the instantaneous rate of change of supply current).

These spikes on the supply voltage line can cause real problems, especially in synchronous circuits. They often cause erroneous switching that is nearly impossible to troubleshoot. The best cure for such problems is prevention.

Figure 15.24 shows the addition of a decoupling capacitor to a totem pole output to eliminate switching spikes. A low-inductance ceramic disc capacitor of about 0.1 μF is placed between the $V_{CC}$ and ground pins of the chip to be decoupled. This capacitor offsets the power line inductance and acts as a low-impedance path to ground for high-frequency noise (i.e., spikes). Since a capacitor is an open circuit for low frequencies, the normal DC supply voltage is not shorted out.

**Figure 15.24**

**Decoupling the Power Supply**

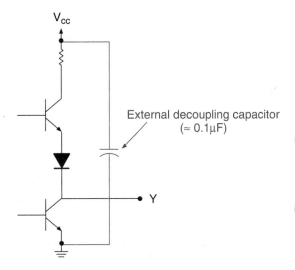

It is important that the capacitor be placed *physically close* to the decoupled chip. Inductance of the power line accumulates with distance, and if the capacitor is far away from the chip (say, at the end of the circuit board), the decoupling effect of the capacitor is lost.

It is not necessary to decouple every chip on a circuit board for designs operating at relatively low frequencies ($\leq 1$ MHz). In such cases, one capacitor for every two ICs is enough. The capacitor should be connected between $V_{CC}$ and ground *of the same chip*, as shown in Figure 15.25.

**Figure 15.25**

**Placement of Decoupling Capacitor (Low-Frequency Designs)**

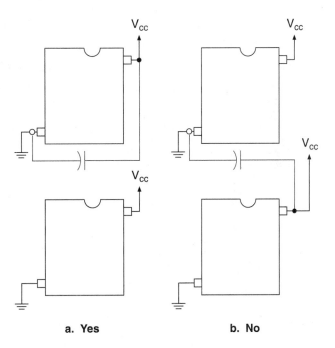

a. Yes                    b. No

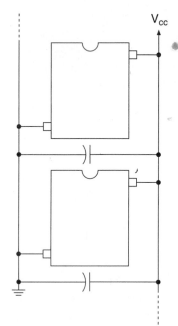

V_cc

For high-frequency designs, use one capacitor per IC, as shown in Figure 15.26. Connect directly to power and ground traces on a printed circuit board, *as close as possible* to the chip being decoupled.

### Connection of Totem Pole Outputs

Totem pole outputs must never be connected together. As shown in Figure 15.27, the problem occurs when two connected outputs are in opposite states.

The active pull-up consisting of $Q_4$, $D_3$, and the 130-$\Omega$ resistor is designed to supply current to about 10 TTL inputs, each having a large input impedance. It will not withstand the current that flows when the output is forced to ground through the LOW output transistor of another gate.

Under this condition about 30 to 55 mA will flow through $Q_{4A}$ and $Q_{3B}$. This exceeds the ratings of the outputs in both the HIGH and LOW state ($I_{OH} = 0.4$ mA; $I_{OL} = 16$ mA) and will cause damage to the outputs over time. The outputs will probably withstand this sort of abuse for several minutes, but eventually will be damaged.

---

**Section Review Problem for Section 15.1c**

15.3. A totem pole output is likely to be damaged when shorted to ground. Why?

---

**Figure 15.26**

**Placement of Decoupling Capacitors (High-Frequency Designs)**

## *Tristate Gates*

> **Tristate output** *An output having three possible states: logic HIGH and LOW, and a high-impedance state, in which the output acts as an open circuit.*

**Figure 15.27**

**Totem Poles Connected Together**

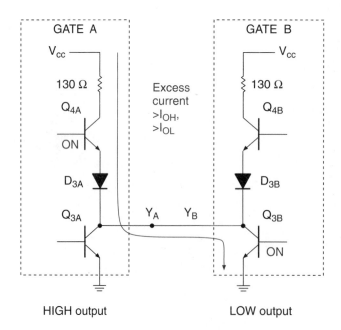

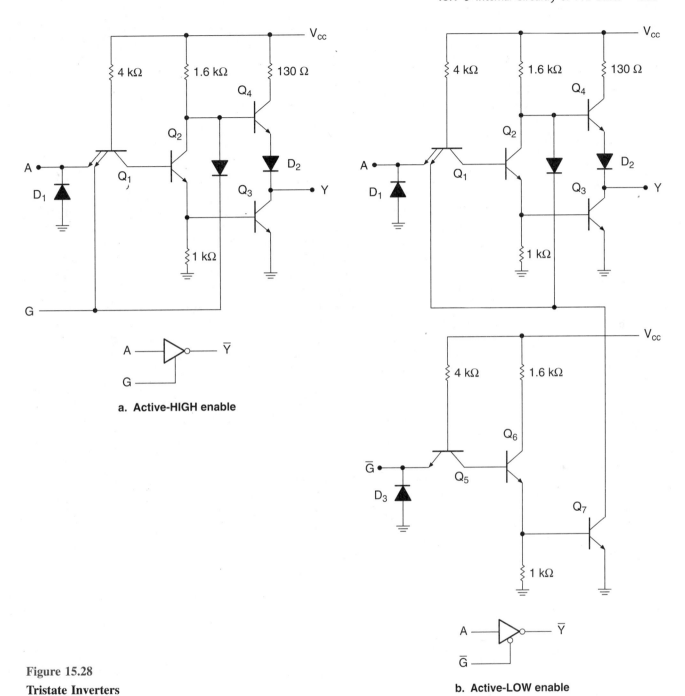

**Figure 15.28**
**Tristate Inverters**

a. Active-HIGH enable

b. Active-LOW enable

Figure 15.28 shows the circuits of two TTL inverters with **tristate outputs.** In addition to the usual binary states of HIGH and LOW, the output of the tristate inverter can also be in a "high-impedance" (Hi-Z) state. This state occurs when both $Q_3$ and $Q_4$ are OFF. The electrical effect is to produce an open circuit at the output, which is neither HIGH nor LOW.

**Table 15.3** Truth Tables of Tristate
Inverters

| $G$ | $A$ | $Y$ | $\overline{G}$ | $A$ | $Y$ |
|---|---|---|---|---|---|
| 0 | 0 | Hi-Z | 0 | 0 | 1 |
| 0 | 1 | Hi-Z | 0 | 1 | 0 |
| 1 | 0 | 1 | 1 | 0 | Hi-Z |
| 1 | 1 | 0 | 1 | 1 | Hi-Z |

The output of a tristate gate combines advantages of a totem pole output and an open-collector output. Like the totem pole output, it has an active pull-up with lower output impedance and faster switching than an open collector. Like the open collector, we can connect several outputs together, provided only one output is active at a time.

Input $G$, the "gating" or "enable" input, controls the gate. When $G$ is active, the gate acts as an ordinary inverter. When inactive, the gate is in the high-impedance state. Table 15.3 summarizes the operation of the tristate inverters in Figure 15.28.

The tristate inverter in Figure 15.28a is enabled by a HIGH at the $G$ input. The circuit is the same as a 7400 NAND gate with two exceptions: (1) an extra diode goes from the base of $Q_4$ to $G$, and (2) $G$ connects directly to one of the emitters of $Q_1$.

When $G = 0$, $Q_1$ acts as though there was a LOW at a NAND gate input. In a 7400 NAND circuit, this causes $Q_2$ and $Q_3$ to be in cutoff mode.

Due to the opposite states of the emitter and collector in $Q_2$, $Q_4$ would normally be ON. Instead, the LOW at $G$ pulls the base of $Q_4$ LOW through the extra diode. Thus, both $Q_3$ and $Q_4$ are OFF.

When $G = 1$, the $G$ emitter of $Q_1$ acts like a HIGH NAND input. By the enable/inhibit rules of a NAND gate, $Y = \overline{A}$. The additional diode prevents the HIGH at $G$ from activating $Q_4$.

The circuit in Figure 15.28b works the same way, except for the opposite sense of the activating input. This opposite active level is achieved by using an open-collector inverter, consisting of $Q_5$, $Q_6$, and $Q_7$, at input $\overline{G}$.

---

**Section Review Problem for Section 15.1d**

15.4. Why is the diode from the base of $Q_4$ necessary in the tristate inverters in Figure 15.28?

---

## Other Basic TTL Gates

Other TTL gates are similar to the NAND and inverter gates we have already examined. A significant variation is the OR/NOR circuit, which has a different input configuration than the AND/NAND/inverter type gates.

### 7402 NOR Gate

Figure 15.29 shows one gate of a 7402 Quadruple 2-input NOR gate package. The difference between this gate and the 7400 NAND gate is the structure of the inputs. The NOR gate does not use the multiple-emitter transistor, but rather an individual transistor ($Q_1$ or $Q_2$) for each input. There are two phase splitters ($Q_3$ and $Q_4$), which are paralleled, emitter-to-emitter and collector-to-collector.

**Figure 15.29**
**7402 NOR Gate Circuit**

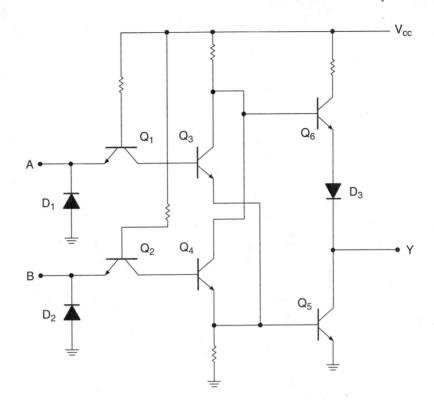

**Table 15.4** 7402 NOR Function and Truth Table

| A | B | $Q_1$ | $Q_2$ | $Q_3$ | $Q_4$ | $Q_5$ | $Q_6$ | Y |
|---|---|-----|-----|-----|-----|-----|-----|---|
| 0 | 0 | ON  | ON  | OFF | OFF | OFF | ON  | 1 |
| 0 | 1 | ON  | OFF | OFF | ON  | ON  | OFF | 0 |
| 1 | 0 | OFF | ON  | ON  | OFF | ON  | OFF | 0 |
| 1 | 1 | OFF | OFF | ON  | ON  | ON  | OFF | 0 |

If either $Q_3$ or $Q_4$ is enabled by a HIGH at its corresponding input, it will turn on $Q_5$, making the output LOW.

If both gate inputs are LOW, both $Q_3$ and $Q_4$ are in cutoff mode, and so is $Q_5$. The output is HIGH through $Q_6$.

Table 15.4 shows the truth table and the states of the transistors for this gate. It is not strictly correct to refer to $Q_1$ and $Q_2$ as being ON or OFF, since there is current flowing in these transistors regardless of whether the inputs are HIGH or LOW. Let us define the ON state of an input transistor as the condition where the base-emitter junction is conducting (LOW input). If the base-collector junction conducts, we will consider the transistor OFF (HIGH input).

## 7408 AND Gate and 7432 OR Gate

It may not be obvious why we would choose to study NAND and NOR gates before AND and OR. After all, AND and OR are the more basic logic functions.

**Figure 15.30**
**7408 AND Gate**

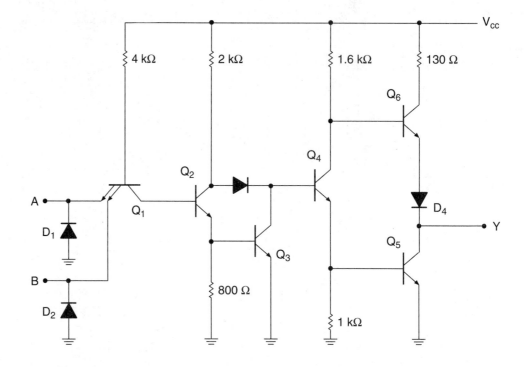

Electrically, it works the other way around. The simplest TTL circuit is the NAND/inverter, followed by the NOR. AND and OR gates are more complex since they are based on the NAND and NOR and require an extra inverter stage.

Figure 15.30 shows the circuit of a 7408 AND gate, and Figure 15.31 shows a 7432 TTL OR gate circuit. Each of these gates is like its NAND/NOR counterpart, except for an additional inverter, implemented by $Q_3$ in the AND gate and $Q_5$ in the OR gate.

Tables 15.5 and 15.6 show the transistor function and truth table for each gate. In keeping with the convention established for the NOR function table, an input transistor with a conducting base-emitter junction is considered ON.

---

**Section Review Problem for Section 15.1e**

15.5. Why are noninverting gates more complex than inverting gates?

---

**Table 15.5**  7408 AND Function and Truth Table

| A | B | $Q_1$ | $Q_2$ | $Q_3$ | $Q_4$ | $Q_5$ | $Q_6$ | Y |
|---|---|-------|-------|-------|-------|-------|-------|---|
| 0 | 0 | ON | OFF | OFF | ON | ON | OFF | 0 |
| 0 | 1 | ON | OFF | OFF | ON | ON | OFF | 0 |
| 1 | 0 | ON | OFF | OFF | ON | ON | OFF | 0 |
| 1 | 1 | OFF | ON | ON | OFF | OFF | ON | 1 |

**Figure 15.31**

**7432 OR Gate**

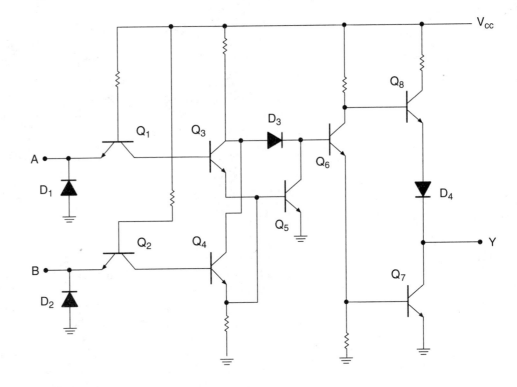

**Table 15.6** 7432 OR Function and Truth Table

| $A$ | $B$ | $Q_1$ | $Q_2$ | $Q_3$ | $Q_4$ | $Q_5$ | $Q_6$ | $Q_7$ | $Q_8$ | $Y$ |
|-----|-----|-------|-------|-------|-------|-------|-------|-------|-------|-----|
| 0 | 0 | ON | ON | OFF | OFF | OFF | ON | ON | OFF | 0 |
| 0 | 1 | ON | OFF | OFF | ON | ON | OFF | OFF | ON | 1 |
| 1 | 0 | OFF | ON | ON | OFF | ON | OFF | OFF | ON | 1 |
| 1 | 1 | OFF | OFF | ON | ON | ON | OFF | OFF | ON | 1 |

## 15.2

# Internal Circuitry of NMOS and CMOS Gates

**MOSFET** *Metal-oxide-semiconductor field effect transistor. A MOSFET has three terminals—gate, source, and drain—which are analogous to the base, emitter, and collector of a bipolar junction transistor.*

**Enhancement-mode MOSFET** *A MOSFET that creates a conduction path (a channel) between its drain and source terminals when the voltage between gate and source exceeds a specified threshold level.*

**Substrate** *The foundation of n- or p-type silicon on which an integrated circuit is built.*

**n-channel enhancement-mode MOSFET** *A MOSFET built on a p-type substrate with n-type drain and source regions. An n-type channel is created in the p-substrate during conduction.*

**p-channel enhancement-mode MOSFET** *A MOSFET built on an n-type substrate with p-type drain and source regions. During conduction, a p-type channel is created in the n-substrate.*

**NMOS** *A logic family based on the switching of n-channel enhancement-mode MOSFETs.*

**CMOS** *A logic family based on the switching of n- and p-channel enhancement-mode MOSFETs.*

All the logic circuits we have examined so far have been based on the switching of bipolar junction transistors. Another major logic family, **CMOS,** is based on the switching of metal-oxide-semiconductor field effect transistors, or **MOSFETs.**

There are two major types of MOSFETs, called depletion-mode and **enhancement-mode MOSFETs.** We will concentrate on the enhancement-mode devices, as they are the type used in the manufacture of digital ICs. Details of the differences between depletion- and enhancement-mode transistors can be found in any good textbook on electronic devices.

MOSFETs can be categorized in another way: as **n-channel** and **p-channel** devices, much as bipolar transistors are classified as NPN or PNP.

CMOS logic is constructed from both n- and p-channel MOSFETs. CMOS ("Complementary MOS") refers to the opposite, or complementary, operation of n- and p-channel transistors.

Another MOS logic family, **NMOS,** is made entirely from n-channel MOSFETs. Since this family requires less space on the silicon **substrate** of an integrated circuit than do other types of logic, it was, in the past, used widely in microprocessors and semiconductor memories. Advances in CMOS technology have made NMOS obsolescent, though not obsolete.

## MOSFET Structure

Figure 15.32 shows the structure and symbol of an n-channel enhancement-mode MOSFET in an integrated circuit. The device is built on a substrate of p-type silicon, which has a deficiency of electrons in its structure. The drain and source regions are "wells" of n-type silicon, which has an excess of electrons. The drain and source are roughly equivalent to the emitter and collector of a bipolar transistor.

The substrate is shown as a terminal with an arrow. The arrow points in for an n-channel device and out for a p-channel device. In nearly all cases, the substrate is shorted to the source terminal. (Some exceptions to this general rule will be examined when we look at circuits of CMOS gates.)

The gate terminal is similar to the base of a bipolar transistor in that it controls the flow of current between the drain and source. The difference is that a MOSFET uses gate *voltage* to control drain current, whereas a bipolar transistor uses base *current* to control collector current.

The gate consists of an insulating layer of silicon dioxide ($SiO_2$) and a layer of metal over the substrate between the drain and source. This gate structure is what gives the MOSFET its name (*metal-oxide-semiconductor* field effect transistor).

**Figure 15.32**
**n-Channel MOSFET**

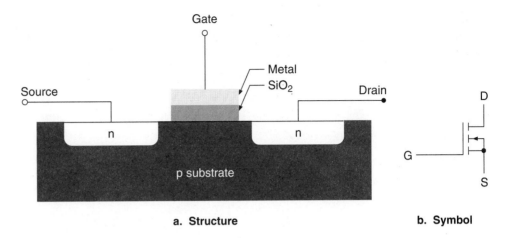

a. Structure                 b. Symbol

---

The oxide layer of the gate structure is subject to damage if excessive voltage (greater than about 100 V) is applied. This especially includes static electricity, or electrostatic discharge (ESD). There are standard precautions for working with MOS devices that should be followed carefully.

Most important are ensuring that MOS devices are stored in antistatic or conducting material, that work surfaces are not likely to generate static, that unused inputs are not left open or floating, that you avoid touching the pins of a MOS device, and that if you must handle a MOS IC, you discharge any static on your person *before* touching it.

A conductive wrist strap with a high series resistance to ground (about 1 MΩ) is often worn to reduce static. The high resistance protects the operator from shock injury in the event of a short circuit.

A list of handling precautions is included in Appendix B.

---

**Section Review Problem for Section 15.2a**

15.6. Why are MOSFET circuits particularly susceptible to static damage?

---

## Bias Requirement for MOS Transistors

**Ohmic region** *The MOSFET equivalent of saturation. When a MOSFET is biased ON, it acts like a relatively low resistance, or "ohmically."*

**n-type inversion layer** *The conducting layer formed between drain and source when an enhancement-mode n-channel MOSFET is biased ON. Also referred to as the channel.*

**Threshold voltage $V_{GS(Th)}$** *The minimum voltage between gate and source of a MOSFET for the formation of the conducting inversion layer (channel).*

**Figure 15.33**

**n-Channel MOSFET in Cutoff Region**

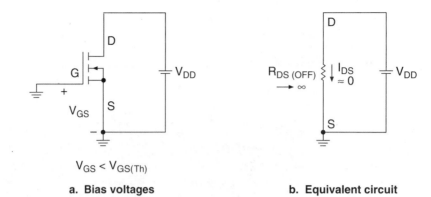

$V_{GS} < V_{GS(Th)}$

**a. Bias voltages**          **b. Equivalent circuit**

When we studied the operation of TTL gate circuits, we discovered that, for the most part, the bipolar transistors in the gates operated either in the saturation or the cutoff regions. In MOS-type gates, we make use of two similar operating regions in the constituent MOSFETs:

1. The cutoff region is the same as that for a bipolar transistor. Under this condition, there is a very high impedance between the drain and source terminals of the MOSFET.

2. The **ohmic region** is analogous to the saturation region of a bipolar transistor. In this state, there is a relatively low resistance between the MOSFET's drain and source.

The MOSFET switches between cutoff and ohmic regions when the voltage between gate and source, $V_{GS}$, is less than or greater than a value called the **threshold voltage.** The abbreviation for this voltage is $V_{GS(Th)}$; its value is between 1 and 5 volts, typically 1.5 V.

Figure 15.33 shows an n-channel MOSFET operating in the cutoff region. The gate-source voltage, $V_{GS}$, is less than $V_{GS(Th)}$. There is no conduction between the drain and source. The resistance, $R_{DS(OFF)}$, between drain and source is very large, typically in the thousands of megohms.

When the value of $V_{GS}$ increases and exceeds the threshold voltage, the MOSFET enters the ohmic region. A conduction channel, called the **n-type inversion layer,** is created in the p-substrate of the transistor, as shown in Figure 15.34. This layer is like an artificially created region of n-type silicon, which allows conduction between the drain and source, provided there is sufficient potential difference between them.

Figure 15.35 shows a MOSFET operating in the ohmic region. $R_{DS(ON)}$, the equivalent resistance of a MOSFET in the ohmic region, is typically around 500 $\Omega$ to 2 k$\Omega$. The drain-source current, $I_{DS}$, is determined by Ohm's Law: $I_{DS} = V_{DD}/R_{DS(ON)}$.

The operation of a p-channel MOSFET is similar, but with polarities reversed. If $V_{GS(Th)}$ is +1.5 V for an n-channel device, an equivalent p-channel MOSFET has a threshold voltage of −1.5 V. $V_{GS} > +1.5$ V turns ON an n-channel transistor; $V_{GS} < −1.5$ V turns ON a p-channel device.

Figure 15.36 summarizes the bias requirements for n- and p-channel enhancement-mode MOSFETs.

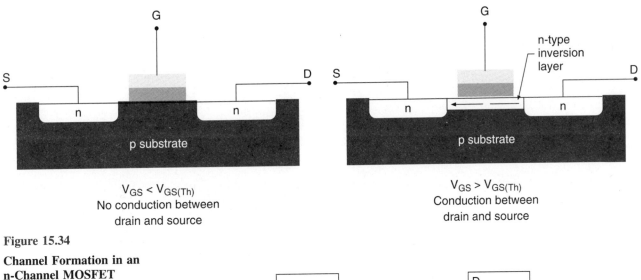

**Figure 15.34**

**Channel Formation in an n-Channel MOSFET**

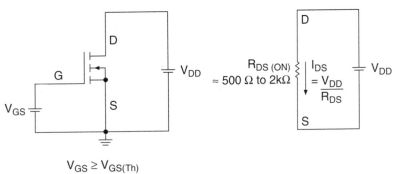

**Figure 15.35**

**n-Channel MOSFET in Ohmic Region**

a. Bias voltages                    b. Equivalent circuit

## NMOS Inverter

Figure 15.37 shows three transistor circuits, all of which act as digital inverters.

Figure 15.37a is the NPN bipolar transistor circuit we examined in Example 15.1. Figure 15.37b shows the same type of circuit using an n-channel MOSFET. Figure 15.37c is an NMOS inverter, that is, an inverter using an n-channel MOSFET as an active pull-up.

Recall that, for Figure 15.37a, when $A$ is HIGH, the transistor is ON, pulling $Y$ to the LOW state. When $A$ is LOW, the transistor is in cutoff mode. $Y$ is pulled HIGH by the collector resistor. The circuit in Figure 15.37b operates in the same way.

No resistor is required on the gate terminal due to its very high impedance.

The circuit in Figure 15.37b is not used in ICs because integrated resistors are relatively bulky. Instead, an NMOS inverter is constructed entirely of n-channel MOSFETs, as shown in Figure 15.37c.

The resistor is replaced by an active load, $Q_2$, an n-channel transistor that is always ON. $Q_2$ is kept ON by making $V_{GS2} \geq V_{GS(Th)}$. The substrate terminal of $Q_2$ is biased separately from the source to ensure correct operation of the device.

The ON resistance, $R_{GS(ON)}$, of a MOSFET depends on the length and width of the channel, which is set during the manufacturing process. The ON resistance of the load

**Figure 15.36**

**Bias Requirements of n- and p-Channel MOSFETs**

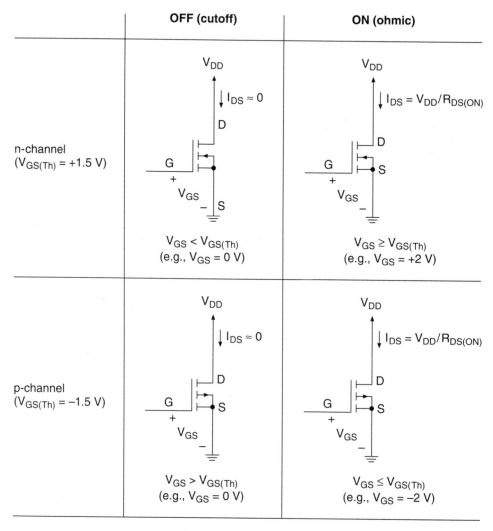

| | OFF (cutoff) | ON (ohmic) |
|---|---|---|
| n-channel ($V_{GS(Th)}$ = +1.5 V) | $I_{DS} \approx 0$ <br> $V_{GS} < V_{GS(Th)}$ <br> (e.g., $V_{GS}$ = 0 V) | $I_{DS} = V_{DD}/R_{DS(ON)}$ <br> $V_{GS} \geq V_{GS(Th)}$ <br> (e.g., $V_{GS}$ = +2 V) |
| p-channel ($V_{GS(Th)}$ = −1.5 V) | $I_{DS} \approx 0$ <br> $V_{GS} > V_{GS(Th)}$ <br> (e.g., $V_{GS}$ = 0 V) | $I_{DS} = V_{DD}/R_{DS(ON)}$ <br> $V_{GS} \leq V_{GS(Th)}$ <br> (e.g., $V_{GS}$ = −2 V) |

**Figure 15.37**

**Transistor Inverters**

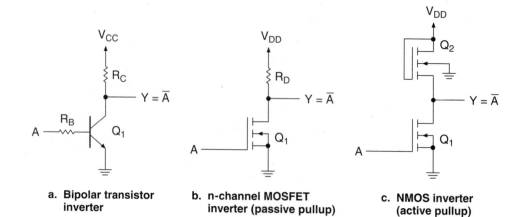

a. Bipolar transistor inverter

b. n-channel MOSFET inverter (passive pullup)

c. NMOS inverter (active pullup)

**Figure 15.38**

**Equivalent Circuits of an NMOS Inverter With LOW and HIGH Inputs**

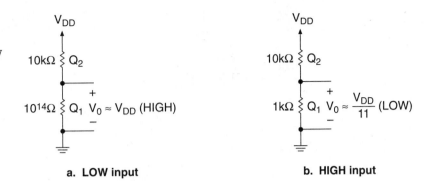

a. LOW input                    b. HIGH input

transistor, $Q_2$ (10 k$\Omega$), is set to be larger than the ON resistance of the switching transistor, $Q_1$ (1 k$\Omega$). This is so that an output LOW can be as close to 0 V as possible.

Figure 15.38 shows how an NMOS inverter acts as a switchable voltage divider. When input is LOW, $V_{GS} < V_{GS(Th)}$ and $Q_1$ acts as a very high impedance ($\sim10^{14}$ $\Omega$). The supply voltage divides across the two transistors, as follows:

$$V_O = \frac{10^{14}\ \Omega}{10^{14}\ \Omega + 10^4\ \Omega} V_{DD} \approx V_{DD}$$

The output is HIGH.

When the input is HIGH, $V_{GS} > V_{GS(Th)}$ and the impedance between the drain and source of $Q_1$ is relatively low. If we assume a ten-to-one ratio between the ON resistances of the two transistors, the output voltage is $V_{DD}/11$ (= 0.454 V for $V_{DD} = 5$ V), which is a logic LOW.

## CMOS Inverter

The LOW-state output voltage of an NMOS device is inevitably somewhat greater than 0 V, since it depends on voltage division across the transistor ON resistances.

CMOS output circuitry improves on this performance. A CMOS output has a HIGH state that approaches $V_{DD}$ and a LOW state that approaches 0 V. This is achieved by actively switching both $V_{DD}$ and ground rather than just ground, as in NMOS devices.

Figure 15.39 shows the circuit of a CMOS inverter, which consists of one n-channel and one p-channel MOSFET.

Recall the bias conditions of the two transistors:

n-channel: threshold voltage, $V_{GS(Th)} \approx +1.5$ V
        ON when $V_{GS} > V_{GS(Th)}$ (e.g., $V_{GS} = V_{DD}$)
        OFF when $V_{GS} < V_{GS(Th)}$ (e.g., $V_{GS} = 0$ V)
p-channel: threshold voltage, $V_{GS(Th)} \approx -1.5$ V
        ON when $V_{GS} < V_{GS(Th)}$ (e.g., $V_{GS} = -V_{DD}$)
        OFF when $V_{GS} > V_{GS(Th)}$ (e.g., $V_{GS} = 0$ V)

The operation of the CMOS inverter, and any other CMOS gate, depends on arranging the bias conditions of each complementary pair of transistors so that they are always in opposite states. Whenever $Q_1$ is ON, $Q_2$ is OFF, and vice versa. Figure 15.40 shows how this is accomplished.

**Figure 15.39**
**CMOS Inverter**

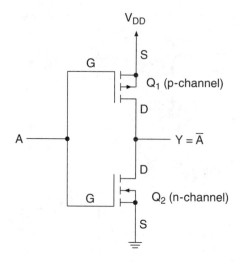

**Figure 15.40**

**Operation of CMOS Inverter**

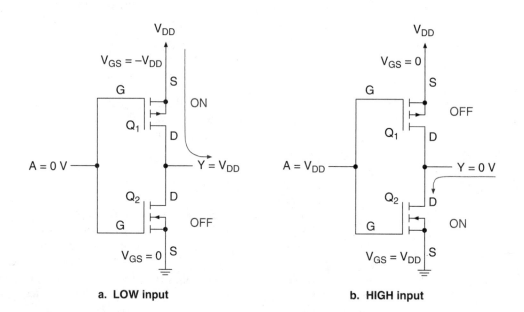

a. **LOW input**          b. **HIGH input**

Assume that a LOW input is at ground potential and that a HIGH input is equal to $V_{DD}$.

When *input A is LOW*, the gate voltage of $Q_2$ is the same as its source voltage; $V_{GS2} = 0$ and $Q_2$ is OFF. This places a high-impedance path between output $Y$ and ground. At the same time, the gate voltage of $Q_1$ is 0 V and its source voltage is $V_{DD}$; $V_{GS1} = V_{G1} - V_{S1} = 0 - V_{DD} = -V_{DD}$. (The p-channel transistor, $Q_1$, is drawn "upside down" to make the complementary pair symmetrical.) $Q_1$ is ON, forming a low-impedance path from $V_{DD}$ to the output $Y$. *Output Y is HIGH.*

When *input A is HIGH*, the gate-source voltage of the n-channel transistor is $V_{DD}$, causing $Q_2$ to turn ON. The gate of $Q_1$ is also at $V_{DD}$. Since the source of the p-channel transistor is at $V_{DD}$, $V_{GS1} = V_{G1} - V_{S1} = V_{DD} - V_{DD} = 0$ V; $Q_1$ is OFF. This combination creates a high impedance between $V_{DD}$ and output $Y$ and a low-impedance path from output $Y$ to ground, as shown in Figure 15.40b. *Output Y is LOW.*

## CMOS NAND/NOR Gates

CMOS NAND and NOR gates are constructed from complementary pairs of MOSFETs. Each MOSFET pair has an n-channel transistor that is turned ON by a HIGH input and a p-channel transistor that is turned ON by a LOW input. The n-channel devices switch the output to ground; the p-channel ones switch the output to $V_{DD}$. NAND and NOR functions are generated by arranging the MOSFET drain-source paths in series (AND) and parallel (OR) configurations.

In Figure 15.41, we see the DeMorgan equivalent forms of a NAND gate. Each form illustrates an aspect of NAND operation that can be described with a brief sentence and implemented by a MOSFET circuit. The combination of forms describes the complete operation of the device.

Figure 15.41a states that both NAND inputs must be HIGH to make the output LOW. A logic HIGH activates an n-channel MOSFET. The gate output is switched to ground by an n-channel MOSFET. Thus, the drain-source paths of two n-channel transistors must be connected in series to make the output LOW under the stated conditions.

Figure 15.41b shows that the NAND output is HIGH if either input is LOW. A p-channel transistor will turn ON when its gate is LOW and will switch a HIGH to the output. A parallel combination of p-channel MOSFETs will satisfy these conditions.

The stated conditions are combined in the CMOS NAND circuit shown in Figure 15.42. Transistors $Q_1$ and $Q_4$ form a complementary pair, as do $Q_2$ and $Q_3$. When $A$ and $B$ are both HIGH, $Q_1$ and $Q_2$ are both OFF, cutting off the connection between $V_{DD}$ and

**Figure 15.41**

**NAND Functions of MOSFETs**

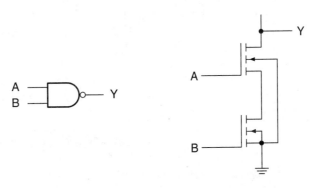

**a. Both inputs HIGH make output LOW**
**(n-channels in series to ground)**

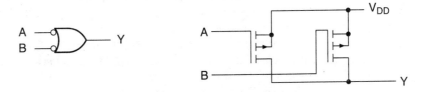

**b. Either input LOW makes output HIGH**
**(p-channels in parallel to V_DD)**

**Figure 15.42**

**CMOS NAND Gate**

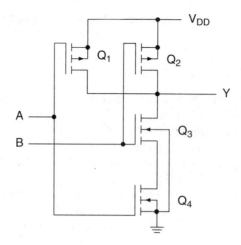

**Table 15.7** Partial CMOS NAND Function and Truth Table

| A | B | $Q_1$ | $Q_2$ | $Q_3$ | $Q_4$ | Y |
|---|---|-------|-------|-------|-------|---|
| 1 | 1 | OFF | OFF | ON | ON | 0 |

**Table 15.8** Partial CMOS NAND Function and Truth Table

| A | B | $Q_1$ | $Q_2$ | $Q_3$ | $Q_4$ | Y |
|---|---|-------|-------|-------|-------|---|
| 0 | 0 | ON | ON | OFF | OFF | 1 |
| 0 | 1 | ON | OFF | ON | OFF | 1 |
| 1 | 0 | OFF | ON | OFF | ON | 1 |

output $Y$. $Q_3$ and $Q_4$ are both ON, supplying a low-impedance path from output $Y$ to ground, making the output LOW. This is shown in the partial truth table in Table 15.7.

When $A$ is LOW and $B$ is HIGH, $Q_1$ is ON. This creates a path from $V_{DD}$ to output $Y$. At the same time, $Q_4$ is OFF. This cuts the $Y$-to-ground path through the n-channel MOSFETs; the series path from output to ground is broken. One parallel path from $V_{DD}$ to output has been established. Output $Y$ is HIGH.

The remaining input combinations also make the output HIGH, as shown in Table 15.8. They do so by breaking the n-channel path from output to ground and enabling one or both p-channel paths from $V_{DD}$ to output.

Each MOSFET in a logic circuit must have its own independent substrate bias. This ensures that the transistor will operate as expected when a logic HIGH or LOW is applied to its gate.

Normally, the substrate of a MOSFET is shorted to its source terminal. If a MOSFET source terminal is isolated from $V_{DD}$ or ground, the substrate must be biased separately. For example, in the NAND gate in Figure 15.42, the substrate of $Q_3$ connects directly to ground.

NOR gates are similar to NANDs in construction. Figure 15.43 shows the DeMorgan equivalent forms of the NOR function and the MOSFET implementations of each aspect of the gate operation.

When either input is HIGH, the output is LOW. This function is implemented by two parallel n-channel MOSFETs. Both inputs must be LOW to make the output HIGH, which implies a series connection of two p-channel transistors. The complete NOR gate circuit is shown in Figure 15.44. (Note that the substrate of $Q_2$ is connected directly to $V_{DD}$ to ensure that it has its own bias voltage.)

As was the case with the NAND circuit, transistors $Q_1$ and $Q_4$ form a complementary MOSFET pair. Transistors $Q_2$ and $Q_3$ form the second pair.

**Figure 15.43**

**NOR Functions of MOSFETs**

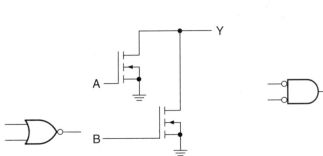
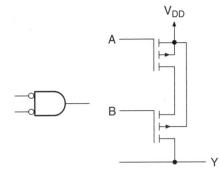

a. **Either input HIGH makes output LOW (n-channels in parallel to ground)**      b. **Both inputs LOW make output HIGH (p-channels in series to $V_{DD}$)**

**Figure 15.44**

**CMOS NOR Gate**

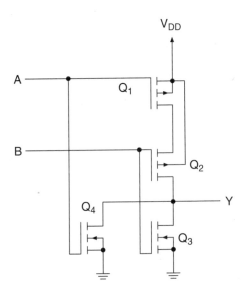

**Table 15.9**  Partial CMOS NOR Function and Truth Table

| A | B | $Q_1$ | $Q_2$ | $Q_3$ | $Q_4$ | Y |
|---|---|-------|-------|-------|-------|---|
| 0 | 0 | ON | ON | OFF | OFF | 1 |

**Table 15.10**  Partial CMOS NOR Function and Truth Table

| A | B | $Q_1$ | $Q_2$ | $Q_3$ | $Q_4$ | Y |
|---|---|-------|-------|-------|-------|---|
| 0 | 1 | ON | OFF | ON | OFF | 0 |
| 1 | 0 | OFF | ON | OFF | ON | 0 |
| 1 | 1 | OFF | OFF | ON | ON | 0 |

When both inputs are LOW, both p-channel transistors are ON. This creates a low-impedance path from $V_{DD}$ to output $Y$. The n-channel transistors, $Q_3$ and $Q_4$, are both OFF. This isolates the output from ground. Output $Y$ is HIGH. Table 15.9 shows the MOSFET states under this condition.

If either input is HIGH, one or both of the p-channel transistors will turn OFF. This action breaks the path from $V_{DD}$ to output $Y$. The complementary n-channel transistor will turn ON. This creates a low-impedance path from output $Y$ to ground. Output $Y$ is LOW. Table 15.10 summarizes the possible input conditions and MOSFET states when the NOR output is LOW.

**Figure 15.45**

**CMOS AND Gate**

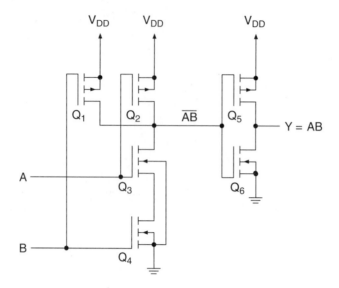

**Figure 15.46**

**CMOS OR Gate**

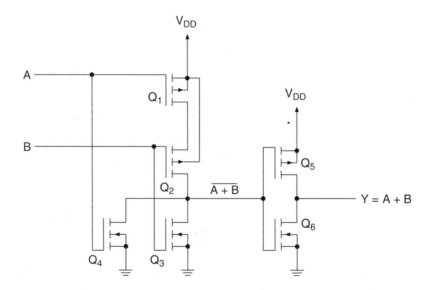

## CMOS AND and OR Gates

Figures 15.45 and 15.46 show the circuits of CMOS AND and OR gates. The AND gate is the same as the NAND circuit, except for the output inverter section constructed from $Q_5$ and $Q_6$. The OR gate is the same as the NOR with an output inverter section.

---

**Section Review Problem for Section 15.2b**

15.7. Why is the source of a p-channel MOSFET connected to $V_{DD}$ in a CMOS gate?

---

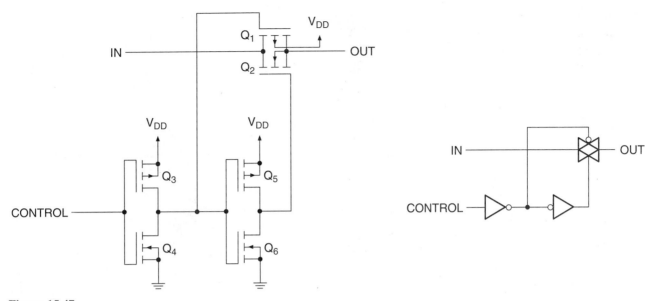

**Figure 15.47**

**CMOS Transmission Gate**

## CMOS Transmission Gate

Figure 15.47 shows the circuit of a CMOS transmission gate. We examined this device briefly in Section 10.6 (Demultiplexers) in connection with the CMOS 4097B Dual 8-channel Multiplexer/Demultiplexer.

A CMOS transmission gate, or analog switch, conducts in both directions. This makes it possible to enable or inhibit time-varying analog signals having both positive and negative values. Conduction takes place between the input and output terminals through MOSFETs $Q_1$ and $Q_2$. Positive current (left to right in the diagram) flows through $Q_2$, and negative current (right to left) flows through $Q_1$. Two inverters, consisting of the $Q_3/Q_4$ and $Q_5/Q_6$ pairs of MOSFETs, control the ON/OFF state of the circuit.

When *CONTROL* = 1, the inverters bias both $Q_1$ and $Q_2$ ON, allowing them to conduct. When *CONTROL* = 0, the circuit inhibits conduction between input and output.

The substrate terminal of $Q_1$ is connected, not to the source terminal of that transistor, but directly to $V_{DD}$, thus providing the correct bias to $Q_1$ in the ON state.

A particular device with this function is the 4066B Quad Analog Switch, whose circuit symbol is shown in Figure 15.48. When the *CONTROL* input is HIGH, analog and digital signals can pass between the bidirectional input terminals.

**Figure 15.48**

**One of Four Analog Switches From 4066B**

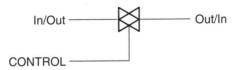

---

**EXAMPLE 15.5**    Figure 15.49 shows a circuit where the analog switches in a 4066B package are used to control the selection and muting of two pairs of speakers in a stereophonic audio system. Briefly explain the circuit operation.

**Figure 15.49**

**Example 15.5**
**4066B Analog Switches as**
**Audio Selectors**

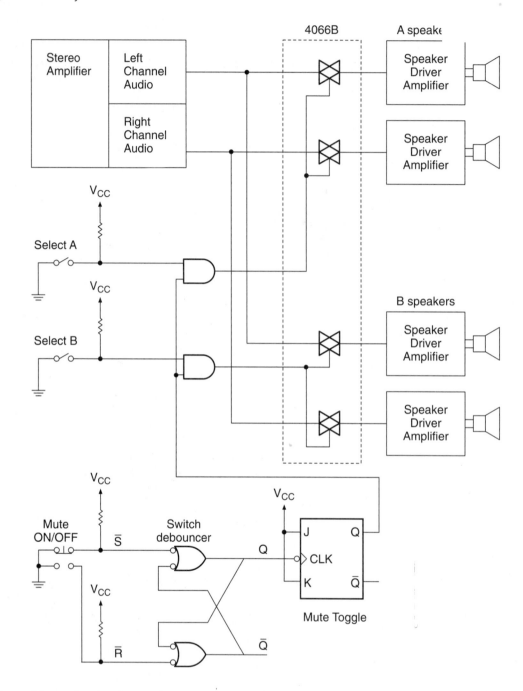

**Solution**   The audio signal to each speaker is passed or blocked by a CMOS transmission gate. The speakers are paired into *A* and *B* groups. Each pair has a left and a right channel speaker. The same logic gate controls both speakers of each group.

The Select *A* switch enables the *A* speakers when it is open (logic HIGH). The Select *B* switch enables the *B* speakers when it is open. The Mute Toggle flip-flop mutes (disables) both sets of speakers when *Q* is LOW. This action inhibits both AND gates, making all transmission gate *CONTROL* inputs LOW. The mute function toggles ON and OFF with each push of the Mute ON/OFF switch.

# 15.3

## TTL and CMOS Variations

All the logic gate circuits shown so far in this chapter have been either standard TTL or CMOS. These early logic families represented the two main standards of logic design for many years, and their influence is still visible in other, more advanced types of logic. The changes that have been made in newer logic families are not fundamental changes in the working concepts, but improvements to the specifications, particularly switching speed and power dissipation.

### TTL Logic Families

**Schottky barrier diode**  *A specialized diode with a forward drop of about +0.4 V.*

**Schottky transistor**  *A bipolar transistor with a Schottky diode across its base-collector junction, which prevents the transistor from going into deep saturation.*

**Schottky TTL**  *A series of unsaturated TTL logic families based on Schottky transistors. Schottky TTL switches faster than standard TTL due to decreased storage time in its transistors.*

**Speed-power product**  *A measure of a logic circuit's efficiency, calculated by multiplying its propagation delay by its power dissipation. Unit: picojoule (pJ)*

Probably the most important development in TTL technology was the introduction, in the early 1970s, of the **Schottky barrier diode** into circuit designs. This made possible the first family of nonsaturated bipolar logic, with its resultant improvement in switching speed.

Figure 15.50 shows a bipolar transistor with a Schottky diode connected across its base and collector and the equivalent circuit symbol of this combination. We call this configuration a **Schottky transistor** and logic devices using such transistors **Schottky TTL.**

Normally the base-collector junction of a saturated bipolar transistor has a drop of about 0.5 volts, as shown in Figure 15.51. The Schottky diode clamps this junction voltage to about 0.4 volts. This keeps the transistor out of deep saturation in its ON state.

**Figure 15.50**
**Schottky Transistor**

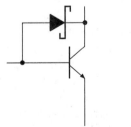 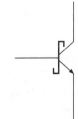

a. Schottky diode-clamped
   base-collector junction

b. Equivalent circuit
   symbol

**Figure 15.51**

**ON-State Operating Voltages of Bipolar Transistors**

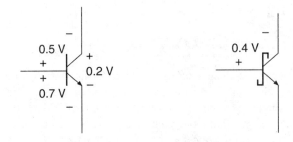

a. **Saturated bipolar transistor**

b. **Base-collector voltage in Schottky transistor**

The base region of the Schottky-clamped transistor holds less charge than does a standard bipolar transistor. Its storage time, the time required to dissipate base charge upon turn-off, is substantially reduced. The transistor can switch faster with the Schottky diode than without.

Figure 15.52 shows the circuits of the 74S00 Schottky and 74LS00 Low-Power Schottky NAND gates. Compare these circuits to each other and to the 7400 standard TTL NAND gate in Figure 15.21.

In the 74S00 circuit, $Q_1$ acts as the input and $Q_2$ as the phase splitter, as in the 7400 gate. The HIGH output circuit consists of $Q_3$ and $Q_4$ connected as a modified Darlington pair. When $Q_2$ is OFF (at least one input is LOW), enough base current flows in $Q_3$ to turn it on. Collector-emitter current in $Q_3$ turns on $Q_4$, making the output HIGH.

When $Q_2$ is ON (both inputs are HIGH), the base of $Q_3$ is pulled LOW, turning it OFF. Sufficient current flows in the base of $Q_5$ to turn it ON. The resultant current through $Q_5$ will turn on $Q_6$, making the output LOW. A similar analysis can be made for the 74LS00 gate.

One difference between the 74S00 and 74LS00 circuits is the size of the resistors; the LS device has larger resistors. Less current flows in the gate circuit. This reduces power dissipation of the chip. The larger resistor values also slow down the switching times of the various transistors by increasing the $RC$ time constants of the circuit elements.

## Speed-Power Product

One measure of logic circuit efficiency is its **speed-power product,** calculated by multiplying switching speed and power dissipation, usually expressed in picojoules (pJ). (The joule is the SI unit of energy. Power is the rate of energy used per unit time.) A major goal of logic circuit design is the reduction of a device's speed-power product.

Table 15.11 shows the propagation delay, supply current, and speed-power product for a NAND gate in six TTL families: standard TTL (7400), Schottky (74S00), Low-Power Schottky (74LS00), Fast TTL (74F00), Advanced Schottky (74AS00), and Advanced Low-Power Schottky (74ALS00).

The speed-power product shown is the worst-case value. This is calculated by multiplying the largest value of $I_{CC}/4$ by the slowest switching speed by 5 volts for each family. We use $I_{CC}/4$ because $I_{CC}$ is specified per chip (four gates).

A faster switching speed results in an overall increase in speed-power product, other factors being equal. For example, the speed-power product of either Advanced Schottky family is lower than that of the LS and S families. However, the ALS series (the slower Advanced Schottky family) has a lower speed-power product than the AS series.

**Figure 15.52**
**Schottky TTL Circuits**

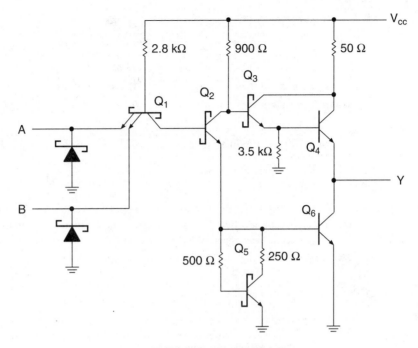

**a. 74S00 Schottky NAND gate**

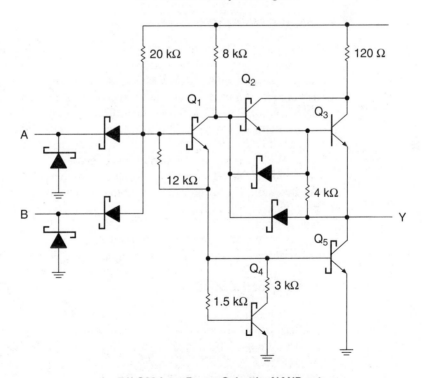

**b. 74LS00 Low-Power Schottky NAND gate**

Table 15.11  TTL Speed and Power Specifications

|  | **7400** | **74LS00** | **74S00** | **74F00** | **74ALS00** | **74AS00** |
|---|---|---|---|---|---|---|
| $t_{pLH}$ **(max)** | 22 ns | 15 ns | 4.5 ns | 6 ns | 11 ns | 4.5 ns |
| $t_{pHL}$ **(max)** | 15 ns | 15 ns | 5 ns | 5.3 ns | 8 ns | 4 ns |
| $I_{CCH}/4$ **(max)** | 2 mA | 0.4 mA | 4 mA | 0.7 mA | 0.21 mA | 0.8 mA |
| $I_{CCL}/4$ **(max)** | 5.5 mA | 1.1 mA | 9 mA | 2.6 mA | 0.75 mA | 4.35 mA |
| **Speed-Power Product (per gate)** | 605 pJ | 82.5 pJ | 225 pJ | 78.0 pJ | 41.25 pJ | 97.9 pJ |

The smaller resistors used to speed up output switching imply a proportional drop in propagation delay (higher speed) but an increased supply current. Power dissipation increases in proportion to the square of the supply current, thus offsetting the effect of the increased switching speed.

## CMOS Logic Families

The CMOS gates we have looked at in this chapter are simpler than most gates actually in use. There are two main families of CMOS devices: Metal-Gate CMOS, and silicon-gate, or High-Speed, CMOS.

### Metal-Gate CMOS

Metal-Gate CMOS is the circuit technology we have examined in previous sections of this chapter. There are two main variations on this type of circuit, designated B-series and UB-series CMOS. Most CMOS gates are B-series; UB-series is available in a limited number of inverting-type gates, such as inverters and 2-, 3-, and 4-input NAND and NOR gates. Figure 15.53 shows the difference in the two configurations.

Figure 15.53b shows one gate from a 4011UB Quadruple 2-input NAND package. Its circuit is the same as the NAND configuration examined in Section 15.2.

The B-series configuration of this circuit has two additional inverter outputs in cascade with the NAND logic. (The same gate becomes an AND when we add a third output inverter.) The inverter configuration is actually an amplifier; extra inverter stages provide additional gain and increase noise margin by allowing the circuit to accept smaller input signals.

CMOS gates are sometimes used in analog applications, such as oscillators. The UB-series gates, with their lower gain, are more desirable for such applications.

### High-Speed CMOS

**High-Speed (silicon-gate) CMOS**  *A CMOS logic family with a smaller device structure and thus higher speed than standard (Metal-Gate) CMOS.*

CMOS has often been considered a nearly ideal family for logic designs, with its high noise immunity, low power consumption, and flexible power supply requirements. Unfortunately, its propagation delay times, typically 10 to 20 times greater than those of equivalent TTL devices, are just not fast enough for use in modern microprocessor-based systems.

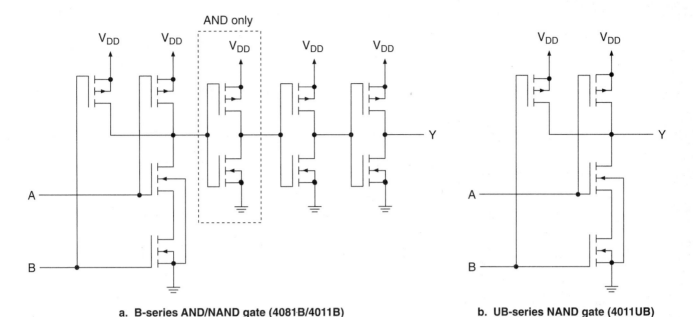

**a. B-series AND/NAND gate (4081·B/4011B)**    **b. UB-series NAND gate (4011UB)**

**Figure 15.53**

**Metal-Gate CMOS Circuits**

**High-Speed CMOS** was developed to address the problem of switching speed, while striving to keep the other advantages of CMOS. This is achieved by using MOS-FETs with a polysilicon material for the gate, rather than metal, as in standard CMOS. Because of advantages gained in this manufacturing process, each transistor is physically smaller and has a lower gate capacitance than metal-gate MOSFETs. Both these factors contribute to a lower propagation delay for the logic gate circuit.

Several subfamilies of High-Speed CMOS are available for various logic and linear applications, designated by the labels 74HCXX, 74HC4XXX, 74HCTXX, and 74HCUXX.

The 74HCXX series duplicates equivalent LSTTL functions in packages having identical pinouts to LSTTL. The 74HC4XXX replaces CMOS functions pin for pin. Both these series have CMOS-equivalent input and output levels, within the power supply limits (2.0 V to 6.0 V) of High-Speed CMOS.

The 74HCTXX devices are designed to be directly compatible with LSTTL devices, and thus have LSTTL-equivalent inputs and CMOS-equivalent outputs.

74HCUXX devices have no output buffers, like the 4000 UB-series standard CMOS devices. The 74HCU devices are used, as are the 4000UB devices, for linear applications such as oscillators and multivibrators.

Table 15.12 shows the relative performance of the various CMOS families. As in TTL, the 2-input NAND gate is used as the standard, except for the HCT and HCU families, where this gate is not available. The quiescent speed-power product of all CMOS families is much smaller than that of any TTL family. The High-Speed CMOS families have propagation delays comparable to those of LSTTL.

The power dissipation of a CMOS device increases directly with frequency. The speed-power product also goes up, with higher frequencies.

Table 15.12 shows CMOS speed-power product for a switching speed of 1 MHz. At these speeds, B-series CMOS has no advantage over the common TTL families in terms of its efficiency. It still has the edge on TTL with respect to noise immunity and power supply flexibility.

Table 15.12 CMOS Speed and Power Specifications

| | 4011B | 4011UB | 74HC00A | 74HCT04 | 74HCU04 |
|---|---|---|---|---|---|
| $t_{pLH}$ (max) | 250 ns | 180 ns | 13 ns | 15 ns | 14 ns |
| $t_{pHL}$ (max) | 250 ns | 180 ns | 13 ns | 17 ns | 14 ns |
| $I_{DD}/4$ (max) | 0.25 μA | 0.25 μA | | | |
| $I_{CC}/4$ (max) (quiescent) | | | 0.25 μA | 0.17 μA | 0.33 μA |
| $V_{DD}$ | 5.0 V | 5.0 V | | | |
| $V_{CC}$ | | | 6.0 V | 5.5 V | 6.0 V |
| Dynamic Power Dissipation (1 MHz) | 1.5 mW | 1.5 mW | 790 μW | 550 μW | 540 μW |
| Speed-Power Product (quiescent; per gate) | 0.31 pJ | 0.23 pJ | 0.02 pJ | 0.016 pJ | 0.028 pJ |
| Speed-Power Product (1 MHz; per gate) | 375 pJ | 270 pJ | 10.3 pJ | 11.3 pJ | 7.6 pJ |

---

**Section Review Problem for Section 15.3**

15.8. Assuming that power dissipation of a 74HC00A NAND gate is directly proportional to its switching frequency, what is the speed-power product of the gate at 2 MHz, 5 MHz, and 10 MHz?

---

## 15.4

# Other Logic Families

TTL and CMOS are the most widely used digital logic families, but they are not the only ones in existence. A number of logic families have been available in the past, such as resistor-transistor logic (RTL), diode-transistor logic (DTL), and high threshold logic (HTL). These families are all obsolete.

Two families still in use are emitter coupled logic (ECL) and BiCMOS. ECL is a high-speed family of unsaturated bipolar logic. BiCMOS is relatively new and represents an attempt to combine the best features of CMOS and TTL circuitry. BiCMOS is so named because its input and logic sections are CMOS circuits and its output is a bipolar circuit.

### Emitter Coupled Logic

Emitter coupled logic (ECL) is the fastest logic family commonly in use today. Its basic switching unit is the bipolar junction transistor operating in the unsaturated mode. Its high speed is the result of not having to switch transistors in and out of saturation.

Several families of ECL are available, the fastest of which have propagation delays around 300 to 500 picoseconds. With comparable rise and fall times, this implies maximum frequencies approaching 1 GHz.

Low-end ECL families typically have propagation delays of 1 nanosecond. Rise and fall times of output waveforms are similar. Maximum frequencies are in the hun-

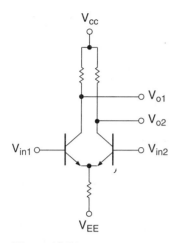

**Figure 15.54**

**Simple Differential Amplifier**

dreds of MHz. Contrast this with LSTTL, which has maximum frequencies of about 30 MHz, or Advanced Schottky TTL, where $f_{max}$ is about 75 MHz.

At the high frequencies available to ECL, wires aren't just wires any more. The residual inductance and capacitance inherent in all conductors start to become significant. If not taken into account, this residual inductance and capacitance will seriously degrade the performance of any ECL circuit.

Designers of high-frequency circuits must use coaxial cables and other transmission line techniques to connect circuit components. Such design information is available from manufacturers of ECL devices.

The basis of ECL is the differential amplifier, shown in its simplest form in Figure 15.54. If the two collector resistors are equal and the transistors have the same gain, the output voltages $V_{o1}$ and $V_{o2}$ will be proportional to the difference $V_{in1} - V_{in2}$. $V_{o1}$ and $V_{o2}$ are 180° out of phase with each other. That is, $V_{o1} = -V_{o2}$.

The lefthand input transistor in Figure 15.54 can be replicated several times to produce an ECL OR/NOR gate. Such a circuit is shown in Figure 15.55. Transistors $Q_1$ to $Q_4$ act as logic inputs. Each interacts with $Q_5$ as part of a differential amplifier. The input voltage to $Q_5$ is supplied by an internal bias network consisting of $Q_6$, the two diodes, and their associated resistors.

The biasing is such that if one or more of the logic inputs is in the HIGH state, $Q_7$ turns ON, making the OR output HIGH. At the same time, $Q_8$ is OFF, making the NOR output LOW.

ECL logic levels are quite different from those in TTL and CMOS levels. A logic HIGH is usually about $-0.9$ V and a LOW is about $-1.7$ V, depending on supply voltages. Recommended power supply voltages are 0 V for $V_{CC}$ and $-5.2$ V for $V_{EE}$.

Noise margin is small for an ECL gate, since the logic levels are so close together. Depending on the family of ECL, typical values range from about 100 mV to 200 mV.

**Figure 15.55**

**Basic ECL OR/NOR Gate**

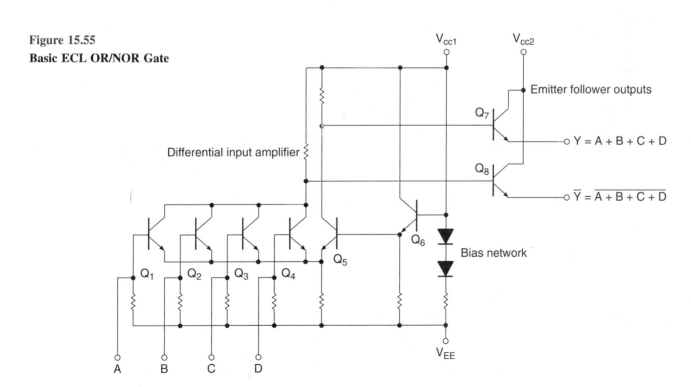

Noise immunity is improved by connecting the system ground to $V_{CC}$. Noise voltage on the $V_{EE}$ line is applied to all inputs equally, including $Q_5$. Because the ECL gate output is proportional to the *difference* of voltages at $Q_5$ and the other inputs, the noise cancels out. This is not the case with noise on the $V_{CC}$ line. Therefore, noise is minimized if the $V_{CC}$ input is grounded.

---

**Section Review Problem for Section 15.4**

15.9. What logic functions are available at the outputs of a basic ECL gate, as shown in Figure 15.55? What is the advantage of having two opposite outputs?

---

## BiCMOS Logic

A BiCMOS device is essentially a High-Speed CMOS device with a Schottky TTL output stage. By combining the two types of logic, we get a device with low power dissipation, high output drive current, and low propagation delay. Noise margins are comparable to those of TTL.

Figure 15.56 shows the data sheet for a 74BC00 BiCMOS NAND gate. Compare the specifications of this device to High-Speed CMOS and FAST TTL.

Relatively few devices are available in this family at present, although the list of available parts will probably increase over time.

Since this family has a high-current output, it is suitable for driving bus lines. Therefore, manufacturers' offerings are concentrated in the basic gates and interface components, such as octal latches and bus drivers.

## ● GLOSSARY

**CMOS** Complementary metal-oxide-semiconductor. A logic family based on the switching of n- and p-channel MOSFETs.

**Cutoff mode** The operating mode of a bipolar transistor when there is no collector current flowing and the path from collector to emitter is effectively an open circuit. In a digital application, a transistor in cutoff mode is considered OFF.

**Enhancement-mode MOSFET** A MOSFET that creates a conduction path (a channel) between its drain and source terminals when the voltage between gate and source exceeds a specified threshold level.

**High-Speed (silicon-gate) CMOS** A CMOS logic family with a smaller device structure and thus higher speed than standard (Metal-Gate) CMOS.

**MOSFET** Metal-oxide-semiconductor field effect transistor. A MOSFET has three terminals—gate, source, and drain—which are analogous to the base, emitter, and collector of a bipolar junction transistor.

**n-channel enhancement-mode MOSFET** A MOSFET built on a p-type substrate with n-type drain and source

regions. An n-type channel is created in the p-substrate during conduction.

**n-type inversion layer** The conducting layer formed between drain and source when an enhancement-mode n-channel MOSFET is biased ON. Also referred to as the channel.

**NMOS** A logic family based on the switching of n-channel enhancement-mode MOSFETs.

**Ohmic region** The MOSFET equivalent of saturation. When a MOSFET is biased ON, it acts like a relatively low resistance, or "ohmically."

**Open-collector output** A TTL output where the collector of the LOW-state output transistor is brought out directly to the output pin. There is no built-in HIGH-state output circuitry, which allows two or more open-collector outputs to be connected without possible damage.

**p-channel enhancement-mode MOSFET** A MOSFET built on an n-type substrate with p-type drain and source regions. During conduction, a p-type channel is created in the n-substrate.

 **MOTOROLA**

**MC74BC00**

QUAD
2-INPUT
NAND GATE

## Advance Information
# Quad 2-Input NAND Gate

The MC74BC00 is a quad 2-input NAND gate utilizing silicon gate Bi-CMOS technology to achieve operating speed equivalent to FAST parts.

All inputs are equipped with protection circuits against static discharge or transient excess voltage.

- High Speed Operation: $t_{pd}$ = 3.8 ns
- High Drivability: $I_{OL}$ = 20 mA
  $I_{OH}$ = -1.0 mA
- Low Power Consumption: $I_{CC}$ (opr) = 3.0 mA (typ)
- ESD Protection Exceeds 2000 V (MIL Standard)
- The Same Pin Connection and the Same Function as FAST (MC74F00)
- Wide Operation Temperature: -40 to +85°C

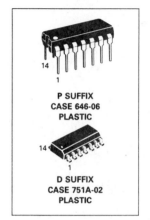

**P SUFFIX
CASE 646-06
PLASTIC**

**D SUFFIX
CASE 751A-02
PLASTIC**

**TRUTH TABLE**

| A | B | Y |
|---|---|---|
| L | L | H |
| L | H | H |
| H | L | H |
| H | H | L |

**LOGIC SYMBOL**

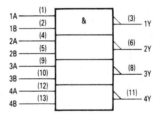

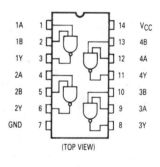

(TOP VIEW)

**INPUT AND OUTPUT EQUIVALENT CIRCUIT**

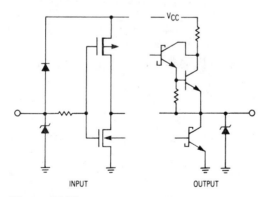

**Figure 15.56**

**74BC00 Data** (Reprinted with permission of Motorola)

*Continued.*

## MC74BC00

### MAXIMUM RATINGS

| Parameter | Symbol | Value | Unit |
|---|---|---|---|
| Supply Voltage | $V_{CC}$ | $-0.5$ to $+7.0$ | V |
| Input Voltage | $V_{IN}$ | $-0.5$ to $V_{CC}+0.5$ | V |
| Output Voltage | $V_{OUT}$ | $-0.5$ to $V_{CC}+0.5$ | V |
| Input Clamp Diode Current | $I_{IK}$ | $\pm 30$ | mA |
| Output Diode Current | $I_{OK}$ | $-30$ | mA |
| Power Dissipation | $P_D$ | 500 ($T_A = +25°C$) | mW |
| Storage Temperature | $T_{stg}$ | $-65$ to $+150$ | °C |

### RECOMMENDED OPERATING CONDITIONS

| Parameter | | Symbol | Min | Typ | Max | Unit |
|---|---|---|---|---|---|---|
| Supply Voltage | | $V_{CC}$ | 4.5 | 5.0 | 5.5 | V |
| Input Voltage | | $V_{IN}$ | 0 | — | $V_{CC}$ | V |
| Output Voltage | | $V_{OUT}$ | 0 | — | $V_{CC}$ | V |
| Output Current | "H" Level | $I_{OH}$ | — | — | $-1.0$ | mA |
| | "L" Level | $I_{OL}$ | — | — | 20 | mA |
| Operating Temperature | | $T_{opr}$ | $-40$ | 25 | 85 | °C |

### DC CHARACTERISTICS (unless otherwise specified, $T_A = -40$ to $+85°C$, $V_{CC} = 4.5$ to 5.5 V)

| Parameter | | Symbol | Test Condition | $V_{CC}$ (V) | Min | Typ* | Max | Unit |
|---|---|---|---|---|---|---|---|---|
| Input Voltage | "H" Level | $V_{IH}$ | | | 2.0 | — | — | V |
| | "L" Level | $V_{IL}$ | | | — | — | 0.8 | V |
| Input Clamp Diode Voltage | | $V_{IK}$ | $I_{IK} = -18$ mA | 4.5 | — | — | $-1.2$ | V |
| Output Voltage | "H" Level | $V_{OH}$ | $I_{OH} = -1.0$ mA | 4.5 | 2.5 | — | — | V |
| | | | $I_{OH} = -1.0$ mA | 4.75 | 2.7 | 3.4 | — | |
| | "L" Level | $V_{OL}$ | $I_{OL} = 20$ mA | 4.5 | — | 0.35 | 0.5 | V |
| Input Current | | $I_I$ | $V_{IN} = V_{CC}$ or GND | 5.5 | — | — | $\pm 1.0$ | $\mu A$ |
| Short Circuit Current (Note 1) | | $I_{OS}$ | | 5.5 | $-60$ | — | $-180$ | mA |
| Static Power Supply Current (total) | | $I_{CCL}$ | $V_{IN} = V_{CC}$ or GND ALL OUTPUT "L" | 5.5 | — | 6.0 | 11 | mA |
| | | $I_{CCH}$ | $V_{IN} = V_{CC}$ or GND ALL OUTPUT "H" | 5.5 | — | — | 10 | $\mu A$ |
| Static Power Supply Current (per bit) (Note 2) | | $\Delta I_{C1}$ | Per Input: $V_{IN} = 0.5$ V Other Input: $V_{CC}$ or GND | | — | — | 1.5 | mA |
| | | $\Delta I_{C2}$ | Per Input: $V_{IN} = V_{CC} - 2.1$ V Other Input: $V_{CC}$ or GND | | — | — | 1.5 | mA |

*All typical values are at $V_{CC} = 5.0$ V, $T_A = 25°C$
(Note 1) Not more than one output should be shorted at a time, nor for more than 1 second.
(Note 2) $\Delta I_{CC}$ specification is the increase in $I_{CCL}$, $I_{CCH}$

### AC CHARACTERISTICS

| Parameter | Symbol | Test Condition | $T_A = +25°C$ $V_{CC} = 5.0$ V | | | $T_A = -40$ to $+85°C$ $V_{CC} = 5.0$ V $\pm 10\%$ | | Unit |
|---|---|---|---|---|---|---|---|---|
| | | | Min | Typ | Max | Min | Max | |
| Propagation Delay | $t_{PHL}$ | $C_L = 50$ pF | 2.5 | 3.7 | 5.7 | 2.5 | 6.7 | ns |
| | $t_{PLH}$ | $R_L = 500$ $\Omega$ | 2.5 | 3.9 | 5.7 | 2.5 | 6.7 | |
| Dynamic Supply Current | $I_{CC}$ | $f = 1.0$ MHz, Output Open | — | 3.0 | 6.0 | — | 7.0 | mA |

**Figure 15.56 (cont.)**

**Phase splitter**   A transistor in a TTL circuit that ensures that the LOW- and HIGH-state output transistors of a totem pole output are always in opposite phase (i.e., one ON, one OFF).

**Saturation mode**   The operating mode of a bipolar transistor when an increase in base current will not cause a further increase in the collector current and the path from collector to emitter is very nearly (but not quite) a short circuit. This is the ON state of a transistor in a digital circuit.

**Schottky barrier diode**   A specialized diode with a forward drop of about $+0.4$ V.

**Schottky transistor**   A bipolar transistor with a Schottky diode across its base-collector junction, which prevents the transistor from going into deep saturation.

**Schottky TTL**   A series of unsaturated TTL logic families based on Schottky transistors. Schottky TTL switches faster than standard TTL due to decreased storage time in its transistors.

**Speed-power product**   A measure of a logic circuit's efficiency, calculated by multiplying its propagation delay by its power dissipation. Unit: picojoule (pJ)

**Storage time**   Time required to transport stored charge away from the base region of a bipolar transistor before it can turn off.

**Substrate**   The foundation of n- or p-type silicon on which an integrated circuit is built.

**Threshold voltage** $V_{GS(Th)}$   The minimum voltage between gate and source of a MOSFET for the formation of the conducting inversion layer (channel).

**Totem pole output**   A type of TTL output with a HIGH and a LOW output transistor, only one of which is active at any time.

**Tristate output**   An output having three possible states: logic HIGH and LOW, and a high-impedance state, in which the output acts as an open circuit.

**Wired-AND**   A connection where open-collector outputs of logic gates are wired together. The logical effect is the ANDing of connected functions.

## ● PROBLEMS

### Section 15.1 Internal Circuitry of TTL Gates

**15.1**   In what logic state is an open TTL input? Why?

**15.2**   Briefly describe the operation of the TTL open-collector inverter shown in Figure 15.5. What is the purpose of the diode?

**15.3**   Briefly explain the operation of a multiple-emitter input transistor used in a TTL NAND gate. Describe how the transistor responds to various combinations of HIGH and LOW inputs.

**15.4**   Draw a wired-AND circuit consisting of three open-collector NAND gates and an output pull-up resistor. The gate inputs are as follows:

Gate 1: Inputs $A$, $B$
Gate 2: Inputs $C$, $D$
Gate 3: Inputs $E$, $F$

Write the Boolean function of the circuit output.

**15.5**   Calculate the minimum value of the pull-up resistor if the circuit drawn in Problem 15.4 is to drive a logic gate having input current $I_{IL} = 0.8$ mA and the NAND gates can sink 12 mA in the LOW output state. (Assume that $V_{OL} = 0.4$ V.)

**15.6**   Draw a circuit consisting only of open-collector gates whose Boolean expression is the product-of-sums expression

$$(A + B)(C + D)(E + F)(G + H).$$

**\*15.7**   Is an open-collector TTL output likely to be damaged if shorted to ground? Why or why not?

**\*15.8**   Is an open-collector TTL output likely to be damaged if shorted to $V_{CC}$? Why or why not?

**15.9**   Draw the totem pole output of a standard TTL gate.

**15.10**   Refer to the TTL NAND gate in Figure 15.21.

**a.** Why are $Q_3$ and $Q_4$ never on at the same time (ideally)?

**b.** How does switching noise originate in a totem pole output? How can the problem be controlled?

**15.11**   Explain briefly why two totem pole outputs should not be connected together.

**15.12**   Derive the truth table of the tristate inverter (active-LOW Enable), shown in Figure 15.28b by

**Figure 15.57**

**Problem 15.13**
**TTL Gate**

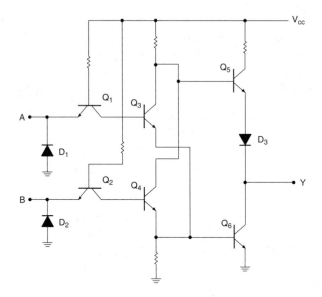

analyzing the states of the transistors under all input conditions.

**15.13** Derive the truth table for the TTL gate shown in Figure 15.57 by analyzing the operation of each of its transistors. What type of gate is it? (i.e., what is its logic function?)

**15.14** Repeat Problem 15.13 for the gate shown in Figure 15.58.

## Section 15.2 Internal Circuitry of NMOS and CMOS Gates

**15.15** State several precautions that should be taken to prevent electrostatic damage to MOSFET circuits.

**15.16** **a.** Draw the circuit symbols for an n-channel and a p-channel enhancement-mode MOSFET.

 **b.** Describe the required bias conditions for each type of MOSFET in the cutoff and ohmic regions.

 **c.** State the approximate channel resistance for a MOSFET in the cutoff and ohmic regions.

**15.17** **a.** Explain why an n-channel MOSFET, rather than a resistor, is used as an active pull-up in an inverter circuit.

 **b.** Why must the two transistors in an NMOS inverter have different channel ON-resistances?

 **c.** Calculate the LOW-state output voltage as a fraction of $V_{DD}$ for the case where both transistors in an NMOS inverter have equal channel ON-resistances.

**15.18** The output voltages of a CMOS gate are closer to 0 V and $V_{DD}$ than are those of an NMOS gate. How is this achieved?

**15.19** Draw the circuit diagram of a CMOS AND gate. Derive the truth table of the gate by analyzing the operation of all the transistors under all possible input conditions.

**15.20** Repeat Problem 15.19 for a CMOS OR gate.

**\*15.21** Figure 15.59 shows a circuit that can switch two analog signals to an automotive speedometer/tachometer. Each sensor produces an analog voltage proportional to its measured quantity. Briefly explain how these analog signals are switched to the display output circuitry.

## Section 15.3 TTL and CMOS Variations

**15.22** Briefly explain how a Schottky barrier diode can improve the performance of a transistor in a TTL circuit.

**Figure 15.58**
**Problem 15.14**
**TTL Gate**

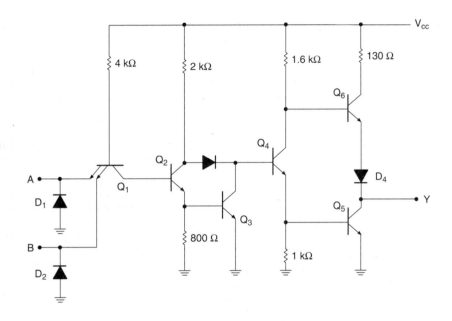

**Figure 15.59**
**Problem 15.21**
**Speedometer/Tachometer**
**Switching Circuit**

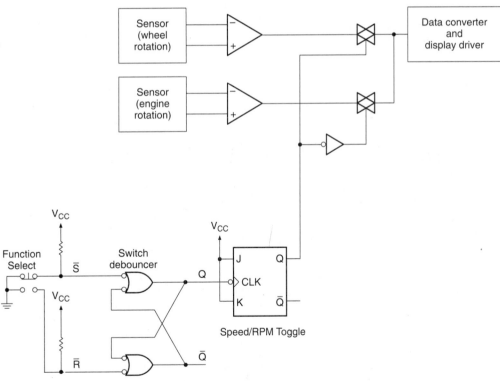

**15.23** Is the speed-power product of a TTL gate affected by the switching frequency of its output? Explain.

**15.24** Use data sheets to calculate the speed-power products of the following gates:

    **a.** 74LS00

    **b.** 74S00

    **c.** 74ALS00

    **d.** 74AS00

    **e.** 74HC00A (quiescent and 10 MHz)

    **f.** 74HCT04

    **g.** 74F00

**15.25** Briefly explain the differences among the following High-Speed CMOS logic families: 74HCXX, 74HC4XXX, 74HCTXX, and 74HCUXX.

**15.26** Assume that the power dissipation of a Metal-Gate or High-Speed CMOS gate increases in proportion to the switching frequency of its output. Calculate the speed-power product of the following gates at 2 MHz, 5 MHz, and 10 MHz:

    **a.** 4011B

    **b.** 74HCT04

    **c.** 74HCU04

## Section 15.4 Other Logic Families

**15.27** State the approximate maximum switching frequencies of an ECL gate. How does this compare to maximum frequencies of other logic families?

**15.28** How is ECL able to switch as fast as it does?

**15.29** State the values of logic HIGH and LOW voltages usually used in ECL devices.

**15.30** What is the basic circuit element of an ECL gate?

**15.31** Briefly explain why ECL has a relatively poor noise immunity, and describe one technique of improving its performance.

**15.32** State one or more advantages of combining CMOS and TTL elements into a single gate, as in the BiCMOS family.

**15.33** Based on the specification for supply voltage, $V_{CC}$, which logic family most closely resembles the BiCMOS family?

**15.34** List the values of $I_{OH}$ and $I_{OL}$ for a 74LS00 NAND, a 74HC00A NAND, and a 74BC00 NAND. Assuming that tristate components with similar specs are available in each family, which family is most suited to driving bus lines? Why?

## ▼ *Answers to Section Review Questions*

### Section 15.1a

**15.1.** a. Provision of logic HIGH when output transistor is OFF b. Limitation of $I_{OL}$ when output transistor is ON

### Section 15.1b

**15.2.** $R_{ext}$ = 611.8 Ω. Minimum standard value: 680 Ω

### Section 15.1c

**15.3.** When the output is HIGH, current flows to ground through a low-impedance path, causing $I_{OH}$ to exceed its rating.

### Section 15.1d

**15.4.** The diode allows the base of $Q_4$ to be pulled LOW through *G*, but will not allow a HIGH at *G* to turn it on. This keeps both output transistors OFF in the high-impedance state and allows them to be in opposite states when the output is enabled.

### Section 15.1e

**15.5.** Noninverting gates are actually double-inverting gates. They require an extra transistor stage to cancel the inversion introduced by NAND or NOR transistor logic.

### Section 15.2a

**15.6.** The thin oxide layer in the gate region can be damaged by overvoltage, such as that caused by electrostatic discharge. If the oxide layer is damaged, it may no longer insulate the gate terminal from the MOSFET substrate, which causes the transistor to malfunction.

### Section 15.2b

**15.7.** It allows complementary operation with an n-channel MOSFET. Specifically, a gate voltage of 0 V turns OFF an n-channel device having a grounded source. The same voltage turns ON the p-channel device whose source is tied to $V_{DD}$. It does so by making the p-channel gate-source voltage more negative than the required threshold.

### Section 15.3

**15.8.** 20.6 pJ, 51.5 pJ, and 103 pJ.

### Section 15.4

**15.9.** OR and NOR. Opposite outputs make both logic levels available without an extra inverter. The inverter would add one gate delay to one of the outputs.

# Interfacing Analog and Digital Circuits

## 16

● CHAPTER OUTLINE

**16.1** Analog and Digital Signals

**16.2** Digital-to-Analog Conversion

**16.3** Analog-to-Digital Conversion

**16.4** Data Acquisition

● CHAPTER OBJECTIVES

$U$pon successful completion of this chapter, you will be able to:

- Define the terms "analog" and "digital" and give examples of each.

- Explain the sampling of an analog signal and the effects of sampling frequency and quantization on the quality of the converted digital signal.

- Draw the block diagram of a generic digital-to-analog converter (DAC) and circuits of a weighted resistor DAC and an R-2R ladder DAC.

- Calculate analog output voltages of a DAC, given a reference voltage and a digital input code.

- Configure an MC1408 integrated circuit DAC for unipolar and bipolar output, and calculate output voltage from known component values, reference voltage, and digital inputs.

- Describe important performance specifications of a digital-to-analog converter.

- Draw the circuit for a flash analog-to-digital converter (ADC) and briefly explain its operation.

- Define "quantization error" and describe its effect on the output of an ADC.

- Explain the basis of the successive approximation ADC, draw its block diagram, and briefly describe its operation.

- Describe the operation of an integrator with constant input voltage.

- Draw the block diagram of a dual slope (integrating) ADC and briefly explain its operation.

- Explain the necessity of a sample and hold circuit in an ADC and its operation.

- Describe data acquisition at an elementary level.

Electronic circuits and signals can be divided into two main categories: analog and digital. Analog signals can vary continuously throughout a defined range. Digital signals take on specific values only, each usually described by a binary number.

Many phenomena in the world around us are analog in nature. Sound, light, heat, position, velocity, acceleration, time, weight, and volume are all analog quantities. Each of these can be represented by a voltage or current in an electronic circuit. This voltage or current is a copy, or analog, of the sound, velocity, or whatever.

We can also represent these physical properties digitally, that is, as a series of numbers, each describing an aspect of the property, such as its magnitude at a particular time. To translate between the physical world and a digital circuit, we must be able to convert analog signals to digital and vice versa.

We will begin by examining some of the factors involved in the conversion between analog and digital signals, including sampling rate, resolution, range, and quantization.

We will then examine circuits for converting digital signals to analog, since these have a fairly standard form. Analog-to-digital conversion has no standard method. We will study several of the most popular: simultaneous (flash) conversion, successive approximation, and dual slope (integrating) conversion.

## 16.1

## Analog and Digital Signals

**Continuous**  *Smoothly connected. An unbroken series of consecutive values with no instantaneous changes.*

**Discrete**  *Separated into distinct segments or pieces. A series of discontinuous values.*

**Analog**  *A way of representing some physical quantity, such as temperature or velocity, by a proportional continuous voltage or current. An analog voltage or current can have any value within a defined range.*

**Digital**  *A way of representing a physical quantity by a series of binary numbers. A digital representation can have only specific discrete values.*

**Analog-to-digital converter**  *A circuit that converts an analog signal at its input to a digital code. (Also called an A-to-D converter, A/D converter, or ADC.)*

**Digital-to-analog converter**  *A circuit that converts a digital code at its input to an analog voltage or current. (Also called a D-to-A converter, D/A converter, or DAC.)*

Electronic circuits are tools to measure and change our environment. Measurement instruments tell us about the physical properties of objects around us. They answer questions such as "How hot is this water?", "How fast is this car going?", and "How many electrons are flowing past this point per second?" These data can correspond to voltages and currents in electronic instruments.

If the internal voltage of an instrument is directly proportional to the quantity being measured, with no breaks in the proportional function, we say that it is an **analog** voltage. Like the property being measured, the voltage can vary continuously throughout a defined range.

For example, sound waves are **continuous** movements in the air. We can plot these movements mathematically as a sum of sine waves of various frequencies. The grooves on a phonographic record are proportional to the sound waves that produce them and mechanically trace the same mathematical functions. When the record is played, the phono cartridge produces a voltage that is also proportional to the original sound waves. This analog audio voltage can be any value between the maximum and minimum voltages of the audio system amplifier.

If an instrument represents a measured quantity as a series of binary numbers, the representation is **digital.** Since the binary numbers in a circuit necessarily have a fixed number of bits, the instrument can represent the measured quantities only as having specific **discrete** values.

A compact disc stores a record of sound waves as a series of binary numbers. Each number represents the amplitude of the sound at a particular time. These numbers are decoded and translated into analog sound waves upon playback. The values of the stored numbers (the encoded sound information) are limited by the number of bits in each stored digital "word."

The main advantage of a digital representation is that it is not subject to the same distortions as an analog signal. Nonideal properties of analog circuits, such as stray inductance and capacitance, amplification limits, and unwanted phase shifts, all degrade an analog signal. Storage techniques, such as magnetic tape, can also introduce distortion due to the nonlinearity of the recording medium.

Digital signals, on the other hand, do not depend on the shape of a waveform to preserve the encoded information. All that is required is to maintain the integrity of the logic HIGHs and LOWs of the digital signal. Digital information can be easily moved around in a circuit and stored in a latch or on some magnetic or optical medium. When the information is required in analog form, the analog quantity is reproduced as a new copy every time it is needed. Each copy is as good as any previous one. Distortions are not introduced between copy generations, as is the case with analog copying techniques, unless the constituent bits themselves are changed.

Digital circuits give us a good way of measuring and evaluating the physical world, with many advantages over analog methods. However, most properties of the physical world are analog. How do we bridge the gap?

We can make these translations with two classes of circuits. An **analog-to-digital converter** accepts an analog voltage or current at its input and produces a corresponding digital code. A **digital-to-analog converter** generates a unique analog voltage or current for every combination of bits at its inputs.

## Sampling an Analog Voltage

**Sample** *An instantaneous measurement of an analog voltage, taken at regular intervals.*

**Sampling frequency** *The number of samples taken per unit time of an analog signal*

> **Quantization** *The number of bits used to represent an analog voltage as a digital number.*
>
> **Resolution** *The difference in analog voltage corresponding to two adjacent digital codes. Analog step size.*

Before we examine actual D/A and A/D converter circuits, we need to look at some of the theoretical issues behind the conversion process. We will look at the concept of **sampling** an analog signal and discover how the **sampling frequency** affects the accuracy of the digital representation. We will also examine **quantization,** or the number of bits in the digital representation of the analog sample, and its effect on the quality of a digital signal.

Figure 16.1 shows a circuit that converts an analog signal to a series of 4-bit digital codes, then back to an analog output. The analog input and output voltages are shown on the two graphs.

There are two main reasons why the output is not a very good copy of the input. First, the number of bits in the digital representation is too low. Second, the input signal is not sampled frequently enough. To help us understand the effect of each of these factors, let us examine the conversion process in more detail.

The analog input signal varies between 0 and 8 volts. This is evenly divided into 16 ranges, each corresponding to a 4-bit digital code (0000 to 1111). We say that the

**Figure 16.1**

**Analog Input and Output Signals**

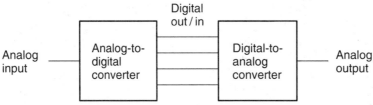

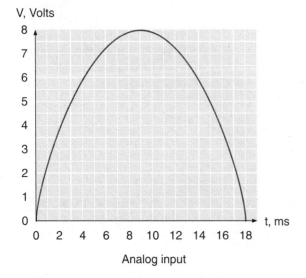

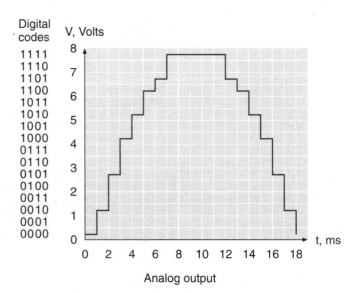

**Table 16.1** 4-bit Digital Codes for 0 to 8 V Analog Range

| Analog Voltage | Digital Code |
|---|---|
| 0.0–0.5 | 0000 |
| 0.5–1.0 | 0001 |
| 1.0–1.5 | 0010 |
| 1.5–2.0 | 0011 |
| 2.0–2.5 | 0100 |
| 2.5–3.0 | 0101 |
| 3.0–3.5 | 0110 |
| 3.5–4.0 | 0111 |
| 4.0–4.5 | 1000 |
| 4.5–5.0 | 1001 |
| 5.0–5.5 | 1010 |
| 5.5–6.0 | 1011 |
| 6.0–6.5 | 1100 |
| 6.5–7.0 | 1101 |
| 7.0–7.5 | 1110 |
| 7.5–8.0 | 1111 |

**Table 16.2** 4-bit Codes for a Sampled Analog Signal

| Time (ms) | Code |
|---|---|
| 0 | 0000 |
| 1 | 0010 |
| 2 | 0101 |
| 3 | 1000 |
| 4 | 1010 |
| 5 | 1100 |
| 6 | 1101 |
| 7 | 1111 |
| 8 | 1111 |
| 9 | 1111 |
| 10 | 1111 |
| 11 | 1111 |
| 12 | 1101 |
| 13 | 1100 |
| 14 | 1010 |
| 15 | 1000 |
| 16 | 0101 |
| 17 | 0010 |
| 18 | 0000 |

signal is quantized into 4 bits. The **resolution,** or analog step size, for a 4-bit quantization is 8 V/16 steps = 0.5 V/step. Table 16.1 shows the codes for each analog range.

The analog input is sampled and converted at the beginning of each time division on the graph. The 4-bit digital code does not change until the next conversion, 1 ms later. This is the same as saying that the system has a sampling frequency of 1 kHz ($f = 1/T = 1/(1 \text{ ms}) = 1$ kHz).

Table 16.2 shows the digital codes for samples taken from $t = 0$ to $t = 18$ ms.

The digital-to-analog converter in Figure 16.1 continuously converts the digital codes to their analog equivalents. Each code produces an analog voltage whose value is the midpoint of the range corresponding to that code.

The A/D converter introduces the greatest inaccuracy at the peak of the analog waveform, where the magnitude of the input voltage changes the least per unit time. There is not sufficient difference between the values of successive analog samples to map them into unique codes. As a result, the output waveform flattens out at the top.

This is the consequence of using a 4-bit quantization, which allows only 16 different analog ranges in the signal. By using more bits, we could divide the analog signal into a greater number of smaller ranges, allowing more accurate conversion of a signal having small changes in amplitude.

Figure 16.2 shows how different levels of quantization affect the accuracy of a digital representation of an analog signal. The analog input is a sine wave, converted to digital codes and back to analog, as in Figure 16.1. The graphs show the analog input and three analog outputs, each of which has been sampled 28 times per cycle, but with different quantizations. The corresponding digital codes range from a maximum negative value of $n$ 0s to a maximum positive value of $n$ 1s for an $n$-bit quantization (e.g., for a 4-bit quantization, maximum negative = 0000, maximum positive = 1111).

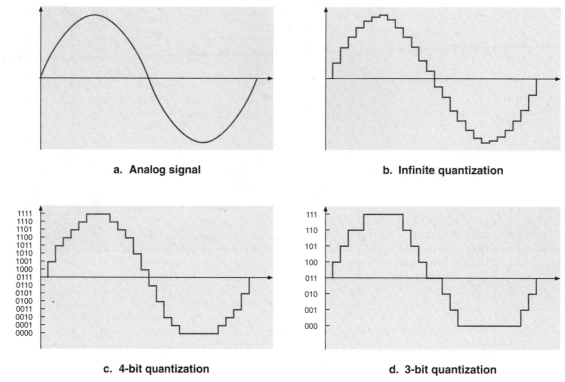

a. Analog signal

b. Infinite quantization

c. 4-bit quantization

d. 3-bit quantization

**Figure 16.2**

**Effect of Quantization**

The first output signal has an infinite number of bits in its quantization. Even the smallest analog change between samples has a unique code. This ideal case is not attainable, since a digital circuit always has a finite number of bits. This does point out that the number of bits in the quantization should be as large as possible.

The 4-bit and 3-bit quantizations in the next two graphs show progressively worse representation of the original signal, especially at the peaks. The change in analog voltage is too small for each sample to have a unique code at these low quantizations.

Figure 16.3 shows how the digital representation of a signal can be improved by increasing its sampling frequency. It shows an analog signal and three analog waveforms resulting from an analog-digital-analog conversion. All waveforms have infinite quantization, but different numbers of samples in the analog-to-digital conversion. As the number of samples decreases, the output waveform becomes a poorer copy of the input.

---

**Section Review Problem for Section 16.1**

16.1. An analog signal has a range of 0 to 24 mV. The range is divided into 32 equal steps for conversion to a series of digital codes. How many bits are in the resultant digital codes? What is the resolution of the A/D converter?

**Figure 16.3**

**Effect of Sampling Frequency**

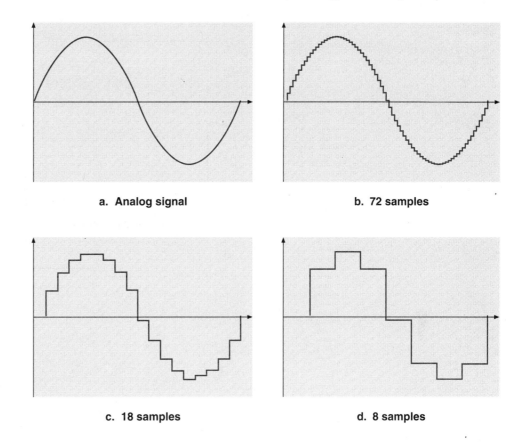

a. **Analog signal**

b. **72 samples**

c. **18 samples**

d. **8 samples**

## 16.2

# Digital-to-Analog Conversion

**Full scale** *The maximum analog reference voltage or current of a digital-to-analog converter.*

Figure 16.4 shows the block diagram of a generalized digital-to-analog converter. Each digital input switches a proportionally weighted current on or off, with the current for the MSB being the largest. The second MSB produces a current half as large. The current generated by the third MSB is one quarter of the MSB current, and so on.

These currents all sum at the operational amplifier's (op amp's) inverting input. The total analog current for an $n$-bit circuit is given by:

$$I_a = \frac{b_{n-1}2^{n-1} + \cdots + b_2 2^2 + b_1 2^1 + b_0 2^0}{2^n} I_{ref}$$

The bit values $b_0$, $b_1$, $\cdots$, $b_n$ can be only 0 or 1. The function of each bit is to include or exclude a term from the general expression.

**Figure 16.4**

**Analysis of a Generalized Digital-to-Analog Converter**

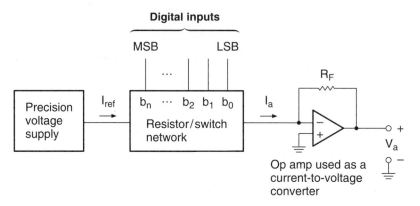

a. **Generic DAC circuit**

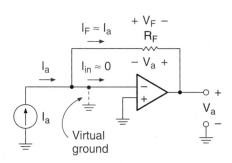

b. **Op amp analysis**

The op amp acts as a current-to-voltage converter. The analysis, illustrated in Figure 16.4b, is the same as for an inverting op amp circuit with a constant input current.

The input impedance of the op amp is the impedance between its inverting (−) and noninverting (+) terminals. This value is very large, on the order of 2 MΩ. If this is large compared to other circuit resistances, we can neglect the op amp input current, $I_{in}$.

This implies that the voltage drop across the input terminals is very small; the inverting and noninverting terminals are at approximately the same voltage. Since the noninverting input is grounded, we can say that the inverting input is "virtually grounded."

Current $I_F$ flows in the feedback loop, through resistor $R_F$. Since $I_a - I_{in} - I_F = 0$ and $I_{in} \approx 0$, then $I_F \approx I_a$. By Ohm's Law, the voltage across $R_F$ is given by $V_F = I_a R_F$. The feedback resistor is connected to the output at one end and to virtual ground at the other. The op amp output voltage is measured with respect to ground. The two voltages are effectively in parallel. Thus, the output voltage is the same as the voltage across the feedback resistor, with a polarity opposite to $V_F$, calculated above.

$$V_a = -V_F = -I_a R_F$$

$$= -\frac{b_{n-1} 2^{n-1} + \cdots + b_2 2^2 + b_1 2^1 + b_0 2^0}{2^n} I_{ref} R_F$$

The range of analog output voltage is set by choosing the appropriate value of $R_F$.

**EXAMPLE 16.1**

Write the expression for analog current, $I_a$, of a 4-bit D/A converter. Calculate values of $I_a$ for input codes $b_3b_2b_1b_0 = 0000, 0001, 1000, 1010,$ and $1111$, if $I_{ref} = 1$ mA.

**Solution**

The analog current of a 4-bit converter is:

$$I_a = \frac{b_3\,2^3 + b_2\,2^2 + 2_1\,b^1 + b_0\,2^0}{2^4}\,I_{ref}$$

$$= \frac{8b_3 + 4b_2 + 2b_1 + b_0}{16}\,(1\text{ mA})$$

$$b_3b_2b_1b_0 = 0000,\ I_a = \frac{(0+0+0+0)\,(1\text{ mA})}{16} = 0$$

$$b_3b_2b_1b_0 = 0001,\ I_a = \frac{(0+0+0+1)\,(1\text{ mA})}{16} = \frac{1\text{ mA}}{16} = 62.5\ \mu\text{A}$$

$$b_3b_2b_1b_0 = 1000,\ I_a = \frac{(8+0+0+0)\,(1\text{ mA})}{16} = \frac{8}{16}\,(1\text{ mA}) = 0.5\text{ mA}$$

$$b_3b_2b_1b_0 = 1010,\ I_a = \frac{(8+0+2+0)\,(1\text{ mA})}{16} = \frac{10}{16}\,(1\text{ mA}) = 0.625\text{ mA}$$

$$b_3b_2b_1b_0 = 1111,\ I_a = \frac{(8+4+2+1)\,(1\text{ mA})}{16} = \frac{15}{16}\,(1\text{ mA}) = 0.9375\text{ mA}$$

Example 16.1 suggests an easy way to calculate D/A analog current. $I_a$ is a fraction of the reference current $I_{ref}$. The denominator of the fraction is $2^n$ for an $n$-bit converter. The numerator is the decimal equivalent of the binary input. For example, for input $b_3b_2b_1b_0 = 0111$, $I_a = (7/16)(I_{ref})$.

Note that when $b_3b_2b_1b_0 = 1111$, the analog current is not the full value of $I_{ref}$, but 15/16 of it. This is one least significant bit less than full scale.

This is true for any D/A converter, regardless of the number of bits. The maximum analog current for a 5-bit converter is 31/32 of full scale. In an 8-bit converter, $I_a$ cannot exceed 255/256 of full scale. This is because the analog value 0 has its own code. An $n$-bit converter has $2^n$ input codes, ranging from 0 to $2^n - 1$.

---

**Section Review Problem for Section 16.2a**

16.2.  Calculate the range of analog voltage of a 4-bit D/A converter having values of $I_{ref} = 1$ mA and $R_F = 10$ kΩ. Repeat the calculation for an 8-bit D/A converter.

---

## Weighted Resistor D/A Converter

Figure 16.5 shows the circuit of a 4-bit weighted resistor D/A converter. The heart of this circuit is a parallel network of binary-weighted resistors. The MSB has a resistor value of $R$. Successive branches have resistor values that double with each bit: $2R$, $4R$, and $8R$. The branch currents decrease by halves with each descending bit value.

**Figure 16.5**

**Weighted Resistor D-to-A Converter**

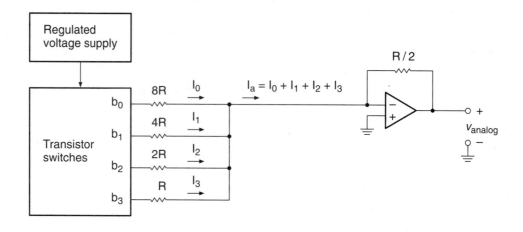

The bit inputs, $b_3$, $b_2$, $b_1$, and $b_0$, are either 0 V or $V_{CC}$. When the corresponding bits are HIGH, the branch currents are:

$$I_3 = V_{CC}/R$$
$$I_2 = V_{CC}/2R$$
$$I_1 = V_{CC}/4R$$
$$I_0 = V_{CC}/8R$$

The sum of branch currents gives us the analog current $I_a$.

$$I_a = \frac{b_3\,V_{CC}}{R} + \frac{b_2\,V_{CC}}{2R} + \frac{b_1\,V_{CC}}{4R} + \frac{b_0\,V_{CC}}{8R}$$

$$= \left[ \frac{b_3}{1} + \frac{b_2}{2} + \frac{b_1}{4} + \frac{b_0}{8} \right] \frac{V_{CC}}{R}$$

We can calculate the analog voltage by Ohm's Law:

$$V_a = -I_a\,R_F = -I_a\,(R/2)$$

$$= -\left[ \frac{b_3}{1} + \frac{b_2}{2} + \frac{b_1}{4} + \frac{b_0}{8} \right] \frac{V_{CC}}{R} \frac{R}{2}$$

$$= -\left[ \frac{b_3}{1} + \frac{b_2}{2} + \frac{b_1}{4} + \frac{b_0}{8} \right] \frac{V_{CC}}{2}$$

$$= -\left[ \frac{b_3}{2} + \frac{b_2}{4} + \frac{b_1}{8} + \frac{b_0}{16} \right] V_{CC}$$

The choice of $R_F = R/2$ makes the analog output a binary fraction of $V_{CC}$.

**EXAMPLE 16.2**

Calculate the analog voltage of a weighted resistor D/A converter when the binary inputs have the following values: $b_3 b_2 b_1 b_0 = 0000, 1000, 1111$. $V_{CC} = 5$ V.

**Solution**

$b_3b_2b_1b_0 = 0000$

$$V_a = -\left[\frac{0}{2} + \frac{0}{4} + \frac{0}{8} + \frac{0}{16}\right] V_{CC} = 0$$

$b_3b_2b_1b_0 = 1000$

$$V_a = -\left[\frac{1}{2} + \frac{0}{4} + \frac{0}{8} + \frac{0}{16}\right] V_{CC} = -\frac{1}{2}(5\text{ V}) = -2.5\text{ V}$$

$b_3b_2b_1b_0 = 1111$

$$V_a = -\left[\frac{1}{2} + \frac{1}{4} + \frac{1}{8} + \frac{1}{16}\right] V_{CC} = -\frac{15}{16}(5\text{ V}) = -4.69\text{ V}$$

The weighted resistor DAC is seldom used in practice. One reason is the wide range of resistor values required for a large number of bits. Another reason is the difficulty in obtaining resistors whose values are sufficiently precise.

A 4-bit converter needs a range of resistors from $R$ to $8R$. If $R = 1$ kΩ, then $8R = 8$ kΩ. An 8-bit DAC must have a range from 1 kΩ to 128 kΩ. Standard value resistors are specified to two significant figures; there is no standard 128-kΩ resistor. We would need to use relatively expensive precision resistors for any value having more than two significant figures.

Another DAC circuit, the R-2R ladder, is more commonly used. It requires only two values of resistance for any number of bits.

---

**Section Review Problem for Section 16.2b**

16.3. The resistor for the MSB of a 12-bit weighted resistor D/A converter is 1 kΩ. What is the resistor value for the LSB?

---

## R-2R Ladder D/A Converter

Figure 16.6 shows the circuit of an R-2R ladder D/A converter. Like the weighted resistor DAC, this circuit produces an analog current that is the sum of binary-weighted currents. An operational amplifier converts the current to a proportional voltage.

The op amp shown is a 34071 High Slew Rate Operational Amplifier. Slew rate is the rate at which the output changes after a step change at the input. If a standard op

**Figure 16.6**

**R-2R Ladder DAC**

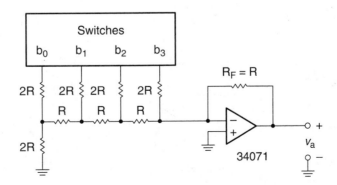

amp (e.g., 741C) is used, the circuit will not accurately reproduce changes introduced by large changes in the digital input.

The method of generating the analog current for an R-2R ladder DAC is a little less obvious than for the weighted resistor DAC. As the name implies, the resistor network is a ladder that has two values of resistance, one of which is twice the other. This circuit is expandable to any number of bits simply by adding one resistor of each value for each bit.

The analog output is a function of the digital input and the value of the op amp feedback resistor. If logic HIGH = $V_{CC}$, logic LOW = 0 V, and $R_F = R$, the analog output is given by:

$$V_a = -\left[\frac{b_3}{2} + \frac{b_2}{4} + \frac{b_1}{8} + \frac{b_0}{16}\right]V_{CC}$$

One way to analyze this circuit is to replace the R-2R ladder with its Thévenin equivalent circuit and treat the circuit as an inverting amplifier. Figure 16.7 shows the equivalent circuit for the input code $b_3b_2b_1b_0 = 1000$.

Figure 16.8a shows the equivalent circuit of the R-2R ladder when $b_3b_2b_1b_0 = 1000$. All LOW bits are grounded, and the HIGH bit connects to $V_{CC}$. We can reduce the network to two resistors by using series and parallel combinations.

The two resistors at the far left of the ladder are in parallel: $2R \parallel 2R = R$. This equivalent resistance is in series with another: $R + R = 2R$. The new resistance is in parallel with yet another: $2R \parallel 2R = R$. We continue this process until we get the simplified circuit shown in Figure 16.8b.

Next, we find the Thévenin equivalent of the simplified circuit. To find $E_{Th}$, calculate the terminal voltage of the circuit, using voltage division.

**Figure 16.7**

**Equivalent Circuit for $b_3b_2b_1b_0 = 1000$**

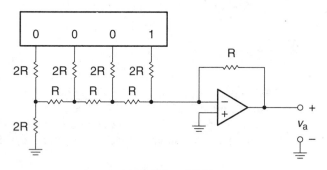

a. $b_3b_2b_1b_0 = 1000$

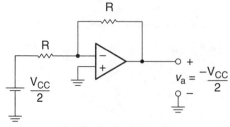

b. Equivalent circuit

**Figure 16.8**

**R-2R Circuit Analysis for $b_3b_2b_1b_0 = 1000$**

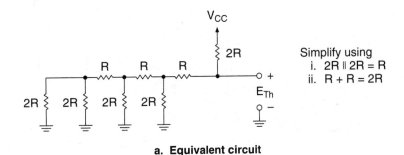

a. Equivalent circuit

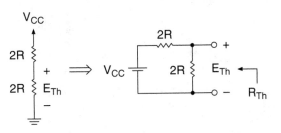

b. Simplified equivalent circuit

c. Thévenin equivalent

$$E_{Th} = \frac{2R}{2R + 2R} V_{CC} = V_{CC}/2$$

$R_{Th}$ is the resistance of the circuit, as measured from the terminals, with the voltage source short-circuited. Its value is that of the two resistors in parallel: $R_{Th} = 2R \parallel 2R = R$.

The value of the Thévenin resistance of the R-2R ladder will always be $R$, regardless of the digital input code. This is because we short-circuit any voltage sources when we make this calculation, which grounds the corresponding bit resistors. The other resistors are already grounded by logic LOWs. We reduce the circuit to a single resistor, $R$, by parallel and series combinations of $R$ and $2R$. Figure 16.9 shows the equivalent circuit.

On the other hand, the value of $E_{Th}$ will be different for each different binary input. It will be the sum of binary fractions of the full-scale output voltage, as previously calculated for the generic DAC.

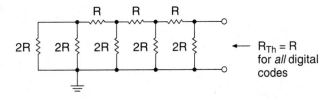

$R_{Th} = ((((((2R \parallel 2R) + R) \parallel 2R) + R) \parallel 2R) + R) \parallel 2R = R$

**Figure 16.9**

**Equivalent Circuit for Calculating $R_{Th}$**

**Figure 16.10**

**Equivalent Circuit of R-2R DAC**

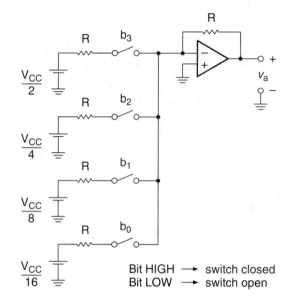

Bit HIGH → switch closed
Bit LOW → switch open

Similar analysis of the R-2R ladder shows that when $b_3b_2b_1b_0 = 0100$, $V_a = -V_{CC}/4$, when $b_3b_2b_1b_0 = 0010$, $V_a = -V_{CC}/8$, and when $b_3b_2b_1b_0 = 0001$, $V_a = -V_{CC}/16$.

If two or more bits in the R-2R ladder are active, each bit acts as a separate voltage source. Analysis becomes much more complicated if we try to solve the network as we did for one active bit.

There is no need to go through a tedious circuit analysis to find the corresponding analog voltage. We can simplify the process greatly by applying the Superposition Theorem. This theorem states that the effect of two or more sources in a network can be determined by calculating the effect of each source separately and adding the results.

The Superposition Theorem suggests a generalized equivalent circuit of the R-2R ladder DAC. This is shown in Figure 16.10. A Thévenin equivalent source and resistance corresponds to each bit. The source and resistance are switched in and out of the circuit, depending on whether or not the corresponding bit is active.

This model is easily expanded. The source for the most significant bit always has the value $V_{CC}/2$. Each source is half the value of the preceding bit. Thus, for a 5-bit circuit, the source for the least significant bit has a value of $V_{CC}/32$. An 8-bit circuit has an LSB equivalent source of $V_{CC}/256$.

---

**Section Review Problem for Section 16.2c**

16.4. Calculate $V_a$ for an 8-bit R-2R ladder DAC when the input code is 10100001.
    Assume that $V_{CC}$ is 10 V.

---

## MC1408 Integrated Circuit D/A Converter

**Multiplying DAC** *A DAC whose output changes linearly with a change in DAC reference voltage.*

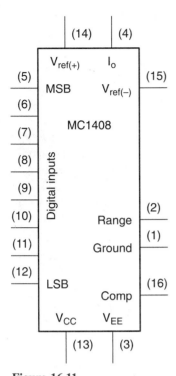

**Figure 16.11**

**MC1408 DAC**

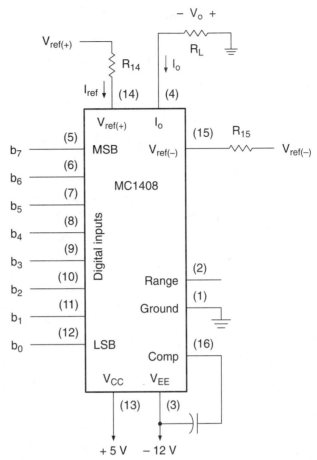

**Figure 16.12**

**MC1408 Configured for Analog Output**

A common and inexpensive DAC is the MC1408 8-bit Multiplying Digital-to-Analog Converter. This device also goes by the designation DAC0808. A logic symbol for this DAC is shown in Figure 16.11.

The output current, $I_o$, flows into pin 4. $I_o$ is a binary fraction of the current flowing into pin 14, as specified by the states of the digital inputs. Other inputs select the range of output voltage and allow for phase compensation.

Figure 16.12 shows the MC1408 in a simple D/A configuration. $R_{14}$ and $R_{15}$ are approximately equal. Pin 14 is approximately at ground potential. This implies:

1. That the DAC reference current can be calculated using only $V_{ref}(+)$ and $R_{14}$ ($I_{ref} = V_{ref}(+)/R_{14}$)

2. That $R_{15}$ is not strictly necessary in the circuit. (It is used primarily to stabilize the circuit against temperature drift.)

The reference voltage *must* be set up so that current flows into pin 14 and out of pin 15. Thus, $V_{ref}(+)$ must be positive with respect to $V_{ref}(-)$. (It is permissible to ground pin 14 if pin 15 is at a negative voltage.)

$I_o$ is given by:

$$I_o = \left[\frac{b_7}{2} + \frac{b_6}{4} + \frac{b_5}{8} + \frac{b_4}{16} + \frac{b_3}{32} + \frac{b_2}{64} + \frac{b_1}{128} + \frac{b_0}{256}\right] \frac{V_{ref}(+)}{R_{14}}$$

Since the output is proportional to $V_{ref}(+)$, we refer to the MC1408 as a **multiplying DAC**.

$I_o$ should not exceed 2 mA. We calculate the output voltage by Ohm's Law: $V_o = -I_o R_L$. The output voltage is negative because current flows from ground into pin 4.

The open pin on the Range input allows the output voltage dropped across $R_L$ to range from $+0.4$ V to $-5.0$ V without damaging the output circuit of the DAC. If the Range input is grounded, the output can range from $+0.4$ to $-0.55$ V. The lower voltage range allows the output to switch about four times faster than it can in the higher range.

▼

**EXAMPLE 16.3**

The DAC circuit in Figure 16.12 has the following component values: $R_{14} = R_{15} = 5.6$ k$\Omega$; $R_L = 3.3$ k$\Omega$. $V_{ref}(+)$ is $+8$ V, and $V_{ref}(-)$ grounded.

Calculate the value of $V_o$ for each of the following input codes: $b_7b_6b_5b_4b_3b_2b_1b_0$ = 00000000, 00000001, 10000000, 10100000, 11111111.

What is the resolution of this DAC?

**Solution**    First, calculate the value of $I_{ref}$.

$$I_{ref} = V_{ref}(+)/R_{14}$$

$$= +8 \text{ V}/5.6 \text{ k}\Omega = 1.43 \text{ mA}$$

Calculate the output current by using the binary fraction for each code. Multiply $-I_o$ by $R_L$ to get the output voltage.

$$b_7b_6b_5b_4b_3b_2b_1b_0 = 00000000$$

$$I_o = 0, V_o = 0$$

$$b_7b_6b_5b_4b_3b_2b_1b_0 = 00000001$$

$$I_o = (1/256) (1.43 \text{ mA}) = 5.58 \text{ }\mu\text{A}$$

$$V_o = -(5.58 \text{ }\mu\text{A})(3.3 \text{ k}\Omega) = -18.4 \text{ mV}$$

$$b_7b_6b_5b_4b_3b_2b_1b_0 = 10000000$$

$$I_o = (1/2) (1.43 \text{ mA}) = 714 \text{ }\mu\text{A}$$

$$V_o = -(714 \text{ }\mu\text{A})(3.3 \text{ k}\Omega) = -2.36 \text{ V}$$

$$b_7b_6b_5b_4b_3b_2b_1b_0 = 10100000$$

$$I_o = = (1/2 + 1/8)(1.43 \text{ mA}) = (5/8)(1.43 \text{ mA}) = 893 \text{ }\mu\text{A}$$

$$V_o = -(893 \text{ }\mu\text{A})(3.3 \text{ k}\Omega) = -2.95 \text{ V}$$

$$b_7b_6b_5b_4b_3b_2b_1b_0 = 11111111$$

$$I_o = (255/256) (1.43 \text{ mA}) = 1.42 \text{ mA}$$

$$V_o = -(1.42 \text{ mA})(3.3 \text{ k}\Omega) = -4.69 \text{ V}$$

Resolution is the same as the output resulting from the LSB: 18.4 mV/step

## Op Amp Buffering of MC1408

The MC1408 DAC will not drive much of a load on its own, particularly when the Range input is grounded. We can use an operational amplifier to increase the output voltage and current. This allows us to select the lower voltage range for faster switching while retaining the ability to drive a reasonable load. The output voltage is limited only by the op amp supply voltages. We use a 34071 High Slew Rate op amp for fast switching.

Figure 16.13 shows such a circuit. $R_{14}$ adjusts the maximum output current of the MC1408 and therefore the maximum output voltage of the op amp. The 0.1-$\mu$F capacitor decouples the +5-V supply. (The manufacturer actually recommends that the +5-V logic supply not be used as a reference voltage. It doesn't matter for a demonstration circuit, but may introduce noise that is unacceptable in a commercial design.) The 75-pF capacitor is for phase compensation.

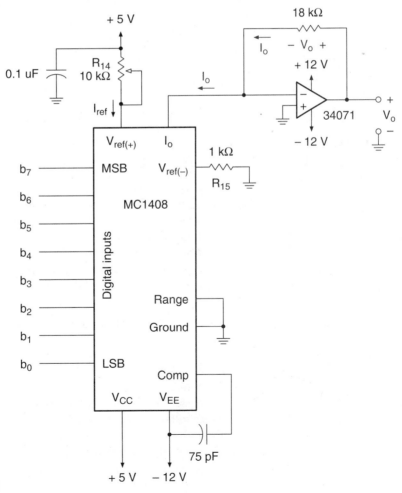

**Figure 16.13**
**DAC With Op Amp Buffering**

$V_o$ is positive because the voltage drop across $R_F$ is positive with respect to the virtual ground at the op amp $(-)$ input. This feedback voltage is in parallel with (i.e., the same as) the output voltage, since both are measured from output to ground.

The output voltage is given by:

$$V_o = \left[ \frac{b_7}{2} + \frac{b_6}{4} + \frac{b_5}{8} + \frac{b_4}{16} + \frac{b_3}{32} + \frac{b_2}{64} + \frac{b_1}{128} + \frac{b_0}{256} \right] \frac{R_F}{R_{14}} V_{ref}$$

$V_o$ can, in theory, be any positive value less than the op amp positive supply $(+12$ V in this case). Any attempt to exceed this voltage makes the op amp saturate. The actual maximum value, if not the same as the op amp's saturation voltage, depends on the values of the two resistors.

**EXAMPLE 16.4**

Refer to the DAC circuit in Figure 16.13. To what value must $R_{14}$ be set to make $V_o = +6$ V for $b_7 b_6 b_5 b_4 b_3 b_2 b_1 b_0 = 10000000$? When $b_7 b_6 b_5 b_4 b_3 b_2 b_1 b_0 = 11111111$ what is the value of $V_o$?

**Solution**

To make $V_o = +6$ V, $I_o = V_o/R_F = 6$ V/18 k$\Omega = 333$ μA.

$$I_o = \tfrac{1}{2}(V_{ref}(+)/R_{14})$$

$$R_{14} = \tfrac{1}{2}(V_{ref}(+)/I_o) = \tfrac{1}{2}(5 \text{ V}/333 \text{ μA})$$

$$= 7.5 \text{ k}\Omega$$

For $b_7 b_6 b_5 b_4 b_3 b_2 b_1 b_0 = 11111111$, $V_o$ is:

$$V_o = (255/256)(18 \text{ k}\Omega/7.5 \text{ k}\Omega)(5 \text{ V}) = 11.953 \text{ V}$$

$$( = \text{full scale} - 1 \text{ LSB})$$

**EXAMPLE 16.5**

Figure 16.14 shows the circuit of an analog ramp (sawtooth) generator built from an MC1408 DAC, an op amp, and an 8-bit synchronous counter. (A ramp generator has numerous analog applications, such as sweep generation in an oscilloscope and frequency sweep in a spectrum analyzer.)

Briefly explain the operation of the circuit and sketch the output waveform. Calculate the step size between analog outputs resulting from adjacent codes. Assume that the DAC is set for $+6$-V output when the input code is 10000000.

Calculate the output sawtooth frequency when the clock is running at 1 MHz.

**Solution**

The 8-bit counter cycles from 00000000 to 11111111 and repeats continuously. This is a total of 256 states.

The DAC output is 0 V for an input code of 00000000 and (12 V $-$ 1 LSB) for a code of 11111111. We know this because a code of 10000000 always gives an output voltage of half the full-scale value (6 V = 12 V/2), and the maximum code gives an output that is one step less than the full-scale voltage. The step size is 12 V/256 steps = 46.9 mV/step. The DAC output advances linearly from 0 to (12 V $-$ 1 LSB) in 256 clock cycles.

Figure 16.15 (see page 710) shows the analog output plotted against the number of input clock cycles. The ramp looks smooth at the scale shown. A section enlarged 32 times shows the analog steps resulting from eight clock pulses.

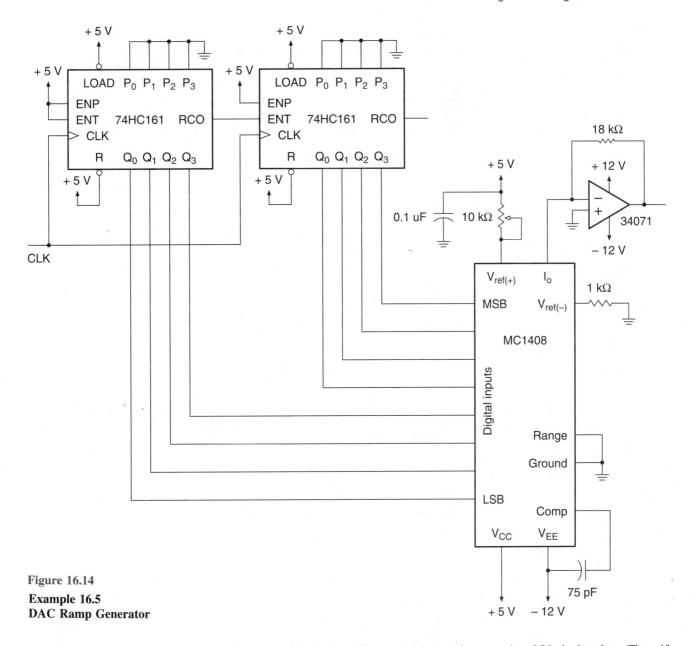

**Figure 16.14**

**Example 16.5**
**DAC Ramp Generator**

One complete cycle of the sawtooth waveform requires 256 clock pulses. Thus, if $f_{CLK} = 1$ MHz, $f_o = 1$ MHz/256 = 3.9 kHz.

(Note that if we do not use a 34071 high slew rate op amp, the sawtooth waveform will not have vertical sides.)

## Bipolar Operation of MC1408

Many analog signals are bipolar, that is, they have both positive and negative values. We can configure the MC1408 to produce a bipolar output voltage. Such a circuit is shown in Figure 16.16.

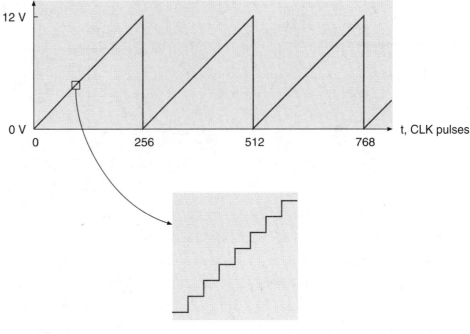

Enlargement (32:1)

We can model the bipolar DAC as shown in Figure 16.16b. The amplitude of the constant-current sink, $I_o$, is set by $V_{ref}(+)$, $R_{14}$, and the binary value of the digital inputs. $I_s$ is determined by Ohm's Law: $I_s = V_{ref}(+)/R_4$.

The output voltage is set by the value of $I_F$:

$$V_o = I_F R'_F = I_F (R_{F1} + R_{F2})$$

By Kirchhoff's Current Law:

$$I_s + I_F - I_o = 0$$

or

$$I_F = I_o - I_s$$

Thus, output voltage is given by:

$$V_o = (I_o - I_s)R_F = I_o R_F - I_s R_F$$

$$= \left[ \frac{b_7}{2} + \frac{b_6}{4} + \frac{b_5}{8} + \frac{b_4}{16} + \frac{b_3}{32} + \frac{b_2}{64} + \frac{b_1}{128} + \frac{b_0}{256} \right] \frac{R_F}{R_{14}} V_{ref} - \frac{R_F}{R_4} V_{ref}$$

How do we understand the circuit operation from this mathematical analysis?

The current sink, $I_o$, is a variable element. The voltage source, $V_{ref}(+)$, remains constant. To satisfy Kirchhoff's Current Law, the feedback current, $I_F$, must vary to the same degree as $I_o$. Depending on the value of $I_o$ with respect to $I_s$, $I_F$ can be positive or negative.

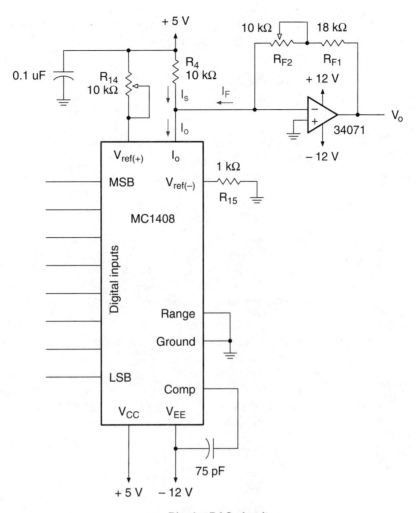

a. Bipolar DAC circuit

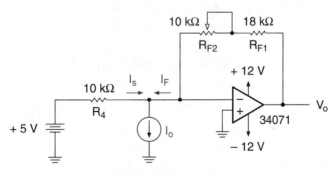

b. Equivalent circuit

Figure 16.16
MC1408 as a Bipolar D/A Converter

We can get some intuitive understanding of the circuit operation by examining several cases of the equation for $V_o$.

**Case 1: $I_o = 0$.** This corresponds to the digital input $b_7b_6b_5b_4b_3b_2b_1b_0 = 00000000$. The output voltage is:

$$V_o = (I_o - I_s)R_F = - I_s R_F$$

This is the maximum negative output voltage.

**Case 2: $0 < I_o < I_s$.** The term $(I_o - I_s)$ is negative, so output voltage is also a negative value.

**Case 3: $I_o = I_s$.** The output is given by:

$$V_o = (I_o - I_s)R_F = 0$$

The digital code for this case could be any value, depending on the setting of $R_{14}$. To make the output symmetrical (i.e., to make maximum positive and maximum negative the same magnitude), set the digital input to 10000000 and adjust $R_{14}$ for 0 V.

**Case 4: $I_o > I_s$.** Since the term $(I_o - I_s)$ is positive, output voltage is positive. The largest value of $I_o$ (and thus the maximum positive output voltage) corresponds to the input code $b_7b_6b_5b_4b_3b_2b_1b_0 = 11111111$.

The magnitude of the maximum positive output voltage of this particular circuit is 2 LSB less than the magnitude of the maximum negative voltage. Specifically, $V_o = (127/128)(R_F/R_4)(V_{ref})$ if $R_4 = 2R_{14}$. The proof of this is left as an exercise.

To summarize:

| Input Code | Output Voltage |
|---|---|
| 00000000 | Maximum negative* |
| 10000000 | 0 V** |
| 11111111 | Maximum positive |

*As adjusted by $R_{F2}$.
**As adjusted by $R_{14}$.

---

**EXAMPLE 16.6**

Calculate the values to which $R_{14}$ and $R_{F2}$ must be set to make the output of the bipolar DAC in Figure 16.19 range from $-12$ V to $(+12$ V $- 2$ LSB$)$. Describe the procedure you would use to set the circuit output as specified.

Confirm that the calculated resistor settings generate the correct values of maximum and minimum output.

**Solution**  Set $R_{14}$ so that the DAC circuit has an output of 0 V when input code is $b_7b_6b_5b_4b_3b_2b_1b_0 = 10000000$. We can calculate the value of $R_{14}$ as follows:

$$\frac{R_F}{2R_{14}} V_{ref} - \frac{R_F}{R_4} V_{ref} = 0$$

The first term is set by the value of the input code. Solving for $R_{14}$, we get:

$$\left[\frac{1}{2R_{14}} - \frac{1}{R_4}\right] R_F V_{ref} = 0$$

$$\frac{1}{2R_{14}} - \frac{1}{R_4} = 0$$

$$\frac{1}{2R_{14}} = \frac{1}{R_4}$$

$$2R_{14} = R_4$$

$$R_{14} = R_4/2 = 10 \text{ k}\Omega/2 = 5 \text{ k}\Omega$$

To set the maximum negative value, set the input code to 00000000 and adjust $R_{F2}$ for $-12$ V. $R_{F2} = R_F - R_{F1}$. Solve the following equation for $R_F$:

$$-\frac{R_F}{R_4} V_{ref} = -12 \text{ V}$$

$$-\frac{R_F}{10 \text{ k}\Omega} (5 \text{ V}) = -12 \text{ V}$$

$$R_F = (12 \text{ V})(10 \text{ k}\Omega)/5 \text{ V} = 24 \text{ k}\Omega$$

$$R_{F2} = 24 \text{ k}\Omega - 18 \text{ k}\Omega = 6 \text{ k}\Omega$$

**Settings**

$R_{14} = R_4/2 = 5 \text{ k}\Omega$ for symmetrical waveform.
$R_{F2} = 6 \text{ k}\Omega$ for output of $\pm 12$ V.

**Check Output Range**

For $b_7b_6b_5b_4b_3b_2b_1b_0 = 00000000$:

$$V_o = \left[\frac{0}{R_{14}} - \frac{1}{R_4}\right] R_F V_{ref} = -\frac{(24 \text{ k}\Omega)(5 \text{ V})}{10 \text{ k}\Omega} = -12 \text{ V}$$

For $b_7b_6b_5b_4b_3b_2b_1b_0 = 11111111$:

$$V_o = \left[\frac{255}{256 R_{14}} - \frac{1}{R_4}\right] R_F V_{ref}$$

$$= \left[\frac{255}{(256)(5 \text{ k}\Omega)} - \frac{1}{10 \text{ k}\Omega}\right] (24 \text{ k}\Omega)(5 \text{ V}) = 11.906 \text{ V}$$

(Note: 12 V $-$ 2 LSB = 12 V $-$ (12 V/128) = 12 V $-$ 94 mV = 11.906 V.)
    The output voltage can be adjusted so that it is absolutely symmetrical, but then the code 10000000 does not correspond to an analog output of 0 V.

---

▲

**Section Review Problem for Section 16.2e**

16.6. Why is the actual maximum value of an 8-bit DAC less than its reference (i.e., its apparent maximum) voltage?

**Figure 16.17**
**DAC Monotonicity**

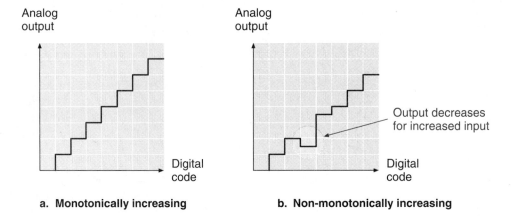

a. **Monotonically increasing**          b. **Non-monotonically increasing**

**Figure 16.18**
**DAC Gain Errors**

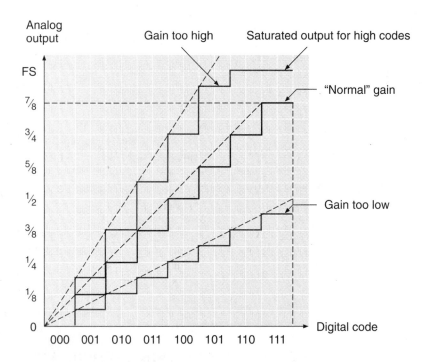

## DAC Performance Specifications

A number of factors affect the performance of a digital-to-analog converter. The major factors are briefly described below.

**Monotonicity.**  The output of a DAC is monotonic if the magnitude of the output voltage increases every time the input code increases. Figure 16.17 shows the output of a DAC that increases monotonically and the output of a DAC that does not.

**Absolute Accuracy.**  This is a measure of DAC output voltage with respect to its expected value.

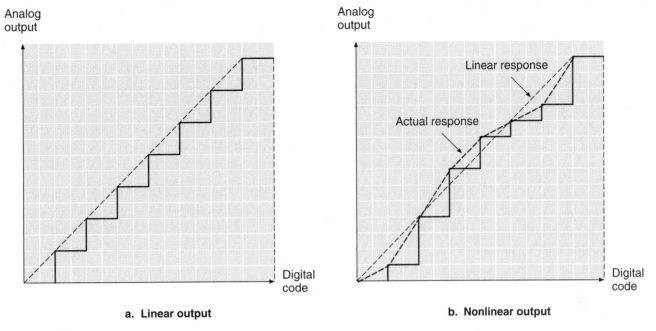

**a. Linear output**                      **b. Nonlinear output**

**Figure 16.19**
**DAC Linearity**

**Relative Accuracy.** Relative accuracy is a more frequently used measurement than absolute accuracy. It measures the deviation of the actual from the ideal output voltage as a fraction of the full-scale voltage. The MC1408 DAC has a relative accuracy of $\pm\frac{1}{2}$ LSB = $\pm$0.195% of full scale.

**Settling Time.** The time required for the outputs to switch and settle to within $\pm\frac{1}{2}$ LSB when the input code switches from all 0s to all 1s. The MC1408 has a settling time of 300 ns for 8-bit accuracy, limiting its output switching frequency to 1/300 ns = 3.33 MHz. Depending on the value of $R_4$, the output resistor, the settling time of the MC1408 may increase to as much as 1.2 $\mu$s when the Range input is open.

**Gain Error.** Gain error primarily affects the high end of the output voltage range. If the gain of a DAC is too high, the output saturates before reaching the maximum output code. Figure 16.18 shows the effect of gain error in a 3-bit DAC. In the high gain response, the last two input codes (110 and 111) produce the same output voltage.

**Linearity Error.** This error is present when the analog output does not follow a straight-line increase with increasing digital input codes. Figure 16.19 shows this error.

**Differential Nonlinearity.** This specification measures the difference between actual and expected step size of a DAC when the input code is changed by 1 LSB. An actual step that is smaller than the expected step can result in a nonmonotonic output.

**Offset Error.** This error occurs when the analog output of a positive-value DAC is not 0 V when the input code is all 0s. Figure 16.20 shows the effect of offset error.

**Figure 16.20**
**DAC Offset Error**

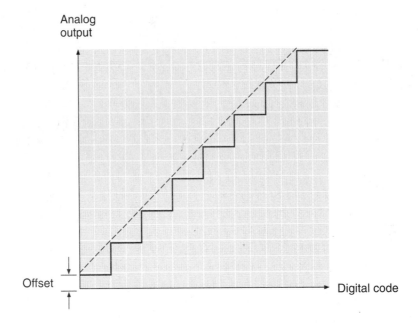

## 16.3

## Analog-to-Digital Conversion

We saw in an earlier section of this chapter that all digital-to-analog converters can be described by a generic form. This is not true of analog-to-digital converters. There are many circuits for converting analog signals to digital codes, each with its own advantages. We will look at several of the most popular.

### Flash A/D Converter

> **Flash converter (or simultaneous converter)**  *An analog-to-digital converter that uses comparators and a priority encoder to produce a digital code.*
>
> **Priority encoder**  *An encoder that will produce a binary output corresponding to the subscript of the highest-priority active input. This is usually defined as the input with the largest subscript.*

Figure 16.21 shows the circuit for a 3-bit **flash analog-to-digital converter.** The circuit consists of a resistive voltage divider, seven analog comparators, a **priority encoder,** and an output latch array.

The voltage divider has a total resistance of $8R$. The resistors are selected to produce seven equally spaced reference voltages ($V_{ref}/16$, $3\ V_{ref}/16$, $5\ V_{ref}/16$, . . . $15\ V_{ref}/16$; each is separated by $V_{ref}/8$). Each reference voltage is fed to the inverting input of a comparator.

**Figure 16.21**

**Flash Converter (ADC)**

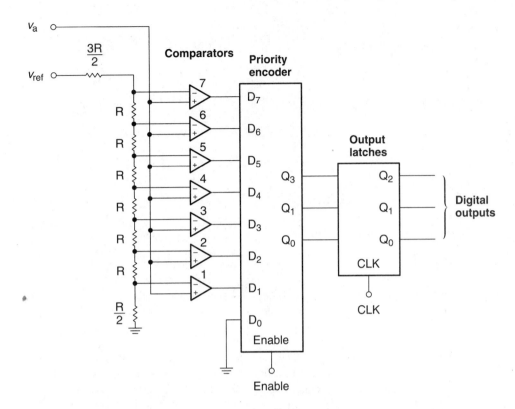

The analog voltage, $V_a$, is applied to the noninverting inputs of all comparators simultaneously. If the analog voltage exceeds the reference voltage of a particular comparator, that comparator switches its output to the HIGH state.

For most analog input values, more than one comparator will have a HIGH output. For example, the reference voltage of comparator 3 is ($5\ V_{ref}/16$). Comparator 4 has a reference voltage of ($7\ V_{ref}/16$). If the analog voltage is in the range ($5\ V_{ref}/16$) $\leq v_a <$ ($7\ V_{ref}/16$), comparators 3, 2, and 1 all have HIGH outputs and comparators 4, 5, 6, and 7 all have LOW outputs.

The priority encoder recognizes that input $D_3$ is the highest-priority active input and produces the digital code 011 at its outputs. The output latches store this value when the *CLK* input is pulsed.

We can regularly sample an analog signal by applying a pulse waveform to the *CLK* input of the latch circuit. The sampling frequency is the same as the clock frequency.

The $D_0$ input of the priority encoder is grounded, rather than connected to a comparator output. No comparator is needed for this input; if $V_a < (V_{ref}/16)$, all comparator outputs are LOW and the resulting digital code is 000.

Figure 16.22 shows the transfer characteristic of the flash ADC with a reference voltage of 8 V. The digital steps are centered on the analog voltages that are whole-number fractions (1/8, 1/4, 3/8, . . . 7/8) of the reference voltage. The transitions are midway between these points. This is why the resistor for the least significant bit is $R/2$, rather than $R$.

The general form of this circuit has $2^n -1$ comparators for an $n$-bit output. For example, an 8-bit flash converter has $2^8 -1 = 255$ comparators. For any large number of bits, the circuit becomes overly complex.

**Figure 16.22**

**Transfer Characteristic of Flash ADC**

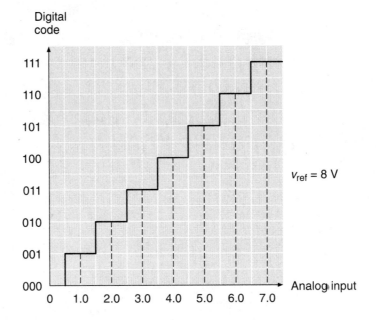

The main advantage of this circuit is its speed. Since the analog input is compared to the threshold values of all possible input codes at one time, conversion occurs in one clock cycle. Clock rates of 25 MHz are attainable with this circuit. Such circuits are used for high-speed applications such as radar and video processing.

## Successive Approximation A/D Converter

**Successive approximation register** *A synchronous circuit used to generate a sequence of closer and closer binary approximations to an analog signal.*

**Quantization error** *Inaccuracy introduced into a digital signal by the inability of a fixed number of bits to represent the exact value of an analog signal.*

Probably the most widely used type of analog-to-digital converter is the successive approximation ADC. The idea behind this type of converter is a technique a computer programmer would call "binary search."

The analog voltage to be converted is a number within a defined range. The search technique works by narrowing down progressively smaller binary fractions of the known range of numbers.

Suppose we know that the analog voltage is a number between 0 and 255, inclusive. We can find the binary value of any randomly chosen number in this range in no more than eight guesses, or approximations, since $2^8 = 256$. Each approximation adds one more bit to the estimated digital value.

The first approximation determines which half of the range the number is in. The second test finds which quarter of the range, the third test which eighth, the fourth test which sixteenth, and so on until we run out of bits.

**EXAMPLE 16.7**    Use a binary search technique to find the value of a number in the range 0 to 255. (The number is 44.)

**Solution**    1. The number must be in the upper or lower half of the range. Cut the range in half: 0–127, 128–255.
Is $x \geq 128$? No. $0 \leq x < 128$.

2. Cut the remaining range in half: 0–63, 64–127.
Is $x \geq 64$? No. $0 \leq x < 64$.

3. Cut the remaining range in half: 0–31, 32–63.
Is $x \geq 32$? Yes. $32 \leq x < 64$.

4. Cut the remaining range in half: 32–47, 48–63.
Is $x \geq 48$? No. $32 \leq x < 48$.

5. Cut the remaining range in half: 32–39, 40–47.
Is $x \geq 40$? Yes. $40 \leq x < 48$.

6. Cut the remaining range in half: 40–43, 44–47.
Is $x \geq 44$? Yes. $44 \leq x < 48$.

7. Cut the remaining range in half: 44–45, 46–47.
Is $x \geq 46$? No. $44 \leq x < 46$.

8. Cut the remaining range in half: 44–45.
Is $x = 45$? No. $x = 44$.

The test criteria for each step in Example 16.7 are phrased so that the answer is always yes or no. (For example, $x \geq 64$? can only be answered yes or no.) Assume that a 1 means yes and a 0 means no. The tests in Example 16.7 give the following sequence of results: 00101100. The decimal equivalent of this binary number is 44, our original value.

A successive approximation ADC such as the one shown in Figure 16.23 applies a similar technique. The circuit has three main components: an analog comparator, a digital-to-analog converter, and a synchronous circuit called a **successive approximation register** (SAR). The SAR is an 8-bit register whose bits can be set and cleared individually, according to a specific control sequence and the logic value at the output of the analog comparator.

When a pulse activates the Start Conversion input, bit $Q_7$ of the SAR is set. This makes the SAR output 10000000. The DAC converts the SAR output to an analog equivalent. When only the MSB is set, this is one half the reference voltage of the DAC.

The DAC output voltage is compared to an analog input voltage. (In effect, the SAR asks, "Is this approximation greater or less than the actual analog voltage?")

If $V_{analog} > V_{DAC}$, the comparator output is HIGH and the MSB remains set. Otherwise, the comparator output is LOW and the MSB is cleared. The process is repeated for all bits.

After all bits have been set or cleared, the End of Conversion (EOC) output generates a pulse. This pulse can be used to load the final digital value into an 8-bit latch.

**Figure 16.23**

**Successive Approximation ADC**

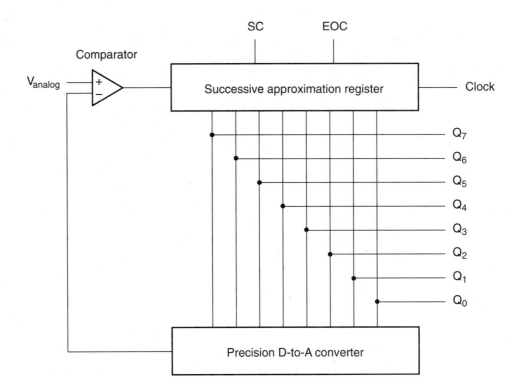

SC = Start Conversion
EOC = End of Conversion

---

**EXAMPLE 16.8**

An 8-bit successive approximation ADC has an analog input voltage of 9.5 V. Describe the steps the circuit performs to generate an 8-bit digital equivalent value if the DAC in the circuit has a reference voltage of 12 V.

**Solution**

Figure 16.24 shows the steps the converter performs to generate the 8-bit digital equivalent of 9.5 V. The conversion process is also summarized in Table 16.3.

The following steps occur for each bit:

1. The bit is set.

2. The digital output is converted to an analog voltage and compared to the actual analog input.

Table 16.3  8-Bit Successive Approximation Conversion

| Bit | New Digital Value | Analog Equivalent | $v_{analog} \geq v_{DAC}$? | Comparator Output | Accumulated Digital Value |
|---|---|---|---|---|---|
| $Q_7$ | 10000000 | 6 V | Yes | 1 | 10000000 |
| $Q_6$ | 11000000 | 9 V | Yes | 1 | 11000000 |
| $Q_5$ | 11100000 | 10.5 V | No | 0 | 11000000 |
| $Q_4$ | 11010000 | 9.75 V | No | 0 | 11000000 |
| $Q_3$ | 11001000 | 9.375 V | Yes | 1 | 11001000 |
| $Q_2$ | 11001100 | 9.5625 V | No | 0 | 11001000 |
| $Q_1$ | 11001010 | 9.46875 V | Yes | 1 | 11001010 |
| $Q_0$ | 11001011 | 9.512625 V | No | 0 | 11001010 |

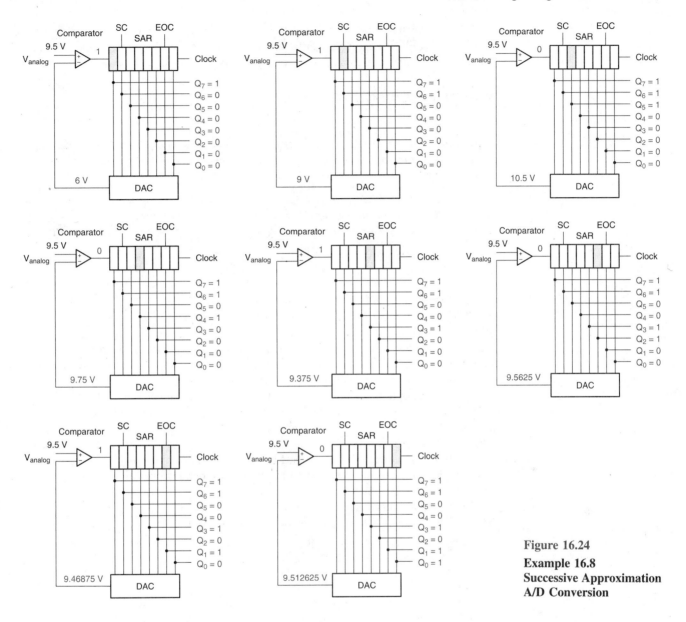

**Figure 16.24**

**Example 16.8**
**Successive Approximation**
**A/D Conversion**

3. If the analog voltage is greater than the DAC output voltage, the bit remains set. Otherwise it is cleared.

There is no exact 8-bit binary value for the analog voltage specified in Example 16.8 (9.5 V). The final answer is within 13 mV, out of 12 V, which is pretty close but not exact. This difference is called **quantization error.**

An advantage of a successive approximation ADC is that the conversion time is always the same, regardless of the analog input voltage. This is not true with all types of analog-to-digital converters. The constant conversion time allows the output to be synchronized so that it can be read at known intervals.

The conversion time can be as few as $(n + 1)$ clock pulses for an $n$-bit device, if a bit is set by a clock edge and cleared asynchronously or by the opposite clock edge. Some SARs require four clock pulses per bit. A typical conversion time is about 30 μs.

## Dual Slope A/D Converter

> **Integrator** *A circuit whose output is the accumulated sum of all previous input values. The integrator's output changes linearly with time when the input voltage is constant.*
>
> **Dual slope ADC** *Also called an integrating ADC. An analog-to-digital converter based on an integrator. The name derives from the fact that during the conversion process the integrator output changes linearly over time, with two different slopes.*

A **dual slope analog-to-digital converter** is based on an **integrator** circuit, such as the one shown in Figure 16.25. The circuit output is proportional to the integral of the input voltage as a function of time. (An integral is the result of the calculus operation of integration.) Integration with respect to time is the summing of instantaneous values of a function over a specified period of time. In other words, the output of an integrator is the accumulated total of all previous values of input voltage.

We can analyze the circuit without calculus under special conditions, such as when the input voltage is constant. An integrator is similar to an inverting amplifier and can be analyzed using similar techniques. Since the input impedance of the op amp is large, there is very little current flowing into its (−) terminal. Ohm's Law thus implies that there is very little voltage difference between the (+) and (−) terminals. Since they are at almost the same potential and the (+) terminal is grounded, we can say that the (−) terminal is "virtually grounded."

If the input voltage is constant, a DC current, $I$, flows in $R$. Since $R$ is connected to the positive terminal of the input voltage source at one end and virtual ground at the other, the entire source voltage drops across the resistor. By Ohm's Law,

$$I = V_{in}/R$$

Since the op amp input impedance is large, most current flows into the capacitor, causing it to charge over time. The current direction defines a polarity for $V_c$, the capacitor voltage.

The op amp output voltage is measured with respect to ground. The capacitor is connected from the op amp output to virtual ground. Therefore, the output voltage, $V_o$, is dropped across the capacitor. Notice that the polarities defined for $V_o$ and $V_c$ are opposite:

$$V_o = -V_c$$

The capacitor voltage is determined by the stored charge, $Q$, and the value of capacitance, $C$:

$$V_c = Q/C$$

**Figure 16.25**
**Integrator**

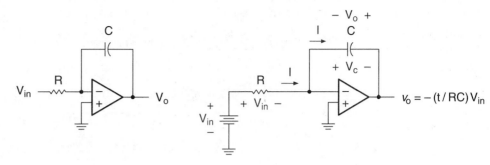

**a. General circuit**                    **b. Constant input voltage**

The current $I$ is the amount of charge flowing past a given point in a fixed time:

$$I = Q/t$$

Thus,

$$V_c = It/C$$

and

$$V_o = -It/C$$

---

Substitute the expression for $I$ into this equation to get

$$V_o = -(t/RC)V_{in}$$

The output of an integrator with a constant input changes linearly with time, with a slope proportional to $V_{in}$ and inversely proportional to $R$ and $C$.

---

This equation describes the *change* in output voltage due to a constant input. When the input goes to 0 V, the capacitor holds its charge (ideally forever; in practice until it leaks away through circuit impedances) and maintains the output voltage at its final value. If a new input voltage is applied, we can use the integrator equation to calculate the change in output, which must then be added to the previous value.

---

**EXAMPLE 16.9**

The integrator circuit of Figure 16.25 has the following component values:

$$C = 0.025 \ \mu F, R = 10 \ k\Omega$$

Sketch the graph of the output voltage if the waveform shown in the graph of Figure 16.26a is applied to the integrator input. The integrator output is originally at 0 V.

**Solution**    We must examine the graph in two sections:

1. From 0 to 3 ms
2. From 3 to 9 ms

**Figure 16.26**

**Example 16.9**
**Integrator Operation**

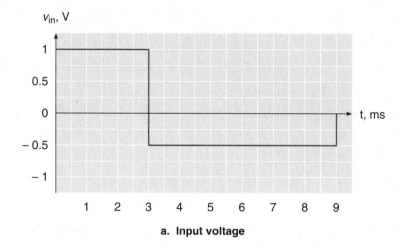

a. **Input voltage**

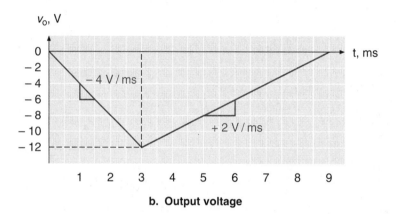

b. **Output voltage**

A different constant input voltage is applied for each section of the graph.

*0 to 3 ms.* The output at 3 ms is given by:

$$v_o(3 \text{ ms}) = v_o(0) - (t/RC) V_{in}$$

$$= 0 \text{ V} - [(3 \text{ ms})/((10 \text{ k}\Omega)(0.025 \text{ μF}))] [1 \text{ V}]$$

$$= -12 \text{ V}$$

The output changes at a rate of −4 V/ms for 3 ms.

*3 to 9 ms.* The output at 9 ms is given by:

$$v_o(9 \text{ ms}) = v_o(3 \text{ ms}) - (t/RC) V_{in}$$

$$= -12 \text{ V} - [(6 \text{ ms})/((10 \text{ k}\Omega)(0.025 \text{ μF}))] [-0.5 \text{ V}]$$

$$= -12 \text{ V} - (-12 \text{ V})$$

$$= 0 \text{ V}$$

The output changes at a rate of +2 V/ms for 6 ms. This cancels the effect of the original input.

**Figure 16.27**

**Dual Slope ADC**

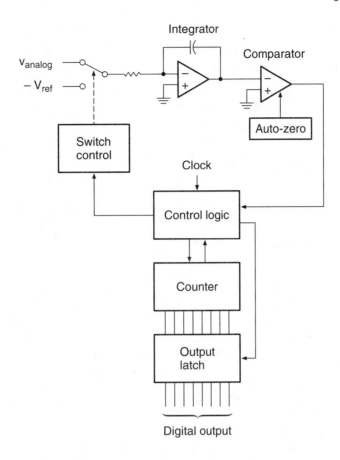

Figure 16.27 shows the block diagram of an 8-bit dual slope analog-to-digital converter. Integrator output voltages for several input values are shown in Figure 16.28. Assume that the integrator has the same $R$ and $C$ values as in Figure 16.25.

1. Before conversion starts, an auto-zero circuit sets the comparator output to 0 V by applying a compensating voltage to the comparator op amp.

2. The input analog voltage causes the integrator output to increase in magnitude, as shown in the left half of Figure 16.28. As soon as this integrator voltage is nonzero, the comparator enables a counter via the control logic.

3. When the counter overflows (i.e., recycles to 00000000), the integrator input is switched from the analog input to $-V_{ref}$.

4. The reference voltage causes the integrator output to move toward 0 V at a known rate, as shown in the right half of Figure 16.28. During this rezeroing time, the counter continues to clock. When the integrator output voltage reaches 0 V, the comparator disables the counter. The digital equivalent of the analog voltage is now contained in the counter.

The reason this works is that in the initial integrating phase, the integrator output operates for a *known time,* producing a final output proportional to the input voltage. In

**Figure 16.28**

**Integrator Outputs for Various Input Voltages**

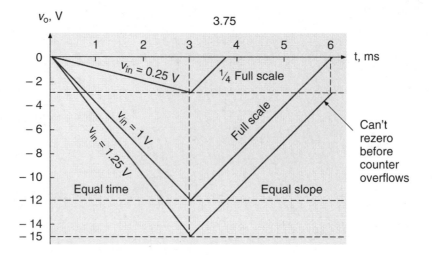

the second phase, the output moves toward zero at a *known rate,* reaching zero in a time proportional to the final voltage of the first phase.

For example, assume that the components of the integrator and the clock rate of the counter are such that a 1-V input corresponds to the full-scale digital output *(FS)*. The integrator output reaches a value of $-12$ V in 3 ms. The time required to rezero the integrator is the same as the initial integrating phase, 3 ms. The counter completes one cycle in the integrating phase and another cycle in the rezeroing phase, so that its final value is 00000000. (Note that this is the result obtained when 1 LSB is added to 11111111.)

If the input voltage is 0.25 volts, the integrator output is $-3$ V after 3 ms (one counter cycle). Since the integrator always rezeros at the same rate (4 V/ms), the rezeroing time is 0.75 ms, or one fourth of a counter cycle (since 12 V/4 = 3 V). The counter has time to reach state 01000000 or $\frac{1}{4}$ *FS.*

If we attempt to measure a voltage beyond that corresponding to full scale, the integrator output cannot rezero within the second counter cycle. Usually, an output pin on the ADC activates to show this condition. Some digital multimeters that use dual slope ADCs show an overvoltage or out-of-range condition by blanking the display, except for a leading digit 1.

One advantage of a dual slope ADC is its accuracy. One particular dual slope ADC is accurate to within $\pm0.05\% \pm 1$ count. This accuracy is balanced against a relatively slow conversion time, in the milliseconds, compared to microseconds for a successive approximation ADC and nanoseconds for a flash converter.

Another advantage is the ability of the integrator to reject noise. If we assume that noise voltage is random, then it will be positive about half the time and negative about half the time. Over time it should average out to zero.

As was alluded to above, a common application of this device is as a voltmeter circuit, where speed is less important than accuracy.

---

**Section Review Problems for Section 16.3**

16.7. Suppose that the dual slope ADC described above (same component values) has an input voltage of 0.375 V (3/8 full scale).

a. What is the slope of the integrator voltage during the integrating phase?

b. What is its slope during the rezeroing phase?

c. How much time elapses during the rezeroing phase?

d. What digital code is contained in the output latch after the conversion is complete?

## Sample and Hold Circuit

**Sample and hold circuit**  *A circuit that samples an analog signal at periodic intervals and holds the sampled value long enough for an ADC to convert it to a digital code.*

For the sake of analysis, we have been assuming that the analog input voltage of any analog-to-digital converter is constant. This is an actual requirement. Most of these circuits will not produce a correct digital code if the analog voltage at the input changes during conversion time.

Unfortunately, most analog signals are not constant. Usually, we want to sample these signals at periodic intervals and generate a series of digital codes that tells us something about the way the input signal is changing over time. A circuit called a **sample and hold circuit** must be used to bridge the gap between a changing analog signal and a requirement for a constant ADC input voltage.

Figure 16.29 shows a basic sample and hold circuit. The voltage followers act as buffers with high input and low output impedances. The transmission gate is enabled during the sampling period, during which it charges the hold capacitor to the current value of the analog signal. During the hold period, the capacitor retains its charge, thus preserving the sampled analog voltage. The high input impedance of the output voltage follower prevents the capacitor from discharging significantly during the hold period.

Figure 16.30 shows how a sample and hold circuit produces a steady series of constant analog voltages for an ADC input. Since these sampled values have yet to be converted to digital codes, they can take on any value within the analog range; they are not yet limited by the number of bits in the quantization.

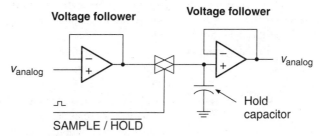

**Figure 16.29**
**Sample and Hold Circuit**

**Figure 16.30**
**Sample and Hold Output**

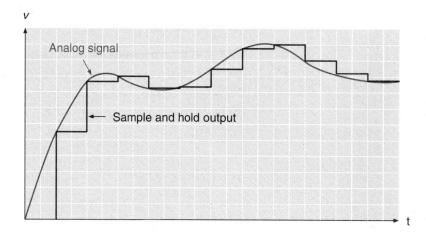

## 16.4

# Data Acquisition

> **Data acquisition network** *A circuit, often used in conjunction with a computer system, that gathers digital data from several analog sources.*
>
> **Parallel port** *A group of parallel inputs and/or outputs to a microcomputer.*

In many industrial and other applications, it is desirable to monitor continuously the values of a number of analog variables in a system. These values can be used to trigger various control responses, set alarms, provide an ongoing record of a process, and so forth. Data are gathered from various monitored points by a **data acquisition network.**

Figure 16.31 shows the block diagram of an 8-channel data acquisition system. It consists of eight analog inputs, an 8-channel *analog* multiplexer, an 8-bit successive approximation ADC, and a microcomputer (MCU). The system provides the MCU with 8-bit data from eight separate analog sources.

The MCU is a very complex device that requires a lot more understanding than we will get from a short examination of the data acquisition network. For now, we need only understand the idea of a **parallel port.**

Inputs and outputs of a microcomputer are organized in groups called "ports." A parallel port is a parallel group of inputs and/or outputs. Often, any particular line can be designated as an input or an output, depending on the status of a bit in an MCU register called the data direction register.

In the system of Figure 16.31, two ports, Port A and Port B, are used. All lines of Port A, $PA_0$ to $PA_7$, are configured as inputs to the MCU. This port gets 8-bit digital values from the ADC output.

Port B is used to control the ADC and the analog MUX. All lines are configured as outputs, except $PB_1$. Lines $PB_2$ to $PB_4$ select an analog channel. The analog value at the MUX output, $Y$, is presented to the ADC input. (A sample and hold circuit could be inserted between the MUX output and the ADC input, if necessary.)

Upon receiving the Start Conversion signal from $PB_0$, the ADC generates an equivalent 8-bit digital code. When conversion is complete, the End of Conversion signal is sent to $PB_1$. This tells the MCU to read the bits at Port A.

**Figure 16.31**

**8-Channel Data Acquisition Network**

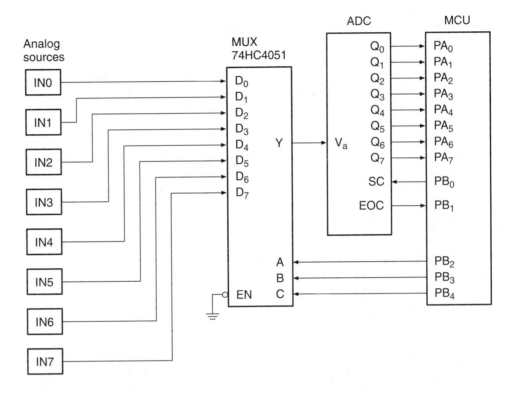

What happens next depends on the application. Typically, the digital data will be stored, compared, or processed in some way. The states of $PB_2$ to $PB_4$ can be changed to select the next analog channel. This continues until all channels have been sampled, converted, and the digital equivalents read by the MCU, which probably has a different place to store data from each channel. After this, the cycle repeats.

## ● GLOSSARY

**Analog** A way of representing some physical quantity, such as temperature or velocity, by a proportional continuous voltage or current. An analog voltage or current can have any value within a defined range.

**Analog-to-digital converter** A circuit that converts an analog signal at its input to a digital code. (Also called an A-to-D converter, A/D converter, or ADC.)

**Continuous** Smoothly connected. An unbroken series of consecutive values with no instantaneous changes.

**Data acquisition network** A circuit, often used in conjunction with a computer system, that gathers digital data from several analog sources.

**Digital** A way of representing a physical quantity by a series of binary numbers. A digital representation can have only specific discrete values.

**Digital-to-analog converter** A circuit that converts a digital code at its input to an analog voltage or current. (Also called a D-to-A converter, D/A converter, or DAC.)

**Discrete** Separated into distinct segments or pieces. A series of discontinuous values.

**Dual slope ADC** Also called an integrating ADC. An analog-to-digital converter based on an integrator. The name derives from the fact that during the conversion process the integrator output changes linearly over time, with two different slopes.

**Flash converter (or simultaneous converter)** An analog-to-digital converter that uses comparators and a priority encoder to produce a digital code.

**Full scale** The maximum analog reference voltage or current of a digital-to-analog converter.

**Integrator** A circuit whose output is the accumulated sum of all previous input values. The integrator's output changes linearly with time when the input voltage is constant.

**Multiplying DAC** A DAC whose output changes linearly with a change in DAC reference voltage.

**Parallel port** A group of parallel inputs and/or outputs to a microcomputer.

**Priority encoder** An encoder that will produce a binary output corresponding to the subscript of the highest-priority active input. This is usually defined as the input with the largest subscript.

**Quantization** The number of bits used to represent an analog voltage as a digital number.

**Quantization error** Inaccuracy introduced into a digital signal by the inability of a fixed number of bits to represent the exact value of an analog signal.

**Resolution** The difference in analog voltage corresponding to two adjacent digital codes. Analog step size.

**Sample** An instantaneous measurement of an analog voltage, taken at regular intervals.

**Sample and hold circuit** A circuit that samples an analog signal at periodic intervals and holds the sampled value long enough for an ADC to convert it to a digital code.

**Sampling frequency** The number of samples taken per unit time of an analog signal.

**Successive approximation register** A synchronous circuit used to generate a sequence of closer and closer binary approximations to an analog signal.

## ● PROBLEMS

### Section 16.1 Analog and Digital Signals

**16.1** An analog signal with a range of 0 to 12 V is converted to a series of 3-bit digital codes. Make a table similar to Table 16.1 showing the analog range for each digital code.

**\*16.2** Sketch the positive half of a sine wave with a peak voltage of 12 V. Assume that this signal will be quantized according to the table constructed in Problem 16.1. Write the digital codes for the points 0, $T/8$, $T/4$, $3T/8$, . . . , $T$ where $T$ is the period of the half sine wave.

**\*16.3** Repeat Problems 16.1 and 16.2 for a 4-bit quantization.

**16.4** Write the 3-bit and 4-bit digital codes for the points 0, $T/16$, $T/8$, $3T/16$, . . . , $T$ for the half sine wave described in Problem 16.2.

**16.5** An analog-to-digital converter divides the range of an analog signal into 64 equal parts. The analog input has a range of 0 to 500 mV. How many bits are there in the resultant digital codes? What is the resolution of the A/D converter?

**16.6** Repeat Problem 16.5 if the analog range is divided into 256 equal parts.

**\*\*16.7** The analog range of a signal is divided into $m$ equal parts, yielding a digital quantization of $n$ bits. If the range is divided into $2m$ parts, how many bits are in the equivalent digital codes? (That is, how many extra bits do we get for each doubling of the number of codes?)

### Section 16.2 Digital-to-Analog Conversion

**16.8**  **a.** Calculate the analog output voltage, $V_a$, for a 4-bit DAC when the input code is 1010.

   **b.** Calculate $V_a$ for an 8-bit DAC when the input code is 10100000.

   **c.** Compare the results of parts a and b. What can you conclude from this comparison?

**16.9**  **a.** Calculate the analog output voltage, $V_a$, for a 4-bit DAC when the input code is 1100.

   **b.** Calculate $V_a$ for an 8-bit DAC when the input code is 11001000.

   **c.** Compare the results of parts a and b. What can you conclude from this comparison? How does this differ from the comparison made in Problem 16.8?

**16.10** Refer to the generalized D/A converter in Figure 16.4. For $I_{ref} = 500$ μA and $R_F = 22$ kΩ,

**Figure 16.32**

**Problem 16.17**
**Waveform**

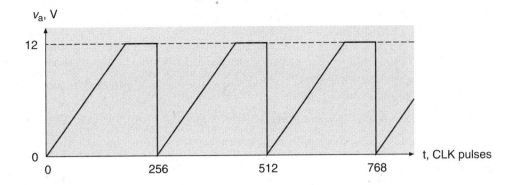

**Figure 16.33**

**Problem 16.18**
**Waveform**

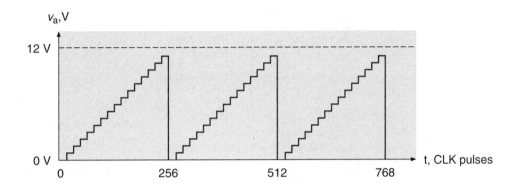

calculate the range of analog output voltage, $V_a$, if the DAC is a 4-bit circuit. Repeat the calculation for an 8-bit DAC.

**16.11** The resistor for the MSB of a 16-bit weighted resistor D/A converter is 1 kΩ. List the resistor values for all bits. What component problem do we encounter when we try to build this circuit?

**16.12** Draw the circuit for an 8-bit R-2R ladder DAC.

**16.13** Calculate the value of $V_a$ of an R-2R ladder DAC when digital inputs are as follows. $V_{CC}$ = 12 V.

     **DCBA**
**a.** 1 1 1 1
**b.** 1 0 1 1
**c.** 0 1 1 0
**d.** 0 0 1 1

**16.14** An MC1408 DAC is configured as shown in Figure 16.12. $R_{14} = R_{15} = 6.8$ kΩ, $V_{ref}$ (+) = +12 V, $V_{ref}$ (−) = ground, and $R_L = 2.2$ kΩ. Calculate the output voltage, $V_a$, for the following digital input codes: 00000000,

00000001, 10000000, 10101010, 11100010, 11111111.

**16.15** Calculate the resolution of the DAC in Problem 16.14.

**16.16** Refer to the positive-value MC1408 DAC in Figure 16.13. To what value should $R_{14}$ be set to make the output voltage +12 V when $b_7b_6b_5b_4b_3b_2b_1b_0 = 11111111$? When $R_{14}$ is set to this value, what is the output voltage for an input code of 10000000? (The answers are *not* the same as in Example 16.4.)

**16.17** The waveform in Figure 16.32 is observed at the output of the DAC ramp generator of Figure 16.14. (Compare this to the proper waveform, found in Figure 16.15.) What is likely to be the problem with the circuit? Can it be easily fixed? How?

**\*16.18** The waveform in Figure 16.33 is observed at the output of the DAC ramp generator in Figure 16.14. What is likely to be the problem with the circuit?

**16.19** Refer to the bipolar DAC circuit in Figure 16.16. Describe how you would adjust the output for a range of $-10$ V to $(+10$ V $- 2$ LSB). Include values of variable components. Calculate the resolution of this circuit.

## Section 16.3 Analog-to-Digital Conversion

**16.20** How many comparators are needed to construct an 8-bit flash converter? Sketch the circuit of this converter. (It is only necessary to show a few of the comparators and indicate how many there are.)

**16.21** Briefly explain the operation of a flash ADC. What is the purpose of the priority encoder? Explain how the latch can be used to synchronize the output to a particular sampling frequency.

**16.22** Why do we choose a value of $R/2$ for the LSB resistor of a flash ADC?

**16.23** An 8-bit successive approximation ADC has a reference voltage of $+16$ V. Describe the conversion sequence for the case where the analog input is 4.75 V. Summarize the steps in Table 16.4. (Refer to Example 16.8.)

**16.24** What is displayed on the seven-segment display in Figure 16.34 when $v_{analog} = 5.25$ V? Assume that the reference voltage is 12 V and that the display can show hex digits.

**16.25** Describe the operation of each part of the successive approximation ADC shown in Figure 16.34 when the analog input changes from 5.25 V to 8.0 V. What is the new number displayed on the seven-segment display?

**16.26** **a.** An 8-bit successive approximation ADC has a reference voltage of 12 V. Calculate the resolution of this ADC.

**b.** The analog input voltage to the ADC in part a is 8 V. Can this input voltage be repre-sented exactly? What digital code represents the closest value to 8 V? What exact analog value does this represent? Calculate the percent error of this conversion.

**\*16.27** An 8-bit dual slope analog-to-digital converter has a reference voltage of 16 V. The integrator component values are: R = 80 k$\Omega$, C = 0.1 $\mu$F. The analog input voltage is 14 V.

Calculate the slope of the integrator voltage during:

**a.** the integrating phase, and

**b.** the rezeroing phase.

**c.** How much time elapses during the rezeroing phase?

(Assume that (1) the integrating and rezeroing time are equal if the integrator output is at full scale, and (2) the reference voltage will rezero the integrator from full scale in exactly one counter cycle.)

**d.** Sketch the integrator output waveform.

**e.** What digital code is contained in the output latch after the conversion is complete?

**\*16.28** Repeat Problem 16.27 if the analog input voltage is 3 V.

**\*\*16.29** Repeat Problem 16.27 if the analog input voltage is 18 V.

**16.30** Make a sketch of a basic sample and hold circuit and briefly explain its operation.

**16.31** Explain why a sample and hold circuit may be needed at the input of an analog-to-digital converter.

## Section 16.4 Data Acquisition

**16.32** Briefly describe the operation of a simple data acquisition network, such as the one shown in Figure 16.31.

| Table 16.4 Table for Problem 16.23 | Bit | New Digital Value | Analog Equivalent | $v_{analog} \geq v_{DAC}$? | Comparator Output | Accumulated Digital Value |
|---|---|---|---|---|---|---|
| | $Q_7$ | | | | | |
| | $Q_6$ | | | | | |
| | $Q_5$ | | | | | |
| | $Q_4$ | | | | | |
| | $Q_3$ | | | | | |
| | $Q_2$ | | | | | |
| | $Q_1$ | | | | | |
| | $Q_0$ | | | | | |

**Figure 16.34**

**Problem 16.24**
**Successive Approximation**
**ADC and Seven-Segment**
**Display**

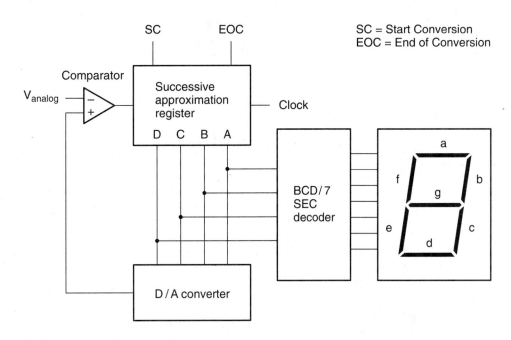

▼ *Answers to Section Review Questions*

**Section 16.1**

**16.1.** 5 bits ($2^5$ = 32). Resolution = 24 mV/32 steps = 0.75 mV/step.

**Section 16.2a**

**16.2.** 4-bit: $I_a$ = 0 to (15/16)(1 mA) = 0 to 0.9375 mA; $V_a = -I_a \ R_F = 0$ to $-9.375$ V  8-bit: $I_a$ = 0 to (255/256)(1 mA) = 0 to 0.9961 mA; $V_a$ = 0 to $-9.961$ V

**Section 16.2b**

**16.3.** 2.048 MΩ.

**Section 16.2c**

**16.4.** $V_a$ = (10 V/2) + (10 V/8) + (10 V/256) = 6.29 V.

**Section 16.2d**

**16.5.** The maximum switching speed is higher if we choose the lower range of output voltage.

**Section 16.2e**

**16.6.** The output 0 V requires its own code. This leaves 255, not 256, codes for the remaining output values. The maximum value of a positive-only output is 255/256 of the reference voltage. A bipolar DAC ranges from $-128/128$ to $+127/128$ of the reference voltage.

**Section 16.3**

**16.7.** a. $-1.5$ V/ms; b. $+4$ V/ms; c. 1.125 ms; d. 01100000.

# Memory

● CHAPTER OUTLINE

**17.1** Basic Memory Concepts
**17.2** Random Access Read/Write Memory (RAM)

**17.3** Read Only Memory (ROM)
**17.4** Sequential Memory: FIFO and LIFO

● CHAPTER OBJECTIVES

Upon successful completion of this chapter, you will be able to:

- Describe basic memory concepts of address and data.

- Explain how latches and flip-flops act as simple memory devices and sketch simple memory systems based on these devices.

- Distinguish between random access read/write memory (RAM) and read only memory (ROM).

- Describe various types of ROM cells and arrays: mask-programmed, fusible-link, and UV erasable.

- Use various types of ROM in simple applications, such as seven-segment decoding and digital function generation.

- Describe the basic configuration and operation of two types of sequential memory: first-in-first-out (FIFO) and last-in-first-out (LIFO).

● INTRODUCTION

In the last few years, memory has become one of the most important topics in digital electronics. This is tied closely to the increasing prominence of cheap and readily available microprocessor chips.

The simplest memory is a device we are already familiar with: the D flip-flop. This device stores a single bit of information as long as is necessary. This simple concept is at the heart of all memory devices.

The other basic concept of memory is the organization of stored data. Bits are stored in locations specified by an "address," a unique number that tells a digital system how to find data that have been previously stored. (By analogy, think of your street address: a unique way to find you and anyone you live with.)

Some memory can be written to and read from in random order; this is called random access read/write memory (RAM). Other memory can be read only: it is called read only memory (ROM). Yet another type of memory, sequential memory, can be read or written only in a specific sequence. There are several variations on all these basic classes.

## 17.1

## Basic Memory Concepts

**Memory**  *A device for storing digital data in such a way that they can be recalled for later use in a digital system.*

**Data**  *Binary digits (0s and 1s) that contain some kind of information. The digital contents of a memory device.*

**Address**  *A number, represented by the binary states of a group of inputs or outputs, uniquely defining the location of data stored in a memory device.*

**Write**  *Store data in a memory device.*

**Read**  *Retrieve data from a memory device.*

**Byte**  *A group of 8 bits.*

**Nibble**  *Half a byte; 4 bits.*

### Address and Data

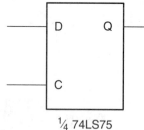

¼ 74LS75

**Figure 17.1**
**D-Type Latch**

A **memory** is a digital device or circuit that can store one or more bits of **data.** The simplest memory device, a D-type latch, shown in Figure 17.1, can store 1 bit. A 0 or 1 is stored in the latch and remains there until changed.

A simple extension of the single D-type latch is an array of latches that can store a group of bits. The 74LS374 octal latch shown in Figure 17.2 will store the 8 bits present at the $D$ inputs when a clock pulse is applied. The data are available at the eight $Q$ outputs simultaneously as long as the $\overline{OC}$ (Output Control) line is held LOW.

A group of 8 bits is called a **byte** and is commonly used as a standard data size in digital and microcomputer systems. It is a convenient size, since one byte can represent one ASCII character plus a parity bit or two hexadecimal digits or two binary-coded decimal digits. (Half a byte is called a **nibble.**) For example,

$$11010111 = \text{ASCII ``W''} + \text{even parity bit } (57_{16})$$

$$01101011 = 6B_{16}$$

$$00111001 = 39_{BCD}$$

We can further extend the concept of memory as an array of latches if we connect four of the 74LS374 octal latches as shown in Figure 17.3.

The forward slash (/) through the input and output data lines and the adjacent number 8 indicate eight parallel lines called the input data bus and output data bus. This notation saves us drawing the lines individually.

**Figure 17.2**

**74LS374 Octal Latch**
(Reprinted by permission of
Texas Instrument)

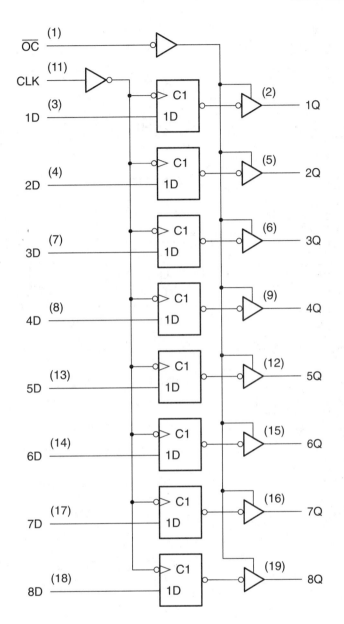

The circuit can store any 8-bit number in one of the four latches. The other latches can hold other 8-bit data at the same time.

Independent decoders steer a clock pulse or Output Control signal to one of the four latches. The input data are connected to all the latch inputs and stored in the latch selected by the states of lines $A_1$ and $A_0$. For example, if $A_1A_0 = 10$ and a clock pulse is applied to the *CLK* input line, the pulse will be steered to the third latch from the top, and the data on the input data lines will be stored there.

We can **read** the contents of any latch by using $Y_1$ and $Y_0$ to select the latch and setting the $\overline{OC}$ input LOW. The output decoders steer the $\overline{OC}$ signal to the selected latch. The output lines from all latches are connected to an output data bus. Only the data from

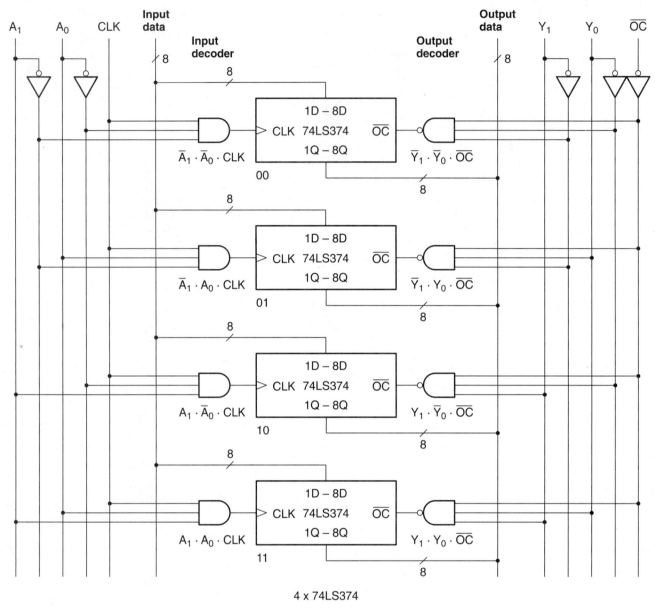

**Figure 17.3**

**4 × 8 Memory From '374 Latches**

the selected latch will be present on the output lines. For example, if $Y_1 Y_0 = 11$ and $\overline{OC} = 0$, the fourth latch is selected and its contents appear on the 8-bit output data bus.

Each latch corresponds to a different binary state of the $A$ or the $Y$ inputs, called an **address.** The number of addresses, or addressable locations, in any memory is specified by the number of possible binary combinations of the address-select inputs, given by $2^n$ for $n$ inputs. The $A$ inputs can address $2^2 = 4$ locations, and the $Y$ inputs can address $2^2 = 4$ locations. These are the same four locations, but the separate input and output decoders allow us simultaneously to **write** data to one location and read from a different one. (There is nothing to prevent us from writing to and reading from the same location at the same time. However, this will make the output data unreliable.) In most memory devices, one set of address lines accesses both the input and output data.

Since the data lines are not decoded but are connected directly to the internal latches, the number of data bits at any address is the same as the number of data inputs and outputs. For the circuit shown in Figure 17.3, there are 8 data bits at each address.

## RAM and ROM

**Random access memory (RAM)** *A type of memory device where data can be accessed in any order, that is, randomly. The term usually refers to random access read/write memory.*

**Read only memory (ROM)** *A type of memory where data are permanently stored and can only be read, not written.*

The memory circuit in Figure 17.3 is one type of **random access memory,** or RAM. Data can be stored in or retrieved from any address at any time. The data can be accessed randomly, without the need to follow a sequence of addresses, as would be necessary in a sequential storage device such as magnetic tape.

RAM has come to mean random access read/write memory, memory that can have its data changed by a Write operation, as well as have its data read. The data in another type of memory, called **read only memory,** or ROM, can also be accessed randomly, although it cannot be changed; there is no Write function; hence the name "read only." Even though both types of memory are random access, we generally do not include ROM in this category.

## Memory Capacity

**b** *Bit.*

**B** *Byte.*

**K** *1024 (= $2^{10}$). Analogous to the metric prefix "k" (kilo-).*

**M** *1,048,576 (= $2^{20}$). Analogous to the metric prefix "M" (mega-).*

The capacity of a memory device is specified by the address and data sizes. The circuit shown in Figure 17.3 has a capacity of $4 \times 8$ bits ("four-by-eight"). This tells us that the memory can store 32 bits, organized in groups of 8 bits at 4 different locations.

For large memories, with capacities of thousands of bits, we can use a shorthand designation, known as the **K,** for large binary numbers. The K is analogous to the metric prefix "kilo-" and is of similar size. $1K = 2^{10} = 1024$. For instance, a memory device might have a capacity of 64 Kb (kilobits), organized as $8K \times 8$. The memory has a capacity of 65,536 ($= 64 \times 1024$) bits organized into groups of 8 bits in 8192 (8K) addressable locations. (A computer system using this memory would see it as having a capacity of 8K bytes, or 8 KB (kilobytes).)

Usually, the range of numbers spanning 1K is expressed as the 1024 numbers from $0_{10}$ to $1023_{10}$ ($0000000000_2$ to $1111111111_2$). This is the full range of numbers that can be expressed by 10 bits. In hexadecimal, the range of numbers spanning 1K is from $000_{16}$ to $3FF_{16}$.

**Figure 17.4**

**Address and Data in an 8K × 8 Memory**

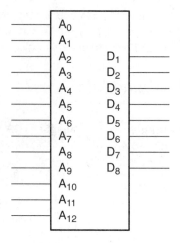

a. Address and data lines

| $D_8$ | | | | | | | $D_1$ | | Addresses | |
|---|---|---|---|---|---|---|---|---|---|---|
| | | | | | | | | | Binary | Hexadecimal |
| 1 | 0 | 1 | 1 | 0 | 1 | 0 | 1 | | 0 0000 0000 0000 | 0000 |
| 0 | 0 | 0 | 1 | 1 | 0 | 1 | 1 | | 0 0000 0000 0001 | 0001 |
| 1 | 1 | 0 | 1 | 0 | 0 | 1 | 1 | | 0 0000 0000 0010 | 0002 |
| 0 | 0 | 0 | 0 | 0 | 1 | 1 | 1 | | 0 0000 0000 0011 | 0003 |
| 0 | 1 | 1 | 1 | 0 | 1 | 1 | 1 | | 0 0000 0000 0100 | 0004 |
| 1 | 0 | 0 | 0 | 1 | 0 | 1 | 0 | | 0 0000 0000 0101 | 0005 |
| 0 | 1 | 0 | 1 | 1 | 1 | 1 | 1 | | 0 0000 0000 0110 | 0006 |

⋮ ⋮ ⋮

| | | | | | | | | | | |
|---|---|---|---|---|---|---|---|---|---|---|
| 1 | 0 | 1 | 0 | 1 | 0 | 1 | 0 | | 1 1111 1111 1101 | 1FFD |
| 0 | 0 | 0 | 1 | 1 | 1 | 1 | 1 | | 1 1111 1111 1110 | 1FFE |
| 1 | 1 | 0 | 0 | 1 | 0 | 1 | 1 | | 1 1111 1111 1111 | 1FFF |

b. Contents (data) and location (address)

The range of numbers spanning 8K can be written in 13 bits ($8 \times 1K = 2^3 \times 2^{10} = 2^{13}$). The addresses in an 8K × 8 memory range from 0000000000000 to 1111111111111, or 0000 to 1FFF in hexadecimal. Thus, a memory device that is organized as 8K × 8 has 13 address lines and 8 data lines.

Figure 17.4 shows the address and data lines of an 8K × 8 memory and a map of its contents. The addresses progress in binary order, but the contents of any location are the last data stored there. Since there is no way to predict what those data are, they are essentially random. For example, in Figure 17.4, the byte at address $0000000000100_2$ ($0004_{16}$) is $01110111_2$ ($77_{16}$). (One can readily see the advantage of using hexadecimal notation.)

▼

**EXAMPLE 17.1**

How many address lines are needed to access all addressable locations in a memory that is organized as 64K × 4? How many data lines are required?

**Solution**    Address lines: $2^n = 64K$

$$64K = 64 \times 1K = 2^6 \times 2^{10} = 2^{16}$$

$$n = 16 \text{ address lines}$$

Data lines: There are 4 data bits for each addressable location. Thus, the memory requires 4 data lines.

▲

## Control Signals

Two memory devices are shown in Figure 17.5. The device in Figure 17.5a is an Am9114 (Advanced Micro Devices) $1K \times 4$ random access read/write memory (RAM). Figure 17.5b shows an Am27C64 $8K \times 8$ erasable programmable read only memory (EPROM). The address lines are designated by $A$ and the data lines by $DQ$. The dual notation $DQ$ indicates that these lines are used for both input $(D)$ and output $(Q)$ data, using the conventional designations of D-type latches. These devices, like most memories, have tristate outputs that allow them to be electrically isolated from the data bus.

In addition to the address and data lines, most memory devices, including those in Figure 17.5, have one or more of the following control signal inputs. (Different manufacturers use different notation, so several alternate designations for each function are listed.)

$\overline{E}$ **(or $\overline{CE}$ or $\overline{CS}$).** $\overline{\text{Enable}}$ (or $\overline{\text{Chip Enable}}$ or $\overline{\text{Chip Select}}$). The memory is enabled when this line is pulled LOW. If this line is HIGH, the memory cannot be written to or read from.

**Figure 17.5**

**Address, Data, and Control Signals**

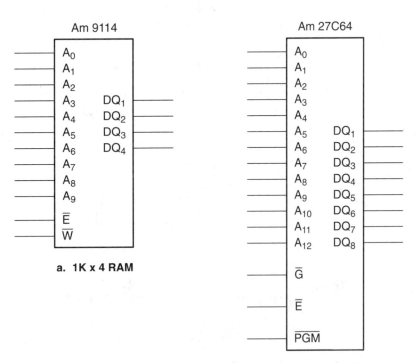

a. **1K x 4 RAM**

b. **1K x 8 ROM (EPROM)**

$\overline{W}$ (or $\overline{WE}$ or $R/\overline{W}$). $\overline{\text{Write}}$ (or $\overline{\text{Write Enable}}$ or Read/$\overline{\text{Write}}$). This input is used to select the Read or Write function when data input and output are on the same lines. When HIGH, this line selects the Read (output) function if the chip is selected. When LOW, the Write (input) function is selected.

$\overline{G}$ (or $\overline{OE}$). $\overline{\text{Gate}}$ (or $\overline{\text{Output Enable}}$). Some memory chips have a separate control to enable their tristate output buffers. When this line is LOW, the output buffers are enabled and the memory can be read. If this line is HIGH, the output buffers are in the high-impedance state. The Chip Select performs this function in devices without Output Enable pins.

The electrical functions of these control signals are illustrated in Figure 17.6.

**Figure 17.6**

**Memory Control Signals**

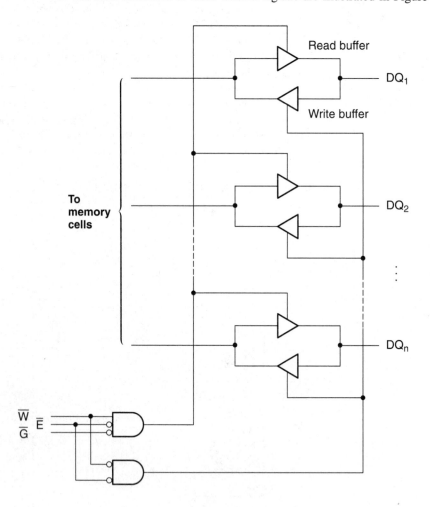

## 17.2

# Random Access Read/Write Memory (RAM)

**Volatile** *A memory is volatile if its stored data are lost when electrical power is lost.*

**Static RAM** *A random access memory that can retain data indefinitely as long as electrical power is available to the chip.*

**Dynamic RAM** *A random access memory that cannot retain data for more than a few milliseconds without being "refreshed."*

**RAM cell** *The smallest storage unit of a RAM, capable of storing 1 bit.*

Random access read/write memory (RAM) is used for temporary storage of large blocks of data. An important characteristic of RAM is that it is **volatile.** It can retain its stored data only as long as power is applied to the memory. When power is lost, so are the data. There are two main RAM configurations: static (SRAM) and dynamic (DRAM).

**Static RAM** (SRAM) consists of arrays of memory cells that are essentially flip-flops. Data can be stored in a static RAM cell and left there indefinitely, as long as power is available to the RAM.

A **dynamic RAM cell** stores a bit as the charged or discharged state of a small capacitor. Since the capacitor can hold its charge for only a few milliseconds, the charge must be restored ("refreshed") regularly. This makes a dynamic RAM (DRAM) system more complicated than SRAM, as it introduces a requirement for memory refresh circuitry.

However, DRAM is sometimes preferable to SRAM due to its large bit capacity. By the time this sees print, 16-Mb DRAMs will probably be available. 1-Mb DRAMs are common, and 4-Mb DRAMs are now commercially available. At the time of this writing, the largest static RAM chips are 1 Mb. 256-Kb SRAMs are readily available, and 64-Kb SRAMs are typically used in system designs.

## Static RAM Cells

The typical static RAM cell consists of at least two transistors that are cross-coupled in a flip-flop arrangement. Other parts of the cell include pull-up circuitry that can be active (transistor switches) or passive (resistors) and some decoding/switching logic. Figure 17.7 shows an SRAM cell in three technologies: bipolar, NMOS, and CMOS.

Each of these cells can store 1 bit of data, a 0 or a 1, as the state of one of the transistors in the cell. The data are available in true or complement form, as the *BIT* and $\overline{BIT}$ outputs of the flip-flop.

All types of SRAM cells operate in more or less the same way. We will analyze the operation of the NMOS cell (Figure 17.7b) and then compare it to the other types.

Transistors $Q_1$ and $Q_2$ are permanently biased ON, making them into pull-up resistors. Channel width and length are chosen to give a resistance of about 1 k$\Omega$. These NMOS load transistors are considered passive pull-ups, as they do not switch on and off.

A bit is stored as $V_{DS3}$, the drain voltage of $Q_3$ with respect to its source. If this voltage is HIGH, the gate of $Q_4$ is HIGH with respect to its source and $Q_4$ is biased ON. This completes a conduction path from the drain of $Q_4$ to its source, making $V_{DS4}$ logic LOW. This LOW is fed back to the gate of $Q_3$, turning it OFF. There is no conduction path between the drain and source of $Q_3$, so $V_{DS3} = V_{DD}$ or logic HIGH. The cell is storing a 1.

This bit can be read by making the *ROW SELECT* line HIGH. This turns $Q_5$ and $Q_6$ ON, which puts the data onto the *BIT* and $\overline{BIT}$ lines where it can be read by other circuitry inside the RAM chip.

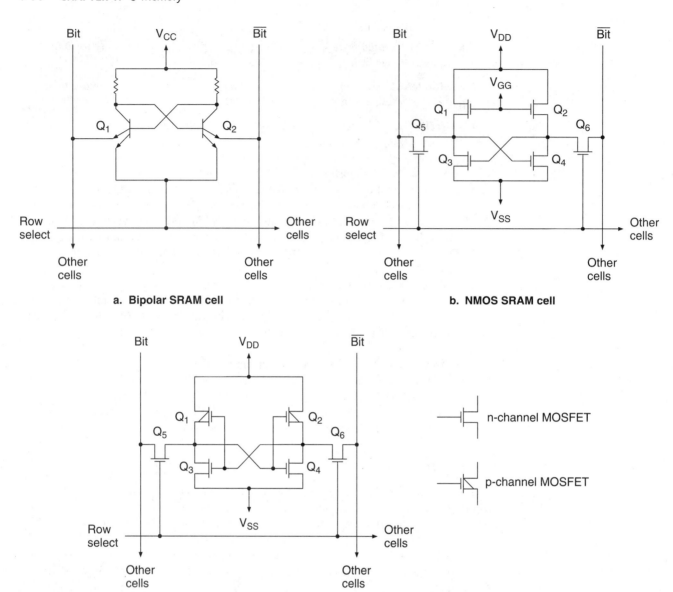

**a. Bipolar SRAM cell**

**b. NMOS SRAM cell**

**c. CMOS SRAM cell**

n-channel MOSFET

p-channel MOSFET

**Figure 17.7**
**SRAM Cells**

To change the cell contents to a 0, we make the *BIT* line LOW and the *ROW SELECT* line HIGH. The *ROW SELECT* line gives access to the cell by turning on $Q_5$ and $Q_6$, completing the conduction path between the *BIT* lines and the flip-flop inputs. The LOW on the *BIT* line pulls the gate of $Q_4$ LOW, turning it OFF. This breaks the conduction path from $Q_4$ drain to source and makes $V_{DS4} = V_{DD}$, a logic HIGH. This HIGH is applied to the gate of $Q_3$, turning it ON. A conduction path is established between $Q_3$ drain and source, pulling the drain of $Q_3$ LOW. The cell now stores a logic 0.

> The contents of an SRAM cell must be changed by introducing a LOW on the *BIT* or the $\overline{BIT}$ line. The data cannot be changed by pulling an input HIGH without pulling the opposite input LOW. If a MOSFET gate is at the LOW state, a HIGH applied to that gate will be pulled down by the LOW level already existing there and will not cause the cell to change state.

The CMOS cell (Figure 17.7c) functions in the same way, except for the actions of $Q_1$ and $Q_2$. $Q_1$ and $Q_3$ are a complementary pair, as are transistors $Q_2$ and $Q_4$. For each of these pairs, when the p-channel transistor is ON, the n-channel is OFF, and vice versa. This arrangement is more energy efficient than the NMOS cell, since there is not the constant current drain associated with the load transistors. Power is consumed primarily during switching between states.

The main design goal of new memory technology is to increase speed and capacity while reducing power consumption and chip area. The NMOS cell has the advantage of being constructed from only one type of component. This makes it possible to manufacture more cells in the same chip area than can be done in either the CMOS or bipolar technologies. NMOS chips, however, are slower than bipolar. New advances in High-Speed CMOS technologies have made possible CMOS memories that are as dense or denser than NMOS and faster. Because of this, NMOS will probably decline in importance over time.

Bipolar SRAMs can be either TTL, as shown in Figure 17.7a, or ECL, which is not shown. Of the two bipolar technologies, ECL is the faster. Historically, all bipolar SRAMs have had the advantage of speed over NMOS and CMOS chips. New CMOS devices are approaching or even exceeding the speeds of TTL.

The bipolar SRAM cell is the least suitable for high-density memory. Both bipolar transistors and resistors are large components compared to a MOSFET. Thus, the bipolar cell is inherently larger than the CMOS or NMOS cell. Bipolar memories are used when a small amount of high-speed memory is required.

The operation of the bipolar SRAM cell is similar to that of the MOSFET cells. In the quiescent state, the *ROW SELECT* line is LOW. In either the Read or the Write mode, the *ROW SELECT* line is HIGH. To change the data in the cell, pull one of the emitters LOW. When the emitter of $Q_1$ goes LOW, the cell contents become 0. When the emitter of $Q_2$ is pulled LOW, the cell contents are 1.

Table 17.1 summarizes the advantages and disadvantages of each of the three technologies.

## Static RAM Cell Arrays

**Word-organized** *A memory is word-organized if one address accesses one word of data.*

**Word** *Data accessed at one addressable location.*

**Word length** *Number of bits in a word.*

Static RAM cell arrays are arranged in a square format, accessible by groups in rows and columns. Each column corresponds to a complementary pair of *BIT* lines and each row to a *ROW SELECT* line, as shown in Figure 17.8.

**Table 17.1**  Memory Technologies

| Memory Technology | Advantages | Disadvantages |
|---|---|---|
| Bipolar | Speed | Complex manufacturing process (resistors, bipolar transistors) High power consumption Limited capacity High cost |
| NMOS | Low cost High density Low power | Low speed |
| CMOS | Lower power than NMOS | More complex than NMOS Low speed (improving) |

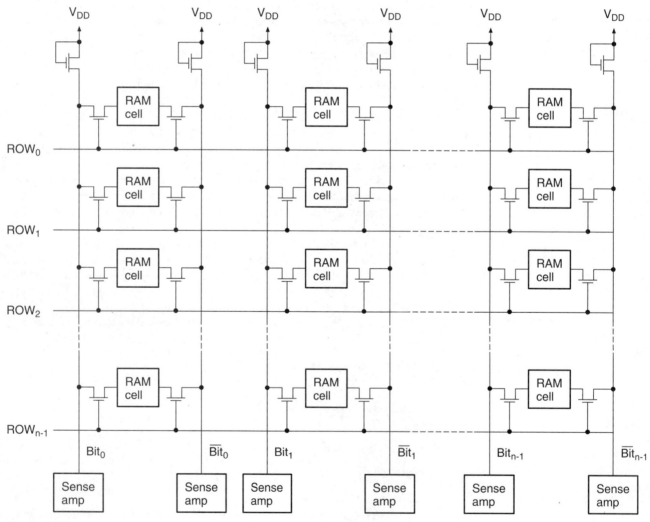

**Figure 17.8**
**SRAM Cell Array**

The column lines have MOSFETs configured as pull-up resistors at one end and a circuit called a sense amplifier at the other. The sense amp is a large RAM cell that amplifies the charge of an active storage cell on the same *BIT* line. Having a larger RAM cell as a sense amp allows the storage cells to be smaller, since each individual cell need not carry the charge required for a logic level output.

Figure 17.9 shows the block diagram of a Motorola MCM6164 8K × 8 Static RAM. The RAM cells are arrayed in a square pattern of 256 rows by 256 columns. The square cell array is used since it is a more efficient layout, in terms of packaging, than a rectangular array of 8192 rows and 8 columns. The square layout requires a different access scheme than the rectangular format.

An internal 8-to-256 line decoder accesses the 256 rows of cells by decoding the eight most significant address lines ($A_{12}$. . .$A_5$).

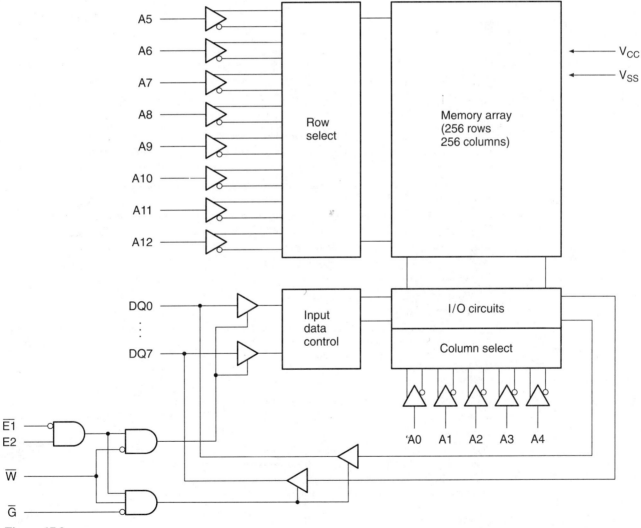

**Figure 17.9**
**6164 8K × 8 Static RAM**

**Figure 17.10**

**Word-Organized SRAM (MCM6164)** *Shown:* **Row 35, Word 20 (Columns 160–167)**

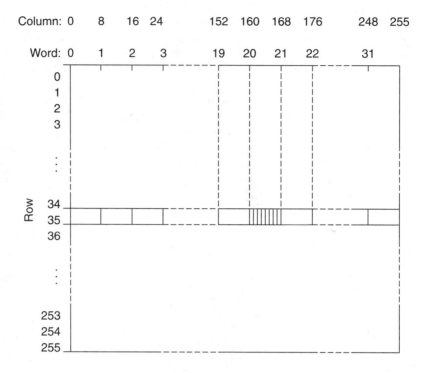

Since the 6164 SRAM is **word-organized** (one address selects one **word**), data are selected in groups of 8 bits. That is, this SRAM has a **word length** of 8 bits. For each row, the 256 columns in the RAM cell array are arranged in 32 groups of 8 cells. Each group of cells (one word) requires a specific 5-bit address ($A_4$. . . $A_0$) to access it uniquely.

Figure 17.10 shows a word location in a 6164 SRAM with respect to its row and column (word) addresses. The word is located at row 35, word 20. The eight most significant address bits specify the row: $A_{12}$. . . $A_5 = 00100011$ ($= 35_{10}$). The five least significant address bits specify the word: $A_4$. . . $A_0 = 10100$ ($= 20_{10}$). (There are eight columns per word, corresponding to columns 160–167; start column = word × 8 = 20 × 8.) The chip address is $A_{12}$. . . $A_0 = 0\ 0100\ 0111\ 0100 = 0474_{16}$ for this particular word.

The $BIT_0$ lines for each word are connected to the output and input buffers corresponding to $DQ_0$. Similarly, the other $BIT$ lines correspond to the remaining $DQ$ lines.

---

**EXAMPLE 17.2**

A Motorola MCM6268 4K × 4 SRAM requires 12 address lines to access each 4-bit word uniquely. The RAM cells are arranged in a 128-row-by-128-column square array. (4K × 4 = ($2^2 × 2^{10}$) × $2^2 = 2^{14} = 2^7 × 2^7 = 128 × 128$.)

The row address lines can access all rows of cells in the array, and the column address lines select a word within each row.

How many row address lines are there? How many column address lines? How many words are in each row of cells? How many bits are there in each word?

**Solution** The 128 rows can be uniquely addressed by 7 row address lines, since $2^7 = 128$. The chip has 12 address lines in total ($2^{12} = 4096$), leaving 5 lines for column addresses.

These 5 lines can address $2^5 = 32$ words of 4 bits each in every row. (32 words $\times$ 4 bits/word = 128 bits; one bit per column.)

---

**Section Review Problem for Section 17.2**

17.1. A Motorola MCM6206 32K $\times$ 8 SRAM has its RAM cells arranged in a square format. How many rows and columns are in the array? How many row and column address lines? How many words per row? How many bits per word?

## Dynamic RAM Cells

**Refresh cycle** *The process that periodically recharges the storage capacitors in a dynamic RAM.*

A dynamic RAM (DRAM) cell consists of a capacitor and an access transistor, as shown in Figure 17.11. A bit is stored in the cell as the charged or discharged state of the capacitor. The bit location is read from or written to by activating the cell MOSFET via the Word Select line, thus connecting the capacitor to the *BIT* line.

**Figure 17.11**

**Dynamic RAM Cell**

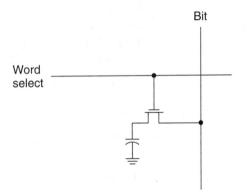

The major disadvantage of dynamic RAM is that the capacitor will eventually discharge by internal leakage current and must be recharged periodically to maintain integrity of the stored data. The recharging of the DRAM cell capacitors, known as refreshing the memory, must be done every 2 to 8 ms, depending on the device.

The **refresh cycle** adds an extra level of complication to the DRAM hardware and also to the timing of the Read and Write cycles, since the memory might have to be refreshed between Read and Write tasks. DRAM timing cycles are much more complicated than the equivalent SRAM cycles.

This inconvenience is offset by the high bit densities of DRAM, which are possible due to the simplicity of the DRAM cell. Up to 4 to 16 megabits of data can be stored on a single 20-pin chip.

## DRAM Cell Arrays

**Bit-organized** *A memory is bit-organized if one address accesses one bit of data.*

**Address multiplexing** *A technique of addressing storage cells in a dynamic RAM that sequentially uses the same inputs for the row address and column address of the cell.*

$\overline{RAS}$ *Row address strobe. A signal used to latch the row address into the decoding circuitry of a dynamic RAM with multiplexed addressing.*

$\overline{CAS}$ *Column address strobe. A signal used to latch the column address into the decoding circuitry of a dynamic RAM with multiplexed addressing.*

Dynamic RAM is often **bit-organized** rather than word-organized. That is, one address will access one bit rather than one word of data. A bit-organized DRAM with a large capacity requires more address lines than a static RAM (e.g., 4 Mb × 1 DRAM requires 22 address lines ($2^{22} = 4{,}194{,}304 = 4M$) and 1 data line to access all cells).

In order to save pins on the IC package, a system of **address multiplexing** is used to specify the address of each cell. Each cell has a row address and a column address, which use the same input pins. Two negative-edge signals called **row address strobe** ($\overline{RAS}$) and **column address strobe** ($\overline{CAS}$) latch the row and column addresses into the DRAM's decoding circuitry. Figure 17.12 shows a simplified block diagram of the row and column addressing circuitry of a 1 Mb × 1 dynamic RAM.

The memory cell array is rectangular, not square. One of the Row Address lines is connected internally to the Column Address decoder, resulting in a 512-row-by-2048-column memory array.

**Figure 17.12**

**Row and Column Decoding in a 1M × 1 Dynamic RAM**

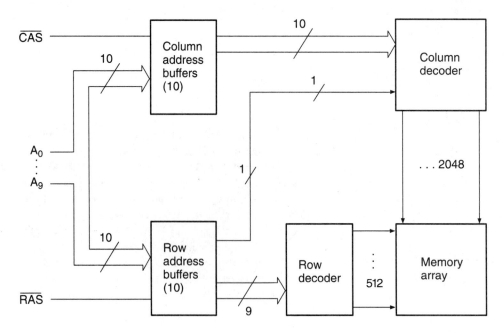

One advantage to the rectangular format is that it cuts the memory refresh time in half, since all the cells are refreshed by accessing the rows in sequence. Fewer rows means a faster refresh cycle. All cells in a row are also refreshed by normal Read and Write operations.

Figure 17.13 shows the block diagram of a Motorola MCM511000A 1 Mb × 1 dynamic RAM. In addition to the row and column decoding circuitry, there are also circuit blocks for data input, data output, sense amplifiers, refresh circuitry, Write Enable control, and an internal Test function.

---

**Section Review Problems for Section 17.2b**

17.2. How many address and data lines are required for the following sizes of dynamic RAM, assuming that each memory cell array is organized in a square format, with common Row and Column Address pins?

a. 256K × 1          d. 1M × 4

b. 256K × 4          e. 4M × 1

c. 1M × 1

---

**Figure 17.13**

**Block Diagram of MCM511000A 1M × 1 Dynamic RAM**
(Reprinted with permission of Motorola)

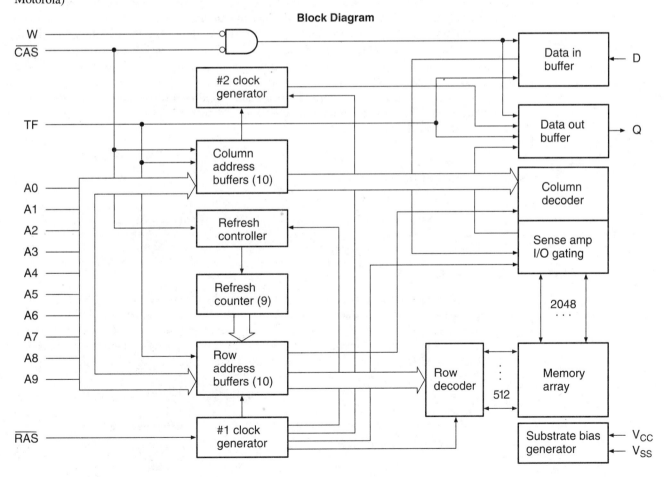

**Block Diagram**

## 17.3

# Read Only Memory (ROM)

> **Hardware** *The electronic circuit of a digital or computer system.*
>
> **Software** *Programming instructions required to make hardware perform specified tasks.*
>
> **Firmware** *Software instructions permanently stored in ROM.*
>
> **Slushware** *Software used by a hardware programmer to store firmware in a ROM or PLD.*

The main advantage of read only memory (ROM) over random access read/write memory (RAM) is that ROM is nonvolatile. It will retain data even when electrical power is lost to the ROM chip. The disadvantage of this is that stored data are difficult or impossible to change.

ROM is used for storing data required for tasks that never or rarely change, such as **software** instructions for a bootstrap loader in a personal computer. (The bootstrap loader—a term derived from the whimsical idea of pulling oneself up by one's bootstraps, that is, starting from nothing—is the software that gives the personal computer its minimum startup information. Generally, it contains the instructions needed to read a magnetic disk containing further operating instructions. This task is always the same for any given machine and is needed every time the machine is turned on, thus making it the ideal candidate for ROM storage.)

Software instructions stored in ROM are called **firmware.** Software instructions used by ROM or PLD programming **hardware** to transfer firmware into ROM or a PLD are called **slushware.**

## *Mask-Programmed ROM*

> **Mask-programmed ROM** *A type of read only memory (ROM) where the stored data are permanently encoded into the memory device during the manufacturing process.*

The most permanent form of read only memory is the **mask-programmed ROM,** where the stored data are manufactured into the memory chip. Due to the inflexibility of this type of ROM and the relatively high cost of development, it is used only for well-developed high-volume applications. However, even though development cost of a mask-programmed ROM is high, volume production is cheaper than for some other types of ROM.

Examples of applications suitable to mask-programmed ROM include:

- BCD-to-seven-segment or ASCII-to-5×7-matrix decoders
- Character generators (decoders that convert ASCII codes into alphanumeric characters on a CRT display)

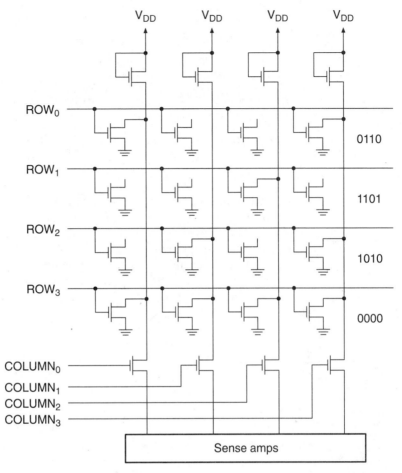

**Figure 17.14**
**Mask-Programmed ROM**

- Function lookup tables (tables corresponding to binary values of trigonometric, exponential, or other functions)

- Special software instructions that must be permanently stored and never changed (firmware)

Figure 17.14 shows a ROM based on a matrix of MOSFETs. Each cell is manufactured with a MOSFET and its gate and source connections. LOWs are programmed by making a connection between the drain of the cell's MOSFET and the corresponding Bit line. When the appropriate Row Select goes HIGH, the MOSFET turns ON, providing a path to ground from the selected Bit line. Cells programmed HIGH have no connection between the MOSFET drain and the Bit line, which thus cannot be pulled LOW when the cell is selected.

These connections can be made by a custom overlay of connections (a mask) on top of the standard-cell layer. The standard-cell-plus-custom-overlay format is cheaper to manufacture than custom cells for each bit, even if many of the MOSFETs are never used.

## Fusible-Link PROM

> **PROM** *Programmable read only memory. A type of ROM whose data are not manufactured into the chip, but are programmed by the user.*
>
> **Fusible-link PROM** *A type of programmable read only memory programmed by blowing selected internal fuses.*
>
> **Burning** *Programming a fusible-link PROM by blowing selected fuses.*

The **fusible-link programmable read only memory** is a ROM that can be programmed in the field by anyone having the proper equipment. The stored data need not be manufactured into the chip. This gives the user more flexibility, allowing him or her to customize a circuit using different versions of the PROM.

The PROM is built like a mask-programmed ROM, with standard cells. Each cell has a small titanium-tungsten (Ti-W) or platinum silicide fuse between the logic element of each cell and its Bit line. Figure 17.15 shows the general layout of a PROM cell matrix.

**Figure 17.15**

**Fusible-Link PROM**

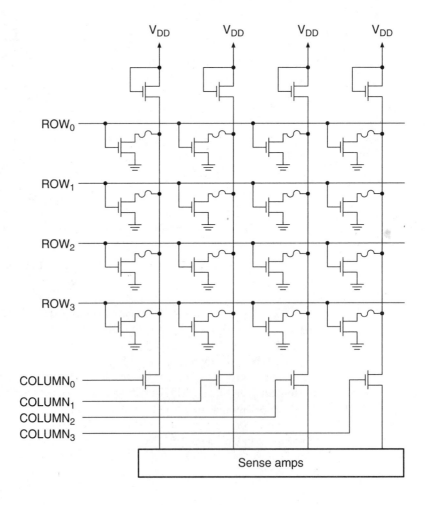

When all fuses in the cell matrix are intact, the stored data are all at logic 0. Any cell can be permanently programmed to a logic 1 by blowing the cell's fuse. This is done by selecting the cell's address and applying a high enough current for a specified time. The programming process is also called **burning** or blasting the PROM.

The main disadvantage of fusible-link PROMs is that the fuse links are relatively large compared to the memory cells, which limits the physical size of the PROM and thus its bit storage capacity. The largest PROMs are in the 32 Kb (4096 × 8) range, with most PROMs being smaller.

A second disadvantage is that, although you can program a PROM for any application you might need, you can't change your mind. Once programmed, a PROM retains its data permanently. This is a comparatively small problem, since PROMs are generally used for finished designs, not for designs undergoing development, and there are other types of PROMs on the market that are erasable.

## Application: PROM Decoder

Fusible-link PROMs are suitable for small-circuit applications, such as a decoder.

How does a read only memory compare to a decoder? A ROM accepts specified bit combinations into a set of inputs (addresses) and responds by placing specific bit combinations on a set of outputs (data). A decoder accepts specified bit combinations (the input code) at its inputs and responds by placing specified logic states (the output code) at its outputs.

A suitably programmed ROM can, for instance, accept any 4-bit binary number at its address inputs, and can respond by placing the proper HIGHs and LOWs on its data outputs required to drive the LED segments of a seven-segment display so that it shows the hexadecimal digit equivalent of the binary input.

**EXAMPLE 17.3**

TBP18S030

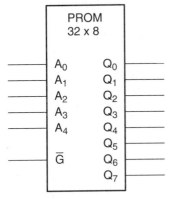

**Figure 17.16**

**Texas Instruments TBP18S030**

A Texas Instruments TBP18S030 fusible-link PROM, shown in Figure 17.16, can store 256 bits, organized as 32 words of 8 bits each. Make a table of addresses and data for a hexadecimal-to-seven-segment decoder using a TBP18S030, assuming that the seven-segment display is common-cathode (each segment turned on by a logic HIGH) and that the digit patterns are as shown in Figure 17.17. Draw the circuit required to implement the decoder after the PROM has been burned.

### Solution

A hexadecimal-to-seven-segment decoder needs 16 different 7-bit codes (1 bit per LED segment for each of the 16 combinations). Four address lines, $A_3 . . . A_0$, will give us all the locations we need. The remaining address line, $A_4$, can be disabled by tying it LOW.

We need only seven data outputs, one for each LED segment. We can use lines $Q_6 . . . Q_0$, assigning segment $a$ to $Q_6$ and segment $g$ to $Q_0$, with the others in sequence in between. Since we don't need $Q_7$, we can leave it unconnected. The internal data for $Q_7$ can be any value, but for simplicity, let us specify that it is 0 for every location.

To display a hex digit, we must illuminate the segments making up that digit by applying HIGHs to the correct LED anodes. For instance, digit 0 is displayed by illuminating segments $a$, $b$, $c$, $d$, $e$, and $f$, which corresponds to a digital output of $Q_7 . . . Q_0 = 01111110$. ($Q_7 = 0$ for all codes.)

The complete data for addresses 00000 to 01111 are shown in Table 17.2. (The hexadecimal equivalents of data $Q_7 . . . Q_0$ are also listed, since these can be used to represent the decoder data when using a commercial PROM burner.)

The circuit for the decoder is shown in Figure 17.18.

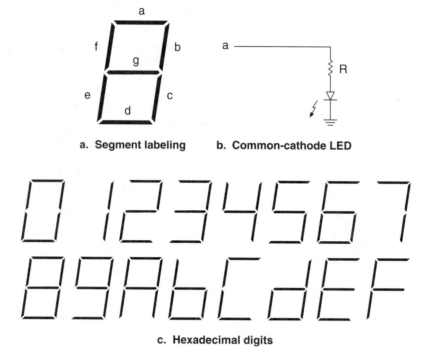

a. Segment labeling    b. Common-cathode LED

c. Hexadecimal digits

Figure 17.17

Example 17.3
Seven-Segment Display

Table 17.2  Addresses and Data for Hexadecimal-to-Seven-Segment Decoder

| Hex Digit | $A_4$ 0 | $A_3$ D | $A_2$ C | $A_1$ B | $A_0$ A | $Q_7$ 0 | $Q_6$ a | $Q_5$ b | $Q_4$ c | $Q_3$ d | $Q_2$ e | $Q_1$ f | $Q_0$ g | Hex Equivalent of Data |
|---|---|---|---|---|---|---|---|---|---|---|---|---|---|---|
| 0 | 0 | 0 | 0 | 0 | 0 | 0 | 1 | 1 | 1 | 1 | 1 | 1 | 0 | 7E |
| 1 | 0 | 0 | 0 | 0 | 1 | 0 | 0 | 1 | 1 | 0 | 0 | 0 | 0 | 30 |
| 2 | 0 | 0 | 0 | 1 | 0 | 0 | 1 | 1 | 0 | 1 | 1 | 0 | 1 | 6D |
| 3 | 0 | 0 | 0 | 1 | 1 | 0 | 1 | 1 | 1 | 1 | 0 | 0 | 1 | 79 |
| 4 | 0 | 0 | 1 | 0 | 0 | 0 | 0 | 1 | 1 | 0 | 0 | 1 | 1 | 33 |
| 5 | 0 | 0 | 1 | 0 | 1 | 0 | 1 | 0 | 1 | 1 | 0 | 1 | 1 | 5B |
| 6 | 0 | 0 | 1 | 1 | 0 | 0 | 1 | 0 | 1 | 1 | 1 | 1 | 1 | 5F |
| 7 | 0 | 0 | 1 | 1 | 1 | 0 | 1 | 1 | 1 | 0 | 0 | 0 | 0 | 70 |
| 8 | 0 | 1 | 0 | 0 | 0 | 0 | 1 | 1 | 1 | 1 | 1 | 1 | 1 | 7F |
| 9 | 0 | 1 | 0 | 0 | 1 | 0 | 1 | 1 | 1 | 1 | 0 | 1 | 1 | 7B |
| A | 0 | 1 | 0 | 1 | 0 | 0 | 1 | 1 | 1 | 0 | 1 | 1 | 1 | 77 |
| B | 0 | 1 | 0 | 1 | 1 | 0 | 0 | 0 | 1 | 1 | 1 | 1 | 1 | 1F |
| C | 0 | 1 | 1 | 0 | 0 | 0 | 1 | 0 | 0 | 1 | 1 | 1 | 0 | 4E |
| D | 0 | 1 | 1 | 0 | 1 | 0 | 0 | 1 | 1 | 1 | 1 | 0 | 1 | 3D |
| E | 0 | 1 | 1 | 1 | 0 | 0 | 1 | 0 | 0 | 1 | 1 | 1 | 1 | 4F |
| F | 0 | 1 | 1 | 1 | 1 | 0 | 1 | 0 | 0 | 0 | 1 | 1 | 1 | 47 |

**Figure 17.18**

**Example 17.3**
**Decoder Connections**

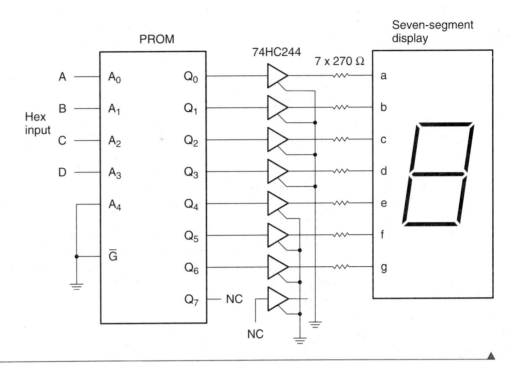

## EPROM

> **EPROM** *Erasable programmable read only memory. A type of ROM that can be programmed ("burned") by the user and erased later, if necessary, by exposing the chip to ultraviolet radiation.*
>
> **FAMOS FET** *Floating-gate avalanche MOSFET. A MOSFET with a second, "floating" gate in which charge can be trapped to change the MOSFET's gate-source threshold voltage.*

Mask-programmed ROM is useful because of its nonvolatility, but it is hard to program and impossible to erase. Fusible-link PROM can be programmed easily, but not erased. **Erasable programmable read only memory** (EPROM) combines the nonvolatility of ROM with the ability to change the internal data if necessary.

This erasability is particularly useful in the development of a ROM-based system. Anyone who has built a complex circuit or written a computer program knows that there is no such thing as getting it right the first time. Modifications can be made easily and cheaply to data stored in an EPROM. Later, when the design is complete, a mask ROM version can be prepared for mass production. Alternatively, if the design will be produced in small numbers, the ROM data can be stored in EPROMs, saving the cost of preparing a mask-programmed ROM.

The basis of the EPROM memory cell is the **FAMOS FET,** whose circuit symbol is shown in Figure 17.19. FAMOS stands for Floating-gate avalanche metal-oxide-semiconductor. ("Avalanche" refers to electron behavior in a semiconductor under certain bias conditions.) This is a MOSFET with a second, or floating, gate that is insulated from the first by a thin oxide layer.

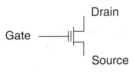

**Figure 17.19**
**FAMOS FET**

The floating gate has no electrical contact with either the first gate or the source and drain terminals. As is the case in a standard MOSFET, conduction between drain and source terminals is effected by the voltage of the gate terminal with respect to the source. If this voltage is above a certain threshold level, the transistor will turn ON, allowing current to flow between drain and source.

In the unprogrammed state the FAMOS transistor's threshold voltage is low enough for the transistor to be turned ON by a 5-V Read signal on the Row Select line. During the programming operation, a relatively high voltage pulse (about 12 V to 25 V, depending on the device) on the Row Select line drives high-energy electrons into the floating gate and traps them there. This raises the threshold voltage of the programmed cell to a level where the cell won't turn ON when selected by a 5-V Read.

The EPROM cells are configured so that an unprogrammed location contains a logic HIGH and the programming signal forces it LOW.

To erase an EPROM, the die (i.e., the silicon chip itself) must be exposed for about 20 to 45 minutes to high-intensity ultraviolet light of a specified wavelength (2537 angstroms) at a distance of 2.5 cm (1 inch). The high-energy photons that make up the UV radiation release the electrons trapped in the floating gate and restore the cell threshold voltages to their unprogrammed levels.

EPROMS are manufactured with a quartz window over the die to allow the UV radiation in. Since both sunlight and fluorescent light contain UV light of the right wavelength to erase the EPROM over time (several days to several years, depending on the intensity of the source), the quartz window should be covered by an opaque label after the EPROM has been programmed.

## EPROM Application: Digital Function Generator

An EPROM can be used as the central component of a digital function generator. Other components in the system include a clock generator, a counter, a digital-to-analog converter, and an output op amp buffer. The portion of the circuit including the last three of these components is shown in Figure 17.20.

The generator can produce the usual analog waveforms—sine, square, triangle, sawtooth—and any other waveforms that you wish to store in the EPROM. A single cycle of each waveform is stored as 256 consecutive 8-bit numbers. For example, the data for one cycle of the sine waveform are stored at addresses $0000_{16}$ to $00FF_{16}$, as shown in hex form in Table 17.3 (see page 760). (FF is maximum positive, 80 is zero, and 00 is maximum negative.) The square wave data are stored at addresses 0100 to 01FF, also shown in Table 17.3. The data for other functions, stored in subsequent 256-byte blocks, are not shown. A full list of the function data and a QuickBASIC program to generate an EPROM record file (Intel format) are included in Appendix C.

The most significant bits of the EPROM address select the waveform function by selecting a block of 256 address. The 8 least significant bits of the EPROM address are connected to an 8-bit (mod-256) counter, which continuously cycles through the 256 selected addresses. A 2764 EPROM (8K × 8) has 13 address lines. After the eight lower lines are accounted for, the remaining five lines can be used to select up to 32 digital functions. With the four binary Function Select switches, we can potentially select 16 functions.

For example, to select the Sine function, inputs $A_{11}$. . . $A_8$, which comprise the most significant digit of the EPROM address, are set to 0000. Thus, the 8-bit counter cycles through addresses 0000–00FF, the location of the sine data. The Square Wave

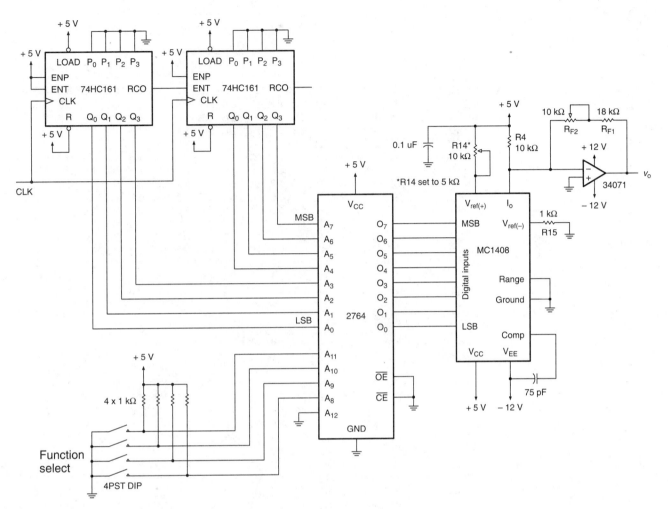

**Figure 17.20**

**Digital Function Generator**

function is selected by setting $A_{11} \ldots A_8$ to 0001, thus selecting the address block 0100–01FF. Other functions can be similarly selected.

The data at each address are sent to the D/A converter (MC1408), which, in combination with the op amp, is configured to produce a bipolar (both positive and negative) output. (We use a 34071 high slew rate op amp so that the generated square waves will have vertical sides.) The circuit generates a continuous waveform by retracing the data points in one 256-byte section of the EPROM over and over.

The DAC/op amp combination produces a maximum negative voltage for a hex input of 00, a 0 V output for an input of 80, and a maximum positive voltage for an input of FF. (You might wish to refer to the section "Bipolar Operation of MC1408" in Chapter 16 for details of the DAC operation.)

You can see from the Sine function data in Table 17.3 that 8 bits are not sufficient to represent each of the 256 steps of a digital sine function as a unique number. The peaks of the waveform are changing too slowly to be represented accurately by an 8-bit quantization, and as a result, the top of the sine wave is flat for several clock pulses. (Mathematically, a sine function is tangential to a horizontal line at its peak. However,

Table 17.3
EPROM Sine and
Square Wave Data

SINE

| Base Address | Byte Addresses | | | | | | | | | | | | | | | |
|---|---|---|---|---|---|---|---|---|---|---|---|---|---|---|---|---|
| | 0 | 1 | 2 | 3 | 4 | 5 | 6 | 7 | 8 | 9 | A | B | C | D | E | F |
| 0000 | 80 | 83 | 86 | 89 | 8C | 8F | 92 | 95 | 98 | 9C | 9F | A2 | A5 | A8 | AB | AE |
| 0010 | B0 | B3 | B6 | B9 | BC | BF | C1 | C4 | C7 | C9 | CC | CE | D1 | D3 | D5 | D8 |
| 0020 | DA | DC | DE | E0 | E2 | E4 | E6 | E8 | EA | EC | ED | EF | F0 | F2 | F3 | F5 |
| 0030 | F6 | F7 | F8 | F9 | FA | FB | FC | FC | FD | FE | FE | FF | FF | FF | FF | FF |
| 0040 | FF | FF | FF | FF | FF | FF | FE | FE | FD | FC | FC | FB | FA | F9 | F8 | F7 |
| 0050 | F6 | F5 | F3 | F2 | F0 | EF | ED | EC | EA | E8 | E6 | E4 | E2 | E0 | DE | DC |
| 0060 | DA | D8 | D5 | D3 | D1 | CE | CC | C9 | C7 | C4 | C1 | BF | BC | B9 | B6 | B3 |
| 0070 | B0 | AE | AB | A8 | A5 | A2 | 9F | 9C | 98 | 95 | 92 | 8F | 8C | 89 | 86 | 83 |
| 0080 | 7F | 7C | 79 | 76 | 73 | 70 | 6D | 6A | 67 | 63 | 60 | 5D | 5A | 57 | 54 | 51 |
| 0090 | 4F | 4C | 49 | 46 | 43 | 40 | 3E | 3B | 38 | 36 | 33 | 31 | 2E | 2C | 2A | 27 |
| 00A0 | 25 | 23 | 21 | 1F | 1D | 1B | 19 | 17 | 15 | 13 | 12 | 10 | 0F | 0D | 0C | 0A |
| 00B0 | 09 | 08 | 07 | 06 | 05 | 04 | 03 | 03 | 02 | 01 | 01 | 00 | 00 | 00 | 00 | 00 |
| 00C0 | 00 | 00 | 00 | 00 | 00 | 00 | 01 | 01 | 02 | 03 | 03 | 04 | 05 | 06 | 07 | 08 |
| 00D0 | 09 | 0A | 0C | 0D | 0F | 10 | 12 | 13 | 15 | 17 | 19 | 1B | 1D | 1F | 21 | 23 |
| 00E0 | 25 | 27 | 2A | 2C | 2E | 31 | 33 | 36 | 38 | 3B | 3E | 40 | 43 | 46 | 49 | 4C |
| 00F0 | 4F | 51 | 54 | 57 | 5A | 5D | 60 | 63 | 67 | 6A | 6D | 70 | 73 | 76 | 79 | 7C |

SQUARE

| Base Address | Byte Addresses | | | | | | | | | | | | | | | |
|---|---|---|---|---|---|---|---|---|---|---|---|---|---|---|---|---|
| | 0 | 1 | 2 | 3 | 4 | 5 | 6 | 7 | 8 | 9 | A | B | C | D | E | F |
| 0100 | FF | FF | FF | FF | FF | FF | FF | FF | FF | FF | FF | FF | FF | FF | FF | FF |
| 0110 | FF | FF | FF | FF | FF | FF | FF | FF | FF | FF | FF | FF | FF | FF | FF | FF |
| 0120 | FF | FF | FF | FF | FF | FF | FF | FF | FF | FF | FF | FF | FF | FF | FF | FF |
| 0130 | FF | FF | FF | FF | FF | FF | FF | FF | FF | FF | FF | FF | FF | FF | FF | FF |
| 0140 | FF | FF | FF | FF | FF | FF | FF | FF | FF | FF | FF | FF | FF | FF | FF | FF |
| 0150 | FF | FF | FF | FF | FF | FF | FF | FF | FF | FF | FF | FF | FF | FF | FF | FF |
| 0160 | FF | FF | FF | FF | FF | FF | FF | FF | FF | FF | FF | FF | FF | FF | FF | FF |
| 0170 | FF | FF | FF | FF | FF | FF | FF | FF | FF | FF | FF | FF | FF | FF | FF | FF |
| 0180 | 00 | 00 | 00 | 00 | 00 | 00 | 00 | 00 | 00 | 00 | 00 | 00 | 00 | 00 | 00 | 00 |
| 0190 | 00 | 00 | 00 | 00 | 00 | 00 | 00 | 00 | 00 | 00 | 00 | 00 | 00 | 00 | 00 | 00 |
| 01A0 | 00 | 00 | 00 | 00 | 00 | 00 | 00 | 00 | 00 | 00 | 00 | 00 | 00 | 00 | 00 | 00 |
| 01B0 | 00 | 00 | 00 | 00 | 00 | 00 | 00 | 00 | 00 | 00 | 00 | 00 | 00 | 00 | 00 | 00 |
| 01C0 | 00 | 00 | 00 | 00 | 00 | 00 | 00 | 00 | 00 | 00 | 00 | 00 | 00 | 00 | 00 | 00 |
| 01D0 | 00 | 00 | 00 | 00 | 00 | 00 | 00 | 00 | 00 | 00 | 00 | 00 | 00 | 00 | 00 | 00 |
| 01E0 | 00 | 00 | 00 | 00 | 00 | 00 | 00 | 00 | 00 | 00 | 00 | 00 | 00 | 00 | 00 | 00 |
| 01F0 | 00 | 00 | 00 | 00 | 00 | 00 | 00 | 00 | 00 | 00 | 00 | 00 | 00 | 00 | 00 | 00 |

since tangential means touching at one point, the flat top is a distortion.) A unique number for each of 256 steps of a sine function needs at least 13 bits,[1] but this requires additional bits on the D/A converter input, and therefore a different DAC and an expanded memory word length.

---

[1]Bits required:
- 360°/256 steps = 1.40625°/step.
- Sine function changes most slowly at peak, so calculate $A \sin(90° \pm 1.40625°)$ to find smallest amplitude change.
- The smallest power-of-2 amplitude, $A$, for which $A \sin(90°) - A \sin(90° \pm 1.40625°) \geq 1$ is 4096.
- The amplitude range $-4096 \leq A \leq 4095$ can be represented by a 13-bit number.

Another drawback of this function generator is that the clock frequency must be 256 times the output waveform frequency (one clock pulse per byte). This means that even audio frequencies require a clock rate that pushes the limit of an inexpensive D/A converter. (20 kHz × 256 = 5.12 MHz; the MC1408 DAC has a settling time of 300 ns, implying an upper limit of 1/300 ns = 3.33 Mhz for 8-bit accuracy. A 3.33 Mhz clock yields an output frequency of 13 kHz.)

## *EEPROM*

> **EEPROM (or E²PROM)**  *Electrically erasable programmable read only memory. A type of read only memory that can be field-programmed and selectively erased while still in a circuit.*

As was discussed in the previous section, EPROMs have the useful property of being erasable. The problem is that they must be removed from the circuit for erasure, and bits cannot be selectively erased; the whole memory cell array is erased as a unit.

**Electrically erasable programmable read only memory** (EEPROM or E²PROM) provides the advantages of EPROM along with the additional benefit of allowing erasure of selected bits while the chip is in the circuit; it combines the read/write properties of RAM with the nonvolatility of ROM. EEPROM is useful for storage of data that need to be changed occasionally, but that must be retained when power is lost to the EEPROM chip. One example is the memory circuit in an electronically tuned car radio that stores the channel numbers of local stations.

Like the UV-erasable EPROM, the memory cell of the EEPROM is based on the FAMOS transistor. Unlike the EPROM, the FAMOS FET is coupled with a standard MOSFET, as shown in Figure 17.21.

The FAMOS FET is programmed in the same way as UV-erasable EPROM: a programming voltage pulse *(V_{PP})* drives high-energy electrons into the floating gate of the FAMOS transistor, where they remain trapped and change the threshold voltage of the transistor. The cell is read by keeping the programming line at 5 V and making the cell's Row Select line HIGH. The FAMOS transistor will or will not turn on, depending on its programmed state.

The FAMOS transistors used in EPROM and EEPROM differ in one important respect. The EEPROM transistor is manufactured with a very thin oxide layer between

**Figure 17.21**

**EEPROM Cell**

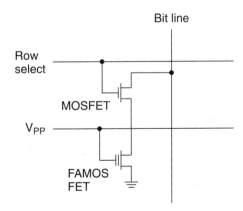

the drain and the upper (nonfloating) gate. This construction allows trapped electrons in the floating gate to be forced out electrically, thus erasing the cell contents.

Given the obvious advantages of EEPROM, why doesn't it replace all other types of memory? There are several reasons:

1. EEPROM has a much slower access time than RAM and is thus not good for high-speed applications.

2. The currently available EEPROMs have significantly smaller bit capacities than commercially available RAM (especially dynamic RAM) and EPROM.

3. Other types of memory, particularly static and dynamic RAM and EPROM, are commercially available in a much wider variety of devices than EEPROM.

## 17.4

## Sequential Memory: FIFO and LIFO

**Sequential memory** *Memory in which the stored data cannot be read or written in random order, but must be addressed in a specific sequence.*

**FIFO** *First-in first-out. A sequential memory in which the stored data can be read only in the order in which it was written.*

**Queue** *A FIFO memory.*

**LIFO** *Last-in first-out. A sequential memory in which the last data written are the first data read.*

**Stack** *A LIFO memory.*

The RAM and ROM devices we have examined up until now have all been random access devices. That is, any data could be read from or written to any sequence of addresses in any order. There is another class of memory in which the data must be accessed in a particular order. Such devices are called **sequential memory.**

There are two main ways of organizing a sequential memory—as a **queue** or as a **stack.** Figure 17.22 shows the arrangement of data in each of these types of memory.

A queue is a **first-in first-out** (FIFO) memory, meaning that the data can be read only in the same order they are written, much as subway cars always come out of a tunnel in the same order they go in.

One common use for FIFO memory is to connect two devices that have different data rates. For instance, a computer can send data to a dot-matrix printer much faster than the printer can use it. To keep the computer from either waiting for the printer to print everything or periodically interrupting the computer's operation to continue the print task, data can be sent in a burst to a FIFO, where the printer can read them as needed. The only proviso is that there must be some logic signal to the computer telling it when the queue is full and not to send more data and another signal to the printer letting it know that there are some data to read from the queue.

Specialty FIFO chips are available from various manufacturers in sizes from about $64 \times 4$ bits up to about $2048 \times 9$ bits.

**Figure 17.22**
**Sequential Memory**

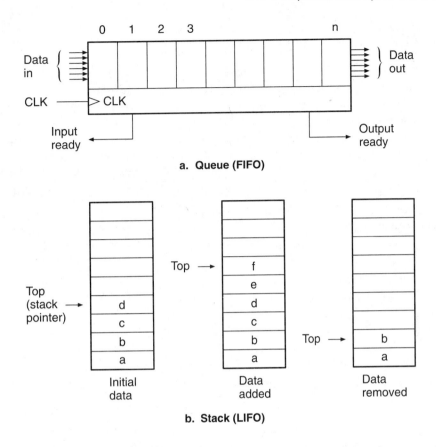

a. **Queue (FIFO)**

b. **Stack (LIFO)**

The **last-in-first-out** (LIFO), or stack, memory configuration, also shown in Figure 17.22, is not available as a special chip, but rather is a way of organizing RAM in a memory system.

The term "stack" derives from the idea of a spring-loaded stack of plates in a cafeteria line. When you put a bunch of plates on the stack, they settle into the recessed storage area. When a plate is removed, the stack springs back slightly and brings the second plate to the top level. (The other plates, of course, all move up a notch.) The top plate is the only one available for removal from the stack, and plates are always removed in reverse order from that in which they were loaded.

Figure 17.22b shows how data are transferred to and from a LIFO memory. A block of addresses in a RAM is designated as a stack, and one or two bytes of data in the RAM store a number called the stack pointer, which is the current address of the top of the stack.

In Figure 17.22, the value of the stack pointer changes with every change of data in the stack, pointing to the last-in data in every case. When data are removed from the stack, the stack pointer is used to locate the data that must be read first. After the Read, the stack pointer is modified to point to the next-out data.

The most common application for LIFO memory is in a computer system. If a program is interrupted during its execution by a demand from the program or some piece of hardware that needs attention, the status of various registers within the computer are stored on a stack and the computer can pay attention to the new demand, which will certainly change its operating state. After the interrupting task is finished, the original operating state of the computer can be taken from the top of the stack and reloaded into the appropriate registers, and the program can resume where it left off.

## ● GLOSSARY

**Address**  A number, represented by the binary states of a group of inputs or outputs, uniquely defining the location of data stored in a memory device.

**Address multiplexing**  A technique of addressing storage cells in a dynamic RAM that sequentially uses the same inputs for the row address and column address of the cell.

**b**  Bit.

**B**  Byte.

**Bit-organized**  A memory is bit-organized if one address accesses one bit of data.

**Burning**  Programming a fusible-link PROM by blowing selected fuses.

**Byte**  A group of 8 bits.

$\overline{CAS}$  Column address strobe. A signal used to latch the column address into the decoding circuitry of a dynamic RAM with multiplexed addressing.

**Data**  Binary digits (0s and 1s) that contain some kind of information. The digital contents of a memory device.

**Dynamic RAM**  A random access memory that cannot retain data for more than a few milliseconds without being "refreshed."

**EEPROM (or E$^2$ PROM)**  Electrically erasable programmable read only memory. A type of read only memory that can be field-programmed and selectively erased while still in a circuit.

**EPROM**  Erasable programmable read only memory. A type of ROM that can be programmed ("burned") by the user and erased later, if necessary, by exposing the chip to ultraviolet radiation.

**FAMOS FET**  Floating-gate avalanche MOSFET. A MOSFET with a second, "floating" gate in which charge can be trapped to change the MOSFET's gate-source threshold voltage.

**FIFO**  First-in first-out. A sequential memory in which the stored data can be read only in the order in which it was written.

**Firmware**  Software instructions permanently stored in ROM.

**Fusible-link PROM**  A type of programmable read only memory programmed by blowing selected internal fuses.

**Hardware**  The electronic circuit of a digital or computer system.

**K**  1024 (= $2^{10}$). Analogous to the metric prefix "k" (kilo).

**LIFO**  Last-in first-out. A sequential memory in which the last data written is the first data read.

**M**  1,048,576 (= $2^{20}$). Analogous to the metric prefix "M" (mega-).

**Mask-programmed ROM**  A type of read only memory (ROM) where the stored data are permanently encoded into the memory device during the manufacturing process.

**Memory**  A device for storing digital data in such a way that it can be recalled for later use in a digital system.

**Nibble**  Half a byte; 4 bits.

**PROM**  Programmable read only memory. A type of ROM whose data are not manufactured into the chip, but are programmed by the user.

**Queue**  A FIFO memory.

**RAM cell**  The smallest storage unit of a RAM, capable of storing 1 bit.

**Random access memory (RAM)**  A type of memory device where data can be accessed in any order, that is, randomly. The term usually refers to random access read/write memory.

$\overline{RAS}$  Row address strobe. A signal used to latch the row address into the decoding circuitry of a dynamic RAM with multiplexed addressing.

**Read**  Retrieve data from a memory device.

**Read only memory (ROM)**  A type of memory where data is permanently stored and can only be read, not written.

**Refresh cycle**  The process that periodically recharges the storage capacitors in a dynamic RAM.

**Sequential memory**  Memory in which the stored data cannot be read or written in random order, but must be addressed in a specific sequence.

**Slushware**  Software used by a hardware programmer to store firmware in a ROM or PLD.

**Software**  Programming instructions required to make hardware perform specified tasks.

**Stack**  A LIFO memory.

**Static RAM**  A random access memory that can retain data indefinitely as long as electrical power is available to the chip.

**Volatile**  A memory is volatile if its stored data are lost when electrical power is lost.

**Word**  Data accessed at one addressable location.

**Word length**  Number of bits in a word.

**Word-organized**  A memory is word-organized if one address accesses one word of data.

**Write**  Store data in a memory device.

# ● PROBLEMS

## Section 17.1 Basic Memory Concepts

**17.1** Explain why it is convenient to have 8 bits in a byte. Give three different examples.

**17.2** How many address lines are necessary to make an $8 \times 8$ memory similar to the $4 \times 8$ memory in Figure 17.3? How many address lines are necessary to make a $16 \times 8$ memory?

**17.3** Briefly explain the difference between RAM and ROM.

**17.4** Calculate the number of address lines and data lines needed to access all stored data in each of the following sizes of memory:

  **a.** $64K \times 8$

  **b.** $128K \times 16$

  **c.** $128K \times 32$

  **d.** $256K \times 16$

Calculate the total bit capacity of each memory.

**17.5** Explain the difference between the Chip Enable $(\overline{E})$ and the Output Enable $(\overline{G})$ control functions in a RAM.

**17.6** Refer to Figure 17.6. Briefly explain the operation of the $\overline{W}$, $\overline{E}$, and $\overline{G}$ functions of the RAM shown.

## Section 17.2 Random Access Read/Write Memory (RAM)

**17.7** Draw the circuit for an NMOS static RAM cell. Label one output *BIT* and the other $\overline{BIT}$.

**17.8** Refer to the NMOS static RAM cell drawn in Problem 17.7. Assume that *BIT* = 1. Describe the operation required to change *BIT* to 0.

**17.9** Briefly explain the purpose of load transistors $Q_1$ and $Q_2$ in an NMOS RAM cell. Why are transistors used for this function rather than resistors?

**17.10** Describe the main difference between a CMOS and an NMOS static RAM cell.

**17.11** Explain how a particular RAM cell is selected from a group of many cells.

**17.12** List the advantages and disadvantages of NMOS, CMOS, and TTL memory technologies.

**17.13** A Motorola MCM6208 $64K \times 4$ SRAM has its RAM cells arranged in a rectangular format (256 rows). The column address lines select a word within each row. How many columns are in the array? How many row and column address lines? How many words per row? How many bits per word?

**17.14** A Motorola MCM6228 $256K \times 4$ SRAM has its RAM cells arranged in a rectangular format (512 rows). The column address lines select a word within each row. How many columns are in the array? How many row and column address lines? How many words per row? How many bits per word?

**17.15** How many address lines are required to access all elements in a $1M \times 1$ dynamic RAM with address multiplexing?

**17.16** What is the capacity of an address-multiplexed DRAM with one more address line than the DRAM referred to in Problem 17.15? With two more address lines?

**17.17** Draw the block diagram of a $1M \times 1$ dynamic RAM, showing Row and Column address latches and decoders, correct number of common address pins, $\overline{RAS}$ and $\overline{CAS}$ lines, correct number of address decoder output lines, and Row/Column dimensions of cell array. (It is not necessary to draw all Row/Column lines; just show how many there are.)

## Section 17.3 Read Only Memory (ROM)

**17.18** Briefly list some of the differences between mask-programmed ROM, fusible-link PROM, UV-erasable EPROM, and EEPROM. List one possible application for each type and say why that type is most suitable.

**17.19** Sketch an array of ROM cells that are encoded with the following 4-bit data: 0001, 0011, 1011, 1111.

**\*17.20** A Texas Instruments TBP18S030 $32 \times 8$ PROM is to be used as a BCD-to-seven-segment decoder with a ripple blanking feature. (See Section 10.1 for a review of this feature.) Ripple Blanking input *(RBI)* and output *(RBO)* are to be active HIGH. If the 4-bit input has a code beyond the BCD range, a dash (segment g) is displayed. A summary of operating conditions is given below.

| BCD Input | *RBI* | *RBO* | Display Status |
|---|---|---|---|
| 0000 to 1001 | 0 | 0 | Decimal equivalent of BCD input |
| 0000 | 1 | 1 | Blank |
| 1001 to 1111 | X | 1 | Segment *g* only |

Make a table of hexadecimal data that could be burned into a fusible-link PROM to implement this decoder for a common-anode LED display. (The LED segments on a common-anode display are illuminated by logic LOW.) Show

addresses for each byte of data. Use the following input and output assignments: $A_4 = RBI$, $A_3A_2A_1A_0 = BCD$ input, $Q_7 = RBO$, $Q_6 =$ segment *a*, $Q_0 =$ segment *g*.

**\*\*17.21** Draw a circuit diagram showing how to connect the PROM decoder of Problem 17.20 to a seven-segment LED display. Account for connections to all address, data, and control lines. Remember that PROM outputs will require buffering and that current must be limited through each LED.

**17.22** Briefly describe the programming and erasing process of a UV-EPROM.

## ▼ Answers to Section Review Questions

### Section 17.2a

**17.1.**
Array size: 512 rows × 512 columns
Row address lines: 9
Column address lines: 6
Words per row: 64
Bits per word: 8

### Section 17.2b

**17.2.** a. 9 address, 1 data; b. 9 address, 4 data; c. 10 address, 1 data; d. 10 address, 4 data; e. 11 address, 1 data. (Note that the devices in parts b and c have an equal number of cells, as do the devices in parts d and e.)

# RAM Devices and Memory Systems

● CHAPTER OUTLINE

**18.1** Static RAM Timing Cycles
**18.2** Static RAM Read and Write Cycles
**18.3** Dynamic RAM Timing Cycles

**18.4** Dynamic RAM Modules
**18.5** Memory Systems

● CHAPTER OBJECTIVES

Upon successful completion of this chapter, you will be able to:

- Sketch timing diagrams of Read and Write timing cycles from static and dynamic RAM.

- Sketch timing diagrams of dynamic RAM refresh cycles.

- Sketch a basic memory system, consisting of several memory devices, an address and a data bus, and address decoding circuitry.

- Represent the location of various memory device addresses on a system memory map.

- Expand memory capacity by parallel busing and MSI decoding.

- Describe how a memory system can be expanded beyond the capacity of the system address bus by using a memory management unit and a paged system of addressing.

- Explain how dynamic RAM is configured into high-capacity memory modules.

● INTRODUCTION

Static and dynamic RAM are generally too complicated for their operation to be understood solely in terms of truth tables. The timing diagram of a memory is often a good summary of device operation. A system of notation has been developed to represent timing diagrams in a standard way. We will examine this notation as it applies to static and dynamic RAM.

Memory devices are usually part of a larger system, including a microprocessor, peripheral devices, and a system of tristate buses. If dynamic RAM is used in such a system, it is often in a memory module of some type.

The capacity of a single memory chip is usually less than the memory capacity of the microprocessor system in which it is used. In order to use the full system capacity, it is necessary to use a method of memory address decoding to select a particular RAM device for a specified portion of system memory.

### 18.1

## Static RAM Timing Cycles

A semiconductor memory is such a complicated device that its operation is best understood by looking at a timing diagram rather than reading a description. Conventional symbols have been developed to describe the various operations of these devices.

Before looking at the timing parameters of a specific SRAM, we will look at the conventional timing diagram symbols used for common memory states and some examples of their use. Transition times are measured between the 50% points of the waveforms, unless stated otherwise.

### Timing Parameters

Times between events in a Read or Write cycle are specified by a standard multiple-subscript notation, developed for "unclassified" signals. (A parallel subscript system exists for signals that are classified as access, hold, valid, and so on.) Timing parameters are written in the form $t_{ABCD}$. $A$ and $C$ are input or output signals. $B$ is the state of signal $A$ at the beginning of interval $t_{ABCD}$. $D$ is the state of signal $C$ at the end of the interval. Thus, when signal $A$ goes to state $B$, signal $C$ goes to state $D$ after an elapsed time $t_{ABCD}$.

$B$ or $D$ can be at any of the following states:

*H:* HIGH state or transition to a HIGH state
*L:* LOW state or transition to a LOW state
*V:* any valid steady-state level
*X:* don't care state
*Z:* high-impedance state

The signals $A$ and $C$ in a static RAM are usually one of the following:

*A:* address input
*D:* data input
*Q:* data output
*W:* write enable
*E:* chip enable
*G:* output enable

For example, $t_{GHQZ}$ is the time elapsed between the Output Enable *(G)* input going HIGH and the data output *(Q)* going to the high-impedance state. The Read (or Write) cycle time is called $t_{AVAV}$, indicating that it is the elapsed time between the beginning and end of one valid address *(V)* on the memory address inputs *(A)* and the beginning of the next valid address. The interval $t_{AVQV}$ is the elapsed time between a valid address *(V)* appearing at the address inputs *(A)* and valid data *(V)* appearing at the data outputs *(Q)*.

## Bus Waveforms

We do not usually show all possible address and data bus combinations on a timing diagram. (Imagine it with even an 8-bit bus, never mind a 32-bit one.) Rather, the generalized bus contents are shown as two parallel lines, with transitions shown as a crossover at the transition point, as illustrated in Figure 18.1. (We have seen similar notation in the simulation waveforms of PLDs.)

If there is no cross-hatching between the parallel lines, the bus contents are at a steady-state level referred to as valid data or address. The interval specified in Figure 18.1, $t_{AVAV}$, is the time required between two valid addresses appearing on the address inputs.

## Compulsory Levels

If an input must be at a certain logic level or an output is guaranteed to be at a certain level, this is shown as a straight horizontal line, as in any other timing diagram. Figure 18.2 shows the $\overline{W}$ input of a memory making two transitions, from HIGH to LOW and back to HIGH. There is no cross-hatching on this line, indicating that the time elapsed between the two transitions, the Write pulse width, $t_{WLWH}$, must occur exactly as shown, within the limits of the timing specification.

## Optional Transitions

An input may be allowed to make a transition within a specified timing interval. An output may be guaranteed to make a transition within a particular time frame. These optional transitions are shown by cross-hatching in the direction of the required transition.

Figure 18.3 shows the relation between the timing of the $\overline{W}$ line and the $\overline{E}$ line of a memory. The hatching on the $\overline{W}$ input implies that $\overline{W}$ must go LOW no later than $t_{ELWL}$ after $\overline{E}$. $\overline{W}$ can go LOW earlier, as long as it is after $\overline{E}$ goes LOW. The interval $t_{WHEH}$ must occur as shown.

Figure 18.1
**Bus Waveforms**

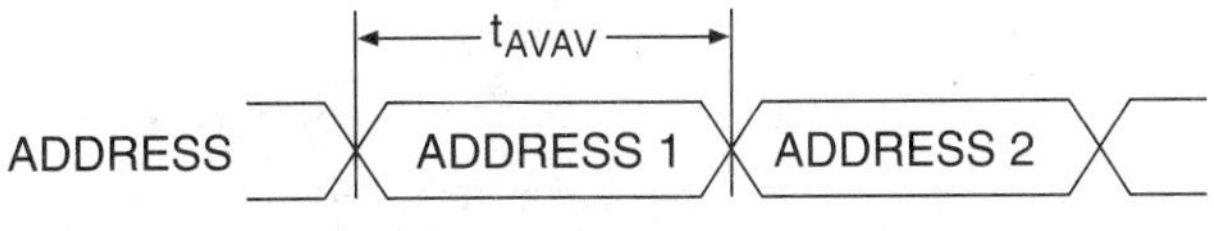

Figure 18.2
**Compulsory Levels**

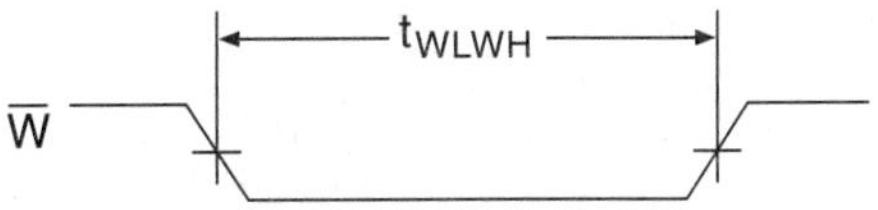

Figure 18.3
**Optional Transitions**

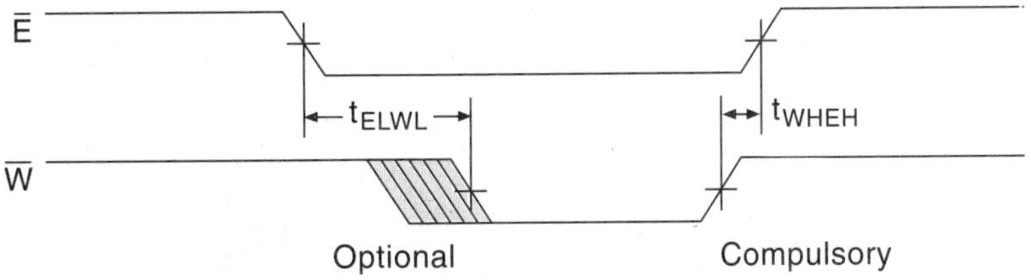

**Figure 18.4**

**Don't Care States**

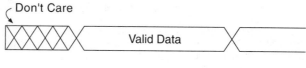

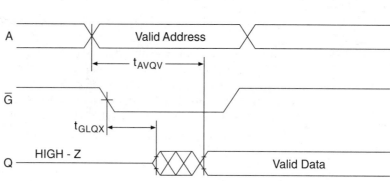

**Figure 18.5**

**High-Impedance States**

## Don't Care States

A don't care or unknown state on an input or output is shown as a double cross-hatch on the waveform. Figure 18.4 shows the transition from a don't care state to valid data on a data bus.

## HIGH-Z

A tristate output in its high-impedance (HIGH-Z) state is shown as a line midway between the HIGH and LOW logic levels. Figure 18.5 shows the output data (Q) lines of a semiconductor memory with transitions from a high-impedance state to a don't care state to a valid data state.

Two timing intervals are shown. The first, $t_{GLQX}$, indicates the elapsed time between the Output Enable (G) line going LOW and the data output (Q) lines making a transition from the high-impedance to the low-impedance don't care (X) state. The second, $t_{AVQV}$, shows the maximum time between a valid address (V) appearing on the address inputs (A) and valid data (V) appearing on the data outputs (Q).

### Section Review Problem for Section 18.1

18.1. Label the timing parameters in the diagram in Figure 18.6, using the same multiple-subscript notation used in the previous section.

**Figure 18.6**

**Timing Diagram for Section Review Problem**

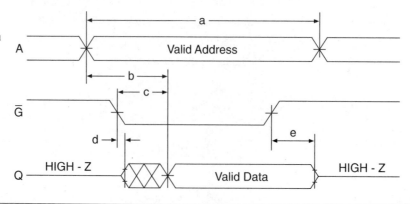

## 18.2

# Static RAM Read and Write Cycles

$F$igures 18.7 through 18.9 show the timing diagrams for the Read, $\overline{W}$-controlled Write, and $E$-controlled Write cycles for an MCM6164 SRAM. As memory devices go, these cycles are pretty straightforward and are a good starting point for understanding the symbols and parameters used in memory cycle timing diagrams.

## SRAM Read Cycle

The Read cycle for the MCM6164 SRAM is shown in the data sheet of Figure 18.7. To read data from a static RAM, all you have to do is address a particular word location, make the $\overline{W}$ line HIGH (Read mode), and activate the Chip Enable and Output Enable inputs ($\overline{E1} = 0$, $E2 = 1$, $\overline{G} = 0$). In an ideal system, the data would be instantaneously available as soon as these conditions were met. In a real system, there are inevitable delays that are specified by the device manufacturer.

Timing parameters specified as minimum values are given from an external point of view, and maximum values are given from an internal point of view, relative to the memory. That is, a parameter specified by a minimum value requires *at least* that amount of time to be allowed between the specified signal transitions. A parameter specified as a maximum value implies that after the initiating transition, the responding transition will occur *no later* than the specified time.

For example, $t_{AXQX}$ (output hold from address change) is 5 ns, minimum, for an MCM6164-45 SRAM. This means that *at least* 5 nanoseconds must be allowed for the data outputs $(Q)$ to release the previous valid data after the address inputs $(A)$ have gone to a new state. ($Q$ is shown as going to the high-impedance state since both $\overline{G}$ and $\overline{E}$ are inactive at this point.)

On the other hand, $t_{GLQV}$ ($\overline{G}$ access time) is 20 ns, maximum. This means that the data outputs $(Q)$ will present valid data *no later* than 20 ns after the Output Enable $(\overline{G})$ input goes LOW; the system can expect to read valid data within this time.

Let us analyze the timing diagram in Figure 18.7 to get an overall view of the MCM6164 Read cycle. We will examine events in the order in which they occur on the $Q$ outputs.

**1.** The Read cycle begins with the SRAM in the Read mode and all enable inputs inactive. $Q$ is in the high-impedance state.

**2.** Inputs $\overline{G}$, $\overline{E1}$, and $E2$ must all be active to switch the data outputs from the high-impedance to the low-impedance $(X)$ state. At least 5 ns must be allowed for this transition if it is controlled by either of the Chip Enable inputs ($t_{E1LQX} = t_{E2HQX} = 5$ ns, minimum). If the transition is controlled by the Output Enable input, no lead time is required ($t_{GLQX} = 0$). At this point, valid data are not yet available on the data outputs, so the $Q$s are in don't care states.

**3.** A transition from a don't care state to valid data appears on the data outputs shortly after the enable inputs are activated and a valid address appears on the address inputs. These times are given by $t_{GLQV}$, $t_{E1LQV}$, $t_{E2HQV}$, and $t_{AVQV}$. All these times are specified as 45 ns, except for $t_{GLQV}$, which is 20 ns. Thus, valid data are available on the data outputs 45 ns after these inputs are activated, if this is done simultaneously. If there is some choice to be made about the order in which to activate these inputs, activate the Output Enable input last, since it is the fastest.

## AC OPERATING CONDITIONS AND CHARACTERISTICS
($V_{CC}$ = 5 V ±10%, $T_A$ = 0 to +70°C, Unless Otherwise Noted)

Input Timing Measurement Reference Level . . . . . . . . . . 1.5 V
Input Pulse Levels . . . . . . . . . . . . . . . . . . . . . . . 0 to 3.0 V
Input Rise/Fall Time . . . . . . . . . . . . . . . . . . . . . . . 5 ns

Output Timing Measurement Reference Level . . . 0.8 V and 2.0 V
Output Load. . . . . . . . . . . . . . . . . . . . . . . . . . . . Figure 1

### READ CYCLE (See Note 1)

| Characteristic | Symbol | Alt Symbol | MCM6164-45 MCM61L64-45 | | MCM6164-55 MCM61L64-55 | | Unit | Notes |
|---|---|---|---|---|---|---|---|---|
| | | | Min | Max | Min | Max | | |
| Read Cycle Time | $t_{AVAV}$ | $t_{RC}$ | 45 | — | 55 | — | ns | — |
| Address Cycle Time | $t_{AVQV}$ | $t_{AA}$ | — | 45 | — | 55 | ns | — |
| $\overline{E1}$ Access Time | $t_{E1LQV}$ | $t_{AC1}$ | — | 45 | — | 55 | ns | — |
| E2 Access Time | $t_{E2HQV}$ | $t_{AC2}$ | — | 45 | — | 55 | ns | — |
| $\overline{G}$ Access Time | $t_{GLQV}$ | $t_{OE}$ | — | 20 | — | 25 | ns | — |
| Output Hold from Address Change | $t_{AXQX}$ | $t_{OH}$ | 5 | — | 5 | — | ns | — |
| Chip Enable to Output Low-Z | $t_{E1LQX}$, $t_{E2HQX}$ | $t_{CLZ}$ | 5 | — | 5 | — | ns | 2, 3 |
| Output Enable to Output Low-Z | $t_{GLQX}$ | $t_{OLZ}$ | 0 | — | 0 | — | ns | 2, 3 |
| Chip Enable to Output High-Z | $t_{E1HQZ}$, $t_{E2LQZ}$ | $t_{CHZ}$ | 0 | 20 | 0 | 20 | ns | 2, 3 |
| Output Enable to Output High-Z | $t_{GHQZ}$ | $t_{OHZ}$ | 0 | 20 | 0 | 20 | ns | 2, 3 |

NOTES:
1. $\overline{W}$ is high at all times for read cycles.
2. All high-Z and low-Z parameters are considered in a high or low impedance state when the output has made a 500 mV transition from the previous steady state voltage.
3. Periodically sampled rather than 100% tested.

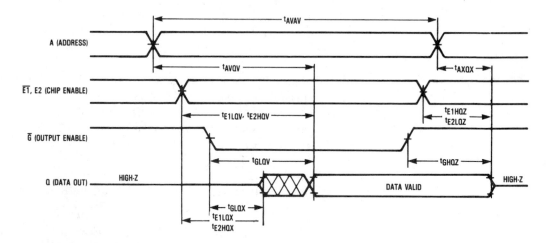

### TIMING LIMITS

The table of timing values shows either a minimum or a maximum limit for each parameter. Input requirements are specified from the external system point of view. Thus, address setup time is shown as a minimum since the system must supply at least that much time (even though most devices do not require it). On the other hand, responses from the memory are specified from the device point of view. Thus, the access time is shown as a maximum since the device never provides data later than that time.

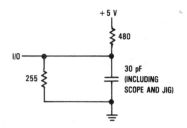

Figure 1. Test Load

Figure 18.7

**Read Cycle MCM6164 Static RAM** (Reprinted with permission of Motorola)

## MCM6164•MCM61L64

**WRITE CYCLE 1 ($\overline{W}$ CONTROLLED)** (See Note 1)

| Characteristic | Symbol | Alt Symbol | MCM6164-45 MCM61L64-45 Min | MCM6164-45 MCM61L64-45 Max | MCM6164-55 MCM61L64-55 Min | MCM6164-55 MCM61L64-55 Max | Unit | Notes |
|---|---|---|---|---|---|---|---|---|
| Write Cycle Time | $t_{AVAV}$ | $t_{WC}$ | 45 | — | 55 | — | ns | — |
| Address Setup Time | $t_{AVWL}$ | $t_{AS}$ | 0 | — | 0 | — | ns | — |
| Address Valid to End of Write | $t_{AVWH}$ | $t_{AW}$ | 40 | — | 50 | — | ns | — |
| Write Pulse Width | $t_{WLWH}$ | $t_{WP}$ | 25 | — | 30 | — | ns | 2 |
| Data Valid to End of Write | $t_{DVWH}$ | $t_{DW}$ | 20 | — | 25 | — | ns | — |
| Data Hold Time | $t_{WHDX}$ | $t_{DH}$ | 0 | — | 0 | — | ns | 3 |
| Write Low to Output in High-Z | $t_{WLQZ}$ | $t_{WHZ}$ | 0 | 20 | 0 | 20 | ns | 4, 5 |
| Write High to Output Low-Z | $t_{WHQX}$ | $t_{OW}$ | 5 | — | 5 | — | ns | 4, 5 |
| Write Recovery Time | $t_{WHAX}$ | $t_{WR}$ | 0 | — | 0 | — | ns | — |

NOTES:
1. A write cycle starts at the latest transition of a low $\overline{E1}$, low $\overline{W}$ or high E2. A write cycle ends at the earliest transition of a high $\overline{E1}$, high $\overline{W}$ or low E2.
2. If $\overline{W}$ goes low coincident with or prior to $\overline{E1}$ low or E2 high then the outputs will remain in a high impedance state.
3. During this time the output pins may be in the output state. Signals of opposite phase to the outputs must not be applied at this time.
4. All high-Z and low-Z parameters are considered in a high or low impedance state when the output has made a 500 mV transition from the previous steady state voltage.
5. Periodically sampled rather than 100% tested.

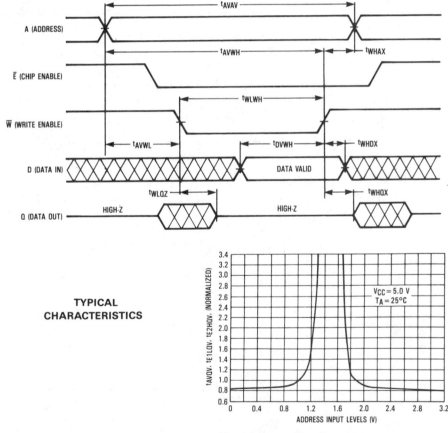

TYPICAL
CHARACTERISTICS

Figure 2. Access Time Versus Address Input Levels

Figure 18.8

**$\overline{W}$-Controlled Write Cycle MCM6164 Static RAM** (Reprinted with permission of Motorola)

773

**WRITE CYCLE 2 (ENABLE CONTROLLED)** (See Notes 1 and 2)

| Characteristic | Symbol | Alt Symbol | MCM6164-45 MCM61L64-45 | | MCM6164-55 MCM61L64-55 | | Unit | Notes |
|---|---|---|---|---|---|---|---|---|
| | | | Min | Max | Min | Max | | |
| Write Cycle Time | $t_{AVAV}$ | $t_{WC}$ | 45 | — | 55 | — | ns | — |
| Address Setup Time | $t_{AVE1L}$ | $t_{AS}$ | 0 | — | 0 | — | ns | — |
| Address Valid to End of Write | $t_{AVE1H}$ | $t_{AW}$ | 40 | — | 50 | — | ns | — |
| Chip Enable to End of Write | $t_{E1LE1H}$ | $t_{CW}$ | 40 | — | 50 | — | ns | 3 |
| Data Valid to End of Write | $t_{DVE1H}$ | $t_{DW}$ | 20 | — | 25 | — | ns | — |
| Data Hold Time | $t_{E1HDX}$ | $t_{DH}$ | 0 | — | 0 | — | ns | 4 |
| Write Recovery Time | $t_{E1HAX}$ | $t_{WR}$ | 0 | — | 0 | — | ns | — |

NOTES:
1. A write cycle starts at the latest transition of a low $\overline{E1}$, low $\overline{W}$ or high E2. A write cycle ends at the earliest transition of a high $\overline{E1}$, high $\overline{W}$ or low E2.
2. $\overline{E1}$ and E2 timings are identical when E2 signals are inverted.
3. If $\overline{W}$ goes low coincident with or prior to $\overline{E1}$ low or E2 high then the outputs will remain in a high impedance state.
4. During this time the output pins may be in the output state. Signals of opposite phase to the outputs must not be applied at this time.

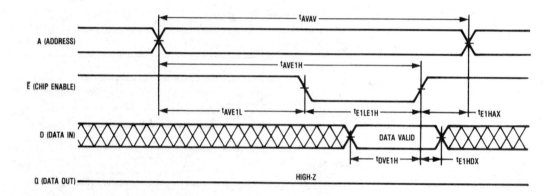

**LOW $V_{CC}$ DATA RETENTION CHARACTERISTICS** ($T_A = 0$ to $+70°C$) (MCM61L64 Only)

| Characteristic | Symbol | Min | Typ | Max | Unit |
|---|---|---|---|---|---|
| $V_{CC}$ for Data Retention ($\overline{E1} \geq V_{CC} - 0.2$ V or E2 $\leq 0.2$ V, $V_{in} \geq V_{CC} - 0.2$ V or $V_{in} \leq 0.2$ V) | $V_{DR}$ | 2.0 | 1.0 | 7.0 | V |
| Data Retention Current ($V_{CC} = 3.0$ V, $\overline{E1} \geq 2.8$ V or E2 $\leq 0.2$ V, $V_{in} \geq 2.8$ V or $V_{in} \leq 0.2$ V) | $I_{CCDR}$ | — | 10 | 30 | $\mu$A |
| Chip Disable to Data Retention Time (see waveform below) | $t_{CDR}$ | 0 | — | — | ns |
| Operation Recovery Time (see waveform below) | $t_{rec}$ | $t_{AVAV}$* | — | — | ns |

*$t_{AVAV}$ = Read Cycle Time

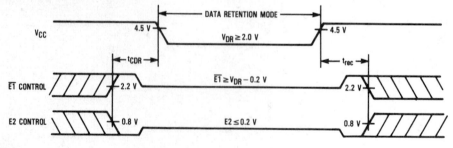

Figure 18.9

**E-Controlled Write Cycle MCM6164 Static RAM** (Reprinted with permission of Motorola)

**4.** When the address is changed, the output will hold its previous value for at least 5 ns ($t_{AXQX}$ = 5 ns, minimum).

**5.** When the chip is disabled by deactivating $\overline{G}$, $\overline{E1}$, or $E2$, the $Q$ outputs will return to the high-impedance state within 20 ns ($t_{GHQZ}$ = $t_{E1HQZ}$ = $t_{E2LQZ}$ = 20 ns, maximum).

---

**Minimum Read Cycle Requirements (MCM6164-45)**

1. Allow 45 ns for completion of Read cycle $(t_{AVAV})$.

2. Allow 5 ns for data output to go to don't care state after address changes $(t_{AXQX})$.

3. Allow 5 ns for Chip Enable inputs to switch data outputs to the low-impedance state $(t_{E1LQX}, t_{E2HQX})$.

---

## SRAM Write Cycles

Two Write cycles are shown in the MCM6164 data sheet. One cycle is controlled by the Write Enable input (Figure 18.8) and the other by the Chip Enable inputs (Figure 18.9).

The main difference is that in the $E$-controlled Write, the Write Enable input is held LOW for the whole cycle. This has the effect of keeping the data outputs, which are not needed for the Write cycle, always in the high-impedance state.

In the $\overline{W}$-controlled cycle, the time limits of the Write pulse (a HIGH-LOW-HIGH pulse on the Write Enable line) must be completely within the limits of the Chip Enable time. If the Output Enable line is active, the data outputs will enter the low-impedance state briefly before and after the Write pulse is applied. (The condition for the output buffers to be enabled is $\overline{G}$ = 0, $\overline{E1}$ = 0, $E2$ = 1, $\overline{W}$ = 1.)

The Chip Enable function is shown as the $\overline{E1}$ line in both Write cycle diagrams. $E2$ has the same timing constraints. Both $\overline{E1}$ and $E2$ must be active to enable the chip.

Both Write cycles shown are for the MCM6164-45 SRAM. Cycles for the MCM6164-55, a slower version of the same device, have the same form but different timing parameters, as specified in the data sheets in Figures 18.7 through 18.9.

### $\overline{W}$-Controlled Write (MCM6164-45)

The $\overline{W}$-controlled Write (Figure 18.8) has a cycle time of 45 ns ($t_{AVAV}$ = 45 ns, minimum). There are five constraints to keep in mind for this cycle, all of them having to do with the Write Enable line:

**1.** The Write Enable line pulses LOW to write data into the selected address. It is permissible for the Write pulse to begin as soon as a valid address is present on the address bus ($t_{AVWL}$ = 0). However, the Write Enable line must remain LOW for at least 40 ns after the beginning of a valid address ($t_{AVWH}$ = 40 ns, minimum).

**2.** The input data *(D)* must be valid for at least 20 ns before the end of the Write pulse ($t_{DVWH}$ = 20 ns, minimum). It is not necessary to hold the data inputs at a valid state after the Write pulse is finished ($t_{WHDX}$ = 0).

**3.** The Write pulse itself must be at least 25 ns long ($t_{WLWH}$ = 25 ns, minimum).

**4.** If the data outputs have been in the low-impedance state, they will switch back to the high-impedance state within 20 ns of $\overline{W}$ going LOW ($t_{WLQZ}$ = 20 ns, maximum). If the Write Enable goes LOW before the Chip Enable inputs are active, the data outputs remain in the high-impedance state. (Data should never be present at both $Q$ and $D$ at the same time.)

**5.** After the Write pulse is finished, 5 ns must be allowed for the data outputs to return to the low-impedance state, if this is required ($t_{WHQX}$ = 5 ns, minimum).

### *E*-Controlled Write (MCM6164-45)

For the *E*-controlled Write (Figure 18.9), the Write Enable line is always LOW. As a result, the data outputs *(Q)* are always in the high-impedance state. If desired, one of the Chip Enable inputs can be held at its active state and the other used to control the Write cycle (e.g., hold *E2* permanently HIGH).

The *E*-controlled Write is 45 ns long for an MCM6164-45 SRAM ($t_{AVAV}$ = 45 ns).

There are several constraints related to the *E* input in this cycle. All parameters are given with respect to the $\overline{E1}$ line. The *E2* waveforms, if used, are the complement of the $\overline{E1}$ waveforms.

**1.** Data are written into the selected location by pulsing the $\overline{E1}$ input LOW. It is permissible to start the Enable pulse as soon as there is a valid address at the address inputs of the chip ($t_{AVEIL}$ = 0), but the Chip Enable line must remain active for at least 40 ns after the beginning of the valid address ($t_{AVEIH}$ = 40 ns, minimum).

**2.** Valid input data *(D)* must be present for at least 20 ns prior to the end of the Enable pulse ($t_{DVEIH}$ = 20 ns, minimum). However, there is no need to hold data valid after the end of the Enable pulse ($t_{EIHDX}$ = 0).

**3.** The Chip Enable pulse must be at least 40 ns long ($t_{EILEIH}$ = 40 ns, minimum).

## 18.3

# Dynamic RAM Timing Cycles

The operation of a dynamic RAM, which requires more types of cycles than a static RAM, is best explained in terms of its timing diagrams. In addition to the normal Read and Write cycles, there are combined Read-Write cycles called Read-Modify-Write and Read-while-Write. Some newer DRAMs have several fast access modes, called fast page mode, nibble mode, and static column decode, each of which can be used for Read, Write, or Read-Write operations. Since each of these fast access modes depends on a different physical design of the dynamic RAM, only one is available on any particular chip.

It is necessary to have some type of Refresh cycle so that the stored data remains valid. Three types of Refresh cycle are $\overline{RAS}$-only Refresh, $\overline{CAS}$ before $\overline{RAS}$ Refresh, and Hidden Refresh.

You might wish to refer back to the block diagram of the address-multiplexed dynamic RAM in Figure 17.12. Recall that in this type of device, the address is split into row and column segments, so that the entire address requires fewer input pins. The signals $\overline{RAS}$ (row address strobe) and $\overline{CAS}$ (column address strobe) latch the row and column addresses into their respective decoders.

**Figure 18.10**

**Normal Read Operation Dynamic RAM ($\overline{W}$ is HIGH)**

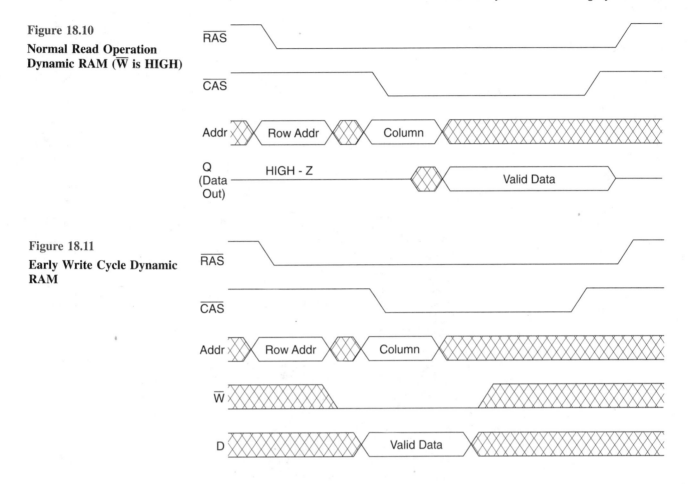

**Figure 18.11**

**Early Write Cycle Dynamic RAM**

## Normal Read Cycle

Figure 18.10 shows the normal Read cycle of a dynamic RAM. Each memory cell has a unique row and column address within the RAM. When the row address is stable on the chip's address inputs, a negative transition on the $\overline{RAS}$ line latches it into the row address decoder. After a short time (about 20 to 50 ns), the column address is placed on the address inputs and latched into the column address decoder by a negative transition on the $\overline{CAS}$ line.

To read data at the selected address, the $\overline{W}$ line must be HIGH for the entire time that the $\overline{CAS}$ line is LOW. (Although there is no requirement for $\overline{W}$ to remain HIGH before and after $\overline{CAS}$ is LOW, it will increase the reliability of operation if the two signals are not coincident.) Data appears on the $Q$ output shortly after these conditions are satisfied and remains there until shortly after the $\overline{CAS}$ line goes back to the HIGH state.

## Write Cycles

The Write cycle shown in Figure 18.11 is referred to as the Early Write cycle, so called because the Write Enable input is active before the $\overline{CAS}$ line. Row and column addresses are selected by the address, $\overline{RAS}$, and $\overline{CAS}$ inputs, as in the Read cycle. The $\overline{W}$ line goes

**Figure 18.12**

**Read-Write Cycle Dynamic RAM**

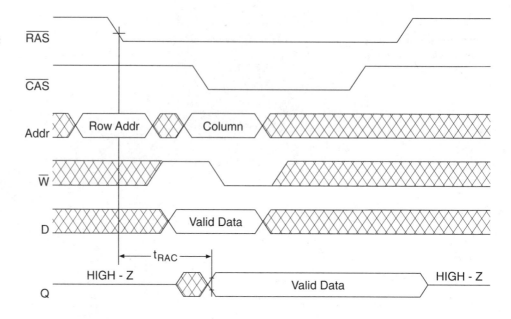

LOW before $\overline{CAS}$ does. Valid data must be present on the data input *(D)* at or before the time $\overline{CAS}$ goes LOW. Since $\overline{W}$ goes LOW before $\overline{CAS}$, the data output *(Q; not shown)* remains in the high-impedance state for the entire cycle.

## Read-Write Cycles

A Read-Write cycle is used to read data from an address and write new data to the same address without having to go through the $\overline{RAS}$ and $\overline{CAS}$ access times more than once. There are two variations on this cycle: Read-Modify-Write and Read-while-Write. The only difference is in the timing of the $\overline{W}$ line with respect to $\overline{RAS}$ and $\overline{CAS}$. The Read-Write cycle of a Motorola MCM511000A 1 M × 1 DRAM is shown in Figure 18.12. *D* and *Q* each have their own pins on this device.

The Read-Modify-Write cycle starts as a normal Read operation. After data appears on the *Q* output, the $\overline{W}$ line goes LOW, writing the data on the *D* input into the selected location. The original output data remains at the *Q* output until $\overline{CAS}$ goes HIGH and is not modified by the Write operation. Only the stored data at the addressed location is changed.

The Read-while-Write cycle is the same as the Read-Modify-Write except that the Write pulse is applied sooner. With the proper timing, the Write pulse can be applied *before* data appears at the *Q* output and new data can be stored in the selected location. Built-in delays inside the dynamic RAM allow this to happen; the output data is already on the way out when the new input data is on the way in.

## Fast Access Cycles

**Page** *A contiguous block of addresses selected by high-order address bits. Low-order bits of the addresses select a word within the page.*

Some of the newer high-capacity (1 Mb and larger) dynamic RAMs have access modes that allow faster Read and Write operations when several related locations are being accessed.

Fast or enhanced page mode allows you to address any sequence of columns in the same row (**page**) of a dynamic RAM with only one pulse on the $\overline{RAS}$ input. In a 1-Mb DRAM, any of 1024 columns in one row can be accessed one after the other, using only the $\overline{CAS}$ line to latch in successive column addresses.

Nibble mode will retrieve any 2, 3, or 4 adjacent bits of data and, if required, cycle through them continuously.

Static column decode mode is similar to fast page mode, except that it requires only one $\overline{RAS}$ and one $\overline{CAS}$ transition, which addresses the first row and column location. Other addresses in the same row are accessed by changing the column address without a $\overline{CAS}$ transition.

Each of the fast access modes requires a memory with a different physical design, so only one of the modes is available on any particular chip. Table 18.1 shows a partial list of these devices by different manufacturers. The devices are not necessarily exact replacements for one another.

Any of the fast access modes can be used for Read, Write, or Read-Write operations or any combination of these. We will examine each of the modes as it applies to a Read operation.

## Fast Page Mode

Figure 18.13 shows the fast-page-mode Read cycle of a Motorola MCM511000A-70 1-Mb DRAM. After the row address is latched into the RAM, any of the 1024 columns in that row can be accessed for Read, Write, or Read-Write operations by holding the $\overline{RAS}$ line LOW and toggling the $\overline{CAS}$ line.

The new column address for each location is given by the state of the address inputs when the $\overline{CAS}$ line makes its negative transition. The data stored at the selected location appears on the $Q$ output a short time after the $\overline{CAS}$ line goes LOW. Data is valid at $Q$ until shortly after $\overline{CAS}$ goes HIGH.

Data in other rows cannot be retrieved without a new negative transition on the $\overline{RAS}$ line.

The first location in a row requires 70 ns $(t_{RAC})$ to access, but any following locations in the same row require only 40 ns $(t_{PC})$. Since we do not need the $\overline{RAS}$ transition for any but the first address, the second and following access times are shorter.

## Nibble Mode

Nibble mode is used to serially address 2, 3, or 4 bits of data with only one negative transition of the $\overline{RAS}$ line. This mode is available with the Motorola MCM511001A-70

Table 18.1  1-Mb DRAMs With Fast Access Modes

| Manufacturer | Fast Page | Nibble | Static Column Decode |
|---|---|---|---|
| Motorola | MCM511000A | MCM511001A | MCM511002A |
| Texas Inst. | TMS4C1024 | TMS4C1025 | TMS4C1027 |
| Fujitsu | MB811000 | MB811001 | — |
| Mitsubishi | M5M4C1000 | M5M4C1001 | M5M4C1002 |
| Toshiba | TC511000 | TC511001 | TC511002 |

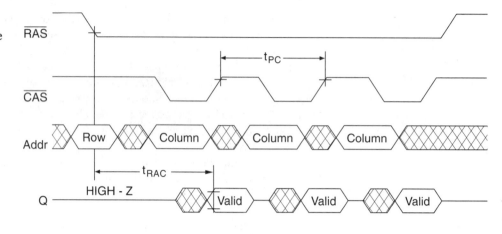

**Figure 18.13**

**Fast-Page-Mode Read Cycle
Dynamic RAM**

$\overline{W}$ is held HIGH

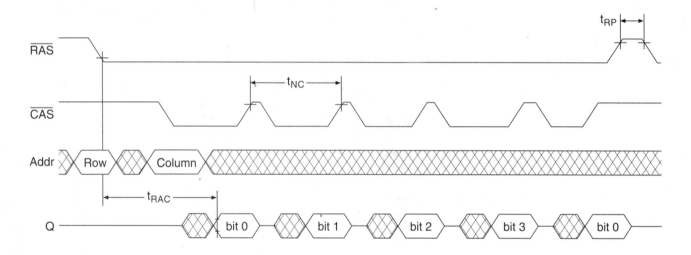

$\overline{W}$ is HIGH

**Figure 18.14**

**Nibble-Mode Read Cycle
Dynamic RAM**

1-Mb dynamic RAM. The nibble-mode Read cycle for this device is shown in Figure 18.14.

Every cell in a 1-Mb dynamic RAM has a unique 20-bit address. Every nibble (group of 4 bits) has a unique 18-bit address ($2^{18} = 256$K; 256K nibbles $\times$ 4 bits/nibble = 1024 Kb = 1 Mb). Each nibble is retrieved by an 18-bit address consisting of the 9 least significant bits of the row and column addresses, bits $A_8. . . A_0$.

A 9-bit row address is latched by a negative transition on the $\overline{RAS}$ line. A 9-bit column address is latched by a negative transition on the $\overline{CAS}$ line. The states of $CA_9$ and $RA_9$, the most significant bits of the column and row addresses, specify which of the 4 possible selected bits will be addressed first. This 2-bit value is stored in an internal counter, which is clocked by pulses on the $\overline{CAS}$ line.

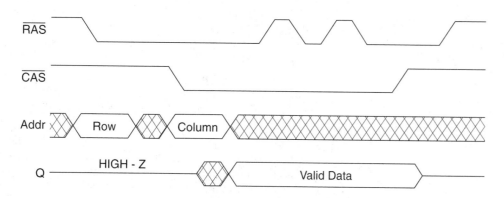

**Figure 18.18**

$\overline{CAS}$ **Before** $\overline{RAS}$ **Refresh**

**Figure 18.19**

**Hidden Refresh (One READ and Two REFRESH Cycles Shown)**

$\overline{CAS}$ goes LOW for 10 ns prior to the negative transition of $\overline{RAS}$ ($t_{CSR}$) and must be held for 30 ns ($t_{CHR}$). Row addresses are generated automatically by an internal counter, and external addresses on pins $A_9$. . . $A_0$ are ignored. The internal counter is clocked by the $\overline{RAS}$ line, allowing $\overline{CAS}$ to remain active during subsequent Refresh cycles, if required. The Refresh cycle ends when $\overline{CAS}$ goes HIGH.

## Hidden Refresh Cycle

Hidden Refresh is a variation of the $\overline{CAS}$ before $\overline{RAS}$ Refresh mode. The cycle is called Hidden Refresh because it can be performed between Read or Write cycles with no externally visible effect. Figure 18.19 shows a Read cycle followed by two Hidden Refresh cycles for an MCM511000A-70 device.

The cycle starts as a normal Read cycle, with the row and column addresses being latched into the RAM in the usual way. The $\overline{RAS}$ line then goes HIGH, and after a short precharge time, goes LOW to initiate a $\overline{CAS}$ before $\overline{RAS}$ Refresh cycle. From this point on, the $\overline{RAS}$ line clocks the internal refresh counter and the external addresses are ignored. Valid data from the initial Read operation will remain on the $Q$ output as long as $\overline{CAS}$ stays LOW. When $\overline{CAS}$ goes HIGH, the Refresh cycle terminates and the RAM is ready for a Read, Write, or Refresh operation.

---

**Figure 18.15**

**Bit Sequence in a Nibble-Mode Dynamic RAM**

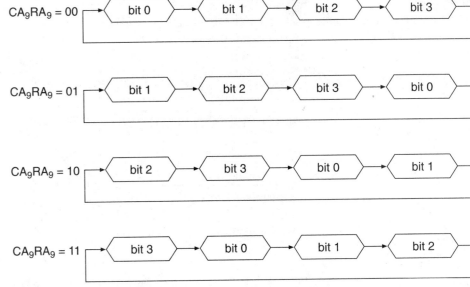

Figure 18.15 shows the sequence in which the selected bits are accessed. For example, if $CA_9RA_9 = 10$, bit 2 will be accessed first. Further pulses on the $\overline{CAS}$ input will access bit 3, bit 0, bit 1, bit 2, and so on until the $\overline{RAS}$ line goes HIGH.

The first access of a bit in nibble mode takes about 70 ns ($t_{RAC}$). Following locations in the same nibble require about 35 ns ($t_{NC}$). To select a new nibble, the $\overline{RAS}$ line must go HIGH for 50 ns ($t_{RP}$) before going LOW again. The new location is accessed 70 ns after that.

## Static Column Decode

Static column decode functions much the same as fast page mode. Both modes allow fast access to all 1024 column addresses in the same row. Static column mode requires only one $\overline{CAS}$ transition; after the first location, column addresses can be changed without a $\overline{CAS}$ pulse, as long as $\overline{RAS}$ and $\overline{CAS}$ are held LOW. This access mode generates less switching noise than fast page mode, since there are no pulses on the $\overline{CAS}$ line after the first column has been selected.

The timing diagram of a static-column-mode Read cycle for a Motorola MCM511002A-70 1-Mb DRAM is shown in Figure 18.16. The first access needs about 70 ns ($t_{RAC}$), and subsequent accesses are about 40 ns ($t_{SC}$).

## Comparison of Fast Access Modes

For a true measure of the relative speed of the various fast access modes, we should compare each mode with a normal cycle for the largest amounts of data they are designed to retrieve. For a 1-Mb dynamic RAM, the page mode and static column mode are designed to select up to 1024 bits of data for each row. The nibble mode can call up to 4 bits of data per $\overline{RAS}$ transition. A normal cycle retrieves 1 bit per row access. Table 18.2 compares the various Read modes when they access 4 bits in the same nibble and 1024 bits in the same row.

**Figure 18.16**

**Static-Column-Decode Read Cycle Dynamic RAM**

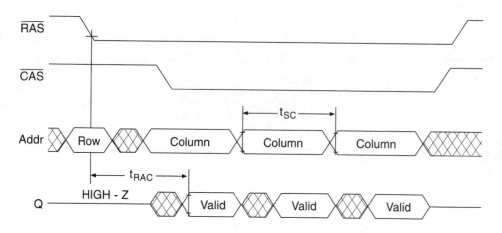

$\overline{W}$ is HIGH

**Table 18.2** Relative Speeds of MCM51100xA Fast Access Modes

|  | Normal | Page | Nibble | Static Column |
|---|---|---|---|---|
| **4 Bits** | 280 ns | 190 ns | 175 ns | 190 ns |
| **Fraction of Normal** | 1.00 | 0.68 | 0.63 | 0.68 |
| **1024 Bits** | 71,680 ns | 40,990 ns | 57,600 ns | 40,990 ns |
| **Fraction of Normal** | 1.00 | 0.57 | 0.80 | 0.57 |

**Normal:**

4-bit Read = $4t_{RAC}$
1024-bit Read = $1024t_{RAC}$

**Page:**

4-bit Read = $t_{RAC} + 3t_{PC}$
1024-bit Read = $t_{RAC} + 1023t_{PC}$

**Nibble:**

4-bit Read = $t_{RAC} + 3t_{NC}$
1024-bit Read = $256(t_{RAC} + 3t_{NC} + t_{RP})$

**Static Column:**

4-bit Read = $t_{RAC} + 3t_{SC}$
1024-bit Read = $t_{RAC} + 1023t_{SC}$

The nibble mode has the fastest access time for a 4-bit Read, but it is limited to those 4 bits for one $\overline{RAS}$ transition. It still maintains an edge over a normal Read cycle for the 1024-bit Read, since it requires a $\overline{RAS}$ pulse for every 4 bits, giving 256 row accesses, as opposed to 1024 needed for the normal Read. The page and static column modes are not as fast as the nibble mode for the 4-bit Read, but clearly show a speed

advantage during the 1024-bit Read operation, since they require only one row access for the entire operation.

## Refresh Cycles

To ensure that stored data remains valid, each cell in a dynamic RAM must be refreshed within a specified period, ranging from 2 to 8 ms. Considering that access times of a DRAM are on the order of several tens of nanoseconds, this isn't too bad. There is a lot of time for memory operations between Refresh cycles (1 ms = 1,000,000 ns).

A row of cells is automatically refreshed when a Read or Write operation is performed on any cell in that row. For a 1 Mb × 1 dynamic RAM, there are 10 bits for row and column addresses. However, the row address is split internally so that 1 bit ($RA_9$) is added to the column address, giving an internal configuration of 512 rows (9-bit address) and 2048 columns (12-bit address).

One row is refreshed in about 130 ns ($t_{RC}$) for a Motorola MCM511000A-70 dynamic RAM.

The 512 rows in the RAM can be refreshed one after the other in an operation called "Burst Refresh." This takes 512 × 130 ns = 66.56 µs for an MCM51100xA-70 DRAM. Read and Write operations are suspended during this time. Out of an 8-ms maximum cycle, 7.93 ms (approximately 99.2% of the cycle) is left over for Read and Write operations.

Memory can also be periodically refreshed in the spaces between Read and Write operations. This is called "distributed Refresh." These modes are controlled by external devices called dynamic memory controllers.

There are several variations on the Refresh cycle. $\overline{RAS}$-only Refresh uses timing signals on the $\overline{RAS}$ line to latch in externally provided row addresses while the $\overline{CAS}$ line is held HIGH. $\overline{CAS}$ before $\overline{RAS}$ Refresh generates internal row addresses when the usual order of address clock signals is reversed. Hidden Refresh also generates internal addresses and refreshes one or more rows of cells while maintaining valid data on the output.

### $\overline{RAS}$-Only Refresh

Figure 18.17 shows the $\overline{RAS}$-only Refresh cycle for a Motorola MCM511000A dynamic RAM. In this cycle, a 9-bit address must be supplied to address inputs $A_8$. . . $\overline{CAS}$ must be held HIGH. $A_9$ and $\overline{W}$ can be in any state. The $\overline{RAS}$ line goes LOW enough to access the selected row (about 70 ns), then HIGH, in preparation for row address.

### $\overline{CAS}$ Before $\overline{RAS}$ Refresh

In the $\overline{CAS}$ before $\overline{RAS}$ Refresh cycle, the usual order of the row and column strobes is reversed. Figure 18.18 shows this cycle for an MCM511000A

**Figure 18.17**

**$\overline{RAS}$-Only Refresh**

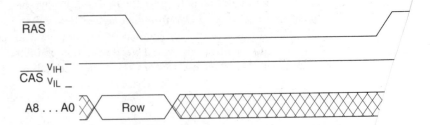

## 18.4

## Dynamic RAM Modules

Dynamic RAM is often bit-organized, rather than word-organized. That is, only one bit at a time is available for a Read or Write operation. This is inconvenient in a memory system, where a data bus is 8 or more bits wide and memory data should be available in parallel groups to fit the data bus. To meet this need, manufacturers have developed dynamic RAM modules. These modules are circuit boards having eight or nine bit-organized dynamic RAM chips with their control and power inputs connected in parallel.

Several types of IC packaging are available for these modules, including DIP (dual in-line package), SIP (single in-line package), ZIP (zig-zag in-line package), and SOJ (small-outline J-lead). Since one goal of a DRAM module is to maximize the amount of memory storage and minimize the space taken by the memory, the most advanced designs use the ZIP and SOJ packaging, which have the smallest packages for the same memory.

Figures 18.20 and 18.21 show the data sheet for a Motorola MCM81000 1M × 8 dynamic RAM module. The module consists of eight MCM511000 1M × 1 DRAMs, connected to make a 1M × 8 memory module. All power, address, and control inputs are connected in parallel, so that one address will select the same memory location on each DRAM chip and one $\overline{RAS}$, $\overline{CAS}$, or $\overline{WRITE}$ signal will activate the appropriate function on all chips. The data lines are independent of one another, giving an 8-bit input or output for every Read or Write.

The MCM81000 memory module has 30 pins, available as pins that can be soldered into a circuit board or as an edge connector, which can fit into a circuit board card slot. The pin form is called a single in-line package (SIP), and the edge connector version is referred to as a single in-line memory module (SIMM). The SIMM package is popular for high-density memory in personal computers.

Texas Instruments has extended the capacity of the DRAM module into the range of 4 to 8 megabytes (4M × 8 bits to 8M × 8 bits) with their memory intensive modules (MIMs). A MIM is a type of SIMM with the memory chips arranged in several blocks, each having a 1M × 8-bit capacity. The input address accesses a particular block and an 8-bit word within that block.

## 18.5

## Memory Systems

**I/O** *Input/output.*

**Bus contention** *The condition that results when two or more devices try to send data to a bus at the same time. Bus contention can damage the output buffers of the devices involved.*

**CPU** *Central processing unit. The central portion of a microcomputer system, which, among other things, controls the flow of signals on the system buses.*

MOTOROLA
## ▪▪ SEMICONDUCTOR ▬▬▬
### TECHNICAL DATA

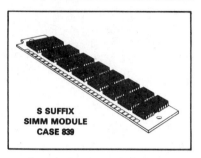

MCM81000

*Product Preview*
## 1M × 8 Bit Dynamic Random Access Memory Module

The MCM81000L and MCM81000S are 8M, dynamic random access memory (DRAM) modules organized as 1,048,576 × 8 bits. The modules are 30-lead single-in-line memory modules (SIMM) or 30-pin single-in-line packages (SIP) consisting of eight MCM511000 DRAMs housed in a 20/26 J-lead small outline package (SOJ) and mounted on a substrate along with a 0.22 μF decoupling capacitor mounted under each DRAM. The MCM511000 is a 1.0μ CMOS high speed, dynamic random access memory organized as 1,048,576 one-bit words and fabricated with CMOS silicon-gate process technology.

- Three-State Data Output
- Early-Write Common I/O Capability
- Fast Page Mode Capability
- TTL-Compatible Inputs and Outputs
- $\overline{RAS}$ Only Refresh
- $\overline{CAS}$ Before $\overline{RAS}$ Refresh
- Hidden Refresh
- 512 Cycle, 8 ms Refresh
- Consists of Eight 1M DRAMs and Eight 0.22 μF Decoupling Capacitors
- Unlatched Data Out at Cycle End Allows Two Dimensional Chip Selection
- Fast Access Time ($t_{RAC}$):
    - MCM81000-80 = 80 ns (Max)
    - MCM81000-10 = 100 ns (Max)
- Low Active Power Dissipation:
    - MCM81000-80 = 3.0 W (Max)
    - MCM81000-10 = 2.6 W (Max)
- Low Standby Power Dissipation:
    - TTL Levels = 88 mW (Max)
    - CMOS Levels = 44 mW (Max)
- $\overline{CAS}$ Control for Eight Common I/O Lines
- Available in Edge Connector (MCM81000S) or Pin Connector (MCM81000L)

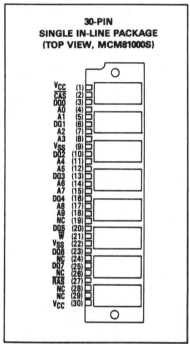

**S SUFFIX
SIMM MODULE
CASE 839**

**30-PIN
SINGLE IN-LINE PACKAGE
(TOP VIEW, MCM81000S)**

| PIN NAMES | |
|---|---|
| A0–A9 . . . . . . . . . . . . . . . Address Inputs | |
| DQ0–DQ7 . . . . . . . . . . . Data Input/Output | |
| $\overline{CAS}$ . . . . . . . . . . Column Address Strobe | |
| $\overline{RAS}$ . . . . . . . . . . . Row Address Strobe | |
| $\overline{W}$ . . . . . . . . . . . . . . Read/Write Input | |
| $V_{CC}$ . . . . . . . . . . . . . . . . . Power (+5 V) | |
| $V_{SS}$ . . . . . . . . . . . . . . . . . . . . Ground | |
| NC . . . . . . . . . . . . . . . No Connection | |

**MCM81000S (SIMM)**

**MCM81000L (SIP)**

**Figure 18.20**

**Data Sheet for MCM81000 DRAM Module** (Reprinted with permission of Motorola)

**Figure 18.21**

**Block Diagram for
MCM81000 DRAM Module**

MCM81000

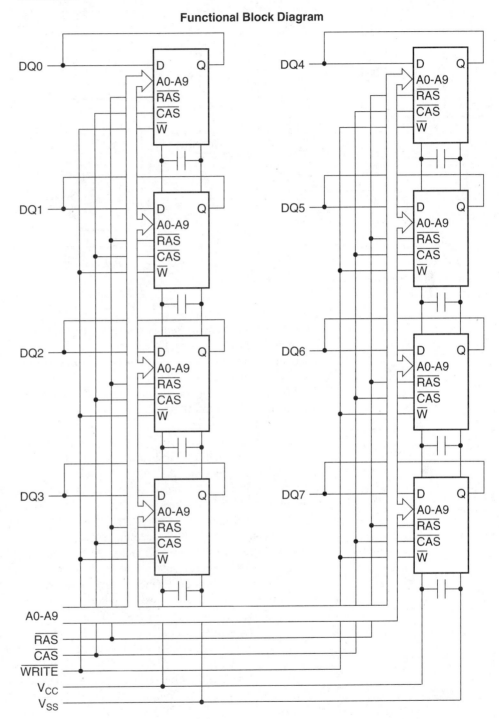

Functional Block Diagram

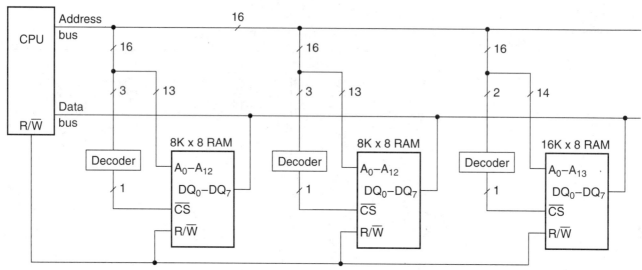

**Figure 18.22**
**Simplified Microcomputer**
**System (Partial)**

Memory devices are seldom used singly. They are usually connected in groups to form a memory system, often controlled by a microprocessor. The $\overline{E}$ (or $\overline{CS}$ or $\overline{CE}$) control inputs of the memory chips can be used to select which chip is to be read from or written to at a particular time.

In such a system, shown in Figure 18.22, the microprocessor has a data bus, an address bus, and a control bus, all of which are connected to memory and various other input and output devices.

(In some microprocessors, notably the Motorola 6800/68000 series, peripheral chips, such as parallel interfaces and serial communications controllers, are treated as if they were part of memory. An output is handled by writing data to an address assigned to an output device. An input is performed by reading from an address assigned to an input device. This is known as "memory-mapped **I/O**" (input/output).)

We will not go into the operational details of a full microcomputer system. As far as we are concerned, the **CPU** (central processing unit) shown in Figure 18.22 is just a source of signals on the data and address buses—an electronic traffic cop, as it were.

If the CPU requires data for some operation, it will send a *READ* and a $\overline{CHIP\ SELECT}$ signal to one of the memory chips, as shown in Figure 18.23a. If the CPU needs some data to be stored, it will send a $\overline{WRITE}$ signal and a $\overline{CHIP\ SELECT}$ to the corresponding device, as shown in Figure 18.23b.

Since the devices in Figure 18.22 are permanently connected, the flow of address and data signals must be controlled so that no two outputs send conflicting data to a bus. As well as making the data uncertain, such action can result in damage to the output buffers of the devices involved. (Recall, from Chapter 15, the effect of opposite logic levels when two totem pole outputs are connected together.)

The condition in which two outputs simultaneously send data to a bus is called **bus contention.** Figure 18.24 shows two CMOS output buffers in contention for a bus. The excess current that flows when opposite logic levels are applied to the bus can damage both outputs.

**Figure 18.23**
**Memory Read and Write**

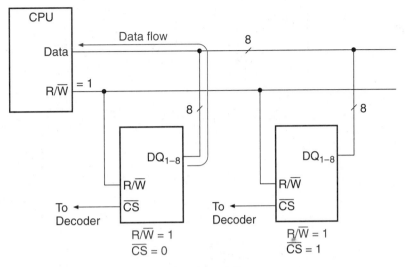

a. Read operation

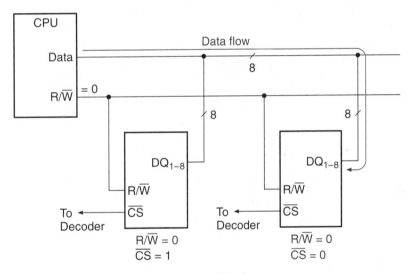

b. Write operation

## Memory Mapping and Address Decoding

**Address decoder**  *A circuit enabling a memory device to be selected by the address bus of a microprocessor-based system.*

**Address space**  *A block of addresses in a microprocessor system.*

**Memory map**  *A diagram showing the total address space of a microprocessor-based system and the placement of various memory devices within that space.*

**Figure 18.24**

**Bus Contention**

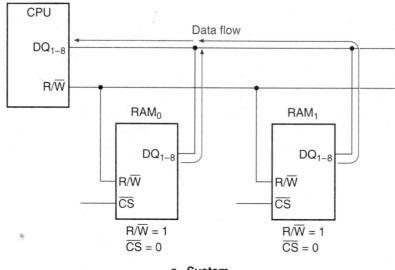

a. System

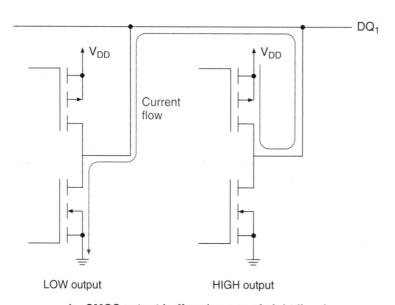

b. CMOS output buffers (on one of eight lines)

Given that no two memories should be accessed at the same time, how does the CPU know not to do this? The circuits that ensure access to only one memory at a time are called **address decoders.**

The address size of a memory chip is usually smaller than that of a microprocessor address bus. For instance, a 64K (16-bit) address bus will probably have several smaller memories, such as 8K × 8 RAMs (13-bit address), 16K × 8 ROMs (14-bit address), or other memories, as well as parallel and serial input and output devices connected to it, each of which must be independently addressable.

One way to ensure independent access is to assign the **address space** of each device to a specific block of addresses within the CPU address space. These assignments

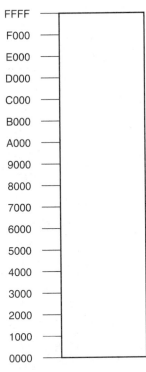

FFFF
F000
E000
D000
C000
B000
A000
9000
8000
7000
6000
5000
4000
3000
2000
1000
0000

**Figure 18.25**

**Memory Map for 64 KB (65,536 Addresses)**

are shown on a diagram called a **memory map,** which is a list of all memory locations addressable by the CPU address bus.

For all of our examples, we will assume that the system address bus has 16 bits. This is the case for many 8-bit microprocessors and microcontrollers. There are many microprocessor systems, such as the Intel 8086, 80286, 80386, and 80486 and the Motorola 68000 series, that employ address buses of 20, 24, or 32 bits, but we will use the smaller address bus because of the simpler examples it generates. The principles are the same, the only difference being the larger address spaces available to the CPUs with the larger address buses.

The range of possible addresses for a microprocessor system with a 16-bit address bus extends from 0000 0000 0000 0000 to 1111 1111 1111 1111 in binary or from 0000 to FFFF in hexadecimal, which corresponds to a 64-KB address space, shown by the memory map in Figure 18.25.

The memory system shown in Figure 18.26 has only one device: an 8K × 8 RAM, which has 13 address lines ($2^{13}$ = 8K). The possible range of internal RAM addresses extends from 0 0000 0000 0000 to 1 1111 1111 1111 in binary, or from 0000 to 1FFF in hexadecimal. Clearly, the address space of the 8K × 8 RAM is not going to take up the entire address space of the CPU. So where do we put it? The easiest place is at the beginning of the CPU address space, as shown in Figure 18.27.

The CPU and RAM address ranges in binary and hexadecimal are:

|  | CPU Address | | | | | RAM Address | | | |
|---|---|---|---|---|---|---|---|---|---|
|  | $A_{15}$ | | | $A_0$ | | $A_{12}$ | | $A_0$ | |
| Start Address | 0000 | 0000 | 0000 | 0000 | (0000) | 0 0000 | 0000 | 0000 | (0000) |
| End Address | 0001 | 1111 | 1111 | 1111 | (1FFF) | 1 1111 | 1111 | 1111 | (1FFF) |

The range of hexadecimal addresses is the same for the CPU and RAM in this case, but the binary addresses are different. They have to be; the RAM does not have as many address lines as the CPU.

**Figure 18.26**

**CPU System With 8K × 8 RAM**

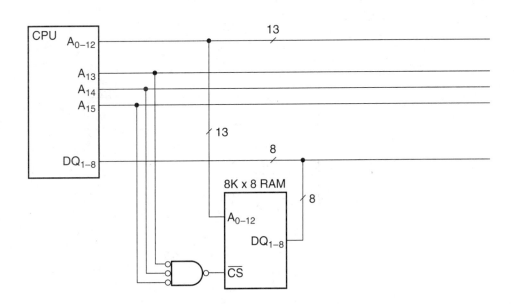

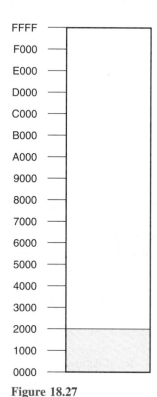

**Figure 18.27**

**8K Memory Space at Beginning of 64K CPU Space**

The three most significant bits on the CPU address bus are all 0 throughout the entire range of addresses shown. This is detected by the address decoder (a DeMorgan-equivalent OR gate) in Figure 18.26. The address decoder places the RAM space within the CPU space by making the $\overline{CHIP\ SELECT}$ of the RAM LOW only when $A_{15}A_{14}A_{13}$ = 000.

What happens to the CPU address space when we add another memory device? Figure 18.28 shows the same system as in Figure 18.26 with another 8K × 8 RAM added. The data bus is not shown.

The address decoders have been chosen to place the first RAM ($RAM_0$) in the same memory space as in Figure 18.27 and the second RAM (labeled $RAM_1$) in the CPU address space from 2000 to 3FFF.

|  | **CPU Address** | | | | | **RAM Address** | | | |
|---|---|---|---|---|---|---|---|---|---|
|  | $A_{15}$ | | | $A_0$ | $A_{12}$ | | | $A_0$ | |
| **RAM₀** | | | | | | | | | |
| Start Address | 0000 | 0000 | 0000 | 0000 | (0000) | 0 | 0000 | 0000 | 0000 | (0000) |
| End Address | 0001 | 1111 | 1111 | 1111 | (1FFF) | 1 | 1111 | 1111 | 1111 | (1FFF) |
| **RAM₁** | | | | | | | | | |
| Start Address | 0010 | 0000 | 0000 | 0000 | (2000) | 0 | 0000 | 0000 | 0000 | (0000) |
| End Address | 0011 | 1111 | 1111 | 1111 | (3FFF) | 1 | 1111 | 1111 | 1111 | (1FFF) |

The memory spaces for $RAM_0$ and $RAM_1$ are shown on the CPU memory map in Figure 18.29.

> *It is extremely important to distinguish between memory address space and CPU address space.* The hexadecimal addresses assigned to $RAM_1$ on the CPU address space are *not* the same as the internal addresses of the RAM. (They can't be; the RAM has only 13 address lines and you need 14 bits to write the binary equivalent of the range from 2000 to 3FFF.) The addresses on the 13 least significant bits of the CPU address bus *are* the same as the internal addresses of $RAM_1$. The upper 3 bits are used to select the chip.

**Figure 18.28**

**System With Two 8K Address Blocks**

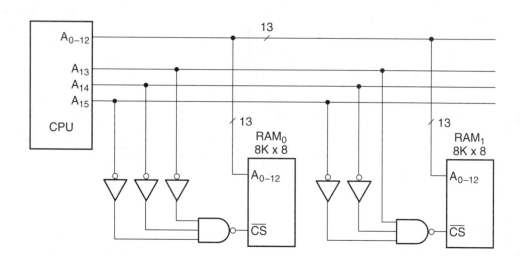

**EXAMPLE 18.1**

What address must appear on the CPU address bus of the system in Figure 18.28 to access location 0F3C in $RAM_0$? What CPU address will access location 0F3C in $RAM_1$?

**Solutions**

$RAM_0$ is selected when $A_{15}A_{14}A_{13} = 000$.

Location 0F3C is selected when address pins $A_{12}. . . A_0$ have the value 0 1111 0011 1100.

CPU address: $A_{15}. . . A_0 = 0000\ 1111\ 0011\ 1100 = 0F3C$.

$RAM_1$ is selected when $A_{15}A_{14}A_{13} = 001$.

Location 0F3C is selected when address pins $A_{12}. . . A_0$ have the value 0 1111 0011 1100.

CPU address: $A_{15}. . . A_0 = 0010\ 1111\ 0011\ 1100 = 2F3C$.

The same internal location is selected in each RAM, but each requires a different CPU address.

**EXAMPLE 18.2**

A microprocessor system with a 16-bit address bus has the following memory devices, with the hexadecimal start addresses as given.

| Device | Capacity | Start Address |
|--------|----------|---------------|
| $ROM_0$ | 8K × 8 | 0000 |
| $ROM_1$ | 16K × 8 | 4000 |
| $RAM_0$ | 16K × 8 | C000 |

Draw a memory map of the system address bus, showing the placement of each device.

**Solution**

Calculate the end addresses of the memory blocks, based on the number of address pins on each device.

$ROM_0$: $8K = 8 \times 1K = 2^3 \times 2^{10}$; 13 address lines.

Internal address range: 0  0000  0000  0000  (0000)
1  1111  1111  1111  (1FFF)

End address = start address + internal range

$= 0000 + 1FFF$

$= 1FFF$

$ROM_1$: $16K = 16 \times 1K = 2^4 \times 2^{10}$; 14 address lines.

Internal address range: 00  0000  0000  0000  (0000)
11  1111  1111  1111  (3FFF)

End address = start address + internal range

$= 4000 + 3FFF$

$= 7FFF$

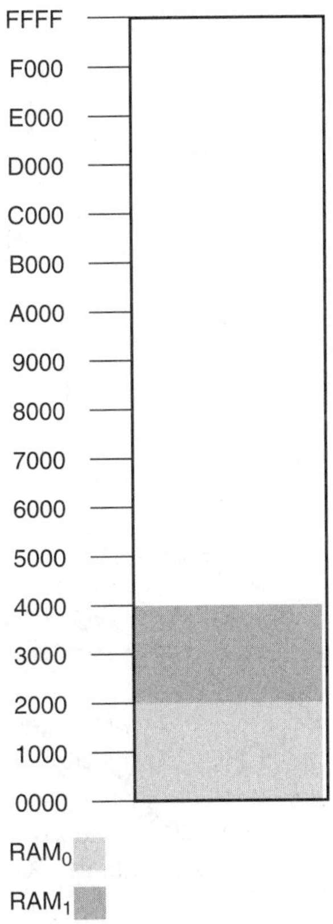

Memory map labels: FFFF, F000, E000, D000, C000, B000, A000, 9000, 8000, 7000, 6000, 5000, 4000, 3000, 2000, 1000, 0000

RAM_0, RAM_1

**Figure 18.29**
**Memory Map for Two 8K × 8 RAMs in 64K-Byte System**

$RAM_0$: $16K = 16 \times 1K = 2^4 \times 2^{10}$; 14 address lines.

Internal address range: 00   0000   0000   0000   (0000)

11   1111   1111   1111   (3FFF)

End address = start address + internal range

= C000 + 3FFF

= FFFF

Figure 18.30 shows the memory map with the three address ranges.

**EXAMPLE 18.3**

Draw the system in Example 18.2, including the CPU address bus, memory devices, and decoding logic.

**Solution**

To draw the decoding logic, we need to find out which states of the high-order bits of the CPU address bus will select each chip.

**CPU Address Range**

$ROM_0$: 000   0   0000   0000   0000   (0000)

000   1   1111   1111   1111   (1FFF)

13 internal address bits + 3 decoding bits
Decoding bits: $A_{15}A_{14}A_{13} = 000$

$ROM_1$: 01   00   0000   0000   0000   (4000)

01   11   1111   1111   1111   (7FFF)

14 internal address bits + 2 decoding bits
Decoding bits: $A_{15}A_{14} = 01$

$RAM_0$: 11   00   0000   0000   0000   (C000)

11   11   1111   1111   1111   (FFFF)

14 internal address bits + 2 decoding bits
Decoding bits: $A_{15}A_{14} = 11$

Figure 18.31 shows the complete system, including address decoding logic.

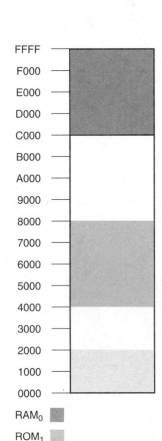

FFFF
F000
E000
D000
C000
B000
A000
9000
8000
7000
6000
5000
4000
3000
2000
1000
0000

$RAM_0$

$ROM_1$

$ROM_0$

**Figure 18.30**
**Example 18.2**
**Memory Map**

## Decoding With MSI Chips

Instead of using SSI gates for address decoding, we can use MSI decoders. Table 18.3 lists some available TTL and CMOS decoders. The symbols of these decoders are shown in Figure 18.32.

Recall the function of an MSI decoder with active-LOW outputs. When the gating (enable) inputs are active, the output selected by the select inputs goes LOW. All other outputs are HIGH. The binary value of the select inputs is the equivalent of the decimal subscript of the selected output. For example, if the gating inputs of a 74HC138 decoder are active and the select inputs have the value $CBA = 110$, output $Y_6$ is selected.

**Figure 18.31**

**Example 18.3
System Block Diagram**

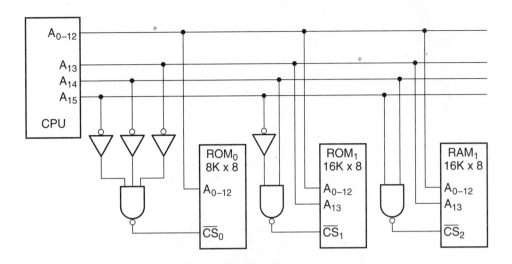

**Table 18.3** MSI Decoders

| Function | TTL | CMOS | High-Speed CMOS |
|---|---|---|---|
| Dual 2-to-4 line decoder | 74LS139 | 4556B | 74HC139A |
| 3-to-8 line decoder | 74LS138 | | 74HC138A |
| 4-to-16 line decoder | 74154 | 4515B | 74HC154 |

**EXAMPLE 18.4**

A microcomputer system with a 16-bit address bus has the following memory assignments.

| Device | Capacity | Start Address |
|---|---|---|
| $RAM_0$ | $8K \times 8$ | 0000 |
| $RAM_1$ | $8K \times 8$ | A000 |

Draw a system memory map and block diagram, using a 74HC138 3-to-8 line decoder as an address decoder.

**Solution**   Find the end addresses of each memory block.

$RAM_0$: $8K = 8 \times 1K = 2^3 \times 2^{10}$; 13 address lines.

Internal address range: 0   0000   0000   0000   (0000)

1   1111   1111   1111   (1FFF)

End address = start address + internal range

= 0000 + 1FFF

= 1FFF

$RAM_1$: End address = start address + internal range

= A000 + 1FFF

= BFFF

These blocks are shown on the memory map in Figure 18.33.

**Figure 18.32**
**MSI Decoders**

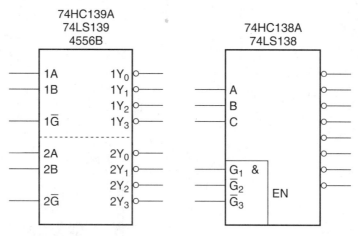

a. Dual 2-to-4 line
   decoder (TTL and CMOS
   are identical)

b. 3-to-8 line decoder

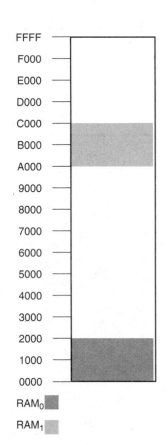

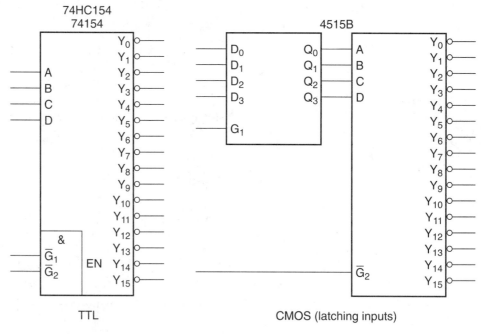

c. 4-to-16 line decoders

Connect CPU address lines $A_{15}$, $A_{14}$, and $A_{13}$ to select inputs $C$, $B$, and $A$ of the 74HC138 and examine the binary values of the CPU addresses for each block. This tells us which decoder outputs to connect to the memory Chip Selects.

**Figure 18.33**

**Example 18.4**
**Memory Map**

$$A_{15} \qquad\qquad\qquad\qquad A_0$$

RAM$_0$:  000  0  0000  0000  0000  (0000)    Decoding: $A_{15}A_{14}A_{13} = 000$
          000  1  1111  1111  1111  (1FFF)    74HC138: Output $Y_0$

**Figure 18.34**

**Example 18.4**
**System Block Diagram**

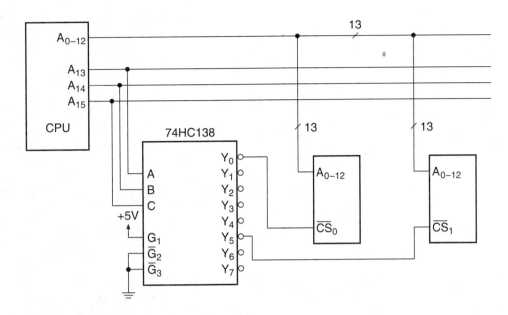

RAM$_1$: 101  0  0000  0000  0000  (A000)   Decoding: $A_{15}A_{14}A_{13} = 101$
        101  1  1111  1111  1111  (BFFF)      74HC138: Output $Y_5$

Figure 18.34 shows the system block diagram.

---

The size of a memory block accessed by one output of an MSI decoder is given by (total address space/$n$) for a decoder with $n$ outputs. (For example, in Example 18.4, one decoder output will access 64K/8 outputs = 8K addresses/output.)

Larger blocks can be decoded by using more than one output. Additional logic is required.

---

**EXAMPLE 18.5**

A microcomputer system with a 16-bit address bus has the following memory assignments.

| Device | Capacity | Start Address |
|--------|----------|---------------|
| ROM$_0$ | 8K × 8 | 2000 |
| ROM$_1$ | 16K × 8 | 4000 |
| ROM$_2$ | 16K × 8 | 8000 |

Draw the address decoding logic for the system, using a 74HC138 3-to-8 line decoder.

**Solution**  Calculate the end addresses of each block.

ROM$_0$: Internal address range (13 bits)  0  0000  0000  0000  (0000)
                                            1  1111  1111  1111  (1FFF)

$$\text{End address} = \text{start address} + \text{internal range}$$

$$= 2000 + 1FFF$$

$$= 3FFF$$

ROM$_1$: Internal address range (14 bits)    00   0000   0000   0000   (0000)

11   1111   1111   1111   (3FFF)

$$\text{End address} = 4000 + 3FFF$$

$$= 7FFF$$

ROM$_2$: Internal address range (14 bits)    00   0000   0000   0000   (0000)

11   1111   1111   1111   (3FFF)

$$\text{End address} = 8000 + 3FFF$$

$$= BFFF$$

One output of a 74HC138 decoder will select an 8K address block. Since the address blocks for ROM$_1$ and ROM$_2$ are each 16K addresses in length, they need two outputs each, which will be ORed together. Connect address lines $A_{15}$, $A_{14}$, and $A_{13}$ to decoder inputs C, B, and A. Analyze the CPU addresses for each block to assign the appropriate decoder outputs.

|  | $A_{15}$ |  |  |  | $A_0$ |  |  |
|---|---|---|---|---|---|---|---|
| ROM$_0$: | 001 | 0 | 0000 | 0000 | 0000 | (2000) | Decoding: $A_{15}A_{14}A_{13} = 001$ |
|  | 001 | 1 | 1111 | 1111 | 1111 | (3FFF) | 74HC138: Output $Y_1$ |
| ROM$_1$: | 01 | 00 | 0000 | 0000 | 0000 | (4000) |  |
|  | 01 | 11 | 1111 | 1111 | 1111 | (7FFF) |  |

*Problem:* A 3-to-8 line decoder requires 3 decoding bits. ROM$_1$ has 14 address inputs, leaving only 2 decoding bits.

*Solution:* Break up the ROM$_1$ block into two 8K subranges and OR the corresponding decoder outputs so that either subrange will select ROM$_1$.

Subranges:

| 010 | 0 | 0000 | 0000 | 0000 | (4000) | Decoding: $A_{15}A_{14}A_{13} = 010$ |
|---|---|---|---|---|---|---|
| 010 | 1 | 1111 | 1111 | 1111 | (5FFF) | 74HC138: Output $Y_2$ |
| 011 | 0 | 0000 | 0000 | 0000 | (6000) | Decoding: $A_{15}A_{14}A_{13} = 011$ |
| 011 | 1 | 1111 | 1111 | 1111 | (7FFF) | 74HC138: Output $Y_3$ |

Note: $A_{13}$ is connected as both an internal address and a decoding bit.

### ROM$_2$

Subranges:

| 100 | 0 | 0000 | 0000 | 0000 | (8000) | Decoding: $A_{15}A_{14}A_{13} = 100$ |
|---|---|---|---|---|---|---|
| 100 | 1 | 1111 | 1111 | 1111 | (9FFF) | 74HC138: Output $Y_4$ |
| 101 | 0 | 0000 | 0000 | 0000 | (A000) | Decoding: $A_{15}A_{14}A_{13} = 101$ |
| 101 | 1 | 1111 | 1111 | 1111 | (BFFF) | 74HC138: Output $Y_5$ |

**Figure 18.35**

**Example 18.5
Decoding Logic**

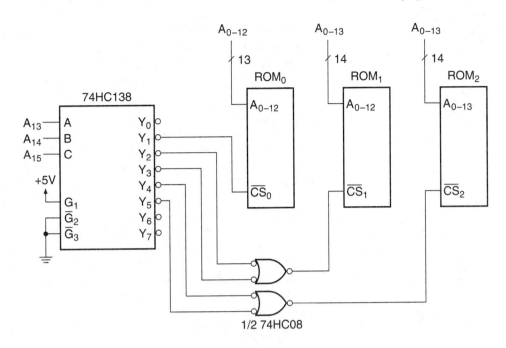

Note: $A_{13}$ is connected as both an internal address and a decoding bit.

The decoding logic is shown in Figure 18.35. To show that either $\overline{Y}_2$ OR $\overline{Y}_3$ activates $\overline{CS}_1$ (all active LOW), we combine the decoder outputs in an AND gate shown in its DeMorgan equivalent form. The same applies to $\overline{Y}_4$, $\overline{Y}_5$, and $\overline{CS}_2$.

---

**Section Review Problem for Section 18.5**

18.2. A microprocessor system with a 16-bit address bus has the following memory assignments:

| Device | Start Address | End Address |
|---|---|---|
| RAM$_0$ | 2000 | 2FFF |
| RAM$_1$ | 4000 | 5FFF |
| RAM$_2$ | 8000 | BFFF |
| RAM$_3$ | C000 | CFFF |
| RAM$_4$ | E000 | FFFF |

How many outputs of a 74154 4-to-16 line decoder are used to decode each address block?

## Memory Expansion

There are several things we might mean when we talk about memory expansion:

**1.** We might mean *horizontal expansion of word length* to fill the complete width of the system data bus. For instance, if we have memory chips with four data input/output lines, we have to know how to configure two chips to fit an 8-bit data bus.

**2.** Memory expansion can also mean adding extra chips to give *more addressable locations* within the limits of the system address bus. If we have a system with four 8K × 8 RAM chips (a total capacity of 32K bytes, or 32 KB, of memory), we need to know how to configure additional chips to give us a larger capacity—say, 64 KB of memory.

**3.** We might mean *expanding memory beyond the limits of the system address bus* by adding memory management hardware and/or software. One system, used in IBM PCs and compatibles, is the LIM (Lotus-Intel-Microsoft) EMS (Expanded Memory Specification) system, which allows 2048 different 16-KB blocks of RAM to be individually mapped into the high addresses of the CPU address space, yielding up to 32 MB of expanded memory. (A related specification is the XMS Extended Memory Specification, which allows an IBM or compatible personal computer to make maximum use of the existing system address bus.)

A single-chip memory expander is the 74LS610 memory mapper, which can expand a system address bus by 8 bits. A system with a 16-bit address bus has an unexpanded capacity of 64 KB, assuming that the entire space is available for user memory. (Usually this is not the case; the system generally reserves ROM space for various startup, maintenance, and supervisory functions.) A '610 memory mapper can expand a 16-bit address bus to 16 MB of address space ($2^{16} \times 2^8 = 2^{24} = 16M = 16,777,216$). The extra address space is allocated in 16 blocks of 1 MB each.

We will examine each of these cases separately.

## Word Length Expansion

To expand the length of a data word in a block of addresses, the address and control lines of all chips in the block are connected in parallel, but the data lines of each chip have different connections to the system data bus.

---

**EXAMPLE 18.6**   A microprocessor system has a 4K × 8 block of RAM assigned to CPU addresses B000 to BFFF. The only available memory devices are two Am2168 4K × 4 RAM chips. Draw a diagram of the system.

**Solution**   The address of the RAM is decoded by a 74154 4-to-16 line decoder. The Chip Select inputs of both chips are connected to output $Y_{11}$ of the decoder. Also, the lower 12 address lines and the Read/Write line are tied to both chips in parallel. CPU data lines $D_0$ to $D_3$ are connected to the data lines of one chip, and the CPU lines $D_4$ to $D_7$ are connected to the data lines of the other chip. The system is shown in Figure 18.36.

---

## Expansion of Memory Capacity Within Existing CPU Space

Most of the previous examples have involved expanding memory capacity while maintaining the same word size at each location. We do this by connecting the data, control, and nondecoding address lines in parallel to all chips and connect each Chip Select to different decoding circuitry.

This type of expansion can be combined with word length expansion, as the following example demonstrates.

**Figure 18.36**
**Example 18.6**
**Word Length Expansion**

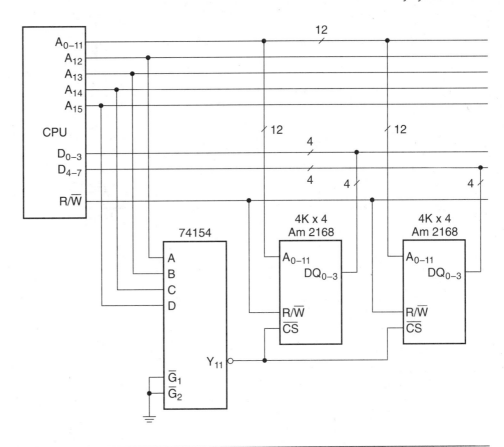

EXAMPLE 18.7     Configure as many Am2168 4K × 4 RAM chips as necessary to make a contiguous 12K × 8 memory space, with a start address of 6000. Assume a 16-bit address bus.

**Solution**     The address space can be broken down into three 4K × 8 blocks. Each block has two chips with a common address space and different data line connections. Use a 74154 decoder, and assign the Chip Selects of each block as follows:

Block1     6000 to 6FFF     Decoder output: $Y_6$
Block2     7000 to 7FFF     Decoder output: $Y_7$
Block3     8000 to 8FFF     Decoder output: $Y_8$

The system is shown in Figure 18.37.

## Expansion of CPU Address Space

**Paging** *Selecting a contiguous block of addresses with the upper bits of the CPU address bus.*

**Bank switching** *Selecting a different group of RAM devices with each combination of upper bits of the CPU address bus.*

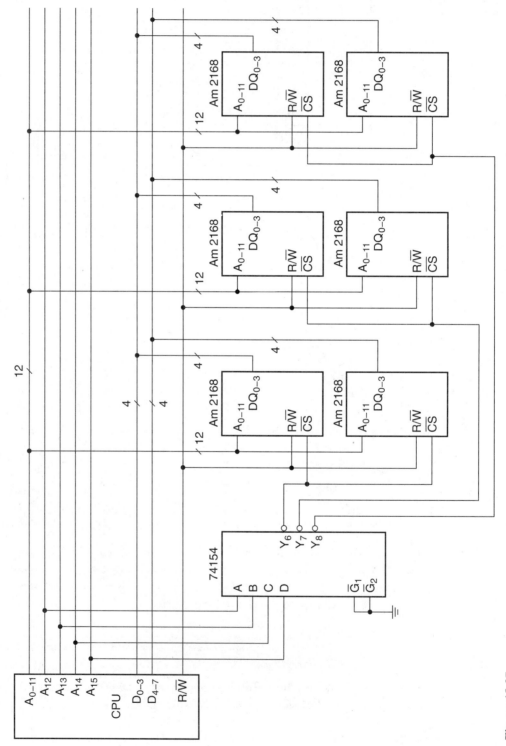

**Figure 18.37**
**Example 18.7**
**12K × 8 RAM From 4K × 4 Chips**

Expanding memory capacity by increasing the size of the CPU address space often involves some kind of **paging** or **bank switching** scheme. The upper bits of the CPU address are used to address a contiguous block of memory, and the lower bits access particular locations within the block. This is not quite the same as decoding the upper bits of the CPU address to access a particular RAM chip, as we have done in our previous examples. The main difference is ease of access to other blocks of memory.

In the various decoding schemes examined so far, if we wanted to access a block of memory other than the one we were looking at, all we would do is put a different address on the CPU bus, and a different block would be decoded by the corresponding logic. Accessing another block is no more difficult than accessing a different location within the same block.

In a paged addressing scheme, only a limited number of pages of memory are available at a time. We can access any location on these pages fairly easily. We can look at another location on the present page by changing the lower bits of the CPU address bus. An address on another page in the selected group can be accessed by using the high-order address bits to select the page and the low-order bits to select the location. But to point to a location on a page not already selected, we first have to call up a new group of pages, which may include some of those previously selected, and lose access to the original pages. This inconvenience is the price we pay for access to a larger overall address space.

Figure 18.38 shows the simplified block diagram of a 74LS610 memory mapper. Other memory management units are generally similar, but differ in details such as the number of bits in the expanded address.

The 4 most significant bits of the CPU address bus select one of 16 mapping registers. Each of these registers contains a 12-bit number loaded in from the 12 least significant bits of the CPU data bus. (This assumes that the system has at least a 12-bit and

**Figure 18.38**

**74LS610 Memory Mapper**

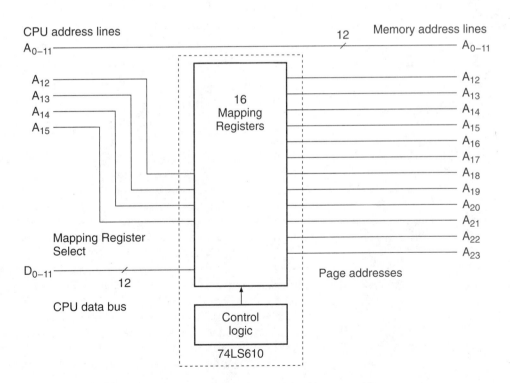

probably a 16-bit data bus. Other similar units, such as the Motorola 6829 memory management unit, are configured for an 8-bit data bus.)

The 12-bit values in the mapping registers become the most significant bits of the memory address bus. We have taken 4 address bits and expanded them to 12, with which we can independently address 4096 ($=2^{12}$) separate pages. Each will have 4096 bytes of memory, since there are 12 low-order address bits, a total of 16 MB. However, there are only 4 high-order CPU address lines to select the mapping registers (pages), so we are limited to accessing 16 at a time, a total of 64 KB of memory (16 pages $\times$ 4 KB/page).

The mapping registers can contain any sixteen 12-bit numbers, each of which selects a page address. For example, suppose the mapping registers of a 74LS610 memory mapper contain the following data:

| Register | | Contents | | | | |
|---|---|---|---|---|---|---|
| $A_{15}$ $A_{12}$ | | $D_{11}$ ($A_{23}$) | | $D_0$ ($A_{12}$) | | |
| 0000 | (0) | 0000 | 0011 | 1111 | | (03F) |
| 0001 | (1) | 0000 | 0111 | 0111 | | (077) |
| 0010 | (2) | 0000 | 0111 | 1111 | | (07F) |
| 0011 | (3) | 0000 | 1011 | 0111 | | (0B7) |
| 0100 | (4) | 0000 | 1011 | 1100 | | (0BC) |
| 0101 | (5) | 0000 | 1011 | 1111 | | (0BF) |
| 0110 | (6) | 0000 | 1111 | 1100 | | (0FC) |
| 0111 | (7) | 0000 | 1111 | 1111 | | (0FF) |
| 1000 | (8) | 0001 | 0011 | 1100 | | (13C) |
| 1001 | (9) | 0001 | 0111 | 1010 | | (17A) |
| 1010 | (A) | 0001 | 0111 | 1100 | | (17C) |
| 1011 | (B) | 0001 | 1011 | 1001 | | (1B9) |
| 1100 | (C) | 0001 | 1011 | 1100 | | (1BC) |
| 1101 | (D) | 0001 | 1111 | 1100 | | (1FC) |
| 1110 | (E) | 1111 | 0111 | 0111 | | (F77) |
| 1111 | (F) | 1111 | 0111 | 1001 | | (F79) |

Each mapping register is addressed by the CPU address lines $A_{15}$. . . $A_{12}$. The contents, $D_{11}$. . . $D_0$, loaded in from the data bus, become the output address lines $A_{23}$. . . $A_{12}$. The hexadecimal page numbers in parentheses are the only ones of the possible 4096 pages that are available at the present time. The partial memory map in Figure 18.39 shows the selected page addresses. (The entire map consists of $64 \times 64$ pages, which is too large to fit on the page and still be legible.)

Each of the selected pages has 4096 bytes, ranging from XXX000 to XXXFFF, where XXX is the page number. For example, if we decide to decode address 13C76D, we must select page 13C with the upper 4 CPU address lines and location 76D with the lower 12 lines. To address page 13C, we select map register 8, which has the contents 13C. Thus, the CPU address 876D maps into memory space 13C76D. Similarly, to access memory location 0B796A, we use CPU address 396A, since the contents of register 3 is 0B7, the memory page number.

If we want to access any location other than one in the 16 pages listed, we must change the contents of the mapping registers.

**Figure 18.39**

**Partial Memory Map of Memory Addressable by 74LS610 Memory Mapper**

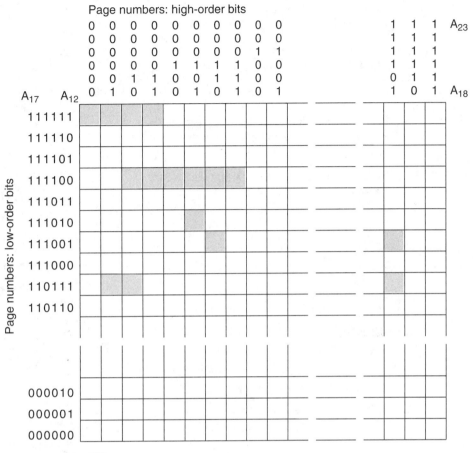

Selected pages

**Address decoder**   A circuit enabling a memory device to be selected by the address bus of a microprocessor-based system.

**Address space**   A block of addresses in a microprocessor system.

**Bank switching**   Selecting a different group of RAM devices with each combination of upper bits of the CPU address bus.

**Bus contention**   The condition that results when two or more devices try to send data to a bus at the same time. Bus contention can damage the output buffers of the devices involved.

**CPU**   Central processing unit. The central portion of a microcomputer system, which, among other things, controls the flow of signals on the system buses.

**I/O**   Input/output.

**Memory map**   A diagram showing the total address space of a microprocessor-based system and the placement of various memory devices within that space.

**Page**   A contiguous block of addresses selected by high-order address bits. Low-order bits of the addresses select a word within the page.

**Paging**   Selecting a contiguous block of addresses with the upper bits of the CPU address bus.

## ● PROBLEMS

### Section 18.1 Static RAM Timing Cycles
### Section 18.2 Static RAM Read and Write Cycles

**18.1**  Figure 18.40 shows a timing diagram for a static RAM. Write the labels for the timing parameters $a$ through $f$, using the multiple-subscript notation described in Section 18.1.

**18.2**  Sketch a partial timing diagram for each of the following timing parameters:

   **a.**  $t_{AVEIL}$

   **b.**  $t_{DVEIH}$

   **c.**  $t_{GHQZ}$

   **d.**  $t_{EILQV}$

### Section 18.3 Dynamic RAM Timing Cycles

**18.3**  Sketch the timing diagram of a normal Read cycle of a dynamic RAM.

**18.4**  Sketch the timing diagram of an early Write cycle of a dynamic RAM.

**18.5**  Briefly explain the difference between a Read-while-Write and a Read-Modify-Write cycle of a dynamic RAM.

**18.6**  Briefly explain the difference between fast page mode and static column decode mode.

**18.7**  List three types of DRAM Refresh cycle. Briefly describe each one.

### Section 18.5 Memory Systems

**18.8**  A microprocessor system with a 16-bit address bus is connected to a 4K $\times$ 8 RAM chip and an 8K $\times$ 8 RAM chip. The 8K address begins at $6000_{16}$. The 4K address block starts at $2000_{16}$.

   Calculate the end address for each block and show address blocks for both memory chips on a 64K memory map.

**18.9**  Draw the memory system of Problem 18.8, including CPU, address bus, memory chips, and decoding logic.

**18.10**  A microprocessor system with a 16-bit address bus has the following memory assignments:

| Memory | Size | Start Address |
|--------|------|---------------|
| $RAM_0$ | 16K | 4000 |
| $RAM_1$ | 8K | 8000 |
| $RAM_2$ | 8K | A000 |

Show the blocks on a 64K memory map.

**Figure 18.40**

**Problem 18.1**

**Timing Diagram**

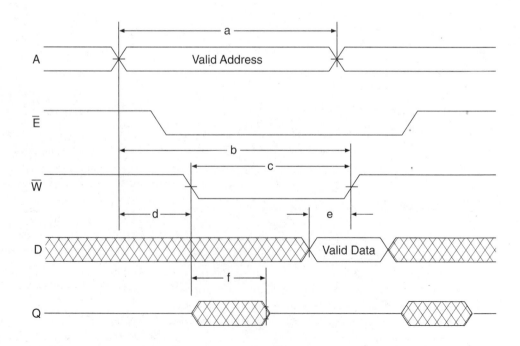

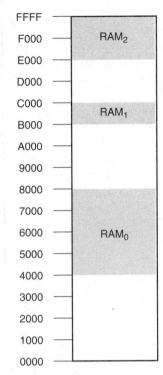

**Figure 18.41**
**Problems 18.12 and 18.13**
**Memory Map**

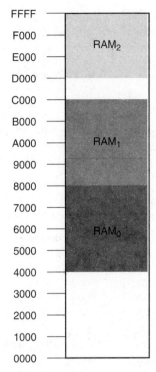

**Figure 18.42**
**Problems 18.14 and 18.15**
**Memory Map**

**18.11**  Draw the memory system described in Problem 18.10.

**18.12**  The memory map of a microcomputer system is shown in Figure 18.41. Make a table of start and end addresses and the size of each block of memory shown in the map.

**18.13**  Draw a circuit block diagram of the memory system represented by the map in Figure 18.41.

**18.14**  Repeat Problem 18.12 for the memory map of Figure 18.42.

**\*18.15**  Draw a circuit block diagram of the memory system represented by the memory map in Figure 18.42.

**18.16**  A microprocessor system with a 16-bit address bus has the following memory assignments:

| Device | Start Address | End Address |
|--------|---------------|-------------|
| $RAM_0$ | 4000 | 5FFF |
| $RAM_1$ | 8000 | BFFF |
| $RAM_2$ | E000 | FFFF |

Calculate the size of each address block.

**18.17**  How many outputs of a 74154 4-to-16 line decoder are needed to decode each address block in the system described in Problem 18.16? (Which ones?) Sketch the decoder connections.

**18.18**  How many outputs of a 74138 3-to-8 line decoder are needed to decode each address block in the system of Problem 18.16? Sketch the decoder connections and any additional logic needed to decode these address blocks.

**18.19**  A 16K × 16 block of memory is assigned to addresses 4000 to 7FFF. Draw a circuit diagram of the system, using MCM6288 16K × 4 static RAMs.

**18.20**  Draw a diagram of the system described in Problem 18.19, using MCM6264 8K × 8 static RAMs.

**\*18.21**  **a.**  Draw a memory map of a system having a 24K memory space, beginning at address 2000 and consisting of contiguous 8K blocks.

**b.** Draw a circuit diagram showing how to implement this system, using as many MCM6264 8K × 8 RAM chips as needed. Also select an MSI decoder for the system and show how to connect its outputs to the RAM.

**18.22** Refer to Figure 18.38. Calculate the address space available to a microprocessor with a 16-bit address bus and a 74LS610 memory mapper. State how much of this space is available at any time.

**18.23** The mapping registers of a 74LS610 memory mapper have the following contents:

| Register | | Contents | | | |
|---|---|---|---|---|---|
| $A_{15}$ | $A_{12}$ | $D_{11}$ ($A_{23}$) | | $D_0$ ($A_{12}$) | |
| 0000 | (0) | 0000 | 0001 | 1001 | (019) |
| 0001 | (1) | 0000 | 0001 | 1111 | (01F) |
| 0010 | (2) | 0000 | 0010 | 1100 | (02C) |
| 0011 | (3) | 0000 | 0011 | 0111 | (037) |
| 0100 | (4) | 0000 | 1011 | 0010 | (0B2) |
| 0101 | (5) | 0000 | 1011 | 1001 | (0B9) |
| 0110 | (6) | 0000 | 1100 | 1100 | (0CC) |
| 0111 | (7) | 0000 | 1101 | 0101 | (0D5) |
| 1000 | (8) | 0010 | 0111 | 0100 | (274) |
| 1001 | (9) | 0011 | 0111 | 1010 | (37A) |
| 1010 | (A) | 1001 | 0101 | 0100 | (954) |
| 1011 | (B) | 1011 | 1111 | 1001 | (BF9) |
| 1100 | (C) | 1101 | 1111 | 1111 | (DFF) |
| 1101 | (D) | 1110 | 0101 | 1000 | (E58) |
| 1110 | (E) | 1110 | 1111 | 0111 | (EF7) |
| 1111 | (F) | 1111 | 1011 | 1101 | (FBD) |

**a.** State the number of bits in each selected RAM address. How do they correspond to the address on the CPU address bus?

**b.** State the selected RAM address (in hexadecimal) for each of the following 16-bit CPU addresses.
  i. 0000 0110 0110 1010
  ii. 0001 0110 0110 1010
  iii. 0010 0110 0110 1010
  iv. 0100 0110 0110 1010
  v. 1010 1000 0000 0110
  vi. 1110 1100 1001 0010
  vi. 1111 1100 1001 0010

## ▼ Answers to Section Review Questions

| Section 18.1 | Section 18.5 |
|---|---|
| **18.1.** | **18.2.** |
| a. $t_{AVAV}$ | RAM$_0$: 1 output |
| b. $t_{AVQV}$ | RAM$_1$: 2 outputs |
| c. $t_{GLQV}$ | RAM$_2$: 4 outputs |
| d. $t_{GLQX}$ | RAM$_3$: 1 output |
| e. $t_{GHQZ}$ | RAM$_4$: 2 outputs |

# Manufacturers' Data Sheets

**A**

## SN54AS00, SN74AS00
### QUADRUPLE 2-INPUT POSITIVE-NAND GATES

**absolute maximum ratings over operating free-air temperature range (unless otherwise noted)**

Supply voltage, $V_{CC}$ . . . . . . . . . . . . . . . . . . . . . . . . . . . . . . . . . . . . . . . . . . . . . . . . . 7 V
Input voltage . . . . . . . . . . . . . . . . . . . . . . . . . . . . . . . . . . . . . . . . . . . . . . . . . . . . . . . . 7 V
Operating free-air temperature range: SN54AS00 . . . . . . . . . . . . . . . . . . . . . . . −55 °C to 125 °C
                                                                  SN74AS00 . . . . . . . . . . . . . . . . . . . . . . . . . 0 °C to 70 °C
Storage temperature range . . . . . . . . . . . . . . . . . . . . . . . . . . . . . . . . . . . . . −65 °C to 150 °C

**recommended operating conditions**

| | | SN54AS00 | | | SN74AS00 | | | UNIT |
|---|---|---|---|---|---|---|---|---|
| | | MIN | NOM | MAX | MIN | NOM | MAX | |
| $V_{CC}$ | Supply voltage | 4.5 | 5 | 5.5 | 4.5 | 5 | 5.5 | V |
| $V_{IH}$ | High-level input voltage | 2 | | | 2 | | | V |
| $V_{IL}$ | Low-level input voltage | | | 0.8 | | | 0.8 | V |
| $I_{OH}$ | High-level output current | | | −2 | | | −2 | mA |
| $I_{OL}$ | Low-level output current | | | 20 | | | 20 | mA |
| $T_A$ | Operating free-air temperature | −55 | | 125 | 0 | | 70 | °C |

**electrical characteristics over recommended operating free-air temperature range (unless otherwise noted)**

| PARAMETER | TEST CONDITIONS | | SN54AS00 | | | SN74AS00 | | | UNIT |
|---|---|---|---|---|---|---|---|---|---|
| | | | MIN | TYP[†] | MAX | MIN | TYP[†] | MAX | |
| $V_{IK}$ | $V_{CC} = 4.5$ V, | $I_I = -18$ mA | | | −1.2 | | | −1.2 | V |
| $V_{OH}$ | $V_{CC} = 4.5$ V to 5.5 V, | $I_{OH} = -2$ mA | $V_{CC}-2$ | | | $V_{CC}-2$ | | | V |
| $V_{OL}$ | $V_{CC} = 4.5$ V, | $I_{OL} = 20$ mA | | 0.35 | 0.5 | | 0.35 | 0.5 | V |
| $I_I$ | $V_{CC} = 5.5$ V, | $V_I = 7$ V | | | 0.1 | | | 0.1 | mA |
| $I_{IH}$ | $V_{CC} = 5.5$ V, | $V_I = 2.7$ V | | | 20 | | | 20 | μA |
| $I_{IL}$ | $V_{CC} = 5.5$ V, | $V_I = 0.4$ V | | | −0.5 | | | −0.5 | mA |
| $I_O$[‡] | $V_{CC} = 5.5$ V, | $V_O = 2.25$ V | −30 | | −112 | −30 | | −112 | mA |
| $I_{CCH}$ | $V_{CC} = 5.5$ V, | $V_I = 0$ V | | 2 | 3.2 | | 2 | 3.2 | mA |
| $I_{CCL}$ | $V_{CC} = 5.5$ V, | $V_I = 4.5$ V | | 10.8 | 17.4 | | 10.8 | 17.4 | mA |

[†]All typical values are at $V_{CC} = 5$ V, $T_A = 25$ °C.
[‡]The output conditions have been chosen to produce a current that closely approximates one half of the true short-circuit output current, $I_{OS}$.

**switching characteristics**

| PARAMETER | FROM (INPUT) | TO (OUTPUT) | $V_{CC} = 4.5$ V to 5.5 V, $C_L = 50$ pF, $R_L = 50$ Ω, $T_A =$ MIN to MAX | | | | UNIT |
|---|---|---|---|---|---|---|---|
| | | | SN54AS00 | | SN74AS00 | | |
| | | | MIN | MAX | MIN | MAX | |
| $t_{PLH}$ | A or B | Y | 1 | 5 | 1 | 4.5 | ns |
| $t_{PHL}$ | A or B | Y | 1 | 5 | 1 | 4 | |

Reprinted by permission of Texas Instruments

## SN54ALS00A, SN74ALS00A
## QUADRUPLE 2-INPUT POSITIVE-NAND GATES

---

### absolute maximum ratings over operating free-air temperature range (unless otherwise noted)

Supply voltage, $V_{CC}$ . . . . . . . . . . . . . . . . . . . . . . . . . . . . . . . . . . . . . . . . . . . . . . . . . . 7 V
Input voltage . . . . . . . . . . . . . . . . . . . . . . . . . . . . . . . . . . . . . . . . . . . . . . . . . . . . . . . . . 7 V
Operating free-air temperature range: SN54ALS00A . . . . . . . . . . . . . . . . . . . . . -55°C to 125°C
                                         SN74ALS00A . . . . . . . . . . . . . . . . . . . . . . . . 0°C to 70°C
Storage temperature range . . . . . . . . . . . . . . . . . . . . . . . . . . . . . . . . . . . . . . . . . . -65°C to 150°C

### recommended operating conditions

| | | SN54ALS00A | | | SN74ALS00A | | | UNIT |
|---|---|---|---|---|---|---|---|---|
| | | MIN | NOM | MAX | MIN | NOM | MAX | |
| $V_{CC}$ | Supply voltage | 4.5 | 5 | 5.5 | 4.5 | 5 | 5.5 | V |
| $V_{IH}$ | High-level input voltage | 2 | | | 2 | | | V |
| $V_{IL}$ | Low-level input voltage | | | 0.7 | | | 0.8 | V |
| $I_{OH}$ | High-level output current | | | -0.4 | | | -0.4 | mA |
| $I_{OL}$ | Low-level output current | | | 4 | | | 8 | mA |
| $T_A$ | Operating free-air temperature | -55 | | 125 | 0 | | 70 | °C |

### electrical characteristics over recommended operating free-air temperature range (unless otherwise noted)

| PARAMETER | TEST CONDITIONS | | SN54ALS00A | | | SN74ALS00A | | | UNIT |
|---|---|---|---|---|---|---|---|---|---|
| | | | MIN | TYP[†] | MAX | MIN | TYP[†] | MAX | |
| $V_{IK}$ | $V_{CC}$ = 4.5 V, | $I_I$ = -18 mA | | | -1.5 | | | -1.5 | V |
| $V_{OH}$ | $V_{CC}$ = 4.5 V to 5.5 V, | $I_{OH}$ = -0.4 mA | $V_{CC}-2$ | | | $V_{CC}-2$ | | | V |
| $V_{OL}$ | $V_{CC}$ = 4.5 V, | $I_{OL}$ = 4 mA | | 0.25 | 0.4 | | 0.25 | 0.4 | V |
| | $V_{CC}$ = 4.5 V, | $I_{OL}$ = 8 mA | | | | | 0.35 | 0.5 | |
| $I_I$ | $V_{CC}$ = 5.5 V, | $V_I$ = 7 V | | | 0.1 | | | 0.1 | mA |
| $I_{IH}$ | $V_{CC}$ = 5.5 V, | $V_I$ = 2.7 V | | | 20 | | | 20 | µA |
| $I_{IL}$ | $V_{CC}$ = 5.5 V, | $V_I$ = 0.4 V | | | -0.1 | | | -0.1 | mA |
| $I_O$[‡] | $V_{CC}$ = 5.5 V, | $V_O$ = 2.25 V | -30 | | -112 | -30 | | -112 | mA |
| $I_{CCH}$ | $V_{CC}$ = 5.5 V, | $V_I$ = 0 V | | 0.5 | 0.85 | | 0.5 | 0.85 | mA |
| $I_{CCL}$ | $V_{CC}$ = 5.5 V, | $V_I$ = 4.5 V | | 1.5 | 3 | | 1.5 | 3 | mA |

[†] All typical values are at $V_{CC}$ = 5 V, $T_A$ = 25°C.
[‡] The output conditions have been chosen to produce a current that closely approximates one half of the true short-circuit output current, $I_{OS}$

### switching characteristics

| PARAMETER | FROM (INPUT) | TO (OUTPUT) | $V_{CC}$ = 5 V, $C_L$ = 50 pF, $R_L$ = 500 Ω, $T_A$ = 25°C | $V_{CC}$ = 4.5 V to 5.5 V, $C_L$ = 50 pF, $R_L$ = 500 Ω, $T_A$ = MIN to MAX | | | | UNIT |
|---|---|---|---|---|---|---|---|---|
| | | | 'ALS00A | SN54ALS00A | | SN74ALS00A | | |
| | | | TYP | MIN | MAX | MIN | MAX | |
| $t_{PLH}$ | A or B | Y | 7 | 3 | 16 | 3 | 11 | ns |
| $t_{PHL}$ | A or B | Y | 5 | 2 | 13 | 2 | 8 | |

**MOTOROLA**

### QUAD 2-INPUT NAND GATE

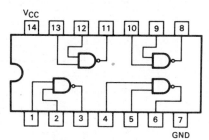

J Suffix — Case 632-08 (Ceramic)
N Suffix — Case 646-06 (Plastic)
D Suffix — Case 751A-02 (SOIC)

**MC54F/74F00**

### QUAD 2-INPUT NAND GATE
FAST™ SCHOTTKY TTL

## GUARANTEED OPERATING RANGES

| SYMBOL | PARAMETER | | MIN | TYP | MAX | UNIT |
|--------|-----------|------|------|-----|-----|------|
| $V_{CC}$ | Supply Voltage | 54, 74 | 4.5 | 5.0 | 5.5 | V |
| $T_A$ | Operating Ambient Temperature Range | 54<br>74 | −55<br>0 | 25<br>25 | 125<br>70 | °C |
| $I_{OH}$ | Output Current — High | 54, 74 | | | −1.0 | mA |
| $I_{OL}$ | Output Current — Low | 54, 74 | | | 20 | mA |

## DC CHARACTERISTICS OVER OPERATING TEMPERATURE RANGE (unless otherwise specified)

| SYMBOL | PARAMETER | | LIMITS | | | UNITS | TEST CONDITIONS | |
|--------|-----------|--|-----|-----|-----|-------|-----------------|--|
| | | | MIN | TYP | MAX | | | |
| $V_{IH}$ | Input HIGH Voltage | | 2.0 | | | V | Guaranteed Input HIGH Voltage | |
| $V_{IL}$ | Input LOW Voltage | | | | 0.8 | V | Guaranteed Input LOW Voltage | |
| $V_{IK}$ | Input Clamp Diode Voltage | | | | −1.2 | V | $V_{CC}$ = MIN, $I_{IN}$ = −18 mA | |
| $V_{OH}$ | Output HIGH Voltage | 54, 74 | 2.5 | | | V | $I_{OH}$ = −1.0 mA | $V_{CC}$ = 4.50 V |
| | | 74 | 2.7 | | | V | $I_{OH}$ = −1.0 mA | $V_{CC}$ = 4.75 V |
| $V_{OL}$ | Output LOW Voltage | | | | 0.5 | V | $I_{OL}$ = 20 mA | $V_{CC}$ = MIN |
| $I_{IH}$ | Input HIGH Current | | | | 20 | μA | $V_{CC}$ = MAX, $V_{IN}$ = 2.7 V | |
| | | | | | 0.1 | mA | $V_{CC}$ = MAX, $V_{IN}$ = 7.0 V | |
| $I_{IL}$ | Input LOW Current | | | | −0.6 | mA | $V_{CC}$ = MAX, $V_{IN}$ = 0.5 V | |
| $I_{OS}$ | Output Short Circuit Current (Note 2) | | −60 | | −150 | mA | $V_{CC}$ = MAX, $V_{OUT}$ = 0 V | |
| $I_{CC}$ | Power Supply Current<br>Total, Output HIGH | | | | 2.8 | mA | $V_{CC}$ = MAX, $V_{IN}$ = GND | |
| | Total, Output LOW | | | | 10.2 | mA | $V_{CC}$ = MAX, $V_{IN}$ = Open | |

NOTES:
1. For conditions shown as MIN or MAX, use the appropiate value specified under recommended operating conditions for the applicable device type.
2. Not more than one output should be shorted at a time, nor for more than 1 second.

## MC54F00/74F00

### AC CHARACTERISTICS

| SYMBOL | PARAMETER | 54/74F<br>$T_A = +25°C$<br>$V_{CC} = +5.0$ V<br>$C_L = 50$ pF | | 54F<br>$T_A = -55°C$ to $+125°C$<br>$V_{CC} = 5.0$ V ± 10%<br>$C_L = 50$ pF | | 74F<br>$T_A = 0°C$ to $70°C$<br>$V_{CC} = 5.0$ V ± 10%<br>$C_L = 50$ pF | | UNITS |
|---|---|---|---|---|---|---|---|---|
| | | MIN | MAX | MIN | MAX | MIN | MAX | |
| $t_{PLH}$ | Propagation Delay | 2.4 | 5.0 | 2.0 | 7.0 | 2.4 | 6.0 | ns |
| $t_{PHL}$ | Propagation Delay | 1.5 | 4.3 | 1.5 | 6.5 | 1.5 | 5.3 | ns |

### AC TEST CIRCUIT

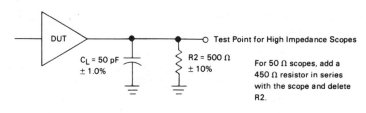

DUT

$C_L = 50$ pF ± 1.0%

R2 = 500 Ω ± 10%

Test Point for High Impedance Scopes

For 50 Ω scopes, add a 450 Ω resistor in series with the scope and delete R2.

**Fig. 1**

J SUFFIX
CERAMIC
CASE 632-08

N SUFFIX
PLASTIC
CASE 646-06

D SUFFIX
SOIC
CASE 751A-02

# MC54/74HC00A

# Quad 2-Input NAND Gate
## High-Performance Silicon-Gate CMOS

The MC54/74HC00A is identical in pinout to the LS00. The device inputs are compatible with standard CMOS outputs; with pullup resistors, they are compatible with LSTTL outputs.

- Output Drive Capability: 10 LSTTL Loads
- Outputs Directly Interface to CMOS, NMOS, and TTL
- Operating Voltage Range: 2.0 to 6.0 V
- Low Input Current: 1.0 $\mu$A
- High Noise Immunity Characteristic of CMOS Devices
- In Compliance with the Requirements Defined by JEDEC Standard No. 7A
- Chip Complexity: 32 FETs or 8 Equivalent Gates
- Improvements over HC00
  - Improved Propagation Delays
  - 50% Lower Quiescent Power
  - Improved Input Noise and Latchup Immunity

## ORDERING INFORMATION

MC74HCXXAN    Plastic
MC54HCXXAJ    Ceramic
MC74HCXXAD    SOIC

$T_A = -55°$ to $125°$C for all packages.
Dimensions in Chapter 6.

## LOGIC DIAGRAM

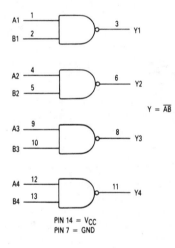

$Y = \overline{AB}$

PIN 14 = $V_{CC}$
PIN 7 = GND

## PIN ASSIGNMENT

| | | | |
|---|---|---|---|
| A1 | 1 • | 14 | $V_{CC}$ |
| B1 | 2 | 13 | B4 |
| Y1 | 3 | 12 | A4 |
| A2 | 4 | 11 | Y4 |
| B2 | 5 | 10 | B3 |
| Y2 | 6 | 9 | A3 |
| GND | 7 | 8 | Y3 |

## FUNCTION TABLE

| Inputs | | Output |
|---|---|---|
| **A** | **B** | **Y** |
| L | L | H |
| L | H | H |
| H | L | H |
| H | H | L |

Reprinted with permission of Motorola

# MC54/74HC00A

## MAXIMUM RATINGS*

| Symbol | Parameter | Value | Unit |
|--------|-----------|-------|------|
| $V_{CC}$ | DC Supply Voltage (Referenced to GND) | $-0.5$ to $+7.0$ | V |
| $V_{in}$ | DC Input Voltage (Referenced to GND) | $-1.5$ to $V_{CC} + 1.5$ | V |
| $V_{out}$ | DC Output Voltage (Referenced to GND) | $-0.5$ to $V_{CC} + 0.5$ | V |
| $I_{in}$ | DC Input Current, per Pin | $\pm 20$ | mA |
| $I_{out}$ | DC Output Current, per Pin | $\pm 25$ | mA |
| $I_{CC}$ | DC Supply Current, $V_{CC}$ and GND Pins | $\pm 50$ | mA |
| $P_D$ | Power Dissipation in Still Air, Plastic or Ceramic DIP        SOIC Package | 750   500 | mW |
| $T_{stg}$ | Storage Temperature | $-65$ to $+150$ | °C |
| $T_L$ | Lead Temperature, 1 mm from Case for 10 Seconds        (Plastic DIP or SOIC Package)        (Ceramic DIP) | 260   300 | °C |

This device contains protection circuitry to guard against damage due to high static voltages or electric fields. However, precautions must be taken to avoid applications of any voltage higher than maximum rated voltages to this high-impedance circuit. For proper operation, $V_{in}$ and $V_{out}$ should be constrained to the range GND $\leq$ ($V_{in}$ or $V_{out}$) $\leq V_{CC}$.

Unused inputs must always be tied to an appropriate logic voltage level (e.g., either GND or $V_{CC}$). Unused outputs must be left open.

*Maximum Ratings are those values beyond which damage to the device may occur.
Functional operation should be restricted to the Recommended Operating Conditions.

## RECOMMENDED OPERATING CONDITIONS

| Symbol | Parameter | | Min | Max | Unit |
|--------|-----------|---|-----|-----|------|
| $V_{CC}$ | DC Supply Voltage (Referenced to GND) | | 2.0 | 6.0 | V |
| $V_{in}$, $V_{out}$ | DC Input Voltage, Output Voltage (Referenced to GND) | | 0 | $V_{CC}$ | V |
| $T_A$ | Operating Temperature, All Package Types | | $-55$ | $+125$ | °C |
| $t_r$, $t_f$ | Input Rise and Fall Time (Figure 1) | $V_{CC} = 2.0$ V   $V_{CC} = 4.5$ V   $V_{CC} = 6.0$ V | 0   0   0 | 1000   500   400 | ns |

## DC ELECTRICAL CHARACTERISTICS (Voltages Referenced to GND)

| Symbol | Parameter | Test Conditions | | $V_{CC}$ V | Guaranteed Limit | | | Unit |
|--------|-----------|-----------------|---|------------|------------------|---|---|------|
| | | | | | 25°C to $-55$°C | $\leq$85°C | $\leq$125°C | |
| $V_{IH}$ | Minimum High-Level Input Voltage | $V_{out} = 0.1$ V or $V_{CC} - 0.1$ V  $|I_{out}| \leq 20\ \mu$A | | 2.0   4.5   6.0 | 1.5   3.15   4.2 | 1.5   3.15   4.2 | 1.5   3.15   4.2 | V |
| $V_{IL}$ | Maximum Low-Level Input Voltage | $V_{out} = 0.1$ V or $V_{CC} - 0.1$ V  $|I_{out}| \leq 20\ \mu$A | | 2.0   4.5   6.0 | 0.5   1.35   1.8 | 0.5   1.35   1.8 | 0.5   1.35   1.8 | V |
| $V_{OH}$ | Minimum High-Level Output Voltage | $V_{in} = V_{IH}$ or $V_{IL}$  $|I_{out}| \leq 20\ \mu$A | | 2.0   4.5   6.0 | 1.9   4.4   5.9 | 1.9   4.4   5.9 | 1.9   4.4   5.9 | V |
| | | $V_{in} = V_{IH}$ or $V_{IL}$ | $|I_{out}| \leq 4.0$ mA  $|I_{out}| \leq 5.2$ mA | 4.5   6.0 | 3.98   5.48 | 3.84   5.34 | 3.7   5.2 | |
| $V_{OL}$ | Maximum Low-Level Output Voltage | $V_{in} = V_{IH}$ or $V_{IL}$  $|I_{out}| \leq 20\ \mu$A | | 2.0   4.5   6.0 | 0.1   0.1   0.1 | 0.1   0.1   0.1 | 0.1   0.1   0.1 | V |
| | | $V_{in} = V_{IH}$ or $V_{IL}$ | $|I_{out}| \leq 4.0$ mA  $|I_{out}| \leq 5.2$ mA | 4.5   6.0 | 0.26   0.26 | 0.33   0.33 | 0.4   0.4 | |
| $I_{in}$ | Maximum Input Leakage Current | $V_{in} = V_{CC}$ or GND | | 6.0 | $\pm 0.1$ | $\pm 1.0$ | $\pm 1.0$ | $\mu$A |
| $I_{CC}$ | Maximum Quiescent Supply Current (per Package) | $V_{in} = V_{CC}$ or GND  $I_{out} = 0\ \mu$A | | 6.0 | 1.0 | 10 | 40 | $\mu$A |

# MC54/74HC00A

**AC ELECTRICAL CHARACTERISTICS** ($C_L$ = 50 pF, Input $t_r$ = $t_f$ = 6.0 ns)

| Symbol | Parameter | $V_{CC}$ V | Guaranteed Limit | | | Unit |
|--------|-----------|------------|------------------|---|---|------|
| | | | 25°C to −55°C | ≤85°C | ≤125°C | |
| $t_{PLH}$, $t_{PHL}$ | Maximum Propagation Delay, Input A or B to Output Y (Figures 1 and 2) | 2.0 4.5 6.0 | 75 15 13 | 95 19 16 | 110 22 19 | ns |
| $t_{TLH}$, $t_{THL}$ | Maximum Output Transition Time, Any Output (Figures 1 and 2) | 2.0 4.5 6.0 | 75 15 13 | 95 19 16 | 110 22 19 | ns |
| $C_{in}$ | Maximum Input Capacitance | — | 10 | 10 | 10 | pF |

| $C_{PD}$ | Power Dissipation Capacitance (Per Gate) Used to determine the no-load dynamic power consumption: $P_D = C_{PD} V_{CC}^2 f + I_{CC} V_{CC}$ | Typical @ 25°C, $V_{CC}$ = 5.0 V | |
|----------|---|---|---|
| | | 22 | pF |

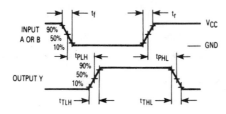

Figure 1. Switching Waveforms

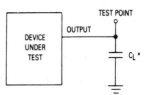

*Includes all probe and jig capacitance.

Figure 2. Test Circuit

**EXPANDED LOGIC DIAGRAM**
**(1/4 of the Device)**

# MC54/74HC02A

J SUFFIX
CERAMIC
CASE 632-08

N SUFFIX
PLASTIC
CASE 646-06

D SUFFIX
SOIC
CASE 751A-02

# Quad 2-Input NOR Gate
## High-Performance Silicon-Gate CMOS

The MC54/74HC02A is identical in pinout to the LS02. The device inputs are compatible with standard CMOS outputs; with pullup resistors, they are compatible with LSTTL outputs.

- Output Drive Capability: 10 LSTTL Loads
- Outputs Directly Interface to CMOS, NMOS, and TTL
- Operating Voltage Range: 2.0 to 6.0 V
- Low Input Current: 1.0 $\mu$A
- High Noise Immunity Characteristic of CMOS Devices
- In Compliance with the Requirements Defined by JEDEC Standard No. 7A
- Chip Complexity: 40 FETs or 10 Equivalent Gates
- Improvements over HC02
  - Improved Propagation Delays
  - 50% Lower Quiescent Power
  - Improved Input Noise and Latchup Immunity

**ORDERING INFORMATION**

| MC74HCXXAN | Plastic |
| MC54HCXXAJ | Ceramic |
| MC74HCXXAD | SOIC |

$T_A = -55°$ to $125°C$ for all packages.
Dimensions in Chapter 6.

**PIN ASSIGNMENT**

| Y1 | 1 ● | 14 | $V_{CC}$ |
| A1 | 2 | 13 | Y4 |
| B1 | 3 | 12 | B4 |
| Y2 | 4 | 11 | A4 |
| A2 | 5 | 10 | Y3 |
| B2 | 6 | 9 | B3 |
| GND | 7 | 8 | A3 |

**LOGIC DIAGRAM**

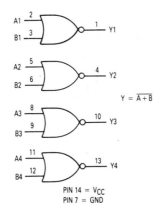

$$Y = \overline{A + B}$$

PIN 14 = $V_{CC}$
PIN 7 = GND

**FUNCTION TABLE**

| Inputs | | Output |
|---|---|---|
| A | B | Y |
| L | L | H |
| L | H | L |
| H | L | L |
| H | H | L |

Reprinted with permission of Motorola

## MC54/74HC02A

### MAXIMUM RATINGS*

| Symbol | Parameter | Value | Unit |
|--------|-----------|-------|------|
| $V_{CC}$ | DC Supply Voltage (Referenced to GND) | $-0.5$ to $+7.0$ | V |
| $V_{in}$ | DC Input Voltage (Referenced to GND) | $-1.5$ to $V_{CC}+1.5$ | V |
| $V_{out}$ | DC Output Voltage (Referenced to GND) | $-0.5$ to $V_{CC}+0.5$ | V |
| $I_{in}$ | DC Input Current, per Pin | $\pm 20$ | mA |
| $I_{out}$ | DC Output Current, per Pin | $\pm 25$ | mA |
| $I_{CC}$ | DC Supply Current, $V_{CC}$ and GND Pins | $\pm 50$ | mA |
| $P_D$ | Power Dissipation in Still Air, Plastic or Ceramic DIP    SOIC Package | 750 500 | mW |
| $T_{stg}$ | Storage Temperature | $-65$ to $+150$ | °C |
| $T_L$ | Lead Temperature, 1 mm from Case for 10 Seconds (Plastic DIP or SOIC Package) (Ceramic DIP) | 260 300 | °C |

This device contains protection circuitry to guard against damage due to high static voltages or electric fields. However, precautions must be taken to avoid applications of any voltage higher than maximum rated voltages to this high-impedance circuit. For proper operation, $V_{in}$ and $V_{out}$ should be constrained to the range GND $\le$ ($V_{in}$ or $V_{out}$) $\le V_{CC}$.

Unused inputs must always be tied to an appropriate logic voltage level (e.g., either GND or $V_{CC}$). Unused outputs must be left open.

*Maximum Ratings are those values beyond which damage to the device may occur.
 Functional operation should be restricted to the Recommended Operating Conditions.

### RECOMMENDED OPERATING CONDITIONS

| Symbol | Parameter | | Min | Max | Unit |
|--------|-----------|---|-----|-----|------|
| $V_{CC}$ | DC Supply Voltage (Referenced to GND) | | 2.0 | 6.0 | V |
| $V_{in}, V_{out}$ | DC Input Voltage, Output Voltage (Referenced to GND) | | 0 | $V_{CC}$ | V |
| $T_A$ | Operating Temperature, All Package Types | | $-55$ | $+125$ | °C |
| $t_r, t_f$ | Input Rise and Fall Time (Figure 1) | $V_{CC} = 2.0$ V $V_{CC} = 4.5$ V $V_{CC} = 6.0$ V | 0 0 0 | 1000 500 400 | ns |

### DC ELECTRICAL CHARACTERISTICS (Voltages Referenced to GND)

| Symbol | Parameter | Test Conditions | | $V_{CC}$ V | Guaranteed Limit | | | Unit |
|--------|-----------|-----------------|---|-----------|------------------|---|---|------|
| | | | | | 25°C to $-55$°C | $\le 85$°C | $\le 125$°C | |
| $V_{IH}$ | Minimum High-Level Input Voltage | $V_{out} = 0.1$ V or $V_{CC} - 0.1$ V $\|I_{out}\| \le 20$ μA | | 2.0 4.5 6.0 | 1.5 3.15 4.2 | 1.5 3.15 4.2 | 1.5 3.15 4.2 | V |
| $V_{IL}$ | Maximum Low-Level Input Voltage | $V_{out} = 0.1$ V or $V_{CC} - 0.1$ V $\|I_{out}\| \le 20$ μA | | 2.0 4.5 6.0 | 0.5 1.35 1.8 | 0.5 1.35 1.8 | 0.5 1.35 1.8 | V |
| $V_{OH}$ | Minimum High-Level Output Voltage | $V_{in} = V_{IH}$ or $V_{IL}$ $\|I_{out}\| \le 20$ μA | | 2.0 4.5 6.0 | 1.9 4.4 5.9 | 1.9 4.4 5.9 | 1.9 4.4 5.9 | V |
| | | $V_{in} = V_{IH}$ or $V_{IL}$ | $\|I_{out}\| \le 4.0$ mA $\|I_{out}\| \le 5.2$ mA | 4.5 6.0 | 3.98 5.48 | 3.84 5.34 | 3.7 5.2 | |
| $V_{OL}$ | Maximum Low-Level Output Voltage | $V_{in} = V_{IH}$ or $V_{IL}$ $\|I_{out}\| \le 20$ μA | | 2.0 4.5 6.0 | 0.1 0.1 0.1 | 0.1 0.1 0.1 | 0.1 0.1 0.1 | V |
| | | $V_{in} = V_{IH}$ or $V_{IL}$ | $\|I_{out}\| \le 4.0$ mA $\|I_{out}\| \le 5.2$ mA | 4.5 6.0 | 0.26 0.26 | 0.33 0.33 | 0.4 0.4 | |
| $I_{in}$ | Maximum Input Leakage Current | $V_{in} = V_{CC}$ or GND | | 6.0 | $\pm 0.1$ | $\pm 1.0$ | $\pm 1.0$ | μA |
| $I_{CC}$ | Maximum Quiescent Supply Current (per Package) | $V_{in} = V_{CC} = $ GND $I_{out} = 0$ μA | | 6.0 | 1.0 | 10 | 40 | μA |

## MC54/74HC02A

**AC ELECTRICAL CHARACTERISTICS** ($C_L$ = 50 pF, Input $t_r$ = $t_f$ = 6.0 ns)

| Symbol | Parameter | $V_{CC}$ V | Guaranteed Limit | | | Unit |
|---|---|---|---|---|---|---|
| | | | 25°C to −55°C | ≤85°C | ≤125°C | |
| $t_{PLH}$, $t_{PHL}$ | Maximum Propagation Delay, Input A or B to Output Y (Figures 1 and 2) | 2.0<br>4.5<br>6.0 | 80<br>16<br>14 | 100<br>20<br>17 | 120<br>24<br>20 | ns |
| $t_{TLH}$, $t_{THL}$ | Maximum Output Transition Time, Any Output (Figures 1 and 2) | 2.0<br>4.5<br>6.0 | 75<br>15<br>13 | 95<br>19<br>16 | 110<br>22<br>19 | ns |
| $C_{in}$ | Maximum Input Capacitance | — | 10 | 10 | 10 | pF |

| Symbol | Parameter | Typical @ 25°C, $V_{CC}$ = 5.0 V | Unit |
|---|---|---|---|
| $C_{PD}$ | Power Dissipation Capacitance (Per Gate)<br>Used to determine the no-load dynamic power consumption:<br>$P_D = C_{PD} V_{CC}^2 f + I_{CC} V_{CC}$ | 22 | pF |

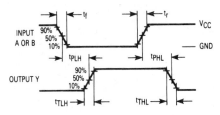

**Figure 1. Switching Waveforms**

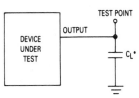

*Includes all probe and jig capacitance.

**Figure 2. Test Circuit**

**EXPANDED LOGIC DIAGRAM**
**(¼ of the Device)**

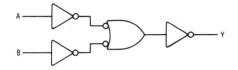

# Hex Inverter with LSTTL-Compatible Inputs
## High-Performance Silicon-Gate CMOS

**MC54/74HCT04A**

**J SUFFIX**
**CERAMIC**
**CASE 632-08**

**N SUFFIX**
**PLASTIC**
**CASE 646-06**

**D SUFFIX**
**SOIC**
**CASE 751A-02**

The MC54/74HCT04A may be used as a level converter for interfacing TTL or NMOS outputs to High-Speed CMOS inputs.

The HCT04A is identical in pinout to the LS04.

- Output Drive Capability: 10 LSTTL Loads
- TTL/NMOS-Compatible Input Levels
- Outputs Directly Interface to CMOS, NMOS, and TTL
- Operating Voltage Range: 4.5 to 5.5 V
- Low Input Current: 1 $\mu$A
- In Compliance with the Requirements Defined by JEDEC Standard No. 7A
- Chip Complexity: 48 FETs or 12 Equivalent Gates

### ORDERING INFORMATION

| MC74HCTXXAN | Plastic |
| MC54HCTXXAJ | Ceramic |
| MC74HCTXXAD | SOIC |

$T_A = -55°$ to $125°C$ for all packages.
Dimensions in Chapter 6.

**LOGIC DIAGRAM**

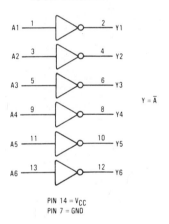

$Y = \overline{A}$

PIN 14 = $V_{CC}$
PIN 7 = GND

### PIN ASSIGNMENT

| | | | |
|---|---|---|---|
| A1 | 1 ● | 14 | $V_{CC}$ |
| Y1 | 2 | 13 | A6 |
| A2 | 3 | 12 | Y6 |
| Y2 | 4 | 11 | A5 |
| A3 | 5 | 10 | Y5 |
| Y3 | 6 | 9 | A4 |
| GND | 7 | 8 | Y4 |

### FUNCTION TABLE

| Inputs A | Outputs Y |
|---|---|
| L | H |
| H | L |

Reprinted with permission of Motorola

## MC54/74HCT04A

### MAXIMUM RATINGS*

| Symbol | Parameter | Value | Unit |
|--------|-----------|-------|------|
| $V_{CC}$ | DC Supply Voltage (Referenced to GND) | $-0.5$ to $+7$ | V |
| $V_{in}$ | DC Input Voltage (Referenced to GND) | $-1.5$ to $V_{CC}+1.5$ | V |
| $V_{out}$ | DC Output Voltage (Referenced to GND) | $-0.5$ to $V_{CC}+0.5$ | V |
| $I_{in}$ | DC Input Current, per Pin | $\pm 20$ | mA |
| $I_{out}$ | DC Output Current, per Pin | $\pm 25$ | mA |
| $I_{CC}$ | DC Supply Current, $V_{CC}$ and GND Pins | $\pm 50$ | mA |
| $P_D$ | Power Dissipation in Still Air, Plastic or Ceramic DIP† / SOIC Package† | 750 / 500 | mW |
| $T_{stg}$ | Storage Temperature | $-65$ to $+150$ | °C |
| $T_L$ | Lead Temperature, 1 mm from Case for 10 Seconds (Plastic DIP or SOIC Package) (Ceramic DIP) | 260 / 300 | °C |

This device contains protection circuitry to guard against damage due to high static voltages or electric fields. However, precautions must be taken to avoid applications of any voltage higher than maximum rated voltages to this high-impedance circuit. For proper operation, $V_{in}$ and $V_{out}$ should be constrained to the range GND $\leq (V_{in}$ or $V_{out}) \leq V_{CC}$.

Unused inputs must always be tied to an appropriate logic voltage level (e.g., either GND or $V_{CC}$). Unused outputs must be left open.

* Maximum Ratings are those values beyond which damage to the device may occur.
 Functional operation should be restricted to the Recommended Operating Conditions.
† Derating — Plastic DIP: $-10$ mW/°C from 65° to 125°C
 Ceramic DIP: $-10$ mW/°C from 100° to 125°C
 SOIC Package: $-7$ mW/°C from 65° to 125°C

### RECOMMENDED OPERATING CONDITIONS

| Symbol | Parameter | Min | Max | Unit |
|--------|-----------|-----|-----|------|
| $V_{CC}$ | DC Supply Voltage (Referenced to GND) | 4.5 | 5.5 | V |
| $V_{in}, V_{out}$ | DC Input Voltage, Output Voltage (Referenced to GND) | 0 | $V_{CC}$ | V |
| $T_A$ | Operating Temperature, All Package Types | $-55$ | $+125$ | °C |
| $t_r, t_f$ | Input Rise and Fall Time (Figure 1) | 0 | 500 | ns |

### DC ELECTRICAL CHARACTERISTICS (Voltages Referenced to GND)

| Symbol | Parameter | Test Conditions | $V_{CC}$ V | Guaranteed Limit 25°C to $-55$°C | Guaranteed Limit $\leq 85$°C | Guaranteed Limit $\leq 125$°C | Unit |
|--------|-----------|-----------------|------------|------|------|------|------|
| $V_{IH}$ | Minimum High-Level Input Voltage | $V_{out}=0.1$ V / $\|I_{out}\| \leq 20 \ \mu A$ | 4.5 / 5.5 | 2 / 2 | 2 / 2 | 2 / 2 | V |
| $V_{IL}$ | Maximum Low-Level Input Voltage | $V_{out}=V_{CC}-0.1$ V / $\|I_{out}\| \leq 20 \ \mu A$ | 4.5 / 5.5 | 0.8 / 0.8 | 0.8 / 0.8 | 0.8 / 0.8 | V |
| $V_{OH}$ | Minimum High-Level Output Voltage | $V_{in}=V_{IL}$ / $\|I_{out}\| \leq 20 \ \mu A$ | 4.5 / 5.5 | 4.4 / 5.4 | 4.4 / 5.4 | 4.4 / 5.4 | V |
| | | $V_{in}=V_{IL}$ / $\|I_{out}\| \leq 4$ mA | 4.5 | 3.98 | 3.84 | 3.7 | |
| $V_{OL}$ | Maximum Low-Level Output Voltage | $V_{in}=V_{IH}$ / $\|I_{out}\| \leq 20 \ \mu A$ | 4.5 / 5.5 | 0.1 / 0.1 | 0.1 / 0.1 | 0.1 / 0.1 | V |
| | | $V_{in}=V_{IH}$ / $\|I_{out}\| \leq 4$ mA | 4.5 | 0.26 | 0.33 | 0.4 | |
| $I_{in}$ | Maximum Input Leakage Current | $V_{in}=V_{CC}$ or GND | 5.5 | $\pm 0.1$ | $\pm 1$ | $\pm 1$ | $\mu A$ |
| $I_{CC}$ | Maximum Quiescent Supply Current (per Package) | $V_{in}=V_{CC}$ or GND / $I_{out}=0 \ \mu A$ | 5.5 | 1 | 10 | 40 | $\mu A$ |

| Symbol | Parameter | Test Conditions | | $\geq -55$°C | 25°C to 125°C | Unit |
|--------|-----------|-----------------|---|------|------|------|
| $\Delta I_{CC}$ | Additional Quiescent Supply Current | $V_{in}=2.4$ V, Any One Input $V_{in}=V_{CC}$ or GND, Other Inputs $I_{out}=0 \ \mu A$ | 5.5 | 2.9 | 2.4 | mA |

NOTES:
1. Information on typical parametric values along with frequency or heavy load considerations can be found in Chapter 4.
2. Total Supply Current = $I_{CC} + \Sigma \Delta I_{CC}$.

# MC54/74HCT04A

**AC ELECTRICAL CHARACTERISTICS** ($V_{CC}$ = 5.0 V ±10%, $C_L$ = 50 pF, Input $t_r$ = $t_f$ = 6 ns)

| Symbol | Parameter | Guaranteed Limit | | | Unit |
|---|---|---|---|---|---|
| | | 25°C to −55°C | ≤85°C | ≤125°C | |
| $t_{PLH}$ | Maximum Propagation Delay, Input A to Output Y (Figures 1 and 2) | 15 | 19 | 22 | ns |
| $t_{PHL}$ | | 17 | 21 | 26 | |
| $t_{TLH}$, $t_{THL}$ | Maximum Output Transition Time, Any Output (Figures 1 and 2) | 15 | 19 | 22 | ns |
| $C_{in}$ | Maximum Input Capacitance | 10 | 10 | 10 | pF |

| | | Typical @ 25°C, $V_{CC}$ = 5 V | |
|---|---|---|---|
| $C_{PD}$ | Power Dissipation Capacitance (Per Inverter) Used to determine the no-load dynamic power consumption: $P_D = C_{PD} V_{CC}^2 f + I_{CC} V_{CC}$ | 22 | pF |

NOTE: For propagation delays with loads other than 50 pF and information on typical parametric values and load considerations, see Chapter 4.

## SWITCHING WAVEFORMS

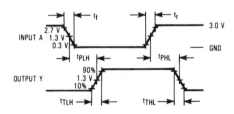

**Figure 1. Switching Waveforms**

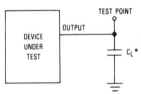

*Includes all probe and jig capacitance.

**Figure 2. Test Circuit**

## EXPANDED LOGIC DIAGRAM
### (1/6 of Device Shown)

A —▷o—▷o—▷o— Y

# Hex Unbuffered Inverter
## High Performance Silicon-Gate CMOS

The MC54/74HCU04 is identical in pinout to the LS04 and the MC14069UB. The device inputs are compatible with standard CMOS outputs; with pullup resistors, they are compatible with LSTTL outputs.

This device consists of six single-stage inverters. These inverters are well suited for use as oscillators, pulse shapers, and in many other applications requiring a high-input impedance amplifier. For digital applications, the HC04 is recommended.

● Output Drive Capability: 10 LSTTL Loads
● Outputs Directly Interface to CMOS, NMOS, and TTL
● Operating Voltage Range: 2 to 6 V; 2.5 to 6 V in Oscillator Configurations
● Low Input Current: 1 μA
● High Noise Immunity Characteristic of CMOS Devices
● In Compliance with the Requirements Defined by JEDEC Standard No. 7A
● Chip Complexity: 12 FETs or 3 Equivalent Gates

## MC54/74HCU04

**J SUFFIX**
**CERAMIC**
**CASE 632-08**

**N SUFFIX**
**PLASTIC**
**CASE 646-06**

**D SUFFIX**
**SOIC**
**CASE 751A-02**

### ORDERING INFORMATION

| | |
|---|---|
| MC74HCUXXN | Plastic |
| MC54HCUXXJ | Ceramic |
| MC74HCUXXD | SOIC |

$T_A = -55°$ to $125°C$ for all packages.
Dimensions in Chapter 6.

### PIN ASSIGNMENT

| | | | |
|---|---|---|---|
| A1 | 1 ● | 14 | $V_{CC}$ |
| Y1 | 2 | 13 | A6 |
| A2 | 3 | 12 | Y6 |
| Y2 | 4 | 11 | A5 |
| A3 | 5 | 10 | Y5 |
| Y3 | 6 | 9 | A4 |
| GND | 7 | 8 | Y4 |

### FUNCTION TABLE

| Inputs A | Outputs Y |
|---|---|
| L | H |
| H | L |

### LOGIC DIAGRAM

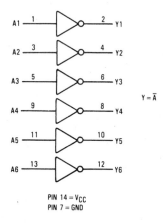

$Y = \overline{A}$

PIN 14 = $V_{CC}$
PIN 7 = GND

Reprinted with permission of Motorola

## MC54/74HCU04

### MAXIMUM RATINGS*

| Symbol | Parameter | Value | Unit |
|---|---|---|---|
| $V_{CC}$ | DC Supply Voltage (Referenced to GND) | −0.5 to +7.0 | V |
| $V_{in}$ | DC Input Voltage (Referenced to GND) | −1.5 to $V_{CC}$ +1.5 | V |
| $V_{out}$ | DC Output Voltage (Referenced to GND) | −0.5 to $V_{CC}$ +0.5 | V |
| $I_{in}$ | DC Input Current, per Pin | ±20 | mA |
| $I_{out}$ | DC Output Current, per Pin | ±25 | mA |
| $I_{CC}$ | DC Supply Current, $V_{CC}$ and GND Pins | ±50 | mA |
| $P_D$ | Power Dissipation in Still Air, Plastic or Ceramic DIP† / SOIC Package† | 750 / 500 | mW |
| $T_{stg}$ | Storage Temperature | −65 to +150 | °C |
| $T_L$ | Lead Temperature, 1 mm from Case for 10 Seconds (Plastic DIP or SOIC Package) (Ceramic DIP) | 260 / 300 | °C |

This device contains protection circuitry to guard against damage due to high static voltages or electric fields. However, precautions must be taken to avoid applications of any voltage higher than maximum rated voltages to this high-impedance circuit. For proper operation, $V_{in}$ and $V_{out}$ should be constrained to the range GND ≤ ($V_{in}$ or $V_{out}$) ≤ $V_{CC}$.

Unused inputs must always be tied to an appropriate logic voltage level (e.g., either GND or $V_{CC}$). Unused outputs must be left open.

*Maximum Ratings are those values beyond which damage to the device may occur.
Functional operation should be restricted to the Recommended Operating Conditions.
†Derating — Plastic DIP: −10 mW/°C from 65° to 125°C
Ceramic DIP: −10 mW/°C from 100° to 125°C
SOIC Package: −7 mW/°C from 65° to 125°C
For high frequency or heavy load considerations, see Chapter 4.

### RECOMMENDED OPERATING CONDITIONS

| Symbol | Parameter | Min | Max | Unit |
|---|---|---|---|---|
| $V_{CC}$ | DC Supply Voltage (Referenced to GND) | 2.0 | 6.0 | V |
| $V_{in}, V_{out}$ | DC Input Voltage, Output Voltage (Referenced to GND) | 0 | $V_{CC}$ | V |
| $T_A$ | Operating Temperature, All Package Types | −55 | +125 | °C |
| $t_r, t_f$ | Input Rise and Fall Time (Figure 1) | — | No Limit | ns |

### DC ELECTRICAL CHARACTERISTICS (Voltages Referenced to GND)

| Symbol | Parameter | Test Conditions | $V_{CC}$ V | Guaranteed Limit 25°C to −55°C | ≤85°C | ≤125°C | Unit |
|---|---|---|---|---|---|---|---|
| $V_{IH}$ | Minimum High-Level Input Voltage | $V_{out}$ = 0.5 V* $|I_{out}|$ ≤ 20 μA | 2.0 / 4.5 / 6.0 | 1.7 / 3.6 / 4.8 | 1.7 / 3.6 / 4.8 | 1.7 / 3.6 / 4.8 | V |
| $V_{IL}$ | Maximum Low-Level Input Voltage | $V_{out}$ = $V_{CC}$ − 0.5 V* $|I_{out}|$ ≤ 20 μA | 2.0 / 4.5 / 6.0 | 0.3 / 0.8 / 1.1 | 0.3 / 0.8 / 1.1 | 0.3 / 0.8 / 1.1 | V |
| $V_{OH}$ | Minimum High-Level Output Voltage | $V_{in}$ = GND $|I_{out}|$ ≤ 20 μA | 2.0 / 4.5 / 6.0 | 1.8 / 4.0 / 5.5 | 1.8 / 4.0 / 5.5 | 1.8 / 4.0 / 5.5 | V |
|  |  | $V_{in}$ = GND | $|I_{out}|$ ≤ 4.0 mA (4.5) / $|I_{out}|$ ≤ 5.2 mA (6.0) | 3.86 / 5.36 | 3.76 / 5.26 | 3.70 / 5.20 |  |
| $V_{OL}$ | Maximum Low-Level Output Voltage | $V_{in}$ = $V_{CC}$ $|I_{out}|$ ≤ 20 μA | 2.0 / 4.5 / 6.0 | 0.2 / 0.5 / 0.5 | 0.2 / 0.5 / 0.5 | 0.2 / 0.5 / 0.5 | V |
|  |  | $V_{in}$ = $V_{CC}$ | $|I_{out}|$ ≤ 4.0 mA (4.5) / $|I_{out}|$ ≤ 5.2 mA (6.0) | 0.32 / 0.32 | 0.37 / 0.37 | 0.40 / 0.40 |  |
| $I_{in}$ | Maximum Input Leakage Current | $V_{in}$ = $V_{CC}$ or GND | 6.0 | ±0.1 | ±1.0 | ±1.0 | μA |
| $I_{CC}$ | Maximum Quiescent Supply Current (per Package) | $V_{in}$ = $V_{CC}$ or GND $I_{out}$ = 0 μA | 6.0 | 2 | 20 | 40 | μA |

NOTE: Information on typical parametric values can be found in Chapter 4.
*For $V_{CC}$ = 2.0 V, $V_{out}$ = 0.2 V or $V_{CC}$ − 0.2 V.

## MC54/74HCU04

**AC ELECTRICAL CHARACTERISTICS** ($C_L = 50$ pF, Input $t_r = t_f = 6$ ns)

| Symbol | Parameter | $V_{CC}$ V | Guaranteed Limit | | | Unit |
|---|---|---|---|---|---|---|
| | | | 25°C to −55°C | ≤ 85°C | ≤ 125°C | |
| $t_{PLH}$, $t_{PHL}$ | Maximum Propagation Delay, Input A to Output Y (Figures 1 and 2) | 2.0 | 80 | 100 | 120 | ns |
| | | 4.5 | 16 | 20 | 24 | |
| | | 6.0 | 14 | 17 | 20 | |
| $t_{TLH}$, $t_{THL}$ | Maximum Output Transition Time, Any Output (Figures 1 and 2) | 2.0 | 75 | 95 | 110 | ns |
| | | 4.5 | 15 | 19 | 22 | |
| | | 6.0 | 13 | 16 | 19 | |
| $C_{in}$ | Maximum Input Capacitance | — | 10 | 10 | 10 | pF |

NOTES:
1. For propagation delays with loads other than 50 pF, see Chapter 4.
2. Information on typical parametric values can be found in Chapter 4.

| $C_{PD}$ | Power Dissipation Capacitance (Per Inverter) Used to determine the no-load dynamic power consumption: $P_D = C_{PD} V_{CC}^2 f + I_{CC} V_{CC}$ For load considerations, see Chapter 4. | Typical @ 25°C, $V_{CC} = 5.0$ V | |
|---|---|---|---|
| | | 15 | pF |

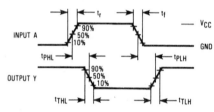

**Figure 1. Switching Waveforms**

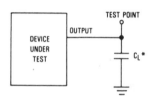

*Includes all probe and jig capacitance.

**Figure 2. Test Circuit**

**LOGIC DETAIL**
**(1/6 of Device Shown)**

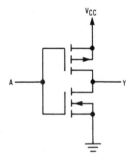

# MC54/74HCU04

## TYPICAL APPLICATIONS

### Crystal Oscillator

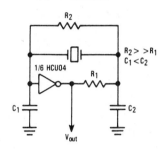

$R_2 >> R_1$
$C_1 < C_2$

### Stable RC Oscillator

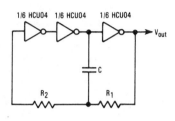

### Schmitt Trigger

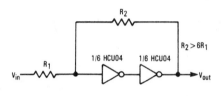

$R_2 > 6R_1$

### High Input Impedance Single-Stage Amplifier
### with a 2 to 6 V Supply Range

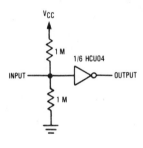

### Multi-Stage Amplifier

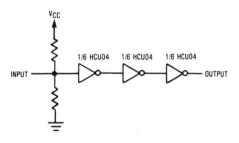

### LED Driver

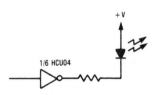

For reduced power supply current, use high-efficiency LEDs
such as the Hewlett-Packard HLMP series or equivalent.

# Quad 2-Input AND Gate
## High-Performance Silicon-Gate CMOS

The MC54/74HC08A is identical in pinout to the LS08. The device inputs are compatible with standard CMOS outputs; with pullup resistors, they are compatible with LSTTL outputs.

● Output Drive Capability: 10 LSTTL Loads
● Outputs Directly Interface to CMOS, NMOS, and TTL
● Operating Voltage Range: 2.0 to 6.0 V
● Low Input Current: 1.0 μA
● High Noise Immunity Characteristic of CMOS Devices
● In Compliance with the Requirements Defined by JEDEC Standard No. 7A
● Chip Complexity: 24 FETs or 6 Equivalent Gates
● Improvements over HC08
  ● Improved Propagation Delays
  ● 50% Lower Quiescent Power
  ● Improved Input Noise and Latchup Immunity

## MC54/74HC08A

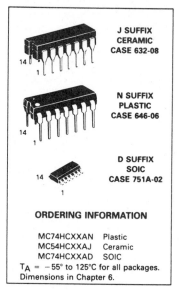

J SUFFIX
CERAMIC
CASE 632-08

N SUFFIX
PLASTIC
CASE 646-06

D SUFFIX
SOIC
CASE 751A-02

### ORDERING INFORMATION

MC74HCXXAN    Plastic
MC54HCXXAJ    Ceramic
MC74HCXXAD    SOIC

$T_A$ = −55° to 125°C for all packages.
Dimensions in Chapter 6.

### PIN ASSIGNMENT

| | | | |
|---|---|---|---|
| A1 | 1 | 14 | $V_{CC}$ |
| B1 | 2 | 13 | B4 |
| Y1 | 3 | 12 | A4 |
| A2 | 4 | 11 | Y4 |
| B2 | 5 | 10 | B3 |
| Y2 | 6 | 9 | A3 |
| GND | 7 | 8 | Y3 |

### FUNCTION TABLE

| Inputs | | Output |
|---|---|---|
| A | B | Y |
| L | L | L |
| L | H | L |
| H | L | L |
| H | H | H |

### LOGIC DIAGRAM

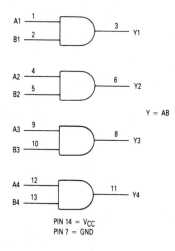

Y = AB

PIN 14 = $V_{CC}$
PIN 7 = GND

Reprinted with permission of Motorola

## MC54/74HC08A

**MAXIMUM RATINGS***

| Symbol | Parameter | Value | Unit |
|--------|-----------|-------|------|
| $V_{CC}$ | DC Supply Voltage (Referenced to GND) | $-0.5$ to $+7.0$ | V |
| $V_{in}$ | DC Input Voltage (Referenced to GND) | $-1.5$ to $V_{CC}+1.5$ | V |
| $V_{out}$ | DC Output Voltage (Referenced to GND) | $-0.5$ to $V_{CC}+0.5$ | V |
| $I_{in}$ | DC Input Current, per Pin | $\pm 20$ | mA |
| $I_{out}$ | DC Output Current, per Pin | $\pm 25$ | mA |
| $I_{CC}$ | DC Supply Current, $V_{CC}$ and GND Pins | $\pm 50$ | mA |
| $P_D$ | Power Dissipation in Still Air, Plastic or Ceramic DIP  SOIC Package | 750 500 | mW |
| $T_{stg}$ | Storage Temperature | $-65$ to $+150$ | °C |
| $T_L$ | Lead Temperature, 1 mm from Case for 10 Seconds  (Plastic DIP or SOIC Package)  (Ceramic DIP) | 260 300 | °C |

*Maximum Ratings are those values beyond which damage to the device may occur.
Functional operation should be restricted to the Recommended Operating Conditions.

This device contains protection circuitry to guard against damage due to high static voltages or electric fields. However, precautions must be taken to avoid applications of any voltage higher than maximum rated voltages to this high-impedance circuit. For proper operation, $V_{in}$ and $V_{out}$ should be constrained to the range GND $\leq$ ($V_{in}$ or $V_{out}$) $\leq V_{CC}$.
Unused inputs must always be tied to an appropriate logic voltage level (e.g., either GND or $V_{CC}$). Unused outputs must be left open.

**RECOMMENDED OPERATING CONDITIONS**

| Symbol | Parameter | | Min | Max | Unit |
|--------|-----------|---|-----|-----|------|
| $V_{CC}$ | DC Supply Voltage (Referenced to GND) | | 2.0 | 6.0 | V |
| $V_{in}, V_{out}$ | DC Input Voltage, Output Voltage (Referenced to GND) | | 0 | $V_{CC}$ | V |
| $T_A$ | Operating Temperature, All Package Types | | $-55$ | $+125$ | °C |
| $t_r, t_f$ | Input Rise and Fall Time (Figure 1) | $V_{CC}=2.0$ V $V_{CC}=4.5$ V $V_{CC}=6.0$ V | 0 0 0 | 1000 500 400 | ns |

**DC ELECTRICAL CHARACTERISTICS** (Voltages Referenced to GND)

| Symbol | Parameter | Test Conditions | | $V_{CC}$ V | Guaranteed Limit | | | Unit |
|--------|-----------|-----------------|---|------------|------------------|---|---|------|
| | | | | | 25°C to $-55$°C | $\leq 85$°C | $\leq 125$°C | |
| $V_{IH}$ | Minimum High-Level Input Voltage | $V_{out}=0.1$ V or $V_{CC}-0.1$ V $|I_{out}|\leq 20$ μA | | 2.0 4.5 6.0 | 1.5 3.15 4.2 | 1.5 3.15 4.2 | 1.5 3.15 4.2 | V |
| $V_{IL}$ | Maximum Low-Level Input Voltage | $V_{out}=0.1$ V or $V_{CC}-0.1$ V $|I_{out}|\leq 20$ μA | | 2.0 4.5 6.0 | 0.5 1.35 1.8 | 0.5 1.35 1.8 | 0.5 1.35 1.8 | V |
| $V_{OH}$ | Minimum High-Level Output Voltage | $V_{in}=V_{IH}$ or $V_{IL}$ $|I_{out}|\leq 20$ μA | | 2.0 4.5 6.0 | 1.9 4.4 5.9 | 1.9 4.4 5.9 | 1.9 4.4 5.9 | V |
| | | $V_{in}=V_{IH}$ or $V_{IL}$ | $|I_{out}|\leq 4.0$ mA $|I_{out}|\leq 5.2$ mA | 4.5 6.0 | 3.98 5.48 | 3.84 5.34 | 3.7 5.2 | |
| $V_{OL}$ | Maximum Low-Level Output Voltage | $V_{in}=V_{IH}$ or $V_{IL}$ $|I_{out}|\leq 20$ μA | | 2.0 4.5 6.0 | 0.1 0.1 0.1 | 0.1 0.1 0.1 | 0.1 0.1 0.1 | V |
| | | $V_{in}=V_{IH}$ or $V_{IL}$ | $|I_{out}|\leq 4.0$ mA $|I_{out}|\leq 5.2$ mA | 4.5 6.0 | 0.26 0.26 | 0.33 0.33 | 0.4 0.4 | V |
| $I_{in}$ | Maximum Input Leakage Current | $V_{in}=V_{CC}$ or GND | | 6.0 | $\pm 0.1$ | $\pm 1.0$ | $\pm 1.0$ | μA |
| $I_{CC}$ | Maximum Quiescent Supply Current (per Package) | $V_{in}=V_{CC}$ or GND $I_{out}=0$ μA | | 6.0 | 1.0 | 10 | 40 | μA |

# MC54/74HC08A

**AC ELECTRICAL CHARACTERISTICS** ($C_L$ = 50 pF, Input $t_r$ = $t_f$ = 6.0 ns)

| Symbol | Parameter | $V_{CC}$ V | Guaranteed Limit | | | Unit |
|--------|-----------|------------|------------------|--------|---------|------|
| | | | 25°C to −55°C | ≤85°C | ≤125°C | |
| $t_{PLH}$, $t_{PHL}$ | Maximum Propagation Delay, Input A or B to Output Y (Figures 1 and 2) | 2.0 4.5 6.0 | 75 15 13 | 95 19 16 | 110 22 19 | ns |
| $t_{TLH}$, $t_{THL}$ | Maximum Output Transition Time, Any Output (Figures 1 and 2) | 2.0 4.5 6.0 | 75 15 13 | 95 19 16 | 110 22 19 | ns |
| $C_{in}$ | Maximum Input Capacitance | — | 10 | 10 | 10 | pF |

| | | Typical @ 25°C, $V_{CC}$ = 5.0 V | |
|--------|-----------|--------------------------------|------|
| $C_{PD}$ | Power Dissipation Capacitance (Per Gate) Used to determine the no-load dynamic power consumption: $P_D = C_{PD} V_{CC}^2 f + I_{CC} V_{CC}$ | 20 | pF |

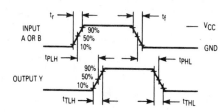

**Figure 1. Switching Waveforms**

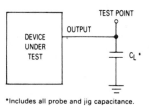

*Includes all probe and jig capacitance.

**Figure 2. Test Circuit**

**EXPANDED LOGIC DIAGRAM**
**(1/4 of the Device)**

**MC54/74HC32A**

# Quad 2-Input OR Gate
## High-Performance Silicon-Gate CMOS

The MC54/74HC32A is identical in pinout to the LS32. The device inputs are compatible with standard CMOS outputs; with pullup resistors, they are compatible with LSTTL outputs.

- Output Drive Capability: 10 LSTTL Loads
- Outputs Directly Interface to CMOS, NMOS, and TTL
- Operating Voltage Range: 2.0 to 6.0 V
- Low Input Current: 1.0 μA
- High Noise Immunity Characteristic of CMOS Devices
- In Compliance with the Requirements Defined by JEDEC Standard No. 7A
- Chip Complexity: 48 FETs or 12 Equivalent Gates
- Improvements over HC32
  - Improved Propagation Delays
  - 50% Lower Quiescent Power
  - Improved Input Noise and Latchup Immunity

**J SUFFIX**
**CERAMIC**
**CASE 632-08**

**N SUFFIX**
**PLASTIC**
**CASE 646-06**

**D SUFFIX**
**SOIC**
**CASE 751A-02**

### ORDERING INFORMATION

MC74HCXXAN   Plastic
MC54HCXXAJ   Ceramic
MC74HCXXAD   SOIC

$T_A$ = −55° to 125°C for all packages.
Dimensions in Chapter 6.

**LOGIC DIAGRAM**

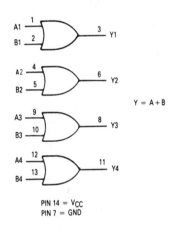

$$Y = A + B$$

PIN 14 = $V_{CC}$
PIN 7 = GND

### PIN ASSIGNMENT

| | | | |
|---|---|---|---|
| A1 | 1 | 14 | $V_{CC}$ |
| B1 | 2 | 13 | B4 |
| Y1 | 3 | 12 | A4 |
| A2 | 4 | 11 | Y4 |
| B2 | 5 | 10 | B3 |
| Y2 | 6 | 9 | A3 |
| GND | 7 | 8 | Y3 |

### FUNCTION TABLE

| Inputs | | Output |
|---|---|---|
| A | B | Y |
| L | L | L |
| L | H | H |
| H | L | H |
| H | H | H |

Reprinted with permission of Motorola

## MC54/74HC32A

**MAXIMUM RATINGS***

| Symbol | Parameter | Value | Unit |
|--------|-----------|-------|------|
| $V_{CC}$ | DC Supply Voltage (Referenced to GND) | −0.5 to +7.0 | V |
| $V_{in}$ | DC Input Voltage (Referenced to GND) | −1.5 to $V_{CC}$ + 1.5 | V |
| $V_{out}$ | DC Output Voltage (Referenced to GND) | −0.5 to $V_{CC}$ + 0.5 | V |
| $I_{in}$ | DC Input Current, per Pin | ± 20 | mA |
| $I_{out}$ | DC Output Current, per Pin | ± 25 | mA |
| $I_{CC}$ | DC Supply Current, $V_{CC}$ and GND Pins | ± 50 | mA |
| $P_D$ | Power Dissipation in Still Air, Plastic or Ceramic DIP<br>SOIC Package | 750<br>500 | mW |
| $T_{stg}$ | Storage Temperature | −65 to +150 | °C |
| $T_L$ | Lead Temperature, 1 mm from Case for 10 Seconds<br>(Plastic DIP or SOIC Package)<br>(Ceramic DIP) | <br>260<br>300 | °C |

*Maximum Ratings are those values beyond which damage to the device may occur.
Functional operation should be restricted to the Recommended Operating Conditions.

This device contains protection circuitry to guard against damage due to high static voltages or electric fields. However, precautions must be taken to avoid applications of any voltage higher than maximum rated voltages to this high-impedance circuit. For proper operation, $V_{in}$ and $V_{out}$ should be constrained to the range GND ≤ ($V_{in}$ or $V_{out}$) ≤ $V_{CC}$.

Unused inputs must always be tied to an appropriate logic voltage level (e.g., either GND or $V_{CC}$). Unused outputs must be left open.

**RECOMMENDED OPERATING CONDITIONS**

| Symbol | Parameter | Min | Max | Unit |
|--------|-----------|-----|-----|------|
| $V_{CC}$ | DC Supply Voltage (Referenced to GND) | 2.0 | 6.0 | V |
| $V_{in}$, $V_{out}$ | DC Input Voltage, Output Voltage (Referenced to GND) | 0 | $V_{CC}$ | V |
| $T_A$ | Operating Temperature, All Package Types | −55 | +125 | °C |
| $t_r$, $t_f$ | Input Rise and Fall Time (Figure 1)      $V_{CC}$ = 2.0 V<br>$V_{CC}$ = 4.5 V<br>$V_{CC}$ = 6.0 V | 0<br>0<br>0 | 1000<br>500<br>400 | ns |

**DC ELECTRICAL CHARACTERISTICS** (Voltages Referenced to GND)

| Symbol | Parameter | Test Conditions | $V_{CC}$ V | Guaranteed Limit | | | Unit |
|--------|-----------|-----------------|-----------|------------------|--|--|------|
| | | | | 25°C to −55°C | ≤85°C | ≤125°C | |
| $V_{IH}$ | Minimum High-Level Input Voltage | $V_{out}$ = 0.1 V or $V_{CC}$ − 0.1 V<br>\|$I_{out}$\|≤20 µA | 2.0<br>4.5<br>6.0 | 1.5<br>3.15<br>4.2 | 1.5<br>3.15<br>4.2 | 1.5<br>3.15<br>4.2 | V |
| $V_{IL}$ | Maximum Low-Level Input Voltage | $V_{out}$ = 0.1 V or $V_{CC}$ − 0.1 V<br>\|$I_{out}$\|≤20 µA | 2.0<br>4.5<br>6.0 | 0.5<br>1.35<br>1.8 | 0.5<br>1.35<br>1.8 | 0.5<br>1.35<br>1.8 | V |
| $V_{OH}$ | Minimum High-Level Output Voltage | $V_{in}$ = $V_{IH}$ or $V_{IL}$<br>\|$I_{out}$\|≤20 µA | 2.0<br>4.5<br>6.0 | 1.9<br>4.4<br>5.9 | 1.9<br>4.4<br>5.9 | 1.9<br>4.4<br>5.9 | V |
| | | $V_{in}$ = $V_{IH}$ or $V_{IL}$      \|$I_{out}$\|≤4.0 mA<br>\|$I_{out}$\|≤5.2 mA | 4.5<br>6.0 | 3.98<br>5.48 | 3.84<br>5.34 | 3.7<br>5.2 | |
| $V_{OL}$ | Maximum Low-Level Output Voltage | $V_{in}$ = $V_{IH}$ or $V_{IL}$<br>\|$I_{out}$\|≤20 µA | 2.0<br>4.5<br>6.0 | 0.1<br>0.1<br>0.1 | 0.1<br>0.1<br>0.1 | 0.1<br>0.1<br>0.1 | V |
| | | $V_{in}$ = $V_{IH}$ or $V_{IL}$      \|$I_{out}$\|≤4.0 mA<br>\|$I_{out}$\|≤5.2 mA | 4.5<br>6.0 | 0.26<br>0.26 | 0.33<br>0.33 | 0.4<br>0.4 | |
| $I_{in}$ | Maximum Input Leakage Current | $V_{in}$ = $V_{CC}$ or GND | 6.0 | ± 0.1 | ± 1.0 | ± 1.0 | µA |
| $I_{CC}$ | Maximum Quiescent Supply Current (per Package) | $V_{in}$ = $V_{CC}$ or GND<br>$I_{out}$ = 0 µA | 6.0 | 1.0 | 10 | 40 | µA |

## MC54/74HC32A

**AC ELECTRICAL CHARACTERISTICS** ($C_L$ = 50 pF, Input $t_r$ = $t_f$ = 6.0 ns)

| Symbol | Parameter | $V_{CC}$ V | Guaranteed Limit | | | Unit |
|---|---|---|---|---|---|---|
| | | | 25°C to −55°C | ≤85°C | ≤125°C | |
| $t_{PLH}$, $t_{PHL}$ | Maximum Propagation Delay, Input A or B to Output Y (Figures 1 and 2) | 2.0 4.5 6.0 | 75 15 13 | 95 19 16 | 110 22 19 | ns |
| $t_{TLH}$, $t_{THL}$ | Maximum Output Transition Time, Any Output (Figures 1 and 2) | 2.0 4.5 6.0 | 75 15 13 | 95 19 16 | 110 22 19 | ns |
| $C_{in}$ | Maximum Input Capacitance | — | 10 | 10 | 10 | pF |

| | | Typical @ 25°C, $V_{CC}$ = 5.0 V | |
|---|---|---|---|
| $C_{PD}$ | Power Dissipation Capacitance (Per Gate) Used to determine the no-load dynamic power consumption: $P_D = C_{PD} V_{CC}^2 f + I_{CC} V_{CC}$ | 20 | pF |

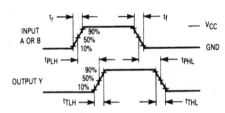

Figure 1. Switching Waveforms

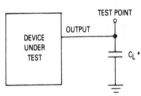

*Includes all probe and jig capacitance.

Figure 2. Test Circuit

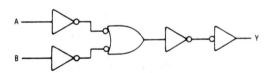

**EXPANDED LOGIC DIAGRAM**
(1/4 of the Device)

# Handling Precautions for CMOS

**MOTOROLA**

## HANDLING AND DESIGN GUIDELINES

### HANDLING PRECAUTIONS

All MOS devices have insulated gates that are subject to voltage breakdown. The gate oxide for Motorola CMOS devices is about 800 Å thick and breaks down at a gate-source potential of about 100 volts. To guard against such a breakdown from static discharge or other voltage transients, the protection network shown in Figure 1 is used on each input to the CMOS device.

Static damaged devices behave in various ways, depending on the severity of the damage. The most severely damaged inputs are the easiest to detect because the input has been completely destroyed and is either shorted to $V_{DD}$, shorted to $V_{SS}$, or open-circuited. The effect is that the device no longer responds to signals present at the damaged input. Less severe cases are more difficult to detect because they show up as intermittent failures or as degraded performance. Another effect of static damage is that the inputs generally have increased leakage currents.

Although the input protection network does provide a great deal of protection, CMOS devices are not immune to large static voltage discharges that can be generated during handling. For example, static voltages generated by a person walking across a waxed floor have been measured in the 4-15 kV range (depending on humidity, surface conditions, etc.). Therefore, the following precautions should be observed:

1. Do not exceed the Maximum Ratings specified by the data sheet.
2. All unused device inputs should be connected to $V_{DD}$ or $V_{SS}$.
3. All low-impedance equipment (pulse generators, etc.) should be connected to CMOS inputs only after the device is powered up. Similarly, this type of equipment should be disconnected before power is turned off.
4. Circuit boards containing CMOS devices are merely extensions of the devices, and the same handling precautions apply. Contacting edge connectors wired directly to device inputs can cause damage. Plastic wrapping should be avoided. When external connections to a PC board are connected to an input of a CMOS device, a resistor should be used in series with the input. This resistor helps limit accidental damage if the PC board is removed and brought into contact with static generating materials. The limiting factor for the series resistor is the added delay. This is caused by the time constant formed by the series resistor and

input capacitance. Note that the maximum input rise and fall times should not be exceeded. In Figure 2, two possible networks are shown using a series resistor to reduce ESD (Electrostatic Discharge) damage. For convenience, an equation for added propagation delay and rise time effects due to series resistance size is given.

5. All CMOS devices should be stored or transported in materials that are antistatic. CMOS devices must not be inserted into conventional plastic "snow", styrofoam, or plastic trays, but should be left in their original container until ready for use.
6. All CMOS devices should be placed on a grounded bench surface and operators should ground themselves prior to handling devices, since a worker can be statically charged with respect to the bench surface. Wrist straps in contact with skin are strongly recommended. See Figure 3 for an example of a typical work station.
7. Nylon or other static generating materials should not come in contact with CMOS devices.
8. If automatic handlers are being used, high levels of static electricity may be generated by the movement of the device, the belts, or the boards. Reduce static build-up by using ionized air blowers or room humidifiers. All parts of machines which come into contact with the top, bottom, or sides of IC packages must be grounded to metal or other conductive material.
9. Cold chambers using $CO_2$ for cooling should be equipped with baffles, and the CMOS devices must be contained on or in conductive material.
10. When lead-straightening or hand-soldering is necessary, provide ground straps for the apparatus used and be sure that soldering ties are grounded.
11. The following steps should be observed during wave solder operations:
    a. The solder pot and conductive conveyor system of the wave soldering machine must be grounded to an earth ground.
    b. The loading and unloading work benches should have conductive tops which are grounded to an earth ground.
    c. Operators must comply with precautions previously explained.
    d. Completed assemblies should be placed in antistatic containers prior to being moved to subsequent stations.

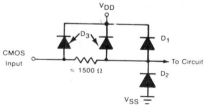

**FIGURE 1 — INPUT PROTECTION NETWORK**

12. The following steps should be observed during board-cleaning operations:
    a. Vapor degreasers and baskets must be grounded to an earth ground.
    b. Brush or spray cleaning should not be used.
    c. Assemblies should be placed into the vapor degreaser immediately upon removal from the antistatic container.
    d. Cleaned assemblies should be placed in antistatic containers immediately after removal from the cleaning basket.
    e. High velocity air movement or application of solvents and coatings should be employed only when assembled printed circuit boards are grounded and a static eliminator is directed at the board.
13. The use of static detection meters for production line surveillance is highly recommended.
14. Equipment specifications should alert users to the presence of CMOS devices and require familiarization with this specification prior to performing any kind of maintenance or replacement of devices or modules.

15. Do not insert or remove CMOS devices from test sockets with power applied. Check all power supplies to be used for testing devices to be certain there are no voltage transients present.
16. Double check test equipment setup for proper polarity of $V_{DD}$ and $V_{SS}$ before conducting parametric or functional testing.
17. Do not recycle shipping rails or trays. Repeated use causes deterioration of their antistatic coating.

### RECOMMENDED FOR READING:

"Total Control of the Static in Your Business"

Available by writing to:
3M Company
Static Control Systems
P.O. Box 2963
Austin, Texas 78769-2963
Or by Calling:
1-800-328-1368

**FIGURE 2 — NETWORKS FOR MINIMIZING ESD AND REDUCING CMOS LATCH UP SUSCEPTIBILITY**

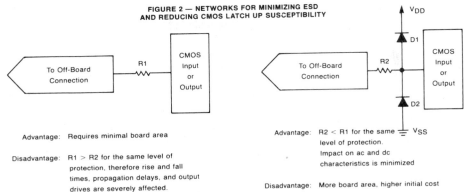

Advantage:   Requires minimal board area

Disadvantage:   R1 > R2 for the same level of protection, therefore rise and fall times, propagation delays, and output drives are severely affected.

Advantage:   R2 < R1 for the same level of protection. Impact on ac and dc characteristics is minimized

Disadvantage:   More board area, higher initial cost

Note: These networks are useful for protecting the following
A   digital inputs and outputs    C   3-state outputs
B   analog inputs and outputs    D   bidirectional (I/O) ports

**PROPAGATION DELAY AND RISE TIME vs. SERIES RESISTANCE**

$$R \approx \frac{t}{C \cdot k}$$

where:
R = the maximum allowable series resistance in ohms
t = the maximum tolerable propagation delay or rise time in seconds
C = the board capacitance plus the driven device's input capacitance in farads
k = 0.7 for propagation delay calculations
k = 2.3 for rise time calculations

**FIGURE 3 — TYPICAL MANUFACTURING WORK STATION**

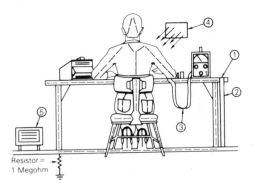

NOTES: 1. 1/16 inch conductive sheet stock covering bench top work area.
2. Ground strap.
3. Wrist strap in contact with skin.
4. Static neutralizer. (Ionized air blower directed at work.) Primarily for use in areas where direct grounding is impractical.
5. Room humidifier. Primarily for use in areas where the relative humidity is less than 45%. Caution: building heating and cooling systems usually dry the air causing the relative humidity inside of buildings to be less than outside humidity.

# EPROM Data for a Digital Function Generator

This appendix includes:

1. A complete set of EPROM data for the EPROM-based digital function generator described in Section 17.4.
2. A program written in Microsoft QuickBASIC to generate an EPROM record file (Intel format).
3. A copy of the generated record file.

## C.1

### EPROM Data

0₀ is the maximum negative voltage of a waveform, FF is maximum positive, and 80 is the zero-crossing point.

SINE

| Base Address | | | | | Byte Addresses | | | | | | | | | | | |
|---|---|---|---|---|---|---|---|---|---|---|---|---|---|---|---|---|
| | 0 | 1 | 2 | 3 | 4 | 5 | 6 | 7 | 8 | 9 | A | B | C | D | E | F |
| 0000 | 80 | 83 | 86 | 89 | 8C | 8F | 92 | 95 | 98 | 9C | 9F | A2 | A5 | A8 | AB | AE |
| 0010 | B0 | B3 | B6 | B9 | BC | BF | C1 | C4 | C7 | C9 | CC | CE | D1 | D3 | D5 | D8 |
| 0020 | DA | DC | DE | E0 | E2 | E4 | E6 | E8 | EA | EC | ED | EF | F0 | F2 | F3 | F5 |
| 0030 | F6 | F7 | F8 | F9 | FA | FB | FC | FC | FD | FE | FE | FF | FF | FF | FF | FF |
| 0040 | FF | FF | FF | FF | FF | FF | FE | FE | FD | FC | FC | FB | FA | F9 | F8 | F7 |
| 0050 | F6 | F5 | F3 | F2 | F0 | EF | ED | EC | EA | E8 | E6 | E4 | E2 | E0 | DE | DC |
| 0060 | DA | D8 | D5 | D3 | D1 | CE | CC | C9 | C7 | C4 | C1 | BF | BC | B9 | B6 | B3 |
| 0070 | B0 | AE | AB | A8 | A5 | A2 | 9F | 9C | 98 | 95 | 92 | 8F | 8C | 89 | 86 | 83 |
| 0080 | 7F | 7C | 79 | 76 | 73 | 70 | 6D | 6A | 67 | 63 | 60 | 5D | 5A | 57 | 54 | 51 |
| 0090 | 4F | 4C | 49 | 46 | 43 | 40 | 3E | 3B | 38 | 36 | 33 | 31 | 2E | 2C | 2A | 27 |
| 00A0 | 25 | 23 | 21 | 1F | 1D | 1B | 19 | 17 | 15 | 13 | 12 | 10 | 0F | 0D | 0C | 0A |
| 00B0 | 09 | 08 | 07 | 06 | 05 | 04 | 03 | 03 | 02 | 01 | 01 | 00 | 00 | 00 | 00 | 00 |
| 00C0 | 00 | 00 | 00 | 00 | 00 | 00 | 01 | 01 | 02 | 03 | 03 | 04 | 05 | 06 | 07 | 08 |
| 00D0 | 09 | 0A | 0C | 0D | 0F | 10 | 12 | 13 | 15 | 17 | 19 | 1B | 1D | 1F | 21 | 23 |
| 00E0 | 25 | 27 | 2A | 2C | 2E | 31 | 33 | 36 | 38 | 3B | 3E | 40 | 43 | 46 | 49 | 4C |
| 00F0 | 4F | 51 | 54 | 57 | 5A | 5D | 60 | 63 | 67 | 6A | 6D | 70 | 73 | 76 | 79 | 7C |

## SQUARE

| Base Address | Byte Addresses | | | | | | | | | | | | | | | |
|---|---|---|---|---|---|---|---|---|---|---|---|---|---|---|---|---|
| | 0 | 1 | 2 | 3 | 4 | 5 | 6 | 7 | 8 | 9 | A | B | C | D | E | F |
| 0100 | FF | FF | FF | FF | FF | FF | FF | FF | FF | FF | FF | FF | FF | FF | FF | FF |
| 0110 | FF | FF | FF | FF | FF | FF | FF | FF | FF | FF | FF | FF | FF | FF | FF | FF |
| 0120 | FF | FF | FF | FF | FF | FF | FF | FF | FF | FF | FF | FF | FF | FF | FF | FF |
| 0130 | FF | FF | FF | FF | FF | FF | FF | FF | FF | FF | FF | FF | FF | FF | FF | FF |
| 0140 | FF | FF | FF | FF | FF | FF | FF | FF | FF | FF | FF | FF | FF | FF | FF | FF |
| 0150 | FF | FF | FF | FF | FF | FF | FF | FF | FF | FF | FF | FF | FF | FF | FF | FF |
| 0160 | FF | FF | FF | FF | FF | FF | FF | FF | FF | FF | FF | FF | FF | FF | FF | FF |
| 0170 | FF | FF | FF | FF | FF | FF | FF | FF | FF | FF | FF | FF | FF | FF | FF | FF |
| 0180 | 00 | 00 | 00 | 00 | 00 | 00 | 00 | 00 | 00 | 00 | 00 | 00 | 00 | 00 | 00 | 00 |
| 0190 | 00 | 00 | 00 | 00 | 00 | 00 | 00 | 00 | 00 | 00 | 00 | 00 | 00 | 00 | 00 | 00 |
| 01A0 | 00 | 00 | 00 | 00 | 00 | 00 | 00 | 00 | 00 | 00 | 00 | 00 | 00 | 00 | 00 | 00 |
| 01B0 | 00 | 00 | 00 | 00 | 00 | 00 | 00 | 00 | 00 | 00 | 00 | 00 | 00 | 00 | 00 | 00 |
| 01C0 | 00 | 00 | 00 | 00 | 00 | 00 | 00 | 00 | 00 | 00 | 00 | 00 | 00 | 00 | 00 | 00 |
| 01D0 | 00 | 00 | 00 | 00 | 00 | 00 | 00 | 00 | 00 | 00 | 00 | 00 | 00 | 00 | 00 | 00 |
| 01E0 | 00 | 00 | 00 | 00 | 00 | 00 | 00 | 00 | 00 | 00 | 00 | 00 | 00 | 00 | 00 | 00 |
| 01F0 | 00 | 00 | 00 | 00 | 00 | 00 | 00 | 00 | 00 | 00 | 00 | 00 | 00 | 00 | 00 | 00 |

## TRIANGLE

| Base Address | Byte Addresses | | | | | | | | | | | | | | | |
|---|---|---|---|---|---|---|---|---|---|---|---|---|---|---|---|---|
| | 0 | 1 | 2 | 3 | 4 | 5 | 6 | 7 | 8 | 9 | A | B | C | D | E | F |
| 0200 | 80 | 82 | 84 | 86 | 88 | 8A | 8C | 8E | 90 | 92 | 94 | 96 | 98 | 9A | 9C | 9E |
| 0210 | A0 | A2 | A4 | A6 | A8 | AA | AC | AE | B0 | B2 | B4 | B6 | B8 | BA | BC | BE |
| 0220 | C0 | C2 | C4 | C6 | C8 | CA | CC | CE | D0 | D2 | D4 | D6 | D8 | DA | DC | DE |
| 0230 | E0 | E2 | E4 | E6 | E8 | EA | EC | EE | F0 | F2 | F4 | F6 | F8 | FA | FC | FE |
| 0240 | FE | FC | FA | F8 | F6 | F4 | F2 | F0 | EE | EC | EA | E8 | E6 | E4 | E2 | E0 |
| 0250 | DE | DC | DA | D8 | D6 | D4 | D2 | D0 | CE | CC | CA | C8 | C6 | C4 | C2 | C0 |
| 0260 | BE | BC | BA | B8 | B6 | B4 | B2 | B0 | AE | AC | AA | A8 | A6 | A4 | A2 | A0 |
| 0270 | 9E | 9C | 9A | 98 | 96 | 94 | 92 | 90 | 8E | 8C | 8A | 88 | 86 | 84 | 82 | 80 |
| 0280 | 7E | 7C | 7A | 78 | 76 | 74 | 72 | 70 | 6E | 6C | 6A | 68 | 66 | 64 | 62 | 60 |
| 0290 | 5E | 5C | 5A | 58 | 56 | 54 | 52 | 50 | 4E | 4C | 4A | 48 | 46 | 44 | 42 | 40 |
| 02A0 | 3E | 3C | 3A | 38 | 36 | 34 | 32 | 30 | 2E | 2C | 2A | 28 | 26 | 24 | 22 | 20 |
| 02B0 | 1E | 1C | 1A | 18 | 16 | 14 | 12 | 10 | 0E | 0C | 0A | 08 | 06 | 04 | 02 | 00 |
| 02C0 | 02 | 04 | 06 | 08 | 0A | 0C | 0E | 10 | 12 | 14 | 16 | 18 | 1A | 1C | 1E | 20 |
| 02D0 | 22 | 24 | 26 | 28 | 2A | 2C | 2E | 30 | 32 | 34 | 36 | 38 | 3A | 3C | 3E | 40 |
| 02E0 | 42 | 44 | 46 | 48 | 4A | 4C | 4E | 50 | 52 | 54 | 56 | 58 | 5A | 5C | 5E | 60 |
| 02F0 | 62 | 64 | 66 | 68 | 6A | 6C | 6E | 70 | 72 | 74 | 76 | 78 | 7A | 7C | 7E | 80 |

## SAWTOOTH

| Base Address | Byte Addresses | | | | | | | | | | | | | | | |
|---|---|---|---|---|---|---|---|---|---|---|---|---|---|---|---|---|
| | 0 | 1 | 2 | 3 | 4 | 5 | 6 | 7 | 8 | 9 | A | B | C | D | E | F |
| 0300 | 00 | 01 | 02 | 03 | 04 | 05 | 06 | 07 | 08 | 09 | 0A | 0B | 0C | 0D | 0E | 0F |
| 0310 | 10 | 11 | 12 | 13 | 14 | 15 | 16 | 17 | 18 | 19 | 1A | 1B | 1C | 1D | 1E | 1F |
| 0320 | 20 | 21 | 22 | 23 | 24 | 25 | 26 | 27 | 28 | 29 | 2A | 2B | 2C | 2D | 2E | 2F |
| 0330 | 30 | 31 | 32 | 33 | 34 | 35 | 36 | 37 | 38 | 39 | 3A | 3B | 3C | 3D | 3E | 3F |
| 0340 | 40 | 41 | 42 | 43 | 44 | 45 | 46 | 47 | 48 | 49 | 4A | 4B | 4C | 4D | 4E | 4F |
| 0350 | 50 | 51 | 52 | 53 | 54 | 55 | 56 | 57 | 58 | 59 | 5A | 5B | 5C | 5D | 5E | 5F |
| 0360 | 60 | 61 | 62 | 63 | 64 | 65 | 66 | 67 | 68 | 69 | 6A | 6B | 6C | 6D | 6E | 6F |
| 0370 | 70 | 71 | 72 | 73 | 74 | 75 | 76 | 77 | 78 | 79 | 7A | 7B | 7C | 7D | 7E | 7F |
| 0380 | 80 | 81 | 82 | 83 | 84 | 85 | 86 | 87 | 88 | 89 | 8A | 8B | 8C | 8D | 8E | 8F |
| 0390 | 90 | 91 | 92 | 93 | 94 | 95 | 96 | 97 | 98 | 99 | 9A | 9B | 9C | 9D | 9E | 9F |
| 03A0 | A0 | A1 | A2 | A3 | A4 | A5 | A6 | A7 | A8 | A9 | AA | AB | AC | AD | AE | AF |
| 03B0 | B0 | B1 | B2 | B3 | B4 | B5 | B6 | B7 | B8 | B9 | BA | BB | BC | BD | BE | BF |
| 03C0 | C0 | C1 | C2 | C3 | C4 | C5 | C6 | C7 | C8 | C9 | CA | CB | CC | CD | CE | CF |
| 03D0 | D0 | D1 | D2 | D3 | D4 | D5 | D6 | D7 | D8 | D9 | DA | DB | DC | DD | DE | DF |
| 03E0 | E0 | E1 | E2 | E3 | E4 | E5 | E6 | E7 | E8 | E9 | EA | EB | EC | ED | EE | EF |
| 03F0 | F0 | F1 | F2 | F3 | F4 | F5 | F6 | F7 | F8 | F9 | FA | FB | FC | FD | FE | FF |

```
PERIODIC EXPONENTIAL

Base                              Byte Addresses
Address    0   1   2   3   4   5   6   7   8   9   A   B   C   D   E   F

0400      00  09  13  1C  24  2D  35  3D  44  4B  52  59  5F  65  6B  71
0410      76  7B  80  85  8A  8E  93  97  9B  9E  A2  A6  A9  AC  B0  B3
0420      B5  B8  BB  BE  C0  C2  C5  C7  C9  CB  CD  CF  D1  D3  D4  D6
0430      D7  D9  DA  DC  DD  DE  E0  E1  E2  E3  E4  E5  E6  E7  E8  E9
0440      EA  EA  EB  EC  ED  ED  EE  EF  EF  F0  F0  F1  F1  F2  F2  F3
0450      F3  F4  F4  F5  F5  F5  F6  F6  F6  F7  F7  F7  F8  F8  F8
0460      F9  F9  F9  F9  F9  FA  FA  FA  FA  FA  FA  FB  FB  FB  FB  FB
0470      FB  FB  FC  FC  FC  FC  FC  FC  FC  FC  FC  FC  FC  FD  FD  FF
0480      FF  F5  EB  E2  DA  D1  C9  C1  BA  B3  AC  A5  9F  99  93  8D
0490      88  83  7E  79  74  70  6B  67  63  60  5C  58  55  52  4E  4B
04A0      49  46  43  40  3E  3C  39  37  35  33  31  2F  2D  2B  2A  28
04B0      27  25  24  22  21  20  1E  1D  1C  1B  1A  19  18  17  16  15
04C0      14  14  13  12  11  11  10  0F  0F  0E  0E  0D  0D  0C  0C  0B
04D0      0B  0A  0A  09  09  09  08  08  08  07  07  07  07  06  06  06
04E0      05  05  05  05  05  04  04  04  04  04  04  03  03  03  03  03
04F0      03  03  02  02  02  02  02  02  02  02  02  02  02  01  01  01


PULSE WAVEFORM WITH PSEUDORANDOM PERIOD

Base                              Byte Addresses
Address    0   1   2   3   4   5   6   7   8   9   A   B   C   D   E   F

0500      FF  FF  FF  FF  FF  FF  FF  FF  FF  FF  FF  FF  FF  FF  FF  FF
0510      FF  FF  FF  FF  FF  80  80  80  80  80  80  80  80  80  80  80
0520      80  80  80  80  80  80  80  80  80  80  80  80  80  80  80  80
0530      80  80  80  80  80  FF  FF  FF  FF  FF  FF  FF  FF  FF  FF  FF
0540      FF  FF  FF  FF  FF  FF  FF  FF  FF  FF  FF  FF  FF  FF  FF  FF
0550      FF  FF  FF  FF  FF  FF  FF  FF  FF  FF  FF  FF  FF  FF  FF  FF
0560      FF  FF  FF  FF  80  80  80  80  80  80  80  80  80  80  80  80
0570      80  80  80  80  80  80  80  80  80  80  80  80  80  80  80  80
0580      80  80  80  80  80  80  80  80  80  80  80  80  80  80  80  80
0590      80  80  80  80  80  80  80  80  80  80  80  80  80  80  80  80
05A0      80  80  80  80  80  80  80  80  80  80  80  80  80  80  FF  FF
05B0      FF  FF  FF  FF  FF  FF  FF  FF  FF  FF  FF  FF  FF  FF  FF  FF
05C0      FF  FF  FF  FF  FF  FF  FF  FF  FF  FF  FF  FF  FF  FF  FF  FF
05D0      FF  FF  FF  FF  FF  FF  FF  FF  FF  FF  FF  FF  FF  FF  FF  FF
05E0      FF  FF  FF  FF  FF  FF  FF  FF  FF  FF  FF  FF  FF  FF  FF  FF
05F0      FF  FF  FF  FF  FF  FF  FF  80  80  80  80  80  80  80  80  80


BIPOLAR WAVEFORM WITH PSEUDORANDOM PERIOD

Base                              Byte Addresses
Address    0   1   2   3   4   5   6   7   8   9   A   B   C   D   E   F

0600      FF  FF  FF  FF  FF  FF  FF  FF  FF  FF  FF  FF  FF  FF  FF  FF
0610      FF  FF  FF  FF  FF  FF  FF  FF  FF  FF  FF  FF  FF  FF  FF  FF
0620      FF  FF  FF  FF  FF  FF  00  00  00  00  00  00  00  00  00  00
0630      00  00  00  00  00  00  00  00  00  00  00  00  00  00  00  00
0640      00  00  00  00  00  00  00  00  00  00  00  00  00  00  00  00
0650      00  00  00  00  00  00  00  00  00  00  00  00  00  00  00  00
0660      00  00  00  00  00  00  00  00  00  00  00  00  00  FF  FF  FF
0670      FF  FF  FF  FF  FF  FF  FF  FF  FF  FF  FF  FF  FF  FF  FF  FF
0680      FF  FF  FF  FF  00  00  00  00  00  00  00  00  00  00  00  00
0690      00  00  00  00  00  00  00  00  00  00  00  00  00  00  00  00
06A0      00  00  00  00  00  00  00  00  00  00  00  00  00  00  00  00
06B0      00  00  FF  FF  FF  FF  FF  FF  FF  FF  FF  FF  FF  FF  FF  FF
06C0      FF  FF  FF  FF  FF  FF  FF  FF  FF  FF  FF  FF  FF  FF  FF  FF
06D0      FF  FF  FF  FF  FF  FF  FF  FF  FF  FF  FF  FF  FF  FF  FF  FF
06E0      FF  FF  FF  FF  FF  FF  FF  FF  FF  FF  FF  FF  FF  FF  FF  FF
06F0      FF  FF  FF  00  00  00  00  00  00  00  00  00  00  00  00  00
```

## C.2

# QuickBASIC Program

This program is also included on the accompanying diskette in file \EPROM\ FUNCTION.BAS.

```
DECLARE SUB PseudoBipolar (Addr%, Ampl%)
DECLARE SUB Exponential (Addr%, Ampl%)
DECLARE SUB PseudoRandom (Addr%, Ampl%)
DEFINT A-Z
DECLARE SUB Sawtooth (Addr, Ampl)
DECLARE SUB triangle (Addr, Ampl)
DECLARE SUB Square (Addr, Ampl)
DECLARE FUNCTION AddrByte (Addr)
DECLARE FUNCTION Chksum (sum)
DECLARE SUB sine (Addr, Ampl)
DECLARE FUNCTION HexString$ (value)
'     This program is to create a hex file in a specific EPROM record format.
'The EPROM is addressed in blocks of 256 bytes (8 address lines) by an 8 bit
'counter to create a digital image of one of several waveform outputs.
'     When the EPROM data is run through a D/A Converter, it creates an analog
'waveform running at the frequency of the counter divided by 256.
'     The waveforms are sine, square, triangle, sawtooth, periodic exponential,
'an aperiodic digital pulse waveform with a pseudorandom period, and a bipolar
'pseudorandom pulse waveform.
'     The record format is as follows. (Spaces are inserted only for clarity.
'The actual record must have NO spaces.)
'
'     :10 0080 00 AF5F67F0602703E0322CFA92007780C3 61
'
'     (:Record Length = 10hex = 16dec)
'     (Address = 0080hex; location in EPROM of first data byte in record)
'     (Record type = 00 = data)
'     (16 data bytes = 32 hex digits)
'     (Checksum; Record Length + Address High byte + Address Low byte
'     + Record type + data bytes + Checksum = 00, after discarding carry)
'
'     An END record is also required, having a similar format, except that
'Record Type = 01 = END and there are no data bytes. Checksum is still
'required.
'     eg. :000FFF01F1
'
OPEN "a:function.hex" FOR OUTPUT AS #1
Fcn = 1
Addr = 0
DO UNTIL Fcn = 8                    'Create hex records for 7
functions.
     FOR Linenum = 1 TO 16     'Each function has 16 lines of data.
          IF Addr < 16 THEN
               Record$ = ":10" + "000" + HEX$(Addr)
          ELSE
               IF Addr < 256 THEN
                    Record$ = ":10" + "00" + HEX$(Addr)
               ELSE
                    Record$ = ":10" + "0" + HEX$(Addr)
               END IF
          END IF
          Record$ = Record$ + "00"
          sum = 16 + AddrByte(Addr)    'Accumulate sum for calculating checksum.
          FOR Byte = 1 TO 16  'Each line has 16 data bytes.
```

```
'
'                         Calculate the waveform amplitude for a particular point (0 to 255)
'                         in the waveform.
                              IF Fcn = 1 THEN
                                  CALL sine(Addr MOD 256, Ampl)
                              ELSEIF Fcn = 2 THEN
                                  CALL Square(Addr MOD 256, Ampl)
                              ELSEIF Fcn = 3 THEN
                                  CALL triangle(Addr MOD 256, Ampl)
                              ELSEIF Fcn = 4 THEN
                                  CALL Sawtooth(Addr MOD 256, Ampl)
                              ELSEIF Fcn = 5 THEN
                                  CALL Exponential(Addr MOD 256, Ampl)
                              ELSEIF Fcn = 6 THEN
                                  CALL PseudoRandom(Addr MOD 256, Ampl)
                              ELSE
                                  CALL PseudoBipolar(Addr MOD 256, Ampl)
                              END IF
'
'
'
'                         Append the waveform amplitude data to the record
'                         and update the checksum accumulator.

                              Record$ = Record$ + HexString$(Ampl)
                              sum = sum + Ampl
                              Addr = Addr + 1
                          NEXT Byte
'
'                     Calculate and append the checksum to the record.
'
                          Record$ = Record$ + HexString$(Chksum(sum))
                          PRINT #1, Record$
                      NEXT Linenum
                      Fcn = Fcn + 1
LOOP
PRINT #1, ":000FFF01F1"
'LPRINT CHR$(12)      Form feed from optional printout. (Remove comment.)
CLOSE
END


FUNCTION AddrByte (Addr)
      IF Addr < 256 THEN
            AddrByte = Addr
      ELSE
            AddrByte = (Addr \ 256) + (Addr MOD 256)
      END IF
END FUNCTION

FUNCTION Chksum (sum)
      IntRem = sum MOD 256
      IF IntRem = 0 THEN
            Chksum = 0
      ELSE
            Chksum = 256 - IntRem
      END IF
END FUNCTION
```

```
SUB Exponential (Addr, Ampl)
    IF Addr < 127 THEN
        Ampl = INT(255 * (1 - EXP(-CSNG(Addr * 2.5 / 64))))
    ELSE
        Ampl = INT(255 * EXP(-CSNG((Addr - 128) * 2.5 / 64)))
    END IF
END SUB

FUNCTION HexString$ (value)
    IF value < 16 THEN
        HexString$ = "0" + HEX$(value)
    ELSEIF value > 255 THEN
        HexString$ = "FF"
    ELSE
        HexString$ = HEX$(value)
    END IF
END FUNCTION

SUB PseudoBipolar (Addr, Ampl) STATIC
'
'    This subroutine generates the digital codes for a bipolar
'pseudorandom digital pulse train.  The idea is to generate pulse max
'and min times of random widths varying between 15 and 75 clock pulses.
'
    IF Addr = 0 THEN              'First pass through the loop?
        AmplFlag = 0             'If so, set timer and amplitude flags.
        TimerFlag = -1
    END IF
    IF TimerFlag THEN
        RANDOMIZE (TIMER)
        Elapsed = INT((60 * RND) + 15)   'Set an elapsed time counter.
        TimerFlag = 0
    END IF
    Ampl = 255 + (255 * AmplFlag)      'Ampl is max or min until timer
    IF Elapsed <= 0 THEN               'runs out.
        TimerFlag = -1
        AmplFlag = NOT AmplFlag        'Invert the output amplitude for the
    END IF                             'next period.
    Elapsed = Elapsed - 1

END SUB

SUB PseudoRandom (Addr, Ampl) STATIC
'
'    This subroutine generates the digital codes for a positive-going
'pseudorandom digital pulse train.  The idea is to generate pulse HIGH
'and LOW times of random widths varying between 15 and 75 clock pulses.
'
    IF Addr = 0 THEN              'First pass through the loop?
        AmplFlag = 0             'If so, set timer and amplitude flags.
        TimerFlag = -1
    END IF
    IF TimerFlag THEN
        RANDOMIZE (TIMER)
        Elapsed = INT((60 * RND) + 15)   'Set an elapsed time counter.
        TimerFlag = 0
    END IF
    Ampl = 255 + (127 * AmplFlag)      'Ampl is HIGH or LOW until timer
    IF Elapsed <= 0 THEN               'runs out.
        TimerFlag = -1
        AmplFlag = NOT AmplFlag        'Invert the output amplitude for the
    END IF                             'next period.
    Elapsed = Elapsed - 1
END SUB
```

```
SUB Sawtooth (Addr, Ampl)
    Ampl = Addr
END SUB

SUB sine (Addr, Ampl)
    Pi# = 3.141592654#
    angle# = (CSNG(Addr) * 2 * Pi#) / 256
    Ampl = INT((SIN(angle#) * 128) + 128)
    IF Ampl > 255 THEN Ampl = 255
END SUB

SUB Square (Addr, Ampl)
    IF Addr < 128 THEN
        Ampl = 255
    ELSE
        Ampl = 0
    END IF
END SUB

SUB triangle (Addr, Ampl)
    IF Addr < 64 THEN
        Ampl = 128 + (2 * Addr)
    ELSEIF Addr >= 64 AND Addr < 192 THEN
        Ampl = 256 - 2 * (Addr - 63)
    ELSE
        Ampl = 2 * (Addr - 191)
    END IF
END SUB
```

## C.3

## Resultant Record File

3) Resultant Record File

```
:10000000808386898C8F9295989C9FA2A5A8ABAE81
:10001000B0B3B6B9BCBFC1C4C7C9CCCED1D3D5D893
:10002000DADCDEE0E2E4E6E8EAECEDEFF0F2F3F54C
:100030000F6F7F8F9FAFBFCFCFDFEFEFFFFFFFFF01
:10004000FFFFFFFFFFFFFEFEFDFCFCFBFAF9F8F7E8
:10005000F6F5F3F2F0EFEDECEAE8E6E4E2E0DEDC00
:10006000DAD8D5D3D1CECCC9C7C4C1BFBCB9B6B319
:10007000B0AEABA8A5A29F9C9895928F8C898683E1
:100080007F7C797673706D6A6763605D5A575451EF
:100090004F4C494643403E3B383633312E2C2A27BD
:1000A0002523211F1D1B1917151312100F0D0C0AE4
:1000B00009080706050403030201010000000000F
:1000C000000000000000000010102030304050607080B
:1000D000090A0C0D0F1012131517191B1D1F2123D0
:1000E00025272A2C2E313336383B3E404346494C97
:1000F0004F5154575A5D6063676A6D707376797CAF
:10010000FFFFFFFFFFFFFFFFFFFFFFFFFFFFFFFF
:10011000FFFFFFFFFFFFFFFFFFFFFFFFFFFFFFFFEF
:10012000FFFFFFFFFFFFFFFFFFFFFFFFFFFFFFFFDF
:10013000FFFFFFFFFFFFFFFFFFFFFFFFFFFFFFFFCF
:10014000FFFFFFFFFFFFFFFFFFFFFFFFFFFFFFFFBF
:10015000FFFFFFFFFFFFFFFFFFFFFFFFFFFFFFFFAF
:10016000FFFFFFFFFFFFFFFFFFFFFFFFFFFFFFFF9F
:10017000FFFFFFFFFFFFFFFFFFFFFFFFFFFFFFFF8F
:10018000000000000000000000000000000000006F
:10019000000000000000000000000000000000005F
:1001A000000000000000000000000000000000004F
:1001B000000000000000000000000000000000003F
:1001C000000000000000000000000000000000002F
```

```
:1001D00000000000000000000000000000000000001F
:1001E00000000000000000000000000000000000000F
:1001F0000000000000000000000000000000000000FF
:100200008082848688A8C8E90929496989A9C9EFE
:10021000A0A2A4A6A8AAACAEB0B2B4B6B8BABCBEEE
:10022000C0C2C4C6C8CACCCED0D2D4D6D8DADCDEDE
:10023000E0E2E4E6E8EAECEEF0F2F4F6F8FAFCFECE
:10024000FEFCFAF8F6F4F2F0EEECEAE8E6E4E2E0BE
:10025000DEDCDAD8D6D4D2D0CECCCAC8C6C4C2C0AE
:10026000BEBCBAB8B6B4B2B0AEACAAA8A6A4A2A09E
:100270009E9C9A98969492908E8C8A88868482808E
:100280007E7C7A78767472706E6C6A68666462607E
:100290005E5C5A58565452504E4C4A48464442406E
:1002A0003E3C3A38363432302E2C2A28262422205E
:1002B0001E1C1A18161412100E0C0A08060402004E
:1002C00000020406080A0C0E10121416181A1C1E201E
:1002D0002224262828A2C2E30323436383A3C3E400E
:1002E00042444648 4A4C4E50525456585A5C5E60FE
:1002F00062646668 6A6C6E70727476787A7C7E80EE
:1003000000010203040506070809 0A0B0C0D0E0F75
:10031000101112131415161718191A1B1C1D1E1F65
:10032000202122232425262728292A2B2C2D2E2F55
:10033000303132333435363738393A3B3C3D3E3F45
:10034000404142434445464748494A4B4C4D4E4F35
:10035000505152535455565758595A5B5C5D5E5F25
:10036000606162636465666768696A6B6C6D6E6F15
:10037000707172737475767778797A7B7C7D7E7F05
:10038000808182838485868788898A8B8C8D8E8FF5
:10039000909192939495969798999A9B9C9D9E9FE5
:1003A000A0A1A2A3A4A5A6A7A8A9AAABACADAEAFD5
:1003B000B0B1B2B3B4B5B6B7B8B9BABBBCBDBEBFC5
:1003C000C0C1C2C3C4C5C6C7C8C9CACBCCCDCECFB5
:1003D000D0D1D2D3D4D5D6D7D8D9DADBDCDDDEDFA5
:1003E000E0E1E2E3E4E5E6E7E8E9EAEBECEDEEEF95
:1003F000F0F1F2F3F4F5F6F7F8F9FAFBFCFDFEFF85
:100400000009131C242D353D444B52595F656B7117
:10041000767B80858A8E93979B9EA2A6A9ACB0B36B
:10042000B5B8BBBEC0C2C5C7C9CBCDCFD1D3D4D65A
:10043000D7D9DADCDDDEE0E1E2E3E4E5E6E7E8E9AE
:10044000EAEAEBECEDEDEEEFEFF0F0F1F1F2F2F3C2
:10045000F3F4F4F5F5F5F6F6F6F7F7F7F7F8F8F83C
:10046000F9F9F9F9F9FAFAFAFAFAFAFBFBFBFBFBEC
:10047000FBFBFCFCFCFCFCFCFCFCFCFCFCFDFDFFAF
:10048000FFF5EBE2DAD1C9C1BAB3ACA59F99938D60
:100490008837E7974706B6763605C5855524E4BED
:1004A00049464340 3E3C39373533312F2D2B2A28DE
:1004B00027252422212 01E1D1C1B1A19181716156A
:1004C000141413121111100F0F0E0E0D0D0C0C0B36
:1004D0000B0A0A090909080808070707070606069C
:1004E000050505050504040404040403030303 03CC
:1004F0000303020202020202020202020201010 1DD
:10050000FFFFFFFFFFFFFFFFFFFFFFFFFFFFFFFFFB
:10051000FFFFFFFFFFFFFFFFFFFFFFFFFFFFFFFFEB
:10052000FFFFFFFFFFFFFFFFFFFFFFFFFFFFFFFFDB
:10053000FFFFFFFFFFFFFFFFFFFFFFFF80808080C7
:1005400080808080808080808080808080808080AB
:10055000808080808080808080808080808080809B
:10056000808080808080808080808080808080808B
:10057000808080FFFFFFFFFFFFFFFFFFFFFFFFFF08
:10058000FFFFFFFFFFFFFFFFFFFFFFFFFFFFFFFF7B
:10059000FFFFFFFFFFFFFFFFFFFFFFFF8080808067
:1005A000808080808080808080808080808080804B
:1005B000808080808080808080808080808080803B
:1005C000808080808080808080808080808080802B
:1005D00080808080808080808080808080FFFFFF9E
:1005E000FFFFFFFFFFFFFFFFFFFFFFFFFFFFFFFF1B
```

```
:1005F000FFFFFFFFFFFFFFFFFFFFFFFFFFFFFFFF0B
:10060000FFFFFFFFFFFFFFFFFFFFFFFFFFFFFFFFFA
:10061000FFFFFFFFFFFFFFFFFFFFF000000000000E4
:10062000000000000000000000000000000000000CA
:10063000000000000000000000000000000000000BA
:1006400000000000FFFFFFFFFFFFFFFFFFFFFFFFB6
:10065000FFFFFFFFFFFFFFFFFFFFFFFFFFFFFFFFAA
:10066000FFFFFFFFFFFFFFFFFFFFF0000000000094
:10067000000000000000000000000000FFFFFFFF7E
:10068000FFFFFFFFFFFFFFFFFFFFFFFFFFFFFFFF7A
:10069000FFFFFFFFFFFFFFFFFFFFFFFFFFFFFFFF6A
:1006A000FFFFFFFFFFFFFFFFFFFFFFFFFFFFFFFF5A
:1006B000FFFFFFFFFFFFFFFFFFFFFFFFFFFFFFFF4A
:1006C000FFFFFFFFFFFFF00000000000000000030
:1006D00000000000000000000000000000000001A
:1006E000000000000000000000000000FFFFFF0D
:1006F000FFFFFFFFFFFFFFFFFFFFFFFFFFFFFFFF0A
:000FFF01F1
```

# Pinouts of Common LSTTL and CMOS Logic Gate ICs

Figures D.1 and D.2 show pin configurations of the most frequently used TTL and CMOS logic gate integrated circuits. High-Speed CMOS pinouts are usually the same as the corresponding function in TTL. (For example, a Quad 2-input NAND in TTL is 74LS00. In High-Speed CMOS it is 74HC00. Both devices have the same pinout.) The exception to this general rule is where the function of the High-Speed CMOS device is not available in TTL, in which case it has the pinout of the corresponding standard CMOS function (e.g., 74HC4002 Dual 4-input NOR gate).

**Figure D.1**
**TTL Devices**

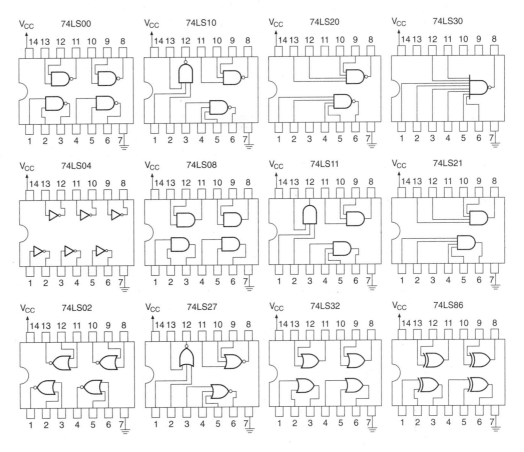

**Figure D.2**

**CMOS Devices**

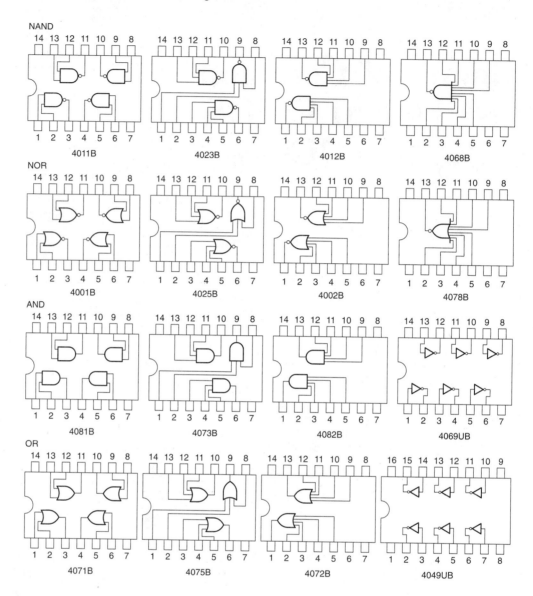

# Answers to Odd Problems

## Chapter 1

**1.1** The analog voltmeter shows the exact value of the measured voltage, although it is very hard or impossible to read beyond the second significant figure. The digital voltmeter shows the exact value up to the third significant figure, but does not display any value at all beyond that.

**1.3** Analog quantities: **a.** Water temperature at the beach; **b.** weight of a bucket of sand; **e.** height of a wave;

Digital quantities: **c.** grains of sand in a bucket; **d.** waves hitting the beach in one hour; **f.** people in a square mile.

Generally, any quantity that can be expressed as "the number of. . ." is digital.

**1.5** **a.** 1101; **b.** 0101; **c.** 1010; **d.** 0001; **e.** 1000

**1.7** 10000, 10001, 10010, 10011, 10100, 10101, 10110, 10111, 11000, 11001, 11010, 11011, 11100, 11101, 11110, 11111

**1.9** **a.** $t_h = 0.5$ μs, $t_l = 2.0$ μs, $T = 2.5$ μs, $f = 400$ kHz, % DC = 20%

**b.** $t_h = 1.0$ μs, $t_l = 4.0$ μs, $T = 5.0$ μs, $f = 200$ kHz, % DC = 20%

**c.** $t_h = 1.5$ μs, $t_l = 6.0$ μs, $T = 7.5$ μs, $f = 133$ kHz, % DC = 20%

These waveforms have different frequencies, but the same duty cycle.

**1.11** See Figure ANS1-11.

**1.13** $t_w = 7.5$ ms
$t_r = 2$ ms
$t_f = 1$ ms

## Chapter 2

**2.1** See Figure ANS2-1. Sixteen different input combinations are possible with this gate.

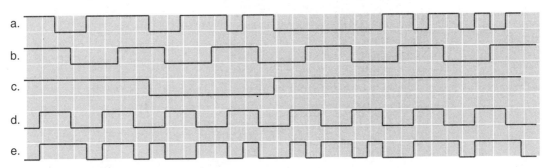

**Figure ANS 1.11**

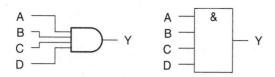

**Distinctive Shape**      **Rectangular Outline**

**Figure ANS 2.1**

**2.3**

| A | B | C | D | Y |
|---|---|---|---|---|
| 0 | 0 | 0 | 0 | 0 |
| 0 | 0 | 0 | 1 | 0 |
| 0 | 0 | 1 | 0 | 0 |
| 0 | 0 | 1 | 1 | 0 |
| 0 | 1 | 0 | 0 | 0 |
| 0 | 1 | 0 | 1 | 0 |
| 0 | 1 | 1 | 0 | 0 |
| 0 | 1 | 1 | 1 | 0 |
| 1 | 0 | 0 | 0 | 0 |
| 1 | 0 | 0 | 1 | 0 |
| 1 | 0 | 1 | 0 | 0 |
| 1 | 0 | 1 | 1 | 0 |
| 1 | 1 | 0 | 0 | 0 |
| 1 | 1 | 0 | 1 | 0 |
| 1 | 1 | 1 | 0 | 0 |
| 1 | 1 | 1 | 1 | 1 |

**2.5** **a.** A: Active HIGH
B: Active HIGH
Y: Active LOW

**b.** A: Active LOW
B: Active LOW
Y: Active LOW

**c.** A: Active LOW
B: Active LOW
C: Active LOW
Y: Active HIGH

**d.** A: Active HIGH
B: Active HIGH
C: Active LOW
D: Active HIGH
Y: Active LOW

**e.** A: Active LOW
Y: Active HIGH

**f.** A: Active LOW
B: Active HIGH
C: Active LOW
D: Active HIGH
Y: Active HIGH

**2.7** **a.**

| A | B | C | D | Y |
|---|---|---|---|---|
| 0 | 0 | 0 | 0 | 1 |
| 0 | 0 | 0 | 1 | 0 |
| 0 | 0 | 1 | 0 | 0 |
| 0 | 0 | 1 | 1 | 0 |
| 0 | 1 | 0 | 0 | 0 |
| 0 | 1 | 0 | 1 | 0 |
| 0 | 1 | 1 | 0 | 0 |
| 0 | 1 | 1 | 1 | 0 |
| 1 | 0 | 0 | 0 | 0 |
| 1 | 0 | 0 | 1 | 0 |
| 1 | 0 | 1 | 0 | 0 |
| 1 | 0 | 1 | 1 | 0 |
| 1 | 1 | 0 | 0 | 0 |
| 1 | 1 | 0 | 1 | 0 |
| 1 | 1 | 1 | 0 | 0 |
| 1 | 1 | 1 | 1 | 0 |

At least one input HIGH makes the output LOW.

**b.** $Y = \overline{A + B + C + D}$

**c.** See Figure ANS2-7.

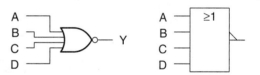

**Distinctive Shape**      **Rectangular Outline**

**Figure ANS 2.7**

**2.9** Y is LOW if A, B, C, D, and E are all HIGH. The truth table for this gate has 32 lines.

**2.11**

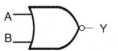

**2.13**

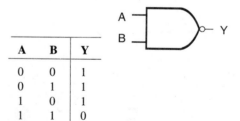

| A | B | Y |
|---|---|---|
| 0 | 0 | 1 |
| 0 | 1 | 1 |
| 1 | 0 | 1 |
| 1 | 1 | 0 |

**2.15**

| A | B | C | Y |
|---|---|---|---|
| 0 | 0 | 0 | 0 |
| 0 | 0 | 1 | 1 |
| 0 | 1 | 0 | 1 |
| 0 | 1 | 1 | 0 |
| 1 | 0 | 0 | 1 |
| 1 | 0 | 1 | 0 |
| 1 | 1 | 0 | 0 |
| 1 | 1 | 1 | 1 |

**2.17 a. and b.**
Figure 2.56a.

| A | B | C | Y |
|---|---|---|---|
| 0 | 0 | 0 | 1 |
| 0 | 0 | 1 | 0 * |
| 0 | 1 | 0 | 0 * |
| 0 | 1 | 1 | 0 * |
| 1 | 0 | 0 | 0 * |
| 1 | 0 | 1 | 0 * |
| 1 | 1 | 0 | 0 * |
| 1 | 1 | 1 | 0 * |

Figure 2.56b.

| A | B | Y |
|---|---|---|
| 0 | 0 | 1 * |
| 0 | 1 | 1 * |
| 1 | 0 | 0 |
| 1 | 1 | 1 * |

Figure 2.56c.

| A | B | C | Y |
|---|---|---|---|
| 0 | 0 | 0 | 1 * |
| 0 | 0 | 1 | 1 * |
| 0 | 1 | 0 | 1 * |
| 0 | 1 | 1 | 1 * |
| 1 | 0 | 0 | 1 * |
| 1 | 0 | 1 | 0 |
| 1 | 1 | 0 | 1 * |
| 1 | 1 | 1 | 1 * |

Figure 2.56d.

| A | B | Y |
|---|---|---|
| 0 | 0 | 1 |
| 0 | 1 | 0 * |
| 1 | 0 | 1 |
| 1 | 1 | 1 |

Figure 2.56e.

| A | B | Y |
|---|---|---|
| 0 | 0 | 0 * |
| 0 | 1 | 1 |
| 1 | 0 | 1 |
| 1 | 1 | 1 |

Figure 2.56f.

| A | B | C | Y |
|---|---|---|---|
| 0 | 0 | 0 | 1 * |
| 0 | 0 | 1 | 0 |
| 0 | 1 | 0 | 0 |
| 0 | 1 | 1 | 0 |
| 1 | 0 | 0 | 0 |
| 1 | 0 | 1 | 0 |
| 1 | 1 | 0 | 0 |
| 1 | 1 | 1 | 0 |

**c.** See Figure ANS2-17c.
**d.** Figure 2.56a.

| A | B | C | Y |
|---|---|---|---|
| 0 | 0 | 0 | 1 * |
| 0 | 0 | 1 | 0 |
| 0 | 1 | 0 | 0 |
| 0 | 1 | 1 | 0 |
| 1 | 0 | 0 | 0 |
| 1 | 0 | 1 | 0 |
| 1 | 1 | 0 | 0 |
| 1 | 1 | 1 | 0 |

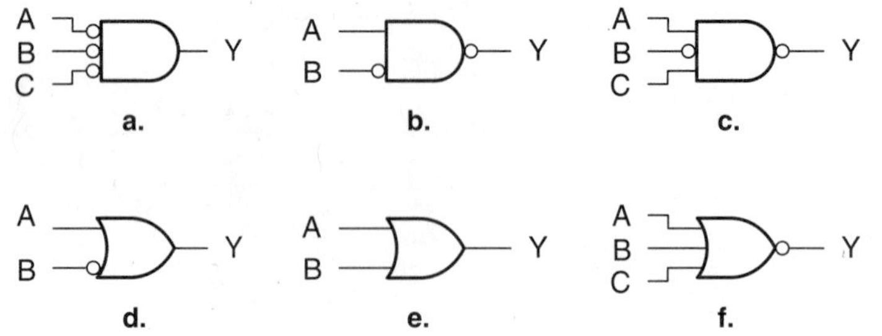

**Figure ANS 2.17c**

Figure 2.56b.

| A | B | Y |
|---|---|---|
| 0 | 0 | 1 |
| 0 | 1 | 1 |
| 1 | 0 | 0 * |
| 1 | 1 | 1 |

Figure 2.56c.

| A | B | C | Y |
|---|---|---|---|
| 0 | 0 | 0 | 1 |
| 0 | 0 | 1 | 1 |
| 0 | 1 | 0 | 1 |
| 0 | 1 | 1 | 1 |
| 1 | 0 | 0 | 1 |
| 1 | 0 | 1 | 0 * |
| 1 | 1 | 0 | 1 |
| 1 | 1 | 1 | 1 |

Figure 2.56d.

| A | B | Y |
|---|---|---|
| 0 | 0 | 1 * |
| 0 | 1 | 0 |
| 1 | 0 | 1 * |
| 1 | 1 | 1 * |

Figure 2.56e.

| A | B | Y |
|---|---|---|
| 0 | 0 | 0 |
| 0 | 1 | 1 * |
| 1 | 0 | 1 * |
| 1 | 1 | 1 * |

Figure 2.56f.

| A | B | C | Y |
|---|---|---|---|
| 0 | 0 | 0 | 1 |
| 0 | 0 | 1 | 0 * |
| 0 | 1 | 0 | 0 * |
| 0 | 1 | 1 | 0 * |
| 1 | 0 | 0 | 0 * |
| 1 | 0 | 1 | 0 * |
| 1 | 1 | 0 | 0 * |
| 1 | 1 | 1 | 0 * |

**2.19**

| A | B | Y |
|---|---|---|
| 0 | 0 | 0 |
| 0 | 1 | 0 |
| 1 | 0 | 0 |
| 1 | 1 | 1 |

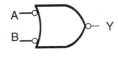

Either input LOW makes the output LOW.

**2.21** See Figure ANS2-21.

Logic circuit 1: Any input LOW makes the output HIGH.

Logic circuit 2: Both inputs HIGH make the output LOW.

The truth table for Logic Circuit 1 has 64 lines.

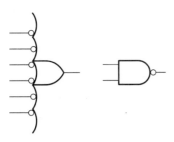

**Logic Circuit 1     Logic Circuit 2**

**Figure ANS 2.21**

The DeMorgan equivalent form shows the function best because the activating inputs go LOW when active.

**2.23** See Figure ANS2-23. The waveform is the complement of the one that would appear at the output of an Exclusive OR gate.

**2.25** See Figure ANS2-25.

**2.27** See Figure ANS2-27.

**2.29** The float switch must produce a logic LOW to make the lamp flash. A HIGH at the float switch would make the OR output always HIGH, thereby inhibiting the logic gate.

**2.31** Transistor-Transistor Logic (TTL) and Complementary Metal-Oxide-Semiconductor (CMOS).

**2.33** 74LS02, 4001B, 74HC02. In each system, the gate function is indicated by the last two or three digits. (e.g. in TTL, ′02 means "NOR" and ′00 means "NAND.")

**2.35** $f = 25$ Hz; DC = 25 %

**2.37** See Figure ANS2-37.

**2.39** The gate output would pulse.

**2.41** Possible faults: Pulsed input open; pullup input open; pulsed input stuck LOW; output stuck HIGH.

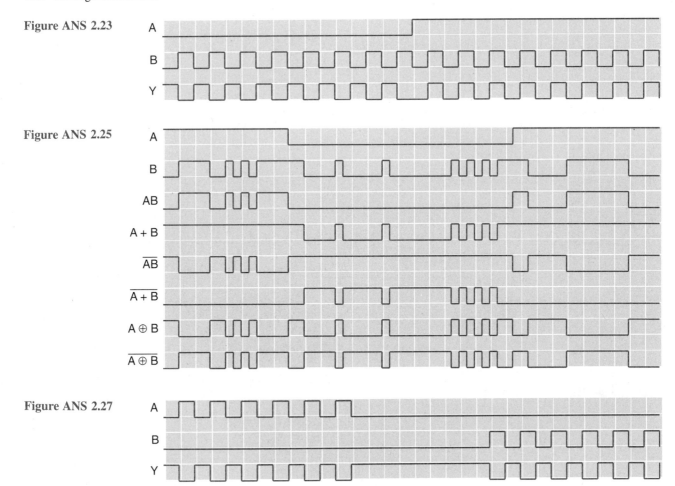

**Figure ANS 2.23**

**Figure ANS 2.25**

**Figure ANS 2.27**

**Figure ANS 2.37**

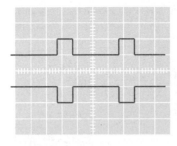

Vertical 5V/div
Horizontal 10ms/div

## Chapter 3

**3.1**  **a.** Y = ABC

**b.** X = PQ + RS

**c.** M = HJKL

**d.** A = W + X + Y + Z

**e.** Y = (A + B)(C + D)

**f.** Y = $\overline{(A + B)(C + D)}$

**g.** Y = $(\overline{A} + \overline{B})(\overline{C} + \overline{D})$

**h.** X = $\overline{P}\,\overline{Q} + \overline{R}\,\overline{S}$

**i.** X = $\overline{\overline{P}\,\overline{Q} + \overline{R}\,\overline{S}}$

**3.3**  Y = ABCDCD

**3.5**  Y = AB + BC + AC

**3.7**  See Figure ANS3-7.

**3.9**  Y = 1, Y = 1

**3.11**  Y = 1

**3.13**  Truth tables for Figures 3.38 to 3.40:

Figure 3.38 a.:

| T | U | V | W | X |
|---|---|---|---|---|
| 0 | 0 | 0 | 0 | 1 |
| 0 | 0 | 0 | 1 | 0 |
| 0 | 0 | 1 | 0 | 1 |
| 0 | 0 | 1 | 1 | 0 |
| 0 | 1 | 0 | 0 | 1 |
| 0 | 1 | 0 | 1 | 0 |
| 0 | 1 | 1 | 0 | 1 |
| 0 | 1 | 1 | 1 | 0 |
| 1 | 0 | 0 | 0 | 1 |
| 1 | 0 | 0 | 1 | 0 |
| 1 | 0 | 1 | 0 | 1 |
| 1 | 0 | 1 | 1 | 0 |
| 1 | 1 | 0 | 0 | 0 |
| 1 | 1 | 0 | 1 | 0 |
| 1 | 1 | 1 | 0 | 1 |
| 1 | 1 | 1 | 1 | 0 |

Figure 3.38 b.:

| H | J | K | L | M |
|---|---|---|---|---|
| 0 | 0 | 0 | 0 | 1 |
| 0 | 0 | 0 | 1 | 1 |
| 0 | 0 | 1 | 0 | 1 |
| 0 | 0 | 1 | 1 | 1 |
| 0 | 1 | 0 | 0 | 1 |
| 0 | 1 | 0 | 1 | 1 |
| 0 | 1 | 1 | 0 | 1 |
| 0 | 1 | 1 | 1 | 1 |
| 1 | 0 | 0 | 0 | 1 |
| 1 | 0 | 0 | 1 | 1 |
| 1 | 0 | 1 | 0 | 1 |
| 1 | 0 | 1 | 1 | 1 |
| 1 | 1 | 0 | 0 | 1 |
| 1 | 1 | 0 | 1 | 0 |
| 1 | 1 | 1 | 0 | 1 |
| 1 | 1 | 1 | 1 | 1 |

Figure 3.38 c.:

| Q | R | S | T | X |
|---|---|---|---|---|
| 0 | 0 | 0 | 0 | 1 |
| 0 | 0 | 0 | 1 | 1 |
| 0 | 0 | 1 | 0 | 1 |
| 0 | 0 | 1 | 1 | 1 |
| 0 | 1 | 0 | 0 | 1 |
| 0 | 1 | 0 | 1 | 0 |
| 0 | 1 | 1 | 0 | 0 |
| 0 | 1 | 1 | 1 | 0 |
| 1 | 0 | 0 | 0 | 1 |
| 1 | 0 | 0 | 1 | 0 |
| 1 | 0 | 1 | 0 | 0 |
| 1 | 0 | 1 | 1 | 0 |
| 1 | 1 | 0 | 0 | 1 |
| 1 | 1 | 0 | 1 | 0 |
| 1 | 1 | 1 | 0 | 0 |
| 1 | 1 | 1 | 1 | 0 |

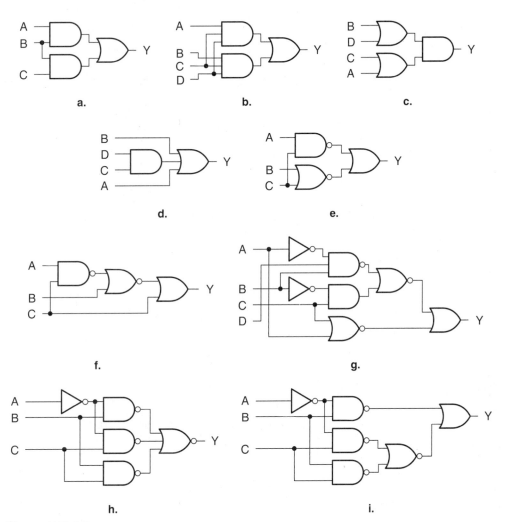

**Figure ANS 3.7**

Figure 3.38 d.:

| Q | R | S | T | X |
|---|---|---|---|---|
| 0 | 0 | 0 | 0 | 1 |
| 0 | 0 | 0 | 1 | 1 |
| 0 | 0 | 1 | 0 | 1 |
| 0 | 0 | 1 | 1 | 1 |
| 0 | 1 | 0 | 0 | 0 |
| 0 | 1 | 0 | 1 | 0 |
| 0 | 1 | 1 | 0 | 0 |
| 0 | 1 | 1 | 1 | 1 |
| 1 | 0 | 0 | 0 | 0 |
| 1 | 0 | 0 | 1 | 0 |
| 1 | 0 | 1 | 0 | 0 |
| 1 | 0 | 1 | 1 | 1 |
| 1 | 1 | 0 | 0 | 0 |
| 1 | 1 | 0 | 1 | 0 |
| 1 | 1 | 1 | 0 | 0 |
| 1 | 1 | 1 | 1 | 1 |

Figure 3.39:

| A | B | C | D | Y |
|---|---|---|---|---|
| 0 | 0 | 0 | 0 | 0 |
| 0 | 0 | 0 | 1 | 0 |
| 0 | 0 | 1 | 0 | 0 |
| 0 | 0 | 1 | 1 | 0 |
| 0 | 1 | 0 | 0 | 0 |
| 0 | 1 | 0 | 1 | 0 |
| 0 | 1 | 1 | 0 | 0 |
| 0 | 1 | 1 | 1 | 0 |
| 1 | 0 | 0 | 0 | 0 |
| 1 | 0 | 0 | 1 | 0 |
| 1 | 0 | 1 | 0 | 0 |
| 1 | 0 | 1 | 1 | 0 |
| 1 | 1 | 0 | 0 | 0 |
| 1 | 1 | 0 | 1 | 0 |
| 1 | 1 | 1 | 0 | 0 |
| 1 | 1 | 1 | 1 | 1 |

Figure 3.40 a.:

| A | B | C | Y |
|---|---|---|---|
| 0 | 0 | 0 | 0 |
| 0 | 0 | 1 | 0 |
| 0 | 1 | 0 | 0 |
| 0 | 1 | 1 | 1 |
| 1 | 0 | 0 | 0 |
| 1 | 0 | 1 | 1 |
| 1 | 1 | 0 | 1 |
| 1 | 1 | 1 | 1 |

Figure 3.40 b.:

| A | B | C | Y |
|---|---|---|---|
| 0 | 0 | 0 | 0 |
| 0 | 0 | 1 | 1 |
| 0 | 1 | 0 | 0 |
| 0 | 1 | 1 | 1 |
| 1 | 0 | 0 | 1 |
| 1 | 0 | 1 | 1 |
| 1 | 1 | 0 | 1 |
| 1 | 1 | 1 | 1 |

**3.15** See Figure ANS3-15.
SOP form: $Y = \overline{A}\,\overline{B}\,\overline{C} + \overline{A}\,\overline{B}\,C + \overline{A}\,B\,\overline{C} + \overline{A}\,B\,C$
POS form: $Y = (\overline{A} + B + C)(\overline{A} + B + \overline{C})(\overline{A} + \overline{B} + C)$
$(\overline{A} + \overline{B} + \overline{C})$

**3.17** See Figure ANS3-17.
SOP form: $Y = \overline{A}\,\overline{B}\,C + \overline{A}\,B\,\overline{C} + A\,\overline{B}\,C + A\,B\,\overline{C} + A\,B\,C$
POS form: $Y = (A + B + C)(A + \overline{B} + \overline{C})$
$(\overline{A} + B + C)$

**3.19** XOR: $(A + B)(\overline{A} + \overline{B})$
XNOR: $(A + \overline{B})(\overline{A} + B)$
See Figure ANS3-19.

**3.21** **a.** $M = PQ + R$

**b.** $M = PQ$

**c.** $S = V + (T + U)$

**d.** $Y = AC + A\overline{B}D$

**e.** $Y = \overline{A\overline{B}D}$

**f.** $P = \overline{Q} + \overline{R} + ST$

**g.** $U = \overline{Y} + X\overline{W} + XZ + WZ$

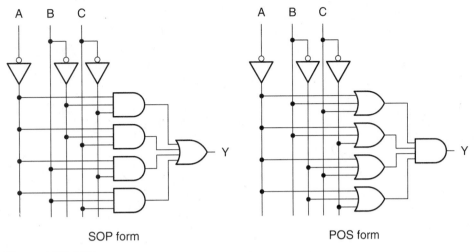

SOP form                              POS form

**Figure ANS 3.15**

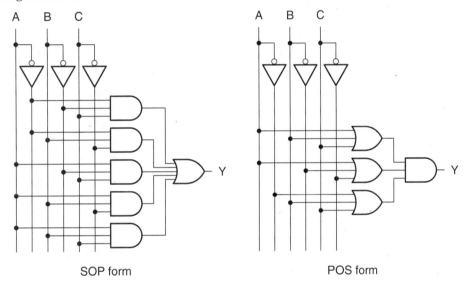

SOP form                              POS form

**Figure ANS 3.17**

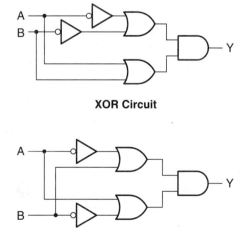

**XOR Circuit**

**XNOR Circuit**

**Figure ANS 3.19**

# Chapter 4

**4.1** Y = C

**4.3** Y = B$\overline{C}$ + AD

**4.5** Y = $\overline{C}$D + $\overline{A}$D + BD + $\overline{A}$BC

**4.7** See Figure ANS4-7 for K-maps.

    **a.** SOP: Y = $\overline{A}$ C + B C; POS: Y = (C)($\overline{A}$ + B)

    **b.** SOP: Y = A B + C; POS: Y = (A + C)(B + C)

    **c.** SOP: Y = A C + B C + $\overline{A}$ $\overline{B}$ $\overline{C}$
    POS: Y = ($\overline{A}$ + C)($\overline{B}$ + C)(A + B + $\overline{C}$)

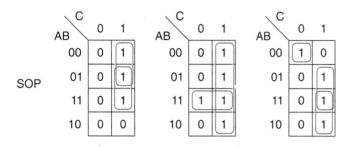

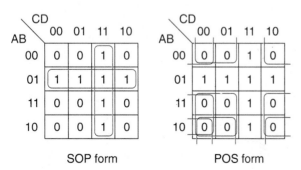

**Figure ANS 4.7**

**4.9** See Figure ANS4-9 for K-maps.
SOP: Y = $\overline{A}$ B + C D
POS: Y = ($\overline{A}$ + C)($\overline{A}$ + D)(B + C)(B + D)

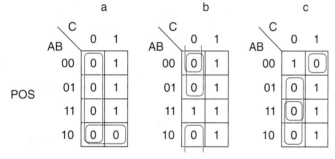

**Figure ANS 4.9**

**4.11** See Figure ANS4-11 for K-maps. (Note that because of the "don't care" states, the groupings in the two maps are not mutually exclusive. This implies that each solution is most efficient for SOP or POS format, but cannot be transformed directly to the other.)
SOP: Y = A D + $\overline{B}$ C
POS: Y = $\overline{B}$(A + C)(C + D)

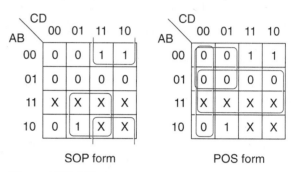

**Figure ANS 4.11**

**4.13** See Figure ANS4-13.
Boolean equation: Y = ($\overline{A}$ + $\overline{B}$ + C)(A + $\overline{C}$)(B + C + $\overline{D}$)

**4.15** See Figure ANS4-15.
Equations:
$E_4 = D_4\overline{D_2} + D_3D_2 + D_3D_1$
$E_3 = \overline{D_3}D_2 + \overline{D_3}D_1 + D_3\overline{D_2}\overline{D_1}$
$E_2 = D_2D_1 + \overline{D_2}\overline{D_1}$
$E_1 = \overline{D_1}$

**4.17** See Figure ANS4-17.
Y = $\overline{A}$ $\overline{B}$ C + A $\overline{B}$ $\overline{C}$ + $\overline{A}$ B $\overline{C}$ + A B C

**4.19** See Figure ANS4-19.
Y = A B $\overline{C}$ + $\overline{A}$ $\overline{B}$ + A $\overline{B}$ $\overline{C}$ + A B C
(Can be further simplified to:
Y = A $\overline{C}$ + $\overline{A}$ $\overline{B}$ + A B)

**4.21** See Figure ANS4-21.

**4.23** See Figure ANS4-23.

**4.25** See Figure ANS4-25.

**4.27** See Figure ANS4-27.

**4.29** See Figure ANS4-29.

**4.31** See Figure ANS4-31.

**Figure ANS 4.13**

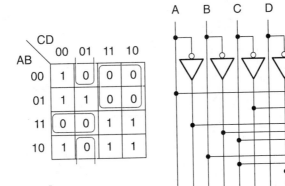

**Figure ANS 4.15**

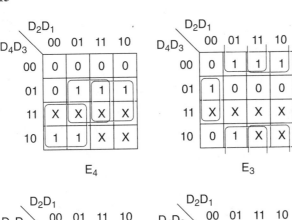

E_4

E_3

E_2

E_1

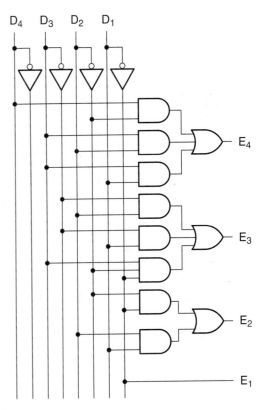

**Figure ANS 4.17**

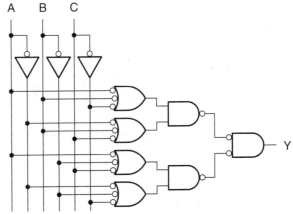

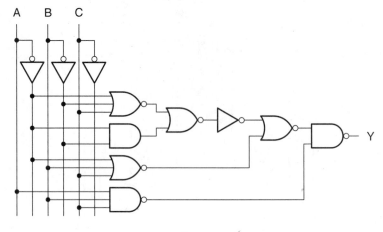

a.

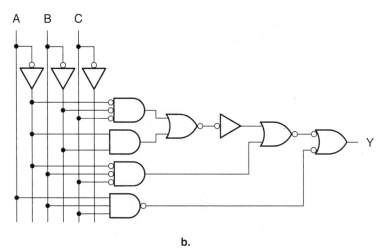

b.

**Figure ANS 4.19**

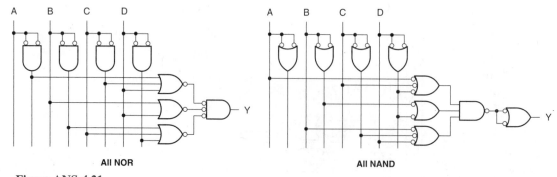

**Figure ANS 4.21**

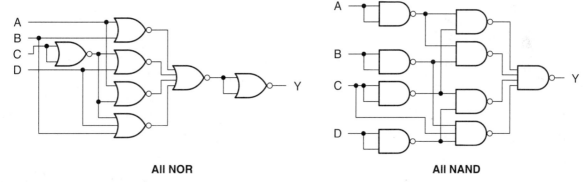

**All NOR**    **All NAND**

**Figure ANS 4.23**

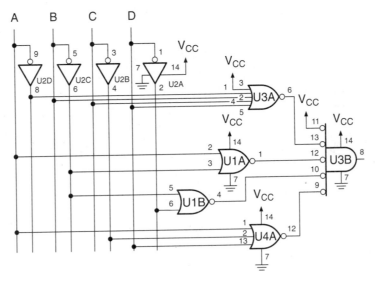

**Unused Gates:**

**Gates:**
74LS02 (QUAD 2-in NOR) U1
74LS04 (HEX INVERTER) U2
74LS25 (DUAL 4-in NOR with ENABLE) U3
74LS27 (TRIPLE 3-in NOR) U4

**Figure ANS 4.25**

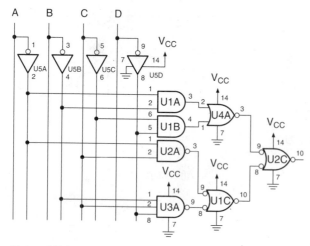

**Unused Gates:**

$V_{DD}$

**Gates:**
U1 4081B (QUAD 2-in AND)
U2 4011B (QUAD 2-in NAND)
U3 4023B (Triple 3-in NAND)
U4 4001B (QUAD 2-in NAND)
U5 4069B (Hex Inverter)

**Figure ANS 4.27**

**4.33** Figure ANS4-33 shows four cases:
1. Door opens, RESET pressed, alarm does not sound.
2. Door opens, RESET not pressed, alarm sounds from 30 seconds after door opens to 60 seconds after door opens.
3. Door opens, but window strip sets off alarm before timer is finished. Alarm does not shut off.
4. Window strip opens, setting off alarm. Alarm does not shut off.
The 200 Hz waveform in each of the diagrams is not shown to scale.

**4.35** The alarm activates on a majority of out-of-range parameters. Each parameter requires a majority of sensors to be out-of-range for alarm indication. Since the alarm is active LOW, it is described by the following Boolean equation:
$$Y = LT + LP + TP$$
where
$$L = L_1L_2 + L_2L_3 + L_1L_3$$
$$T = T_1T_2 + T_2T_3 + T_1T_3$$
$$P = P_1P_2 + P_2P_3 + P_1P_3$$
See Figure ANS4-35 for the circuit diagram.

**4.37** U3-6 stuck LOW, U3-5 open circuit, or U2-4/U3-5 connection stuck HIGH.

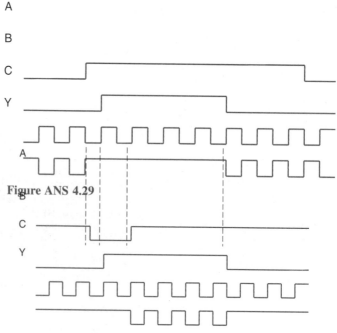

**Figure ANS 4.29**

**Figure ANS 4.31**

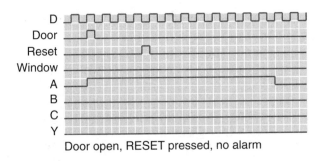

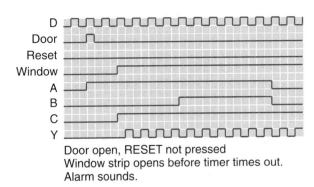

Door open, RESET not pressed
Window strip opens before timer times out.
Alarm sounds.

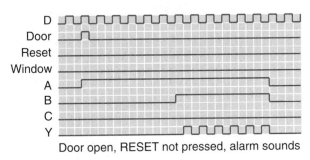

Door open, RESET not pressed, alarm sounds

**Figure ANS 4.33**

Door remains closed
Window strip opens
Alarm sounds.

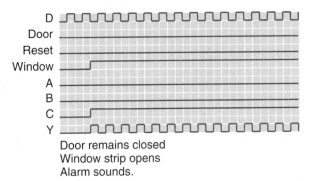

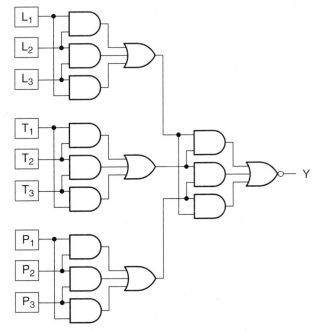

**Figure ANS 4.35**

## Chapter 5

**5.1**  $8^4 = 4096$, $8^3 = 512$, $8^2 = 64$, $8^1 = 8$, $8^0 = 1$

**5.3**  101101, 101110, 101111, 110000, 110001, 110010, 110011, 110100, 110101

**5.5**  **a.** 13,  **b.** 53,  **c.** 3163,  **d.** 2922,  **e.** 90,
**f.** 1023,  **g.** 1024

**5.7**  **a.** 0.625,  **b.** 0.375,  **c.** 0.8125

**5.9**  ⅓

**5.11**  **a.** 0.11,  **b.** 0.101,  **c.** 0.0011,
**d.** 0.10$\overline{1001}$,  **e.** 1.11,  **f.** 11.11$\overline{1100}$,
**g.** 1000011.1101011100001. . . (nonrepeating)

**5.13**  9F7, 9F8, 9F9, 9FA, 9FB, 9FC, 9FD, 9FE, 9FF, A00, A01, A02, A03

**5.15**  **a.** 2C5,  **b.** 761,  **c.** FFF,  **d.** 1000,  **e.** 2790,
**f.** 7D00,  **g.** 8000

**5.17**  **a.** 5E86,  **b.** B6A,  **c.** C5B,  **d.** 6BC4,  **e.** 15785,
**f.** 198B7,  **g.** 28000

**5.19**  367, 370, 371, 372, 373, 374, 375, 376, 377, 400

**5.21**  **a.** 100,  **b.** 31,  **c.** 175,  **d.** 261,  **e.** 310,  **f.** 10000

**5.23 a.** 110 010 100, **b.** 101 010, **c.** 1 111 111, **d.** 10 000 000, **e.** 1 111 101 000, **f.** 1 111 111 111, **g.** 10 000 000 000

The number in part d. ($200_8$) is next in sequence after the number in part c ($177_8$). (i.e. d. is one greater than c.). The same is true of f. and g. ($1777_8$ and $2000_8$.)

**5.25**

| Decimal | True Binary | Gray Code |
|---------|-------------|-----------|
| 0 | 00000 | 00000 |
| 1 | 00001 | 00001 |
| 2 | 00010 | 00011 |
| 3 | 00011 | 00010 |
| 4 | 00100 | 00110 |
| 5 | 00101 | 00111 |
| 6 | 00110 | 00101 |
| 7 | 00111 | 00100 |
| 8 | 01000 | 01100 |
| 9 | 01001 | 01101 |
| 10 | 01010 | 01111 |
| 11 | 01011 | 01110 |
| 12 | 01100 | 01010 |
| 13 | 01101 | 01011 |
| 14 | 01110 | 01001 |
| 15 | 01111 | 01000 |
| 16 | 10000 | 11000 |
| 17 | 10001 | 11001 |
| 18 | 10010 | 11011 |
| 19 | 10011 | 11010 |
| 20 | 10100 | 11110 |
| 21 | 10101 | 11111 |
| 22 | 10110 | 11101 |
| 23 | 10111 | 11100 |
| 24 | 11000 | 10100 |
| 25 | 11001 | 10101 |
| 26 | 11010 | 10111 |
| 27 | 11011 | 10110 |
| 28 | 11100 | 10010 |
| 29 | 11101 | 10011 |
| 30 | 11110 | 10001 |
| 31 | 11111 | 10000 |

**5.27** 31 30 25 20 6F 66 66 20 70 75 72 63 68 61 73 65 73 20 6F 76 65 72 20 24 35 30 2E 20 28 4D 6F 6E 64 61 79 20 6F 6E 6C 79 29

**Chapter 6**

**6.1** See Figure ANS6-1.

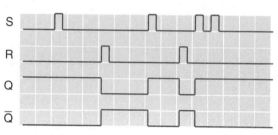

**Figure ANS 6.1**

**6.3** See Figure ANS6-3.

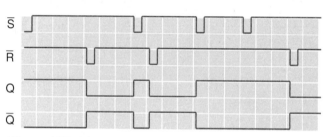

**Figure ANS 6.3**

**6.5** See Figure ANS6-5.

$\overline{S} = 0, \overline{R} = 0$;  Forbidden state (Latch SET and RESET inputs activate simultaneously.)

$\overline{S} = 0, \overline{R} = 1$  SET input active (Q = 1).

$\overline{S} = 1, \overline{R} = 0$  RESET input active (Q = 0).

$\overline{S} = 1, \overline{R} = 1$  No change (Neither SET nor RESET input is active.)

**Figure ANS 6.5**

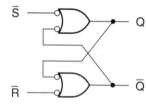

**6.7** See Figure ANS6-7.

**6.9 a.** See Figure ANS6-9a.

**b.** $\overline{S} = \overline{R} = 0$ is forbidden for a NAND latch because, in this state, we are attempting to set and reset the latch at the same time. If both inputs go to the no change state ($\overline{S} = \overline{R} = 1$) at the same time, the final output state is not predictable.

**c.** i) Reset;  ii) Set;  iii) unknown.

**6.11** See Figure ANS6-11.

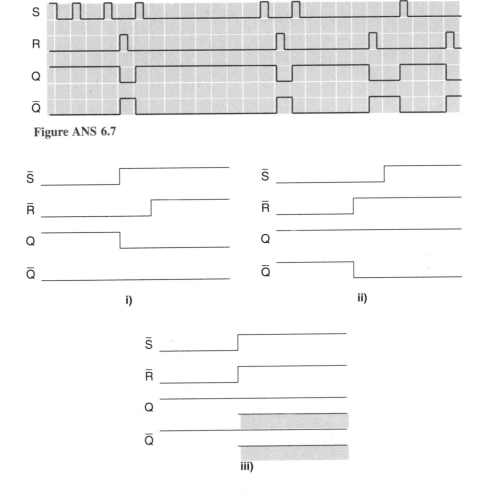

**Figure ANS 6.7**

i)

ii)

iii)

**Figure ANS 6.9a**

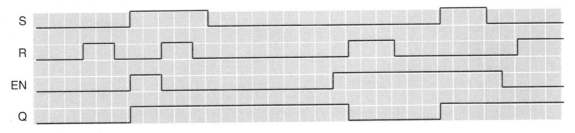

**Figure ANS 6.11**

**6.13** See Figure ANS6-13.

**6.15** See Figure ANS6-15.

**6.17** See Figure ANS6-17.

**6.19** See Figure ANS6-19 for the $Q_1$ and $Q_2$ waveforms. In the first four changes $Q_2$ changes before $Q_1$. This is be-

cause a change occurs on S or R about halfway through the HIGH half-cycle of the EN/CLK waveform. Since the gated latch ENABLE input is active HIGH, the change on $Q_1$ occurs immediately. Changes on $Q_2$, however, must wait for the next positive edge on EN/CLK.

**6.21** See Figure ANS6-21.

Figure ANS 6.13

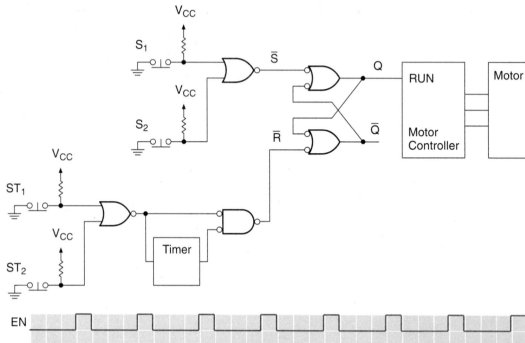

Figure ANS 6.15

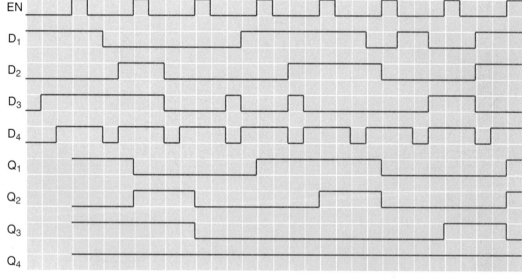

Figure ANS 6.17

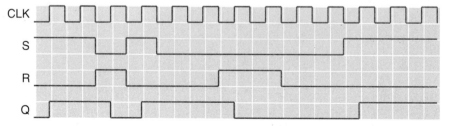

**Figure ANS 6.19**

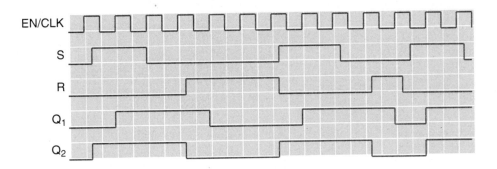

**Figure ANS 6.21**

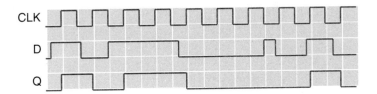

**6.23**  See Figure ANS6-23.

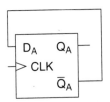

**Figure ANS 6.23**

**6.25**  See Figure ANS6-25.

**Figure ANS 6.25**

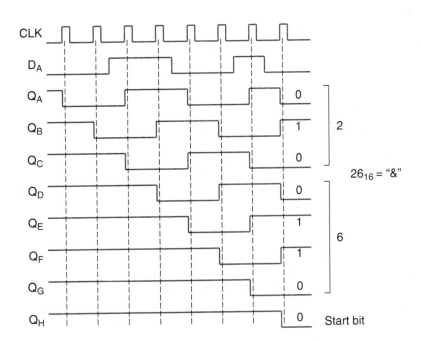

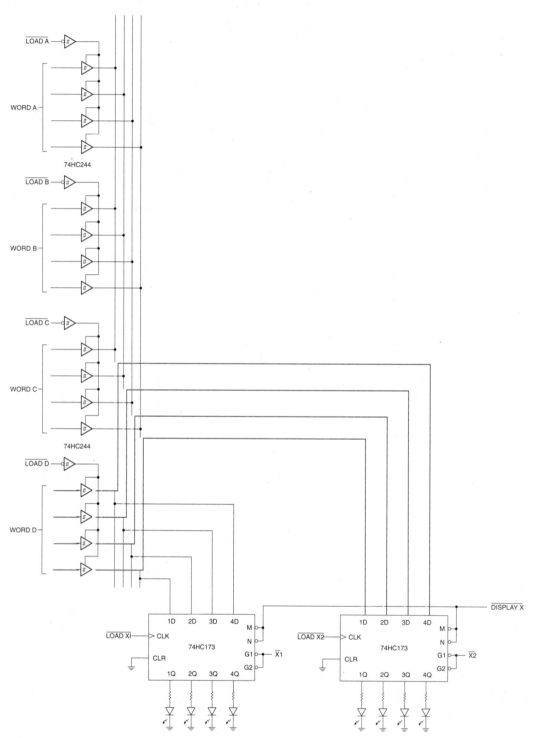

**Figure ANS 6.29**

**6.27**

| | Latch A | Latch B |
|---|---|---|
| 1 | 01100001 | -------- |
| 2 | 01100001 | 00101010 |
| 3 | 10111000 | 00101010 |
| 4 | 10111000 | 11000101 |
| 5 | 00011001 | 11000101 |
| 6 | 00011001 | 00000010 |

New data enter the selected latch as soon as LOAD DATA becomes HIGH. The data are stored in the latch as soon as LOAD DATA goes LOW.

**6.29a.**

| Step | Action | Source | Destination | Output |
|---|---|---|---|---|
| 1 | Transfer | D | X1 | --------- |
| 2 | Transfer | B | X2 | --------- |
| 3 | Display | - | -- | 1111 1010 |
| 4 | Transfer | C | X1 | --------- |
| 5 | Transfer | D | X2 | --------- |
| 6 | Display | - | -- | 1110 1111 |
| 7 | Transfer | A | X1 | --------- |
| 8 | Transfer | A | X2 | --------- |
| 9 | Display | - | -- | 0011 0011 |

    **b.** See Figure ANS6-29.

**6.31**  The state S=R=1 is forbidden in the SR flip-flop. The state J=K=1 is the toggle state of the JK flip-flop.

**6.33**  See Figure ANS6-33.

**6.35**  The circuit timing diagram is shown in Figure ANS6-35. The flip-flop outputs progress through the following series of states.

| $Q_C$ | $Q_B$ | $Q_A$ |
|---|---|---|
| 1 | 1 | 1 |
| 1 | 1 | 0 |
| 1 | 0 | 1 |
| 1 | 0 | 0 |
| 0 | 1 | 1 |
| 0 | 1 | 0 |
| 0 | 0 | 1 |
| 0 | 0 | 0 |

The flip-flop outputs count downward in binary sequence from 111 to 000 and then repeat the sequence.

**6.37**  See Figure ANS6-37.

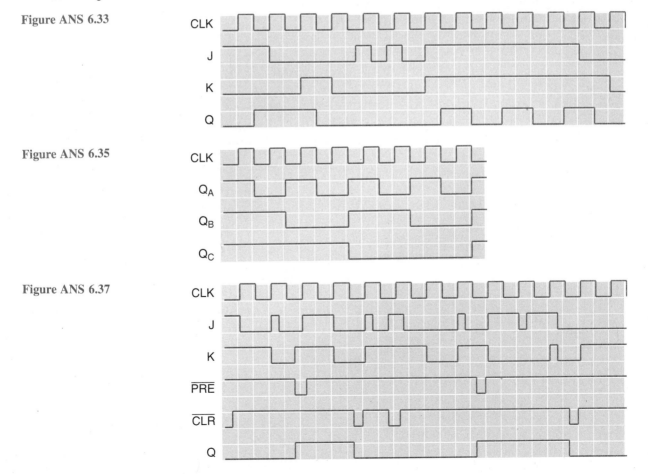

**Figure ANS 6.33**  CLK J K Q

**Figure ANS 6.35**  CLK $Q_A$ $Q_B$ $Q_C$

**Figure ANS 6.37**  CLK J K $\overline{PRE}$ $\overline{CLR}$ Q

**6.39**

| Device | Setup time ($t_{su}$) | Hold time ($t_h$) |
|---|---|---|
| 74LS74A | 20 ns | 5 ns |
| 74HC76 | 20 ns | 3 ns |
| 74LS76A | 20 ns | 0 ns |
| 74LS107A | 20 ns | 0 ns |
| 74ALS112A | 22 ns | 0 ns |
| 74HC112 | 20 ns | 3 ns |

**6.41** The timing parameters are illustrated in Figure 6.55. Table 6.12 gives values of the parameters for the 74LS107A and 74HC107 flip-flops. Specific to-scale timing diagrams are shown in Figure ANS6-41.

**6.43** $\overline{CLR}$ pulse width: $t_w = 16$ ns
Recovery time: $t_{rec} = 20$ ns
Propagation delay from $\overline{CLR}$: $t_{pHL} = 8$ ns, $t_{pLH} = 8$ ns

**6.45** CLK is stuck HIGH or LOW. CLK is open circuited. $\overline{OC}$ is stuck HIGH. D inputs are all stuck LOW.

**6.47** Preset and Clear outputs are stuck HIGH.

## Chapter 7

**7.1** A nonretriggerable monostable ignores all trigger pulses that occur during the output pulse width time, $t_w$. A retriggerable monostable accepts new trigger pulses during $t_w$. The resulting pulse times out $t_w$ after the last trigger pulse.

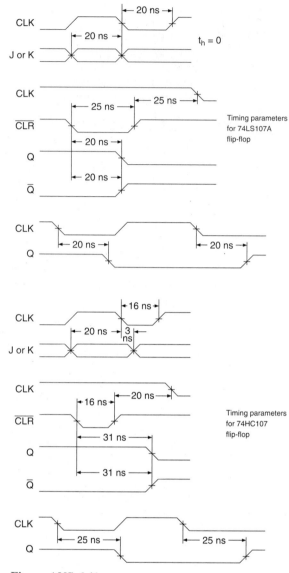

**Figure ANS 6.41**

**7.3**   0.002 μF

**7.5**   See Figure 7.7 for the circuit.
Calculated values: $R_1$ = 37.9 kΩ, $R_2$ = 25.2 kΩ
Closest standard values: $R_1$ = 39 kΩ, $R_2$ = 27 kΩ

**7.7**   R = 39 kΩ; $t_w$ = 1.07 s

**7.9**   $R_a$ = 2.2 kΩ, $R_b$ = 4.7 kΩ, f = 124 kHz, DC = 59.48%

**7.11**  When $R_a$ and $R_b$ are interchanged, the value of $t_l$ of the output waveform changes, but not $t_h$. This is because the timing capacitor charges through two resistors ($R_a + R_b$), but discharges through $R_b$ only. In Problem 7.10a, $t_l$ is greater than in 7.10b, since $R_b$ is larger in the first case. Thus, in 7.10a, the frequency and duty cycle are both smaller than in 7.10b.

**7.13**  Duty cycle is unaffected by different values of the timing capacitor. Frequency changes in inverse proportion to the value of C. The capacitor value in Problem 7.10c is one tenth the value of that in 7.10d. Thus, the frequency defined by the component values in Problem 7.10c is ten times greater than the frequency defined in 7.10d.

**7.15**  f = 756 Hz, DC = 21.3%
The timing components are shown in Figure ANS7-15.

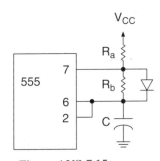

**Figure ANS 7.15**

## Chapter 8

**8.1**   See Figure 8.1.

**8.3**   See Figure ANS8-3.

**8.5**   Inputs: D, EN
Latch outputs: Q, NOTQ
Feedbacks: STEERQ, STEERNQ
Equations:   STEERQ = /(Q * EN)
            STEERNQ = /(Q * /EN)
                Q = /(STEERQ * NOTQ)
             NOTQ = /(STEERNQ * Q)

**8.7**   2421 inputs: $Y_4Y_3Y_2Y_1$
BCD outputs: $D_4D_3D_2D_1$
Equations: $D_4 = Y_3Y_2$
$D_3 = Y_4\overline{Y}_3 + Y_3\overline{Y}_2$
$D_2 = Y_4\overline{Y}_2 + Y_4Y_2$
$D_1 = Y_1$

**8.9**   The required circuit is shown in Figure ANS8-9. The circuit includes pin numbers for an 85C224 implementation. Outputs must be combinational because the chip does not have enough pins for a registered output. Since all pins are used up by input and output functions, there is nothing left for the CLOCK pin.
Equations:
F4 = A4*/S1*/S0 + B4*/S1*S0 + C4*S1*/
S0 + D4*S1*S0

F3 = A3*/S1*/S0 + B3*/S1*S0 + C3*S1*/
S0 + D3*S1*S0

F2 = A2*/S1*/S0 + B2*/S1*S0 + C2*S1*/
S0 + D2*S1*S0

F1 = A1*/S1*/S0 + B1*/S1*S0 + C1*S1*/
S0 + D1*S1*S0

**8.11**  This source file and the corresponding JEDEC file are found on the instructor's diskette in files \INTRO \GLATCH.PDS and \INTRO\GLATCH.JED.
Title       Gated D Latch
Pattern     pds
Revision    1.0
Author      R.Dueck
Company     Seneca College
Date            September 1, 1993


CHIP   glatch   85C220

PIN D       ;Inputs: D, EN
PIN EN

PIN Q       ; Latch outputs: Q, NOTQ
PIN NOTQ

PIN STEERQ   ; Feedbacks: STEERQ, STEERNQ
PIN STEERNQ

EQUATIONS

STEERQ = /(Q * EN)
STEERNQ = /(Q * /EN)
Q = /(STEERQ * NOTQ)
NOTQ = /(STEERNQ * Q)

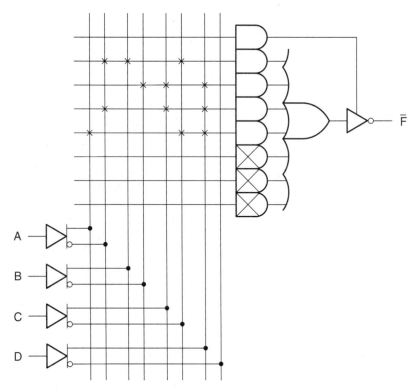

**Figure ANS 8.3**

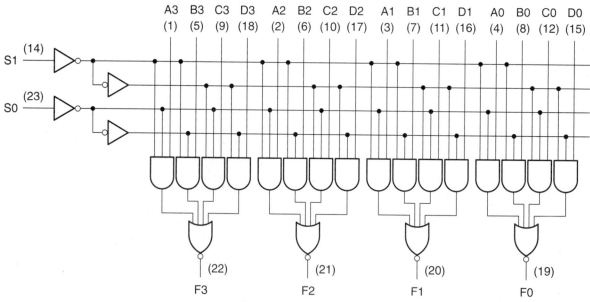

**Figure ANS 8.9**

**8.13a.** Boolean equations for a BCD-to-Excess-3 Decoder:

E4 = D4 + D3*D2 + D3*D1

E3 = D3*/D2*/D1 + /D3*D2 + /D3*D1

E2 = /D2*/D1 + D2*D1

E1 = /D1

**b.** The PLDasm source file and JEDEC file are included on the instructor's diskette in files \INTRO \BCD2XS3.PDS and \INTRO\BCD2XS3.JED.

**8.15** The Boolean equations for the dual decoder are given below. There are two sets of code inputs, IA4-IA1 and IB4-IB1, and two sets of code outputs, OA4-OA1 and OB4-OB1. The equations are the same for each set. In each equation, the first line is for the BCD-to-Excess-3 function and the second line for the Excess-3-to-BCD function. The PLDasm source and JEDEC files are found on the instructor's diskette as files \INTRO\DUALCODE.PDS and \INTRO\DUALCODE.JED.

OA4 = (IA4 + IA3*IA2 + IA3*IA1)*/CODE__A
    + (IA4*IA3 + IA4*IA2*IA1)*CODE__A

OA3 = (IA3*/IA2*/IA1 + /IA3*IA2 + /IA3*IA1)
    */CODE__A
    + (IA3*IA2*IA1 + /IA3*/IA2
    + /IA3*/IA1)*CODE__A

OA2 = (/IA2*/IA1 + IA2*IA1)*/CODE__A
    + (/IA2*IA1 + IA2*/IA1)*CODE__A

OA1 = /IA1

OB4 = (IB4 + IB3*IB2 + IB3*IB1)*/CODE__B
    + (IB4*IB3 + IB4*IB2*IB1)*CODE__B

OB3 = (IB3*/IB2*/IB1 + /IB3*IB2 + /IB3*IB1)
    */CODE__B
    + (IB3*IB2*IB1 + /IB3*/IB2
    + /IB3*/IB1)*CODE__B

OB2 = (/IB2*/IB1 + IB2*IB1)*/CODE__B
    + (/IB2*IB1 + IB2*/IB1)*CODE__B

OB1 = /IB1

# Chapter 9

**9.1** **a.** 11111; **b.** 100000; **c.** 11110; **d.** 101010;
**e.** 101100; **f.** 1100100

**9.3**

|  | Decimal | True Magnitude | 1's Complement | 2's Complement |
|---|---|---|---|---|
| **a.** | $-110$ | 11101110 | 10010001 | 10010010 |
| **b.** | 67 | 01000011 | 01000011 | 01000011 |
| **c.** | $-54$ | 10110110 | 11001001 | 11001010 |
| **d.** | $-93$ | 11011101 | 10100010 | 10100011 |
| **e.** | 0 | 00000000 | 00000000 | 00000000 |
| **f.** | $-1$ | 10000001 | 11111110 | 11111111 |
| **g.** | 127 | 01111111 | 01111111 | 01111111 |
| **h.** | $-127$ | 11111111 | 10000000 | 10000001 |

**9.5** Largest: $01111111_2 = +127_{10}$; smallest: $10000000_2 = -128_{10}$

**9.7** Overflow results in an 8-bit signed addition if the sum is outside the range $-128 \leq \text{sum} \leq +127$. The sums in parts a. and f. do not generate an overflow. The sums in parts b., c., d., and e. do.

**9.9** **a.** 3D; **b.** 120; **c.** B1A; **d.** FFF; **e.** 2A7F

**9.11**

| A | B | $C_0$ | E |
|---|---|---|---|
| 0 | 0 | 0 | 0 |
| 0 | 1 | 0 | 1 |
| 1 | 0 | 0 | 1 |
| 1 | 1 | 1 | 0 |

$C_0 = AB$

$\Sigma = \overline{A}B + A\overline{B} = A \oplus B$

The half adder circuit is shown in Figure 9.2.

**9.13** See Figures 9.6 and 9.7 for the full adder circuits.

**9.15** **a.** $0100 + 1001 = 1101$

$(4_{10} + 9_{10} = 13_{10})$

$A_1 = 0, B_1 = 1, C_0 = 0;$    $C_1 = 0, \Sigma_1 = 1$

$A_2 = 0, B_2 = 0, C_1 = 0;$    $C_2 = 0, \Sigma_2 = 0$

$A_3 = 1, B_3 = 0, C_2 = 0;$    $C_3 = 0, \Sigma_3 = 1$

$A_4 = 0, B_4 = 1, C_3 = 0;$    $C_4 = 0, \Sigma_4 = 1$

Binary Equivalent: $C_4\Sigma_4\Sigma_3\Sigma_2\Sigma_1 = 01101$

**b.** $1010 + 0110 = 10000$

$(10_{10} + 6_{10} = 16_{10})$

$A_1 = 0, B_1 = 0, C_0 = 0;$    $C_1 = 0, \Sigma_1 = 0$

$A_2 = 1, B_2 = 1, C_1 = 0;$    $C_2 = 1, \Sigma_2 = 0$

$A_3 = 0, B_3 = 1, C_2 = 1;$    $C_3 = 1, \Sigma_3 = 0$

$A_4 = 1, B_4 = 0, C_3 = 1;$    $C_4 = 1, \Sigma_4 = 0$

Binary Equivalent: $C_4\Sigma_4\Sigma_3\Sigma_2\Sigma_1 = 10000$

**c.** $0101 + 1101 = 10010$

$(5_{10} + 13_{10} = 18_{10})$

$A_1 = 1, B_1 = 1, C_0 = 0;$    $C_1 = 1, \Sigma_1 = 0$

$A_2 = 0, B_2 = 0, C_1 = 1;$    $C_2 = 0, \Sigma_2 = 1$

$A_3 = 1, B_3 = 1, C_2 = 0;$    $C_3 = 1, \Sigma_3 = 0$

$A_4 = 0, B_4 = 1, C_3 = 1;$    $C_4 = 1, \Sigma_4 = 0$

Binary Equivalent: $C_4\Sigma_4\Sigma_3\Sigma_2\Sigma_1 = 10010$

**d.** $1111 + 0111 = 10110$

$(15_{10} + 7_{10} = 22_{10})$

$A_1 = 1, B_1 = 1, C_0 = 0;$    $C_1 = 1, \Sigma_1 = 0$

$A_2 = 1, B_2 = 1, C_1 = 1;$    $C_2 = 1, \Sigma_2 = 1$

$A_3 = 1, B_3 = 1, C_2 = 1;$    $C_3 = 1, \Sigma_3 = 1$

$A_4 = 1, B_4 = 0, C_3 = 1;$    $C_4 = 1, \Sigma_4 = 0$

Binary Equivalent: $C_4\Sigma_4\Sigma_3\Sigma_2\Sigma_1 = 10110$

**9.17** $C_3 = A_3B_3 + A_2B_2(A_3 + B_3) + A_1B_1(A_3 + B_3)$
$(A_2 + B_2) + C_0(A_3 + B_3)(A_2 + B_2)(A_1 + B_1)$

**9.19** See Figure 9.16. The inverters convert the B inputs to the 1's complement form by complementing each bit. The input carry is forced HIGH, thus adding 1 to the sum of A and the 1's complement of B. This is equivalent to adding A and the 2's complement of B.

**9.21** See Figure 9.20.

| $S_A$ | $S_B$ | $S_\Sigma$ | V |
|-----|-----|-----|---|
| 0 | 0 | 0 | 0 |
| 0 | 0 | 1 | 1 |
| 0 | 1 | 0 | 0 |
| 0 | 1 | 1 | 0 |
| 1 | 0 | 0 | 0 |
| 1 | 0 | 1 | 0 |
| 1 | 1 | 0 | 1 |
| 1 | 1 | 1 | 0 |

$S_A$ = sign bit of A
$S_B$ = sign bit of B
$S_\Sigma$ = sign bit of the sum
V = overflow indication

$$V = \overline{S}_A \overline{S}_B S_\Sigma + S_A S_B \overline{S}_\Sigma$$

**9.23**

| Accumulator | | B Register | | Adder/Subtractor | | | | | |
|---|---|---|---|---|---|---|---|---|---|
| $A_7$ | $A_0$ | $B_7$ | $B_0$ | $\Sigma_7$ | $\Sigma_0$ | $\overline{L}_A$ | $\overline{L}_B$ | $S_U$ | $\overline{E}_U$ |
| 00000000 | | 00000000 | | 00000000 | | 1 | 1 | 0 | 1 |
| 00111001 | | 00000000 | | 00111001 | | 0 | 1 | 0 | 1 |
| 00111001 | | 10010010 | | 11001011 | | 1 | 0 | 0 | 1 |
| 11001011 | | 10010010 | | 11001011 | | 0 | 1 | 0 | 0 |
| 11001011 | | 01110110 | | 01010101 | | 1 | 0 | 1 | 1 |
| 01010101 | | 01110110 | | 01010101 | | 0 | 1 | 1 | 0 |

**9.25** 1999; $3\frac{1}{2}$ digits.

**9.27** $C_4 = C_4' + \Sigma_4' \Sigma_3' + \Sigma_4' \Sigma_2'$
where $C_4$ = BCD carry,
$C_4'$ = carry generated by the binary sum of BCD digits, $\Sigma_2'$, $\Sigma_3'$, and $\Sigma_4'$ are sum bits generated by the binary sum of BCD digits.

**9.29** The circuit will display $3\frac{1}{2}$ digits.

**9.31** The target device for the 8-bit ripple-carry adder must be an 85C090 EPLD since this application requires 17 inputs, 9 outputs, and 7 feedback pins. The PLDasm files, using the same method as Example 9.24, are found on the instructor's diskette in files \ARITH \ADD8RIPL.PDS and \ARITH\ADD8RIPL.JED.

# Chapter 10

**10.1** 1100, 0001, 1111; $Y = D C \overline{B} \overline{A}$,
$Y = \overline{D} \overline{C} \overline{B} A$, $Y = D C B A$

**10.3** See Figure ANS10-3.

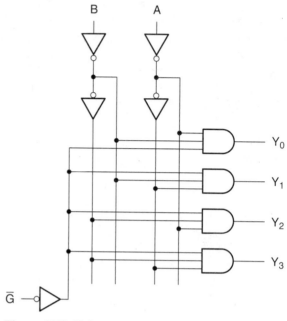

**Figure ANS 10.3**

**10.5** **a.** 32; **b.** 64; **c.** 256; $m = 2^n$.

**10.7** See Figure ANS10-7.

**10.9** See Figure ANS10-9.

**10.11**

| D | C | B | A | a | b | c | d | e | f | g |
|---|---|---|---|---|---|---|---|---|---|---|
| 0 | 0 | 0 | 0 | 0 | 0 | 0 | 0 | 0 | 0 | 1 |
| 0 | 0 | 0 | 1 | 1 | 0 | 0 | 1 | 1 | 1 | 1 |
| 0 | 0 | 1 | 0 | 0 | 0 | 1 | 0 | 0 | 1 | 0 |
| 0 | 0 | 1 | 1 | 0 | 0 | 0 | 0 | 1 | 1 | 0 |
| 0 | 1 | 0 | 0 | 1 | 0 | 0 | 1 | 1 | 0 | 0 |
| 0 | 1 | 0 | 1 | 0 | 1 | 0 | 0 | 1 | 0 | 0 |
| 0 | 1 | 1 | 0 | 1 | 1 | 0 | 0 | 0 | 0 | 0 |
| 0 | 1 | 1 | 1 | 0 | 0 | 0 | 1 | 1 | 1 | 1 |
| 1 | 0 | 0 | 0 | 0 | 0 | 0 | 0 | 0 | 0 | 0 |
| 1 | 0 | 0 | 1 | 0 | 0 | 0 | 1 | 1 | 0 | 0 |

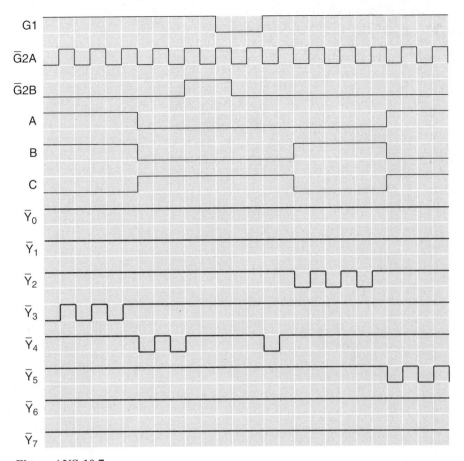

**Figure ANS 10.7**

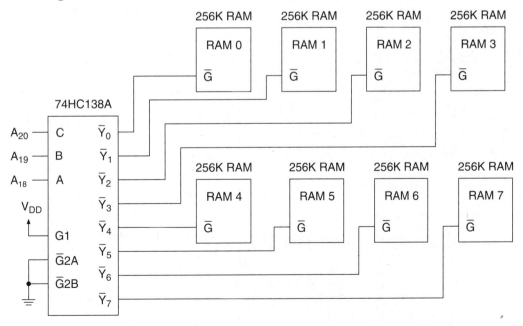

**Figure ANS 10.9**

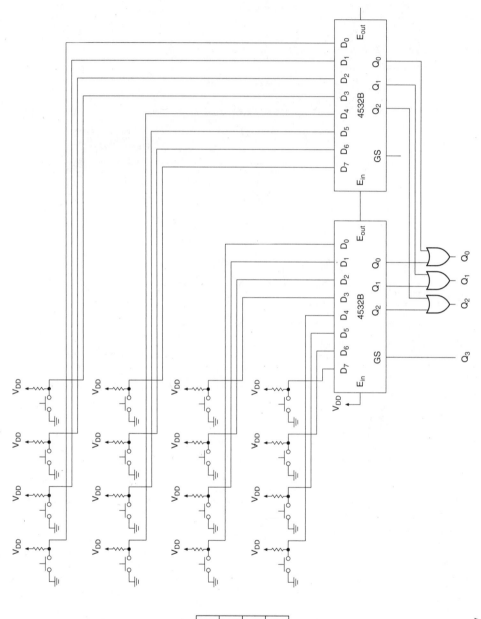

**Figure ANS 10.17**

**10.13**  $a = (\overline{C\ \overline{A}})(\overline{D\ B})(\overline{\overline{D}\ \overline{C}\ \overline{B}\ A})$

$b = (\overline{C\ \overline{B}\ A})(\overline{C\ B\ \overline{A}})(\overline{D\ \overline{B}})$

$c = (\overline{\overline{C}\ B\ \overline{A}})(\overline{D\ C})$

$d = (\overline{\overline{C}\ \overline{B}\ A})(\overline{C\ \overline{B}\ \overline{A}})(\overline{C\ B\ A})$

$e = (\overline{C\ \overline{B}})(A)$

$f = (\overline{\overline{D}\ \overline{C}\ A})(\overline{C\ B})(\overline{B\ A})$

$g = (\overline{\overline{D}\ \overline{C}\ \overline{B}})(\overline{C\ B\ A})$

**10.15**  1000, 1001, 1001

**10.17**  See Figure ANS10-17.

**10.19**  Inputs 4 and 12 are shorted.

**10.21**  See Figure ANS10-21.

Designate the encoders 2, 1, and 0 where 2 is the most significant.

**c.**  $D_{16}$ highest active input: This is the $D_0$ input of the Encoder 2. Thus, $Q_2Q_1Q_0 = 000$. Since an encoding input is active, $GS_2 = 1$, $E_{out2} = 0$. $E_{in1} = 0$, disabling Encoders 1 and 0 . GS = 0 for Encoder 0 and 1. Output code = $GS_2GS_1Q_2Q_1Q_0 = 10000$.

**d.**  $D_{15}$ highest active input: Since no input is active on Encoder 2, $GS_2 = 0$ and $E_{out2} = 1$, enabling Encoder 1. Selected input is $D_7$ on Encoder 1. $Q_2Q_1Q_0 = 111$. $GS_1$ $(Q_4) = 1$. Thus $GS_2GS_1Q_2Q_1Q_0 = 01111$.

**e.**  $D_8$ highest active input: Similar to case d., except $Q_2Q_1Q_0 = 000$. $GS_2GS_1Q_2Q_1Q_0 = 01000$.

**f.**  $D_7$ highest active input: No codes selected on Encoders 2 or 1. Therefore, the cascading chain of $E_{in}$ and $E_{out}$ enables Encoder 0. $GS_0$ active, but it is not part of the 5-bit code output. Selected input generates code $Q_2Q_1Q_0 = 111$. $GS_2GS_1Q_2Q_1Q_0 = 00111$.

**10.23**  See Figure ANS10-23. Each AND-shaped gate compares one bit of A to the B bit in the same position. The gate output is HIGH if $A_n < B_n$ (i.e. if A = 0 AND B = 1). Each XNOR gate produces a HIGH if $A_n = B_n$. These basic conditions are combined in the AND and OR gates at the circuit output. (The complete explanation is similar to that given for the (A > B) function in the main text for this section.)

**10.25**

(A = B)

$= (\overline{A_5 \oplus B_5})(\overline{A_4 \oplus B_4})(\overline{A_3 \oplus B_3})(\overline{A_2 \oplus B_2})(\overline{A_1 \oplus B_1})(\overline{A_0 \oplus B_0})$

(A < B)

$= \overline{A}_5 B_5 + (\overline{A_5 \oplus B_5})\overline{A}_4 B_4 + (\overline{A_5 \oplus B_5})(\overline{A_4 \oplus B_4})\overline{A}_3 B_3$

$+ (\overline{A_5 \oplus B_5})(\overline{A_4 \oplus B_4})(\overline{A_3 \oplus B_3})\overline{A}_2 B_2$

$+ (\overline{A_5 \oplus B_5})(\overline{A_4 \oplus B_4})(\overline{A_3 \oplus B_3})(\overline{A_2 \oplus B_2})\overline{A}_1 B_1$

$+ (\overline{A_5 \oplus B_5})(\overline{A_4 \oplus B_4})(\overline{A_3 \oplus B_3})(\overline{A_2 \oplus B_2})(\overline{A_1 \oplus B_1})\overline{A}_0 B_0$

(A > B)

$= A_5 \overline{B}_5 + (\overline{A_5 \oplus B_5})A_4 \overline{B}_4 + (\overline{A_5 \oplus B_5})(\overline{A_4 \oplus B_4})A_3 \overline{B}_3$

$+ (\overline{A_5 \oplus B_5})(\overline{A_4 \oplus B_4})(\overline{A_3 \oplus B_3})A_2 \overline{B}_2$

$+ (\overline{A_5 \oplus B_5})(\overline{A_4 \oplus B_4})(\overline{A_3 \oplus B_3})(\overline{A_2 \oplus B_2})A_1 \overline{B}_1$

$+ (\overline{A_5 \oplus B_5})(\overline{A_4 \oplus B_4})(\overline{A_3 \oplus B_3})(\overline{A_2 \oplus B_2})(\overline{A_1 \oplus B_1})A_0 \overline{B}_0$

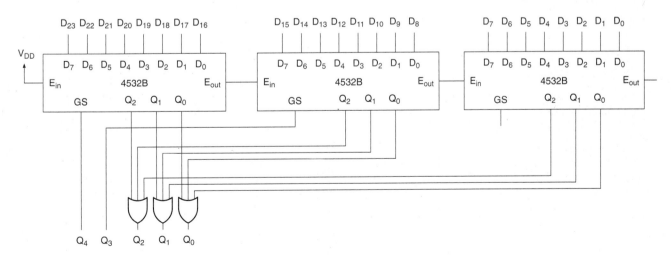

**Figure ANS 10.21**

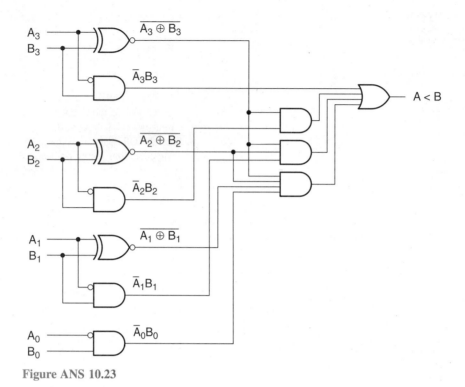

**Figure ANS 10.23**

**10.27**  (A < B)= 1; (A = B) = 0; (A > B) = 0

**10.29**  (A < B) = 1; (A = B) = 0; (A > B) = 0. In the previous comparator stage the A bits are less than the B bits.

**10.31**  See Figure ANS10-31.

**10.33**  **a.** ODD;   **b.** EVEN;   **c.** ODD;
**d.** EVEN;   **e.** ODD

**10.35**  See Figure ANS10-35.

**10.37**  See Figure ANS10-37.

**10.39**  See Figure ANS10-39.

**10.41**  See Figure ANS10-41.

**10.43**  See Figure ANS10-43.

**10.45**  See Figure 10-58a.

**10.47**  00110011

**10.49**  Circuit: $Q_D$ connects to select input D instead of G. MUX should be permanently enabled by active enable inputs. Timing diagram is same as 10.60b. except that ASCII "m" = 6D.

**10.51**  See Figure ANS10-51.

**10.53**  The diagram of the 74HC139A DMUX/Decoder is shown in Figure 10.4.

**10.55**  The circuit is similar to Figure 10.63. The 4-bit counter drives inputs DCBA. One enable input is grounded. The other receives the multiplexed data.

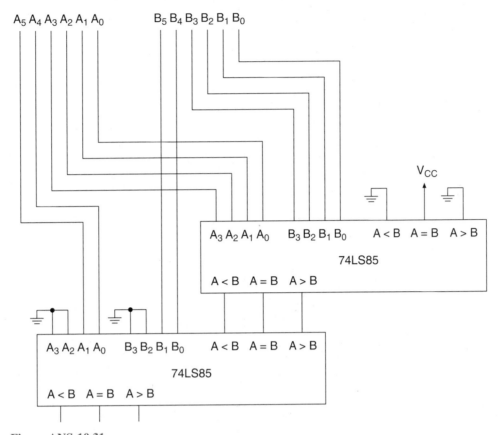

**Figure ANS 10.31**

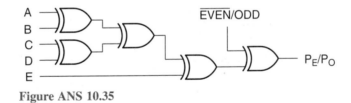

**Figure ANS 10.35**

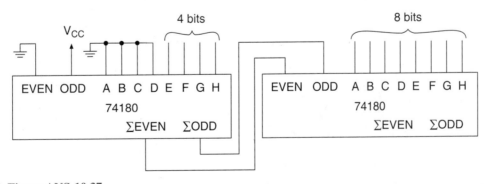

**Figure ANS 10.37**

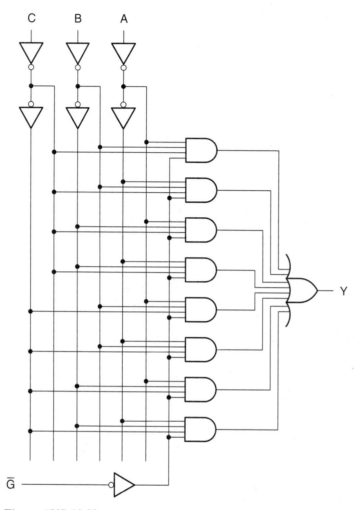

**Figure ANS 10.39**

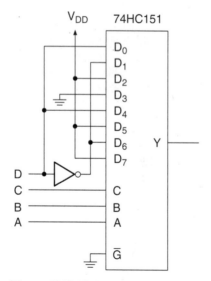

**Figure ANS 10.41**

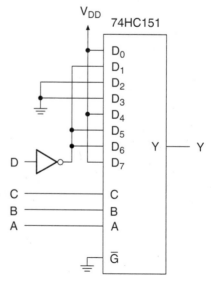

**Figure ANS 10.43**

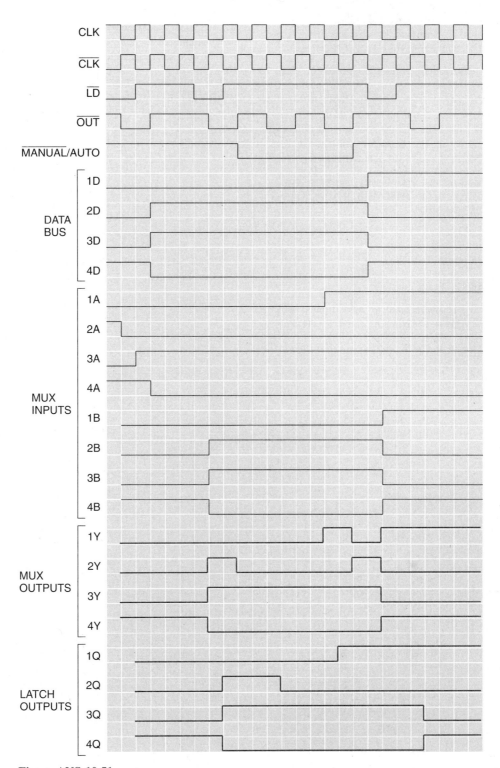

**Figure ANS 10.51**

**10.59** The truth table for each digit is similar to that listed in Table 10.6. The PLDasm source and JEDEC files are included on the instructor's diskette in files MSI\312DCDR.PDS and MSI\312DCDR.JED.

**10.61** The EQUATIONS section of the PLDasm source file is: EQUATIONS
AEQB = /(A7:+:B7)*/(A6:+:B6)*/(A5:+:B5)*/(A4:+:B4)
* /(A3:+:B3)*/(A2:+:B2)*/(A1:+:B1)*/(A0:+:B0)
The complete source file and the corresponding JEDEC file are contained on the instructor's diskette in files \MSI\8BITCMP.PDS and MSI\8BITCMP.JED.

**10.63** The PLDasm source and JEDEC files are located on the instructor's diskette in files \MSI\MUXDMUX8.PDS and \MSI\MUXDMUX8.JED

## Chapter 11

**11.1** See Figure ANS11-1.

**11.3** See Figure ANS11-3.

**11.5** 6 flip-flops; 6 bits are required to count from 0 to 63 (000000 to 111111). Maximum count = $111111_2$ = $63_{10}$.

**11.7** 64; 128; 256

**11.9** See Figure ANS11-9 for state diagram and timing diagram.
Count sequence:

| $Q_D$ | $Q_C$ | $Q_B$ | $Q_A$ | |
|---|---|---|---|---|
| 1 | 0 | 0 | 1 | |
| 1 | 0 | 0 | 0 | |
| 0 | 1 | 1 | 1 | |
| 0 | 1 | 1 | 0 | |
| 0 | 1 | 0 | 1 | |
| 0 | 1 | 0 | 0 | |
| 0 | 0 | 1 | 1 | |
| 0 | 0 | 1 | 0 | |
| 0 | 0 | 0 | 1 | |
| 0 | 0 | 0 | 0 | (recycle) |

**11.11** 4.8 kHz; 24 kHz; the $Q_B$ and $Q_C$ waveforms do not fit neatly into the mod-10 count cycle as they are both LOW at the beginning *and* the end of the cycle.

**11.13** The circuit is like Figure 11.10, but with one less flip-flop. Maximum count: $1111_2$ (= $15_{10}$). Modulus: 16

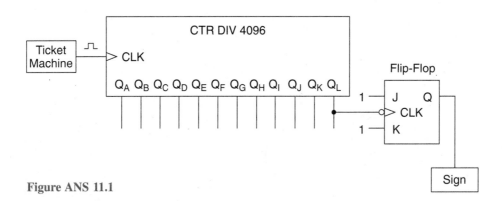

**Figure ANS 11.1**

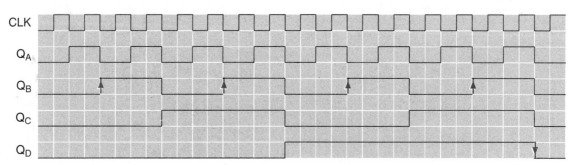

**Figure ANS 11.3**

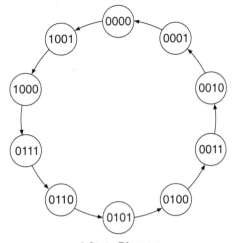

a) State Diagram

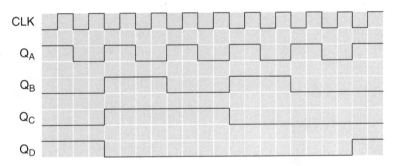

b) Timing Diagram

**Figure ANS 11.9**

**11.15** [See Figure ANS11-15.]

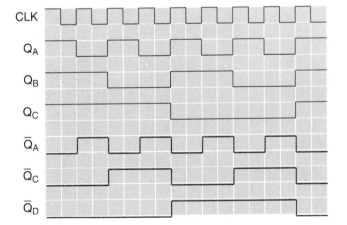

**Figure ANS 11.15**

**11.17** See Figure ANS11-17.

**11.19** The output waveforms are shifted by one half clock cycle relative to the CLK input.

**11.21** Count sequence:

| $Q_D$ | $Q_C$ | $Q_B$ | $Q_A$ | |
|---|---|---|---|---|
| 0 | 0 | 0 | 0 | |
| 0 | 0 | 0 | 1 | |
| 0 | 0 | 1 | 0 | |
| 0 | 0 | 1 | 1 | |
| 0 | 1 | 0 | 0 | |
| 0 | 1 | 0 | 1 | |
| 0 | 1 | 1 | 0 | |
| 0 | 1 | 1 | 1 | |
| 1 | 0 | 0 | 0 | |
| 1 | 0 | 0 | 1 | |
| 1 | 0 | 1 | 0 | |
| 1 | 0 | 1 | 1 | |
| (1 | 1 | 0 | 0) | (recycle state) |

See Figure ANS11-21 for the circuit timing diagram.

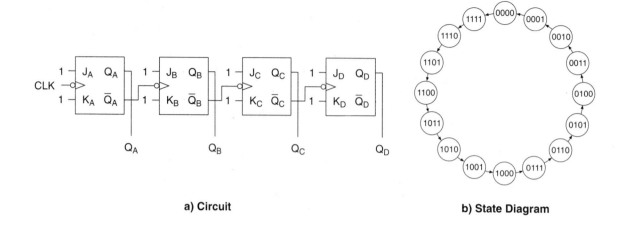

a) Circuit

b) State Diagram

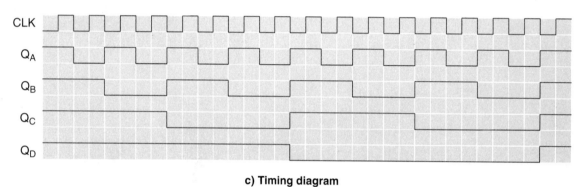

c) Timing diagram

**Figure ANS 11.17**

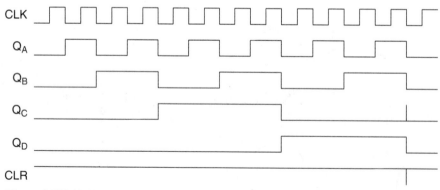

**Figure ANS 11.21**

**11.23** Count sequence:

| $Q_D$ | $Q_C$ | $Q_B$ | $Q_A$ | |
|---|---|---|---|---|
| 0 | 0 | 1 | 1 | |
| 0 | 1 | 0 | 0 | |
| 0 | 1 | 0 | 1 | |
| 0 | 1 | 1 | 0 | |
| 0 | 1 | 1 | 1 | |
| 1 | 0 | 0 | 0 | |
| 1 | 0 | 0 | 1 | |
| 1 | 0 | 1 | 0 | |
| 1 | 0 | 1 | 1 | |
| 1 | 1 | 0 | 0 | |
| (1 | 1 | 0 | 1) | (recycle state) |

See Figure ANS11-23 for the circuit, state diagram and timing diagram.

**11.25** Use a 4-input NAND gate connected to all four counter outputs.

**11.27** See Figure ANS11-27.

**11.29** The maximum frequency has been exceeded because when all outputs change states, the most significant bit output does not change within one clock period.

**11.31** 14.8 MHz

**11.33** See Figure 11.29.

**11.35** Add two flip-flops for outputs $Q_E$ and $Q_F$. Add two AND gates to combine the previous flip-flop output with the previous AND gate output. The AND output connects to J and K of the next flip-flop.

**11.37** $J_A = K_A = 1$
$J_B = K_B = \overline{Q}_A$
$J_C = K_C = \overline{Q}_A \cdot \overline{Q}_B$
$J_D = K_D = \overline{Q}_A \cdot \overline{Q}_B \cdot \overline{Q}_C$
$J_E = K_E = \overline{Q}_A \cdot \overline{Q}_B \cdot \overline{Q}_C \cdot \overline{Q}_D$
$J_F = K_F = \overline{Q}_A \cdot \overline{Q}_B \cdot \overline{Q}_C \cdot \overline{Q}_D \cdot \overline{Q}_E$
$J_G = K_G = \overline{Q}_A \cdot \overline{Q}_B \cdot \overline{Q}_C \cdot \overline{Q}_D \cdot \overline{Q}_E \cdot \overline{Q}_F$
$J_H = K_H = \overline{Q}_A \cdot \overline{Q}_B \cdot \overline{Q}_C \cdot \overline{Q}_D \cdot \overline{Q}_E \cdot \overline{Q}_F \cdot \overline{Q}_G$

**11.39** Boolean Equations: $J_A = K_A = 1$; $J_B = \overline{Q}_C Q_A$,
$K_B = Q_A$; $J_C = Q_B Q_A$, $K_C = Q_A$
modulus = 6; unused states: $110 \rightarrow 111$; $111 \rightarrow 000$.

**11.41** Boolean Equations: $J_A = \overline{Q}_B + \overline{Q}_D$, $K_A = 1$;
$J_B = Q_A$, $K_B = Q_A + Q_D$;
$J_C = K_C = Q_B Q_A$
$J_D = Q_C Q_B Q_A$, $K_D = Q_B$

modulus = 11;
unused states: $1011 \rightarrow 0100$; $1100 \rightarrow 1101$;
$1101 \rightarrow 1110$; $1110 \rightarrow 0100$;
$1111 \rightarrow 0000$.

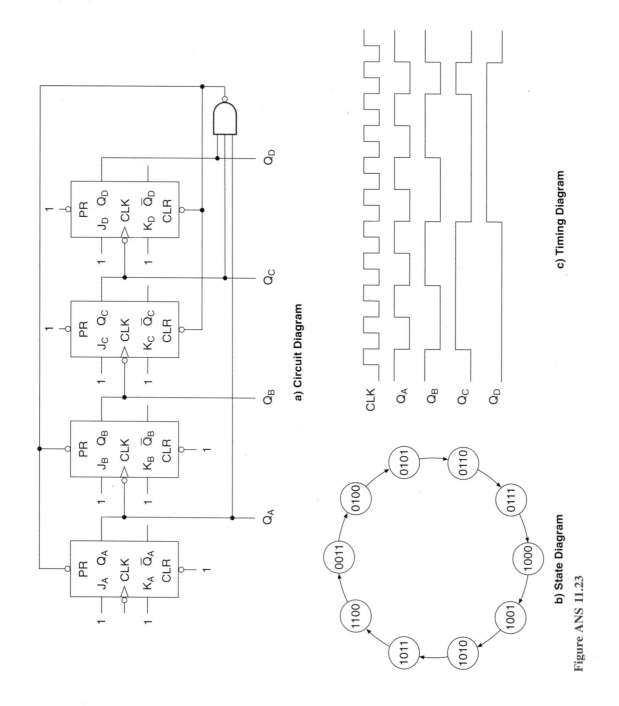

a) Circuit Diagram

b) State Diagram

c) Timing Diagram

**Figure ANS 11.23**

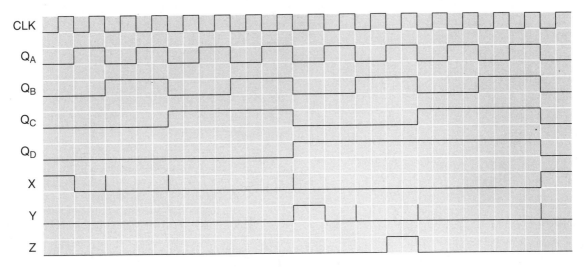

**a) Timing Diagram, showing glitches**

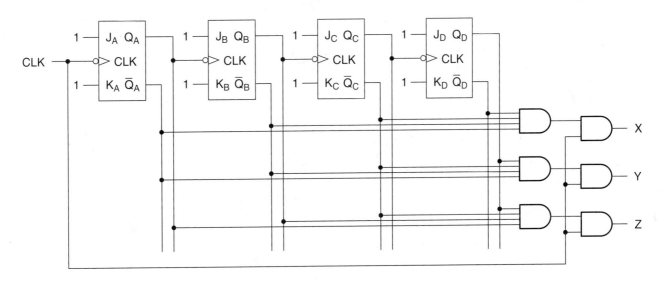

**b) Modified circuit**

**Figure ANS 11.27**

## Chapter 12

**12.1**  **a.** 10;  **b.** 12;  **c.** 16

**12.3**  See Figure 12-2 in the text.

**12.5**  See Figure ANS12-5.

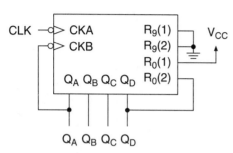

**Figure ANS 12.5**

**12.7**  256

**12.9**  60 seconds

**12.11**  See Figure ANS12-11.

**12.13**  [See Figure 12-18 in the text.]

**12.15**  [See Figure 12-17 in the text.]

**12.17**  Synchronous LOAD transfers the parallel input values to the counter flip-flops when the counter is clocked. Asynchronous LOAD makes the transfer immediately, without waiting for the clock. Figure 12.11 shows an example of each.

**12.19**  The required sampling interval is 1 μs. This is the same period as the pulse generated by the reset monostable in the orginal design. During the reset time, the frequency counter is not functional. To eliminate this problem, the reset pulse must be shortened as much as possible, preferably to at least 1/100 of the sampling interval. (This implies a reset pulse not longer than about 10 ns, which may not be possible.)

**12.21**  The circuit is the same as Figure 12.20, with DCBA = 1100.

**12.23**  90

**12.25**  The circuit is the same as Figure 12.49, with the following values, from most to least significant, at the parallel inputs: 10101111. (The MSB is on the far right in Figure 12.49.)

**12.27**  The 74LS160 counter is identical to the 74LS161, except for the modulus. The 74LS160 is mod-10 and the 74LS161 in mod-16. The description of the '160 is similar to that of the '161 given in the text. The main difference is that RCO activates when the counter output is 1001. This is indicated by the notation 3CT = 9 at the RCO output, indicating a dependency on the CLK and the output count.

**12.29**  The outputs of a Moore machine depend only on the present states of its internal flip-flops. The outputs of a Mealy machine depend on its flip flop states and the states of the inputs to the combinational part of the circuit.

**12.33**  See Figure ANS12-33.

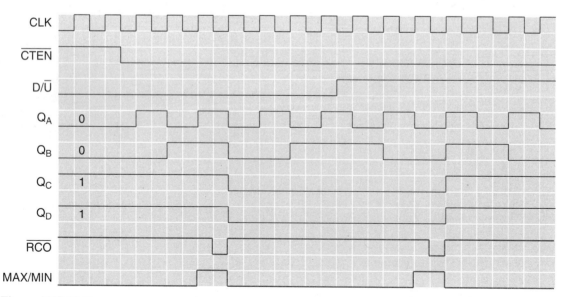

**Figure ANS 12.11**

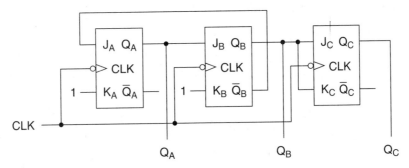

**Figure ANS 12.33**

**12.35** Configure the PLD as two separated BCD counters, one for the Ones digit and and one for the Tens. The Tens digit is synchronously enabled by the Ones when the latter equals 1001.

The complete source file and JEDEC file for this application are found on the instructor's diskette in files MSICTR\2DIGBCD.PDS and MSICTR\2DIGBCD.JED.

**Chapter 13**

**13.1**  See Figure ANS13-1.

**13.3**  See Figure ANS13-3.

**13.5**  See Figure ANS13-5.

**13.7**  See Figure ANS13-7.

**13.9**  See Figure ANS13-9.

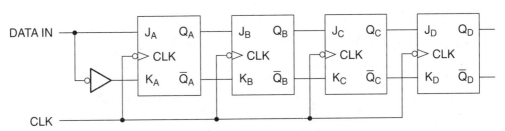

**Figure ANS 13.1**

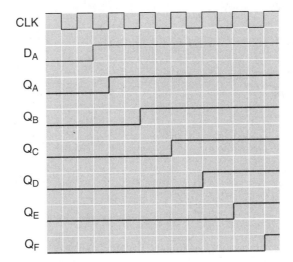

**Figure ANS 13.3**

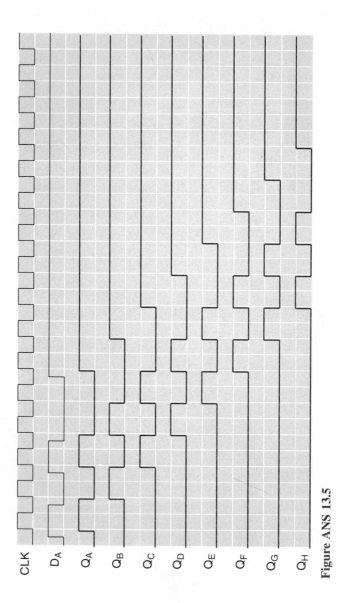

**Figure ANS 13.5**

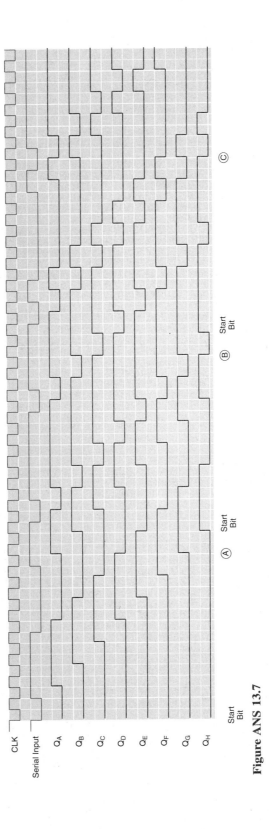

**Figure ANS 13.7**

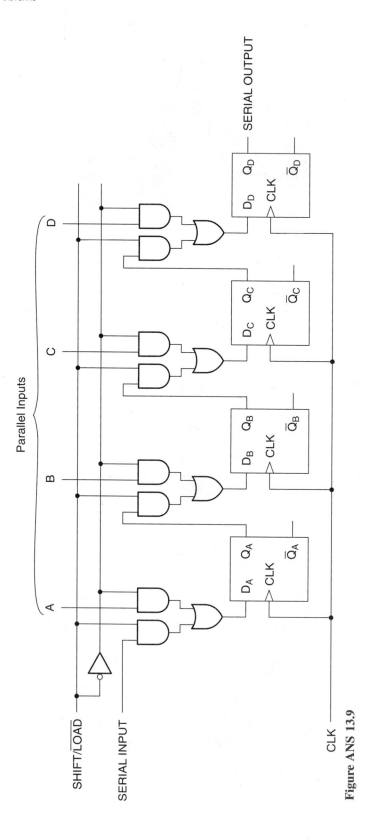

**Figure ANS 13.9**

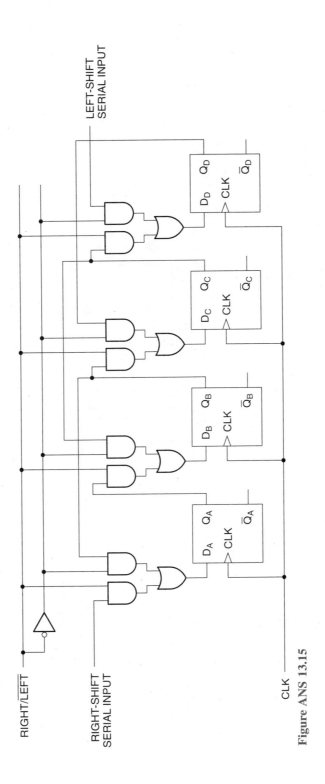

**Figure ANS 13.15**

**13.11**  The shift register must be of such a size that no 1s get shifted out of the register during a multiply or divide operation.

**13.13**  Logic 0.

**13.15**  See Figure ANS13-15.

**13.17**  See Figure ANS13-17.

**13.19**  See Figure ANS13-19.

**13.21**  See Figure ANS13-21.

**13.23**  See Figure ANS13-23. This figure shows the circuit for the 9-state Johnson counter. The tenth decoding gate resets the D flip-flops.

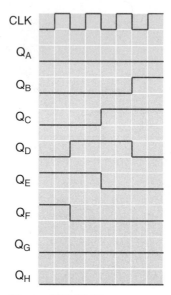

**Figure ANS 13.17**

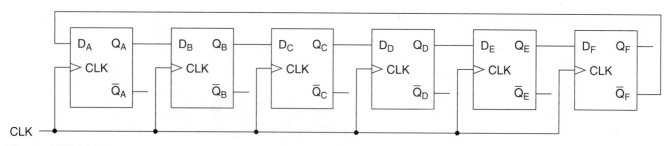

**Figure ANS 13.19**

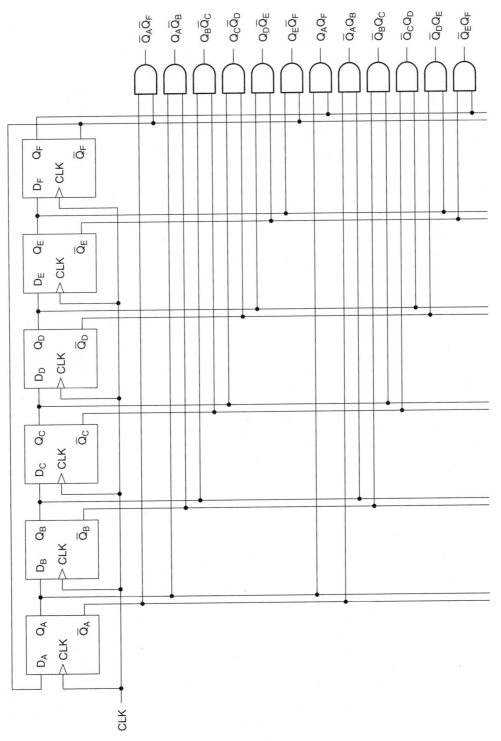

**Figure ANS 13.21**

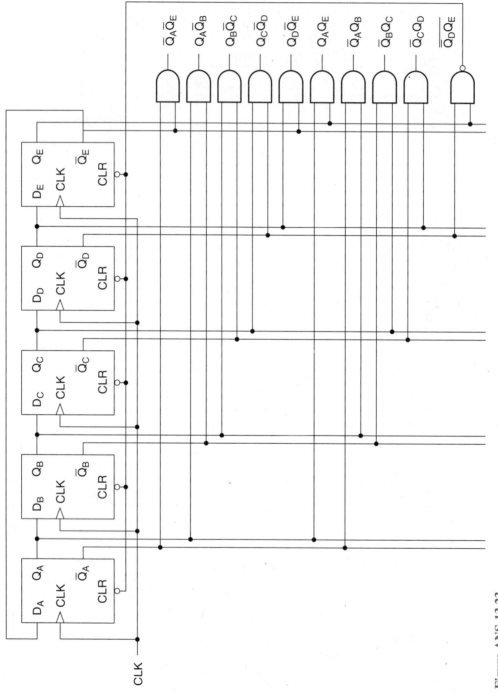

**Figure ANS 13.23**

## Chapter 14

**14.1** TTL: advantages—relatively high speed, high current driving capability; disadvantages—high power consumption, rigid power supply requirements. CMOS: advantages—low power consumption, high noise immunity, flexible power supply requirements; disadvantages—low output current, slow switching speed (except for High-Speed CMOS). ECL: advantages—high speed; disadvantages—high susceptibility to noise, high power consumption.

**14.3** TTL: $t_{pHL} = 12$ ns, $t_{pLH} = 10$ ns; CMOS: $t_{pHL} = 200$ ns, $t_{pLH} = 220$ ns

**14.5** **a.** See Figure ANS14-5.

   **b.** 500 ns **c.** 250 ns

**14.7** $t_{p1} = t_{pLH1} + t_{pLH3} + t_{pLH9}$
$t_{p3} = t_{pHL1} + t_{pHL3} + t_{pHL9}$
OR
$t_{p3} = t_{pHL4,5} + t_{pHL6,7} + t_{pHL8} + t_{pHL9}$

**14.9** $n = 4$

**14.11** $n = 8$

**14.15** **a.** $P_d = 5$ μW; **b.** $P_d = 32.5$ μW; **c.** $P_d = 5.5$ mW

**14.17** **a.** $V_{nH} = 1.45$ V ; $V_{nL} = 1.45$ V

   **b.** [$V_{nH} = 4.35$ V ; $V_{nL} = 3.95$ V]

   **c.** $V_{nH} = 0.7$ V ; $V_{nL} = 0.3$ V

   **d.** $V_{nH} = 0.7$ V ; $V_{nL} = 0.3$ V

   **e.** $V_{nH} = 1.0$ V ; $V_{nL} = 0.3$ V

   **f.** $V_{nH} = 1.0$ V ; $V_{nL} = 0.3$ V

   **g.** $V_{nH} = 0.65$ V ; $V_{nL} = 0.95$ V

   **h.** $V_{nH} = 1.7$ V ; $V_{nL} = 0.4$ V

**14.19** $n = 1$

**14.21** The inputs of a 74HCT series gate are voltage compatible with LSTTL outputs. This is not the case for 4000B or 74HC series gates.

**14.23** [See Figure ANS14-23.]

**14.25** See Figure ANS14-25.

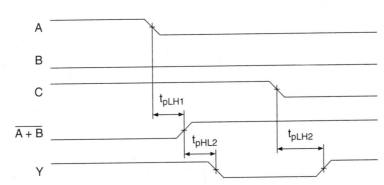

**Figure ANS 14.5**

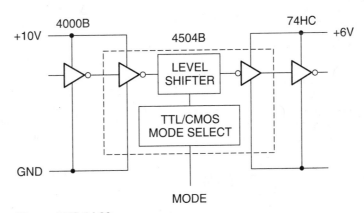

**Figure ANS 14.23**

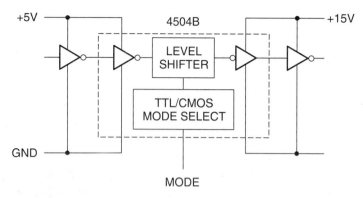

NOTE THAT A 4050B BUFFER CANNOT
INTERFACE A SOURCE AND LOAD WITH
THE SPECIFIED SUPPLY VOLTAGES.

**Figure ANS 14.25**

## Chapter 15

**15.1** Logic HIGH. An open input provides no low impedance path from input to ground.

**15.3** The multiple-emitter transistor input acts as a LOW if any emitter is LOW, due to the low impedance path to ground it provides. The transistor acts as a HIGH only if all inputs are HIGH or open.

**15.5** Calculated: 389 $\Omega$; Standard value: 390 $\Omega$

**15.7** No. No current will flow in the output transistor since its collector and emitter are both at ground potential.

**15.9** See Figure 15.21b.

**15.11** If one output is HIGH and the other is LOW, current in excess of the maximum safe value will flow from the HIGH output to the LOW and both outputs will be damaged over a period of several minutes.

**15.13**

| A | B | $Q_1$ | $Q_2$ | $Q_3$ | $Q_4$ | $Q_5$ | $Q_6$ | Y |
|---|---|---|---|---|---|---|---|---|
| 0 | 0 | ON | ON | OFF | OFF | ON | OFF | 1 |
| 0 | 1 | ON | OFF | OFF | ON | OFF | ON | 0 |
| 1 | 0 | OFF | ON | ON | OFF | OFF | ON | 0 |
| 1 | 1 | OFF | OFF | ON | ON | OFF | ON | 0 |

**15.15** Store MOS devices in antistatic or conducting material. Work only on an antistatic work surface and wear a conductive wrist strap. Connect all unused device inputs to power or ground. Do not touch the pins of the MOS device, if possible.

**15.17** **a.** An NMOS transistor is smaller than an integrated resistor. **b.** Equal values of ON-resistance give an output voltage of $V_{DD}/2$ for logic LOW. Different values are chosen to provide a more favorable voltage divider ratio. **c.** $V_{DD}/2$

**15.19** See Figure 15.45.

| A | B | $Q_1$ | $Q_2$ | $Q_3$ | $Q_4$ | $Q_5$ | $Q_6$ | Y |
|---|---|---|---|---|---|---|---|---|
| 0 | 0 | ON | ON | OFF | OFF | OFF | ON | 0 |
| 0 | 1 | OFF | ON | OFF | ON | OFF | ON | 0 |
| 1 | 0 | ON | OFF | ON | OFF | OFF | ON | 0 |
| 1 | 1 | OFF | OFF | ON | ON | ON | OFF | 1 |

**15.21** The state of flip-flop output Q selects which signal is switched to the data converter/display driver by enabling one of the CMOS transmission gates. When Q = 1, the wheel rotation sensor is selected. Q = 0 selects the engine rotation sensor.

**15.23** No. TTL power dissipation, and therefore the speed-power product, depends on the logic state of the device outputs, not on frequency.

**15.25** 74HCXX: pin replacement for TTL device; CMOS-compatible inputs; TTL-compatible outputs. 74HC4XXX: pin replacement for CMOS device; CMOS-compatible inputs; TTL-compatible outputs. 74HCTXX: pin replacement for TTL device; TTL-compatible inputs; TTL-compatible outputs. 74HCUXX: unbuffered CMOS outputs.

**15.27** 1 GHz (ECL); 30 MHz (LSTTL); 75 MHz (ASTTL)

**15.29** HIGH: −0.9 V; LOW: −1.7 V

**15.31** ECL noise immunity is relative poor because the HIGH and LOW voltages are close together, compared to other logic families. This can be improved by tying the common-mode ($V_{CC}$) terminals to ground.

**15.33** High-Speed CMOS.

**Chapter 16**

**16.1**

| Analog Voltage | Code |
|---|---|
| 0.0 V – 1.5 V | 000 |
| 1.5 V – 3.0 V | 001 |
| 3.0 V – 4.5 V | 010 |
| 4.5 V – 6.0 V | 011 |
| 6.0 V – 7.5 V | 100 |
| 7.5 V – 9.0 V | 101 |
| 9.0 V – 10.5 V | 110 |
| 10.5 V – 12.0 V | 111 |

**16.3**

| Analog Voltage | Code |
|---|---|
| 0.00 V – 0.75 V | 0000 |
| 0.75 V – 1.50 V | 0001 |
| 1.50 V – 2.25 V | 0010 |
| 2.25 V – 3.00 V | 0011 |
| 3.00 V – 3.75 V | 0100 |
| 3.75 V – 4.50 V | 0101 |
| 4.50 V – 5.25 V | 0110 |
| 5.25 V – 6.00 V | 0111 |
| 6.00 V – 6.75 V | 1000 |
| 6.75 V – 7.50 V | 1001 |
| 7.50 V – 8.25 V | 1010 |
| 8.25 V – 9.00 V | 1011 |
| 9.00 V – 9.75 V | 1100 |
| 9.75 V – 10.50 V | 1101 |
| 10.50 V – 11.25 V | 1110 |
| 11.25 V – 12.00 V | 1111 |

| Fraction of T | Sine Voltage | Digital Code |
|---|---|---|
| 0 | 0 V | 0000 |
| T/8 | 4.59 V | 0110 |
| T/4 | 8.48 V | 1011 |
| 3T/8 | 11.09 V | 1110 |
| T/2 | 12.00 V | 1111 |
| 5T/8 | 11.09 V | 1110 |
| 3T/4 | 8.48 V | 1011 |
| 7T/8 | 4.59 V | 0110 |
| T | 0 V | 0000 |

**16.5** 6 bits. 7.8125 mV.

**16.7** n + 1 (One extra bit for each doubling of the number of codes.)

**16.9** **a.** 0.75 $V_{ref}$ **b.** 0.78125 $V_{ref}$ **c.** A 4-bit and 8-bit quantization of the same analog voltage are the same in the first four bits (as in Problem 16.8).

**16.11** From most to least significant bits: 1 kΩ, 2 kΩ, 4 kΩ, 8 kΩ, 16 kΩ, 32 kΩ, 64 kΩ, 128 kΩ, 256 kΩ, 512 kΩ, 1024 kΩ, 2048 kΩ, 4096 kΩ, 8192 kΩ, 16,384 kΩ,

32,768 kΩ. All resistors greater than 64 kΩ are specified to three or more significant figures. These values, which are necessary to maintain conversion accuracy, are not available as commercial components.

**16.13** **a.** 11.25 V; **b.** 8.25 V; **c.** 4.8 V; **d.** 2.25 V

**16.15** 15.16 mV

**16.17** The op amp is saturating on high input codes. This is caused by too high a value of feedback resistance in the op amp circuit or a reference current that is too large.

**16.19** The procedure is described in Example 16.8. $R_{14} = 5$ kΩ, $R_{F2} = 2$ kΩ.

**16.21** The priority encoder converts the highest active comparator voltage to a digital code. The enable input of the latch can be pulsed with a waveform having the same frequency as the sampling frequency.

**16.23**

| Bit | New Digital Value | Analog Equivalent | $V_{analog} \geq V_{DAC}$? | Comparator Output | Accumulated Digital Value |
|---|---|---|---|---|---|
| $Q_7$ | 10000000 | 8 V | No | 0 | 00000000 |
| $Q_6$ | 01000000 | 4 V | Yes | 1 | 01000000 |
| $Q_5$ | 01100000 | 5.5 V | No | 0 | 01000000 |
| $Q_4$ | 01010000 | 3.75 V | Yes | 1 | 01010000 |
| $Q_3$ | 01011000 | 4.125 V | Yes | 1 | 01011000 |
| $Q_2$ | 01011100 | 4.3125 V | Yes | 1 | 01011100 |
| $Q_1$ | 01011110 | 4.40625 V | Yes | 1 | 01011110 |
| $Q_0$ | 01011111 | 4.453125 V | Yes | 1 | 01011111 |

Final answer: 01011111.

**16.25** The new hex digit is A.

**16.27** **a.** −1.75 V/ms; **b.** +2 V/ms; **c.** Integrating phase: 8 ms, Rezeroing phase: 14/16 of FS: 7 ms; **e.** 11100000

**16.29** **a.** −2.25 V/ms; **b.** +2 V/ms; Integrating *and* rezeroing phases: 8 ms (can't rezero); **e.** 11111111.

**16.31** The analog signal may be changing at a faster rate than the conversion cycle of the ADC. To perform the conversion accurately, the analog voltage must be held constant for at least the conversion time.

**Chapter 17**

**17.1** 8 bits can represent two BCD digits (1001 0101 = 95), two hex digits (1110 0111 = E7), or one ASCII character (01000010 = "A").

**17.3** Data can be written to and read from RAM. Data can only be read from ROM.

**17.5** Chip Enable must be active to allow a Read or a Write function to occur in a memory device. Output Enable is the enabling input of a tristate read (output) buffer in a memory device.

**17.7** See Figure 17.7b.

**17.9** The load transistors are MOSFETs that are always biased in the ON state. Since their channel ON resistance is a constant value, they act as pullup resistors. This function is performed by MOSFETs because they take up less chip space and are easier to manufacture than integrated resistors.

**17.11** A selected RAM cell is at the junction of an active ROW line and an active COLUMN line in a rectangular matrix of cells.

**17.13** Array size: 256 rows × 1024 columns
Row address lines: 8
Column address lines: 10
Words per row: 256
Bits per word: 4

**17.15** 10 multiplexed address lines.

**17.17** See Figure 17.12.

**17.19** See Figure ANS17-19.

**17.21** See Figure 17.18.

**Figure ANS 17.19**

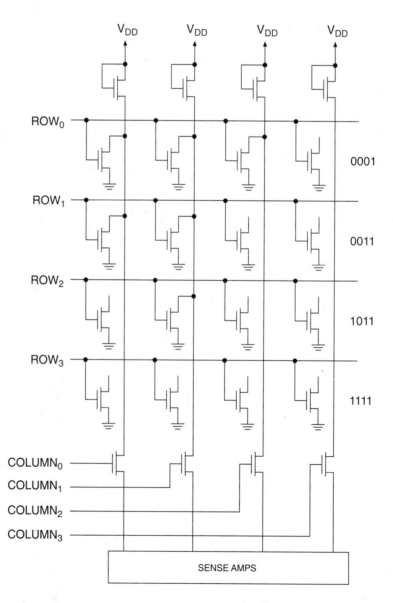

## Chapter 18

**18.1**  **a:** $t_{AVAV}$;  **b:** $t_{AVWH}$;  **c:** $t_{WLWH}$;  **d:** $t_{AVWL}$;
**e:** $t_{DVWH}$;  **f:** $t_{WLQZ}$

**18.3**  See Figure 18.10.

**18.5**  Both cycles read the contents of a memory location, then modify it with a write pulse. The Read-While-Write cycle performs the write sooner than the Read-Modify-Write cycle.

**18.7**  $\overline{RAS}$-only refresh—Uses the Row Address Strobe to clock in individual row address to be refreshed.

$\overline{CAS}$ before $\overline{RAS}$ refresh—The Column Address Strobe goes LOW, followed by Row Address Strobe. External addresses are ignored and the $\overline{RAS}$ line acts as a clock to an internal row counter used only for the refresh cycle.

Hidden Refresh—Refresh cycles occur between the individual Read and Write cycles of the RAM.

**18.9**  See Figure ANS18-9.

**18.11**  See Figure ANS18-11.

**18.13**  See Figure ANS18-13.

**18.15**  See Figure ANS18-15.

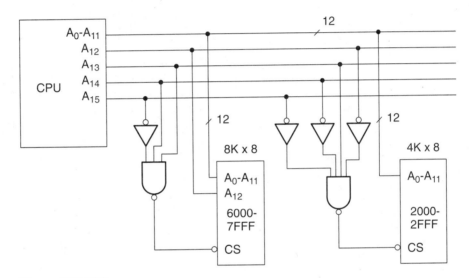

**Figure ANS 18.9**

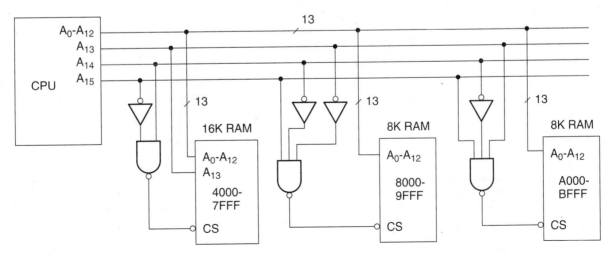

**Figure ANS 18.11**

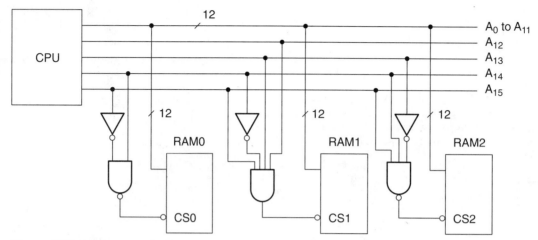

**Figure ANS 18.13**

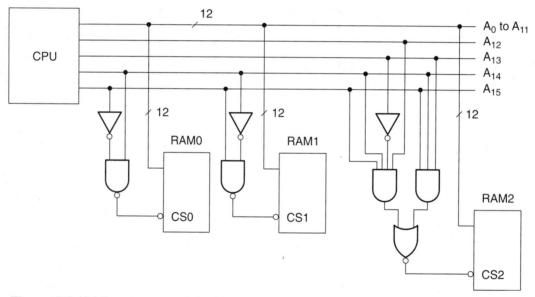

**Figure ANS 18.15**

**18.17** $RAM_0$: outputs $Y_4$ and $Y_5$; $RAM_1$: outputs $Y_8$, $Y_9$, $Y_{10}$, and $Y_{11}$; $RAM_2$: outputs $Y_{14}$ and $Y_{15}$. The decoder connections are similar to those in Figure 18-35.

**18.19** See Figure ANS18-19.

**18.21** See Figure ANS18-21.

**18.23 a.** 24 bits. The upper 12 bits of each address are contained in the mapping resgisters of the memory map-

per. The registers are accessed by the upper 4 bits of the CPU address bus. The CPU address bus generates the lower 12 bits of each address directly.

**b.** i. 01966A;  ii. 01F66A;  iii. 02C66A;
    iv. 0B266A;  v. 954806;  vi. EF7C92;
    vii. FBDC92

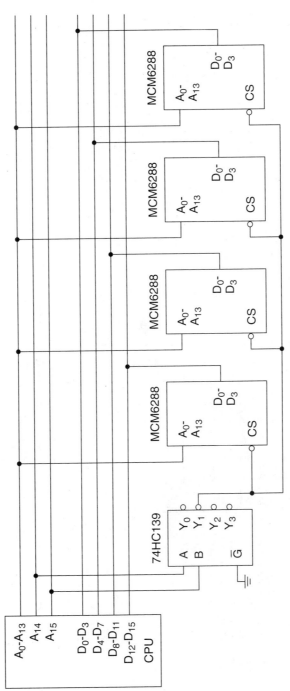

**Figure ANS 18.19**

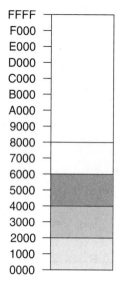

**Figure ANS 18.21a**

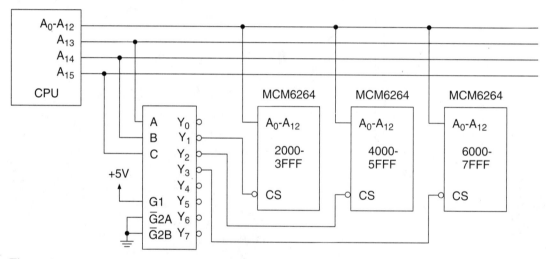

**Figure ANS 18.21b**

ABEL (Date I/O), 301
Absolute accuracy (of D/A converter), 714
Accumulator, 352
Active-HIGH terminal, 23
Active levels, 23–24
    of AND functions, 23
    of OR functions, 23
Active-LOW terminal, 23
Adder/subtractor with overflow indication, 364–65
Address, 736, 738
Address decoder, 789, 790
Address multiplexing, 750
Address space, 789, 790
Adjacent cell (in Karnaugh map), 109
Alphanumeric code, 172, 173
Amplitude, 12, 13
Analog and digital signals, 692–93
    analog-to-digital conversion, 716
        dual slope A/D converter, 722–27
        flash A/D converter, 716–18
        sample and hold circuit, 727–28
        successive approximation A/D converter, 718–22
    data acquisition, 728–29
    digital-to-analog conversion, 697–99
        bipolar operation of MC1408, 709–13
        DAC performance specifications, 714–15
        MC1408 integrated circuit D/A converter, 704–7
        op amp buffering of MC1408, 707–9
        R-2R ladder D/A converter, 701–4
        weighted resistor D/A converter, 699–701
    sampling an analog voltage, 693–97
Analog representation, 2, 692, 693
Analog-to-digital conversion, 716
    dual slope A/D converter, 722–27
    flash A/D converter, 716–18
    sample and hold circuit, 727–28
    successive approximation A/D converter, 718–22
Analog-to-digital converter, 692, 693
AND dependency, 525, 526
AND function, 20–21, 22
    active levels of, 23
    available SSI gates, 136
    Boolean expression of, 73–74
AND gate, 22
    Boolean expression for, 65
    enable and inhibit properties of, 38–39

AND/OR gates, 674–75
Aperiodic waveforms, 11–12
Architecture cell, 283, 286
Arithmetic logic unit (ALU), 352
ASCII, 172
ASCII code, 172–74
Associative property of addition, 86
Associative property of multiplication, 86
Astable multivibrator, 555 timer as, 253–56
Astable multivibrator, 242–43
Asynchronous counters, 463–65
    maximum frequency of, 479–81
    truncated sequence counters, 468–72
    UP counters and DOWN counters, 465–68
Asynchronous inputs (preset and clear), 215–18
Asynchronous MSI counters, 505–9
Audio reproduction, digital versus analog, 3

Base, of number system, 158
b (bit), 739
B (byte), 739
BCD adders, 357–58
    carry output, 358–59
    implementation of, 365–66
    multiple-digit BCD adders, 359–61
    sum correction, 359
BCD codes, 170–71
    8421 code, 171
    excess-3 code, 171–72
BCD priority encoders, 399–400
BCD-to-seven-segment decoders, 386
    decoder, 388–89
    display, 386–87
    MSI seven-segment decoders, 389–94
BiCMOS logic, internal circuitry of, 684
Bidirectional shift registers, 577–80
    74LS194A universal shift register, 580–83
Bidirectional synchronous counters, 509–13
Binary adders and subtractors, 337
    half and full adders, 337–38
        Boolean algebra method, 338–42
        Karnaugh map method, 338
    parallel binary adder/subtractor, 342–46
Binary-coded decimal (BCD), 170–71
Binary counter, 460
Binary inputs, 6–9

Binary numbers, 159
  counting in binary, 159–60
  decimal-to-binary conversion, 160
    repeated division by 2, 161–62
    sum of powers of 2, 160–61
  fractional binary numbers, 163
    fractional-decimal-to-fractional-binary conversion, 163–64
Binary number system, 5
  binary inputs, 6–9
  positional notation, 5–6
Binary point, 163
Binary-to-decimal conversion, 159
Bipolar transistors as logic devices, 638–41
Biquinary sequence, 505, 506
Bistable multivibrator, 240
Bit, 5
  Least significant (LSB), 6–7
  most significant (MSB), 6–7
Bit-organized memory, 750
Boolean algebra, 20, 64
  combinational logic and, 64–65
  rules of, 97
  theorems of, 85
    commutative, associative, and distributive properties, 85–87
    double-variable theorems, 91–96
    single-variable theorems, 87–88: double inversion, 91; $x$ and/or 0/1, 88–89; $x$ and/or $x/\bar{x}$, 89–90
Boolean expressions, 20
  evaluating, 73–74
  from logic diagrams, 65–67
  from truth tables, 78–84
  logic diagrams from, 69–73
  maximum SOP and POS simplification of, 104–8
    DeMorgan equivalent gates, 125–128
    Karnaugh map method of, 109–25
  order of precedence in, 67–69
  truth tables from, 77–78
Boolean function generator, 418–21
Boolean variable, 20
Borrow, 324
Borrowing rules, 325
Breadboard, 42, 43
Bubble, 21
Buffer, 21, 22
Burning, 754
Bus contention, 785, 788
Bus form, 79
Bus waveforms, 769
Byte, 736

Carry, 323
  end-around, 328
  fast, 346–47
  ripple, 342, 343
Carry bit, 323
Carry output, 358–59
Cascade, 342
Cascading MSI counters, 516–19
Case shift, 172, 174
Cell (in PLD), 270, 272
Cell (in Karnaugh map), 109
Ceramic leaded chip carrier (CLCC), 42, 43
Chip, 42
Clear, 215
CLOCK, 198
Clock generator, 47, 48
CMOS, 42, 602

CMOS B-series gates, 621
CMOS logic, propagation delay for, 604
CMOS logic family, 664
CMOS logic gates, 136
  B-series, 610–11
  handling precautions for, 832–33
  input and output parameters of, 627
  interfacing with TTL logic gates, 627–31
  internal circuitry of, 663–64
    AND and OR gates, 674–75
    bias requirements for MOS transistors, 665–67
    inverter, 669–71
    metal-gate CMOS, 680
    MOSFET structure, 664–65
    NAND/NOR gates, 671–74
    NMOS inverter, 667–69
    transmission gate, 675–76
  LSTTL driving 74HC or 4000B CMOS, 629
  LSTTL driving 74HCT CMOS, 629
  pinouts for, 844
  power dissipation in, 620–23
  speed and power specifications, 682
CMOS transmission gate, 429–30
Coincidence gate (XNOR), 28–29
Column address strobe ($\overline{CAS}$), 750
Combinational logic, Boolean algebra and, 64–65
Combinational logic circuits, 64, 104. *See also* Boolean expressions
  designing from word problems, 141–43
  PLDs designed as, 432–41
  practical circuit implementation, 135–39
  propagation delay in, 607–11
  pulsed operation of, 138–41
  troubleshooting, 143–46
Comment, 304
Common-anode display, 386, 387
Common-cathode display, 386, 387
Common control block, 525
Commutative property of addition, 86
Commutative property of multiplication, 86
Comparator, 247, 248
Compile, 301, 302
Complementary metal-oxide semiconductor. *See* CMOS
Complement form, 37, 39
Continuous values, 2, 692, 693
Control dependency, 525, 526
Control gating, 208
Control signals, 741–42
Counters, 422, 456
  asynchronous, 463–72
  asynchronous MSI counters, 503–9
  decoding, 472–73
    maximum frequency of asynchronous counter, 479–81
    propagation delay, glitches, and strobing, 473–78
  digital, 456–63
  with nonstandard sequences, 535–40
  synchronous, 481–94
  synchronous presettable MSI counters, 509–24
Count sequence, 456
Count-sequence table, 460, 461
$C_{PD}$ (internal capacitance of high-speed CMOS device), 615
CPU (central processing unit), 785, 788
CUPL (Logical Devices), 301
Cutoff mode, 639, 640

Data, 736–39
Data acquisition, 728–29
Data acquisition network, 728

Data bus, 352
Data inputs, 415, 416
Decimal-to-binary conversion, 160–62
Decoder, 472–73
   Decoders, 374–75
     BCD-to-seven-segment, 386–94
     MSI, 377
       2- to 4-line (74HC139A), 377–80
       3- to 8-line (74HC138A), 381, 382
       4- to 16-line (74HC154), 381–85
     single-gate, 375–77
DeMorgan equivalent forms, 33–35
   changing gate to, 35–37
DeMorgan equivalent gates, simplification of Boolean expressions
     with, 125–28
DeMorgan's theorems, 33–35, 80
Demultiplexers, 428
   CMOS analog multiplexer/demultiplexer, 429–31
   demultiplexing a TDM signal, 428–29
Dependency notation, 525
Development software (for PLDs), 295–99
   software vendors, 299–301
   using PLDasm, 301–17
Difference, 324
Differential nonlinearity, 715
Digital arithmetic, 322
   hexadecimal arithmetic, 334–36
   representing signed binary numbers, 325–27
   signed binary arithmetic, 327–34
   unsigned binary arithmetic, 323–25
Digital arithmetic circuits
   BCD adders, 357–61
   binary adders and subtractors, 337–52
   parallel binary adder/subtractor, 342–46
   programmable logic implementation of, 361
     adder/subtractor with overflow indication, 364–65
     BCD adder, 365–66
     parallel adder, 361–64
   register arithmetic circuits, 352–56
Digital counters, 456–58
   basic concepts, 458–59
     count sequence table and timing diagram, 460–63
     number of bits and maximum modulus,
       460
     state diagram, 459
Digital electronics, 1–2
Digital function generator, 758–61
   EPROM data for, 834–36
   QuickBASIC program, 837–40
   record file, 840–42
Digital logic gates
   74HC00A, 44
   74HC02A, 44
   74HC04A, 44
   74HC11, 44
   74LS00, 44
   74LS02, 44
   74LS04, 44
   74LS11, 44
   4001B, 44
   4011B, 44
   4069UB, 44
   4073B, 44
Digital logic levels, 4
Digital representation, 2, 692, 693
Digital signal, 37. *See also* Analog and digital signals
Digital signal processor (DSP), 374, 377

Digital-to-analog conversion, 697–99
   DAC performance specifications, 714–15
   MC1408 integrated circuit D/A converter, 704–7
     bipolar operation of, 709–13
     op amp buffering of, 707–9
   R-2R ladder D/A converter, 701–4
   weighted resistor D/A converter, 699–701
Digital-to-analog converter, 692, 693
Digital waveforms, 9–10
   aperiodic, 11–12
   periodic, 10–11
   pulse, 12–15
DIP (dual in-line package), 42, 291–92
Disable, 37
Disassemble, 301, 302
Discrete values, 2, 692, 693
Display
   common-anode, 386, 387
   common-cathode, 386, 387
   seven-segment, 386–87
Distributive properties, 86–87
Don't care states, 120–23, 770
DOWN counter, 459
DRAM. *See* Dynamic RAM (DRAM)
Driving gate, 611, 612
Dual in-line package (DIP). *See* DIP
Dual slope ADC, 722
Dual slope A/D converter, 722–27
Dynamic RAM (DRAM), 743
   timing cycles, 776–77
     fast access cycles, 778–79
     fast access cycles: comparison of fast access modes, 781–83; fast
       page mode, 779; nibble mode, 779–81; static column decode, 781
     normal read cycle, 777
     read-write cycles, 778
     refresh cycles, 783: $\overline{CAS}$ before $\overline{RAS}$ refresh, 783–84; hidden
       refresh cycle, 784; $\overline{RAS}$-only refresh, 783)
     write cycles, 777–78
Dynamic RAM (DRAM) cell arrays, 750–51
Dynamic RAM (DRAM) cells, 749
Dynamic RAM (DRAM) modules, 785, 787–89

ECL (emitter coupled logic), 602
Edge, 12
   falling, 12
   leading, 12, 14
   of pulse waveform, 198
   rising, 12
   trailing, 12, 14
Edge-sensitive (edge-triggered), 198
Edge-triggered D flip-flops, 198, 201–12
Edge-triggered enabling, 198
Edge-triggered JK flip-flops, 212–15
   asynchronous inputs, 215–18
Edge-triggered SR flip-flops, 198–201
EEPROM (electrically erasable programmable read only memory),
     761–62
85C22V10, 292–95
85C060/090, 295
85C220/224, 292
8421 code, 171
Electronics, digital versus analog, 2–3
Emitter-coupled logic, circuitry of, 682–84
ENABLE, 37, 191–92
Encoders, 394
   priority, 394–96
     BCD priority encoders, 399–400

Encoders *(Continued)*
    MSI priority encoders, 397–99
    priority encoder circuit, 396–97
End-around carry, 328
Enhancement-mode MOFSET, 663, 664
EPROM (erasable programmable read only memory), 757–61
Erasable programmable logic devices (EPLD), 291–92
  85C22V10, 292–95
  85C060/090, 295
  85C220/224, 292
EVEN parity, 408, 409
Excess-3 code, 171
Excitation table, 528, 530
Exclusive NOR gate. *See* XNOR gate
Exclusive OR gate. *See* XOR gate
External pull-up resistor, 651–52

Falling edge, 12
Fall time ($t_f$), 12, 13
FAMOS FET (floating-gate avalanche MOSFET), 757–58, 761
Fanout, 611–15
Fast carry, 346–47
Feynman, Richard, 46–47
FIFO (first-in first-out) memory, 762
File extension, 301, 302
Firmware, 752
555 timer, 247–48
  as astable multivibrator, 253–56
    calculating timing component values, 256–58
    minimum value of $R_a$, 259–60
  internal configuration, 248
    output section, 249–50
    SR latch, 249
    voltage divider and comparators, 248–49
  as monostable multivibrator, 250–53
Flash A/D converter, 716–18
Flash converter, 716
Flip-flop excitation tables, 530
Flip-flops, 198–99
  edge-triggered D flip-flops, 198, 201–3
  edge-triggered D flip-flops
    applications of, 203: parallel data transfer, 205–7; serial-to-
      parallel conversion, 203–5; tristate buffering in parallel data
      transfer, 208–12
    control gating, 208
  edge-triggered JK flip-flops, 212–15
  edge-triggered SR flip-flops, 198–201
Floating logic state, 47, 48
Flow control, 310, 312
4000B CMOS devices, 629
  power dissipation in, 620–23
4011B gate, 682
4011UB gate, 682
4067B 16-channel MUX/DMUX, 417
4097B dual 8-channel MUX/DMUX, 417, 430–31
4532B priority encoder, 397–99
4538B IC monostable multivibrator, 243
4585B magnitude comparator, 404
FPLA (field programmable logic array), 264, 268
Frequency counter, 519–21
Full adder, 337–38
Full-sequence counters, 460, 482–87
Functions, active levels of, 23
Fusible-link PROM, 754–57

Gain error, 715
GAL16V8, 283–87

GAL22V10, 287–91
GAL (generic array logic), 283
  GAL16V8, 283–87
  GAL22V10, 287–91
Gated D latch, 192–95
Gated latches, 190
  gated SR latch, 191–92
  74HC75 4-bit transparent latch, 195–97
  transparent latch (gated D latch), 192–95
Gated SR latch, 191–92
Gating, level of, 69–73
Generic array logic. *See* GAL
Glitch, 468, 470, 473–78
Global architecture cell, 283, 286
Graphics shell, 301, 302
Gray code, 172

Half adder, 337–38
Hardware, 752
Hexadecimal arithmetic, 334
  hex addition, 334–35
  hex subtraction, 335–36
Hexadecimal numbers, 165
  conversions between hexadecimal and binary, 167–68
  counting in hexadecimal, 165
  decimal-to-hexadecimal conversion, 166
    repeated division by 16, 167
    sum of weighted hexadecimal digits, 166–67
  hexadecimal-to-decimal conversion, 166
High-current driver, 650–51
High-impedance state, 208, 209
High-speed CMOS devices, power dissipation in, 620, 623–24
High-speed CMOS logic family, 680–82
HIGH-Z state, 770
Hold time ($t_h$), 218

$I_{CCH}$ (TTL supply current with all outputs HIGH), 615
$I_{CCL}$ (TTL supply current with all outputs LOW), 615
$I_{CC}$ (total TTL supply current), 615
IC monostable multivibrators, 243–47
$I_{DD}$ (CMOS supply current under static conditions), 615
Ideal pulse, 12
IEEE/ANSI notation, 525
  74LS91 serial-in-serial-out shift register, 595
  74LS93 and 74LS90 asynchronous counters, 525–26
  74LS95B parallel-access shift register, 596
  74LS161A/163A synchronous presettable counters, 526–27
  74LS164 serial-in-parallel-out shift register, 595–96
  74LS191 bidirectional synchronous presettable counter, 527–28
  74LS194A universal shift register, 596–97
IEEE/ANSI Standard 91–1984, 21, 22
Inhibit, 37
In phase, 37, 38
Input, stuck HIGH or LOW, 53–54
Input line, 264, 265
Input line number, 270, 272
Inputs
  data, 415, 416
  select, 415, 416
Integrated circuit (IC), 41, 42
Integrated circuit logic gates, 41–44
Integrator, 722
Interconnection dependency, 525, 526
Internal logic circuit, fault in, 52
Internal short circuit, 54
Inverter, 21
  CMOS, 669–71

NMOS, 667–69
TTL, 641-45
$I_{OH}$ current measurement, 611, 613–15
I/O (input/output), 785, 788
$I_{OL}$ current measurement, 611, 613–15
$I_T$ (sum of static and dynamic supply currents), 615

Johnson counters, 583, 589–91
   modulus and decoding, 591–95

K (1024), 739
Karnaugh map method of simplification, 109–10
   conditions for maximum simplification, 116–17
   don't care states, 120–23
   grouping cells along outside edges, 112–13
   loading K-map from truth table, 113–14
   multiple groups, 114–15
   overlapping groups, 115–16
   POS simplification, 123–25
   three- and four-variable map, 111–12
   two-variable map, 110–11
   using K-maps for partially simplified circuits, 117–20

Latch circuits, troubleshooting, 221–23
Latches, 180–83
   gated, 190–97
   NAND/NOR, 183–90
Leading edge, 12, 14
Least significant bit (LSB), 6–7
Level of gating, 69–73
Level-sensitive enabling, 198
LIFO (last-in first out) memory, 762
Linearity error (of D/A converter), 715
Load gate, 611, 612
Local architecture cell, 283, 286
Logic
   negative, 4
   positive, 4
   programmable. *See* Programmable logic
Logical product, 22
Logical sum, 22, 23
Logic circuits. *See* Combinational logic circuits
Logic diagrams, 64
   Boolean expressions from, 65–67
   from Boolean expressions, 69–73
   truth tables from, 74–77
Logic families, 136
   BiCMOS logic, 684
   CMOS, 136, 680
   emitter coupled logic
     (ECL), 682-84
   high-speed CMOS, 136,
     680-82
   LSTTL, 136
   TTL, 677-80
Logic functions
   active levels, 23–24
   basic, 20–21
     AND, 20–21, 22
     NOT, 20–22
     OR, 22–23
   derived, 25
     exclusive OR and exclusive NOR, 27–29
     NAND and NOR, 25–29
Logic gate network, 64
Logic gates, 20
   AND, 22, 38–39

changing to DeMorgan equivalent, 35–37, 65
CMOS, 627–30
electrical characteristics of, 602–3
enable and inhibit properties of, 37
   AND and OR, 38–39
   NAND and NOR, 39–40
   XOR and XNOR, 40–41
generalized multiple-input, 32
integrated circuit, 41–44
NAND, 27
NOR, 27
OR, 22–23, 38–39
possible faults in, 52–54
truth table for generalized gate, 29–31
XNOR (coincidence), 28–29
XOR, 27–28
Logic level, 1–2, 4
   compulsory, 769
   floating, 47, 48
   HIGH, 4
   LOW, 4
Logic probe, 47, 48
Logic pulser, 47, 48
LSTTL devices
   CMOS driving LSTTL, 628–29
   LSTTL driving 74HB or 4000B CMOS, 629
   LSTTL driving 74HCT CMOS, 629
LSTTL logic family, 136
   propagation delay for, 604
LSTTL logic gates, pinouts for, 843

M (1,048,576), 739
Magnitude bits, 325, 326
Magnitude comparators, 401–8
Mask-programmed ROM, 752–53
Maximum modulus ($m_{max}/$), 460
Maxterm, 78, 80
MC54/74HC00A quadruple 2-input NAND gates, 813–15
MC54/74HC02A quadruple 2-input NOR gates, 816–18
MC54/74HC08A quadruple 2-input AND gates, 826–28
MC54/74HC32A quadruple 2-input OR gates, 829–31
MC54/74HCT04A hex inverter with LSTTL-compatible inputs, 819–21
MC54/74HCU04 hex unbuffered inverter, 822–25
MC54F/74F00 quadruple 2-input NAND gates, 811–12
MCM6164–45
   E-controlled write cycle, 775–76
   minimum read cycle requirements, 775
   $\overline{W}$-controlled write cycle, 775–76
MCM6164/MCM6L64 static RAM, 772–74
MCM81000 DRAM memory module, 786–87
Mealy-type state machine, 528, 529
Memory
   basic concepts, 736
     address and data, 736–39
   bit-organized, 750
   control signals, 741–42
   expansion of, 799–800
     expansion of CPU address space, 801–5
     expansion of memory capacity within existing CPU space, 800–1
     word length expansion, 800
   FIFO, 762
   LIFO, 762
   memory capacity, 739–41
   random access memory (RAM), 739, 741–52
   read only memory (ROM), 739

sequential, 762–63
volatile, 742, 743
word-organized, 745, 748
Memory map, 789, 791
Memory systems, 785, 788–89
  decoding with MSI chips, 794–99
  memory expansion, 799–800
    expansion of CPU address space, 801–5
    expansion of memory capacity within existing CPU space, 800–1
    word length expansion, 800
  memory mapping and address decoding, 789–94
Metal-gate CMOS, 680
Minterm, 78, 79
Minuend, 324
Mode dependency, 426, 525
Modulo-$n$ (or mod-$n$) counters, 458, 459, 521–23
Modulus, 458, 459
Monostable multivibrators, 240–42
  555 timer as, 250–53
  IC monostables, 243–47
  nonretriggerable, 241
  retriggerable, 241
Monotonicity, 714
Moore-type state machine, 528–29, 541
MOSFET (metal-oxide-semiconductor field effect transistor), 663, 664–65
MOS transistors, bias requirements for, 665–67
Most significant bit (MSB), 6–7
MSI (medium-scale integration) devices, 374
  counters, 503–24
  decoders, 377
    decoding with, 794–99
  multiplexers, 417
  priority encoders, 397–99
  serial shift registers, 568–69, 571–72
  shift registers, 568–69, 571–72
Multichannel data selection, 426–28
Multichip mod-$n$ counters, 523–24
Multiple-digit BCD adders, 359–61
Multiple-input gates, generalized, 32
Multiplexers, 270, 272, 415–17
  Boolean function generator, 418–21
  MSI, 417
  multichannel data selection, 426–28
  single-channel data selection, 417–18
  time-dependent multiplexer applications, 422
    parallel-to-serial conversion, 425–26
    time-division multiplexing, 422–24
    waveform generation, 424–25
Multivibrators, 239–40
  astable, 242–43
  bistable, 240
  monostable, 240–42

NAND function, 25–26
  active level of, 26
  available SSI gates, 136
NAND gates, 20–21, 25
  enable and inhibit properties of, 39–40
  multiple-input, 27
  universal property of, 128
    all-NAND forms, 128–29
NAND latch
  as switch debouncer, 188–90
NAND/NOR gates, 671–74
NOR latches, 183–84
  R = 0, S = 0 (no change condition), 184–85

S = 0, R = 1 (reset condition), 185–86
S = 1, R = 0 (set condition), 186
S = 1, R = 1 (forbidden condition), 186–88
n-channel enhancement mode MOFSET, 664
Negative logic, 4
Nested statements, 310, 312
Nibble, 736
9's complement, 171, 172
NMOS inverter, 667–69
NMOS logic family, 664
NMOS logic gates
  internal circuitry of, 663–64
    bias requirements for MOS transistors, 665–67
    inverter, 667–69
    MOFSET structure, 664–65
Noise, 624
  switching, 655–58
Noise margin, 624–27
NOR function, 25–26
  active level of, 26
  available SSI gates, 136
  exclusive, 27–29
NOR gates
  enable and inhibit properties of, 39–40
  multiple-input, 27
  universal property of, 128
    all-NOR forms, 129–35
NOT function, 20–22
  Boolean expression of, 73–74
n-type inversion layer, 665, 666

Octal numbers, 168
  conversions between octal and binary, octal and decimal, 169–70
  counting in octal, 168–69
Octet (in Karnaugh map), 109, 112
ODD parity, 408, 409
Offset error, 715
Ohmic region, 665, 666
1's complement, 325, 326–27
One-shot multivibrator. *See* monostable multivibrator
OPAL (National Semiconductor), 301
Open circuit, 52
Open-collector applications, 647
  high-current driver, 650–51
  value of external pull-up resistor, 651–52
  wired-AND, 647–50
Open-collector output, 641
Operand, 323
Order of precedence, in Boolean expressions, 67–69
OR function, 22–23
  active levels of, 23
  available SSI gates, 136
  Boolean expression of, 73–74
OR gate, 22–23
  enable and inhibit properties of, 38–39
Oscilloscope, 47, 48–52
OTP (one-time programmable), 291–92
Out of phase, 37, 40
Output
  open-collector, 641
  stuck HIGH or LOW, 53–54
  totem pole, 653-55
  tristate, 658, 659
Output logic macrocell (OLMC), 283
Overflow, 331–33
Overflow detection, 350–52

Pair (in Karnaugh map), 109, 111
PAL16L8 PAL circuit, 270–71
PAL20P8 PAL circuit, 274–76
PALASM, 299–301
PAL (programmable array logic), 264, 268–69
    fuse matrix and combinational outputs, 270–73
    outputs with programmable polarity, 273–77
PAL (programmable array logic) devices with registered outputs, 277–82
Parallel adder, PLD implementation of, 361–64
Parallel binary adder/subtractor, 342–46, 349–50
    2's complement subtractor, 347–49
    fast carry, 346–47
    overflow detection, 350–52
Parallel carry, 346–47
Parallel data transfer, 205–7
    tristate buffering in, 208–12
Parallel loading, 562
Parallel-load shift registers
    74LS95B, 571
    74LS165A, 569–71
Parallel-load shift registers, 569
    applications of, 572
        arithmetic operations, 574–76
        serial data transmission, 572–74
Parallel port, 728
Parallel shifting, 562
Parallel-to-serial conversion, 425–26
Parallel transmission, 203
Parity, 408, 409
Parity bit, 408, 409
Parity generators and checkers, 408–15
p-channel enhancement mode MOFSET, 664
Periodic waveforms, 10–11
Phase
    in, 37, 38
    out of, 37, 40
Phase splitter, 653
Plastic leaded chip carrier (PLCC), 42, 43
PLCC (plastic leaded chip carrier), 291–92
PLD. *See* Programmable logic device
PLDshell Plus/PLDasm (Intel), 301–2
    sequential logic applications, 542–53
    source file, 302–3
        declaration section, 303
        design section, 303–10
        simulation section, 310–17
Polarity, programmable, 273–77
Positional notation, 5–6, 158–59
Positive logic, 4
POS. *See* Product-of-sums (POS)
Power dissipation, 615–16
    in 4000B CMOS devices, 620–23
    in high-speed CMOS devices, 620, 623–24
    in TTL devices, 616–20
Power supplies, interfacing devices with different, 629–31
Preset, 215
Presettable counter, 509
Presettable MSI synchronous counters, 513–16
Priority encoder, 394–96
Product line, 264, 265
Product line first cell number, 270, 272
Product-of-sums (POS), 79, 80
    maximum simplification of, 104–8, 123–25
Product term, 78, 79
Programmable logic
    development software, 295–317
    FPLA, 268

Generic array logic (GAL), 283–91
    introduction to, 264–66
    PAL, 268–69
        PAL devices with registered outputs, 277–82
        PAL fuse matrix and combinational outputs, 270–73
        PAL outputs with programmable polarity, 273–77
    PLD architectures, 266
    PROM, 267–68
    UV-erasable programmable logic devices (EPLD), 291–95
Programmable logic devices (PLDs), 264, 265
    adder/subtractor with overflow indication, 364–65
    architectures, 266
    BCD adder, 365–66
    combinational logic applications of, 432–41
    parallel adder, 361–64
    sequential logic applications of, 4-bit counter (state table method), 548–53
    sequential logic applications of, 542–44
        two 8-bit counters (Boolean equation method), 544–48
PROM (programmable read only memory), 264, 266, 267–68, 754
    PROM decoder, 755–57
Propagation delay, 218, 219, 473–78, 603–7
    in logic circuits, 607–11
Pulse, 12–14
Pulse waveform, 12–15, 37
Pulse width ($t_w$), 12, 13, 218

Quad (in Karnaugh map), 109, 111
Qualifying symbol, 21, 22
Quantization, 694
Quantization error, 718, 721
Quasistable state, 240
Queue, 762

Radix point, 163
RAM. *See* Random access read/write memory (RAM)
Random access read/write memory (RAM), 739, 742–43
    dynamic RAM (DRAM) cells, 749
        DRAM cell arrays, 750–51
    dynamic RAM (DRAM) modules, 785, 787–89
    dynamic RAM (DRAM) timing cycles, 776–84
    static RAM (SRAM) cells, 743–45
        static RAM cell arrays, 745–49
    static RAM (SRAM) timing cycles, 768–71
Random access read/write memory (RAM) cell, 743
Read (from memory), 736, 737
Read only memory (ROM), 739, 752
    EEPROM, 761–62
    EPROM, 757–61
    fusible-link PROM, 754–57
    Mask-programmed ROM, 752–53
Recovery time ($t_{rec}$), 218, 219
Recycle, 456
Recycle state, 468, 470
Refresh cycle, 749
Register, 277, 278, 352
Register arithmetic circuits, 352–56
Registered output, 277, 278
Relative accuracy, 715
RESET, 180
Resolution, 694, 695
Ring counters, 583
    modulus and decoding, 586–89
Ripple blanking, 389, 393
Ripple carry, 342, 343
Ripple counter, 463, 464
Rise time ($t_r$), 12, 13

Volatile (memory), 742, 743
$V_{OL}$ (voltage measured at output when output is LOW), 624

Waveform generation, 424–25
Windowed CERDIP, 291
Wired-AND connection, 647–50
Word, 745, 748

Word length, 745, 748
Word problems, designing logic circuits from, 141–43
Write (to memory), 736, 738

XNOR (exclusive NOR/coincidence gate), 28–29, 40–41, 84–85
XOR (exclusive OR function), 27–28, 40–41, 84–85